BK 621.38 C3970
OPTICAL FIBRE COMMUNICATION
/CE
C1980 49.50 FV
3000 581302 30011
St. Louis Community College
S0-BST-382

WITHDRAWN

OPTICAL FIBRE COMMUNICATION

OPTICAL FIBRE COMMUNICATION

Technical Staff of CSELT
(Centro Studi e Laboratori Telecomunicazioni)

Torino (Turin), Italy

McGraw-Hill Book Company

*New York St. Louis San Francisco Auckland
Bogotá Hamburg Johannesburg London Madrid
Mexico Montreal New Delhi Panama Paris
São Paulo Singapore Sydney Tokyo Toronto*

Library of Congress Cataloging in Publication Data

Centro studi e laboratori telecomunicazioni, Turin.
Optical fibre communication.

Includes bibliographical references and index.
1. Optical communications. 2. Fiber optics.
I. Title.
TK5103.59.C46 1981 621.38'0414 80-22667
ISBN 0-07-014882-1

First published in the United States 1981 by McGraw-Hill Book Company, division of McGraw-Hill, Inc.

2 3 4 5 6 7 8 9 VBVB 8987654

CONTENTS

PREFACE

After a dozen books, published to date on optical fibre communication and related subjects, this is one of the first, to my knowledge, which attempts an in-depth coverage of both the fundamental (i.e. theoretical and technological) and the practical (i.e. design, measurement and implementation) aspects of this new transmission medium, the optical fibre, in telecommunications. This accounts for the book's 900 pages, 500 figures and 800 bibliographic references and for the number of highly qualified CSELT researchers, who have each been working for years on a particular subject covered in the book.

Even if it might be implied that in a so rapidly evolving field any book may be made obsolete in few years, we decided to proceed with the preparation of an extended treatise on optical fibre communication because we realised that optical fibre art was moving from research into industry and operating companies (not only in the telecommunication field) and therefore scores of engineers may soon need a tool for understanding, designing, producing, selling, installing and operating optical fibre systems.

As happened in the past with rapidly evolving technologies (I may only quote as an example semiconductor technology, where a new industrial process was announced something like every six months, especially in the sixties), industry is bound to rely on quasi-freezed-on technologies and components for maybe four or five years, and only after that period a new generation of systems may be devised.

As regards optical fibre communication, the first generation may be that dealt with in this book, mainly based on operation in the 0.8-0.9 μm wavelength region; the second generation may be, for instance, based on the exploitation of the so called " second window ", i.e. the 1.3-1.6 μm wavelength region, whereas the third genera-

tion may be based on exploitation of highly coherent sources, monomode fibres, etc. As a consequence, there is no doubt that, possibly every five years, a book like the one we are presenting today will have to be reviewed and reissued, and I hope that we will be willing to do so, satisfactorily.

I say " I hope " because it took two long years of intense work of senior CSELT researchers to write this book and it will take much effort, to be subtracted from our daily research activity, for a conscious review of a book of this size.

On the other hand, CSELT is in a good position to observe the whole telecommunication field, as it is the central research laboratory of the STET Group, who runs in Italy the telecommunication services through three operating Companies, SIP, Telespazio and Italcable, in addition to providing equipment production through SIT-SIEMENS, Italtel Selenia and Elsag, semiconductor products through SGS-ATES and installation services through SIRTI.

The book is divided into five parts, i.e.:

Part one: The transmission medium
Part two: Sources and detectors
Part three: Cables and connections
Part four: Systems
Part five: Integrated optics.

These headings are more or less standard in all the conferences on optical fibre communications around the world. The contents of each part, subdivided into a suitable number of chapters, have been chosen to reflect the present state-of-the-art, with particular emphasis on components and techniques, which are likely to be used in practical equipment and systems in the near future. Perhaps two aspects have not been treated with due attention to practice: one is that of part five, integrated optics, for the reason that applications are not numerous at present. In addition, we decided not to specifically illustrate the many field experiments (I think they number more than 200 to date) organized all around the world, even if most of our " COS " experiments performed in Italy (COS1 and COS2 in Turin, COS3 in Rome) have constituted the background for much of the practical guidelines given throughout the book.

The authors are people who " know because they work not because they read ". Most of them are known to colleagues attending the most important international conferences on optical communication, particularly the European Conferences on Optical Communication (ECOC), which have been organized with substantial contribution and support from CSELT.

I should like to explicitly mention: Dr. Bruno Costa, Dr. Pietro Di Vita, Dr. Umberto Rossi, Dr. Bruno Sordo, Dr. Giuseppe Cocito, Dr. Giacomo Roba, Dr. Paolo Vergnano, Dr. Giorgio Randone, Dr. Pier Luigi Carni, Dr. Giuseppe Galliano, Dr. Federico Tosco, Dr. Feliciano Esposto, Dr. Emilio Vezzoni, Dr. Angelo Luvison, Dr. Agostino Moncalvo, Dr. Luigi Sacchi, Dr. Alfredo Fausone, Dr. Vittorio Ghergia, Dr. Antonio Scudellari, who have written the fifteen chapters of the book; however, many other researchers from CSELT have valuably contributed in various ways, whom I hereby acknowledge all together.

We are also indebted to the many Organisations and authors who allowed the publication of figures and data taken from their works; their names are reported under "Acknowledgments", at the end of the book, besides being specifically mentioned in the pertinent figures and/or bibliographic references.

We have also tried in this book to avoid the discomfort to the reader of sweeping through a collection of papers written by different authors, without seeing a true connection, common style, good cross-referencing, standardisation of symbols, uniform bibliographic references, etc. throughout the book. I wish to express to Dr. Achille Lanza my appreciation for his hard work in amalgamating all the papers collected from the authors.

We are not sure that our effort will produce a reasonable satisfaction of the readers: here is the result, submitted to my "dear colleagues and chers amis", the telecommunication engineers, with my best auspices that it will help them in their daily and fascinating work.

B. Catania

CSELT Managing Director

ABOUT THE AUTHORS

Pier Luigi CARNI was born in Caravaggio (Bergamo) in 1948. He received his Doctorate in Electronic Engineering from the Polytechnic of Milano in 1972.

In 1974 he joined CSELT, where he performed research on characterization and application of opto-electronic devices. In 1978 he transferred to SIRTI, an Associate Company of CSELT, where he is currently engaged in optical fibre system installation.

Pier Luigi Carni is coauthor of a paper presented at the Fourth European Conference on Optical Communication (ECOC), Genova, 1978, where he was Secretary of both the Technical and the European Management Committee.

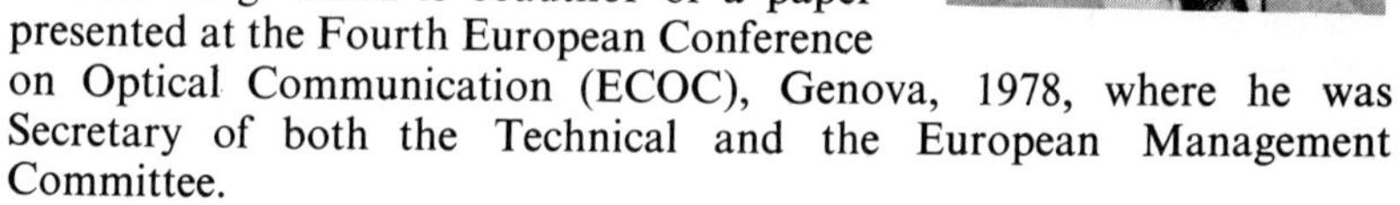

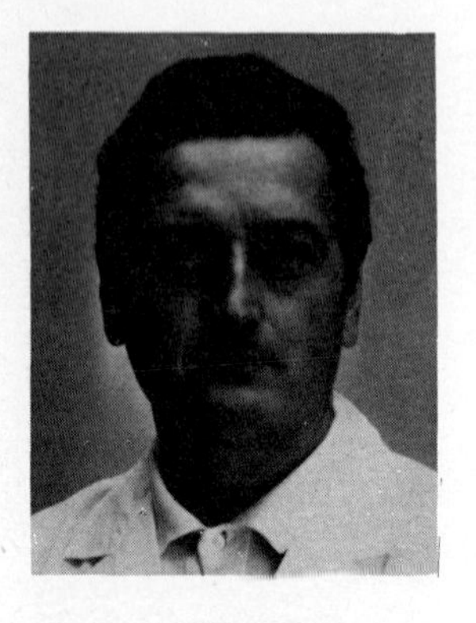

Giuseppe COCITO was born in Alba (Cuneo) in 1930. He received his Doctorate in Physics from the University of Torino in 1958.

He began his career at the Anti-Submarine Warfare Applications Dept. of NATO-Marina Militare Italiana (La Spezia) where he was mainly involved in mathematical problems, computer programming, information theory and random signal processing. From 1967 to 1973 he was Associate Professor of Astronomy at the University of Torino. From 1973 to 1978 he worked at CSELT, where he was engaged in research on optical communications with particular reference to the problems of fibre drawing and connection. He is now a professional consultant.

Giuseppe Cocito holds 10 international patents and is author or coauthor of some 15 papers. He has lectured on fibre technology and coupling at the Scuola Superiore Guglielmo Reiss Romoli (SSGRR).

Bruno COSTA was born in Varapodio (Reggio Calabria) in 1946. He received his Doctorate in Physics from the University of Torino in 1969, where he spent one post-graduate year as an experimental researcher in the field of elementary particle physics.

In 1971 he joined CSELT where, after an initial period devoted to various theoretical investigations, he was engaged in research on optical fibres, with special regard to the physics and transmission properties of fibres. He is presently head of the Laser Section.

Bruno Costa holds an international patent and is author or coauthor of 20 papers, one of which he was invited to present at the Fifth European Conference on Optical Communication (ECOC), Amsterdam, 1979. He has lectured on optical fibre communication at the Scuola Superiore Guglielmo Reiss Romoli (SSGRR).

Pietro DI VITA was born in Trapani in 1949. He received his Doctorate in Physics from the University of Torino in 1971, where he spent a short post-graduate period working in the field of optical models in elementary particle physics.

In 1972 he joined CSELT, where he has been working on theoretical aspects of propagation and measurement methods in optical waveguides.

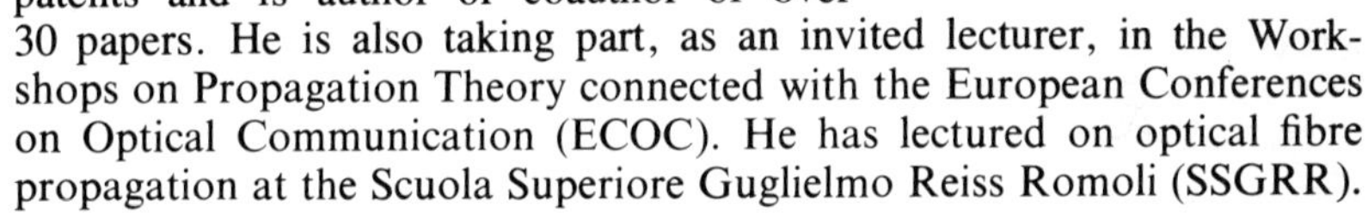

Pietro Di Vita holds 10 international patents and is author or coauthor of over 30 papers. He is also taking part, as an invited lecturer, in the Workshops on Propagation Theory connected with the European Conferences on Optical Communication (ECOC). He has lectured on optical fibre propagation at the Scuola Superiore Guglielmo Reiss Romoli (SSGRR).

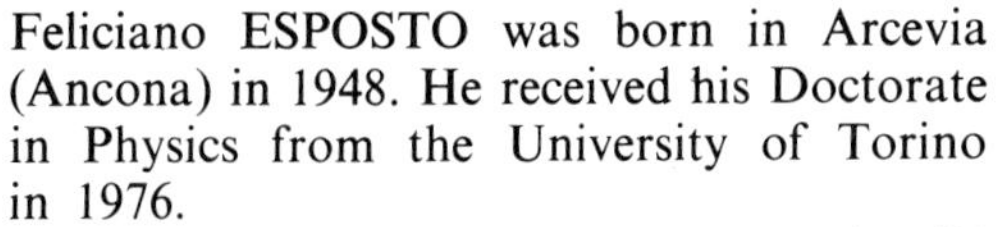

Feliciano ESPOSTO was born in Arcevia (Ancona) in 1948. He received his Doctorate in Physics from the University of Torino in 1976.

In 1969 he joined CSELT, where he did research on optics and optical communications, particularly holography and coupling problems related to the use of optical fibres. Since mid 1979 he is with SIP, the Italian Telephone Operating Company, which, like CSELT, belongs to the STET Group. He is currently engaged in the follow-up of the COS-3 experiment in Rome. He is coauthor of some 10 papers. He has lectured on optical fibre coupling at the Scuola Superiore Guglielmo Reiss Romoli (SSGRR).

Alfredo FAUSONE was born in Ivrea (Torino) in 1943. He obtained his Doctorate in Electronic Engineering from the Polytechnic of Torino in 1968. From 1969 to 1972 he was with Olivetti, involved in computer system development and, in particular, the use of « hybrid » integrated circuit technology.

He joined CSELT in 1972, where he has been involved in high-speed digital transmission system research both over coaxial cables and, recently, over optical fibres. Since January 1979, he holds the responsibility for the Advanced Techniques Section.

He has 4 international patents and is coauthor of 3 papers. He has lectured on fibre optic system design at the Scuola Superiore Guglielmo Reiss Romoli (SSGRR).

Giuseppe GALLIANO was born in Torino in 1947. He received his Doctorate in Physics in 1971 from the University of Torino, where he spent one post-graduate year in conducing research on elementary particle physics.

He joined CSELT in 1972, where he was engaged in works on analogue transmission equipment and on symmetric and coaxial cables to be employed for telephone and cable television systems. At present, his main activity is in the testing and characterization of optical fibre cables both in laboratory and in the field.

Giuseppe Galliano is coauthor of two papers.

Vittorio GHERGIA was born in Milano in 1939. He received his Doctorate in Electronic Engineering from the Polytechnic of Torino in 1965.

After joining CSELT in 1966, he specialised in thin film, thick film and semiconductor technologies. Since 1975 he holds the responsibility for the Material and Component Section, which deals with environmental testing, component reliability, chemical and physical analyses, and carries out material research on opto-electronic components in the 1.3 μm wavelength region. Since 1978 he has also been teaching Electronic Technology at the Polytechnic of Torino.

Vittorio Ghergia is author or coauthor of some 20 papers. He has lectured on components and materials at the Scuola Superiore Guglielmo Reiss Romoli (SSGRR).

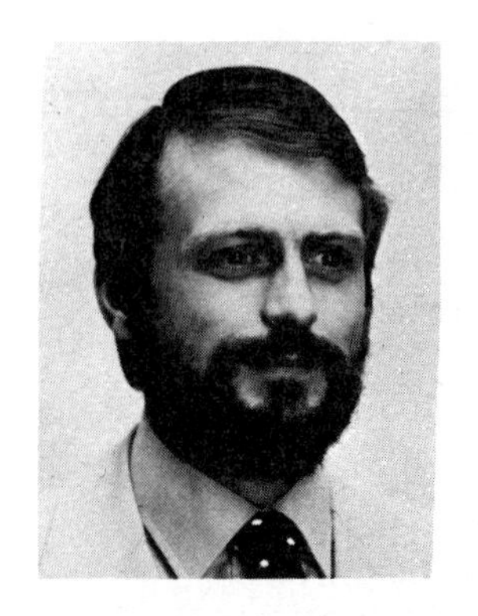

Angelo LUVISON was born in Torino in 1944. He received his Doctorate in Electronic Engineering from the Polytechnic of Torino in 1969.

Since then, he has been working at CSELT, where his research activities have included adaptive data receivers, performance evaluation of digital communication systems, optical fibre communication theory and distributed processing. In 1972, he was one of the recipients of the Siemens Award for the best telecommunication paper presented at the XX Convegno Internazionale delle Comunicazioni, Genova. Since 1975 he has also been teaching Information and Transmission Theory at the University of Torino. At present he holds the responsibility for the Scientific Section dealing with Communication Theory.

Angelo Luvison has 6 international patents and is author or coauthor of over 35 papers. He was invited to present two of these papers: one at the Second European Conference on Optical Fibre Communication (ECOFC), Paris, 1976, and the other at the International Symposium on Circuits and Systems (ISCAS), Tokyo, 1979. He has lectured on various topics in statistical communication theory at the Scuola Superiore Guglielmo Reiss Romoli (SSGRR).

He is the Secretary/Treasurer of the IEEE North-Italy Section.

Agostino MONCALVO was born in Novi Ligure (Alessandria) in 1947. He received his Doctorate in Electronic Engineering from the Polytechnic of Torino in 1971. Since then he has been working at CSELT in the field of communication systems, dealing particularly with equalizers, signal processing, line encoding, and various other topics connected with analogue and digital optical communication systems. In 1971 he was made an Assistant Professor of Electronic Measurements at the Polytechnic of Torino. He was later made Assistant Professor of Circuits and Electromagnetic Fields.

He is coauthor of 4 papers, one of which was invited at the Fourth European Conference on Optical Communication (ECOC), Genova, 1978.

Giorgio RANDONE was born in Trieste in 1939. He received his Doctorate in Physics from the University of Milano in 1962.

From 1962 to 1975 he worked at CISE (Milano), and his research activities ranged from basic solid state physics to GaAs material technology and device development. In 1967, on leave from CISE, he spent a year as Associate Researcher at the University of North Carolina. He joined CSELT in 1976, where he has been involved in the development of semiconductor light sources and detectors for optical fibre communication systems. At present he holds the responsibility for the Opto-electronic Section.

Giorgio Randone is author or coauthor of more than 20 scientific and technical papers. He is taking part in the Injection Laser Workshop connected with the European Conferences on Optical Communication (ECOC). He has lectured on opto-electronic technology and devices at the Scuola Superiore Guglielmo Reiss Romoli (SSGRR).

Giacomo ROBA was born in Cogoleto (Genova) in 1951. He received his Doctorate in Physics from the University of Genova in 1976.

Since then he has been working at CSELT, where he is currently doing research on optical fibre technology. He has contributed to the design and construction of fibre preform production and fibre drawing equipment. In particular, he has studied the automation of these processes together with several improvements in impurity elimination.

Giacomo Roba has applied for one patent and has lectured on fibre technology at the Scuola Superiore Guglielmo Reiss Romoli (SSGRR).

Umberto ROSSI was born in Prato (Firenze) in 1950. He received his Doctorate in Electronic Engineering at the University of Bologna in 1974.

In the following year he received a government scholarship and went on working at the University of Bologna in mode coupling mechanisms in optical waveguides. Since 1976 he has been with CSELT, where he deals with various theoretical aspects of propagation and measurement methods in optical waveguides.

Umberto Rossi is coauthor of 10 papers. He is a member of AEI.

Luigi SACCHI was born in Torino in 1940. He received his Doctorate in Electronic Engineering from the Polytechnic of Torino in 1964. In 1965 he followed a specialization course in telephony at the Polytechnic of Torino. He joined CSELT in 1965, where he has been engaged in research work on digital transmission systems on coaxial cables and optical fibres. He was head of the Digital Transmission Section until 1979. At present he is head of the Planning and Coordination Office. Since 1968, he has participated in the work of international standardization bodies, such as CCITT and CEPT, where he acts as representative of the Italian Administration in various working parties dealing with PCM systems and digital lines.

From 1970 to 1976 he taught as an Assistant Professor at the Polytechnic of Torino in the course on Telephone Transmission. In 1975 he was one of the recipients of the Cassa di Risparmio Award for the best telecommunication paper presented at the XXIII Convegno Internazionale delle Comunicazioni, Genova.

Luigi Sacchi holds two international patents and is author or coauthor of some 20 papers, one of which was invited at the International Symposium on Circuits and Systems (ISCAS), Tokyo, 1979. He has also lectured on digital transmission over coaxial and optical fibre cables at the Scuola Superiore Guglielmo Reiss Romoli (SSGRR).

Antonio SCUDELLARI was born in Ravenna in 1946 and received his Doctorate in Electronic Engineering from the University of Bologna in 1971. In 1971-1972 he was an Assistant Professor of Circuit Theory at the University of Bologna.

He joined CSELT in 1972 and was associated with the Centro Onde Millimetriche, Bologna, where he took part in a research programme on circular waveguide telecommunication systems. Then he performed his military service as a technical officer in the inspection of avionic products. When he returned in 1974, he was assigned to the Microwave Section of CSELT in Torino, where he has been working on the design of passive and active microwave microstrip and hybrid circuits. He is also engaged in exploratory studies concerning integrated optical devices for optical fibre communication systems. Antonio Scudellari is author of 3 papers.

Bruno SORDO was born in Dogliani (Cuneo) in 1948. He received his Doctorate in Physics from the University of Torino in 1972.

Since then he has been working at CSELT, where he is currently doing research on optical fibre communication measurement techniques. In particular, he has been engaged in the investigation on refractive index profile measurements by means of several methods (near-field reflection, refracted near-field) and on attenuation and pulse dispersion measurements, backscattering technique, etc. Bruno Sordo holds an international patent and is coauthor of some 10 papers.

Federico TOSCO was born in Torino in 1941. He received his Doctorate in Electronic Engineering from the Polytechnic of Torino in 1964.

In 1965 he followed a specialization course in telephony at the Polytechnic of Torino. He joined CSELT in same year and has been engaged in research work on FDM systems, coaxial and optical cables, video telephone and television transmission over cables. He was responsible for the COS-2 project, where CSELT, Pirelli, SIP and SIRTI cooperated in September 1977 in the installation of about 5 km of optical cables in ducts between two exchanges of the SIP telephone network of Torino. At present he is head of the Cable Transmission Division, which comprises about 80 researchers.

Since 1968 he has participated in the work of international standardization bodies, such as CCITT, IEC, CEPT, where he acts as representative of the Italian Administration in various working parties dealing with analogue transmission systems, coaxial and optical cables.

From 1970 to 1975 he taught as an Assistant Professor at the Polytechnic of Torino, in the course on telephone transmission, and from 1975 to 1977 as a Professor in the same course.

Federico Tosco is author or coauthor of some 20 papers. He has lectured on analogue transmission systems, on symmetric and coaxial cables, and on optical cables at the Scuola Superiore Guglielmo Reiss Romoli (SSGRR).

Paolo VERGNANO was born in Chieri (Torino) in 1949. He received his Doctorate in Chemical Engineering from the Polytechnic of Torino in 1973.

He spent a short post-graduate period conducting research in the field of the material exchange reactor with the Industrial Chemistry Institute of the same Polytechnic. In 1975 he joined CSELT, where he was involved in research on optical fibre technology. In 1979 he transferred to SIRTI, an Associate Company of CSELT, where his activity includes various system aspects of optical fibre communication.

Emilio VEZZONI was born in Rivarolo del Re (Cremona) in 1952. He received his Doctorate in Electronic Engineering from the University of Bologna in 1976.

Since 1977, he has been working at CSELT, engaged in the characterization of thin film parameters, fibre-to-fibre and source-to-fibre coupling. He is currently doing research on the characterization of opto-electronic devices, particularly for high-capacity optical fibre communication systems.

Emilio Vezzoni is coauthor of 5 papers. He is a member of AEI, and has lectured on opto-electronic devices at the Scuola Superiore Guglielmo Reiss Romoli (SSGRR).

ACKNOWLEDGMENTS

The authors are particularly indebted to the many Institutes, Industries, Publishing Houses, Technical Magazines and individual authors, who have given us permission to publish figures and/or data from publications issued in their name. Their list is given in the following, unless otherwise acknowledged in each chapter. Advanced Technology Publications Inc.; The American Ceramic Society; The American Institute of Mining, Metallurgical, and Petroleum Engineers, Inc.; The American Institute of Physics; The American Physical Society; American Telephone and Telegraph Company; Associazione Elettrotecnica ed Elettronica Italiana; Bell Telephone Laboratories, Incorporated; Centre National d'Etudes des Télécommunications (CNET); Chapman and Hall Ltd.; Comité du Colloque International sur les Transmissions par Fibres Optiques; Corning Glass Works, N. Y.; The Electrical Communication Laboratories, NTT; General Telephone & Electronics Corporation; Hirzel Verlag, Stuttgart; Hitachi Cable Ltd., Tokyo; The Institute of Electrical and Electronics Engineers, Inc.; The Institute of Electronics and Communication Engineers of Japan; The Institute of Electrical Engineers of Japan; The Institution of Electrical Engineers; International Telephone and Telegraph Corporation; Istituto Internazionale delle Comunicazioni, Genova; McGraw-Hill, Inc.; Naval Air Engineering Center, Lakehurst; Nippon Electric Co., Ltd.; Nippon Telegraph & Telephone Public Corporation; North-Holland Publishing Company; Optical Society of America; N. V. Philips' Gloeilampenfabrieken, Eindhoven, The Netherlands; The Publication Board, Japanese Journal of Applied Phisics; RCA Corporation; Tamburini editore S.p.A., Milano; Telephony Publishing Corp.; URSI, Bruxelles; VDE-Verlag GmbH.

LIST OF ABBREVIATIONS

AES
Auger electron spectrometry

AMI
Alternate mark inversion

APD
Avalanche photodiode

ATMP
Adjust-to-maximize power

CATV
Cable television

CCD
Charge-coupled device

CD
Core diameter

CGW
Corning glass works

CMI
Coded mark inversion

COBRA
Commutateur optique binaire rapide

CVD
Chemical vapour deposition

CW
Continuous wave

DBR
Distributed Bragg reflector

DFB
Distributed feedback

DFE
Decision-feedback equalizer

DH
Double heterostructure

DPO
Digital-processing oscilloscope

DSB
Double sideband

e.m.
Electromagnetic

EMI
Electromagnetic interference

EMP
Electromagnetic pollution

EO
Electro-optic

FDM
Frequency division multiplex

FET
Field-effect transistor

FM
Frequency modulation

FRP
Fibre reinforced plastic rod

FWHM
Full width half maximum

GQR
Gauss quadrature rule

HPDE
High density polyethylene

IIR
Integrated interferometric reflector

IM
Intensity modulation

IO
Integrated optics

ISI
Intersymbol interference

IVPO
Inside vapour phase oxidation

JFET
Junction field effect transistor

LAP
Al-laminated polyethylene

LD
Laser diode

LED
Light-emitting diode

LEED
Low energy electron diffraction

LOC
Large optical cavity

LPE
Liquid phase epitaxy

MBE
Molecular beam epitaxy

MCVD
Modified chemical vapour deposition

MFC
Mass flow controller

MO
Magneto-optic

MOSFET
Metal-oxide-semiconductor field-effect transistor

MMSE
Minimum mean-square error

MTBF
Mean time between failure

NA
Numerical aperture

NEP
Noise-equivalent power

NRA
Nuclear reactions analysis

NRZ
Non-return to zero

OD
Outer diameter

OVPO
Outside vapour phase oxidation

PAM
Pulse amplitude modulation

PCM
Pulse code modulation

PCVD
Plasma-activated CVD

PEP
Perfluoronated ethylene-propylene

PEPT
Polyethylene-terephthalate

PF
Parabolic-index fibre

PFM
Pulse frequency modulation

PIX
Proton induced X-rays

PLL
Phase-lock loop

PM
Phase modulation

PPM
Pulse position modulation

PVC
Polyvinylchloride

RAM
Random access memory

RBS
Rutherford back-scattering

RHEED
Reflected high energy electron diffractometer

r.m.s.
Root mean square

RV
Random variable

SCH
Separate confinement hetero-junction

SEM
Scanning electron microscopy

SF
Step-index fibre

SIMS
Secondary ion mass spectrometry

SNR
Signal-to-noise ratio

TDL
Tapped-delay line

TE
Transverse electric

TLC
Telecommunication

TM
Transverse magnetic

VA
Viterbi algorithm

VCO
Voltage controlled oscillator

VPE
Vapour phase epitaxy

WDM
Wavelength division multiplexing

WKBJ
Mathematical method used in physics, from: G. Wentzel, H. A. Kramers, L. Brillouin and J. Jeffreys

XMA
X-ray microanalysis

XRD
X-ray diffractometry

XRT
X-ray topography

OPTICAL FIBRE COMMUNICATION

HISTORICAL REMARKS

by Bruno Costa

The use of light for transmitting signals dates back far in time, but one of the first proposals for its use in the transmission of information in the modern sense of the word may be traced back to the 1870's [1]. It was, however, only after the invention [2] and implementation [3] of the laser that a definite research effort was carried out in this field. The availability of a coherent, monochromatic optical source stimulated the exploration of the potential of optical communications which, due to the very high frequency of the carrier ($\gtrsim 10^{14}$ Hz), would allow a very large amount of information to be transmitted. Light modulation and detection fundamentals were therefore studied. The first experiments were carried out letting the laser beam freely irradiate through the atmosphere, in very much the same way as the transmission of radio waves [4]. Apart from special applications such as transmission through empty space, this approach soon showed various drawbacks, i.e. system unreliability due to precipitations (rain, snow), fog, atmospheric turbulence. Light propagation in a protected environment was therefore suggested, and various solutions, including gas-filled pipes, tubes with focusing lenses inside, and so on were tried. A feasibility test was also carried out using a lens waveguide, nearly 1 km long [5]. Apart from the technical difficulties involved, the economic benefits of such systems lay with the transmission of huge bandwidths, which were not required by any foreseeable application. The use of glass fibres as a guiding medium soon appeared very attractive: size, weight, handling ease, flexibility and cost were clear advantages of such small dielectric waveguides in comparison with the above-mentioned systems.

Glass fibres guide the light through multiple internal reflections. It is very likely that the conduction of light through transparent cylinders was well-known to ancient glassblowers, but the earliest recorded scientific demonstration of this phenomenon was given by John Tyndall at the Royal Society, in England, in 1870. His famous demonstration consisted in illuminating a vessel of water and showing that, when a stream of water was allowed to flow through a hole in the vessel, light was conducted along the curved path of the stream.

The first complete theoretical analysis of electromagnetic propagation in dielectric cylinders was performed by Hondros and Debye [6] in 1910. However, it was only in the 1950's that optical fibres began to find practical application, mainly in the field of image transmission along flexible bundles (" fiberscopes "). In this period the idea of glass-coating glass fibres was developed [7]. After this, glass fibres found increasingly widespread use in optical devices (field flatteners), in instrumentation (coupling of plates for image intensifiers), photoelectronic devices, data processing, and photocopying systems, as well as in medicine (gastroscopy, bronchoscopy, etc.). However, in spite of the potential advantages of optical fibres in optical communications, it was not until 1966 that their use in this field was proposed. The reason for this was that the attenuation exhibited by glass fibres available at that time was in the range of thousands of dB/km. This allowed transmission only over short distances. In addition, light sources and detectors were not at all compatible in size and ruggedness with optical fibres.

In 1966, Kao and Hockam of Standard Telecommunication Laboratories in England, pointed out in a famous paper [8] that the attenuation found in glasses employed for optical fibres was not a basic property of the material, but was produced by the presence of impurities, mainly metallic ions. Since the intrinsic material loss, essentially determined by Rayleigh type scattering, which decreases as the fourth power of wavelength, is very low, reduction of the impurity content would allow attainment of much lower losses than those typically encountered. After that, research in this field started with a programme, sponsored by the British Post Office, for glass purification and study of fibre transmission problems. Other organizations in the USA, Japan and Germany soon became involved in this research area. The monomode fibre was first considered as a suitable medium for transmitting large amounts of information. Multimode fibres were soon taken into consideration also and, in 1968, an analysis of the refractive index grading necessary for minimizing modal dispersion was carried out [9]. On the attenuation side, it was recognized that an attenuation of 20 dB/km for glass fibres was the limit beyond which practical application to long-distance transmission became feasible. It appeared in 1970 that available fibres were not in a position to meet this goal [10]. It was, however, in the same year that Corning Glass Works succeeded in fabricating single-mode fibres, hundreds of meters long, with losses under 20 dB/km [11]. The technique consisted in depositing a thin layer of very pure, doped silica material inside a fused silica tube, which formed the basis for the modern low-loss fibre-production technology. This result was a real breakthrough, and was followed by a very strong research effort in this very promising field in many other industrialized countries (Italy, France, Netherlands, Australia). An important step which supported the view that very low losses could be attained, providing at the same time a valuable tool for experimentation, was the development of liquid-core fibres [12]. These were obtained by filling a quartz capillary tube with properly chosen liquids with low transmission loss in the visible and near-infrared region. Losses below 8 dB/km in the near-infrared were obtained [13]. First long-distance system demonstrations were carried out using such fibres (analogue transmission of a TV signal over a few kilometers). In 1972 further substantial progress was announced, again by Corning Glass, namely the achievement of 4 dB/km minimum loss in a high silica core multimode optical fibre [14]. This result made liquid core fibres obsolete.

The new possibilities offered by optical fibres stimulated research toward optical sources and detectors compatible in size and reliability, with low power consumption. Semiconductor light emitters and solid-state detectors (photodiodes) appeared the most promising devices. First results on laser action in a GaAs junction, cooled to 77 °K, dated back to 1962 [15, 16], and the first demonstration of incoherent emission from *pn* junction was announced in 1963 [17, 18]. In 1968 the first double heterostructure laser was announced [19], but it was not until 1970 that c.w. operation at room temperature was achieved [20]. The life of these devices was not longer than a few hours. This result was obtained by using a structure consisting of alternate layers of GaAs and $Al_xGa_{1-x}As$ (double heterostructure, or DHS, lasers), which resulted in low threshold current. Since then, spectacular progress has been achieved from the viewpoint of reliability and life: in 1973, devices with a life of more than 1000 hr. were announced [21]. Mass survival of stripe geometry DHS lasers for longer than 7000 hours was reported in 1977 [22]. At the present moment (1979), devices with projected lifetimes in excess of 100,000 hours are available on the market. As far as incoherent emitters (LED's) are concerned, an important step was performed in 1971, when C. A. Burrus, of Bell Labs, developed small area (50 μm diameter), head emitting, high radiance, DHS LED's, particularly suitable for coupling to optical fibres [23].

Coming back to fibres, a substantially new achievement took place in 1976. Japanese researchers of NTT and Fujikura cables succeeded in fabricating a fibre, with very low OH content and a minimum loss of 0.47 ± 0.1 dB/km, very close to the intrinsic material loss (Rayleigh scattering) [24]. The only possible improvement at this stage was the exploitation of longer wavelengths, and, in fact, an attenuation as low as 0.2 dB/km has recently been reported for a monomode fibre at 1.55 μm, which seems the minimum attainable loss for doped silica fibres [25]. Together with very low losses, silica fibres exhibit another very attractive feature in the long wavelength region: as was theoretically predicted in 1975 [26], and experimentally verified later [27, 28], a wavelength exists, near 1.3 μm, where the material dispersion of doped silica falls to zero. This means that very large bandwidths can be obtained, as material dispersion is an ultimately limiting factor of the information carrying capacity of optical fibres.

These two features — very low loss, very low dispersion — disclosed exciting new possibilities for long distance, high bit rate transmission on optical fibres. As a consequence, research was stimulated for the development of sources and detectors suitable for the new wavelength region. As regards light sources, many different approaches were undertaken, using various combinations of III-V binary, ternary and quaternary compounds [29, 31]. The most successful seems to have been the GaInAsP/InP combination. Thanks to ease of lattice matching, structural simplicity and better heat conductivity, room temperature c.w. operation of lasers with lives longer than 2000 hours have been quickly obtained [32]. LED's have been fabricated as well [33], and both devices are becoming commercially available. As far as detectors are concerned, silicon pin diodes and avalanche photodiodes with very good characteristics have been developed throughout these years without any particular fabrication difficulty. However, they are not usable beyond 1 μm; germanium has good responsivity up to 1.6 μm, but suffers from excessive dark current and avalanche noise. Ternary alloys such as GaAsSb [34] and InGaAs [35, 36] have been studied, but a low noise detector satisfactory at 1.3 μm is still expected.

As the various components needed for an optical fibre system — light sources, detectors, low loss fibres and fibre cables, splices and connectors — were developed, laboratory demonstration of transmission systems and field trials were carried out in many countries. A good summary of the main field trials can be found in ref. [37]. The best results in terms of repeaterless long links at high bit rate transmission are related to the exploitation of the low loss, low dispersion properties of fibres in the long wavelength region: a 100 Mbit/s repeaterless transmission over 63 km of optical fibre has been performed [38], using an LED with peak wavelength at 1.27 μm. As a consequence of the use of laser sources, a 100 km span between repeaters is obtainable.

First complete optical links are already in operation on board military ships and aircrafts. Also, more complex experiments, including special services for selected customers, are under test [39].

The future of optical fibre communication looks very bright [40]. There is no longer any serious obstacle to the introduction of such systems for the various applications which have been envisaged from the beginning: data buses, voice and video communication on vehicles and within buildings, computer links, CATV broadcasting, public telecommunication network, and so on.

This is essentially due to the strikingly fast progress which has characterized this research field, no doubt due to the extremely attractive features of optical fibres as a transmission medium. It can be said that in many cases results have gone beyond the more optimistic forecasts. Further developments are related to the use of monomode fibres for long distance, very high bit rate applications and to the introduction of integrated optic devices for small, rugged and compact light communication circuits.

However, new possible materials are now being considered, at least on a speculative basis, ~~to~~ see whether new classes of components can be obtained with even better performance than those now available. Attenuations less than 0.001 dB/km are predicted for materials such as TlBr [41] and $ZnCl_2$ [42] in the wavelength region between 3.5 - 5.5 μm. This means that there is still room for continuing research in this exciting field.

REFERENCES

[1] R. Kompfner: *Optics at Bell Laboratories - Optical communications*, Appl. Opt., Vol. 11, No. 11 (1972), pp. 2412-2425.

[2] A. L. Shawlow and C. H. Townes: *Infrared and optical masers*, Phys. Rev. Vol. 12 (December 1958), pp. 1940-1948.

[3] T. H. Maiman: *Stimulated optical radiation in ruby*, Nature, No. 187, (August 1960), pp. 493-494.

[4] J. R. Kerr, P. J. Titterton, A. R. Kraemer and C. R. Cooke: *Atmospheric optical communications systems*, Proc. IEEE, Vol. 58, No. 10 (1970), pp. 1691-1709.

[5] J. B. Christian, G. Goubau, and J. W. Mink: *Further investigations with an optical beam waveguide for long distance transmission*, IEEE Trans. Micr. Theory and Tech., Vol. MTT-15, No. 4 (1967), pp. 216-219.

[6] D. Hondros and P. Debye: *Elektromagnetische Wellen an dielektrischen Drahten*, Ann. Physik, Vol. 32 (1910), pp. 465-470.

[7] N. S. Kapany et al.: J. Opt. Soc. Am., Vol. 49 (1959), pp. 779.

[8] K. C. Kao and G. A. Hockam: *Dielectric fibre surface waveguides for optical frequencies*, Proc. IEEE, Vol. 113 (1966), pp. 1151-1158.

[9] S. Kawakami and J. Nishizawa: *An optical waveguide with the optimum distribution with reference to waveform distortion*, IEEE Trans. Micr. Theory and Tech., Vol. MTT-16, No. 10 (1968), pp. 814-818.

[10] N. Lindgren: *Optical communications. A decade of preparation*, Proc. IEEE, Vol. 58, No. 10 (1970), pp. 1410-1418.

[11] F. P. Kapron and D. B. Keck: *Radiation losses in glass optical waveguides*, Trunk Telecommunications by Guided Waves, London, (September 29-October 2, 1970).

[12] J. Stone: *Optical transmission in liquid-core quartz fibers*, Appl. Phys. Lett., Vol. 20, No. 7 (1972), pp. 237-240.

[13] G. J. Ogilvie and R. J. Esdaile: *Transmission loss of tetrachloroethylene-filled liquid-core-fibre light guide*, Electron. Lett., Vol. 8, No. 22 (1972), pp. 533-534.

[14] R. D. Maurer: *First European Electro-optics Market and Technology*, Geneva, (September 12-15, 1972).

[15] A. N. Hall, G. E. Fenner, T. D. Kingsley, T. J. Soltyo and R. O. Carlson: *Coherent light emission from* GaAs *junctions*, Phys. Rev. Lett., Vol. 9, (November 1962), pp. 366-378.

[16] H. I. Nathan, W. P. Dumbe, G. Burns, F. H. Dill and G. J. Lasher: *Stimulated emission of radiation from* GaAs *p-n junctions*, Appl. Phys. Lett., Vol. 1, (November 1962), pp. 62-64.

[17] W. T. Matzen: *Semiconductor single crystal circuit development*, Report No. ASD-TOR-63-281 (March 1963).

[18] R. H. Rediker: *Infrared and visible light emission from forward biased p-n junctions*, Solid State Design (August 1963), pp. 19-28.

[19] W. F. Kasonocky, R. Cornely and I. J. Hegyi: *Multilayer* GaAs *injection laser*, IEEE J. Quantum Electron., Vol. QE-4, No. 4, (1968), pp. 176-179.

[20] I. Hayashi, M. B. Panish, P. W. Foy and S. Sumelay: *Junction lasers which operate continuously at room temperature*, Appl. Phys. Lett., Vol. 17, No. 3 (1970), pp. 109-111.

[21] R. L. Hartman, J. C. Dymeut, C. J. Hwang and H. Kuhn: *Continuous operation of* $Ga_xAl_{1-x}As$ *double heterostructure lasers with* 30 °C *half-lives exceeding* 1000 h, Appl. Phys. Lett., Vol. 23 (August 1973), pp. 181-183.

[22] A. R. Goodwin, J. R. Peters, M. Pion and W. O. Bourne: GaAs *lasers with consistently low degradation rates at room temperature*, Appl. Phys. Lett., Vol. 30, No. 2 (1977), pp. 110-113.

[23] C. A. Burrus and B. I. Miller: *Small area, double heterostructure aluminum gallium arsenide electroluminescent diode sources for optical fibre transmission lines*, Opt. Commun., Vol. 4, No. 12 (1971), pp. 307-309.

[24] M. Horiguchi and H. Osanai: *Spectral losses of low - OH - content optical fibres*, Electron. Lett., Vol. 12, No. 12 (1976), pp. 310-312.

[25] T. Miyashita, T. Miya and M. Nakahara: *An ultimate low loss single mode fiber at* 1.55 μm, Optical Fiber Communication, Washington D.C., (March 6-8, 1979).

[26] D. N. Payne and W. A. Gambling: *Zero material dispersion in optical fibres*, Electron. Lett., Vol. 11, No. 8 (1975), pp. 176-178.

[27] D. N. Payne and A. H. Hartog: *Determination of the wavelength of zero material dispersion in optical fibres by pulse-delay measurements*, Electron. Lett., Vol. 13, No. 21 (1977), pp. 627-628.

[28] L. G. Cohen and Chinlon Lin: *Pulse delay measurements in the zero material dispersion wavelength region for optical fibres*, Appl. Opt., Vol. 16, No. 12 (1977), pp. 3136-3140.

[29] J. J. Hsieh: *Room temperature operation of* GaInAsP/InP *double heterostructure diode laser emitting at* 1.1 μm, Appl. Phys. Lett., Vol. 28, No. 5 (1976), pp. 283-285.

[30] R. E. Nahory et al.: *Continuous operation of* 1.0 μm *wavelength* $GaAs_{1-x}Sb_x/Al_yGa_{1-y}As_{1-x}Sb_x$ *double-heterostructure injection lasers at room temperature*, Appl. Phys. Lett., Vol. 28, No. 1 (1976), pp. 19-23.

[31] C. J. Nuese, G. H. Olsen and M. Ettemberg: *Vapor-grown cw room-temperature* $GaAs/In_yGa_{1-y}P$ *lasers*, Appl. Phys. Lett., Vol. 29, No. 1 (1976), pp. 54-56.

[32] C. C. Shen, J. J. Hsieh and T. A. Lind: 1500 h *continuous cw operation of double-heterostructure* GaInAsP/InP *lasers*, Appl. Phys. Lett., Vol. 30 (April 1977), pp. 353-354.

[33] A. G. Dentai, T. P. Lee, G. A. Burrus and E. Buehler: *Small area high radiance* InGaAsP *cw LED's emitting at* 1.30 μm, Electron. Lett., Vol. 13 (August 1977), pp. 484-485.

[34] T. P. Pearsall, R. E. Nahory and M. A. Pollack: *Impact ionization rates for electrons and holes in* $GaAs_{1-x}Sb_x$ *alloys*, Appl. Phys. Lett., Vol. 28, No. 7 (1976), pp. 403-405.

[35] T. P. Pearsall, R. E. Nahory and M. A. Pollack: *Impact ionization coefficients for electrons and holes in* $In_{0.14}Ga_{0.86}As$, Appl. Phys. Lett., Vol. 27, No. 6 (1975), pp. 330-332.

[36] G. E. Stillman, C. M. Wolfe, A. G. Foyt and W. T. Lindley: *Schottky barrier* $In_xGa_{1-x}As$ *alloy avalanche photodiodes for* 1.06 μm, Appl. Phys. Lett., Vol. 24, No. 1 (1974), pp. 8-10.

[37] R. L. Gallawa, J. E. Midwinter and S. Shimada: *Survey of world-wide optical waveguide systems*, Optical Fiber Communication, Washington D.C., (March 6-8, 1979).

[38] T. Ito et al.: *Transmission experiments in the* 1.2-1.6 μm *wavelength region using graded-index optical-fiber cables*, Ibidem.

[39] M. Kawahata and M. R. Finley: *Optical visual information systems - some first results*, Ibidem.

[40] B. Catania: *Optical fibre communication through time and space*, Keynote address at Fourth European Conference on Optical Communication, Genoa, (September 12-15, 1978), Supplement to Conference Proceedings, pp. 1-9.

[41] D. A. Pinnow, A. L. Gentile, A. G. Standlee, A. J. Timper and L. M. Hobrock: *Polycrystalline fiber optical waveguides for infrared transmission*, Appl. Phys. Lett., Vol. 33, No. 1 (1978), pp. 28-29.

[42] L. G. Van Uitert and S. H. Wemple: $ZnCl_2$ *glass: a potential ultralow-loss optical fiber material*, Ibidem, pp. 57-59.

PART I
THE TRANSMISSION MEDIUM

CHAPTER 1

THE OPTICAL FIBRE

by Bruno Costa

Optical fibres are dielectric waveguide structures that are used to confine and guide the light. This chapter is a résumé of the main characteristics of optical fibres intended for use in telecommunications. Other chapters discuss the theory of optical fibres, their fabrication and characterization techniques.

1.1 Fibre types

Optical fibres essentially consist of an inner dielectric material called « core », surrounded by another dielectric with smaller refractive index referred to as « cladding ». As regards the geometry, by far the most common and, in practice, the only fibre type currently adopted is the cylindrical circular cross-section fibre. In fact, the associated circular symmetry results in ease of fabrication and handling and allows simpler theoretical treatment. Proposals for triangular-cored fibres [1, 2] have never found practical applications.

In Fig. 1 the cross-section and typical refractive index profile of various proposed optical fibre types are shown. Typical dimensions of the most common fibres are also indicated. Figure 1*a* represents the simplest waveguide structure, consisting of a dielectric rod; the cladding is air. Since the change in refractive index is abrupt, such a fibre is referred to as a « step index » fibre. Suitable materials for the rod are dielectrics transparent in the visible and near infrared region of the electromagnetic spectrum (0.5-1.5 μm wavelength) so that the choice is restricted to glasses or fused silica. Plastic may also be used, but, in general, higher losses result.

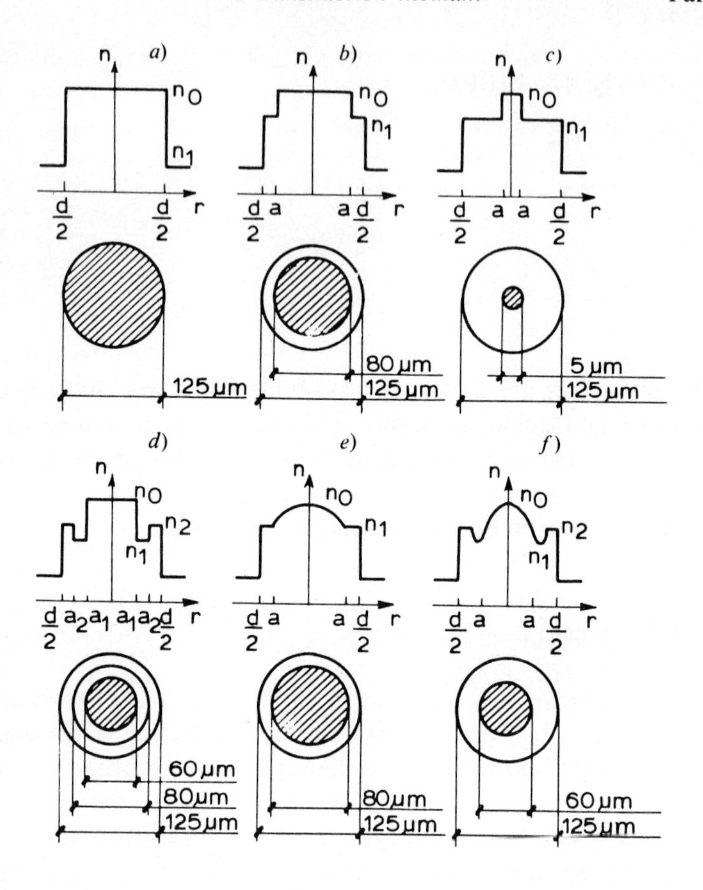

The field of the guided waves in the structure shown extends to the surrounding medium in the form of an evanescent wave (see Section 2.7.2). Therefore, any support for the fibre or surface contamination perturbs light propagation and leads to intolerably high losses, making the use of such a structure very impractical.

For practical use the waveguide configuration in Fig. 1*b* has been adopted. The fibre core, of refractive index n_0, is surrounded by a cladding of index n_1 made of a material with suitably low attenuation. The evanescent field in the cladding can therefore be trasmitted with a minimum of perturbation; by proper choice of cladding thickness, the field at the outer surface becomes negligible, and the fibre can be supported and handled without degrading its transmission characteristics. A residual problem was once constituted by the core-cladding interface, that was subject to contamination and fabrication irregularities resulting in a high loss for the power travelling at the boundary region. However, modern fabrication techniques, such as Chemical Vapour Deposition (CVD, see Section 4.3) have completely resolved the problem. The previously mentioned materials may be employed for the fibre core, provided a material of smaller refractive index can be found for the cladding. This is not an easy task in the case of fused silica, which, being the purest obtainable material and therefore that having the lowest loss, is a very attractive fibre core constituent. The problem has been overcome by using, as a core material, fused silica doped with suitable oxides (GeO_2, P_2O_5) which lead to increased refractive index without degrading the transmission quality. A different approach consists in using high purity fused silica rods as the core, and transparent plastic (such as hexafluoropropane and vinyl fluoride copolymer [3], perfluoronated ethylene-propylene (FEP) [4], or silicone [5]) for the cladding. Ease of fabrication and low cost are attractive features of this solution. A different concept, which is now of historical interest only, is to use a liquid as a core inside a capillary glass tube. Although never very practical, the fabrication of such a fibre [6-8] was very important, as it demonstrated the possibility of achieving very low losses (see « Historical Remarks »).

The most important parameters for specifying optical fibre properties are: core radius, a, the numerical aperture, briefly referred to as NA, which will be herein defined as $\Delta = \sqrt{n_0^2 - n_1^2}$ (n_0 = core index, n_1 = cladding index), and is related to the maximum acceptance angle for rays entering the fibre, and a characteristic parameter, called V, and defined as $V = (2\pi a/\lambda)\sqrt{n_0^2 - n_1^2}$ (λ being the light wavelength in vacuo).

The step index structures described support, in general, many propagating modes (Section 2.7), their number being approximately equal to $V^2/2$ and therefore proportional to a^2 and to the numerical aperture squared. As each mode travels with a different speed, a high number of modes results in large signal distortion and therefore severe bandwidth limitation. This is a drawback of the step index fibre in the context of telecommunication systems.

Practical fibres have therefore $\Delta \ll 1$ (typically $\Delta \leqslant 0.2$). Also, if the fibre core radius a is reduced, the fibre shown in Fig. 1*c* results, which propagates a single mode. This kind of fibre provides the ultimate limit of bandwidth achievable in dielectric waveguides. Although favourable from the viewpoint of bandwidth, a very small refractive index difference (and core radius) leads to poor light gathering efficiency and optical power confinement. This is the reason for the proposal of the waveguide structures shown in Fig. 1*d*. They are called « *W*-type fibres » with reference to their index profile shape [9-12]. The core is surrounded by a double cladding, the inner cladding have a refractive index n_1 smaller than the refractive index, n_2, of the outer cladding.

It can be shown that in such a waveguide the bandwidth is almost equal to that of a singly clad fibre with refractive index difference $n_0 - n_2$, but the power confinement is much better thanks to the presence of the intermediate layer.

The monomode version of this fibre has the advantage of a core radius appreciably larger than that of a singly clad fibre with the same core and outer cladding refractive index. It is also claimed that the waveguide dispersion characteristics of such a fibre can be adjusted so as to compensate for material dispersion (see below, Section 1.2.2).

The impairing effect of differential mode delay on bandwidth can be greatly reduced by suitably varying the refractive index in the fibre cross-section, so as to develop a nearly parabolic index profile with a maximum on the fibre axis. In this way, all modes have nearly the same group velocity. Such a fibre is shown in Fig. 1*e*. It was first developed by an ion exchange technique [13], but now the CVD technique allows a better control of index profile shape, which is obtained in the desired form by changing the dopant level as a function of radius.

A particular class of index profiles which is of great interest for its simplicity and ability to faithfully describe currently produced optical waveguides is given by the following relation:

$$n^2(r) = n^2(0) \left[1 - \Delta_1 \left(\frac{r}{a} \right)^{\alpha} \right] \tag{1}$$

where $\Delta_1 = \Delta^2/n^2(0)$, r is a radial coordinate with its origin at the core centre. The value of $\alpha = 2$ corresponds to the parabolic profile and leads to very small time dispersion.

A possible improvement should be achievable by the fibre type shown in Fig. 1*g* [14, 15], which is a kind of *W*-type fibre with a graded index rather than a step index profile.

The last two fibre types are the most widely produced at the present stage, and appear to be the main contenders for use in long-distance, high bit rate applications. For particular, less demanding applications, step index fibres are also of interest.

1.2 Fibre transmission properties

The transmission properties which are of chief interest for telecommunication use are attenuation and dispersion. A theoretical analysis of these subjects is given in Chapter 2, while the experimental determination of corresponding parameters is discussed in Chapter 3.

In the following sections, we shall review the main factors influencing attenuation and dispersion, with emphasis on the best state-of-the-art results and a look at possible future improvements.

1.2.1 *Attenuation*

There are several mechanisms that lead to transmission losses in fibre waveguides. These are: 1) material absorption, 2) material scattering, 3) waveguide scattering, and 4) leaky modes and fibre design.

Besides these, which can be termed intrinsic losses, factors external to the fibre can also contribute to the total loss. Mechanical deformations, which cause bending and microbending will be discussed, as well as radiation environment.

a) Intrinsic factors

Material absorption. Pure glasses and fused silica, which at the present stage appear to be the most suitable materials for fibre fabrication, exhibit two main intrinsic absorption mechanisms for frequencies in the optical and near optical range. A fundamental absorption band occurs at ultraviolet wavelengths, due to stimulation of electron transitions. At lower energies interactions of photons with molecular vibrations produce absorption bands at wavelengths above 6 μm. In Fig. 2 the absorption spectra of fused

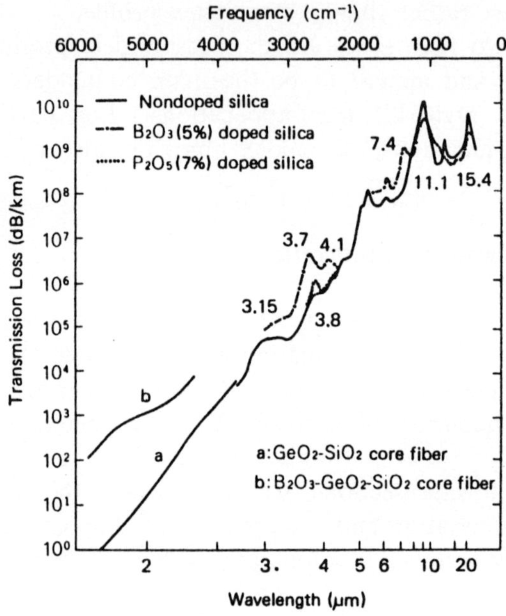

FIG. 2 – Absorption spectra for nondoped fused silica, doped silicas and fibres. (After [17]).

silica, and doped silica used for fibre core, are shown in the infrared region [16]. It has been pointed out [17] that the absorption characteristics in this region are markedly influenced by the kind of dopant. Between these two regions a considerable wavelength interval exists, in the visible and near infrared range, where pure material should not have intrinsic absorption. It has been, however, emphasized that local electric field on a microscopic scale induces broadening on excitation electronic levels, producing a tail on the ultraviolet absorption band [18]. The behaviour of the corresponding loss in a spectral region from about 0.2 to 1.2 μm can be described by an exponential

$$\alpha_{abs} \simeq \exp\left[-\frac{E-E_g}{\Delta E}\right] \tag{2}$$

where α_{abs} is the loss coefficient, E the photon energy, E_g the energy gap in the material, and ΔE is a parameter which is constant for each material. It has been shown [19], on the basis of the analysis of low loss fibres data, that Eq. (2) can be written as a function of wavelength:

$$\alpha_{abs} = c \times 10^{1/\lambda} \tag{3}$$

where c is a constant to be determined and λ the wavelength in μm. In [19] c has been evaluated to be 0.049 for pure fused silica. The corresponding attenuation curve is shown in Fig. 3 (the figure also includes the intrinsic scattering loss, which will be discussed later).

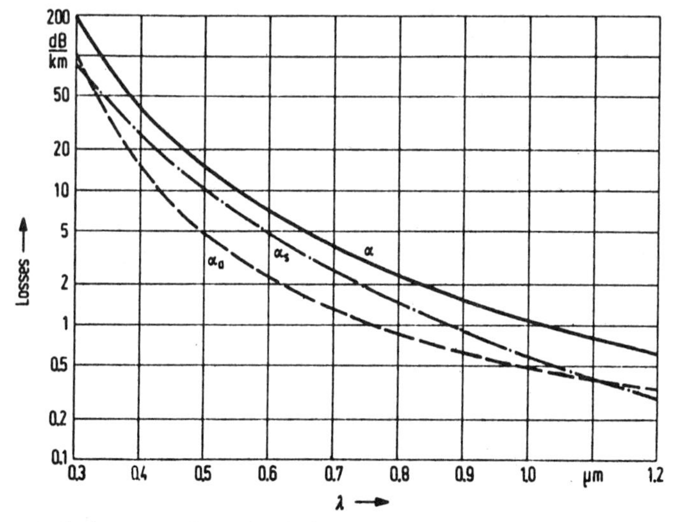

FIG. 3 – Intrinsic attenuation α, intrinsic scattering α_s, and intrinsic absorption α_a of pure silica against wavelength, calculated from Eq. (3). (After [19]).

The effect of the addition of dopant on intrinsic absorption in the shorter wavelength range has not been investigated in detail. Experimental results seem to show that the addition of dopant materials has very little influence on total loss, as is demonstrated by the attainment, in both P_2O_5 and GeO_2 doped silica fibres [16], of losses very close to or even lower than those theoretically predicted for pure fused silica. This is clearly visible in Fig. 4,

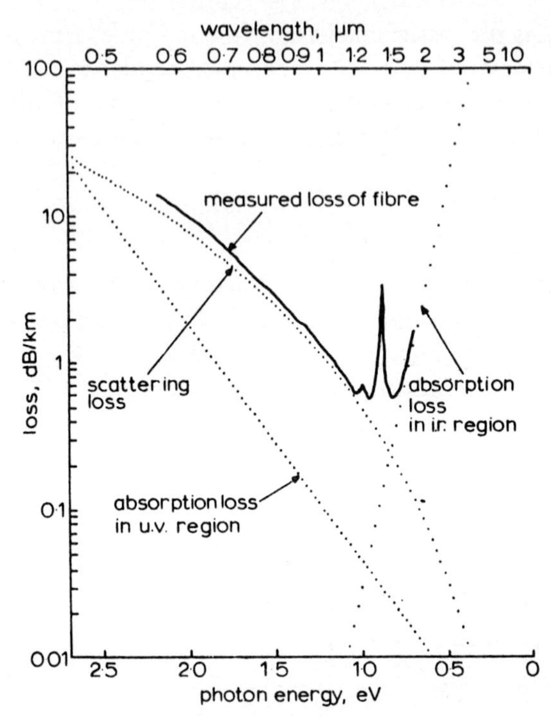

FIG. 4 – Spectral loss curve of germania-doped silica glass core fibre separated into inherent u.v. absorption loss, scattering loss and inherent i.r. absorption loss. (After [16]).

where the attenuation curve of a very low loss germania doped optical fibre is shown, with the estimated scattering and absorption components [16].

From Figs. 3, 4 it appears that the ultraviolet absorption tail contributes for only a small amount of the total attenuation, particularly in the red and infrared region. The main source of absorption loss in this spectral range is due to the presence of impurities in the form of metallic ions with electronic transitions in the 0.5-1 μm region. These are transition metals of the first series (Sc, Ti, V, Cr, Mn, Fe, Co, Ni, Cu). The corresponding absorption bands show a broad peak, whose height depends upon the impurity concentration and state of oxidation. In Fig. 5 absorption spectra for Cr^{2+}, Cu^{2+} and Fe^{3+} are shown [20]. As the peak absorption wavelength and width of the band depend on the glass composition, the represented curves are indicative.

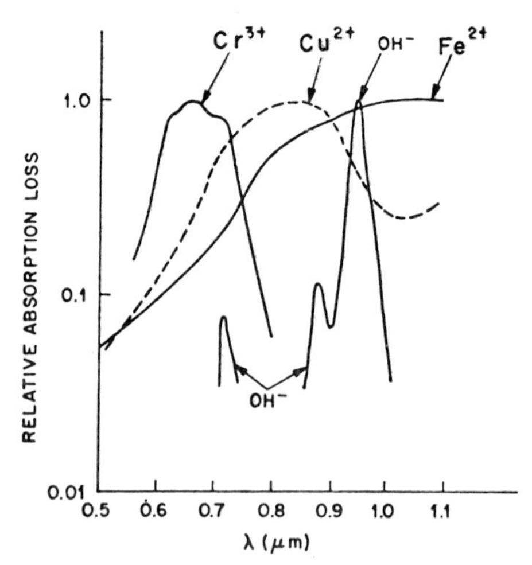

FIG. 5 – Relative absorption loss versus wavelength for certain ions in glass. (After [20]).

The different ion effect in different glasses can be seen in Table I of Section 4.2.3. Here, in Table I, the concentration of the metal ions of Fig. 5 needed for 1 dB/km attenuation is reported. It appears that extremely pure materials are required in order to keep fibre losses at a tolerable level. This means that an impurity control as good as that achieved in semiconductor technology must be obtained.

TABLE I

Ion	Concentration for 1 dB/km Peak Loss in Glass (fractions by weight)
OH^-	1.25 parts in 10^6
Cu^{2+}	2.5 parts in 10^9
Fe^{2+}	1 part in 10^9
Cr^{3+}	1 part in 10^9

This is the reason why low loss fibre waveguides have been consistently produced only after the adoption of such techniques as chemical vapour deposition, derived from semiconductor manufacturing technology.

In addition to transition metal ions, two other absorption causes are present in fibres. One is the vibration absorption associated with hydroxyl ions, OH^-. The fundamental vibration occurs at 2.73 μm with overtones at 1.37, 0.95, 0.72 μm (it is common for overtones not to be exact harmonics of the fundamental) and smaller bands corresponding to the combination frequencies from the various overtones of the hydroxyl ion vibration and the vibration of fused silica [21]. In Figs. 6*a* [22], 6*b* [17] and Table I the absorption spectra of OH^- and the concentration for 1 dB/km loss at 950 nm are shown.

The second effect is absorption arising from defects in the material. These defects are usually produced by high energy irradiation of the glass [23], but may also be induced by the fibre drawing [24]. The wavelength of peak absorption is around 630 nm.

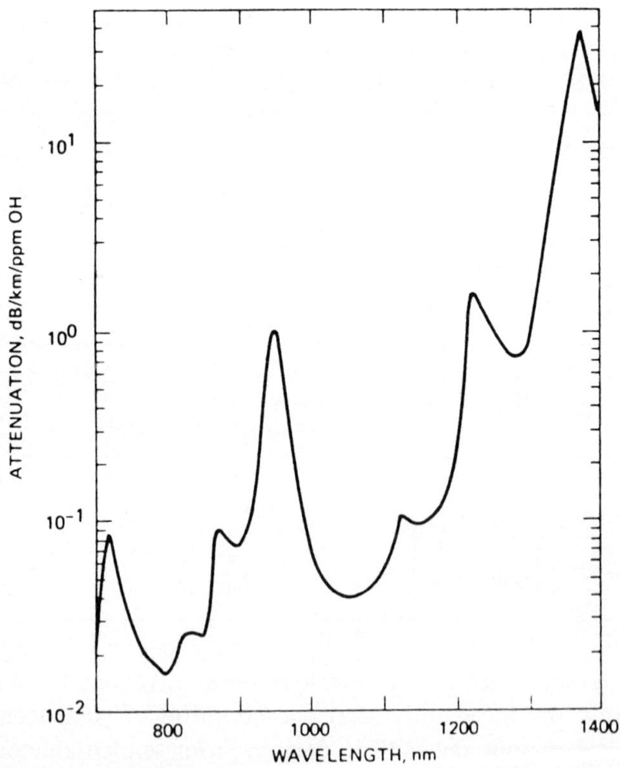

FIG. 6*a*) – Spectral attenuation in db/km/ppm of OH in fused silica. (After [22]).

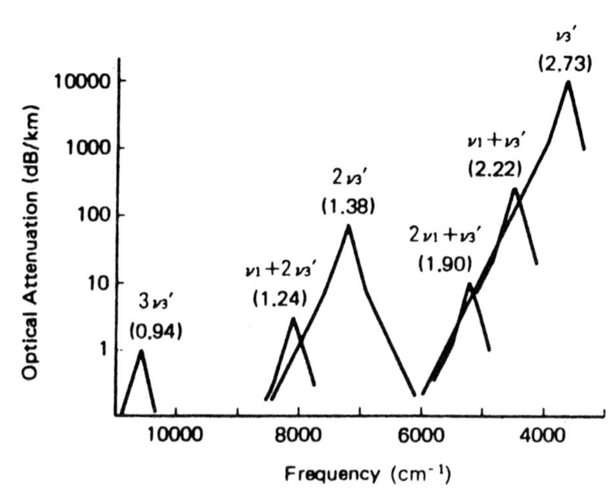

FIG. 6*b*) – Infrared absorption due to water (1 ppm) in silica glass. (After [17]).

It has also been shown that such a defect arises in silica fibres with low OH– content, but is absent in silica fibres with high OH– content.

Material scattering. There is a series of scattering mechanisms contributing to loss which arise from the material properties. They include Rayleigh scattering, Mie scattering, stimulated Raman scattering, and stimulated Brillouin scattering.

Rayleigh scattering is caused by variations in refractive index which occur over distances which are small compared with the light wavelength. Such variations may be generated by thermal fluctuations, compositional fluctuations, phase separations, etc.

Rayleigh scattering is believed to be the fundamental lower limit of attenuation in glass. If only thermally induced fluctuations are considered, the corresponding loss coefficient, α_s, is given by the following relation:

$$\alpha_s = \frac{8}{3}\frac{\pi^3}{\lambda^4} n^8 p^2 kT\beta_T \tag{4}$$

where λ is the optical wavelength, n is the index of refraction, p the photoelastic coefficient, β_T the isothermal compressibility of the material, k is Boltzmann's constant, T the absolute temperature. It must be stressed that, while for liquids T is the ambient temperature, in glasses the random structure is not determined by ambient temperature, but by its « fictive temperature » [18] which is the temperature at which the glass, if heated, would come into thermodynamic equilibrium, a condition for (4) to hold. This temperature is related to the glass softening point.

A very interesting feature of (4) is the λ^{-4} dependence of the attenuation coefficient, which means that Rayleigh scattering decreases rapidly toward longer wavelengths. In Fig. 7 the loss due

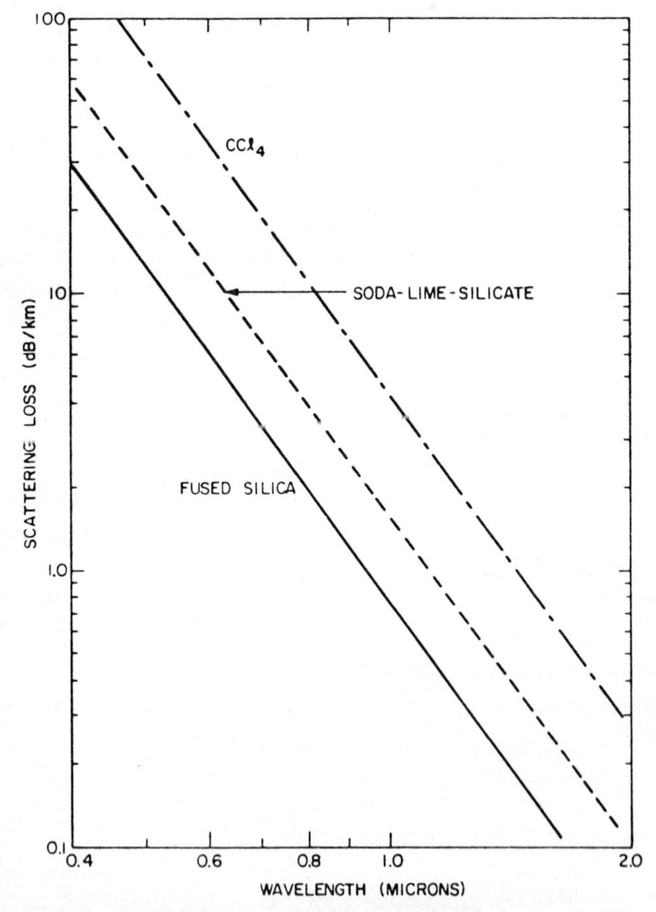

FIG. 7 – Intrinsic scattering loss versus optical wavelength for liquid CCl_4 and vitreous fused silica and soda-lime-silicate. (After [18]).

to Rayleigh scattering is shown [18]. It agrees with curves reported in Fig. 3 [19] for fused silica.

The addition of a dopant affects intrinsic scattering in two ways. On one hand, the fictive temperature is decreased, which tends to lower the scattering loss; on the other hand, variations of dopant concentration constitute an additional scattering source, still following a λ^{-4} law.

According to existing data it seems that phosphorus doping does not greatly affect intrinsic scattering loss, whereas addition of GeO_2 leads to a certain amount of loss increase (a few tens per cent).

According to the previous discussion of Eq. (4), glasses with a lower softening point than fused silica should have lower Rayleigh scattering loss and therefore lower ultimate attenuation limits than fused silica fibres. Due to the more difficult impurity level control in their fabrication, absorption losses in such materials are at present too high to allow exploitation of their smaller intrinsic loss. In addition glasses suffer from compositional variations contributing to Rayleigh scattering.

Mie scattering is caused by unhomogeneities comparable in size to the wavelength, resulting in predominantly forward scatter.

Stimulated Raman and Brillouin scattering are non-linear effects, arising when a threshold power-density level is reached. As a result a wavelength shift is caused by the non-linear interaction between a travelling wave and the material. These effects pose an upper limit to the power level that can be transmitted in a fibre.

Waveguide scattering. Variations in core diameter and peak index may cause transfer of energy from one mode to another. If the transfer is to radiation modes a signal loss results.

Such a loss may be particularly severe if the spatial frequency of geometrical perturbation corresponds to the coupling length between modes [25]. Fortunately experience has shown that the fabrication techniques actually employed do not lead to the most critical correlation lengths, so that a good diameter control ($\sim 1\%$) is enough to assure negligible waveguide induced scattering losses.

Leaky modes and waveguide design. In an optical fibre, modes can be excited which are neither refracted nor completely guided. These are referred to as leaky modes (see Section 2.1.1). These modes radiate according to a phenomenon similar to tunnelling through a potential barrier in quantum mechanical systems; the loss can be very small, but also very high for strongly tunnelling modes. This fact can lead to a correspondingly higher attenuation of the optical waveguide. In practice, however, leaky modes constitute only a few per cent of propagation modes.

A different effect may be produced by the fibre design. As already mentioned the power in a fibre is not totally confined within the core, but a certain fraction travels in the cladding. This fraction is attenuated according to the cladding loss rather than core loss. This means that the fibre must be designed, so that a small amount of power is propagated in the cladding or a cladding material is to be used with losses comparable to those of the core. If the field extends to the cladding-outer medium boundary, propagation can be adversely affected by the presence of lossy coatings, usually applied to strengthen the fibre and to minimize cross-talk between adjacent fibres.

b) External factors

– *Mechanical deformation.* Practical use of the optical fibre makes bending unavoidable, and this in turn, induces radiation loss. Bound energy may radiate; this, in the framework of geometric optics, may be interpreted as refraction of rays which, due to the curvature, are no longer under conditions of total internal reflection. Theoretical calculations have been performed [26, 27] which show that the attenuation coefficient depends exponentially on the radius of curvature (see also Chapter 2). This means that an increase of a factor of two of the radius around a critical value can change the loss from negligible to prohibitive values.

When fibres are cabled or wrapped on a drum, small irregularities in the supporting surfaces cause what is usually called « microbending » which may be a serious source of loss through the mechanism just described. In order to minimize this effect a large NA and/or large cladding-core diameter ratio is required [28]. In addition fibre coatings and cable buffering must be carefully chosen [29].

– *Radiation effects.* Optical waveguide attenuation may be strongly affected by exposure to nuclear radiation. This is due essentially to ionizing effects which cause light-absorbing defect centres in the material [30]. A compendium of the radiation-induced loss in several types of fibres, irradiated with different cobalt 60 doses, is shown in Fig. 8, taken from ref. [31]. Shown

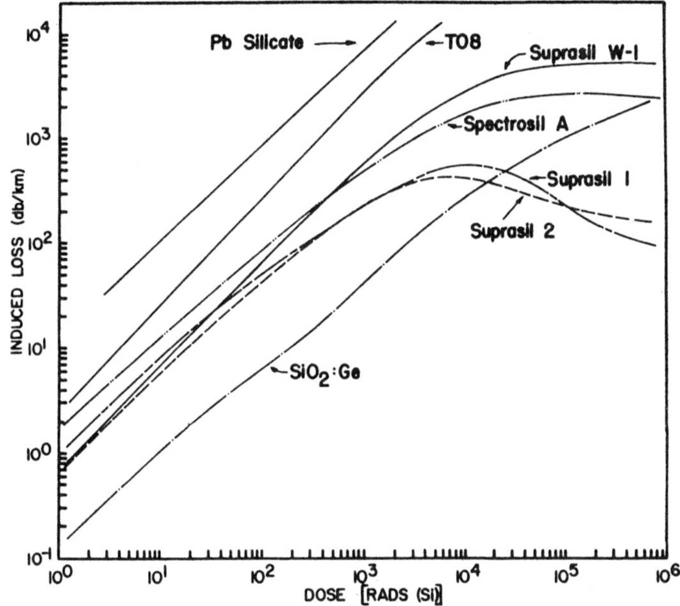

FIG. 8 – Radiation-induced loss versus dose at 820 nm during ^{60}Co irradiation. (After [31]).

are: a compound glass fibre (Pb silicate); plastic clad fibres with different core materials (TO8 is quartz; Suprasil W-1, Spectrosil A, Suprasil 1, and Suprasil 2 are high purity synthetic fused silica); and germanium doped silica core fibres (phosphorus doped silica and germanium doped phosphosilicate core have very similar characteristics). Many interesting observations can be made on the basis of the data shown. First there is a correlation between the level of radiation induced loss and the intrinsic material loss. Fibres with higher intrinsic loss have a higher radiation sensitivity, which is probably due to the higher impurity concentration. Second, polymer coated fused silica fibres show a saturating behaviour such that after a certain dose their response becomes extremely non linear. Suprasil W-1 and Spectrosil A have a typical saturation region after 10^4 rad. In the case of Suprasil 1 and 2 a loss peak is reached, after which induced loss decreases with a further dose. This fact has previously led to erroneous conclusions, because data obtained at high dose levels were extrapolated back to lower doses, obtaining for polymer coated fibres radiation resistances much higher than they actually were.

The other fibres show a fairly linear behaviour along the whole dose range examined.

Another point of interest is the wavelength dependence of radiation damage. Considering a germanium doped silica fibre, codoped with boron and phosphorus, it turns out that there is a minimum of induced loss at 1.05 μm wavelength, as shown in Fig. 9 [32]. According to [30] the damage at short wavelengths

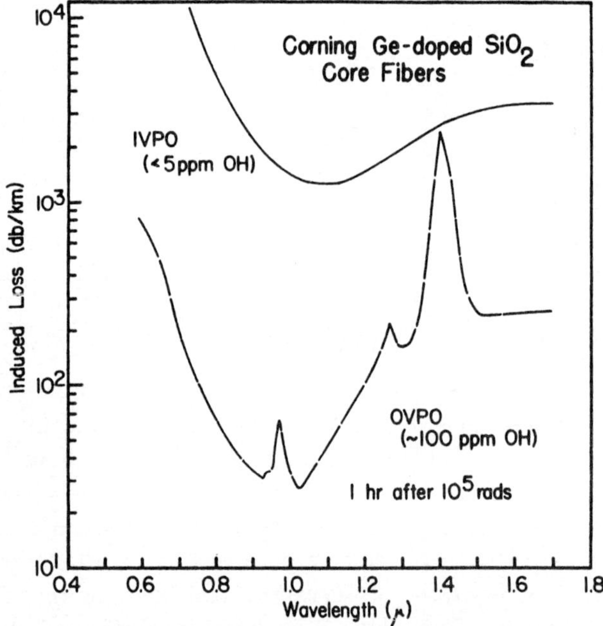

FIG. 9 – Radiation-induced optical absorption spectra. Both fibre cores are codoped with *B*; the IVPO fibre contains *P* as well. (IVPO and OVPO are different CVD techniques (see chapter 4)). (After [32]).

is dominated by an intense absorption in the UV, with a tail extending into infrared. The increase after 1.05 suggests a broad band absorption above 1.7 μm, possibly caused by a change in the glass structure which results in an induced vibrational spectrum. In addition, fibres with low OH^- content are much more sensitive to radiation than fibres with higher OH^- content. When fibres with pure silica core and borosilicate cladding are considered this behaviour reverses at longer wavelengths, so that the damage in dry silica fibres is less than in wet fibres near 1.3 μm, as shown in Fig. 10. Moreover, pure silica core fibres show substantially less damage than doped silica fibres.

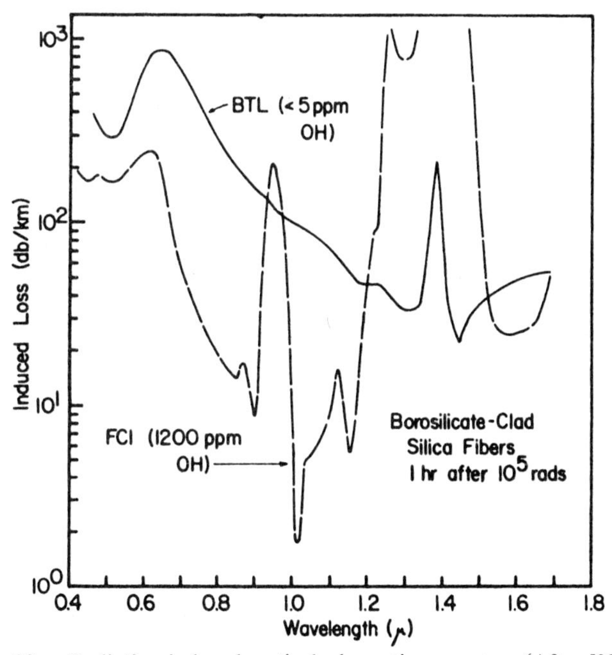

FIG. 10 – Radiation-induced optical absorption spectra. (After [32]).

Finally, the radiation induced-loss is not a static phenomenon, but has a certain time evolution. Time resolved measurements with pulsed electron irradiation at 3700 rad dose have shown transient absorption several thousandths of dB/km high in SiO_2:GeO_2 core fibres due to a broad band absorption in the u.v. Fig. 11 shows the induced absorption as a function of wavelength [32] which illustrates the permanent damage.

After irradiation, a certain amount of recovery is observed.

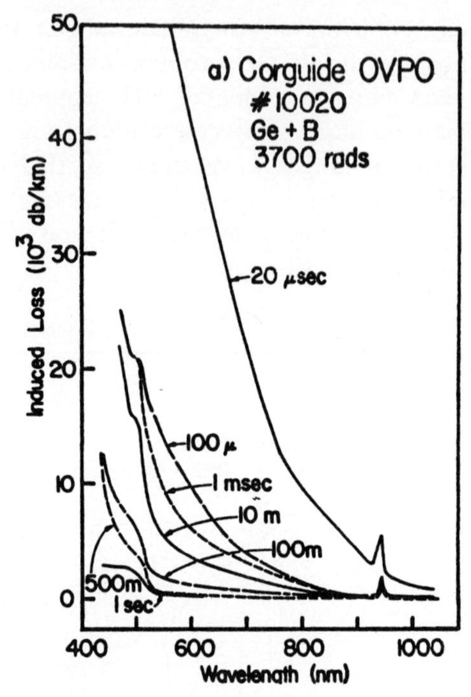

FIG. 11 – Radiation-induced optical absorption spectra as a function of time following a pulsed electron irradiation. Times denoted by « m » indicates millisec. (After [32]).

The extent of the recovery seems to depend upon the purity of the silica; it can be seen from Fig. 12 [31] that the TO8 fibre has a loss substantially higher than that of synthetic silica, and little recovery is also shown by the Ge-doped silica fibre. Two months after irradiation the situation is as shown in Fig. 13.

As a conclusion, fibres which are likely to be used in a nuclear environment should be carefully chosen in order to minimize radiation induced effects, also taking into account the operating wavelength range.

State-of-the-art

Doped silica fibres obtained by the CVD technique have shown the lowest attained losses of any other kind of fibre. The control on the described loss sources may be so good as to produce the results shown in Fig. 14. In it are represented three

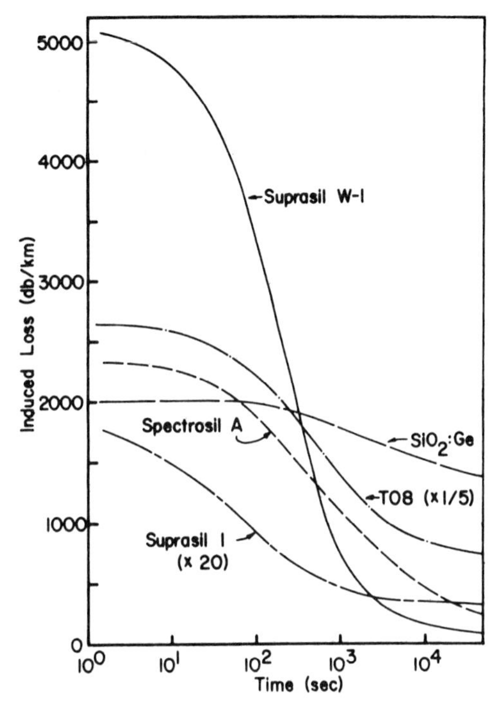

FIG. 12 – Decay of the radiation-induced optical loss at 820 nm following ^{60}Co irradiation. (After [31]).

fibres of different compositions, which are the most widely used compositions at present, with the lowest reported losses. The best result is a 0.46 dB/km total attenuation at a wavelength of 1.55 μm for the germania-doped silica fibre. The phosphorus doped fibre has a minimum of 0.47 dB/km at 1.27 μm. It seems that the fundamental limit has nearly been reached in these fibres, where the Rayleigh scattering accounts for a 70-80% of the total loss, as can be seen in Fig. 4. This fact is particularly evident from Fig. 15 [33], in which the attenuation curve is plotted on semi-logarithmic paper. Nearly all the experimental points fall on a λ^{-4} curve, with the exception of a small OH^- peak at 950 nm and absorption due to colour centres toward smaller wavelengths.

Even more strikingly, a 0.2 dB/km loss has been achieved for a monomode fibre at 1.55 μm wavelength, as shown in Fig. 16 [34].

1.2.2 *Fibre information capacity*

The information capacity of a fibre, i.e. the maximum amount of information which can be transmitted per unit time, is limited by signal distortion which results from pulse spreading. Signal distortion is caused by three main factors: modal dispersion, arising from differences among the group velocities of the various modes,

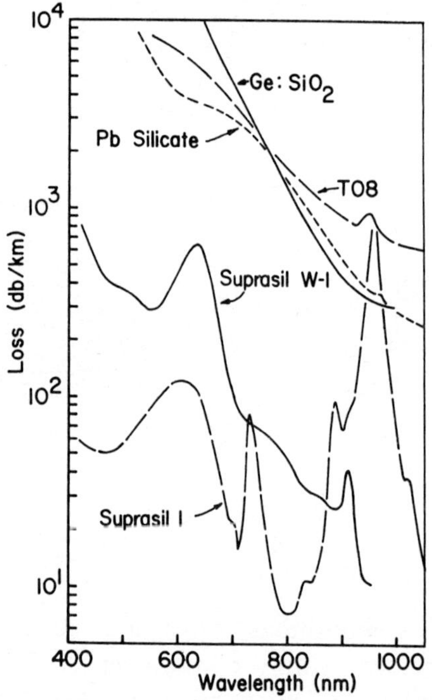

FIG. 13 – Optical spectra of the fibres taken 2 months after irradiation. (After [31]).

material dispersion and waveguide dispersion. In addition, the bandwidth of a certain length of fibre is also determined by the presence of imperfections, or by microbending effects, which cause mode mixing and mode filtering resulting in a sub-linear dependence of bandwidth on length.

A theoretical analysis of these effects is carried out in Chapter 2. Let us summarize here the main results, both theoretical and experimental, related to the most common fibre types.

In a multimode step index fibre the predominant contribution

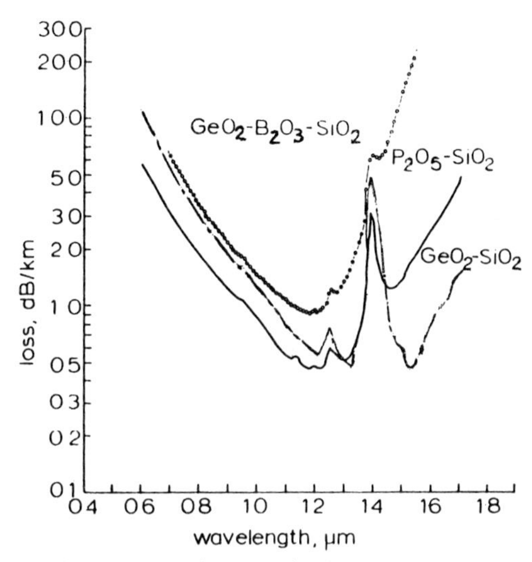

FIG. 14 – Spectral loss curves of germania doped borosilicate glass core fibre, phosphosilicate glass core fibre and germania doped silica glass core fibre whose lengths are 1.2, 1.1 and 0.8 km, respectively. (After [16]).

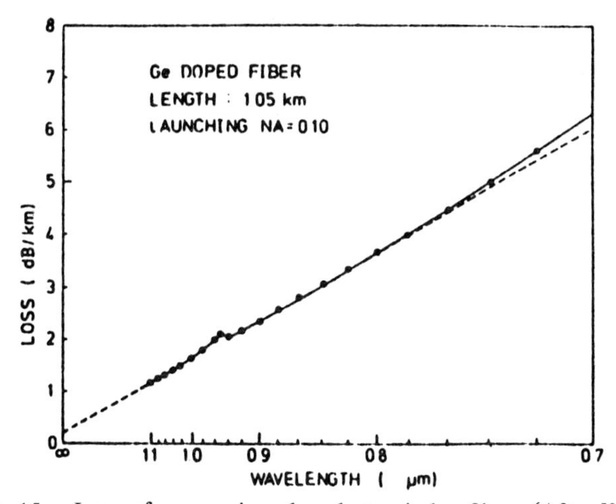

FIG. 15 – Loss of germanium-doped step index fibre. (After [33]).

to signal distortion is given by difference of group delay among the various modes, usually referred to as « modal dispersion ». Pulse broadening is proportional to the squared numerical aperture of the fibre, according to the following equation, which can be directly deduced from ray tracing:

$$\tau_{st} = L \frac{\Delta^2}{2n_0 c} \tag{5}$$

where τ_{st} is the time spread, L the fibre length, c the speed of light in vacuo.

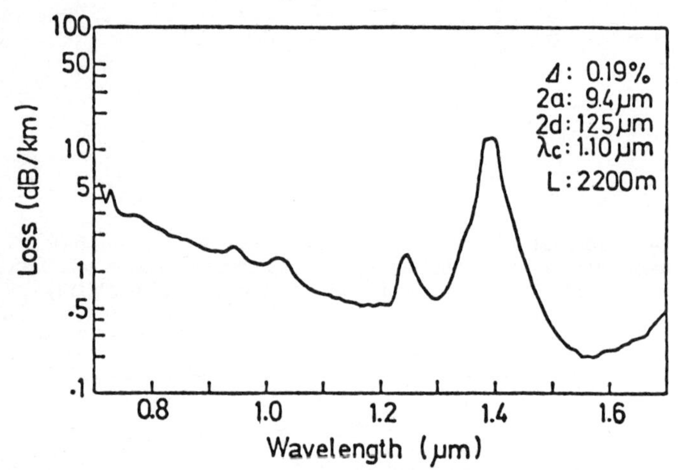

FIG. 16 – Loss spectrum of a single mode fibre core; germanium doped silica cladding. λ_c is cut-off wavelength. (After [34]).

Distortion can therefore be reduced by reducing the core-cladding index difference; this reduction adversely affects coupling efficiency, power confinement, and bending induced losses, so that it is not possible to go beyond certain practical limits. Typical NA's are of the order of 0.2. For a doped silica fibre [$n_0 \simeq 1.47$] this corresponds to a pulse broadening of approximately 48 ns/km. Although this is a worst case estimate, as it does not take into account phenomena like mode mixing and preferential mode attenuation, which tend to increase the bandwidth, it appears that in such fibres a rather severe limitation is imposed on the transmittable information rate.

It has been recognized since 1968 that a proper grading of the refractive index, that is a monotonic decrease of the index from the centre to the core periphery, could greatly enhance the bandwidth, reducing the modal time dispersion.

The underlying mechanism can be schematized as follows, using a ray optics model: rays which travel along the fibre axis have the shortest geometrical path, but traverse a region of high refractive index; rays inclined with respect to the axis have a longer geometrical path but through a less dense medium, i.e. a medium with lower refractive index. Therefore, if the index profile is properly shaped, nearly equal optical paths and, consequently, time delays, can be obtained.

The particular class of index profiles expressed by (1) can well describe many profiles of practical interest, allowing, at the same time, simple analytical treatment. For the parabolic profile ($\alpha=2$) very good equalization is produced and maximum delay spread is given by:

$$\tau_{gr} = L \frac{\Delta^4}{8n_0^3 c}. \tag{6}$$

Comparing (6) with (5) one can easily see that a graded index fibre with parabolic profile produces a time dispersion $4n_0^2/\Delta^2$ times smaller than that of a step fibre with the same NA, i.e. a more than 100 times smaller effect for representative values of n and NA. Since the refractive index is frequency dependent and due to the different dispersive properties of the core and the cladding, the optimum α value is wavelength dependent and is slightly different from 2. (See Section 2.5.1).

In Fig. 17 the effect of fibre index profile on r.m.s. pulse width is shown, considering various sources of different spectral width [35].

This implies that a fibre with profile optimized for a given wavelength is no longer optimized at another wavelength. According to presently available data it seems that germania doped silica fibres are more sensitive to wavelength changes than phosphorous doped fibres. In Fig. 18 α_{opt} as a function of λ is shown for different material compositions [36].

Multiple α profiles have also been suggested for obtaining a large bandwidth over an extended spectral range, thus avoiding the use of special material compositions which may not satisfy other system requirements [37].

Multiple α profiles are obtained by letting different portions of

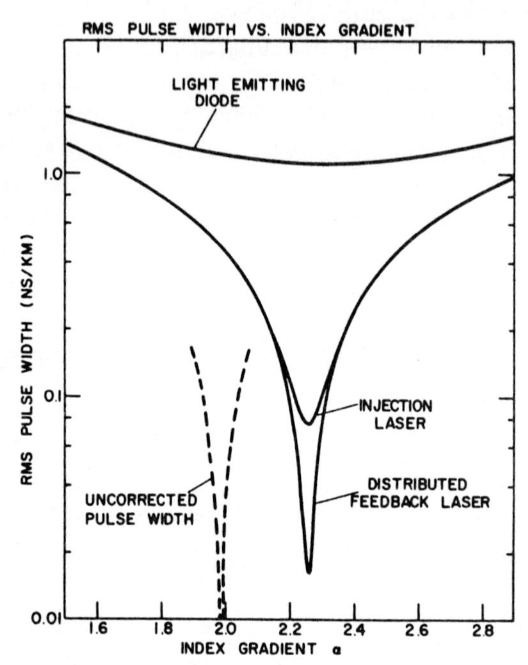

FIG. 17 – Assuming equal power in all modes, the rms pulse width is shown as a function of α for three different sources, all operating at 0.9 μm. The sources are taken to be an LED, a gallium arsenide injection laser, and a distributed feedback laser having rms spectral widths of 150 Å, 10 Å, and 2 Å, respectively. The dashed curve shows the pulse width that would be predicted if all material dispersion effects were neglected. (After [35]).

the refractive index curve as a function of radius be described by different values of α. These additional degrees of freedom can be exploited to obtain low dispersion over an extended range of wavelengths, or at several different wavelengths, as shown in Fig. 19 [38].

Other profiles, different from α-type, have been studied in order to further reduce modal time spread [39, 40]. In particular, nonlinear profile dispersion, i.e. the fact that in principle the dispersive properties of core material vary along a radial coordinate due to the different dopant concentrations, has been considered as an aid for optimization [41]. It has been shown that modal dispersion can be reduced to less than 30 ps/km with an optimized profile [15]. However, this requires control over the 7th decimal digit of refractive index, which appears beyond the capability of present technology. It appears that a more viable way to obtain very high bandwidths is to use monomode fibres.

The kind of dispersion now discussed can be termed « inter-modal » dispersion, as it shows itself among different modes.

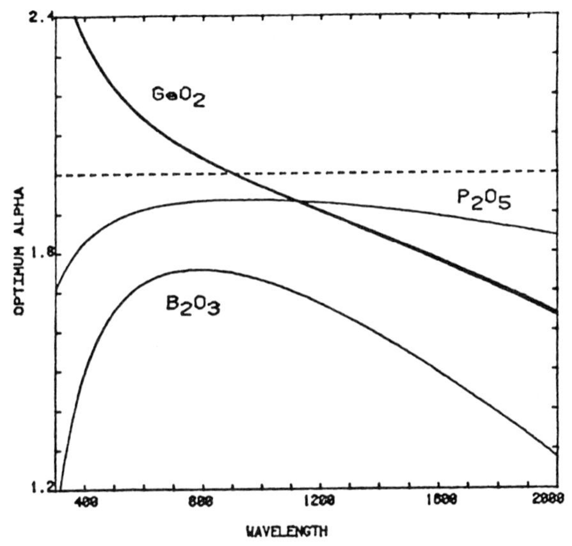

FIG. 18 – Optimum profile for the three dopants shown. The GeO_2 curve comprises results for three GeO_2 concentrations which all lie within the thickness of a line. (After [36]).

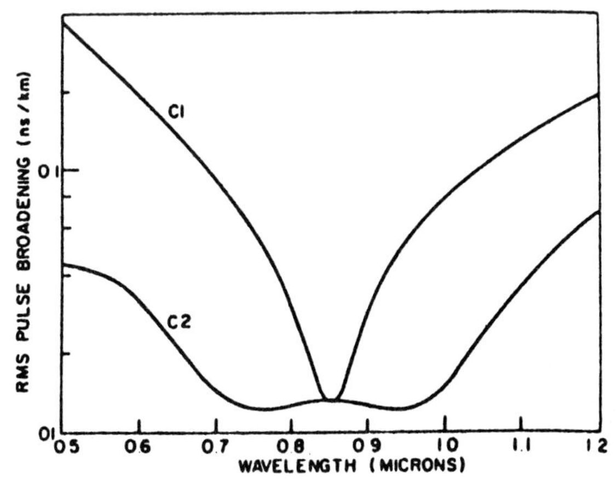

FIG. 19 – Curve C_1 shows the pulse dispersion versus wavelength for a single-α profile designed for minimum dispersion at 0.85 μm. Curve C_2 is for the double-α profile designed for minimal dispersion over a spectral range centered at 0.85 μm. (After [38]).

Waveguide and material dispersion can be termed « intramodal », as they affect each mode. Therefore they are also significant for single mode fibres, and are, in fact, the factors limiting the information carrying capacity of single mode fibres.

Waveguide dispersion [42, 43] is mainly due to the wavelength dependence of the « V » number. It is usually negligible for all modes, except those close to cut-off which, on the other hand, suffer very high losses.

Material dispersion arises in connection with the wavelength dependence of the group velocity of a mode due to the dispersive properties of the material, i.e. the change of its refractive index with frequency. Pulse spreading due to material dispersion can be shown to be given by (see also Section 3.6)

$$\tau = \frac{L}{c} \frac{\Delta\lambda}{\lambda} \lambda^2 \frac{d^2n}{d\lambda^2} \qquad (7)$$

where $\Delta\lambda/\lambda$ is the relative spectral width of the source (assumed $\ll 1$) and $\lambda^2(\mathrm{d}^2n/\mathrm{d}\lambda^2)$ characterizes the dispersion of core material.

It has been shown theoretically [44], and experimentally verified [45], that this term vanishes at a certain wavelength, around $\lambda = 1.27$ μm. The exact wavelength of « zero-material-dispersion » depends on fibre composition, as shown in Fig. 20*a*, *b* [45, 46].

For monomode fibre minimum dispersion is not at the zero material dispersion wavelength, but at a wavelength at which waveguide and material dispersion compensate each other, as shown in Fig. 21 [46].

It must be stressed, however, that these two dispersion terms do not simply add, but interfere in a complicated way [42].

At the wavelength of minimum dispersion the bandwidth of monomode fibres used with narrow bandwidth sources can be extremely high ($>$ 100 GHz/km).

The region of zero material dispersion is also a region of very low loss, due to wavelength dependence of Rayleigh scattering, as already seen. It is therefore envisaged that future long-distance systems will work mainly in the 1.3 μm wavelength region [47].

In addition to these elements, other factors influence the bandwidth of a fibre. They arise mainly from imperfections in the fibre structure, either geometrical or compositional, and from externally induced discontinuities. Such discontinuities can be splices or

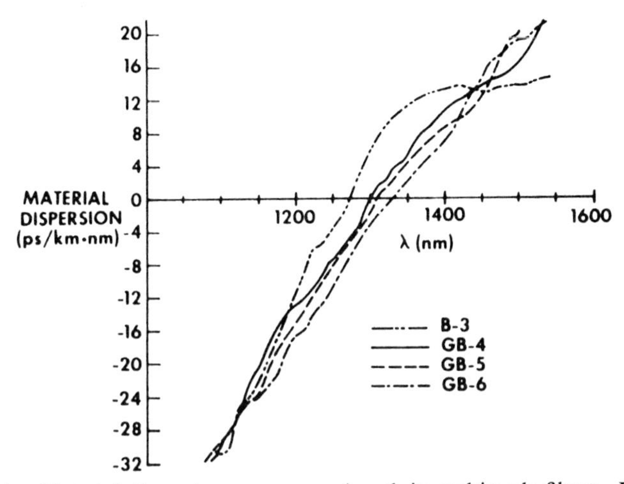

FIG. 20*a*) – Material dispersion versus wavelength in multimode fibres. Fibre B-3 has a graded borosilicate core. The other fibre cores have a small amount of B_2O_3 and a larger graded GeO_2 concentration. The dopant concentrations were increased from fibre GB-4 to GB-6. (After [46]).

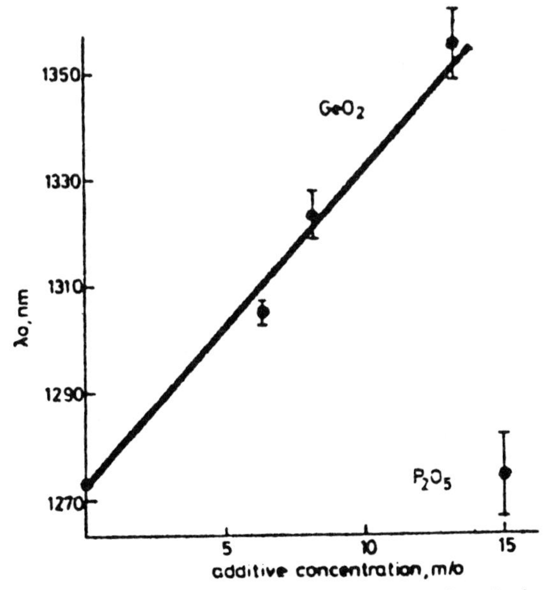

FIG. 20*b*) – Effect of concentration of additive on wavelength λ_0 of zero of material dispersion. The value for O m/o is calculated from the data for pure silica. Note that the 6.4 m/o GeO_2 fibre also contains P_2O_5. (After [45]).

stress-induced microbending, which can arise, for instance, during the cabling process.

All these effects produce coupling, i.e. power exchange among modes, either to radiative modes, determining a loss, or to different guided modes, producing mode mixing. As a consequence, mode dispersion is no longer linearly dependent on length, and it approaches, in extreme cases, a square root length dependence.

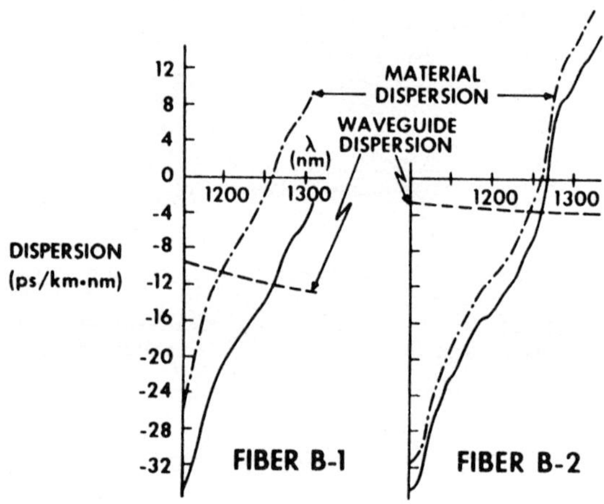

FIG. 21 – Single-mode fibre dispersion (——) versus wavelength. Waveguide dispersion (– – –) is subtracted to obtain material dispersion (–·–·–). Fibre B-1 causes more waveguide dispersion than fibre B-2. (After [46]).

Mode filtering, produced by selective attenuation of certain modes which may be lossy in nature (high order modes, leaky modes) is another factor that tends to increase the bandwidth.

A further recently discovered phenomenon occurs when fibres with overcompensated and undercompensated refractive index profiles are spliced in succession [48]. Overcompensation means that high order modes travel faster than the fundamental, and undercompensation means the opposite. Clearly two such fibres spliced together produce a certain amount of equalization of time delays, further contributing to a bandwidth increase.

As a result of all the factors discussed the prediction of length dependence of bandwidth in optical fibres links is extremely difficult. When modal dispersion is the predominant factor and mode coupling occurs, a square root dependence is often encountered. In the case of well optimized fibres a power law with 0.85 exponent has been measured [49]. In fact, in this case modal dispersion is comparable with material dispersion which, following a strictly linear law [50], will prevail over long distances.

1.3. Mechanical properties

Mechanical properties of optical fibres, such as Young's modulus, breaking strength and extensibility, are very important for a successful application to telecommunication systems. In fact the fibre must possess sufficient strength to withstand the stresses to which it is subjected during the construction of optical cables. In addition, it is desirable that it have a long survival time in an installed system, which requires that multikilometer lengths of fibre survive short-term tensile loading up to 7×10^2 N/mm^2, and sustain lower static loading for a period of 20 years or longer [51].

The intrinsic strength of silica fibres is very high, with theoretical predictions of 1.8×10^4 N mm^{-2} [52] and reported measured strength of 1.4×10^4 N mm^{-2} [53]: these can be found on virgin, as-pulled fibres. Thereafter a rapid degradation occurs, essentially due to surface defects and flaws induced by contact with other solid surfaces, or atmospheric chemical attack. Under tensile loading the stress at the crack tip increases many times over the value of the externally applied stress because of leverage introduced by the long thin crack geometry. As a consequence the bonding strength is exceeded and the crack propagates. This effect is known as stress corrosion, and the stress level at which failure takes place is called the fatigue limit.

Therefore the breaking stress of a fibre depends on a number of variables such as temperature, humidity, glass composition, and thermal history, so that, in effect, one cannot speak of a specific fibre strength, and its determination must be performed on a statistical basis.

In order to obtain high strength fibres three main factors are of importance: *a*) high bulk and surface quality; *b*) clean furnace environment; *c*) high drawing temperature (or low draw force).

Improvement of furnace characteristics has led to the results shown in Fig. 22 [54]. The cumulative percentage of failures on

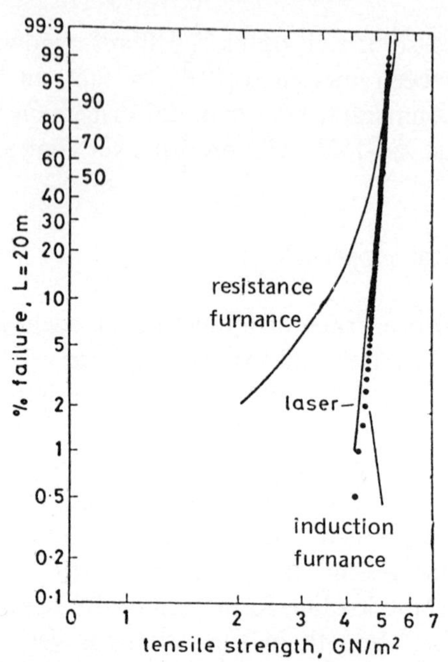

FIG. 22 – Weibull probability of failure for 20 m gauge length fibres using direct tensile measurement with a strain rate of 0.5 min^{-1}. 1 km length drawn with a graphite resistance furnace; 2 km length drawn with a CO_2 laser; 4 km length drawn with a zirconia induction furnace. (After [54]).

fibre samples as a function of tensile strength is shown for fibres drawn with different furnaces or with a CO_2 laser.

Fibre strength can also be improved by applying an outside surface-compression force, for instance in the form of a compressive cladding [55]. In this way no corrosion takes place until tensile forces exceed surface compression.

Further details will be given in Chapter 3. Mechanical properties of optical fibres can be improved by the application of a proper coating just after drawing. This is discussed in Chapter 4.

REFERENCES

[1] R. B. Dyott, C. R. Day and M. C. Brair: *Glass fibre waveguide with a triangular core*, Electron. Lett., Vol. 9, No. 13 (1913), pp. 288-290.

[2] D. R. Tynes: *Partially cladded triangular-cored glass optical fibers and lasers*, J. Opt. Soc. Am., Vol. 64, No. 11 (1974), pp. 1415-1423.

[3] Y. Suzuki and H. Kashiwagi: *Polymer-clad, fused-silica optical fiber*, Appl. Opt., Vol. 13, No. 3 (1974), pp. 1-2.

[4] P. Kaiser, A. C. Hart Jr. and L. L. Blyler Jr.: *Low-loss FEP-clad silica fibers*, Appl. Opt., Vol. 14, No. 1 (1975), pp. 156-162.

[5] S. Tanaka, K. Inada, T. Akimoto and M. Kojima: *Silicone clad fused silica core fiber*, Electron. Lett., Vol. 11, No. 7 (1975).

[6] J. Stone: *Optical transmission in liquid-core quartz fibers*, Appl. Phys. Lett., Vol. 20 (Apr. 1972), pp. 239-240.

[7] G. J. Ogilvie, R. J. Esdaile and G. P. Kidd: *Transmission loss of tetrachloroethylene-filled liquid-core-fibre light guide*, Electron. Lett., Vol. 8, No. 21 (1972), pp. 532-534.

[8] W. A. Gambling, D. N. Payne and H. Matsumura: *Gigahertz bandwidths in multimode, liquid core, optical fibre waveguide*, Opt. Commun., Vol. 6, No. 12 (1976), pp. 317-322.

[9] S. Kawakami and S. Nishida: *Anomalous dispersion of new double clad optical fibre*, Electron. Lett., Vol. 10, No. 4 (1974), pp. 38-40.

[10] S. Kawakami and S. Nishida: *Characteristics of a double clad optical fiber with a low-index inner cladding*, IEEE J. Quantum Electron., Vol. QE-10, No. 12 (1974), pp. 879-887.

[11] K. Furuya and Y. Suematsu: *Mode dependent radiation losses of dielectric waveguides with external higher-index layers*, Electron. and Commun. in Japan, Vol. 57-C, No. 11 (1974), pp. 101-107.

[12] S. Onoda, T. Tanaka and M. Sumi: *W fiber design considerations*, Appl. Opt., Vol. 15, No. 8 (1976), pp. 1930-1935.

[13] T. Uchida, M. Furukawa, I. Kitano, K. Koizumi and H. Hastsumura: *A light focusing fiber guide*, IEEE J. Quantum Electron. (Abstract), Vol. QE-5 (1969), pp. 331.

[14] K. Petermann: *The design of W-fibres with graded index core*, AEÜ, Vol. 29, No. 11 (1975), pp. 485-487.

[15] T. Okoshi: *Optimum profile design of optical fibers and related requirements for profile measurements and control*, 1977 International Conference on Integrated Optics and Optical Fiber Communication, Tokyo (July 18-20, 1977), pp. 391-394.

[16] H. Osanai, T. Shioda, T. Mariyama, S. Araki, M. Horiguchi, T. Izawa and H. Takata: *Effect of dopants on transmission loss of low-OH-content optical fibres*, Electron. Lett., Vol. 12, No. 21 (1976), pp. 549-550.

[17] S. Kobayashi, N. Shibata, S. Shibata and T. Izawa: *Characteristics of optical fibers in the infrared wavelength region*, Rev. Electr. Commun. Lab., Vol. 26, No. 3-4 (1978), pp. 453-467.

[18] D. A. Pinnow, T. C. Rich, F. W. Ostermayer Jr. and M. Di Domenico Jr.: *Fundamental optical attenuation limits in the liquid and glassy state with application to fiber optical waveguide materials*, Appl. Phys. Lett., Vol. 22, No. 10 (1973), pp. 527-529.

[19] W. Heitmann: *Intrinsic attenuation in pure and doped silica for fibre optical waveguides*, NTZ, Vol. 30, No. 6 (1977), pp. 503-506.

[20] S. E. Miller, E. A. J. Marcatili and Tingye Li: *Research toward optical-fiber transmission systems*, Proc. IEEE, Vol. 61, No. 12 (1973), pp. 1703-1751.

[21] D. B. Keck, R. D. Maurer and P. C. Schultz: *On the ultimate lower limit of attenuation in glass optical waveguides*, Appl. Phys. Lett., Vol. 22, No. 7 (1973), pp. 307-309.

[22] M. K. Barnoski and S. D. Personick: *Measurements in fiber optics*, Proc. IEEE, Vol. 66, No. 4 (1978), pp. 429-441.

[23] G. H. Sigel and B. D. Evans: *Effects of ionizing radiation on transmission of optical fibres*, Appl. Phys. Lett., Vol. 24, No. 9 (1974), pp. 419-412.

[24] P. Kaiser: *Drawing induced coloration in vitreous silica fibers*, J. Opt. Soc. Am., Vol. 64, No. 4 (1974), pp. 475-481.

[25] D. Marcuse: *Radiation losses of dielectric waveguides in terms of the power spectrum of the wall distortion function*, Bell. Syst. Tech. J., Vol. 8, No. 10 (1969), pp. 3233-3342.

[26] E. A. J. Marcatili: *Bends in optical dielectric guides*, Bell Syst. Tech. J., Vol. 48 (1969), pp. 2103-2132.

[27] D. Marcuse: *Curvature loss formula for optical fibers*, J. Opt. Soc. Am., Vol. 66, No. 2 (1976), pp. 216-220.

[28] R. Olshansky: *Distortion losses in cabled optical fibers*, Appl. Opt., Vol. 14, No. 1 (1975), pp. 20-21.

[29] D. Gloge: *Optical-fiber packaging and its influence on fiber straightness and loss*, Bell Syst. Tech. J., Vol. 54, No. 2 (1975), pp. 245-262.

[30] E. J. Friebele, G. H. Sigel and D. L. Griscom: *Drawing induced defect centers in silica core fiber optics*, Second European Conference on Optical Fibre Communication, Paris (September 27-30, 1976), pp. 63-68.

[31] E. J. Friebele, G. H. Sigel Jr. and M. E. Gingerich: *Enhanced low dose radiation sensitivity of fused silica and high silica core fiber optic waveguide*, Third European Conference on Optical Communication, Munich (September 14-16, 1977), pp. 72-74.

[32] E. J. Friebele, G. H. Sigel and M. E. Gingerich: *Spectral dependence of radiation damage in fiber optic waveguides in the 0.4-1.7 μm region*, Fourth European Conference on Optical Communication, Genoa (September 12-15, 1978), pp. 80-87.

[33] K. Inada: *A new graphical method relating to optical fibre attenuation*, Opt. Commun., Vol. 19, No. 3 (1976), pp. 437-439.

[34] T. Miaslika, T. Miya and M. Nakakara: *An ultimate low loss single mode fiber at* 1.55 μm, Optical Fiber Communication, Washington, D.C. (March 6-8, 1979).

[35] R. Olshanshy and D. B. Keck: *Pulse broadening in graded-index optical fibers*, Appl. Opt., Vol. 15, No. 2 (1976), pp. 483-491.

[36] F. M. E. Sladen, D. N. Payne and M. J. Adams: *Profile dispersion measurements for optical fibres over the wavelength range* 350 nm *to* 1900 nm, Fourth European Conference on Optical Communication, Genoa (September 12-15, 1978), pp. 48-57.

[37] R. Olshansky: *Optical waveguides with low pulse dispersion over an extended range*, Electron. Lett., Vol. 24, No. 11 (1978), pp. 330-331.

[38] R. Olshansky: *Multiple α-index profiles*, Appl. Opt., Vol. 18, No. 5 (1979), pp. 683-698.

[39] S. Geckeler: *Group delay in graded-index fibres with non power-law refractive profiles*, Electron. Lett., Vol. 13, No. 1 (1977), pp. 29-31.

[40] J. A. Arnaud: *Optimum profiles for dispersive multimode fibres*, Opt. Quantum Electron., Vol. 9 (1977), pp. 111-119.

[41] S. Geckeler: *Nonlinear profile dispersion aids optimization of graded index fibres*, Electron. Lett., Vol. 13, No. 15 (1977), pp. 440-442.

[42] D. Gloge: *Dispersion in weakly guiding fibers*, Appl. Opt., Vol. 10, No. 10 (1971), pp. 2442-2445.

[43] B. Costa and P. Di Vita: *Group velocity of modes and pulse distortion in dielectric optical waveguides*, Opto-electron., Vol. 5 (1973), pp. 439-456.

[44] D. N. Payne and W. A. Gambling: *Zero material dispersion in optical fibres*, Electron. Lett., Vol. 11, No. 8 (1975), pp. 176-178.

[45] D. N. Payne and A. M. Hartog: *Determination of the wavelength of zero material dispersion in optical fibres by pulse-delay measurements*, Electron. Lett., Vol. 13, No. 21 (1977), pp. 627-628.

[46] L. G. Cohen and Chinlon Lin: *Pulse delay measurements in the zero material dispersion wavelength region for optical fibers*, Appl. Opt., Vol. 16, No. 12 (1977), pp. 3136-3139.

[47] T. Ito, K. Nagakawa, S. Shimada, K. Ishihara, Y. Ohmori and K. Sugiyama: *Transmission experiments in the* 1.2-1.6 μm *wavelength region using graded-index optical-fiber cables*, Optical Fiber Communication, Washington D.C. (March 6-8, 1979), pp. 6-8.

[48] M. Eve, A. Hartog, R. Kashyap and D. N. Payne: *Wavelength dependence of light propagation in long fibre links*, Fourth European Conference on Optical Communication, Genoa (September 12-15, 1978), pp. 58-63.

[49] T. Tarifuji and M. Ikeda: *Pulse circulation measurement of transmission characteristic in long optical fiber*, Appl. Opt., Vol. 16, No. 8 (1977), pp. 2175-2179.

[50] B. Sordo, F. Esposto and B. Costa: *Experimental study of modal and material dispersion in spliced optical fibre*, Fourth European Conference on Optical Communication, Genoa (September 12-15, 1978), pp. 71-79.

[51] R. Olshansky and R. D. Maurer: *Tensile strength and fatigue of optical fibers*, J. Appl. Phys., Vol. 47, No. 10 (1976), pp. 4497-4498.

[52] D. G. Holloway: *The physical properties of glass,* London, Wyckebam Publication (1973), p. 220.

[53] A. S. Tetelman and A. J. McEvily: *Fracture of structural material,* Wiley, New York (1967).

[54] F. V. Di Marcello and A. C. Hart: *Furnace-drawn silica fibres with tensile strengths* $>$ 3.5 GN/m^2 (500 kp.s.i) *in* 1 km *lengths*, Electron. Lett., Vol. 14, No. 18 (1978), p. 578.

[55] M. J. Maklad and C. K. Kao: *Fatigue characteristics of chemical-vapor-deposited optical fibers having surface compression,* Optical Fiber Communication, Washington, D.C. (March 6–8, 1979), Paper TuC3.

CHAPTER 2

LIGHT PROPAGATION THEORY IN OPTICAL FIBRES

by Pietro Di Vita and Umberto Rossi

2.0 Introduction

The theoretical study of the propagation characteristics of light in an optical fibre with any radial refractive index distribution may be carried out using two different approaches, namely geometrical and modal or electromagnetic. A geometrical analysis can provide well-approximated results particularly for multimode fibres. The greatest advantages of this theory are the shorter calculation times and a more immediate physical interpretation of results compared to the electromagnetic approach. On the other hand, the electromagnetic approach is the only one which can provide well-approximated results in certain cases. These include monomode or few-mode fibres which must be dealt with using the electromagnetic theory, as must all problems involving coherence or interference phenomena. Moreover, it must be stressed that a purely geometrical approach (in its rigorous formulation) cannot describe some typical undulatory phenomena such as intrinsic losses of leaky skew rays and losses due to a dissipative cladding. On the other hand, using the rigorous electromagnetic theory in these cases does not always lead to satisfactory results because in general the equations involved are not analytically soluble and calculation times for numerical results are too long. Good results in these cases are obtained by the quasi-classical approach, which gives a correction of geometrical theory, including the effects to the first order of wavelength, and allows the above mentioned undulatory phenomena to be taken into account.

In general, one can say that many of the problems encountered in the study of the propagation through practical multimode optical fibres can be tackled using the geometrical approach, as shown in the following; on the other hand problems related to special kinds of fibres (e.g. monomode) can only be handled using an electromagnetic approach.

This chapter is therefore divided into two parts: *A*) « Geometrical Approach » and *B*) « Modal Approach »; each one of these two analyses has its own field of application.

In the part devoted to geometrical analysis, a general formalism is presented which allows most of the practical problems to be dealt with in an original manner; then launching and coupling problems are discussed followed by a study of the power distribution of the fibre using a new formalism. In three separate sections, power attenuation (excluding scattering), pulse distortion and light scattering inside the fibre (scattering is considered separately because of the different formalism required) are also discussed.

In the part devoted to the modal approach, the analysis starts from Maxwell equations, which can be rigorously solved only if the core refractive index is assumed uniform (« step-index » fibres). In other cases solutions can be obtained using various kinds of approximations, the most widely known of which is the WKBJ method; the quantization of the electromagnetic field can be obtained through the characteristic equation. Then in the following sections the problems of optical power flow and pulse distortion are discussed; methods are given for fibres with any refractive index profile, but details can only be supplied for step- and parabolic-index fibres, which are the most widely analyzed in the literature. Finally, power losses and mode-coupling problems are considered, and the differences between modal and geometrical approaches are illustrated.

The number of approximations is the least possible, according to the evidence and a simple interpretation of the obtained results (both by formulae and plots). Optical fibres will be considered with an inner cylindrical core (of radius a) with an arbitrary recfractive index profile $n(r)$ (r is the radial coordinate), surrounded by a cladding with uniform refractive index n_1 which usually assumes the value $n(a)$.

A. Geometrical approach

2.1 General formalism

The ray approach to propagation in optical fibres has been fully discussed in technical literature: starting from step-index fibres [1-3] the method has been applied to fibres with different refractive index profiles [4-6]. By means of geometrical optics nearly all the problems concerning optical fibres have been investigated including pulse distortion [7-10], power coupling [11, 12] and scattering [13].

The fundamental hypothesis adopted to justify the use of a geometrical approach in the propagation theory of optical fibres is that the electromagnetic field inside the fibre may be expanded in a set of local plane waves of the form [14-16]:

$$\bar{A}(\mathbf{r})\cdot \exp\,[2\pi j S(\mathbf{r})/\lambda] \tag{1}$$

in which the amplitude function $\bar{A}$ and the phase function S depend on the position vector $\mathbf{r}$ (λ is the free space wavelength). When the phase function $S(\mathbf{r})$ is known the path of each ray inside the fibre can be determined; in fact:

$$\nabla S = n\mathbf{t} \tag{2}$$

where $\mathbf{t}$ is the unit vector tangent to the ray path at every point and n is the local refractive index. Since the phase fronts $S(\mathbf{r}) =$ $=$ const. advance in the direction of $\nabla S(\mathbf{r})$, an immediate interpretation of each local plane wave in terms of a light ray arises.

From Eq. (2) the eikonal equation can be carried out:

$$|\nabla S|^2 = n^2(\mathbf{r})\,. \tag{3}$$

Such an equation allows, at least in principle, the determination of the phase function $S(\mathbf{r})$ from the knowledge of the spatial distribution of the refractive index $n(\mathbf{r})$, by means of a quadrature process.

In optical fibres the refractive index depends only on the radial coordinate r; if a system of cylindrical coordinates (r, ψ, z) is chosen so that the z-axis coincides with the fibre axis (Fig. 1), Eq. (3) can be given the following formal solution:

$$S(r, \psi, z) = \pm\int^{r} P(r)dr + h\psi + kz\,, \tag{4}$$

where:

$$P(r) \equiv \sqrt{n^2(r) - k^2 - h^2/r^2} \tag{5}$$

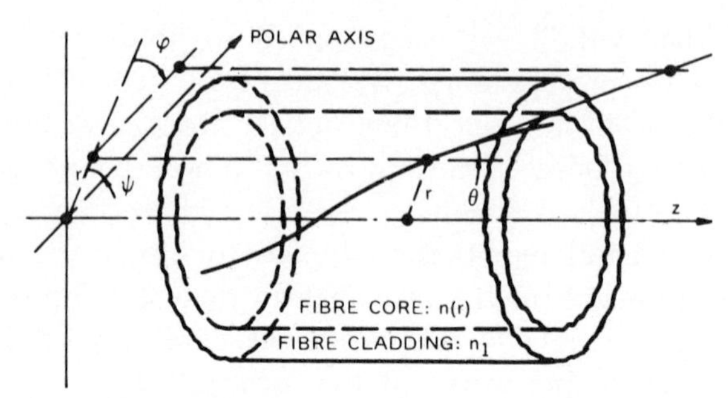

FIG. 1 – Position and angular coordinates of a light ray in the fibre.

and h and k are integration constants that have been determined for every ray by initial launching conditions, and remain constant at every point along the ray path. From a physical point of view P, h and k are proportional to the three components (radial, azimuthal and longitudinal) of the wave vector of the local plane wave:

$$\mathbf{K}(r) = \left(\pm \frac{2\pi P}{\lambda}, \frac{2\pi}{\lambda}\frac{h}{r}, \frac{2\pi}{\lambda}k\right).$$

2.1.1 *Ray congruences*

Indicating as $r = r(z)$, $\psi = \psi(z)$, $z = z$ the parametric equations of a ray in the fibre, from Eq. (2) the following differential equations for $r(z)$ and $\psi(z)$ can be obtained:

$$k^2\dot{r}^2 = P^2(r); \qquad kr^2\dot{\psi} = h \tag{6}$$

(the point denotes the derivative with respect to z). It can be seen from Eqs. (6) that, keeping h and k fixed, there is a double infinity of rays satisfying the above mentioned equations. In the following it will be very useful to consider all these rays (with the same value of h and k) collectively, since they have the same properties: in fact one ray may be transformed into any other ray of this set simply by a rotation around and a translation along the fibre axis. This set of rays (which is called a *ray congruence*) in modal theory corresponds to a mode [15, 16].

Considering the angular polar coordinates of the ray θ and φ (in which θ is the angle of the ray with the fibre axis and φ is the azimuth of the ray with respect to the radius vector (see Fig. 1)) the expressions for h and k can be carried out. In fact, owing to the definition of θ and φ, from simple trigonometric considerations it follows that $\dot{r} = \mathrm{tg}\,\theta \cos\varphi$ and $r\dot{\psi} = \mathrm{tg}\,\theta \sin\varphi$. From Eqs. (5) and (6) then the following explicit expressions for h and k can be obtained:

$$h = n(r) r \sin\theta \sin\varphi; \qquad k = n(r) \cos\theta\,. \tag{7}$$

These expressions allow the determination of the ray congruence (h, k) to which each ray belongs (when, for instance, initial launching conditions are known).

Every ray congruence (h, k) is composed of real rays (which in wave optics correspond to oscillating waves) where $P^2(r) > 0$, and complex rays (which correspond to evanescent waves) where $P^2(r) < 0$ (see Eqs. (1), (4) and (6*a*)). The cylindrical surfaces, coaxial to the fibre axis whose radius r_c is such that $P^2(r_c) = 0$, take the name of *caustic* surfaces which separate the real ray regions from the complex ray regions.

In a congruence of *guided* rays, real rays should not be present in the cladding to avoid any intrinsic radiation loss. In our case this condition means that for such congruences $k \geqslant n_1$. This leads to the following guidance limit condition for a given (h, k) congruence:

$$k = n_1. \tag{8}$$

Then a guided ray congruence is usually characterized by the presence of two caustics in the core: in the inner region delimited by them there are real rays and in the outer regions there are complex rays. If the guidance condition $(k \geqslant n_1)$ is not fulfilled we may have two kinds of ray congruences with intrinsic radiation loss: leaky ray congruences and radiated ray congruences. The loss mechanisms of these two kinds of congruences are very different.

A *leaky* (according to Snyder's definition of tunnelling leakage [17]) ray congruence, in addition to the two caustics in the core, has a third caustic in the cladding beyond which there are real rays. In such congruences the intrinsic radiation loss is due to the leakage of rays from the core through the evanescent field (complex rays) which is present between the second caustic in the core and the caustic in the cladding. Such a phenomenon

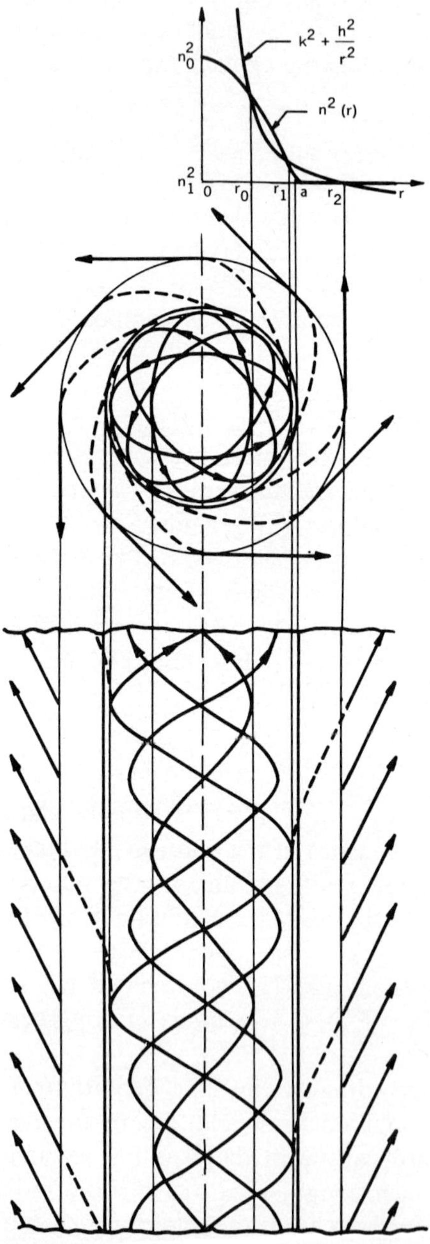

FIG. 2 – Paths of some rays of a leaky congruence in a parabolic-index fibre. Solid lines represent real rays, dashed lines represent complex rays (i.e. evanescent waves).

can be understood completely only by means of an undulatory analysis since it is the optical equivalent of the tunnel effect in quantum mechanics [18, 19]. Then the guidance condition of leaky congruences, which assures the existence of the third caustic in the cladding is $P^2(a) \leqslant 0$ (and at the same time $k < n_1$). This leads to the following guidance limit condition for each leaky (h, k) congruence:

$$P^2(a) = 0 \,. \tag{9}$$

In Fig. 2 the paths of a leaky congruence (for a parabolic-index fibre) are shown: solid lines represent real rays, dashed lines represent complex rays (evanescent field).

Radiated ray congruences have only one caustic in the core beyond which there are real rays. The intrinsic loss of power in these congruences takes place simply through the refraction of rays in the cladding and it is much more drastic in comparison with that of leaky congruences. The condition with which these radiated ray congruences are obtained is of course: $P^2(a) \geqslant 0$.

This can be better understood by referring to Fig. 3*a*, where the refractive index profile $n(r)$ is shown, together with three kinds of ray congruences, identified by three different (h, k) pairs.

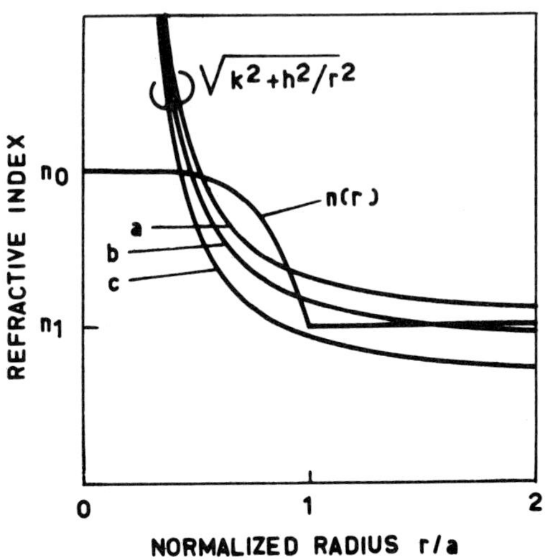

FIG. 3 *a*) – (h, k) congruences on the $(r, n(r))$ plane.

The caustics correspond to the intersections of each congruence curve with the refractive index profile. It is therefore evident that curve *a*) represents a guided congruence (two caustics in the core), curve *b*) a leaky congruence (two caustics in the core and one in the cladding), and curve *c*) a radiated congruence (only one caustic in the core).

The various ray congruences are represented in the (h, k) plane shown in Fig. 3*b* (for a parabolic-index fibre). The congruence domain is limited on the right by the line which represents circular

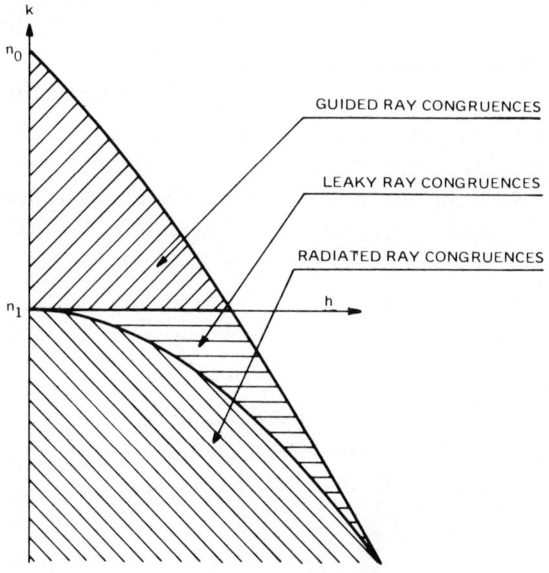

FIG. 3 *b*) – Plane (h, k) in which the guided, leaky and radiated congruences are represented.

helicoidal rays (beyond which no ray exists). This line depends on the index profile of the fibre and has the following parametric equations:

$$\begin{cases} h = \sqrt{-\dfrac{r^3}{2}\dfrac{dn^2}{dr}}\,, \\ k = \sqrt{n^2(r) + \dfrac{r}{2}\dfrac{dn^2}{dr}}\,. \end{cases} \tag{10}$$

Guided congruences are separated from leaky congruences by the straight line of Eq. (8) and these leaky congruences are separated from radiated ones by the ellipse of Eq. (9).

In practice radiated ray congruences are ignored in the calculations of the power present in the fibre, owing to their nearly immediate extinction. Leaky ray congruences, on the other hand, all ought strictly to be taken into account and weighted with their own intrinsic loss coefficient by the optical tunnel effect [19]. In fact, many leaky congruences have very low losses and still carry a considerable power after several kilometres of fibre. Anyway, in the following, calculations will be performed in two ways: either excluding leaky rays (this is the case in which such ray congruences are scarcely present in the fibre because of different mechanisms that may considerably reduce their contribution [19]) or including them (this is the case of short lengths of fibre where a large volume of leaky rays may still be present).

2.1.2 *Numerical aperture of the fibre*

The local numerical aperture seen by a ray which impinges on the input fibre surface is defined as:

$$A(r) \equiv n(r) \sin \theta_M \tag{11}$$

(where θ_M is the maximum θ value for a ray to be guided), and can be evaluated in two ways: either excluding or including leaky rays. In the first case, starting from the limit guidance condition (8), through (7) we have:

$$A(r) = \sqrt{n^2(r) - n_1^2}\,. \tag{12}$$

In the second case, starting from the limit guidance condition (9), we get:

$$A(r, \varphi) = \min \left\{ \sqrt{\frac{n^2(r) - n_1^2}{1 - \frac{r^2}{a^2} \sin^2 \varphi}},\; n(r) \right\}. \tag{13}$$

Equations (11)-(13) are the basis for the evaluation of the most interesting parameters in the propagation theory of optical fibres by

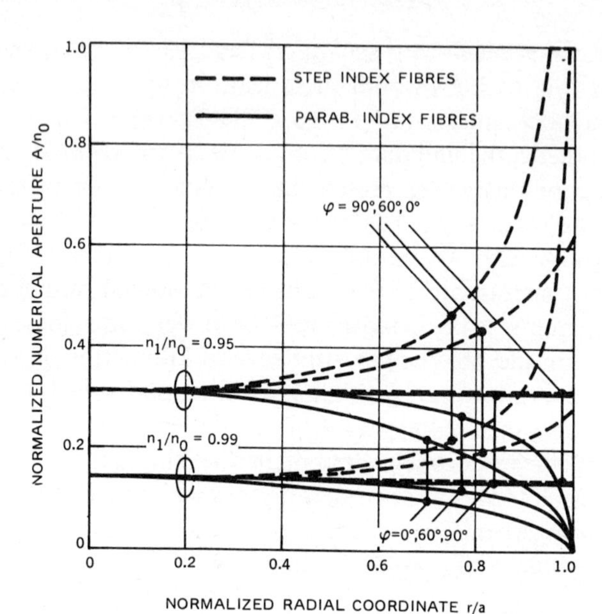

FIG. 4 – Numerical aperture $A(r, \varphi)$ (normalized to n_0) of parabolic- and step-index fibres against r, for different values of φ and two values of n_1/n_0.

a geometrical approach. In Fig. 4 the behaviour of local numerical aperture (normalized to $n_0 = n(0)$) of Eq. (13) is represented for parabolic-index fibres (solid lines) and step-index fibres (dashed lines), and for different values of the azimuthal angle φ. For $\varphi = 0$ we have the local numerical aperture of Eq. (12), which excludes leaky rays.

2.2 Launching and coupling efficiency

The problem of optical power injection into the fibre is one of the most important in optical fibre systems. In fact a lot of power is lost in the coupling region between source and fibre or in the joint region between two different fibres; consequently, the definition and a theoretical evaluation of parameters that can measure the quality of launching and coupling of optical power is of vital importance. In the following a formalism will be introduced for the problem of source-fibre coupling: afterwards, fibre-fibre coupling will be considered by means of a similar formalism.

Geometrical optics allows the evaluation of the desired parameters when the source radiance distribution and the relative geometry between source and fibre are known. We indicate by $R'(r', \psi', \theta', \varphi')$ the radiance distribution of the source in a medium of refractive index n' (e.g. air: $n'=1$) as a function of position (r', ψ') and angular (θ', φ') coordinates of an emitted ray (these coordinates are similar to the fibre coordinates shown in Fig. 1 and are referred to the source axis). Owing to the presence of an optical source, a radiance distribution $R(r, \psi, \theta, \varphi)$ will be obtained on the fibre input surface (in the fibre-medium, i.e. a medium with refractive index $n(r)$). Source coordinates $(r', \psi', \theta', \varphi')$ may be correlated to fibre coordinate $(r, \psi, \theta, \varphi)$ through a geometric transformation (**T**) which describes the relative source-fibre geometry and media or devices interposed, according to the following formal relation:

$$(r', \psi', \theta', \varphi') = (\mathbf{T}|r, \psi, \theta, \varphi). \tag{14}$$

The explicit form of the transformation **T** can be obtained by means of purely geometrical considerations: it is extensively performed in [11, 20, 21]. Then R may be deduced from R', through the radiance conservation law [22]:

$$R(r, \psi, \theta, \varphi) = \frac{n^2(r)}{n'^2} \cdot R'(\mathbf{T}|r, \psi, \theta, \varphi). \tag{15}$$

Note that, if the interposed media introduce some loss, the second side of Eq. (15) should be multiplied by the relative transmittivity $T(r, \psi, \theta, \varphi)$. Knowing $R(r, \psi, \theta, \varphi)$ through Eq. (15), optical power collected and guided by the fibre can be obtained according to:

$$W_0 = \int_0^a r\,dr \int_0^{2\pi} d\psi \int_0^{2\pi} d\varphi \int_0^{\theta_M} R(r, \psi, \theta, \varphi) \sin\theta \cos\theta \, d\theta \tag{16}$$

(where θ_M is defined by Eq. (11) and may be deduced from Eq. (12) or (13)). On the other hand, the power emitted by the source can be expressed as:

$$W' = \iint_S r'\,dr'\,d\psi' \int_0^{2\pi} d\varphi' \int_0^{\pi/2} R'(r', \psi', \theta', \varphi') \sin\theta' \cos\theta' \, d\theta' \tag{17}$$

(where S represents the source surface).

Launching efficiency Λ, which evaluates the quality of coupling, is defined as:

$$\Lambda = W_0/W' \tag{18}$$

and can be deduced from Eqs. (16) and (17). It will be evaluated both in the case of direct coupling (when source and fibre input surfaces are coaxially joined together) and in the presence of coupling errors.

2.2.1 *Direct coupling*

It is known that when the surface of the source is larger than the fibre core cross-section, the best coupling (which assures the maximum of guided power) is direct [22]; otherwise an optical device which improves the coupling may be designed, but in the absence of such a device, the best coupling is still direct.

In the case of direct coupling the transformation (14) is very simple (through Snell's law):

$$r' = r\,, \qquad \psi' = \psi\,, \qquad n' \sin\theta' = n(r)\cdot\sin\theta\,, \qquad \varphi' = \varphi\,. \tag{19}$$

Now let us consider the following radiance distribution of the source:

$$R'(\theta') = R_0\cdot\cos^y\theta'; \qquad y \geqslant 0 \tag{20}$$

with this law it is possible to describe most of the sources used for optical fibres and particularly light emitting diodes (LED's); for $y = 0$ in fact we have the Lambertian source, while increasing y the source becomes more and more directional until it reaches (for $y \to \infty$) the unidirectional distribution. If we consider a circular shaped source (radius b), we have the following expression of launching efficiency for direct coupling [12]:

$$\Lambda = \frac{T}{\pi b^2}\int_0^{a'} r\,dr \int_0^{2\pi} [1 - \cos^{y+2}\theta'_M]\,d\varphi\,, \tag{21}$$

where T is the transmittivity at the fibre input (that in practice can be considered independent of r and φ [12]), a' is the minimum between a and b, and θ'_M is the maximum acceptance angle related to θ_M, defined by Eq. (11), through Snell's law expressed by Eq. (19*c*). Then θ'_M may be deduced from the expressions (12) or (13) of local numerical aperture according to the exclusion or the inclusion of leaky rays (but always $0 \leqslant \theta'_M \leqslant \pi/2$). Excluding the leaky ray contribution, θ'_M ceases to depend on φ and Eq. (21) may be simplified. In Figs. 5 and 6 Λ, deduced from Eq. (21), is given for the following family of refractive index profiles [23]:

$$n^2(r) - n_1^2 = \Delta^2[1 - (r/a)^x]\,, \qquad \text{for } r \leqslant a \text{ and } x > 0\,, \tag{22}$$

where:

$$\Delta \equiv \sqrt{n_0^2 - n_1^2} \tag{23}$$

represents the value of local numerical aperture on fibre axis: for $x = \infty$ we have the step-index fibre, while, as x decreases, the profile becomes smoother and smoother and for $x = 2$ we have the parabolic-index fibre. In Figs. 5 and 6 the dashed lines include the contribution of leaky rays, solid lines exclude it; moreover it is assumed that $T = 1$. Figure 5 shows Λ versus the exponent x

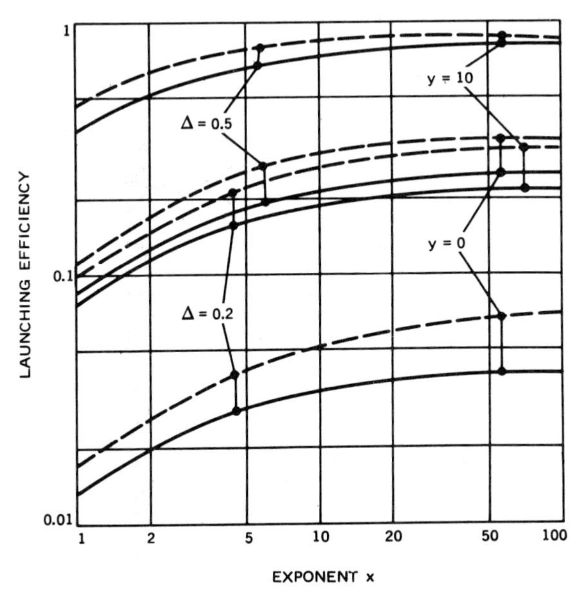

FIG. 5 – Launching efficiency of fibres with different values of Δ, excited by sources with different values of y against the exponent x of the profile family (22). Solid lines exclude leaky ray contribution, dashed lines include it.

of Eq. (22) for different values of Δ and y of the source. Launching efficiency increases with the increase of x, Δ (more guiding fibre) and y (more directional source). In Fig. 5, b is assumed coincident to a while in Fig. 6 Δ is plotted against the ratio b/a

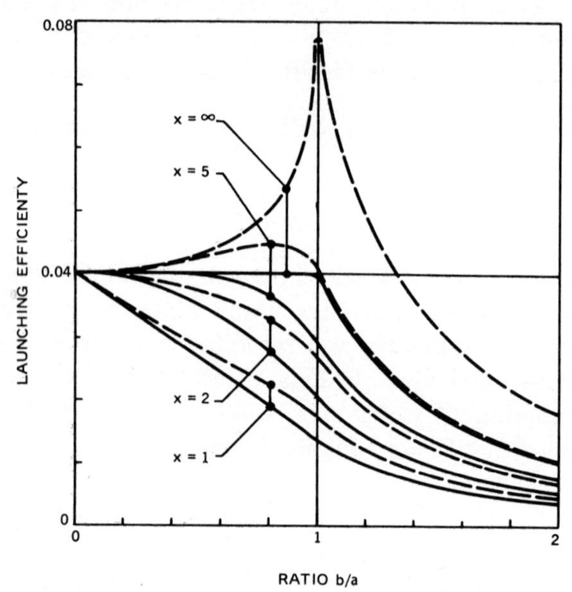

FIG. 6 – Launching efficiency of fibres with different values of exponent x and $\Delta = 0.2$ against the ratio b/a (Lambertian source radius/fibre core radius). Solid lines exclude leaky ray contribution, dashed lines include it.

for different values of x; the source is assumed Lambertian ($y = 0$) and $\Delta = 0.2$. These figures show that for graded-index fibres launching efficiency (excluding leaky rays) decreases with the increase of b while for step-index fibres it remains constant (for $b < a$): this is due to the fact that in graded-index fibres the acceptance decreases towards the edge of the core (see Eq. (12) and Fig. 4). Moreover, these lines are proportional (for $b \leq a$) to $n^2(b)$; such an effect is typical of the profile family (22). The peak shown by launching efficiency (including leaky rays) for $b \simeq a$ is typical of step- or quasi-step-index fibres and is due to the increase of the acceptance of leaky rays towards the edge of the core (see Eq. (13) and Fig. 4).

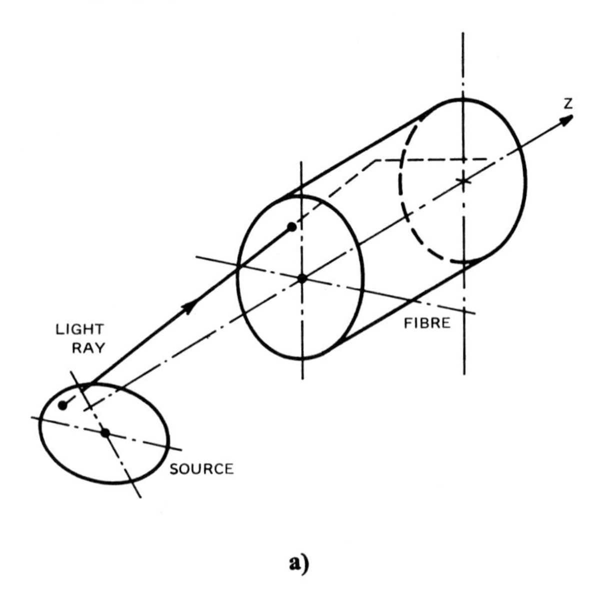

a)

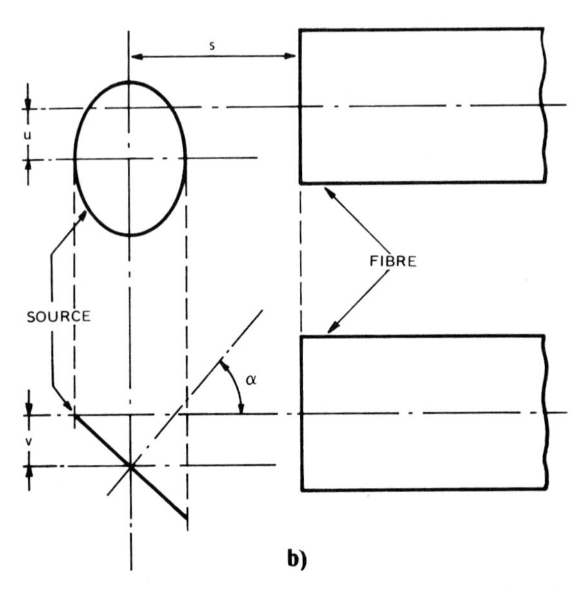

b)

FIG. 7 – Geometry of source-fibre coupling errors: *a*) perspective view, *b*) orthogonal projection.

2.2.2 *Coupling errors*

In practice, direct coupling conditions can rarely be achieved; it thus appears very important to analyze the behaviour of launching efficiency against the various coupling error parameters (that evaluate quantitatively every deviation from the direct coupling condition). There are three fundamental kinds of coupling errors (Fig. 7):

a) *separation*: if source and fibre input surfaces have the same axis, but are separated by a gap s;

b) (*lateral*) *displacement*: if the axes of the two surfaces are parallel but separated by a distance d;

c) (*angular*) *misalignment*: if these two axes form a certain angle α.

In practice, these three error configurations may be simultaneously present; in such a case we have to take into account two kinds of possible displacements (see Fig. 7*b*): u, along the axis around which the emitting surface is rotated by α, and v, along

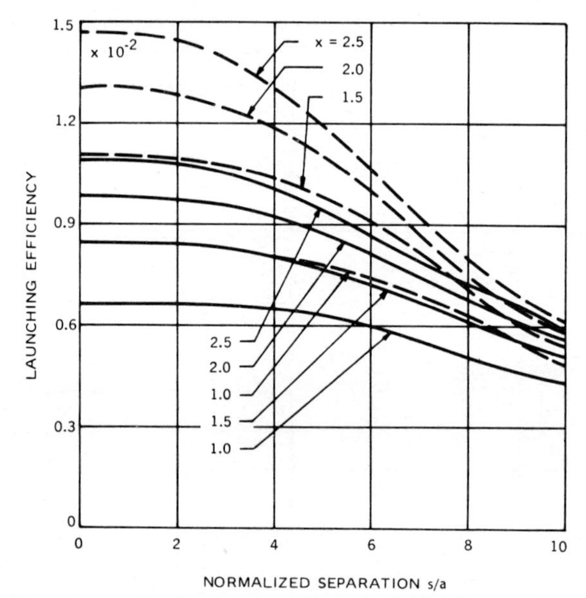

FIG. 8 *a*) – Launching efficiency of fibres with different values of exponent x and $\Delta = 0.14$ against the separation s (normalized to a) between (Lambertian) source and fibre ($u = v = \alpha = 0$). Solid lines exclude leaky ray contribution, dashed lines include it. (After [12]. Reprinted with permission).

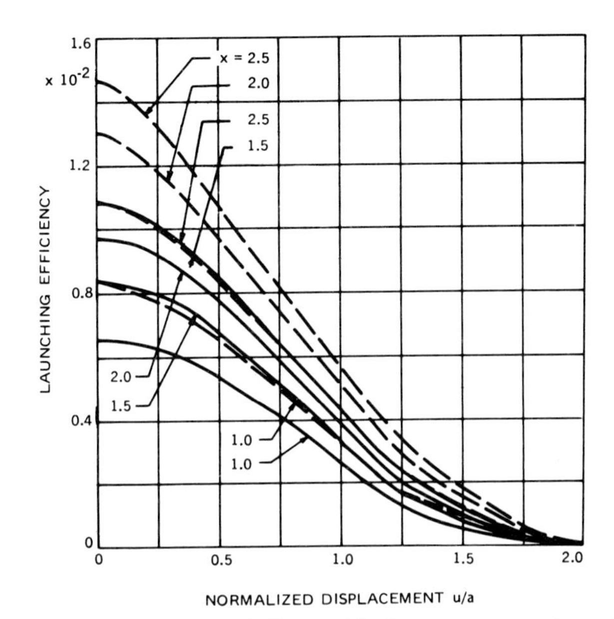

FIG. 8 *b*) – Lauching efficiency of fibres with the same parameters of Fig. 8*a* against the displacement u (normalized to a) between (Lambertian) source and fibre axes ($s = a/2$, $v = \alpha = 0$). Solid lines exclude leaky ray contribution, dashed lines include it. (After [12]. Reprinted with permission).

the axis perpendicular both to the u-axis and to the fibre axis; it follows that $u^2 + v^2 = d^2$, d being the distance between the two axes.

In general transformation (14), which depends on various error parameters: $\mathbf{T} = \mathbf{T}(s, u, v, \alpha)$, is very cumbersome and its explicit form is shown in Refs. [11, 20, 21]. Some numerical results of launching efficiency versus error parameters are shown in Figs. 8. In these figures launching efficiency is given against s, u and α respectively, for some graded-index fibres with $\Delta = 0.14$. In the solid lines the contribution of leaky rays is excluded, in the dashed lines it is included. As one can see, the most drastic coupling losses are caused by displacement, while the least drastic ones are caused by misalignment. Moreover in Fig. 8*a*, particularly for parabolic-index fibres and excluding leaky rays, a certain flattening of losses is evident near $s \simeq 0$. This effect is attributable to a phenomenon of the « virtual lengthening » of graded index fibres, due to the special shape of the acceptance cone [12].

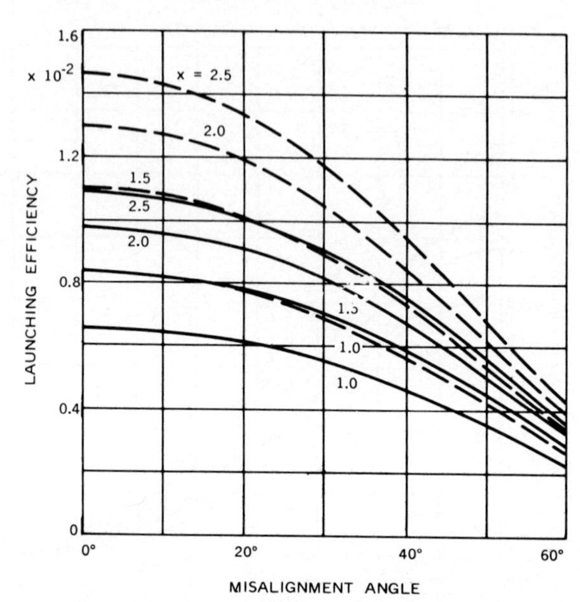

FIG. 8 *c*) – Launching efficiency of fibres with the same parameters of Fig. 8*a* against the misalignment angle α between (Lambertian) source and fibre axes ($s = a$, $u = v = 0$). Solid lines exclude leaky ray contribution, dashed lines include it. (After [12]. Reprinted with permission).

2.2.3 *Fibre-fibre coupling*

The theory developed for the evaluation of losses between source and fibre can be employed for the study of fibre-fibre coupling [21]. It can be demonstrated that if two identical weakly guiding fibres (that is with small difference between cladding and axial core refractive indices—hypothesis fulfilled in many actual cases) are coupled, coupling loss depends only on the following normalized parameters [24]:

$$\left\{\begin{array}{l} \bar{s} = \dfrac{s}{a} \cdot \dfrac{\Delta}{\sqrt{n'^2 - \Delta^2}}\,; \\[2ex] \bar{u} = \dfrac{u}{a}; \qquad \bar{v} = \dfrac{v}{a}; \qquad \left(\bar{d} = \dfrac{d}{a}\right); \\[2ex] \bar{\alpha} = \dfrac{n'}{\Delta} \cdot \sin \alpha\,. \end{array}\right. \tag{24}$$

Some numerical results are shown in Figs. 9 and 10, where curves at constant launching efficiency are given in the $(\bar{s}, \bar{v})$ plane,

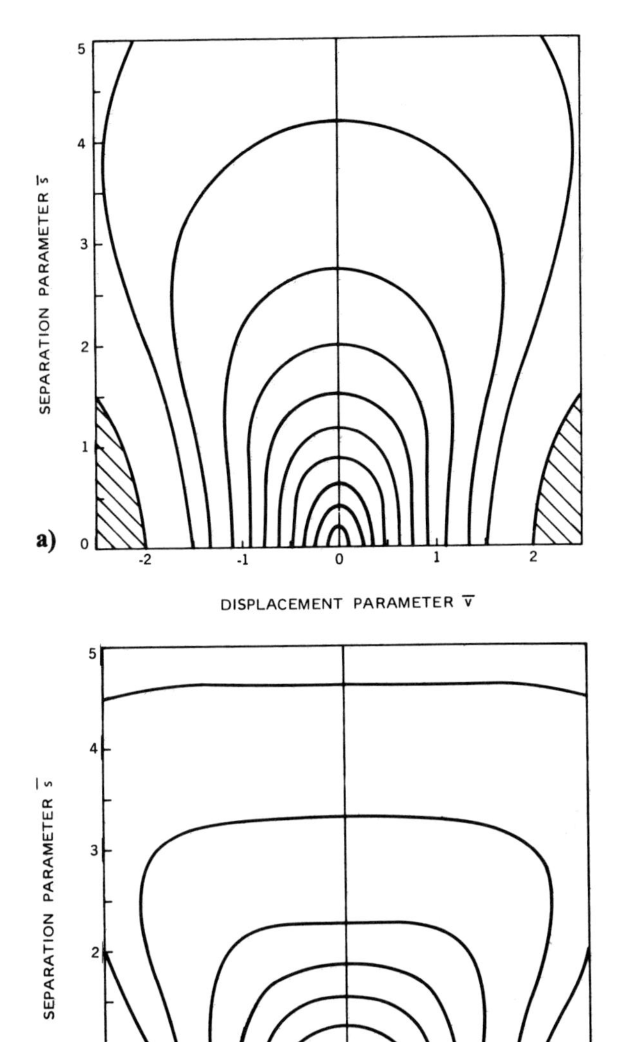

FIG. 9 – Fibre-fibre coupling: curves at constant launching efficiency in the separation-displacement $(\bar{s}, \bar{v})$ plane, *a*) for parabolic-index fibres, *b*) for step-index fibres, with $u = \alpha = 0$ and excluding leaky rays. The curves, starting from the outermost one, hold, for the following values of launching efficiency: 0.05. 0.1, 0.2, 0.3, 0.4, 0.5, 0.6, 0.7, 0.8, 0.9. The straight line represents a symmetry axis; in the shaded regions launching efficiency vanishes.

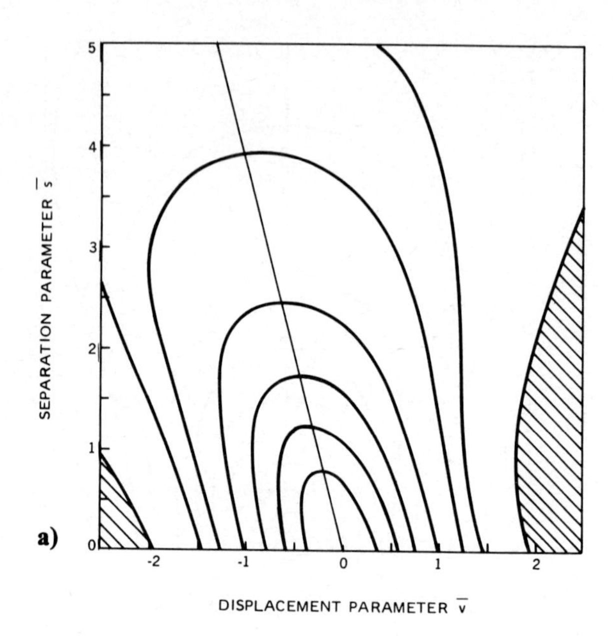

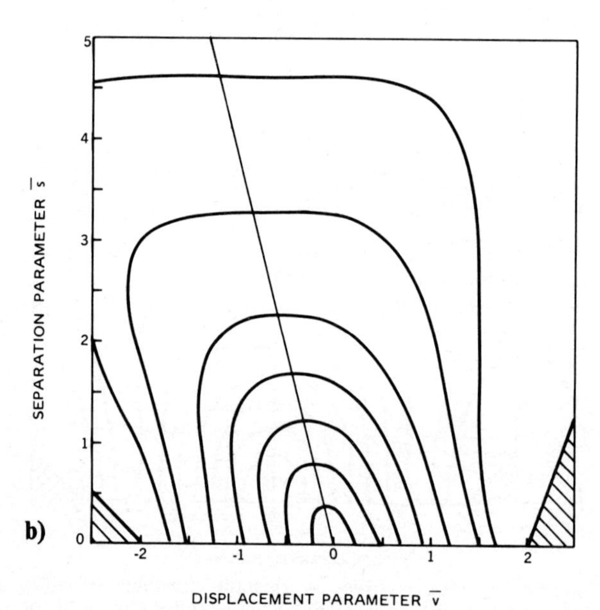

FIG. 10 – Fibre-fibre coupling: the same as Fig. 9: *a*) for parabolic-index, *b*) for step-index fibres, but in the presence of a misalignment: $\bar{\alpha} = 0.5$ ($u = 0$), excluding leaky rays. The curves hold for the same values of launching efficiency as Fig. 9 but the greatest values are not present.

ignoring the leaky ray contribution, and considering equal and uniformly excited fibres. Figures 9 (*a*: parabolic-index fibres; *b*: step-index fibres) hold in the absence of misalignment ($\bar{\alpha} = 0$), while Figs. 10 (*a*: parabolic-index fibres; *b*: step-index fibres) hold in the presence of a misalignment with $\bar{\alpha} = 0.5$. These plots may supply a useful tool both for the evaluation of launching efficiency in fibre-fibre splices and for the design of fibre-fibre joints and connectors.

It can be seen that for small error parameter values a linear launching efficiency decay takes place when the emitting fibre is uniformly excited; in this case the following expression in the presence of a single coupling error can be obtained [21, 24]:

$$\Lambda(\bar{e}) = 1 - k_e \cdot \bar{e} \qquad (\bar{e} \ll 1) \tag{25}$$

where $\bar{e}$ can be any normalized error parameter (24) and k_e is a suitable coefficient, shown in Table I both for parabolic-(PF) and step-(SF) index fibres. It can be observed that for small coupling errors losses are greater in parabolic- than in step-index fibres and for displacement and misalignment than for separation.

TABLE I

Coefficient k_e.

k_e	PF	SF
k_s	$\frac{1}{2} = 0.5$	$\frac{4}{3\pi} \simeq 0.4244$
k_u, k_v k_α, k_d	$\frac{8}{3\pi} \simeq 0.8488$	$\frac{2}{\pi} \simeq 0.6366$

However, knowing the behaviour of losses in the presence of only a single coupling error is not very useful, because normally different coupling errors are simultaneously present in a general offset configuration. Hence, it appears to be very important to know the composition laws between two or more coupling errors. The discussion refers only to fibres of usual interest, namely parabolic- and step-index weakly guiding fibres. However since small index profile variations (e.g. a dip on fibre axis or small departures from parabolic or step profile) do not produce appreciable differences in the results, they can still be considered valid also in quasi-parabolic- or quasi-step-index fibres (which include most cases).

It is possible to obtain non-trivial rigorous laws of composition of coupling errors (valid for every possible value of error parameters) only in the u-α composition ($s = v = 0$): in fact $\bar{u}$ and $\bar{\alpha}$ give rise to a quadratic superposition in parabolic-index fibres [25] while, if small error parameter values are considered, $\bar{u}$ and $\bar{\alpha}$ give rise to a linear superposition in step-index fibres [21, 24].

As regards the remaining coupling error compositions, which cannot be easily investigated by analytical methods, the study can be performed by best fitting a large number of numerical results. The two by two composition in general can be expressed simply as follows:

$$\Lambda(\bar{e}_1, \bar{e}_2) = 1 - [(k_{e_1}\bar{e}_1)^n + (k_{e_2}\bar{e}_2)^n]^{1/n} \tag{26}$$

the values of n which give the best approximation are listed in Table II both for parabolic-(PF) and step-(SF) index fibres.

TABLE II

Exponent n.

$e_1 - e_2$	PF	SF
$s - \alpha$ $u - \alpha$	2	1
$s - u$ $s - d$	2.55	2.52

It must be stressed that relations (25)-(26) hold only if a uniform power distribution on the emitting face of the first fibre is considered. If non-uniform power distributions are taken into account, the evaluation of Eq. (16) becomes very cumbersome.

Experimental results indicate that a parabolic dependence on each error parameter arises and that the most critical parameter is again the lateral displacement [25]. Thus, the following discussion concerns this kind of error. We have generalised some theoretical calculations from parabolic-index profiles [26] to any profile fibres and considered both the power accepted by the receiving fibre (launching efficiency: Λ_A) and the power detected at the output of the receiving fibre (launching efficiency: Λ_B) [27]. In this second case, the additional attenuation induced by the splice in the receiving fibre (due to a greater excitation of lossy congruences) must be taken into account.

Assuming that the power distribution among the ray congruences of the emitting fibre depends only on k: $\varrho_e(k)$ and $\varrho_0(k)$ being the steady-state power distribution (see Sections 2.6.2 and 2.10.2); when $\varrho_e(n_1) = 0$ the linear decay in Λ_A ceases and the parabolic decay arises. For $u \ll a$ we have obtained the following expansions:

$$\Lambda_A \simeq \begin{cases} 1 + \dfrac{u}{\pi} \dfrac{\varrho_e(n_1) \cdot \int_0^a [n^2(r) - n_1^2]\, dr}{\int_0^a \varrho_e(n(r)) \cdot dn^2/dr \cdot r^2\, dr}; & \text{for } \varrho_e(n_1) \neq 0; \\ 1 + \dfrac{u^2}{8n_1} \dfrac{\varrho_e'(n_1) \cdot F(a)}{\int_0^a \varrho_e(n(r)) \cdot dn^2/dr \cdot r^2\, dr}; & \text{for } \varrho_e(n_1) = 0; \end{cases} \tag{27a}$$

$$\Lambda_B \simeq 1 - \frac{u^2}{8} \frac{\int_0^a \varrho_e'(n(r)) \cdot \varrho_0'(n(r)) \cdot [F(r)/n^2(r)] \cdot dn^2/dr \cdot dr}{\int_0^a \varrho_e(n(r)) \cdot \varrho_0(n(r)) \cdot dn^2/dr \cdot r^2\, dr}; \tag{27b}$$

where: $\varrho'(k) \equiv d\varrho/dk$, and $F(r) \equiv \int_0^r (dn^2/dr)^2 r\, dr$.

For actual power distributions $\varrho_e(k)$ and $\varrho_0(k)$ the decay coefficient of $(u/a)^2$ takes values around 1 for Λ_A and around 2 for Λ_B; this is in agreement with experimental results [25].

Consequently, losses are more drastic in the case of uniform power distribution in the emitting fibre, so that expressions (25)-(26) may be considered as the upper limit of joint losses. More details on this subject will be given in Part III, Chapter 2.

As a matter of fact, more realistic predictions of joint losses, should also consider mismatch in optical fibre parameters [27].

2.3 Power distribution in the fibre

It is of the utmost importance to know the spatial and angular distribution of power inside a fibre since information can be deduced not only about the refractive index profile of the fibre [28-31] but also (as will be shown in the next sections) about power attenuation and pulse distortion. In this section near- and far-field intensity distributions will be analyzed.

2.3.1 *Distribution in the (h, k) plane*

It may often be more useful for the following calculations to start from the power distribution in the (h, k) plane (e.g. see Fig. 3*b*) rather than from the radiance distribution $R(r, \psi, \theta, \varphi)$. If W_0 (expressed by Eq. (16)) is the power launched into the fibre and $\varrho_0(h, k)$ the power distribution in the (h, k) plane at launching $(z = 0)$, then:

$$W_0 = \iint_D \varrho_0(h, k)\, dh\, dk \tag{28}$$

where D is the domain of the congruences accepted by the fibre in the (h, k) plane (e.g. see Fig. 3). This domain D may be limited either by the straight line of Eq. (8) or by the ellipse of Eq. (9), depending on the exclusion or inclusion of the contribution of leaky congruences. Equating Eqs. (16) and (28) we can obtain the form of $\varrho_0(h, k)$ for which purpose it is necessary to make in Eq. (16) a change of variables from $(r, \psi, \theta, \varphi)$ to (h, k, r, ψ). Taking into account that such a change of variables is ruled by Eq. (7), summing on r and ψ we obtain the following expression for $\varrho_0(h, k)$:

$$\varrho_0(h, k) = 4 \cdot k \int_{r_0}^{r_1} \frac{dr}{P(r) \cdot n^2(r)} \int_0^{2\pi} d\psi\, R(r, \psi, \theta(k, r), \varphi(h, k, r))\,. \tag{29}$$

where $\theta(k, r)$ and $\varphi(h, k, r)$ are the inverse functions of (7).

Expression (29) of $\varrho_0(h, k)$ for Lambertian sources assumes the following simple expression:

$$\varrho_0(h, k) = \frac{8\pi}{n'^2} R_0 l(h, k)\,, \tag{30}$$

where:

$$l(h, k) \equiv k \int_{r_0}^{r_1} \frac{dr}{P(r)}, \tag{31}$$

represents the half-period of a ray of any (h, k) congruence along the z-axis (this can be easily verified by integrating Eq. (6*a*)). For Lambertian sources it is possible to obtain the following compact expressions of W_0 by means of Eqs. (28), (30) and (31), for every kind of refractive index profile:

$$W_0 = \frac{2\pi^2}{n'^2} R_0 \int_0^a [n^2(r) - n_1^2]\, r\, dr \tag{32}$$

$$W_0 = \frac{2\pi^2}{n'^2} R_0 \int_0^a \frac{n^2(r) - n_1^2}{\sqrt{1 - r^2/a^2}}\, r\, dr\,. \tag{33}$$

In Eq. (32) leaky rays are excluded, in Eq. (33) they are included.

Moreover, if the power density $\varrho_0(h, k)$ is known for $z = 0$, it is possible to obtain the power density $\varrho(h, k, z)$ for every value of $z \geqslant 0$ simply through:

$$\varrho(h, k, z) = \varrho_0(h, k) \exp\,[-\gamma(h, k) \cdot z] \tag{34}$$

where $\gamma(h, k)$ is the attenuation coefficient for every (h, k) congruence; its explicit form will be shown in the following sections. From Eq. (34) the following expression of the distribution of power along the z-axis can be derived:

$$W(z) = \iint_D \varrho(h, k, z)\, dh\, dk\,. \tag{35}$$

Through these quantities it is possible to obtain rigorous expressions for the radial and angular distribution of power in the fibre.

2.3.2 *Near-field intensity*

The near-field intensity of the fibre is given by the radial distribution of power. More generally, considering a light beam emitted by a generic source, the near-field region is limited to the region in which the beam width does not depart significantly from source dimensions. The near-field intensity may be carried out by summing all the contributions of the radial power distribution of each ray congruence. The normalized (*i.e.* its integral over r is unity) radial density of power of each congruence has the following form [19]:

$$w(r, h, k) = \frac{k}{P(r) \cdot l(h, k)}; \qquad \text{for } r_0 < r < r_1 \tag{36}$$

this expression may be easily understood observing that this radial power distribution should be inversely proportional to $dr/dz = \dot{r}$ (see Eq. (6*a*)). Then the whole radial distribution of power normalized to unity is given by the sum of contributions given by Eq. (36), each one weighted by its density $\varrho(h, k, z)$, according to:

$$w(r, z) = \frac{1}{W(z)} \iint\limits_{D,(P^2>0)} w(r, h, k) \cdot \varrho(h, k, z) \, dh \, dk \, . \tag{37}$$

In the case of the Lambertian source Eq. (37) assumes the following expression:

$$w(r, z) = \frac{R_0}{W(z)} \cdot \frac{8\pi}{n'^2} \iint\limits_{D,(P^2>0)} \frac{\exp[-\gamma(h, k) \cdot z]}{P(r)} k \, dh \, dk \tag{38}$$

and for $\gamma \cdot z = 0$ we obtain the following analytical expressions:

$$w(r) = r \frac{R_0}{W_0} \frac{2\pi^2}{n'^2} [n^2(r) - n_1^2] \, , \tag{39}$$

$$w(r) = r \frac{R_0}{W_0} \frac{2\pi^2}{n'^2} \left[\frac{n^2(r) - n_1^2}{\sqrt{1 - r^2/a^2}} \right] . \tag{40}$$

In Eq. (39) leaky rays are excluded, in Eq. (40) they are included. These two equations (apart from the metric factor r) supply near-field intensity for ideal fibres excited by a Lambertian source [23, 31] and are the basis of a method for the experimental evaluation of the index profile in optical fibres [28-31]. This method exploits the fact that near-field intensity is proportional to $n^2(r)$ (see Eq. (39)). However, it must be stressed that this proportionality in practice may be frustrated owing to the presence of leaky rays (see Eq. (40)), to the different attenuation of various congruences (see Eq. (38)), or to the source which is not strictly Lambertian (see Eq. (37)). Then in the various cases it is important to evaluate the actual influence of these phenomena in order to decide which of Eqs. (37)-(40) should be used for the interpretation of results. Several different techniques have been proposed for correlating the near-field power distribution to the refractive index profile [30, 32] and avoiding correction factors, but a simple and reliable method, valid also for monomode or few mode fibres, has yet to be implemented [33]. Experimental details on the correlation between refractive index profile and near-field measurements will be given in Chapter 3. As an example, showing the influence of leaky congruences, the near-field intensity obtained from Eqs. (39) and (40) is given in Fig. 11 for fibres with different values of profile exponent x of Eq. (22).

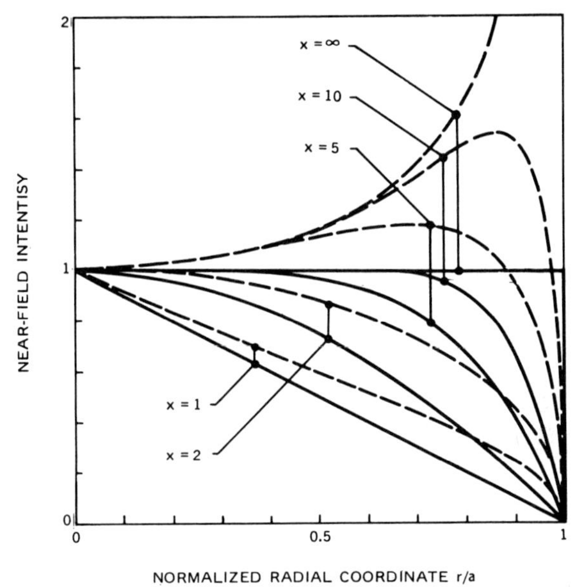

FIG. 11 – Plot of near-field intensity of fibres with different values of exponent x. Solid lines exclude leaky ray contribution, dashed lines include it.

2.3.3 *Far-field intensity*

The far-field intensity of the fibre is given by the angular distribution of power at the output of the fibre (in the outer medium of refractive index n'). Considering again a light beam emitted by a generic source, the far-field region can be considered as the region in which the beam wavefront is almost spherical and the geometrical dimensions of source can be ignored with respect to the beam width. As in the case of near-field intensity, the angular distribution may be obtained as the weighted superposition of normalized contributions of various congruences $w'(\theta', h, k)$. Taking into account that for every (h, k) congruence k, r and θ' (the angle of the ray with the fibre axis in the outer medium) are correlated by the following expression (after Eqs. (7*b*) and (19*c*)):

$$n^2(r) = k^2 + n'^2 \sin^2 \theta' \tag{41}$$

and performing a change of variables according to Eq. (41) from r to θ' in Eq. (36) (to obtain the unitary integral), we obtain the following expression of the normalized angular power distribution for every congruence:

$$w'(\theta', h, k) = n'^2 \sin 2\theta' \cdot w(\nu(k^2 + n'^2 \sin^2 \theta'), h, k) \cdot \\ \cdot |\nu'(k^2 + n'^2 \sin^2 \theta')|; \quad P^2 > 0 \tag{42}$$

where $\nu(n^2)$ is the inverse function of $n^2(r)$ and $\nu'(n^2)$ its derivative with respect to the argument. The normalized (to unity) angular power distribution at the output of the fibre is given by the following superposition (through Eq. (42)):

$$w'(\theta', z) = \frac{1}{W(z)} \iint\limits_{D,(P^2>0)} w'(\theta', h, k) \cdot \varrho(h, k, z)\, dh\, dk\,. \tag{43}$$

For Lambertian sources Eq. (43) assumes the following more explicit form:

$$w'(\theta', z) = \sin 2\theta' \cdot \\ \cdot \frac{R_0}{W(z)} 8\pi \iint\limits_{D,(P^2>0)} \frac{\exp[-\gamma(h, k) \cdot z] \cdot |\nu'(k^2 + n'^2 \sin^2 \theta')| \cdot k\, dh\, dk}{\sqrt{n'^2 \sin^2 \theta' - h^2/\nu^2(k^2 + n'^2 \sin^2 \theta')}}, \tag{44}$$

and for $\gamma \cdot z = 0$ it is possible to obtain from Eq. (44) the following analytical expression which holds for every refractive index profile and excludes leaky congruences:

$$w'(\theta', z) = \sin 2\theta' \cdot \frac{\pi^2 R_0}{W_0} \cdot \nu^2(n_1^2 + n'^2 \sin^2 \theta'), \\ \text{for } |n' \sin \theta'| < \Delta\,. \tag{45}$$

These expressions (apart from the metric factor $\sin 2\theta'$) give directly far-field intensity. In Fig. 12 this intensity is shown in

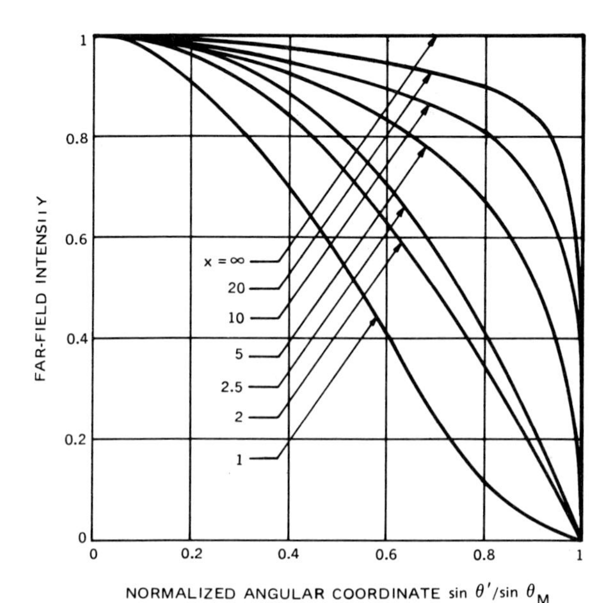

FIG. 12 – Plot of far-field intensity of fibres with different values of exponent x of the profile family (22). Leaky ray contribution is not considered.

the case of a Lambertian source excluding leaky rays and for some values of profile exponent x of Eq. (22). These results are in good agreement with results given in the literature [23]. The obtained results (Eqs. (43) and (44)) may become useful in evaluating the transmission characteristics of an optical fibre. In fact, from a direct measurement of far-field intensity and from a comparison of such quantities, obtained for different values of length z, one can obtain useful information on the evolution of optical power inside the fibre itself, particularly as regards the amount of scattering power which is present in the fibre, and the amount of inhomogeneities, impurities and imperfections which may induce scattering (see also next sections).

2.4 Power attenuation

Equations (34) and (35) supply the longitudinal evolution of power density ϱ and power W in the presence of losses which do not convert power from one ray congruence into another.

The case of power conversion (caused by the scattering of light) is discussed in Section 2.6. These Eqs. (34) and (35) may be rewritten in the following way:

$$W(z) = \iint_D \varrho_0(h, k) \exp[-\gamma(h, k)\cdot z]\, dh\, dk\,. \tag{46}$$

It can be observed that if the attenuation coefficient $\gamma(h, k)$ were the same for every congruence, the longitudinal distribution of power would be exponential. As this is not true for most cases, then an averaged attenuation coefficient has to be introduced:

$$\bar{\gamma}(z) = \frac{1}{W(z)} \frac{dW(z)}{dz} = \frac{1}{W(z)} \iint_D \gamma(h, k)\varrho(h, k, z)\, dh\, dk\,. \tag{47}$$

The z-dependence of the attenuation coefficient $\bar{\gamma}$ is due to the different attenuation suffered by the various (h, k) congruences. In general, as $z \to \infty$, $\bar{\gamma}(z)$ approaches a constant, because, while z increases, the most attenuated congruences give a more and more negligible contribution to the propagating power; then, when all but the least attenuated congruence can be ignored, we have:

$$\lim_{z\to\infty} \bar{\gamma}(z) = \min_{(h,k)\in D} \{\gamma(h, k)\}\,. \tag{48}$$

As regards some kinds of loss, the limit (48) of $\bar{\gamma}(z)$ may be practically reached for actual values of fibre length.

The attenuation coefficient in most kinds of loss is inversely proportional to the half period of the ray along the z-axis, $l(h, k)$; then the whole attenuation coefficient $\gamma(h, k)$ can be split into two different terms: $l(h, k)$ which accounts for the different optical path of various congruences, and $g(h, k)$ which accounts for the different way each congruence is affected by power attenuation:

$$\gamma(h, k) = g(h, k)/l(h, k) \tag{49}$$

in these cases and for the Lambertian source, Eq. (47) may be simplified as follows:

$$\bar{\gamma}(z) = \frac{R_0}{W(z)} \cdot \frac{8\pi}{n'^2} \iint_D g(h, k) \exp[-\gamma(h, k)\cdot z]\, dh\, dk\,. \tag{50}$$

If different causes of attenuation are simultaneously present in the fibre, it is possible to add the single attenuation coefficients of each congruence (provided that the loss tangents are small [34]). Now we are going to analyze separately the most usual causes of loss in fibres (apart from scattering which will be considered further on).

2.4.1 *Core absorption*

Let us consider a fibre with a dissipative core and let $\alpha(r)$ be the attenuation constant of a plane wave in a medium with the same composition as the layer which is at a distance r from the fibre axis (and whose refractive index is $n(r)$). For instance, in the case of fibres produced by CVD methods, $\alpha(r)$ depends not only on the attenuation constant for absorption both of silica and dopants, but also on the concentration of dopants which affects the refractive index value and gives the dependence on r. As a first approximation, the following behaviour for $\alpha(r)$ will be considered

$$\alpha(r) = \alpha_0 \cdot n(r)/n(0) . \tag{51}$$

Anyway, for any behaviour of $\alpha(r)$, it is possible to deduce the following expression of the attenuation coefficient of a congruence (h, k) due to core absorption [19]:

$$\gamma_0(h, k) = \frac{1}{l(h, k)} \int_{r_0}^{r_1} \frac{\alpha(r) \cdot n(r)}{P(r)} \, dr . \tag{52}$$

Equation (52) describes the coefficient $\gamma_0(h, k)$ as being proportional to the length of the geometric path of each ray of the congruence, weighted point-to-point with $\alpha(r)$. This is evident if we assume in Eq. (52) $\alpha(r)$ constant and then:

$$\gamma_0(h, k) = \alpha \cdot L(h, k) \tag{53}$$

$L(h, k)$ is the geometric path length along a unit length of fibre.

In Fig. 13 the behaviour of $W(z)$ (in arbitrary units), for $\alpha(r)$ expressed by Eq. (51), against the attenuation parameter $\alpha_0 z$ (in Np) both for parabolic- and step-index fibres is shown for

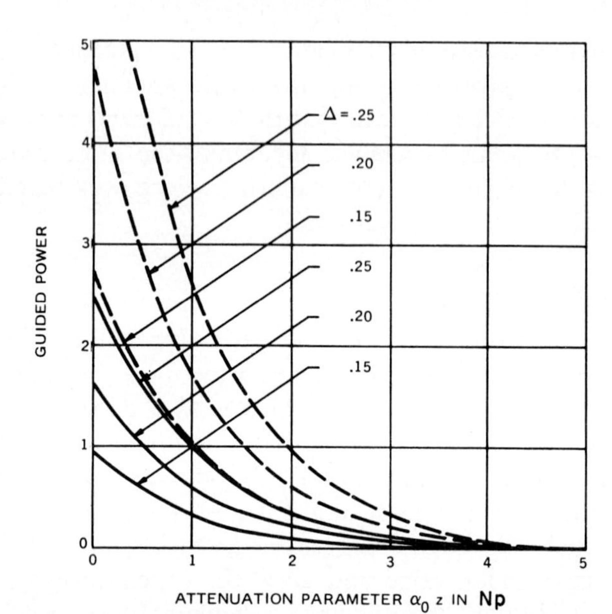

FIG. 13 – Guided power (including leaky rays) in parabolic- (solid lines) and step- (dashed lines) index fibres with dissipative core against the attenuation parameter: $\alpha_0 z$ for different values of Δ and a Lambertian source.

different values of Δ. One can see that, particularly for weakly guiding and parabolic-index fibres, the curves follow a nearly exponential behaviour; this is due to the fact that in these cases $\gamma_0(h, k)$ is practically constant for the different congruences. For these kinds of loss and Lambertian sources the following simpler expression of $\bar{\gamma}_0(0)$ (excluding leaky rays) may be deduced from Eq. (50):

$$\bar{\gamma}_0(0) = 2\int_0^a \alpha(r)n(r)[n(r) - n_1]r\,dr \bigg/ \int_0^a [n^2(r) - n_1^2]\,r\,dr \qquad (54)$$

and as shown in Fig. 13 for weakly guiding and parabolic-index fibres, $\bar{\gamma}(z)$ (for $z > 0$) may be well-approximated with its initial value $\bar{\gamma}_0(0)$ given by Eq. (54).

2.4.2 *Cladding absorption*

Geometrical optics is obtained from wave optics in the limit $\lambda \to 0$; in such an approximation the evanescent field (described by complex rays) vanishes and the whole power is considered confined in the region of real rays. Since, for guided congruences, such rays are confined inside the core, a rigorous geometrical theory cannot describe the loss phenomenon due to the presence of a dissipative cladding. In fact, such a phenomenon acts by attenuating the part of the evanescent field that propagates through the cladding. With the help of a quasi-classical approach (i.e. considering a non-rigorous geometrical theory) it is possible to explain very well such an effect [35] and to obtain the following expression of the relative attenuation coefficient for each congruence [19]:

$$\gamma_1(h, k) = \frac{1}{l(h, k)} \frac{\lambda}{2\pi} \frac{\alpha_1 n_1}{|P^2(a)|} \exp\left(-\frac{4\pi}{\lambda} \int_{r_1}^{a} |P(r)| dr\right), \tag{55}$$

(where α_1 is the attenuation constant of a plane wave in the cladding-medium). To confirm the undulatory origin of such an effect, one can see that $\gamma_1(h, k)$ vanishes when $\lambda \to 0$.

Such a phenomenon is very selective on various congruences: the congruences nearest to the guidance limit condition are much more attenuated than the others. Moreover, we obtain appreciable attenuations of power only for very large values of α_1. This is shown in Fig. 14, where the guided power $W(z)$ (in arbitrary units and for the same fibres as in Fig. 13) is given against the attenuation parameter $\alpha_1 z$ (in Np). One can see that the phenomenon has a greater effect on step-index than on parabolic-index fibres. In absolute for the same values of Δ and a, when $\alpha_1 z \geqslant 100$ Np, step-index fibres contain less residual power than parabolic ones. This is due to the fact that, since the second caustic in step-index fibres always coincides with the core-cladding interface, nearly all the congruences of such fibres have an evanescent field in the cladding greater than that of parabolic fibre congruences. Moreover one can observe that, also for very large values, of $\alpha_1 z$, parabolic-index fibres suffer small losses owing to the presence of a dissipative cladding; this means that for these fibres lower quality materials can be used for cladding.

2.4.3 *Intrinsic attenuation of leaky rays*

This phenomenon is due to the optical tunnel effect [19]. The leaky congruences, as we have already said, are radiated by

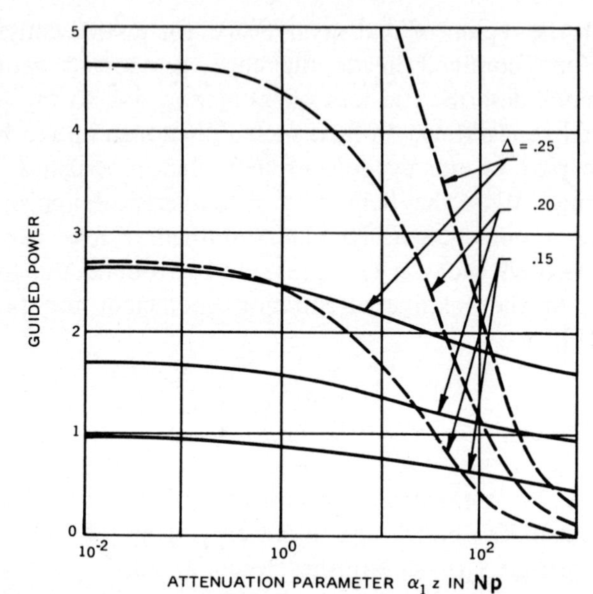

FIG. 14 – Guided power (including leaky rays) in parabolic- (solid lines) and step- (dashed lines) index fibres with dissipative cladding against the attenuation parameter: $\alpha_1 z$ for different values of Δ and a Lambertian source. It is assumed: $\lambda = 0.9\ \mu m$ and $a = 40\ \mu m$.

means of a mechanism that may be very slow, because the power leakage may take place only through the evanescent field (complex rays) which is present between the second caustic in the core and the caustic in the cladding [17]. Thus also this phenomenon cannot be predicted by a rigorous geometrical theory: in fact, such a theory does not explain any power transmission between these two caustics. Also in this case the phenomenon may be explained by means of the quasi-classical approach, and it is possible to obtain the following expression of the relative attenuation coefficient of each congruence [19]:

$$\gamma_l(h, k) = \frac{1}{2l(h, k)} \cdot \frac{\exp[-2L]}{(1 + \exp[-2L]/4)^2};$$

$$\left(L = \frac{2\pi}{\lambda} \int_{r_1}^{r_2} |P(r)| dr\right). \qquad (56)$$

Also in this case, to confirm the typical undulatory character of the phenomenon, $\gamma_l(h, k)$ vanishes for $\lambda \to 0$ (limit of rigorous geometrical optics). Moreover, this coefficient is highly dependent on the distance between the two caustics of radii r_1 and r_2 and increases when the two caustics approach each other. Then losses due to optical tunnel effect are larger in step-index fibres (where $r_1 = a$ always) than in parabolic-index fibres (where $r_1 \leqslant a$).

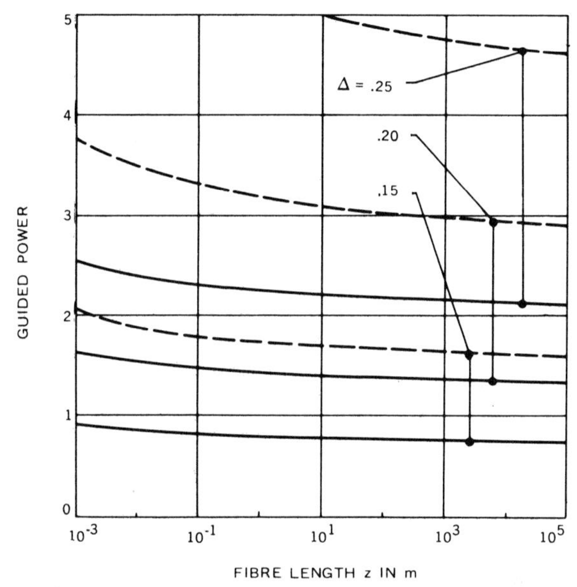

FIG. 15 – Guided power (including leaky rays) in parabolic-(solid lines) and step-(dashed lines) index fibres (considering the intrinsic loss of leaky rays, by optical tunnel effect) against the fibre length z for different values of Δ and a Lambertian source. It is assumed $\lambda = 0.9$ μm, $a = 40$ μm.

This is shown in Fig. 15, where the residual guided power $W(z)$ is shown against the length z both for parabolic- and step-index fibres with different values of Δ, in the presence of the optical tunnel effect only (the units of $W(z)$ and other quantites are the same as Fig. 13). Most leaky power loss occurs in the first portion of the fibre, but about 30% of leaky power may still be present in the fibre also after 10^4 m of fibre length [19]. It is also important to point out that while this optical tunnel effect (which acts even in fibres without imperfections or impurities) attenuates only leaky rays, there are in addition other loss phenomena. These are caused by either fibre imperfections or impurities (particularly the presence of a dissipative cladding [19]) that, even if attenuating all rays, may attenuate chiefly leaky rays and considerably reduce their contribution. Finally, the influence of non-circularity of core cross-section on selective additional extinction of leaky rays is debated [36, 37]. The arguments of Ref. [37] seem convincing, but the severe attenuation mechanism of leaky rays observed by Costa and Sordo [29] might also be due to a certain amount of mode coupling [19] and birefringence (in their transient region) induced by mechanical stresses in addition to non-circularity itself.

2.4.4 *Interface reflection losses*

In the case of step-index fibres the presence of an imperfect core-cladding interface may cause both absorption and light scattering. This phenomenon cannot be analyzed like a volume loss (as the previous ones), but the relative coefficient is proportional to the number of reflections per unit length of fibre and depends on the reflectivity ϱ' of the interface according to [19]:

$$\gamma_r(h, k) = \frac{|\ln \varrho'|}{2l(h, k)} = \frac{|\ln \varrho'| \cdot \operatorname{tg} \theta}{2\sqrt{a^2 - r^2 \sin^2 \varphi}} . \tag{57}$$

The behaviour of guided power in the presence of this kind of loss is illustrated in Fig. 16 against the attenuation parameter $|\ln \varrho'| z$ (in metres) for step-index fibres with various values of Δ and a. Losses are smaller for fibres with a larger core radius (because there are less reflections at the interface). However, losses begin to be very drastic from $|\ln \varrho'| z \simeq (10^3 \div 10^4)$ m.

2.5 Pulse distortion

The evaluation of the bandwidth of an optical fibre is of great importance in characterizing the associated communication systems and can be calculated through the analysis of the distortion suffered by a power pulse launched into the fibre after a given fibre length.

If an infinitely narrow optical power pulse (i.e. with a Dirac-delta shape in time) is launched into the fibre, the pulse detected

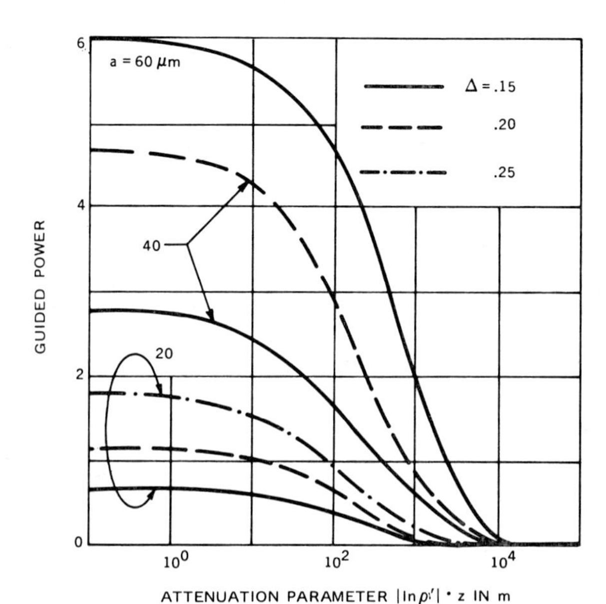

FIG. 16 – Guided power (including leaky rays) in step-index fibres with an imperfect core-cladding interface against the attenuation parameter: $|\ln \varrho'| z$ for different values of a and Δ and a Lambertian source.

at the output has a finite width; the relative broadening in multimode fibres is essentially due to different travelling times along the fibre of rays of various congruences. The travelling time (or group delay) of each ray depends on the length of the optical path of the rays in the fibre and on the material dispersion (inhomogeneous dispersion [38]) of the fibre. Let us consider a ray of a congruence (h, k); its travelling time through the unit length of fibre can be expressed as follows [38-40]:

$$\tau(h, k) = \frac{1}{c \cdot l(h, k)} \int_{r_0}^{r_1} \frac{N(r) \cdot n(r)}{P(r)} \, dr \,, \tag{58}$$

where c is the light speed in free space and $N(r)$ is the group refractive index at the point of the fibre at a distance r from its axis, defined according to:

$$N(r) \equiv n(r) - \lambda \frac{\partial n(r)}{\partial \lambda} \,. \tag{59}$$

In Eq. (59), we have assumed a dependence of n, and then of N, also on λ: $n(r) = n(r, \lambda)$, $N(r) = N(r, \lambda)$. Then in Eq. (58) a dependence of τ on λ is also present: $\tau(h, k) = \tau(h, k, \lambda)$; this dependence is obtained directly through $n(r, \lambda)$ and $N(r, \lambda)$ and indirectly it is also contained in P, l, r_0 and r_1 because all these quantities depend on the form of $n(r)$ $(= n(r, \lambda))$.

Now let us imagine that a Dirac pulse (whose optical energy is normalized to unity): $\delta(t)$ is launched at the fibre input end; we assume temporarily that we have (even if it is not possible, owing to the uncertainty principle) a monochromatic source: at the output of a length z of fibre we have the following normalized pulse response:

$$w(t) = \frac{1}{W(z)} \iint_D \varrho(h, k, z) \cdot \delta(t - z \cdot \tau(h, k)) \, dh \, dk \,. \tag{60}$$

In Eq. (60) a dependence on λ: $w(t) = w(t, \lambda)$ is also understood according to Eqs. (58) and (59). Then if we have a real source that launches into the fibre a temporal pulse described by a function of time t and wavelength λ, $W_\lambda(t)$, the pulse response of the fibre can be obtained from Eq. (60), according to the following expression:

$$\begin{aligned} W(t) &= \int_{\Delta\lambda} d\lambda \int_{\Delta t'} W_\lambda(t') \, w(t - t', \lambda) \, dt' = \\ &= \int_{\Delta\lambda} d\lambda \iint_D \varrho(h, k, z) \cdot W_\lambda(t - z \cdot \tau(h, k, \lambda)) \, dh \, dk \end{aligned} \tag{61}$$

$\Delta\lambda$ represents the wavelength domain of the launched pulse (its spectral bandwidth) and $\Delta t'$ its time « window width », that is the time interval which includes the whole pulse; D is the congruence domain (see Fig. 3*b*). Equations (60) and (61) allow the form of the fibre output pulse to be obtained for any kind of excitation and loss (except the scattering loss that will be analyzed in the next section).

Now we are going to analyze the various elements that contribute to affect the fibre response to the Dirac pulse, and finally some useful formulae for calculating the pulse width will be given.

2.5.1 *Refractive index profile*

Pulse broadening is highly dependent on the refractive index profile of the fibre. When the index profile is not a step one

($x \to \infty$ in Eq. (22)), but is smoothed (for $r \simeq a$), a certain equalization phenomenon begins to take place inside the fibre among the optical paths of various rays, and this leads to a less evident broadening of the output pulse.

Such an effect may be illustrated using the family of refractive index profiles of Eq. (22). We call Δt the window width of the output pulse, t_0 the travelling time for the axial ray that, ignoring the material dispersion, has the following expression:

$$t_0 = n_0 \cdot z/c \tag{62}$$

and $\bar{\varepsilon}$ the following ratio of refractive indices:

$$\bar{\varepsilon} \equiv n_1/n_0 \,. \tag{63}$$

Using these notations we obtain the form $\Delta t/t_0$ (ignoring the contribution of leaky rays but for any difference between n_0 and n_1). Through Eq. (58) we have that, for this profile family, τ depends only on k according to:

$$\tau_x(k) = \frac{1}{c} \frac{x n_0^2 + 2k^2}{(x+2)k} \,. \tag{64}$$

One can see that for $x \geqslant 2$ the axial ray ($k = n_0$) is the fastest, while for $x \leqslant 2\bar{\varepsilon}$ it is the slowest. The congruence rays at guidance limit condition ($k = n_1$: excluding leaky rays) are the slowest for $x \geqslant 2\bar{\varepsilon}$ and the fastest for $x \leqslant 2\bar{\varepsilon}^2$. For $2\bar{\varepsilon}^2 \leqslant x \leqslant 2$ the fastest rays are those belonging to congruences with a value of k intermediate between n_0 and n_1 and given by:

$$k = n_0 \sqrt{x/2} \,. \tag{65}$$

Taking into account these considerations we can obtain the following expressions of $\Delta t/t_0$:

$$\frac{\Delta t}{t_0} = \begin{cases} \delta_1(x) - \delta_2(x)\,, & \text{for } x \leqslant 2\bar{\varepsilon}^2 \\ \delta_1(x)\,, & \text{for } 2\bar{\varepsilon}^2 \leqslant x \leqslant 2\bar{\varepsilon} \\ \delta_2(x)\,, & \text{for } 2\bar{\varepsilon} \leqslant x \leqslant 2 \\ \delta_2(x) - \delta_1(x)\,, & \text{for } x \geqslant 2 \end{cases} \tag{66}$$

where:

$$\delta_1(x) \equiv \frac{(\sqrt{2} - \sqrt{x})^2}{x + 2}, \qquad \delta_2(x) \equiv \frac{(\bar{\varepsilon}\sqrt{2} - \sqrt{x})^2}{(x + 2)\,\bar{\varepsilon}}. \tag{67}$$

Figure 17 shows that where $\Delta t/t_0$ is given against x, the dependence of pulse broadening on the kind of profile is particularly strong in the neighbourhood of $x = 2\bar{\varepsilon}$. At this point in fact

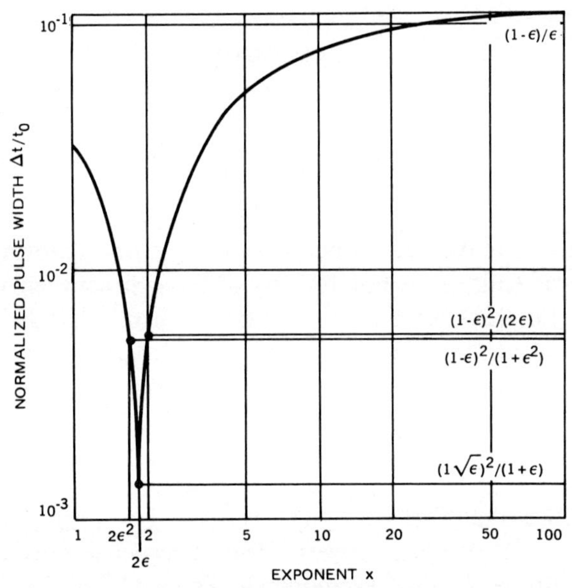

FIG. 17 – Relative broadening $\Delta t/t_0$ of Dirac pulse, in fibres excited by a Lambertian source, against the exponent x. (This figure holds for $\bar{\varepsilon} = 0.9$).

we have the following pulse broadening minimum:

$$\frac{\Delta t}{t_0} = \frac{(1 - \sqrt{\bar{\varepsilon}})^2}{1 + \bar{\varepsilon}} \tag{68}$$

in which the curve presents a cuspidal point: then also small fluctuations of x around $2\bar{\varepsilon}$ may cause a considerable removal from minimum broadening. For small differences between the refractive indices n_0 and n_1 ($\bar{\varepsilon} \simeq 1$) Eqs. (64-68) agree with the results already obtained in the literature [23].

As regards refractive index profiles which do not follow Eq. (22), several methods have recently been proposed to investigate their bandwidth performance. In fact, the refractive index profile may exhibit a dip on the fibre axis or be non-circular. In such cases, it is possible to correlate the refractive index profile to an equivalent profile, following Eq. (22) [41], which has the same pulse response.

Another interesting approach to the problem of optimizing bandwidth performance of graded-index optical fibres has recently been proposed [42-44]. It consists in introducing a further degree of freedom in the refractive index profile, in order to impose further design performance. In particular, the refractive index profile can be expressed as follows:

$$n^2(r) - n_1^2 = \sum_1^N {}_i \, \Delta_i^2 \left[1 - \left(\frac{r}{a}\right)^{x_i}\right] \tag{69}$$

where $r < a$; x_i, $\Delta_i > 0$ and numerical aperture Δ of Eq. (23) can now be expressed as

$$\Delta = \sqrt{n_0^2 - n_1^2} = \sqrt{\sum_1^N {}_i \, \Delta_i^2}\,. \tag{70}$$

In this way the bandwidth characteristic of the fibre can be greatly improved since the shape of the dependence of pulse broadening versus wavelength can be highly modified. In fact, it is possible to design an optical fibre whose r.m.s. pulse broadening versus wavelength presents two points of minimum modal dispersion (allowing the use of the same fibre at two different wavelengths with the same performance). Moreover, the r.m.s. pulse broadening may present zero slope at the point of minimum dispersion (reducing the stringent tolerances on the shape of refractive index profile due to the cuspid of Fig. 17). It seems possible, by means of this technique, to obtain an optical fibre presenting a region of minimum modal dispersion over an extended range of wavelengths (from 0.7 to 1.0 μm) [44].

2.5.2 *Influence of leaky rays*

In general the fibre travelling time of leaky rays is either greater (for instance when $x > 2\bar{\varepsilon}$ in profiles of Eq. (22)) or smaller (for

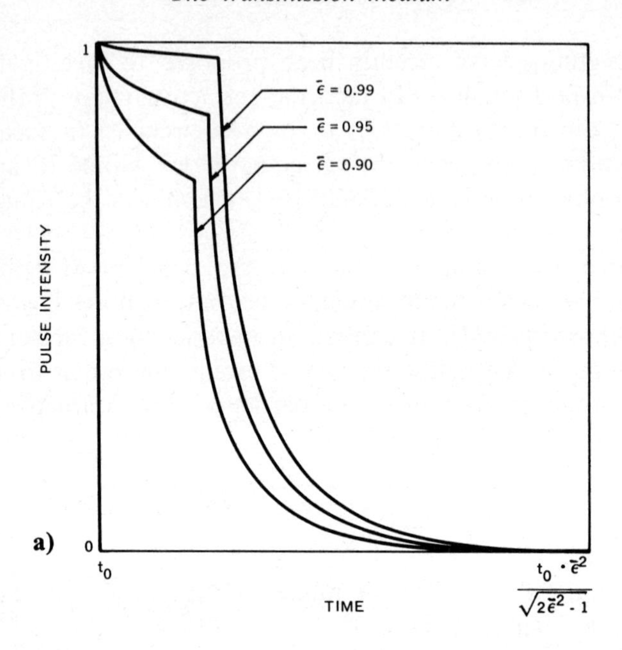

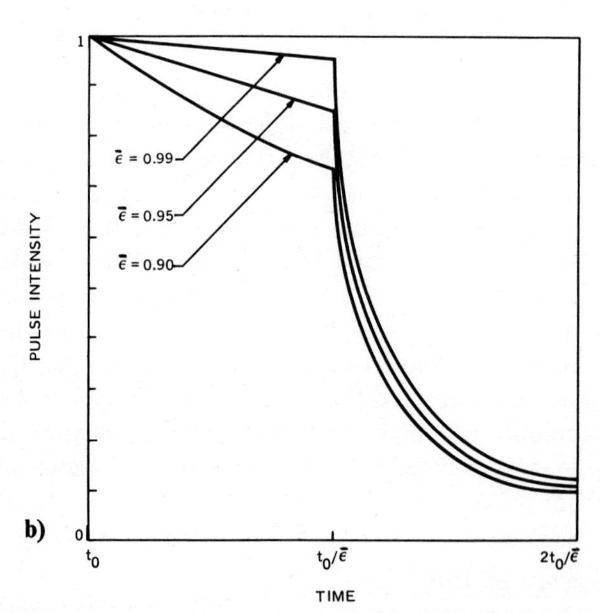

FIG. 18 – Fibre response to Dirac pulse, excited by a Lambertian source, including leaky ray contribution, for parabolic-(*a*) and step-(*b*) index fibres and for different values of $\bar{\varepsilon}$.

instance when $x < 2\bar{\varepsilon}^2$ in profiles of Eq. (22)) than travelling times of all other rays: this fact causes a greater broadening of the output pulse [9, 45].

This effect is shown in Figs. 18 where the shape of fibre output pulses is illustrated (ignoring material dispersion) for parabolic-index fibres (*a*) and for step-index fibres (*b*) excited by a Lambertian source and for various values of $\bar{\varepsilon}$. One can see that leaky rays produce a tail in the pulse (after the discontinuity of the slope of the curves) that in the case of step-index fibres may be unlimited. This effect is greater for more strongly guiding fibres (for smaller values of $\bar{\varepsilon}$), particularly in step-index fibres, because in this case the leaky power is concentrated mostly in shorter times.

2.5.3 *Core and cladding absorption*

Both core and cladding absorption may contribute to a smaller broadening of the fibre output pulse [9, 46]. In fact these kinds of attenuation are, in general, selective for congruences with a greater group delay. Such effects are shown in Figs. 19 and 20

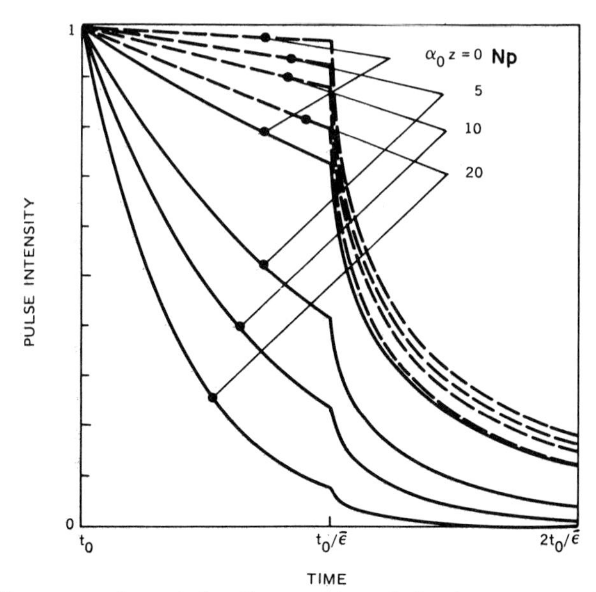

FIG. 19 – Response of step-index fibres, with a dissipative core, to Dirac pulse, excited by a Lambertian source, for different values of $\bar{\varepsilon}$ and of attenuation parameter: $\alpha_0 z$. Leaky ray contribution is included.

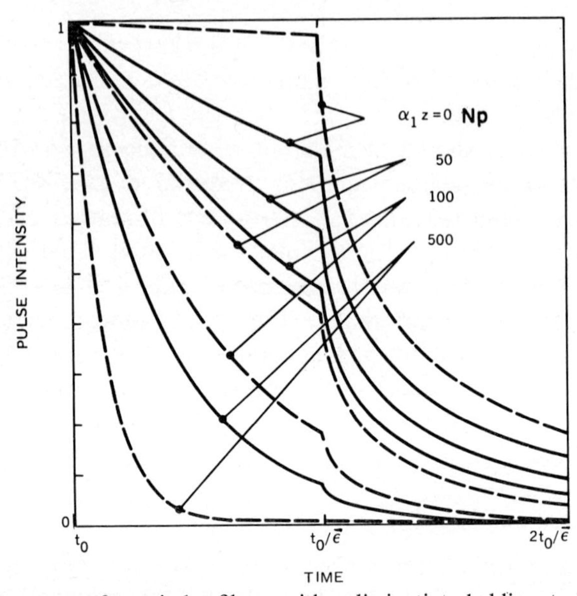

FIG. 20 – Response of step-index fibres, with a dissipative cladding, to Dirac pulse, excited by a Lambertian source, for different values of $\bar{\varepsilon}$ and of attenuation parameter: $\alpha_1 z$. Leaky ray contribution is included and it is assumed $\lambda = 0.9$ μm and $a = 40$ μm.

in the case of step-index fibres (for which they are more considerable). In these figures the fibre output pulse shape is given for different values of attenuation parameters: $\alpha_0 z$ (for core absorption: Fig. 19) and $\alpha_1 z$ (for cladding absorption: Fig. 20) and for $\bar{\varepsilon} = 0.99$ (dashed lines) and $\bar{\varepsilon} = 0.9$ (solid lines). On the whole the pulse narrows as attenuation values increase; however, core absorption in weakly guiding fibres does not produce appreciable effects.

2.5.4 *Source radiance and Material dispersion*

The radiance distribution of the source may also affect the pulse shape at the fibre output. In fact, a non-Lambertian source excites to a different extent the various congruences and this may be reflected in the output pulse shape. However, it must be stressed that in practice a source more directional than the Lambertian one produces appreciable effects only for strongly guiding fibres: for instance for sources whose radiance may be expressed by Eq. (20) with $y = 10$ the output pulse width in a step-index fibre reduces to about a half only for $\bar{\varepsilon} = 0.9$ [9].

Another very important quantity that can affect fibre pulse response is the spectral distribution of source radiance, particularly for low distortion fibres. In fact, since the group delay of each ray depends on λ: $\tau(h, k, \lambda)$, through the material dispersion of the fibre, a spectral spread of radiation emitted by the source gives rise to a temporal spread of each component (due to each ray congruence) of the output pulse. Then the actual pulse shape at the output is given by the convolution expressed in Eq. (61) and in general it is larger than a monochromatic pulse response. In practice this further broadening of the pulse response when semiconductor laser sources are used is appreciable only for quasi-parabolic index fibres, while, when LED sources are used, it may be negligible only for step-index fibres. This is due to the different spectral widths of these sources and to the different amount of distortion in parabolic- and step-index fibres.

It must be emphasized that, also in the case of sources with very narrow spectral width, material dispersion gives rise to effects that may strongly influence pulse distortion. In fact, in Eq. (58) we can see that $\tau(h, k)$ contains the group refractive index $N(r)$ that is highly dependent on the inhomogeneous dispersion of the fibre [38]. This effect for fibres with an index profile given by (22) can be numerically approximated by a translation of the profile exponent x: i.e. the pulse response of a fibre with material dispersion and index profile of exponent x_0 is well-approximated by the pulse response of a fibre without material dispersion but with a profile exponent $x_0 - \delta$ [38, 47]. δ depends on the magnitude of material dispersion but usually is smaller than 0.5. Then such an effect is practically negligible for step-index fibres.

Anyway, for a correct evaluation of the generalized expression for the modal delay (58), material dispersion must be taken into account in all the cases when pulse broadening by modal or waveguide dispersion is comparable (multimode graded-index fibres with nearly optimal refractive index profile) or negligible (monomode fibres with $\lambda \leqslant 1\ \mu$m) with respect to material dispersion. Experimental data of material dispersion can be achieved for instance by measurements on bulk specimens, and generally aim at an evaluation of refractive index versus λ according to the Sellmeier dispersion expansion [48]. More details on this subject will be given in Chapter 3. Here we point out that a reliable evaluation of dispersion properties is of great importance when operating in the neighbourhood of optimal profiles in order to determine accurately the point of zero material dispersion [49]. In this region, in fact, the information capacity of the optical fibre can grow up to very large values, since material and modal dispersion may cancel each other out. Thus, the overall dispersion of the fibre should be limited only by residual second-order chromatic effects [50].

2.5.5 *Pulse width*

In every case a very good evaluation of the broadening of a Dirac pulse in the unit length of fibre is given by its standard deviation:

$$\Delta\tau \equiv \sqrt{\langle\tau^2\rangle - \langle\tau\rangle^2}; \tag{71}$$

where:

$$\langle\tau\rangle \equiv \frac{1}{W(z)} \iint_D \varrho(h, k, z) \cdot \tau(h, k)\, dh\, dk\,, \tag{72}$$

$$\langle\tau^2\rangle \equiv \frac{1}{W(z)} \iint_D \varrho(h, k, z) \cdot \tau^2(h, k)\, dh\, dk\,. \tag{73}$$

In the case of fibres with uniform losses, excited by Lambertian sources, a good approximation of Eq. (71) can be given by:

$$\Delta\tau \simeq 2|\langle\tau\rangle - \tau_0| \tag{74}$$

where $\tau_0 = N(0)/c$ is the travelling time of an axial ray in the unit length of fibre. Equation (74) is valid when τ_0 is an extremal travelling time (this hypothesis holds apart from refractive index profiles close to the minimum dispersion ones). In Eq. (74) $\langle\tau\rangle$ is given by Eq. (72), but for uniformly excited fibres the following simpler expression can be used:

$$\langle\tau\rangle = \frac{2}{c} \int_0^a N(r) n(r) [n(r) - n_1]\, r\, dr \Big/ \int_0^a [n^2(r) - n_1^2]\, r\, dr\,. \tag{75}$$

2.6 Light scattering

Light scattering in optical fibres should be dealt with in a special way because it causes not only power losses but also power conversions from one congruence into another. Then in this case formula (46) no longer represents the power distribution along the z-axis.

Also the temporal power distribution may be considerably modified by the presence of power conversions among guided congruences and we can often notice a phenomenon of equalization of optical paths of various rays that can lead to a narrower pulse response than in the absence of scattering [51].

The phenomenon of light scattering in optical fibres should be analyzed statistically. In general, we must distinguish two cases, for which two different formalisms must be used: the case of rare events (of scattering) and the case of frequent events.

In the case of rare events, the optical power is subjected on average to a very small number of scatterings in the given length of fibre, so that the probability that the light will not undergo any scattering may be significant This case is associated to wide angle scattering because the cases of rare events of narrow angle scattering do not have in practice any appreciable effect either on loss or on power conversion. Then we may include among these cases of rare events: Rayleigh scattering (the ultimate cause of loss in the fibre) and light scattering due to inhomogeneities (such as micro-bubbles, micro-crystals, micro-fractures, etc.) which in a first approximation may be considered as an isotropic scattering.

In the case of frequent events, a large number of scatterings occur in a given length of fibre, so that the probability that the light will not undergo any scattering is definitely negligible. This case is associated to narrow angle scattering because the cases of frequent events of wide angle scattering lead to enormous attenuations (due to the considerable power lost at each scattering) which in practice make them of no interest. Then we may consider among these cases of frequent events the scattering due to small geometric or optic imperfections of the fibre, such as micro-bending, smooth fluctuations of the core radius, or of the numerical aperture Δ or of the refractive index profile of the fibre.

The cases described so far are extreme cases, and, even if in practice they are both normal, it should be pointed out that intermediate situations may exist. In the following the two different formalisms by which the two cases can be described will be outlined.

2.6.1 *Rare events of scattering*

Let P_n be the distribution of probability that the whole optical power, along the fibre length z, undergoes n scattering; making the realistic assumption that scattering is a Poisson process, we obtain [13]:

$$P_n = \exp[-\alpha_s z] \frac{(\alpha_s z)^n}{n!} \tag{76}$$

where α_s is the attenuation coefficient by scattering of core material that is related to the mean free path l_c of a single photon between two consecutive scatterings by:

$$\alpha_s = 1/l_c \tag{77}$$

Calling $\varrho_n(h, k, z)$ the power distribution among various congruences at the output of the fibre when assuming that in the length z the whole optical power undergoes n scatterings, the actual distribution of output power $\varrho(h, k, z)$ can be evaluated according to the following weighted superposition of various contributions:

$$\varrho(h, k, z) = \\ = \sum_0^\infty {}_n P_n \cdot \varrho_n(h, k, z) = \exp[-\alpha_s z] \sum_0^\infty {}_n \frac{(\alpha_s \cdot z)^n}{n!} \varrho_n(h, k, z). \tag{78}$$

Now the contributions $\varrho_n(h, k, z)$ must be determined; to this purpose let us define the scattering function $F(h, k; h', k')$ which is the density of probability that, owing to a single scattering event, a ray is converted from the ray congruence (h, k) to the ray congruence (h', k'). This function may be carried out from the differential cross-section of the single scattering centre [13] and should fulfil the following normalization relation:

$$\iint_{D \cup R} F(h, k; h', k')\, dh'\, dk' = 1 \tag{79}$$

where R represents in the (h, k) plane the set of ray congruences which are not considered as guided (see Fig. 3*b*). Assuming that all the optical power undergoes n scatterings, if $\varrho_j^{(n)}(h, k)$ is the power distribution after the j-th scattering and $\varrho_{j-1}^{\prime(n)}(h, k)$ the power distribution before it, owing to the way $F(h, k; h', k')$ has been defined, we have:

$$\varrho_j^{(n)}(h', k') = \iint_D F(h, k; h', k')\, \varrho_{j-1}^{\prime(n)}(h, k)\, dh\, dk\,; \quad j = 1, 2, \ldots, n\,. \tag{80}$$

In its turn $\varrho_{j-1}^{\prime(n)}(h, k)$ may be deduced from the power distribution $\varrho_{j-1}^{(n)}(h, k)$ after the $(j-1)$-th scattering; taking into account the loss (owing to other causes) between two consecutive scattering centres, if $\gamma(h, k)$ is the relative attenuation coefficient and z/n the average distance between these two centres, we have:

$$\varrho_{j-1}^{\prime(n)}(h, k) = \varrho_{j-1}^{(n)}(h, k) \exp\left[-\gamma(h, k)\cdot z/n\right]. \tag{81}$$

Starting from the input distribution: $\varrho_0(h, k) = \varrho_0^{(n)}(h, k)$, and using Eqs. (80) and (81) iteratively the output searched distribution $\varrho_n(h, k, z) = \varrho_n^{(n)}(h, k)$ can be obtained.

Since this formalism is usually applied to wide angle scatterings (in which large losses occur at every event), one can see that the first contributions ($n \leqslant 5$) are sufficient to characterize the phenomenon in the cases of practical interest. Such a formalism has been applied to step index fibres [13, 52] and, for Rayleigh scattering only, the following expression of the power attenuation coefficient in weakly guiding fibres can be demonstrated:

$$\gamma_s = \alpha_s(1 + 3\bar{\varepsilon})/4\,. \tag{82}$$

Such a formalism may be used also to derive the pulse response of the fibre. In fact, analogously to Eq. (78), this pulse response can be written as the weighted superposition of various contributions:

$$W(t) = \sum_0^\infty {}_n\, P_n \cdot W_n(t)\,. \tag{83}$$

$W_n(t)$ may be obtained iteratively as in the case of $\varrho_n(h, k, z)$ [13]. The response to Dirac pulses of weakly guiding fibres, excited by Lambertian sources in the presence of wide angle scattering, in practice does not differ from the response of a dissipative fibre. This is due to the very small contributions of terms with $n \geqslant 1$

in the superposition (83) [52], and then in this case the formalism may be reduced to the formalism of absorption losses (which has been dealt with in preceding sections).

As regards strongly guiding step-index fibres, the response to a Dirac pulse shows a bump for times which are intermediate between the minimum and the maximum travelling time of the fibre. However, this bump is so modest that it does not lead to a less broadened pulse. This effect is shown in Fig. 21 where

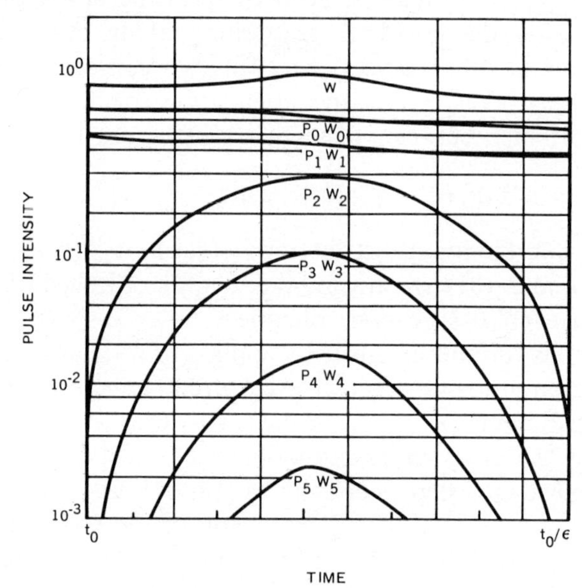

FIG. 21 – Response $W(t)$ of step-index fibres, with Rayleigh scattering, to Dirac pulse, excited by a Lambertian source, together with contribution $(P_n \cdot W_n(t))$ to $W(t)$ up to $n = 5$. Leaky ray contribution is escluded and it is assumed $\alpha_s z = 4$ Np and $\bar{\varepsilon} = 0.9$.

(excluding the contribution of leaky rays) the pulse response $W(t)$ and the first five weighted contributions $P_n \cdot W_n(t)$ of Eq. (83) are shown in the presence of Rayleigh scattering with $\alpha_s z =$ $= 4$ Np in a step-index fibre with $\bar{\varepsilon} = 0.9$ excited by a Lambertian source. A certain equalization of the optical path is reached in the contributions $P_n \cdot W_n(t)$ so that they have a narrower width; but, owing to the large loss at each scattering event, they become negligible while n increases, and contribute to the output pulse $W(t)$ only with a modest bump.

2.6.2 *Frequent events of scattering*

The preceding formalism may be simplified in the case of frequent scattering events if it is possible to assume that a continuous transformation of the distribution $\varrho(h, k, z)$ occurs along the fibre owing to the scattering.

In this case a different scattering function $f(h, k; h', k')$ must be considered, which represents the density of probability that (in the unit length of fibre and owing to a single scattering event), a ray is converted from a ray congruence (h, k) into the ray congruence (h', k'). Also this scattering function may be deduced from the differential cross-section of the single scattering centre; considering that the following normalization relation holds:

$$\iint_{D \cup R} f(h, k; h', k')\, dh'\, dk' = \alpha_s(h, k) = 1/l_c(h, k) \tag{84}$$

where the mean free path l_c (and then α_s) may depend on h and k; for instance, for scattering due to an imperfect core-cladding interface in step-index fibres we have approximately: $l_c(h, k) \simeq l(h, k)$ (see Eq. (31)). Owing to the way $f(h, k; h', k')$ is defined, the following transport equation (in the sense of Boltzmann transport equation [53]), which rules the continuous evolution of $\varrho(h, k, z)$, can be easily demonstrated:

$$\frac{\partial}{\partial z}\varrho(h, k, z) = -[\alpha_s(h, k) + \gamma(h, k)]\cdot\varrho(h, k, z) + \\ + \iint_D f(h', k'; h, k)\varrho(h', k', z)\, dh'\, dk' \tag{85}$$

($\gamma(h, k)$ represents the attenuation coefficient due to other causes).

An expression describing the temporal evolution of power distribution may also be deduced from Eq. (85). To this purpose we should include the variable t in $\varrho(h, k, z)$. In this way a spatial-temporal distribution: $\varrho(h, k, z, t)$ can be obtained, which, in the absence of scattering, has the following expression (that may be deduced from Eq. (61)):

$$\varrho(h, k, z, t) = \varrho(h, k, z)\cdot W_\lambda(t - z\cdot\tau(h, k)) \tag{86}$$

$\varrho(h, k, z, t)$ supplies the spatial-temporal distribution of power of a given (h, k) ray congruence at a certain fibre length z; the integration over h and k gives directly the pulse shape at the output of a fibre length z. Such a spatial-temporal distribution in the presence of scattering, in which a continuous transformation of power distribution can be assumed, no longer has the expression (86) but must fulfil a transport equation similar to Eq. (85) where the differential operator $\partial/\partial z$ is replaced by: $\partial/\partial z + \tau(h, k) \cdot \cdot \partial/\partial t$. This last operator takes into account the group delay of rays of each congruence between two consecutive scatterings (which are considered infinitely close to each other): $dt = \tau(h, k)\, dz$.

By means of this new transport equation the pulse response of the fibre in presence of scattering may be analyzed. In particular the case of frequent narrow-angle scattering modifies considerably the pulse shape and the output pulse may be narrower than the pulse in the absence of scattering, owing to the equalization of the actual paths of all rays. This is shown in Fig. 22 where for a step-index fibre with $\bar{\varepsilon} = 0.9$ the responses to a Dirac pulse launched by a Lambertian source are illustrated.

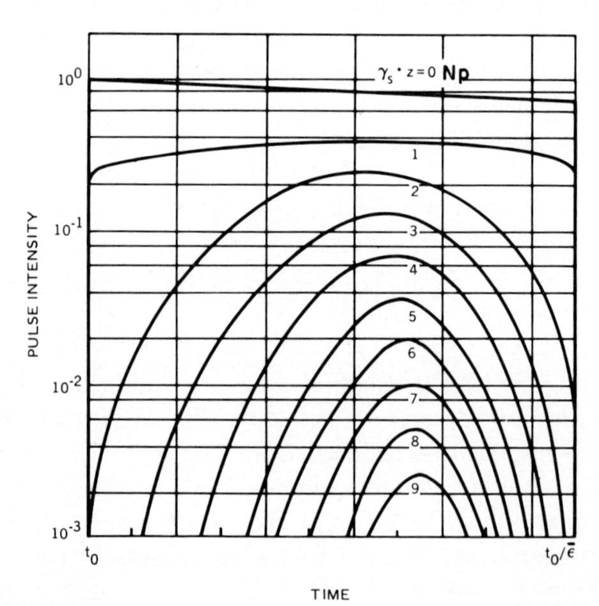

FIG. 22 – Response of step-index fibres, with narrow-angle scattering, to Dirac pulse, excited by a Lambertian source, for different values of attenuation parameter $\alpha_s z$ and $\bar{\varepsilon} = 0.9$. Leaky ray contribution is excluded.

The various curves are given for different values of the whole attenuation due to narrow-angle scattering: $\gamma_s z$ (in Np). Leaky rays are excluded. It is interesting to observe that while $\gamma_s z$ increases, the width of pulse response becomes narrower and narrower with respect to the pulse width in the absence of scattering and the pulse approaches a Gaussian shape. Moreover, to confirm the effect of full conversion of power among the various congruences, we may observe an increase of the average travelling time along the unit length of fibre. Finally, we can observe that there is a particular value of the attenuation $\gamma_s z$ (in this case between 1 and 2 Np) beyond which the power conversion reaches the steady state: only for greater values of $\gamma_s z$ does an effective reduced broadening of the pulse take place. Moreover, beyond this point the behaviour of pulse width versus the fibre length is no longer expressed by a linear law, but the square-root law arises. Such a change of law, predicted by a modal method [51], was also observed experimentally [54, 55].

2.6.3 *Micro-bending*

One of the most important causes of frequent events of scattering is microbending.

Optical fibres employed as a transmission medium in telecommunication systems will have to be assembled in cables in order to prevent their accidental damage and to provide additional tensile strength. In general, small imperfections or slight roughness in the mechanical structure will result in contact forces between the fibre and the supporting surface; random lateral deviations of the fibre axis will arise, which may produce quite significant losses. If the supporting structure is assumed to exert on the fibre a contact lateral force $F(z)$ per unit length, the lateral displacement $y(z)$ produced in this way is related to $F(z)$ according to [56, 57]:

$$EI \frac{d^4 y(z)}{dz^4} = F(z) \tag{87}$$

where E is the Young's modulus and I the moment of inertia of the fibre cross-section.

Integrating Eq. (87) yields an explicit expression of the lateral displacement of the fibre axis; however, this approach must be applied to a statistical model to evaluate the effects of small random perturbations.

The evolution of optical power flow can be considered using the same formalism as for narrow-angle scattering, at least for multimode fibres; in fact, micro-bending gives rise to geometric fluctuations of the fibre axis direction: $y(z)$ derived from Eq. (87) specifies (e.g. through its spatial transform) the scattering function of Eq. (85) which governs the continuous spatial and temporal evolution of optical power inside the fibre. Thus also in this case the same observations made about narrow angle scattering can be repeated, as regards optical power attenuation and reduced pulse broadening, which have been summarized in Fig. 22. However, it must be stressed that the micro-bending phenomenon, owing to its random mechanical origin, will cause losses which may be quite drastic [57], while it can hardly be used to produce intentional mode coupling in order to reduce pulse broadening. There may be some undulatory aspects in this phenomenon as in the case of narrow angle scattering in general; they are considered by means of the modal approach (see 2.10).

B. Modal approach

2.7 Propagation modes in optical fibres

Starting from Maxwell equations the propagation of electromagnetic waves inside a dielectric cylindrical waveguide of arbitrary refractive index profile is analyzed. Expressions of field components are given and the different discrete field configurations (propagation modes) are illustrated.

2.7.1 *Electromagnetic field equations*

The expression of electromagnetic (e.m.) field components can be derived starting from Maxwell equations which, dealing with dielectric waveguides, can be expressed in the following stationary form (having assumed a temporal dependence according to $\exp(j\omega t)$ — ω: angular frequency):

$$\begin{cases} \nabla \times \mathbf{E} = -j\omega \mathbf{B}, & (\nabla \cdot \mathbf{B} = 0), \\ \nabla \times \mathbf{H} = j\omega \mathbf{D}, & (\nabla \cdot \mathbf{D} = 0). \end{cases} \tag{88}$$

The electric and magnetic flux density vectors **D** and **B** are linearly related to electric and magnetic field vectors **E** and **H** according to:

$$\mathbf{D} = \varepsilon(r) \cdot \mathbf{E}; \qquad \mathbf{B} = \mu_0 \cdot \mathbf{H} \tag{89}$$

where the electric permittivity $\varepsilon(r)$ (which has been considered to depend on the radial coordinate only) is related to the index profile of the fibre (of core radius a) through:

$$\varepsilon(r) = \begin{cases} n^2(r) \cdot \varepsilon_0, & \text{for } r \leqslant a \\ n_1^2 \cdot \varepsilon_0, & \text{for } r > a \end{cases} \tag{90}$$

(ε_0 and μ_0 are the electric permittivity and magnetic permeability of free space).

Considering a system of cylindrical polar coordinates (r, ψ, z) coaxial to the fibre, the following dependence of field vectors on ψ and z can be factorized:

$$\begin{cases} \mathbf{E}(r, \psi, z) = \mathbf{E}(r) \cdot \exp[-j(\beta z + \nu\psi)], \\ \mathbf{H}(r, \psi, z) = \mathbf{H}(r) \cdot \exp[-j(\beta z + \nu\psi)], \end{cases} \tag{91}$$

where the axial propagation constant β should be positive so that only progressive waves are considered, and the azimuthal quantum number ν should be an integer to allow the periodicity of fields versus ψ. The factorization of Eqs. (91) is due to the dependence of the refractive index on the radial coordinate only.

Introducing the following homogeneous field vectors:

$$\mathbf{h}(r) \equiv j\omega\mu_0 \mathbf{H}(r), \qquad \mathbf{e}(r) \equiv \beta \mathbf{E}(r) \tag{92}$$

it can be seen from Eqs. (88) and (89) that the radial components of field vectors (h_r and e_r) may be derived through the azimuthal (h_ψ and e_ψ) and axial (h_z and e_z) components of field vectors according to:

$$\begin{cases} h_r = -je_\varphi + j\dfrac{\nu}{\beta r}e_z\,, \\ j\dfrac{k_0^2}{\beta}e_r = \beta h_\varphi - \dfrac{\nu}{r}h_z\,, \end{cases} \tag{93}$$

where the local wave number has been introduced:

$$k_0(r) \equiv \omega\cdot\sqrt{\varepsilon(r)\cdot\mu_0} = \frac{2\pi}{\lambda}n(r) \qquad (\lambda\text{-free space wavelength}). \tag{94}$$

Thus the following linear system of four differential equations of the first-order for e_z, h_z, e_φ and h_φ can be derived:

$$\begin{cases} e_z' = \dfrac{\beta^2}{k_0^2}\cdot\dfrac{\nu}{r}\,h_z + \dfrac{\beta}{k_0^2}\,\varkappa^2 h_\varphi\,, \\ h_z' = \dfrac{\nu}{r}\,e_z + \dfrac{\varkappa^2}{\beta}\,e_\varphi\,, \\ e_\varphi' = -\left(k_0^2 - \dfrac{\nu^2}{r^2}\right)\dfrac{\beta}{k_0^2}\,h_z - \dfrac{1}{r}\,e_\varphi - \dfrac{\nu}{r}\dfrac{\beta^2}{k_0^2}\,h_\varphi\,, \\ h_\varphi' = -\left(k_0^2 - \dfrac{\nu^2}{r^2}\right)\dfrac{1}{\beta}\,e_z - \dfrac{\nu}{r}\,e_\varphi - \dfrac{1}{r}\,h_\varphi\,. \end{cases} \tag{95}$$

The apex indicates derivative with respect to r, and the local transverse propagation constant

$$\varkappa(r) \equiv \sqrt{k_0^2(r) - \beta^2} \tag{96}$$

has been introduced. These equations lead in general to a fourth-order differential equation for each component; at this stage calculations become very cumbersome and some reasonable approximations would be welcome. The following condition, that generally holds for standard telecommunication fibres:

$$\max\{|k_0'/k_0|\} \ll \max\{|e_z'/e_z|\},\ \max\{|e_\varphi'/e_\varphi|\} \tag{97}$$

permits us to split the system (95) into two systems of two differential equations of the first-order. In fact, defining the following two field combinations:

$$p_{\pm} \equiv \frac{k_0}{\beta} e_z \pm h_z\,, \qquad q_{\pm} \equiv \frac{k_0}{\beta} e_{\psi} \pm h_{\psi} \tag{98}$$

such that:

$$\begin{cases} e_z = (p_+ + p_-)\cdot\beta/(2k_0) \\ h_z = (p_+ - p_-)/2 \end{cases} \qquad \begin{cases} e_{\psi} = (q_+ + q_-)\cdot\beta/(2k_0) \\ h_{\psi} = (q_+ - q_-)/2 \end{cases} \tag{99}$$

Eqs. (95) become:

$$\begin{cases} \pm\, p'_{\pm} = \dfrac{\nu}{r}\dfrac{\beta}{k_0}\, p_{\pm} + \dfrac{\varkappa^2}{k_0}\, q_{\pm}\,, \\ \mp\, q'_{\pm} = \dfrac{1}{k_0}\left(k_0^2 - \dfrac{\nu^2}{r^2}\right) p_{\pm} + \left(\dfrac{\nu}{r}\dfrac{\beta}{k_0} \pm \dfrac{1}{r}\right) q_{\pm}\,. \end{cases} \tag{100}$$

Another field combination can be defined so that the radial components of field vectors can be deduced from Eqs. (93):

$$s_{\pm} \equiv \frac{k_0}{\beta} e_r \pm h_r = \pm \frac{j}{k_0}\left(\frac{\nu}{r} p_{\pm} - \beta q_{\pm}\right) \tag{101}$$

such that:

$$e_r = (s_+ + s_-)\beta/(2k_0)\,, \qquad h_r = (s_+ - s_-)/2\,. \tag{102}$$

Owing to the linearity of Eqs. (100) and (101) all the quantities p, q and s contain an arbitrary multiplicative constant; thus we can split the electromagnetic waves propagating along the fibre into two families by alternatively vanishing (if possible) the multiplicative coefficients of p_-, q_-, s_- or the coefficients of p_+, q_+, s_+. These two families of electromagnetic waves can be identified with those which are usually called in the literature the *EH* and *HE* modes [58]. They represent independent regimes of propagating waves. Thus we have (eliminating the subscript $\pm$ and for $\nu \neq 0$):

$$\begin{cases} e_z = \beta\cdot p/(2k_0)\,, \\ h_z = \pm\, p/2\,, \end{cases} \qquad \begin{cases} e_{\psi} = \beta\cdot q/(2k_0)\,, \\ h_{\psi} = \pm\, q/2\,, \end{cases} \qquad \begin{cases} e_r = \beta\cdot s/(2k_0)\,, \\ h_r = \pm\, s/2\,. \end{cases} \tag{103}$$

In these and the following expressions the upper sign holds for *EH* modes and the lower sign for *HE* modes. p, q, and s should fulfill Eqs. (100) and (101); thus a knowledge of only one component of the e.m. field is sufficient to obtain all the others. In particular the z-component p can be shown to fulfill the simplest second-order differential equation, that is (from Eqs. (100)):

$$p'' + \left(\frac{1}{r} - 2\frac{\varkappa'}{\varkappa}\right)p' + \left(\varkappa^2 - \frac{\nu^2}{r^2} \pm 2\frac{\varkappa'}{\varkappa}\frac{\nu}{r}\frac{\beta}{k_0}\right)p = 0\,. \tag{104}$$

Thus Eq. (104) is very important, representing the starting point for any further development of the theory. If weakly guiding optical fibres are considered (which usually is assumed in fibre optics) further simplifications can be made. In this approximation we have:

$$|\beta - k_0| \ll k_0\,. \tag{105}$$

It can be demonstrated [59] that, in monotonic profiles, this condition necessarily implies condition (97). Moreover it can be shown that under condition (105) we have:

$$\varkappa^2 \ll \beta^2, \qquad \max\{p\} \ll \max\{q\}\,. \tag{106}$$

Thus Eqs. (100) and (101) become respectively:

$$\begin{cases} \pm p' = \dfrac{\nu}{r}p + \dfrac{\varkappa^2}{\beta}q\,, \\[2ex] \mp q' = \beta p + \dfrac{\nu \pm 1}{r}q\,, \end{cases} \tag{107}$$

$$s = \pm jq\,, \tag{108}$$

and Eqs. (103):

$$e_z = \pm h_z = p/2; \qquad e_\varphi = \pm h_\varphi = q/2 \; (= \pm je_r = jh_r)\,. \tag{109}$$

Equation (104) for p remains substantially unchanged (only the factor β/k_0 can be approximated with unity), but the second-order differential equation for q can be greatly simplified and we have (from Eqs. (107)):

$$q'' + \frac{1}{r}q' + \left[\varkappa^2 - \left(\frac{\nu \pm 1}{r}\right)^2\right]q = 0\,. \tag{110}$$

Any transversal component of the e.m. field (q or s) fulfills Eq. (110) which is simpler than Eq. (104) for p since it does not contain any derivative of $\varkappa$.

2.7.2 *Solution of field equations*

A rigorous solution of exact field Eqs. (95) is possible only in fibres with uniform refractive index. In fact in this case conditions (97) hold for any value of radius r and Eqs. (100), (104) are exactly equivalent to Eqs. (95). Thus Eq. (104) becomes a Bessel equation and the axial field components can be expressed as follows:

$$p(r) = A \cdot J_\nu(\varkappa r) + B \cdot Y_\nu(\varkappa r) \tag{111}$$

where J_ν and Y_ν are the Bessel function of the first and second kind respectively of order ν [60]. If we consider the e.m. field inside the fibre core (with uniform refractive index n_0), the ratio of constants B/A of Eq. (111) should vanish when $r = 0$ in order to eliminate the divergence of power flow through the fibre and $\varkappa^2$ should be positive (that is: $\beta^2 < [(2\pi/\lambda)\,n_0]^2$) in order to obtain guided modes (without any leakage). If we consider the field in the cladding of the fibre, $B/A = j$ should be imposed: (that is Eq. (111) leads to the Hankel function of the first kind [60]) in order to obtain either guided modes with an exponential radial decay for $r \to \infty$ (if $\varkappa^2 < 0$) or radiation modes with outgoing waves (if $\varkappa^2 > 0$). So for guided modes:

$$p(r) = \begin{cases} J_\nu(\varkappa r)\,, & r < a\,, \\ K_\nu(\gamma r)\,, & r > a\,. \end{cases} \tag{112}$$

Having introduced the modified Bessel function of the second kind K_ν [60] and the cladding transverse propagation constant γ:

$$\gamma \equiv \sqrt{\beta^2 - \left(\frac{2\pi}{\lambda} n_1\right)^2}\,, \qquad (\gamma^2 = -\varkappa^2 > 0)\,. \tag{113}$$

Equations (100a) and (101) allow us to obtain from Eq. (112) the expressions of the transverse components (q and s) of the e.m. field. These expressions can be found in the literature [61, 62]; but now we shall restrict ourselves to the (usual) case of weakly guiding fibres. In this case (from Eq. (107a)), the following expressions of transverse field components can be obtained:

$$q(r) = \begin{cases} -\dfrac{\beta}{\varkappa} J_{\nu\pm1}(\varkappa r)\,, & r < a\,, \\ \pm\dfrac{\beta}{\gamma} K_{\nu\pm1}(\gamma r)\,, & r > a\,. \end{cases} \tag{114}$$

We have seen that solutions of wave equations for step-index fibres are available but for several graded-index fibres rigorous solutions cannot be obtained. The e.m. field in the cladding can still be described by means of modified Bessel functions of the second kind (Eqs. (112*b*) and (114*b*)) but the field inside the core can be expressed rigorously only in weakly guiding fibres with parabolic-index profiles ($n^2(r) = n_0^2 - (\Delta \cdot r/a)^2$; Δ-axial numerical aperture, see Eq. (23)). In this case it can be demonstrated [63] that a solution of Eq. (110) for transverse field component q, which is regular for $r \to 0$, is:

$$q(r) = r^{\nu\pm1} \cdot \exp\left(-\frac{V}{2}\frac{r^2}{a^2}\right) \cdot$$
$$\cdot M\left(\frac{\nu \pm 1 + 1}{2} - \frac{(n_0 \cdot 2\pi/\lambda)^2 - \beta^2}{4V} a^2,\ \nu \pm 1 + 1,\ V\frac{r^2}{a^2}\right) \tag{115}$$

where M is the Kummer confluent hypergeometric function [60] and V the normalized frequency, defined as:

$$V \equiv \frac{2\pi a}{\lambda} \Delta\,. \tag{116}$$

The remaining components can be derived from Eqs. (107*b*) and (108). Thus, for general graded-index profiles, approximate solutions of wave equations must be introduced. Many kinds of approximations have been proposed in the literature [23, 35], [64-69] (most of which are borrowed from mathematical methods of quantum mechanics [70]): *perturbation theory*, which consists in an expansion of the e.m. field as a series of the perturbation parameter, which represents the slight deviation of the index profile from that for which the solution is known [71, 72]; *evanescent wave analysis*, in which the field is expanded by means of a polynomial: the relative sequence generally diverges but a good approximation is often obtained by means of a truncation before the terms that begin to increase [73, 74]; *variational method*, which permits a field approximation by minimizing a functional (which should vanish for the exact expressions of the field) [66]. The most widely used approximation, however, is the *quasi-classical analysis* [18] (also known as WKBJ method, after those who first used it: G. Wentzel, L. Brillouin, H. A. Kramers, and J. Jeffreys) which will be extensively described in the following. This formalism combines a great simplicity with a good degree of approximation for field expressions; moreover it gives results which may be easily matched with geometrical optics and allow a deep physical understanding of propagation phenomena. The quasi-classical analysis consists substantially of an expansion of the field (at the first order of k_0) in the form:

$$q(r) = \bar{A}(r) \cdot \exp\,[j\bar{S}(r)]\,, \tag{117}$$

which expresses the important fact that, for large values of k_0, the field may be described by local plane waves of amplitude $\bar{A}(r)$ and phase $\bar{S}(r) - \beta z - \nu\psi$ (compare with Eqs. (1) and (91)). The zero-th order of this expansion supplies the geometrical optic analysis.

It may be convenient to express the differential equation of the transverse field (110) by means of the quantities h, k and $P(r)$ defined in Section 2.1. These quantities can be related to the various propagation constants as follows:

$$\beta = \frac{2\pi}{\lambda}\,k; \quad \nu \pm 1 = \frac{2\pi}{\lambda}\,h; \quad \varkappa^2(r) - \left(\frac{\nu \pm 1}{r}\right)^2 = \left[\frac{2\pi}{\lambda}\,P(r)\right]^2, \tag{118}$$

thus Eq. (110) becomes:

$$q'' + \frac{q'}{r} + \left[\frac{2\pi}{\lambda} P(r)\right]^2 q = 0 \tag{119}$$

WKBJ solutions of this equation in the form (117) are:

$$q(r) = \begin{cases} (rP)^{-1/2} \cdot \exp\left(\dfrac{2\pi j}{\lambda} \displaystyle\int^{r} P(r')\,dr'\right), \\[2ex] (rP)^{-1/2} \cdot \exp\left(-\dfrac{2\pi j}{\lambda} \displaystyle\int^{r} P(r')\,dr'\right). \end{cases} \tag{120}$$

It is easy to verify (by substituting expression (120) in Eq. (119)) that the WKBJ method is valid when the following condition is fulfilled:

$$|P^2| \gg \frac{1}{2}\frac{P''}{P} - \frac{3}{4}\left(\frac{P'}{P}\right)^2 . \tag{121}$$

This happens when multimode fibres are considered and $P \neq 0$. Thus, solutions (120) fail for $P \simeq 0$; this happens in general for two values of r corresponding to the two caustics; around these values of r, $P(r)$ may be expanded as a linear function and the solutions of this new differential equation (the Airy functions Ai and Bi [60]) can be used to interpolate WKBJ solutions. Thus the following expressions of q in the fibre core (considering two caustics of radii r_0 and r_1 in the core) can be provided:

$$q(r) = \begin{cases} \dfrac{1}{\sqrt{r|P|}} \exp\left(-\dfrac{2\pi}{\lambda}\displaystyle\int_r^{r_0} |P|\, dr\right); & 0 \leqslant r < r_0 \\[2ex] 2\sqrt{\dfrac{\pi}{rP}}\left[\dfrac{3\pi}{\lambda}\displaystyle\int_{r_0}^{r} P\, dr\right]^{\frac{1}{6}} \cdot Ai\left(-\left[\dfrac{3\pi}{\lambda}\displaystyle\int_{r_0}^{r} P\, dr\right]^{\frac{2}{3}}\right); & r \simeq r_0 \\[2ex] \dfrac{1}{\sqrt{rP}} \cos\left(\dfrac{2\pi}{\lambda}\displaystyle\int_{r_0}^{r} P\, dr - \dfrac{\pi}{4}\right); & r_0 < r < r_1 \\[2ex] 2\sqrt{\dfrac{\pi}{rP}}\left[\dfrac{3\pi}{\lambda}\displaystyle\int_{r}^{r_1} P\, dr\right]^{\frac{1}{6}} \cdot \left\{\sin S \cdot Ai\left(-\left[\dfrac{3\pi}{\lambda}\displaystyle\int_{r}^{r_1} P\, dr\right]^{\frac{2}{3}}\right) + \right. & \\ \qquad \left. + \cos S \cdot Bi\left(-\left[\dfrac{3\pi}{\lambda}\displaystyle\int_{r}^{r_1} P\, dr\right]^{\frac{2}{3}}\right)\right\}; & r \simeq r_1 \\[2ex] \dfrac{1}{\sqrt{r|P|}}\left[\sin S \cdot \exp\left(-\dfrac{2\pi}{\lambda}\displaystyle\int_{r_1}^{r} |P|\, dr\right) + \right. & \\ \qquad \left. + 2\cos S \cdot \exp\left(\dfrac{2\pi}{\lambda}\displaystyle\int_{r_1}^{r} |P|\, dr\right)\right]; & r_1 < r \leqslant a \end{cases} \tag{122}$$

where S is defined according to:

$$S = \frac{2\pi}{\lambda} \int_{r_0}^{r_1} P(r)\, dr\,. \tag{123}$$

Equations (107*b*) and (108) allow us to obtain from q (Eq. (122)) the expressions of the other field components: p and s.

2.7.3 *Characteristic equation*

When Eqs. (107) have been solved, the field components p, q (and hence e_z, h_z, e_φ, h_φ through Eqs. (109)) can be obtained apart from a multiplicative constant. Thus, if (p, q) are solutions of Eqs. (107), the normalized e.m. field components can be expressed as:

$$e_z = \pm h_z = A \cdot p; \qquad e_\varphi = \pm h_\varphi = B \cdot q \; (= \pm je_r = jh_r); \tag{124}$$

where A and B may be, *a priori*, arbitrary constants. But they can be determined by means of the boundary conditions; there are two kinds of boundary conditions: *a*) tangential components of the e.m. field (that is, e_z, h_z, e_φ, h_φ) should be continuous at the core-cladding interface [14]; *b*) e.m. field components should be matched at the input face of the fibre with the external fields (launching conditions). As regards conditions (*a*) we have (the subscripts 0 or 1 indicate, in this section, quantities regarding the core or the cladding respectively):

$$\begin{cases} A_0 p_0(a) = A_1 p_1(a)\,, \\ B_0\, q_0(a) = B_1\, q_1(a)\,, \end{cases} \tag{125}$$

as can be easily verified from Eq. (124) imposing the continuity of $e_z(h_z)$ and $e_\varphi(h_\varphi)$ respectively. Moreover, from Eq. (107*a*) we have:

$$\begin{cases} \pm A_0 p_0'(a) = \dfrac{\nu}{a} A_0 p_0(a) + \dfrac{\varkappa_0^2(a)}{\beta} B_0\, q_0(a)\,, \\[2ex] \pm A_1 p_1'(a) = \dfrac{\nu}{a} A_1 p_1(a) + \dfrac{\varkappa_1^2(a)}{\beta} B_1\, q_1(a)\,. \end{cases} \tag{126}$$

Equations (125) and (126) form a homogeneous system of four

equations in the four unknown quantities A_0, A_1, B_0, and B_1. This system admits nontrivial solutions only if the determinant of the coefficients vanishes. In this case three among the four constants can be written as functions of the fourth and this last constant may be obtained from the knowledge of launching conditions (*b*) (e.g. expanding the field at the input face of the fibre as a superposition of the various field configurations—modes—of the fibre). Thus, vanishing the determinant of the system (125), (126) we have:

$$\frac{1}{\varkappa_0^2(a)}\left(\pm\frac{p_0'(a)}{p_0(a)}-\frac{\nu}{a}\right)=\frac{1}{\varkappa_1^2(a)}\left(\pm\frac{p_1'(a)}{p_1(a)}-\frac{\nu}{a}\right). \tag{127}$$

Equation (127) gives the values of the allowable axial propagation constants β's of the field configurations as supported by the fibre structure for every ν. Only a discrete set of values of β satisfies Eq. (127), that is, for a given value of ν only a discrete number (e.g. labelled by $\mu = 1, 2, \ldots$) of field configurations (modes) can be supported by the fibre. Thus, the modes are called $EH_{\nu\mu}$ or $HE_{\nu\mu}$ according to the choice of the upper or lower sign in Eq. (127) (and in the previous ones), and then the possible values of β are $\beta_{\nu\mu}^{\pm}$ (solutions of Eq. (127)). Thus Eq. (127) is very important for the identification of the properties of the mode and is called a *characteristic equation*. The condition expressed through this equation can be expressed saying that the quantity:

$$\frac{1}{\varkappa^2(r)}\left(\pm\frac{p'(r)}{p(r)}-\frac{\nu}{r}\right) \tag{128}$$

must be continuous at the core-cladding interface. From Eqs. (107) we can obtain the following equivalent and simpler forms of characteristic equation:

$$\frac{q_0(a)}{p_0(a)}=\frac{q_1(a)}{p_1(a)}; \qquad \frac{q_0'(a)}{q_0(a)}=\frac{q_1'(a)}{q_1(a)}, \tag{129}$$

that is, the quantities:

$$\frac{q(r)}{p(r)} \quad \text{and} \quad \frac{q'(r)}{q(r)} \tag{130}$$

must be continuous at the core-cladding interface.

These considerations hold, obviously, for weakly guiding fibres; the characteristic equation for fibres with any refractive index difference has been derived and studied only for step-index fibres [62, 75]. In the following we will give explicit expressions of the characteristic equation for weakly guiding fibres on the ground of the expressions of the field component q previously deduced. For step-index fibres from Eqs. (114) we have:

$$\frac{J_{\nu\pm1}(u)}{uJ_\nu(u)} \pm \frac{K_{\nu\pm1}(w)}{wK_\nu(w)} = 0 \tag{131}$$

having defined the following normalized propagation constants:

$$u \equiv a\cdot\varkappa(0) = a\sqrt{k_0^2(0) - \beta^2}; \qquad w \equiv a\cdot\gamma = \sqrt{V^2 - u^2}\,. \tag{132}$$

On the other hand, the characteristic equation for parabolic-index fibres may be derived from Eqs. (114b) and (115):

$$2(\nu\pm1)\frac{M\left(\dfrac{\nu\pm1+1}{2}-\dfrac{u^2}{4V},\ \nu\pm1,\ V\right)}{M\left(\dfrac{\nu\pm1+1}{2}-\dfrac{u^2}{4V},\ \nu\pm1+1,\ V\right)} - (\nu\pm1+V) = \frac{wK'_{\nu\pm1}(w)}{K_{\nu\pm1}(w)}\,. \tag{133}$$

A detailed study of this equation is given in Ref. [63]. Finally, for general graded-index fibres a simple characteristic equation can be obtained from Eqs. (114b) and (122) as follows (for $\nu \gg 1$) [76]:

$$\sqrt{(\nu\pm1)^2 + w^2}\,\frac{2e^T - \tan S}{2e^T + \tan S} = \frac{wK'_{\nu\pm1}(w)}{K_{\nu\pm1}(w)} \tag{134}$$

where S is defined in Eq. (123) and T is given by:

$$T \equiv 2\int_{r_1}^{a}\sqrt{\frac{(\nu\pm1)^2}{r^2} - \varkappa^2(r)}\;dr\,. \tag{135}$$

When $\varkappa\cdot a \ll V$, this characteristic equation can be simplified as follows [23]:

$$S = \mu - \tfrac{1}{2}\,. \tag{136}$$

The study of the characteristic equation for general graded-index fibres is complicated and, at present, an expression of this equation, which holds also for few mode fibres ($\nu \simeq 1$; that is, in the field where the modal approach is preferred), is not known. Thus, the geometrical approach previously described may be considered as sufficiently valid (at least for many-mode fibres), while some undulatory effects [68, 77] may be taken into account within the geometrical theory, considering the quantization conditions (136) and (118*b*). This simply means replacing in Part A of the present Chapter the integrations over h and k with sums on ν and μ.

Now we must comment on the behaviour of the field for $\nu = 0$. First of all, this kind of wave is the only one with circular symmetry (the fields, Eq. (91), do not depend on ψ). In this case we can see from Eqs. (100), (104) and (110) that solutions p_+ and q_+ are identical to p_- and q_-; so we can arrange the multiplicative constants of $p_\pm$ in such a way that either $e_z = 0$ ($h_z = p$) or $h_z = 0$ ($e_z = p$). In the former case we have transverse electric modes ($HE_{0\mu} \equiv TE_{0\mu}$) and from Eqs. (93) and (95) we deduce that: $e_z = e_\psi = h_r = 0$; in the latter case transverse magnetic modes ($EH_{0\mu} \equiv TM_{0\mu}$) are obtained with $h_z = h_\psi = e_r = 0$. It can be shown that only for $\nu = 0$ can *TE* and *TM* modes be obtained, because the boundary conditions (125) at a dielectric interface cannot be satisfied when $\nu \neq 0$ with single *TE* or *TM* waves only [61]. Thus for $\nu \neq 0$ the two families of modes $EH_{\nu\mu}$ and $HE_{\nu\mu}$ are hybrid in the sense that both longitudinal components E_z and H_z do not vanish identically. As will be shown in the case of step-index fibres, the fundamental mode (that does not have cut-off frequency) is not a *TE* or a *TM* mode but the HE_{11} mode. The previous statements represent substantial differences with respect to metallic waveguides [78], and basically are due to the fact that in dielectric waveguides the field components E_z, E_ψ and H_r do not vanish for $r = a$ but (as seen) another kind of boundary condition must be imposed. Moreover, another noticeable difference is the operating condition. While metallic waveguides are commonly used in a monomodal configuration (only one propagating mode, that usually in circular metallic waveguides is not the fundamental one); optical fibres are commonly used either in the strong multimodal configuration (including the fundamental mode propagation) or in the monomodal configuration on the fundamental HE_{11} mode.

At this point it is possible to examine how the mode can be identified in the fibre. Firstly, let us analyze the physical meaning of the two quantum numbers ν and μ. Since in practice a mode is present with a superposition of both angular functions $\exp(j\nu\psi)$ and $\exp(-j\nu\psi)$, its azimuthal behaviour is described by the function $\cos(\nu(\psi - \psi_0))$. This means that in a complete turn around the axis the field of the mode changes sign 2ν times; that is the power density (that is proportional to the square of the field: $\cos^2(\nu(\psi - \psi_0))$ will exibit 2ν maxima. Correspondingly, (as will be shown in the next section) μ represents the number of maxima in the radial direction (from 0 to a).

So a $EH_{\nu\mu}$ or $HE_{\nu\mu}$ mode can be easily identified from an inspection of its near-field intensity: half of the number of azimuthal maxima is ν and the number of maxima in the radial direction is μ (see also Section 3.9 and Fig. 107). The $HE_{1\mu}$ modes are the only ones with a maximum on axis in the near-field intensity. The $TE_{0\mu}$ and $TM_{0\mu}$ modes are the only ones with no azimuthal maximum in the power (that is, as already said, with circular symmetry).

2.7.4 *Characteristic function of modes*

In this section a suitable method for solving characteristic equations is studied. It can be seen that these equations, in general, define implicitly a function $u = f(w)$ for each mode of the waveguide (e.g. see Eqs. (131) or (133) after the elimination of V through Eq. (132*b*)). This function is called a *characteristic function* because all propagation characteristics of a mode can be derived from it. If the approximation of weakly guiding fibres is introduced, it can be verified that characteristic equations are the same for $EH_{\nu\mu}$ and $HE_{\nu+2,\mu}$ modes ($HE_{1,\mu}$ modes are singular). This degeneration implies only that for these sets of modes the propagation constants are in practice the same though the structure of the field is different. In fact, the azimuthal behaviour of all field components and the radial behaviour of the axial components are considerably different for these sets of modes. Only the radial behaviour of the transverse component is the same. The fact that these last components are dominant (see Es. (106*b*)) explains the degeneracy in the macroscopic quantities of the two sets of modes. Thus, indicating by $f^{\pm}_{\nu\mu}(w)$ the characteristic function of an $EH_{\nu\mu}$ (+) or an $HE_{\nu\mu}$ (−) mode the following relations hold (for $\nu = 0$, $TE_{0\mu}$ and $TM_{0\mu}$ modes are considered):

$$f^{+}_{0\mu}(w) = f^{-}_{0\mu}(w) = f^{-}_{2\mu}(w); \qquad f^{+}_{\nu\mu}(w) = f^{-}_{\nu+2,\mu}(w)\,. \tag{137}$$

The following properties of characteristic functions for step-index fibres (Eq. (131)) can be demonstrated [62, 75]:

1) They are positively defined and increasing for each real $w \geqslant 0$.
2) They are limited and (calling $j_{\nu\mu}$ the μ-th zero of the Bessel function $J_\nu(u)$) the minimum takes the value:

$$f^{\pm}_{\nu\mu}(0) = \begin{cases} j_{1,\mu-1}\,, & \text{(for } HE_{1\mu} \text{ modes } - j_{10} = 0) \\ j_{(\nu-1)\pm 1,\mu}\,, & \text{(for all other modes)} \end{cases} \tag{138}$$

   and the asymptotic ($w \to \infty$) value is:

$$f^{\pm}_{\nu\mu}(\infty) = j_{\nu\pm 1,\mu}\,. \tag{139}$$

3) The derivative $f^{\pm\prime}_{\nu\mu}(w)$ with respect to the argument is always positive; it vanishes for $w \to \infty$, while for $w \to 0$ it takes the values:

$$f^{\pm\prime}_{\nu\mu}(w) = \begin{cases} +\infty\,, & \text{(for } HE_{1\mu} \text{ modes)} \\ 0\,, & \text{(for all other modes)}\,. \end{cases} \tag{140}$$

   Thus for all the modes (except $HE_{1\mu}$ modes) there is an inflection point at a certain $w > 0$.
4) The following asymptotic expressions can be deduced:
   For $w \to 0$:

$$f^{\pm}_{\nu\mu}(w) = \begin{cases} \sqrt{f^{\pm 2}_{\nu\mu}(0) + \dfrac{w^2}{(\nu-1)\pm 1}} & (EH_{\nu\mu},\ HE_{\nu\mu};\ \nu > 2)\,, \\ \sqrt{f^{\pm 2}_{0\mu}(0) + (1 - 2\ln w)\,w^2} & (TE_{0\mu},\ TM_{0\mu},\ HE_{2\mu})\,, \\ \sqrt{f^{-2}_{1\mu}(0) - \dfrac{1}{\ln w}}; & (HE_{1\mu};\ \mu > 1)\,, \\ \sqrt{-2/\ln w}; & (HE_{11})\,. \end{cases} \tag{141}$$

For $w \to \infty$:

$$f^{\pm}_{\nu\mu}(w) = f^{\pm}_{\nu\mu}(\infty) \cdot \exp\left[-1/w\right] . \qquad (142)$$

Similar properties hold for any refractive index profile with some modifications to particular values. For example, for parabolic-index fibres, the following values can be deduced: $f^{\pm}_{\nu\mu}(0)$ are the roots (u) of the equation (from Eqs. (132*b*) and (133)):

$$2(\nu \pm 1)\, M\left(\frac{\nu \pm 1 + 1}{2} - \frac{u}{4},\ \nu \pm 1,\ u\right) = $$
$$= uM\left(\frac{\nu \pm 1 + 1}{2} - \frac{u}{4},\ \nu \pm 1 + 1,\ u\right) \qquad (143)$$

while for $w \to \infty$ we have:

$$f^{\pm}_{\nu\mu}(w) = \sqrt{[4\mu + 2(\nu \pm 1) - 2]\, w} . \qquad (144)$$

That is the characteristic function squared shows an horizontal asymptote for $w \to \infty$ in step-index fibres and an oblique asymptote in parabolic-index fibres. In Figs. 23 the characteristic functions for some low order modes are plotted (for step- and parabolic-index weakly guiding fibres respectively). These figures can be used for a graphic determination of the propagation constants of the modes. In fact, intersecting these curves with the circle of radius V represented by

$$u^2 + w^2 = V^2 ; \qquad (145)$$

the coordinates of the intersection point (for every curve) directly supply w and u respectively (for the relative mode) and thus β (and also $\varkappa$ and γ) can be determined. However, it must be emphasized that for a given value of V only a finite number of curves have a real intersection with the circle (145); that is, only a finite number of modes may propagate in a certain fibre excited by a source of a certain frequency. Other modes are said to be below « cut-off ». Figs. 23 clearly show that the guidance condition for a certain mode (that is, the existence of the above-mentioned intersection) is $V \geqslant f^{\pm}_{\nu\mu}(0)$ and then the guidance limit

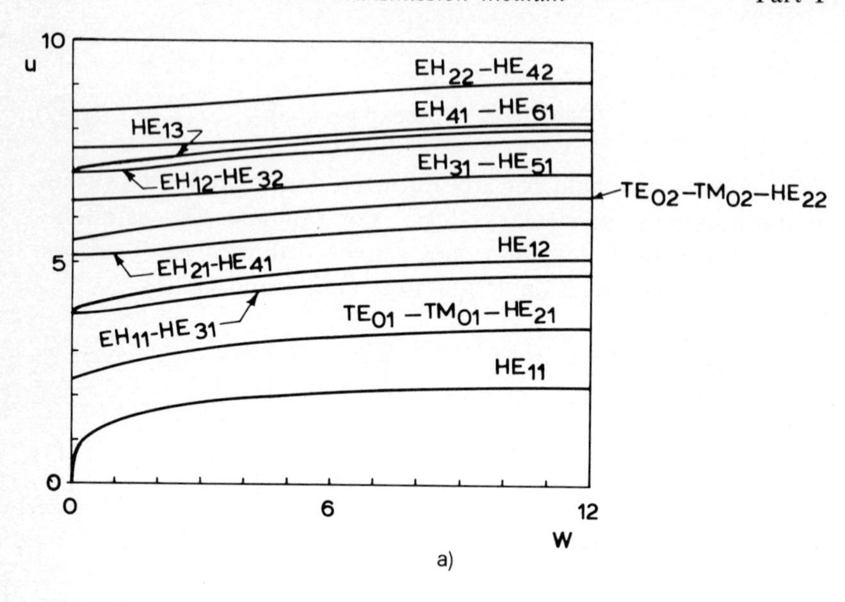

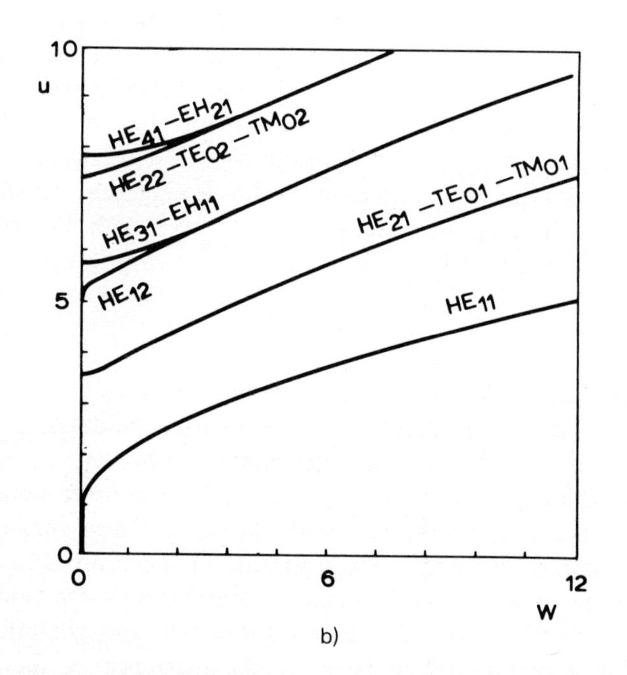

FIG. 23 – Plot of characteristic functions $u = f(w)$ for the lowest-order modes in weakly guiding fibres: *a*) with step profile; *b*) with parabolic profile.

condition or « cut-off » condition for each mode is:

$$V = V_{co} \equiv f^{\pm}_{\nu\mu}(0) \,. \tag{146}$$

Cut-off V-values are reported in Table III for a few lower order modes respectively for step- and parabolic-index fibres.

TABLE III

Cut-off V-values for some low-order modes in step- and parabolic-index fibres.

Modes	V_{co} (step)	V_{co} (parab.)	Modes	V_{co} (step)	V_{co} (parab.)
HE_{11}	0	0	EH_{12}, HE_{32}	7.016	9.645
TE_{01}, TM_{01}, HE_{21}	2.405	3.518	EH_{41}, HE_{61}	7.588	11.938
EH_{12}	3.832	5.068	EH_{22}, HE_{42}	8.417	11.760
EH_{11}, HE_{31}	3.832	5.744	TE_{03}, TM_{03}, HE_{23}	8.654	11.424
EH_{21}, HE_{41}	5.136	7.848	EH_{51}, HE_{71}	8.771	13.959
TE_{02}, TM_{02}, HE_{22}	5.520	7.451	EH_{32}, HE_{52}	9.761	13.833
EH_{31}, HE_{51}	6.380	9.904	EH_{61}, HE_{81}	9.936	15.972
HE_{13}	7.016	9.158			

It can be seen that there is a unique mode that can always propagate (the HE_{11} mode for which $V_{co} = 0$). When the fundamental HE_{11} mode is the only one propagating (this can be accomplished, for instance, designing a step-index waveguide and the source in such a way that $V < 2.405$), the fibre is called *monomode*. Otherwise, the fibre is called *multimode*. Cut-off condition for a given mode is $w = 0$ ($V = u$). On the other hand, when for a certain mode $w \to \infty$ (that is: $w \gg u$; $w \simeq V$), the mode is called « far from cut-off ». Thus asymptotic expressions (141)

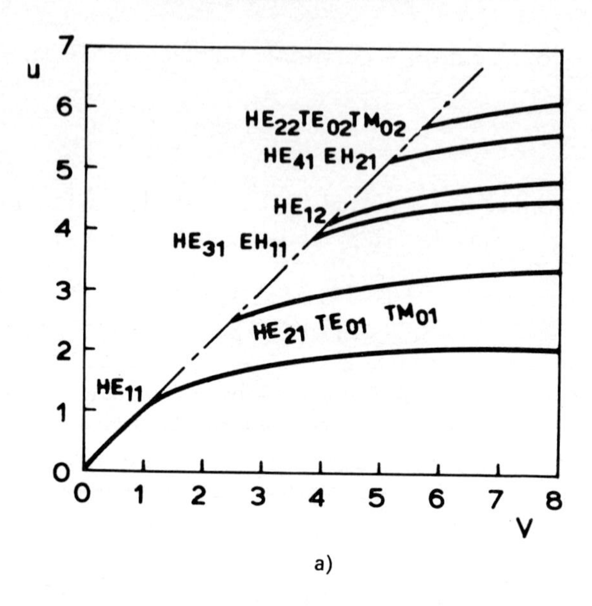

a)

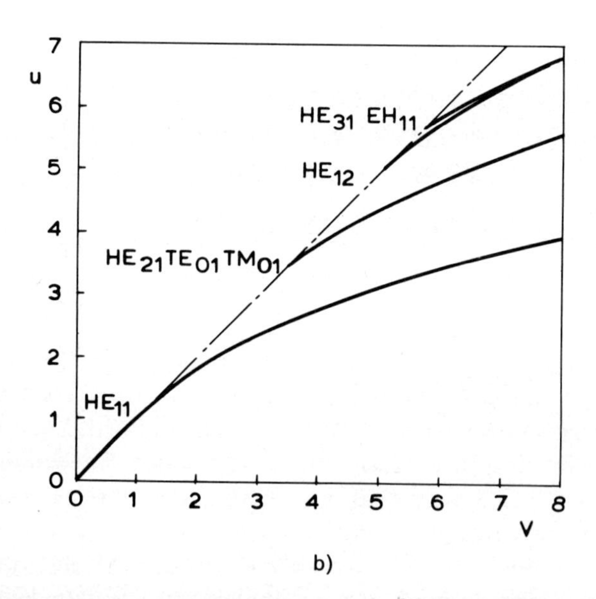

b)

FIG. 24 – Plot of characteristic functions $u = u(V)$ for the lowest-order modes in weakly guiding fibres: *a*) with step profile; *b*) with parabolic profile (After [62]).

and (142) or (144) are expressions of the characteristic function respectively near and far from cut-off.

When a « many-mode » fibre (strongly multimode: $V \gg 1$) is considered, most of the modes (particularly the lower order ones) are certainly far from cut-off; that is, Eqs. (142) or (144) can be used for these modes.

Another characteristic function (instead of $u = f(w)$) is often used [79-81], that is $u = u(V)$. This function can be easily derived from $f(w)$ considering Eq. (145), and the relative properties of $u(V)$ can be easily derived from corresponding properties of $f(w)$. These characteristic functions are plotted in Fig. 24 for lowest order modes in step- and parabolic-index fibres: these figures show that a graphical solution can be obtained intersecting the curves with the vertical line corresponding to the operating value of V: the ordinates of these intersections give directly the u-values for each guided mode (Eq. (145) supplies w).

2.7.5 *Number of modes in a multimode fibre*

Let us consider now a multimode waveguide and evaluate the number N of modes it can support. Imagine that this waveguide is excited with a Lambertian source of radiance R_0 in air (ignoring the reflectivity at the input face). In this case all the modes are excited and optical power is uniformly distributed over all of them [82]; thus, power in each mode: $\bar{w} = W/N$ is independent of the mode.

The power W can be expressed in this case as the product:

$$W = R_0 v(D) \tag{147}$$

where $v(D)$ is the volume of the phase space $(r, \psi, \theta, \varphi)$ domain D of the waveguide defined as follows:

$$v(D) \equiv \iint_S r\, dr\, d\psi \int_0^{2\pi} d\varphi \int_0^{\theta_M(r,\varphi)} \sin\theta \cos\theta\, d\theta\,, \tag{148}$$

where S is the cross-section of the waveguide (θ_M is defined as in Eq. (11)); for circularly symmetric waveguides we have (A is the local numerical aperture):

$$v(D) = \pi \int_0^a r\,dr \int_0^{2\pi} d\varphi\, A^2(r, \varphi) \tag{149}$$

and excluding tunnelling leaky rays we have:

$$v(D) = 2\pi^2 \int_0^a A^2(r)\, r\, dr\,. \tag{150}$$

Each mode occupies a certain constant volume $\bar{v}$ in the phase space $(r, \psi, \theta, \varphi)$; thus the number N of modes is:

$$N = \frac{v(D)}{\bar{v}}, \qquad (\bar{w} = R_0 \cdot \bar{v})\,. \tag{151}$$

It can be demonstrated through the analysis of the case of a step-index fibre [83] (or by means of the uncertainty principle) that:

$$\bar{v} = \frac{\lambda^2}{2}; \qquad \bar{w} = \frac{R_0\,\lambda^2}{2}\,. \tag{152}$$

Actually it should be $\bar{v} = \lambda^2/4$ but for waveguides of circular symmetry a degeneration on the polarization state of the modes occurs. Thus each mode that is being considered should be split into two modes and then the actual $\bar{v}$ of the mode is twice the calculated $\bar{v}$. Through Eqs. (151) and (148) (or (149) or (150)), the number of modes for any many-mode waveguide can be obtained considering that $\bar{v} = \lambda^2/4$ (or $\lambda^2/2$, for circular waveguides); that is for guided modes of circular fibres:

$$N = \frac{4\pi^2}{\lambda^2} \int_0^a A^2(r)\, r\, dr; \tag{153}$$

as special cases we have:

$$N = \begin{cases} \dfrac{V^2}{2} & \text{for step-index fibres}\,, \\[2ex] \dfrac{V^2}{4} & \text{for parabolic-index fibres}\,. \end{cases}$$

2.8 Optical power of modes

It is of considerable importance to know the optical power content of various modes since, as we shall see, it not only affects attenuation and excitation of modes but also their group velocity and dispersion.

2.8.1 *Power flow*

As is known, the density of power flow **P** represents (with its own direction) the amount of power which flows through the unit surface orthogonal to it. In our case **P** is given by the real part of the complex Poynting vector **S**:

$$\mathbf{P} = Re\,\mathbf{S} = Re\,\frac{\mathbf{E}\times\mathbf{H}^*}{2}\,. \tag{154}$$

For guided modes it can be shown that the radial component of **P** vanishes while the azimuthal component is proportional to ν. However, the axial component of **P** is of the greatest importance for the characterization of optical fibres and for weakly guiding fibres it takes the following expression (as can be deduced from Eqs. (91), (92), (124) and (154)):

$$P_z(r) = \frac{|B\cdot q(r)|^2}{\beta\omega\mu_0}\,. \tag{155}$$

Thus, since q fulfills the same differential Eq. (110) both for $EH_{\nu\mu}$ and $HE_{\nu+2,\mu}$ modes, this pair of modes have the same axial power flow distributions. Moreover, it can be shown that the quantum number μ represents the number of maxima in P_z for $0\leqslant r\leqslant a$. This fact can be confirmed in the case of step-index fibres for which P_z is (after Eq. (114)):

$$P_z(r) = \begin{cases} B_0^2\,\dfrac{\beta}{\varkappa^2\,\omega\mu_0}\cdot J_{\nu\pm1}^2(\varkappa r)\,, & r\leqslant a\,,\\[2ex] B_1^2\,\dfrac{\beta}{\gamma^2\,\omega\mu_0}\cdot K_{\nu\pm1}^2(\gamma r)\,, & r\geqslant a\,,\end{cases} \tag{156}$$

and Figs. 25 show the behaviour of $P_z(r)$ (normalized to its maximum) for some lowest order modes in step- and parabolic- index fibres in far from cut-off condition ($P_z(r) = 0$, for $r > a$). Thus,

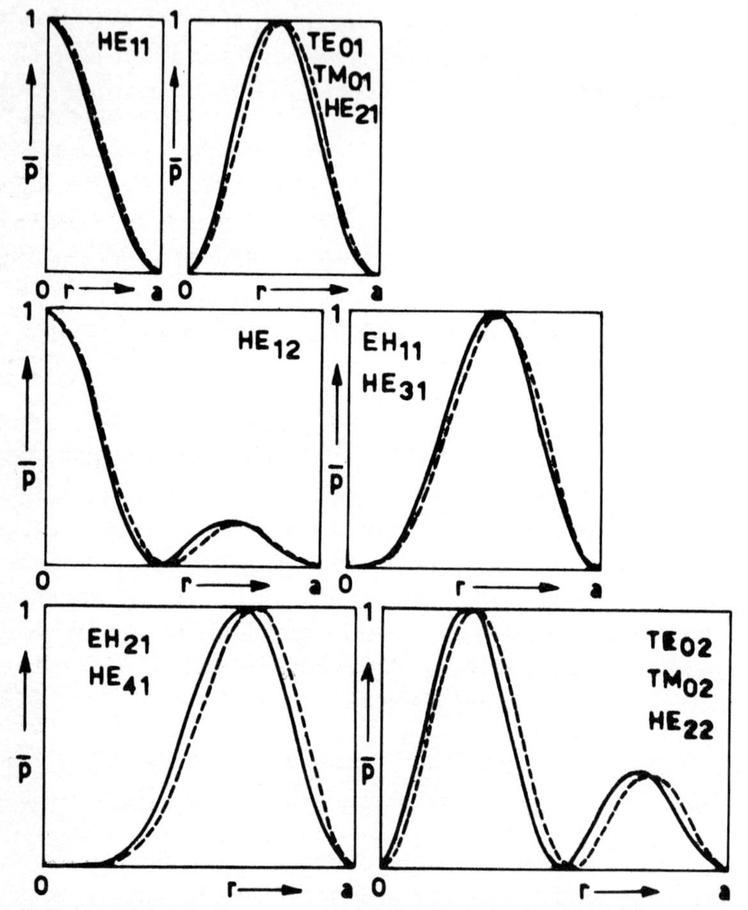

FIG. 25 - Radial behaviour of normalized optical power flow density P_z for the 12 lowest-order modes in a weakly guiding optical fibre (solid lines: parabolic profiles; dashed lines: step profiles - after [62]).

as these figures clearly show, Eq. (155) supplies the near-field distribution of every mode inside the fibre and permits us to recognize each mode starting from near-field measurements [84, 85]. As regards general graded-index fibres, expressions of $P_z(\boldsymbol{r})$ can be obtained from Eq. (154) using expressions (122) for q; and in the limit of $\lambda \to 0$, such expressions lead to:

$$P_z(r) = \frac{B_0^2}{\beta \omega \mu_0} \cdot \frac{1}{r\,P(r)} \tag{157}$$

which agrees with the results of geometrical optics regarding the power distribution of a ray congruence (compare with Eq. (36)).

2.8.2 *Total power of modes*

The total power carried by a mode can be obtained after an integration of $P_z(r)$ over the fibre cross-section. This total power $\bar{w}$ can be split into the two contributions $\bar{w}_0$ and $\bar{w}_1$, representing optical power flows through the core and cladding respectively:

$$\bar{w}_0 \equiv \int_0^{2\pi} d\psi \int_0^a P_z \cdot r\,dr, \qquad \bar{w}_1 \equiv \int_0^{2\pi} d\psi \int_a^\infty P_z \cdot r\,dr, \tag{158}$$

$$\bar{w} \equiv \bar{w}_0 + \bar{w}_1 = \int_0^{2\pi} d\psi \int_0^\infty P_z \cdot r\,dr = 2\pi \int_0^\infty P_z(r)\,r\,dr. \tag{159}$$

For step-index fibres it results (after Eq. (156)):

$$\bar{w}_0 = \pi a^2 \frac{B_0^2\,\beta a^2}{u^2\,\omega\mu_0}\,[J_{\nu\pm1}^2(u) - J_\nu(u) \cdot J_{\nu\pm2}(u)], \tag{160}$$

$$\bar{w}_1 = \pi\,\mathrm{a}^2 \frac{B_1^2\,\beta a^2}{w^2\,\omega\mu_0}\,[K_\nu(w)\,K_{\nu\pm2}(w) - K_{\nu\pm1}^2(w)], \tag{161}$$

and (the ratio B_1/B_0 is given by Eq. (125*b*)):

$$\bar{w} = -\pi a^2 \frac{B_0^2\,\beta a^2}{\omega\mu_0} \frac{V^2}{u^2 w^2} J_\nu(u)\,J_{\nu\pm2}(u). \tag{162}$$

Thus $\bar{w}$ vanishes when the mode is at cut-off (see Eq. (138)). Far from cut-off (most of the modes in a many-mode fibre are in this condition) it results:

$$\bar{w} = \pi\,\mathrm{a}^2 \frac{B_0^2\,\beta a^2}{u^2\,\omega\mu_0} J_\nu^2(u) \qquad (u = j_{\nu\pm1,\mu}). \tag{163}$$

On the other hand, it has been shown that in this case and when a source with constant radiance is considered, $\bar{w}$ is simply given by Eq. (152). Thus, Eqs. (152) and (163) allow the determination of the residual coefficient B_0 for each mode. This is an example of resolution of the problem of launching of modes in the optical fibre.

A quantity that we will prove to be very important is the ratio:

$$e \equiv \frac{\bar{w}_0}{\bar{w}_1} \tag{164}$$

which represents the relative amount of power inside the core and which, for step-index fibres, has the following explicit expression (as can be deduced from Eqs. (160), (161)):

$$e = \frac{(w^2/u^2)\, K_\nu(w) \cdot K_{\nu\pm 2}(w) + K^2_{\nu\pm 1}(w)}{K_\nu(w) \cdot K_{\nu\pm 2}(w) - K^2_{\nu\pm 1}(w)} \,. \tag{165}$$

The following simple relation connecting the parameter e and the characteristic function of each mode holds (and is valid for any difference in refractive index) [75]:

$$e \cdot f(w) \cdot \frac{df(w)}{dw} - w = 0 \,. \tag{166}$$

Equation (166) constitutes a differential equation (together with (165)) for $f(w)$; on the other hand, it allows us to obtain for the parameter e all the expansions which have already been obtained for the characteristic functions. In particular at cut-off we have:

$$(w = 0): \quad e = \begin{cases} \nu\,, & \text{for } EH_{\nu\mu},\ HE_{\nu+2\,\mu} \text{ modes } (\nu \geqslant 1) \\ 0\,, & \text{for the other modes} \end{cases} \tag{167}$$

while in conditions far from cut-off the parameter e diverges according to:

$$(V \gg u): \quad e \simeq \frac{V^3}{f^2(\infty)} \exp\left[-\frac{1}{V}\right] . \tag{168}$$

Thus, the fraction of power flowing through the core is minimum at cut-off (but not vanishing at all) and maximum far from cut-off. This can be verified by an inspection of Fig. 26 where the ratio e is given for a step-index weakly guiding fibre against the normalized frequency V and for the lowest order modes. This

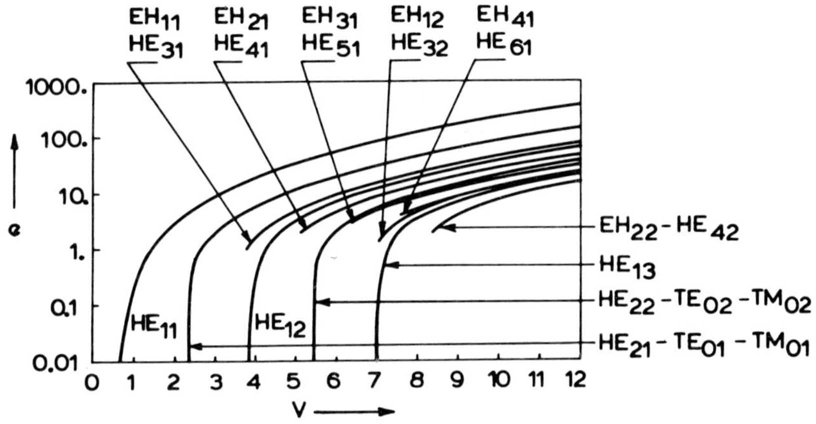

FIG. 26 – Plot of parameter e for the lowest-order modes in a weakly guiding step-index optical fibre.

ratio gives a measure of the confinement of the field inside the core: for $e > 100$ the field may be considered in practice inside the core and geometrical optics is able to supply good approximations.

2.9 Pulse distortion

The evaluation of the behaviour of pulse broadening through the fibre is quite important for determining the transmission characteristics of an optical fibre communication link. Several reasons contribute to broaden the pulse but with a good approximation they can be divided into two families [47]: « intermodal » contribution, due to different group delays of various modes (this for multimode fibres has been essentially dealt with by the preceding ray approach) and « intramodal » contribution. Intramodal broadening acts within each mode: it is due to the fact that the time of flight of the mode along the fibre generally depends on the wavelength. Since the source has a finite spectral width, the pulse carried by the single mode broadens. This kind of broadening represents the ultimate lower limit of pulse distortion.

Let us call τ_g the unitary group delay and D the dispersion for a given mode:

$$\tau_g = \left(\frac{d\beta}{d\omega}\right)_{\omega_0}, \qquad D = \left(\frac{d^2\beta}{d\omega^2}\right)_{\omega_0} \tag{169}$$

(where ω_0 is the central angular frequency of emitted optical power). Assuming the launched pulse to have a Gaussian distribution both on frequency and time, the optical power carried by the single mode spreads (owing to intramodal dispersion) according to:

$$\Delta t(z) = \sqrt{\Delta t_0^2 + \sigma^2 D^2 z^2} \tag{170}$$

where Δt_0 is the width of the launched pulse and σ its spectral width (on ω axis). For sufficiently large z-values Δt depends linearly on z.

Equation (170) is a very good approximation even if Gaussian distributions are not strictly fulfilled. This equation holds for each mode; in the case of monomode fibres it gives directly the actual pulse dispersion because intermodal dispersion is absent. For multimode fibres, intramodal pulse distortion becomes dominant when a strong equalization between group delays of various modes takes place (e.g., for fibres with optimal profile, or for optically or opto-electronically well-equalized fibres [47, 86, 87]). In this case, the actual intramodal contribution to pulse broadening Δt_{tot} can be obtained with a good approximation by averaging all the Δt's (Eq. (170)) of each mode weighted with its power content. For strongly multimode fibres such an average can be represented by the following integral over the ray congruences:

$$\Delta t_{\text{tot}} = \frac{1}{w(z)} \iint \Delta t(h, k)\, \varrho(h, k, z)\, dh\, dk \tag{171}$$

where the dependence of Δt (Eq. (170)) on the congruence (mode) variables h and k has been illustrated. If D can be considered practically constant for every mode, Eq. (170) directly gives the total pulse broadening.

On the other hand, if the intermodal contribution to pulse

distortion is dominant, the pulse broadening is given simply by:

$$\Delta t_{\text{tot}} = \Delta\tau_g \cdot z \tag{172}$$

where $\Delta\tau_g$ is the r.m.s. of the group delay (169*a*) over all modes. For many-mode fibres the summation over the modes can be replaced by an integration on ray congruences and Eq. (172) leads to geometrical optics results expressed by Eqs. (71)-(73).

In general, when both intermodal and intramodal dispersions contribute to pulse broadening the actual width of the pulse must be deduced through an analysis of the output pulse shape. In the reliable hypothesis of Gaussian pulses, this shape is:

$$W(t) = \sum_{1}^{N}{}_{i} \frac{\bar{w}_i \cdot \exp\left[-\gamma_i \cdot z\right]}{2\pi(\Delta t_0^2 + D_i^2 \cdot \sigma^2 \cdot z^2)} \exp\left[-\frac{(t-\tau_{gi}\cdot z)^2}{\Delta t_0^2 + D_i^2 \sigma^2 z^2}\right], \tag{173}$$

having labelled the mode quantities by the subscript i which enumerates all guided modes from the lowest to the highest order one up to N of Eq. (153). $\bar{w}_i$ is the total power carried by each mode, γ_i is the total power attenuation coefficient of the i-th mode, which will be discussed later on. Equation (173) shows that in general, the pulse shape is given through a superposition of N different single pulses; however, in practice, these pulses are often overlapped in such a way that it is difficult to single out each mode contribution.

Now it is necessary to find appropriate expressions of unitary group delay τ_g and dispersion D for each mode (Eqs. (169)).

2.9.1 *Group delay*

The group delay in the unit length of fibre is defined according to Eq. (169*a*). Now, starting from this definition applied to step-index fibres, a more general expression for every refractive index profile can be obtained. From Eqs. (96), (113), (169*a*) and through the characteristic equation, it is possible to demonstrate the following equation involving τ_g and the characteristic function $f(w)$ [75], which holds for any difference of refractive index values:

$$\left[f(w)\frac{df(w)}{dw} + w\right]\tau_g = \tau_{0g}\cdot w + \tau_{1g}\cdot f(w)\frac{df(w)}{dw}, \tag{174}$$

where τ_{0g} and τ_{1g} are local group delays in core and cladding defined according to:

$$\tau_{ig} \equiv \frac{n_i}{c} \cdot \frac{2\pi n_i}{\lambda\beta}; \qquad i = 0, 1; \tag{175}$$

for non-dispersive media, and [88]:

$$\tau_{ig} \equiv \frac{N_i}{c} \cdot \frac{2\pi n_i}{\lambda\beta}, \qquad i = 0, 1; \tag{176}$$

for dispersive media (N_i being the group refractive index defined according to Eq. (59)).

Equation (174) allows the evaluation of the group delay of every mode starting from the properties of the characteristic equation.

Considering Eq. (166) which relates the characteristic equation properties to the power ratio between core and cladding e (Eq. (164)) the following important relation can be obtained

$$\tau_g = \sum_{0}^{n} {}_i \, \tau_{ig} w_i \qquad (n = 1), \tag{177}$$

where w_i is the fraction of power which flows in the i-th medium of the fibre:

$$w_i = \frac{\bar{W}_i}{\bar{W}}. \tag{178}$$

Equation (177) can be given an interesting physical meaning: let us consider the energy stored in the unit length of fibre E, defined as the integral of electromagnetic energy density U [89]:

$$U = \frac{1}{4}\left[\frac{\partial(\omega\varepsilon)}{\partial\omega} \mathbf{E}\cdot\mathbf{E}^* + \frac{\partial(\omega\mu)}{\partial\omega} \mathbf{H}\cdot\mathbf{H}^*\right] \tag{179}$$

over the fibre cross-section:

$$E = \int_0^{2\pi} d\psi \int_0^{\infty} Ur \, dr\,. \tag{180}$$

Such a parameter may be split into its two contributions E_0 and E_1 of the core and cladding respectively:

$$E_0 = \int_0^{2\pi} d\psi \int_0^a U r\, dr\,, \qquad E_1 = \int_0^{2\pi} d\psi \int_a^\infty U r\, dr\,. \tag{181}$$

Thus, owing to the fact that the radial power flow vanishes, the energy conservation law states that the electromagnetic energy contained in the fibre length element dz: $E \cdot dz$ should equate the energy flow $\bar{w} \cdot dt$ through the fibre cross-section in a time interval $dt = \tau_g \cdot dz$; thus,

$$E = \tau_g \cdot \bar{w}; \qquad E_i = \tau_{ig} \cdot \bar{w}_i; \qquad (i = 0, 1)\,. \tag{182}$$

Equations (182*b*) may be obtained from the energy conservation law through core or cladding cross-section only and considering that there is no radial power flow. Equations (182*a*) can be demonstrated thoroughly from Maxwell equations [90].

Thus, taking into account that $E = \sum_i E_i$, we can obtain the relation (177). This indicates that the unit group delay of a mode can simply be evaluated averaging local group delay (defined through (175) or (176)) weighted with the power flow in each medium.

If a fibre with several (say n) coaxial layers is considered, it can be easily demonstrated that relation (177) still holds, with the suitable value of n (see [34]). Note that in this case τ_{ig}, w_i, $\bar{w}_i$, and E_i are still defined through Eqs. (176), (178), (182*b*) but now: $i = 0, 1, \dots, n$.

All this formalism can be extended from a several-layer fibre to a graded-index fibre; in fact, in the latter case it is sufficient to give Eq. (177) the following integral expression:

$$\tau_g = \int_0^\infty \tau(r)\, w(r)\, dr \tag{183}$$

where $w(r)\,dr$ represents the fraction of power which flows in the elementary layer $(r, r+dr)$ and $\tau(r)$ is the local group delay of this layer, which takes (extending Eq. (176)) the following expression:

$$\tau(r) = \frac{N(r)}{c} \cdot \frac{2\pi n(r)}{\lambda\beta} = \frac{n(r)\,N(r)}{c^2}\,\frac{\omega}{\beta}. \tag{184}$$

It can be observed that for a many-mode fibre $w(r)$ takes the expression (36) because in practice it does not vanish only in the region $r_0 \leqslant r \leqslant r_1$; taking this into account, from Eqs. (183) and (184), expression (58) can be derived (we must emphasize that in Eq. (58) a ray congruence corresponds to a mode). Equations (182) are a very important result, which holds for waveguides of arbitrary index profile and arbitrary cross-section, both dielectric and metallic. This equation permits us to obtain the group delay of each mode directly through the field distributions.

In our case of isotropic and dielectric media, U of Eq. (179) can be written simply (after Eq. (90)):

$$U = \frac{\varepsilon_0}{2} n(r) \cdot N(r) |\mathbf{E}|^2. \tag{185}$$

Thus from Eq. (182*a*) we have (after Eq. (91)):

$$\tau_g = \frac{\varepsilon_0}{2} \frac{\int_0^\infty n(r) \cdot N(r) \cdot [E_r \cdot E_r^* + E_\psi \cdot E_\psi^* + E_z \cdot E_z^*]\, r\, dr}{\int_0^\infty Re(E_r H_\psi^* - E_\psi H_r^*)\, r\, dr} \tag{186}$$

and for weakly guiding fibres we have (after Eqs. (92), (106*b*) and (109)):

$$\tau_g = \frac{\omega}{\beta c^2} \int_0^\infty n(r) \cdot N(r) q^2(r) r\, dr \Big/ \int_0^\infty q^2(r) r\, dr. \tag{187}$$

Equation (187) agrees with Eq. (183) and gives the explicit expression of $w(r)$:

$$w(r) = q^2(r) \cdot r \Big/ \int_0^\infty q^2(r)\, r\, dr. \tag{188}$$

These expressions may be used extensively in order to obtain group delay for every mode of arbitrary graded-index fibre (e.g.

for parabolic profile expressions (115), (114*b*) of q can be used in Eq. (187)).

Considering again step-index fibres, in Fig. 27 the group delay

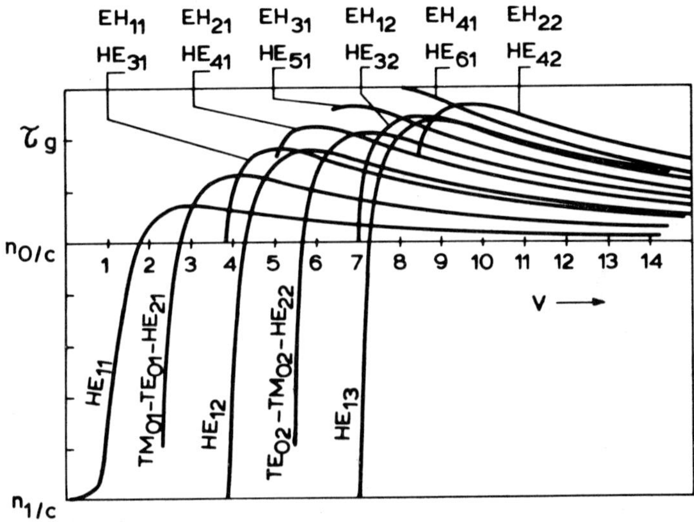

FIG. 27 – Normalized group delay τ_g against normalized frequency V for the lowest-order modes in a weakly guiding step-index optical fibre.

τ_g is given for a weakly guiding (non-dispersive material) fibre for the lowest order modes. It can be observed that a point of maximum τ_g exists for many modes; particularly for the fundamental HE_{11} mode this point is at $V \simeq 3$; as we shall see in the next section this point can be of great importance for minimizing pulse distortion. The curves representing τ_g lie between the two cut-off values of τ_{0g} $(= n_0^2/cn_1)$ and τ_{1g} $(= n_1/c)$; starting from the cut-off value:

$$\tau_g(\text{cut-off}) = \frac{\nu}{\nu+1}\cdot\frac{n_0^2}{cn_1} + \frac{1}{\nu+1}\cdot\frac{n_1}{c} \qquad (\text{for } EH_{\nu\mu},\ EH_{\nu+2,\mu})$$

they approach asymptotically n_0/c as V increases.

2.9.2 *Dispersion*

Dispersion, defined according to Eq. (169*b*), takes the following general expressions for each mode (after Eqs. (177) and

Eqs. (183)):

$$D = (\beta\omega c^2)^{-1} \sum_0^n {}_i \, w_i \frac{\partial(\omega^3 n_i \cdot \partial n_i/\partial\omega)}{\partial\omega} + \tau_g \left(\frac{1}{\omega} - \frac{\tau_g}{\beta}\right) + \sum_0^n {}_i \frac{\partial w_i}{\partial\omega} \tau_{ig}; \quad (189)$$

for multi-layer step-index fibres and:

$$D = (\beta\omega c^2)^{-1} \int_0^\infty w(r) \frac{\partial(\omega^3 n(r) \cdot \partial n(r)/\partial\omega)}{\partial\omega} dr + \tau_g \left(\frac{1}{\omega} - \frac{\tau_g}{\beta}\right) + \\ + \int_0^\infty \frac{\partial w(r)}{\partial\omega} \tau(r) dr; \quad (190)$$

for graded-index fibres.

As one can see, from Eqs. (189) and (190), dispersion results in an overlapping of the two different causes: material dispersion (i.e. the fact that refractive index depends on the frequency) and guide dispersion (i.e. the group delay of each mode generally depends on the frequency in an intrinsic way and apart from the dispersion of the media). With presently used materials (Silica, Ge, P or B doped) when sources of wavelength $\leqslant 1.1$ μm are used, guide dispersion is almost negligible with respect to material dispersion [47, 91]. Thus, in Eqs. (189) and (190) the first term becomes dominant and often practically independent of the mode. The second term is negligible far from cut-off (e.g. in many-mode fibres). However, when sources emitting optical power in a wavelength range $\lambda \geqslant 1.3$ μm are used, material dispersion may vanish [49, 92, 93]; in this case guide dispersion represented by the last two terms of Eqs. (189) and (190) becomes dominant and, for step-index fibres, the dispersion becomes proportional to the slope of curves in Fig. 27. In particular, we can observe that for monomode step-index fibres there is a point of *minimum dispersion* for $V \simeq 3$. This point may be translated on the V-axis (e.g. up to a V-value less than 2.405; that is, the maximum V-value in order to obtain the monomode condition) suitably choosing the operating wavelength around that of minimum material dispersion [94, 95]. In this case one may exploit certain compensation between material and guide dispersion which may occur only around these wavelengths. However; for general many-mode fibres each mode shows a point of minimum dispersion near cut-off and far from cut-off guide dispersion decreases as V^{-3} [75].

The obtained results allow the evaluation of pulse dispersion in various common cases; however, much work is still required to clarify the different effects on pulse dispersion and to determine the actual pulse response and the best condition for pulse broadening minimization in a given situation.

2.10 Power losses

There are many causes which produce power attenuation of a single mode in an optical fibre; these causes have been examined in Section 2.4 by a geometrical analysis, and can be summarized as: core and cladding absorption, optical tunnel effect, imperfect-interface losses; moreover, there are scattering losses. This last phenomenon is considered apart in a sub-section; as regards other causes, formulae expressed in Section 2.4 are still valid for every mode with some adjustments.

2.10.1 *Absorption*

If a fibre is composed of dissipative materials, modal power loss can be well-approximated (for small loss tangents) by the average of the loss coefficient α_i of each fibre layer weighted with the power flow through that layer [34]:

$$\gamma = \sum_{0}^{n}{}_{i} \, \alpha_i \, w_i \, ; \tag{191}$$

for a graded-index fibre the sum is replaced by an integration over r:

$$\gamma = \int_0^\infty \alpha(r) \, w(r) \, dr \, , \tag{192}$$

where (analogously to Eq. (183)) $w(r)$ is given by Eq. (188). Thus from these equations either Eq. (52) or (55) can be obtained when multimode optical fibres are considered.

Imperfect interface losses can be considered from a modal point of view through Eqs. (191) or (192) but now $\alpha(r)$ must be considered to be vanishing everywhere except for $r \simeq a$ [96]; for multimode optical fibres Eq. (57) still holds.

Optical tunnel effect, from a modal point of view, is due to the below cut-off condition of the mode [17]. Thus in the field expressions (91) β has an imaginary part which accounts for the loss of leaky modes. For multimode fibres the attenuation constant which rises from the imaginary part of β perfectly agrees with Eq. (56) [19, 35].

2.10.2 *Mode coupling*

A multimode fibre exhibits power coupling between different modes caused by perturbations in the geometry and in the optical properties of the waveguide. These perturbations may arise from fabrication processes, which cause random fluctuations of geometrical and optical parameters, or from external stresses which cause deviations from the straight geometry of the fibre (e.g., microbending).

The spatial-temporal evolution of optical power in a fibre is described by the coupled power equation [97] which is the modal equivalent of Eq. (85) enlarged to consider also the temporal dependence of $\bar{w}_{\nu\mu} = \bar{w}_{\nu\mu}(z, t)$:

$$\frac{\partial \bar{w}_{\nu\mu}}{\partial z} + \tau_{g_{\nu\mu}} \frac{\partial \bar{w}_{\nu\mu}}{\partial t} = -\gamma_{\nu\mu}\, \bar{w}_{\nu\mu} + \sum_{\nu'\mu'} C_{\nu\nu'\mu\mu'}(\bar{w}_{\nu'\mu'} - \bar{w}_{\nu\mu}) \qquad (193)$$

where $\bar{w}_{\nu\mu}$ is the power flow in the mode, $\tau_{g_{\nu\mu}}$ its group delay time and $\gamma_{\nu\mu}$ its total attenuation coefficient; $C_{\nu\nu'\mu\mu'}$ is the power coupling coefficient which represents the fraction of optical power converted from the mode $EH_{\nu'\mu'}$ (or $HE_{\nu'\mu'}$) to the mode $EH_{\nu\mu}$ (or $HE_{\nu'\mu'}$) in the unit length of fibre.

The coupling coefficient matrix $C_{\nu\nu'\mu\mu'}$ is symmetric; if coupling only between modes with nearly equal propagation constants is assumed, the non-zero elements $C_{\nu\nu'\mu\mu'}$ are only those in the proximity of the diagonal ($\nu \simeq \nu'$, $\mu \simeq \mu'$). This hypothesis of « nearest neighbour » coupling [97] can be explained considering that an exchange of optical power between two modes takes place only if one Fourier component of the spectrum of coupling function (i.e., of the perturbation from the ideal geometry) equals the separation between propagation constants of the two modes. Since most usual perturbations do not extend in spatial frequency by more than a few inverse decimetres, not all the modes which flow in the fibre can couple one another, but only the « nearest » ones. This can be observed from length-dependent broadening of far-field angular distribution, which confirms that optical power is gradually transferred from lower-order to higher-order modes [98].

The phenomenon of mode coupling in multimode fibres produces a smaller increase in pulse broadening after a certain length of fibre. In fact, power transfer produces a certain equalization among various group delays and, when the exchange process has reached its steady state, the width of the impulse response increases only as the square root of fibre length [51].

A second effect is produced by the exchange of power among modes which are subject to different attenuation: higher order modes can also couple with some low order radiative modes (i.e., modes which are just below cut-off condition), and this produces an excess loss which can also reach very high levels.

A quantitative evaluation of excess loss and of reduced pulse broadening (these two quantities depend strictly on each other) has been carried out in the literature [97, 99], where simplifying assumptions are made about perturbation power spectrum, without however reducing the validity of the results. Anyhow, the numerical integration of Eq. (193) requires considerable computer time if the number of modes propagating along the fibre is high. In this case Eq. (193) can be transformed into a simpler differential diffusion equation assuming that the discrete spectrum of modes can be replaced by a continuous spectrum [99-101].

The diffusion equation, obtained intially considering only one mode parameter [100], has been given recently the following more rigorous expression [101]:

$$\frac{\partial \bar{w}(\nu, \mu)}{\partial z} + \tau_g(\nu, \mu) \frac{\partial \bar{w}(\nu, \mu)}{\partial t} = - \gamma(\nu, \mu) \cdot \bar{w}(\nu, \mu) + + \left[\alpha_0 \frac{\partial^2 \bar{w}(\nu, \mu)}{\partial \mu^2} + \alpha_1 \frac{\partial^2 \bar{w}(\nu, \mu)}{\partial \mu \, \partial \nu} + \alpha_2 \frac{\partial^2 \bar{w}(\nu, \mu)}{\partial \nu_2} \right], \tag{194}$$

having regarded both mode parameters: ν and μ as continuous variables ($\alpha_i(\nu, \mu)$, i = 0, 1, 2, are suitable scattering coefficients).

In both Eqs. (193) and (194) the dependence on z can be removed expressing $\bar{w}$ as a superposition of different eigenfunctions exponentially decaying according to: $\exp(-\alpha_i z)$ (with: $0 < \alpha_0 < \alpha_1 < \dots$).

The eigenfunction relative to the lowest eigenvalue α_0 is the only one surviving after an enough long length of fibre and thus represents the *steady-state power distribution*, while α_0 is the *steady-state attenuation constant*. The length $z_s \equiv (\alpha_1 - \alpha_0)^{-1}$ is a good estimate of the *steady-state length*, that is the length of fibre after which the steady-state can be considered reached. Another property of z_s concerns the pulse broadening; in fact for $z > z_s$ it depends on the fibre length z no longer linearly but according to the square root law: $\sqrt{z}$.

The shape and the loss constant of steady-state distribution are determined by the index profile, by the nature and the intensity of the perturbation, but are quite independent of initial launching conditions. According to Ref. [99], but under the restrictive hypothesis of considering a single mode-parameter, all the eigensolutions of Eqs. (194) may be expressed in terms of Bessel functions of suitable order and argument. More rigorous, but more complicated solutions of Eq. (194) are discussed in Ref. [101].

2.11 Conclusion

The aim of this chapter was to supply a general view of problems and methods in the theory of propagation in optical fibres. The most important subjects have been extensively discussed and the most significant results illustrated both by way of formulae and figures in as general a way as possible. The chapter is not a mere review. There are different original points, particularly in the introduction of both geometrical and modal formalisms. These have been made as general as possible for easy understanding. For the sake of simplicity, detailed description of most of the various effects has been given in the part devoted to the geometrical approach, while in the part devoted to the modal approach only the phenomena not previously examined have been discussed in detail. There are many unsolved problems at present, namely, the full characterization of few-mode fibres with generic profile, the behaviour of dispersion joined with a better understanding of the interplay of various effects, a more realistic scattering theory, etc. Much theoretical research is being conducted in various centres; however, closer collaboration with experimental, technological, and systems groups should certainly improve the situation [102-105].

We have not exhausted all the problems in the theory of propagation. The reader interested in some particular aspect can be referred to various texts which may deal with certain of these problems in greater detail [3, 59, 61, 85, 97, 106-108]. We only hope to have supplied the practical grounding to enable the reader to investigate more deeply certain of the theoretical problems in fibre optics.

REFERENCES

[1] R. J. Potter: *Transmission properties of optical fibers*, J. Opt. Soc. Am., Vol. 51 (1961), pp. 1079-1089.

[2] N. S. Kapany: *Fiber Optics - Principles and Applications*, Academic Press, New York (1967).

[3] M. P. Lisitsa, L. I. Berezhinskii and M. Ya. Valakh: *Fiber Optics*, Israel Progr. for Scient. Transl., New York (1972).

[4] A. Fletcher *et al.*: *Solutions of two optical problems*, Proc. Roy. Soc., Vol. 223 (1954), pp. 216-225.

[5] E. G. Rawson *et al.*: *Analysis of refractive index distributions in cylindrical, graded-index glass rods (GRIN rods) used as image relays*, Appl. Opt., Vol. 9 (1970), pp. 753-759.

[6] A. Cozannet *et al.*: *Étude de la propagation de la lumière dans les fibres optiques à gradient d'indice*, Ann. Télécommun., Vol. 29, (1974), pp. 219-226.

[7] J. P. Dakin *et al.*: *Theory of dispersion in lossless multimode optical fibres*, Opt. Commun., Vol. 7 (1973), pp. 1-5.

[8] R. Bouillie *et al.*: *On the pulse broadening in dielectric multimode waveguides*, Opto-electron., Vol. 5 (1973), pp. 457-477.

[9] F. Albertin *et al.*: *Geometrical theory of energy launching and pulse distortion in dielectric optical waveguides*, Opto-electron., Vol. 6 (1974), pp. 369-386.

[10] R. Bouillie *et al.*: *Ray delay in gradient waveguides with arbitrary symmetric refractive profile*, Appl. Opt., Vol. 13 (1964), pp. 1045-1049.

[11] P. Di Vita and R. Vannucci: *Geometrical theory of coupling errors in dielectric optical waveguides*, Opt. Commun., Vol. 14 (1975), pp. 139-144.

[12] P. Di Vita and R. Vannucci: *Multimode optical waveguides with graded refractive index: theory of power launching*, Appl. Opt., Vol. 15 (1976), pp. 2765-2772.

[13] P. Di Vita and R. Vannucci: *Geometrical theory of light scattering in dielectric optical waveguides*, Alta Freq., Vol. 43 (1974), pp. 789-796.

[14] M. Born and E. Wolf: *Principles of Optics*, Pergamon Press, Oxford (1964).

[15] S. J. Maurer and L. B. Felsen: *Ray-optical techniques for guided waves*, Proc. IEEE, Vol. 55 (1967), pp. 1718-1729.

[16] L. B. Felsen: *Rays and modes in optical fibres*, Electron. Lett., Vol. 10 (1974), pp. 95-96.

[17] A. W. Snyder: *Leaky-ray theory of optical waveguides of circular cross section*, Appl. Phys., Vol. 4 (1974), pp. 273-298.

[18] B. G. Levich: *Theoretical Physics, Vol. 3* (*Quantum Mechanics*), North-Holland, Amsterdam (1973).

[19] P. Di Vita and R. Vannucci: *Loss mechanisms of leaky skew rays in optical fibres*, Opt. Quantum Electron., Vol. 9 (1977), pp. 177-188.

[20] G. Cocito *et al.*: *Theory and experiments on LED/Fibre coupling*, Second European Conference on Optical Fibre Communication, Paris (September 27-30, 1976), pp. 281-291.

[21] P. Di Vita and U. Rossi: *Evaluation of coupling efficiency in joints between optical fibres*, Alta Freq., Vol. 47 (1978), pp. 414-423.

[22] P. Di Vita and R. Vannucci: *The « Radiance law » in radiation transfer processes*, Appl. Phys., Vol. 7 (1975), pp. 249-255.

[23] D. Gloge and E. A. J. Marcatili: *Multimode theory of graded-core fibers*, Bell. Syst. Tech. J., Vol. 52 (1973), pp. 1563-1578.

[24] P. Di Vita and U. Rossi: *Theory of power coupling between multimode optical fibres*, Opt. Quantum Electron., Vol. 10 (1978), pp. 107-117.

[25] G. Cocito *et al.*: *Fibre to fibre coupling with three different fibre core diameters*, CSELT Rapporti Tecnici, Vol. 6 (1978), pp. 113-119.

[26] D. Gloge: *Offset and tilt loss in optical fiber splices*, Bell Syst. Tech. J., Vol. 55 (1976), pp. 905-916.

[27] P. Di Vita and U. Rossi: *Realistic evaluation of coupling loss between different optical fibres*, J. Opt. Commun., Vol. 1 (1980), pp. 26-32.

[28] B. Costa and B. Sordo: *Measurement of refractive index profile in optical fibres*, XXIII Congresso Internazionale per l'Elettronica, Rome (March 22-24, 1976), pp. 405-411.

[29] B. Costa and B. Sordo: *Measurements of refractive index profile in optical fibres: comparison between different techniques*, Second European Conference on Optical Fibre Communication, Paris (September 27-30, 1976), pp. 81-86.

[30] J. A. Arnaud and R. M. Derosier: *Novel technique for measuring the index profile of optical fibers*, Bell. Syst. Tech. J., Vol. 55 (1976), pp. 1489-1508.

[31] F. M. E. Sladen *et al.*: *Determination of optical fibre refractive index profiles by a near-field scanning technique*, Appl. Phys. Lett., Vol. 28 (1976), pp. 255-258.

[32] M. Eve: *Near-field and far-field power distributions in a multimode graded fibre*, Opt. Quantum Electron., Vol. 9 (1977), pp. 459-464.

[33] W. J. Stewart: *A new technique for measuring the refractive index profiles of graded optical fibres*, 1977 International Conference on Integrated Optics and Optical Fiber Communication, Tokyo (July 12-20, 1977), pp. 395-398.

[34] R. Roberts: *Propagation characteristics of multimode dielectric waveguides at optical frequencies,* Conference on Trunk Telecommunications by Guided Waves, London (September 29-October 2, 1970), pp. 39-44.

[35] K. Petermann: *The mode attenuation in general graded core multimode fibres*, A.E.Ü., Vol. 29 (1975), pp. 345-348.

[36] K. Petermann: *Leaky mode behaviour of optical fibres with noncircular symmetric refractive index profile*, A.E.Ü., Vol. 31 (1977), pp. 201-204.

[37] A. Ankiewicz: *Ray theory of graded non-circular optical fibres*, Opt. Quantum Electron, Vol. 11 (1979), pp. 197-203.

[38] J. A. Arnaud: *Pulse broadening in multimode optical fibres*, Bell, Syst. Tech. J., Vol. 54 (1975), pp. 1179-1205.

[39] P. Di Vita: *Derivation of power attenuation and pulse distortion in multimode graded index optical fibres*, Prima Riunione Nazionale di Elettromagnetismo Applicato, L'Aquila, (Italy) (June 24-25, 1976), pp. 135-138.

[40] C. C. Timmermann: *Signalübertragung mit vielwelligen Gradientenfasern*, Dr.-Ing. Dissertation Tech. Univ. Braunschweig (1975).

[41] K. Petermann: *A general condition for the delay equalization in multimode optical fibres*, Fourth European Conference on Optical Communication, Genoa (September 12-15, 1978), pp. 281-287.

[42] R. Olshansky: *Optical waveguides with low pulse dispersion over an extended spectral range*, Electron. Lett., Vol. 14 (1978), pp. 330-331.

[43] R. Olshansky: *Multiple-α index profiles*, Appl. Opt., Vol. 18 (1979), pp. 683-689.

[44] R. Olshansky: *Multiple-α index profiles.* Optical Fiber Communication, Washington, D.C. (March 6-8, 1979), pp. 96-98.

[45] S. Geckeler: *Dispersion in optical waveguides with graded refractive index*, Electron. Lett., Vol. 11 (1974), pp. 139-140.

[46] C. G. Someda: *A coupled-mode analysis of imperfect optical dielectric waveguides*, Alta Freq., Vol. 43 (1974), pp. 781-788.

[47] R. Olshansky and D. B. Keck: *Pulse broadening in graded-index optical fibres*, Appl. Opt., Vol. 15 (1976), pp. 483-491.

[48] F. A. Jenkins and H. E. White: *Fundamentals of Optics*, McGraw-Hill, New York (1957).

[49] D. N. Payne and W. A. Gambling: *Zero material dispersion in optical fibres*, Electron. Lett., Vol. 11 (1975), pp. 176-178.

[50] F. P. Kapron: *Maximum information capacity of fibre-optic waveguides*, Electron. Lett., Vol. 13 (1977), pp. 96-97.

[51] S. D. Personick: *Time dispersion in dielectric waveguides*, Bell. Syst. Tech. J., Vol. 50 (1971), pp. 843-859.

[52] P. Di Vita and R. Vannucci: *Theory of light scattering in multimode optical fibres for telecommunications*, XXIII Congresso Internazionale per l'Elettronica, Rome (March 22-24, 1976), pp. 375-382.

[53] K. Huang: *Statistical Mechanics*, J. Wiley & S., New York, (1963).

[54] E. L. Chinnock *et al.*: *The length dependence of pulse spreading in the CGW-Bell-10 optical fibre*, Proc. IEEE, Vol. 61 (1973), pp. 1499-1500.

[55] D. B. Keck: *Observation of externally controlled mode coupling in optical waveguides*, Proc. IEEE, Vol. 62 (1974), pp. 649-650.

[56] R. Olshansky: *Distortion losses in cabled optical fibres*, Appl. Opt., Vol. 14 (1975), pp. 20-21.

[57] D. Gloge: *Optical-fiber packaging and its influence on fiber straightness and loss*, Bell Syst. Tech. J., Vol. 54 (1975), pp. 245-262.

[58] E. Snitzer: *Cylindrical dielectric waveguide modes*, J. Opt. Soc. Am., Vol. 51 (1961), pp. 491-498.

[59] H.-G. Unger: *Planar Optical Waveguides and Fibres*, Clarendon, Press, Oxford (1977).

[60] M. Abramowitz and I. Stegun: *Handbook of Mathematical Functions*, Dover, New York (1953).

[61] D. Marcuse: *Light Transmission Optics*, Van Nostrand-Reinhold, New York (1972).

[62] P. Di Vita: *Teoria elettromagnetica della propagazione in fibre ottiche ideali*, LXXV Riunione AEI, Rome (September 15-21, 1974), B4.

[63] R. Yamada *et al.*: *Guided waves along an optical fiber with parabolic-index profile*, J. Opt. Soc. Am., Vol. 67 (1977), pp. 96-103.

[64] C. N. Kurtz and W. Streifer: *Guided waves in inhomogeneous focusing media - Part I and II*, IEEE Trans. MTT, Vol. 17 (1996), pp. 11-15 and 250-253.

[65] H. Kirchoff: *Wave propagation along radially inhomogeneous glass fibres*, A.E.Ü., Vol. 27 (1973), pp. 13-18.

[66] T. Okoshi and K. Okamoto: *Analysis of wave propagation in inhomogeneous optical fibres using a variational method*, IEEE Trans. MTT, Vol. 22 (1974), pp. 938-945 and: *Analysis of wave propagation in optical fibres having core with α-power low refractive index distribution and uniform cladding*, ibid., Vol. 24 (1976), pp. 416-421.

[67] C. N. Kurtz: *Scalar and vector mode relations in gradient-index light guides*, J. Opt. Soc. Am., Vol. 65 (1975), pp. 1235-1240.

[68] E. Bianciardi and V. Rizzoli: *Propagation in graded-core fibres: a unified numerical description*, Opt. Quantum Electron., Vol. 9 (1977), pp. 121-133.

[69] O. Leminger: *Ein Näherungs-verfahren zur Berechnung der geführten Moden in vielwelligen Gradientenfasern*, A.E.Ü., Vol. 32 (1978), pp. 353-356.

[70] P. M. Morse and H. Feshbach: *Methods of Theoretical Physics*, New York (1953), Part 1 and Part 2.

[71] R. A. Sammut and A. K. Ghatak: *Perturbation theory of optical fibres with power-law core profile*, Opt. Quantum Electron., Vol. 10 (1978), pp. 475-482.

[72] E. Khular *et al.*: *Propagation characteristics of mean square law optical fibres. A perturbation analysis*, Optik, Vol. 50 (1978), pp. 111-120.

[73] L. B. Felsen: *Evanescent waves*, J. Opt. Soc. Am., Vol. 66 (1976), pp. 751-760.

[74] S. Choudary and L. B. Felsen: *Guided modes in graded-index optical fibres*, J. Opt. Soc. Am., Vol. 67 (1977), pp. 1192-1196.

[75] B. Costa and P. Di Vita: *Group velocity of modes and pulse distortion in dielectric optical waveguides*, Opto-electronics, Vol. 5 (1973), pp. 439-456.

[76] L. Brunetti *et al.*: *Una nuova formulazione dell'equazione caratteristica per fibre ottiche a profilo graduale*, Alta Freq., Vol. 47 (1978), pp. 131-134.

[77] M. J. Adams *et al.*: *Resolution limit of the near-field scanning technique*, Third European Conference on Optical Communication, Munich (September 14-16, 1977), pp. 25-27.

[78] S. Ramo, J. R. Whinnery, and T. Van Duzer: *Fields and Waves in Communication Electronics*, J. Wiley & S., New York, (1965).

[79] A. W. Snyder: *Asymptotic expressions for eigenfunctions and eigenvalues of a dielectric or optical waveguide*, IEEE Trans. MTT, Vol. 17 (1969), pp. 1130-1132.

[80] G. Biernson and D. J. Kinsley: *Generalized plots of mode patterns in a cylindrical dielectric waveguide applied to retinal cones*, IEEE Trans. MTT, Vol. 13 (1965), pp. 345-356.

[81] D. Gloge: *Weakly guiding fibers,* Appl. Opt., Vol. 10 (1971), pp. 2252-2258.

[82] A. W. Snyder and C. Pask: *Incoherent illumination of an optical fiber*, J. Opt. Soc. Am., Vol. 63 (1973), pp. 806-812.

[83] C. Pask *et al.*: *Number of modes of optical waveguides*, J. Opt. Soc. Am., Vol. 65 (1975), pp. 356-357.

[84] E. Snitzer and H. Osterberg: *Observed dielectric waveguide modes in the visible spectrum*, J. Opt. Soc. Am., Vol. 51 (1961), pp. 499-505.

[85] N. S. Kapany and J. J. Burke: *Optical Waveguides*, Academic Press, New York (1972).

[86] D. Gloge: *Fibre delay equalization by carrier drift in the detector*, Opto-electronics, Vol. 5 (1973), pp. 345-350.

[87] P. Di Vita and R. Vannucci: *Optical equalizers for multimode optical fibres*, XXIII Congresso Internazionale per l'Elettronica, Rome (March 22-24, 1976), pp. 258-266.

[88] P. Di Vita: unpublished work (1973).

[89] L. Landau and E. Lifchitz: *Electrodynamique des Milieux Continus*, Editions MIR, Moscow (1969).

[90] H. A. Haus and H. Kogelnik: *Electromagnetic momentum and momentum flow in dielectric waveguides*, J. Opt. Soc. Am., Vol. 66 (1976), pp. 320-327.

[91] K. Jürgensen: *Dispersion-optimized optical single-mode glass fiber waveguides*, Appl. Opt., Vol. 14 (1975), pp. 163-168.

[92] L. G. Cohen and C. Lin: *Pulse delay measurements in the zero material dispersion wavelength region for optical fibers,* Appl. Opt., Vol. 16 (1977), pp. 3136-3139.

[93] D. N. Payne and A. H. Hartog: *Determination of the wavelength of zero material dispersion in optical fibres by pulse delay measurements*, Third European Conference on Optical Communication, Munich (September 14-16, 1977), Post-deadline paper; NTZ, Vol. 31, (1978), pp. 130-132, and Electron. Lett., Vol. 13 (1977), pp. 627-629.

[94] L. Smith and E. Snitzer: *Dispersion minimization in dielectric waveguides*, Appl. Opt., Vol. 12 (1973), pp. 1592-1599.

[95] K. Jürgensen: *Comment on: Dispersion minimization in dielectric waveguides*, Appl. Opt., Vol. 13 (1974), pp. 1289-1290; L. Smith and E. Snitzer: *Authors' reply to comments on: dispersion minimization in dielectric waveguides*, ibidem, p. 1290.

[96] D. Gloge: *Propagation effects in optical fibers*, IEEE Trans. MTT, Vol. 23 (1975), pp. 106-120.

[97] D. Marcuse: *Theory of Dielectric Optical Waveguides*, Academic Press, New York (1974).

[98] W. A. Gambling *et al.*: *Mode conversion coefficients in optical fibres*, Appl. Opt., Vol. 14 (1975), pp. 1538-1542.

[99] R. Olshansky: *Mode coupling effects in graded-index optical fibers*, Appl. Opt., Vol. 14 (1975), pp. 935-945.

[100] D. Gloge: *Optical power flow in multimode fibers*, Bell Syst. Tech. J., Vol. 51 (1972), pp. 1767-1783.

[101] J. A. Arnaud and M. Leberre-Rousseau: *Théorie de la propagation dans les fibres multimodales déformées aléatoirement: I. Théorie générale,* Ann. Télécomm. Vol. 35 (1980), pp. 61-73.

[102] J. D. Love *et al.*: *Report on the International Workshop on Optical Waveguide Theory, 21-23 September, 1976, Lannion, France*, Opt. Quantum Electron., Vol. 9 (1977), pp. 84-86.

[103] C. G. Someda and K. Petermann: *International Workshop on Optical Waveguide Theory*, NTZ, Vol. 31 (1978), pp. 119-123.

[104] L. B. Felsen: *Theory of optical waveguides: report on the 3rd International Workshop*, Opt. Quantum Electron., Vol. 11 (1979), pp. 283-286.

[105] J. A. Arnaud: *Optical waveguide theory* (Report on the 4th International Workshop on Optical Waveguide Theory), Opt. Quantum Electron., Vol. 12 (1980), pp. 187-191.

[106] W. B. Allan: *Fibre Optics: Theory and Practice,* Plenum Press, London (1973).

[107] J. A. Arnaud: *Beam and Fiber Optics,* Academic Press, New York (1976).

[108] M. S. Sodha and A. K. Ghatak: *Inhomogeneous Optical Waveguides,* Plenum Press, New York (1977).

CHAPTER 3

FIBRE CHARACTERIZATION

by Bruno Costa and Bruno Sordo

3.1 Introduction

The measurement of optical waveguide properties is important from many standpoints. One is the verification of theoretical models, or the investigation of waveguide behaviour in different conditions; very precise and sophisticated experimental techniques are usually required to perform such measurements. Another is related to the requirements of systems designers, who are mainly concerned with transmission properties such as attenuation and bandwidth at specific wavelengths and with regard to their behaviour as a function of length. Fibre testing in conditions similar to those to be encountered in practical situations is of particular interest. Finally there is a need for fibre manufacturers to measure those parameters (refractive index profile, size, mechanical characteristics, besides, of course, transmission properties) which may allow a check of the quality of the manufacturing process and help in identifying the critical parameters to be controlled.

We shall consider in this chapter the following measurement items:

1) Attenuation: *a*) total loss;
   b) scatter loss;
   c) absorption loss;
   d) differential mode attenuation.
2) Fault location.
3) Bandwidth, pulse dispersion, group delay.
4) Material dispersion.
5) Refractive index profile, numerical aperture.

6) Coupling efficiency
7) Mode distribution.
8) Splice loss.
9) Size, geometrical parameters.
10) Mechanical characteristics.

After a brief survey of fibre preparation and handling, the commonly used techniques for the measurement of the above-listed properties will be presented, with a discussion of the main problems associated both with measuring principles and with experimental apparatus.

3.2 Fibre preparation and handling

In order to perform the great majority of measurements it is required that the fibre end faces be flat, smooth and perpendicular to the fibre axis; moreover, accurate positioning of the fibre is needed. The first problem was initially overcome by means of grinding and polishing processes [1] or by insertion of the fibre end into suitably designed index matching cells [2, 3], i.e. cells filled with a liquid whose refractive index is close to that of the fibre core. An example of such a cell is shown in Fig. 1, which

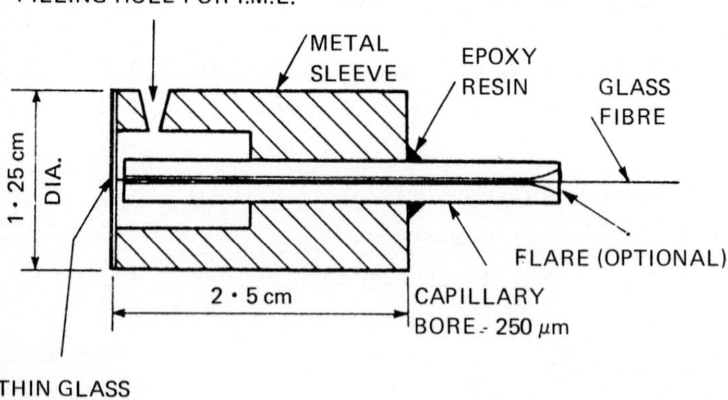

FIG. 1 – An index matching cell for coupling light into fibres (After [3]).

illustrates the small hole for fibre insertion, the flat transparent cover, and the chamber in which liquid is poured. These methods have been nearly abandoned because they are either time-consuming or inconvenient to use. Techniques now in common use consist essentially in creating a local stress on the fibre surface and pulling the fibre with a proper tension. One such method consists in scoring the fibre by means of a blade (diamond or tungsten carbide) to initiate the break and applying a tension; a theoretical analysis shows [4] that a proper stress distribution across the fibre can produce a mirror-like surface on the entire fibre cross-section. The correct stress distribution can be obtained by bending the fibre to an appropriate curvature radius, which

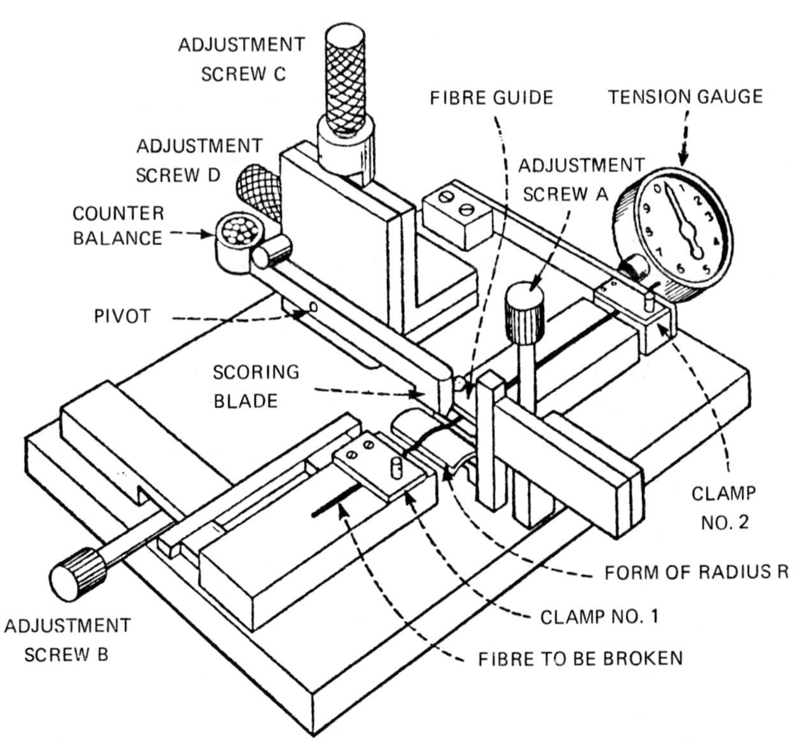

FIG. 2 – A semi-schematic view of a fibre breaking machine (After [4]).

depends on fibre diameter and Young's modulus. For instance, for silica fibres with 125 μm overall diameter, a bend radius of 5.7 cm and a pulling tension in the range 125 to 175 g were found to produce good and reproducible results. A schematic of a tool performing the operations needed for end preparation is shown in Fig. 2. From this basic design handtools have been developed [5-7]: an example is shown in Fig. 3.

FIG. 3 – Fibre breaker (hand type) (After [5]).

It must be noted, however, that according to the authors' experience, (confirmed by a quantitative analysis by other researchers [8]), scribing and pulling by hand can produce better results than machine operation. Typical examples of the surface quality obtained in this way are shown in Figs. 4*a*, *b* representing respectively a normal microscope photograph of a fibre cross-section and a picture taken with an interferometric microscope, where deviations from planarity correspond to deviations from straightness of the fringes.

Other methods for obtaining good ends make use of electrical sparks [9-12] or thermal shocks [13, 14] to create a weakened zone where the fibre is fractured by pulling; an example of an end obtained by thermal shock, i.e., by heating the proper zone of the fibre with a flame, is shown in the electron microscope photograph of Fig. 5. In [14] a hot wire is used to cut the fibre; a schematic of the procedure is represented in Fig. 6. The last class of techniques has the advantage of not requiring a mechanical

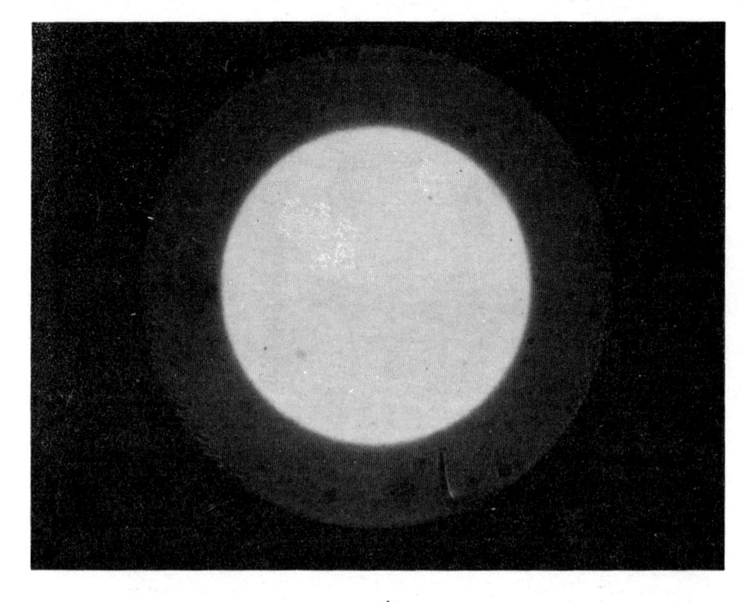

a)

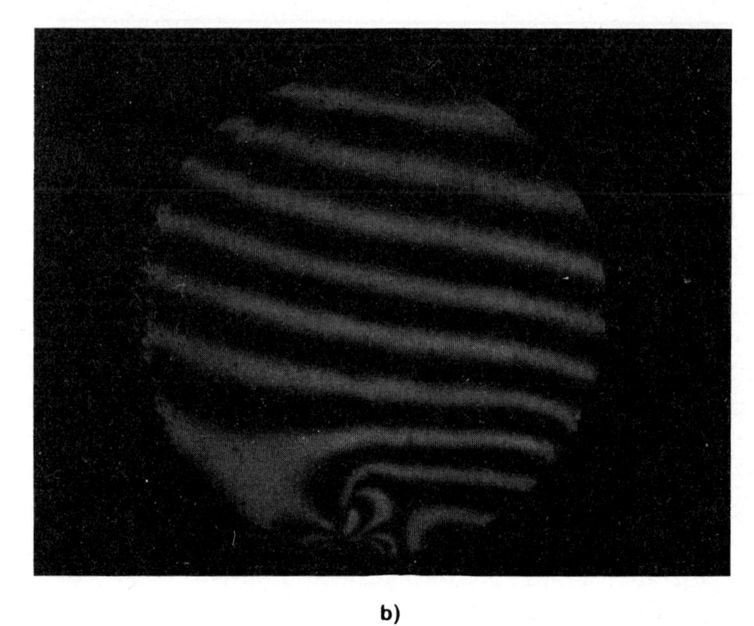

b)

FIG. 4 – Photographs of a fibre cross-section: *a*) normal microscope; *b*) interferometric microscope.

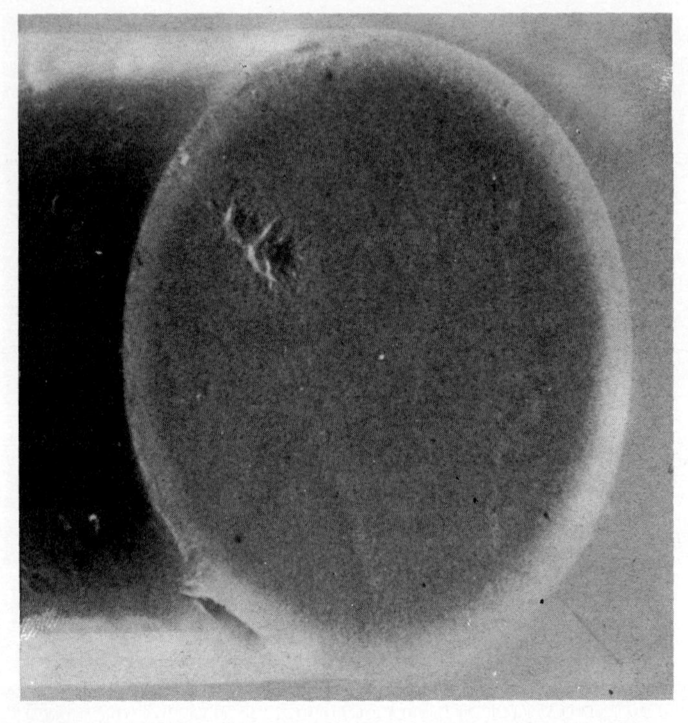

FIG. 5 – Electron microscope photograph of a fibre end obtained by thermal shock.

scoring action, but involves somewhat more cumbersome devices for the operation.

In general, the methods described work very well on ordinary CVD fabricated optical fibres. However, it has been pointed out that fibres with large numerical aperture (i.e. with very high dopant concentrations) break irregularly if the score-and-pull method is used [15]. This is attributed to internal stresses resulting from different properties of the core and the cladding. To eliminate this effect, a new technique consisting of a combination of mechanical scoring and hot wire heating has been implemented [15]. A nichrome wire is heated by electrical current, raising the fibre temperature to ~ 850 °C; the wire itself is used to raise the fibre with respect to the clamps, producing an appropriate bend. Then the fibre is scored and pulled in the ordinary way, and good ends result.

In order to perform measurements on fibres it is usually required that the fibre ends be precisely positioned and fixed. The

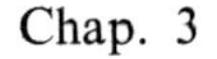

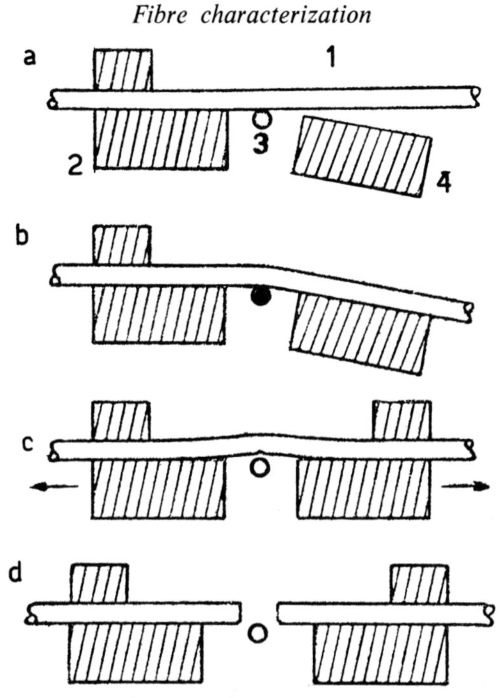

FIG. 6 – Cutting procedure. (After [14]): *a*) Fibre 1, supported by clamp 2, touches wire 3; *b*) The wire is heated and the free end of the fibre contacts second support 4; *c*) Clamp 4 is lifted and is used for pulling; *d*) Fracture finished.

fibre can be conveniently held by inserting it in *V*-shaped grooves or, alternatively, in guiding cylinders, and clamping it by means of an elastic element (e.g., a spring). More sophisticated approaches are also used. One of these consists of a vacuum chuck, shown in Fig. 7. The slot at the groove bottom may be evacuated, so that the fibre is held by atmospheric pressure [16]. Another approach [17], a possible configuration of which is shown in Fig. 8, employs a non-uniform electric field to hold and position the dielectric fibre. Two electrically insulated metal edges present a mechanical reference for the fibre: when a voltage (a few hundred volts) is applied, a non-uniform electric field is obtained which causes an electrostatic force pressing the fibre against the metal edges.

To bring the fibre into the right position (e.g. in the focal point of a lens) it must be mounted on micromanipulators, with positioning resolution of a few microns acting on three orthogonal axes. Angular setting may also be useful, at least around two orthogonal axes, in order to align the fibre axis with the light beam axis.

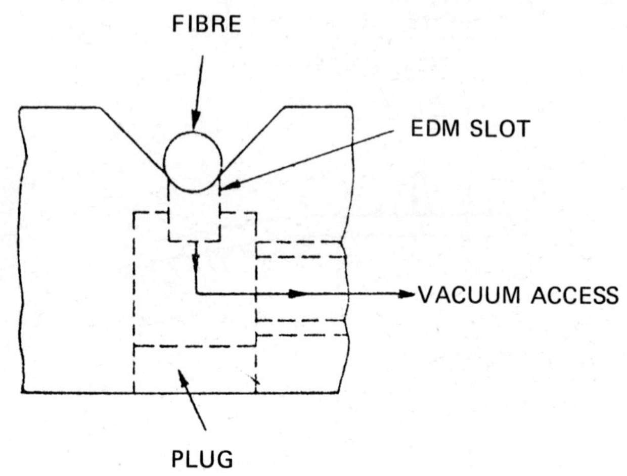

FIG. 7 – Schematic of a vacuum chuck. (After [16]).

Rotatory tables or gimbal mounts are generally used for this purpose. On the other hand, if monomode fibres having a core diameter in the micron range are to be tested, piezo-electrically driven translation stages may be required to achieve a suitable positioning resolution, which may be well below 1 μm.

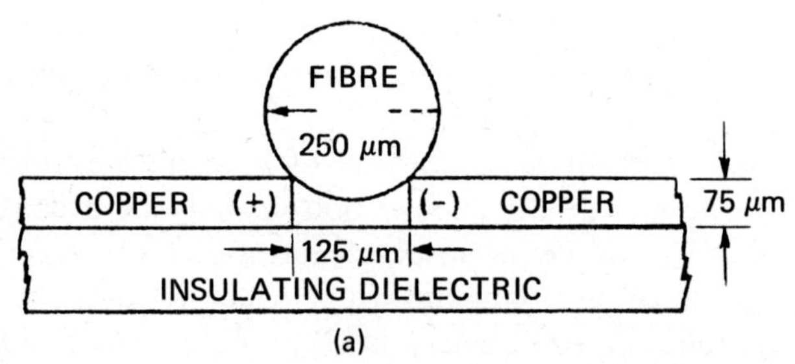

FIG. 8 – Schematic of an electrostatic fibre holder. (After [17]).

3.3 Attenuation measurements

3.3.1 *General discussion*

Attenuation is a basic parameter for optical fibres, as it may limit the maximum achievable length of a communication link. Knowledge of this parameter enables one to know how much an optical signal has been weakened after traversing a known length of fibre. An operational definition of attenuation may be outlined as follows. For a given wavelength λ the light intensity at a distanze z along the fibre is in general given by

$$P(z, \lambda) = P(0, \lambda) \exp \left[-\int_0^z \gamma(\lambda, z) \, dz \right] \tag{1}$$

where $P(0, \lambda)$ is the optical power launched into the guide and $\gamma(\lambda, z)$ is the attenuation coefficient per unit length which may be, in general, position dependent. In practice one is usually interested in an average attenuation coefficient that may be defined as

$$\bar{\gamma}(\lambda) = \frac{1}{z} \int_0^z \gamma(\lambda, z) \, dz \, . \tag{2}$$

In accordance (1) becomes

$$P(z, \lambda) = P(0, \lambda) \exp \left[- z\bar{\gamma}(\lambda) \right] . \tag{3}$$

If the transmitted power is measured for two different lengths z_1 and z_2, $\bar{\gamma}(\lambda)$ can be easily calculated from (3)

$$\bar{\gamma}(\lambda) = \frac{1}{z_2 - z_1} \ln \frac{P(z_1, \lambda)}{P(z_2, \lambda)} \, . \tag{4}$$

The attenuation is usually expressed in terms of dB/unit length as $\bar{\alpha}(\lambda)$, which is given by

$$\bar{\alpha}(\lambda) = 10 \log e \cdot \bar{\gamma}(\lambda) = \frac{1}{z_2 - z_1} 10 \log \frac{P(z_1, \lambda)}{P(z_2, \lambda)} \, . \tag{5}$$

Although an attenuation measurement based on the definition given in (4) or (5) is in principle very simple to implement, it poses some problems of both a conceptual and practical nature when considered in more detail. The conceptual problem arises when multimode fibres are to be measured because each mode propagates, in general, with a different attenuation constant, as discussed in Section 2.4. This means that the attenuation value that is measured depends on the particular modes that are excited and on their relative intensity, or in practical terms, on the particular optical source and the way radiation is injected into the fibre. This problem is not encountered if the measurement is performed when propagation in the fibre is in the so-called « steady state », that is, in an equilibrium condition in which the relative power content of the various modes becomes constant: in this situation all modes have the same loss coefficient. This occurs after a certain length when high loss modes have been eliminated and when the amount of power transferred from one mode to the other modes through mode coupling is compensated by an equal amount received in the opposite direction. A particularly clear example of the consequences of the effects discussed

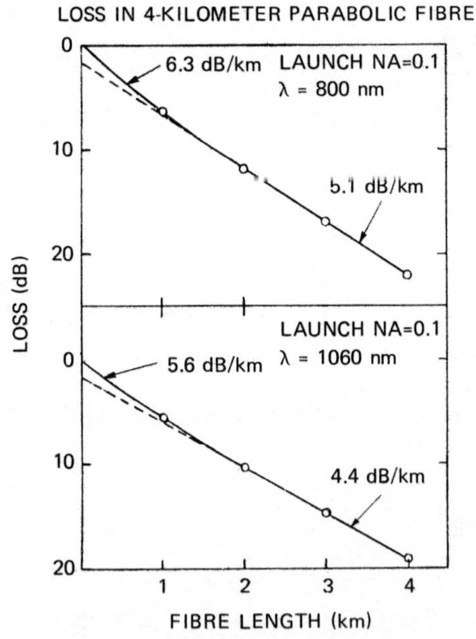

FIG. 9 – Results of attenuation measurements as a function of fibre length with different launch NA. (After [18]).

above is shown in Fig. 9, where the results of attenuation measurements are represented as a function of length in a 4 km long graded index fibre [18]. The loss per unit length is given by the slope of the curve. A higher, length dependent specific loss is measured near the input end: this can be interpreted as an effect of the presence of high loss modes which are being gradually suppressed. A steady state appears to be reached during the second kilometre, after which a constant loss is achieved.

Mode coupling (i.e., power transfer among different modes; also see Chapter 2) that is responsible for these effects is caused by perturbations of the waveguide structure; localized non-homogeneities of the dielectric constant, irregularities at the core-cladding boundary, microbending, bending, etc.; its strength determines the equilibrium length after which steady state propagation settles. This suggests a possible way of performing a measurement that is independent of the particular excitation conditions: it is sufficient to establish at the input end of the fibre a power distribution similar to the steady state power distribution. This can be accomplished either by suitably shaping the input light beam, or by using mode filters or mode scramblers. These techniques will be discussed in the next section.

A practical problem arises in the measuring procedure. As can be seen from Eq. (5), the optical power must be measured at two points. The input end of the fibre is not suitable because this would imply the knowledge of the optical power coupled into the fibre, which is rather difficult to evaluate with the accuracy required for a precise measurement. Therefore, the fibre should be cut at a certain length z_1, thus wasting a portion of fibre. This is very inconvenient, especially for cabled fibres and, in particular, for installed optical cables and field measurements. In view of such difficulties non-destructive techniques have been developed, which will be described later.

Finally, the attenuation of a fibre is not only determined by its intrinsic properties, but also by environmental condition such as microbending, bending, and mechanical stresses. Care must be taken to avoid such conditions; for instance, when fibres are wound on reels for ease of handling, a reasonably large drum diameter and loose wrapping are required.

3.3.2 *Steady state mode exciters*

As previously said, launching an equilibrium distribution of modes would simplify the interpretation of measurements, especially if the results are to be applied to long haul systems. By such techniques the measured attenuation value is very likely to be the most correct one. Three main methods have been proposed to produce such an equilibrium distribution.

The first method consists in inserting a long fibre between the optical source and the fibre under measurement, so that the effective optical source is the long fibre at the output end of which an equilibrium distribution should be reached. This method seems inconvenient in several respects: *a*) it requires the use of a long fibre ($\geqslant$ 1 km) in order to settle steady state propagation conditions; *b*) the steady state pattern of the illuminating fibre might differ from the corresponding pattern for the fibre to be tested; *c*) the reference fibre acts as a filter with respect to the wavelength, so that certain wavelength regions may be strongly attenuated.

The second method consists in inserting a short length of fibre ($\sim$ 1 m) subjected to strong mechanical perturbations causing a heavy mode coupling such that an equilibrium power distribution is obtained after a very short length. Such devices are called « mode scramblers » or « mode mixers ». In order to test their effectiveness two methods can be employed. One is to measure the far field radiation pattern (i.e. the optical power distribution in a meridional plane as a function of the angle with respect to the fibre axis) and verify that it is insensitive to changes in the launching conditions. As the far field is related to the mode distribution (see Section 3.9), this should indicate that the scrambling action has been effective. However, this test does not completely assure the attainment of a steady state distribution because, for instance, a particular mode scrambler can excite preferentially high order modes, as has been shown in [19]. A better test is to check both the insensitivity to launching conditions and the similarity between the far field of the mode scrambler and the far field from a long length of fibre similar to that used for the mode scrambler. An example of this kind of measurement is shown in Fig. 10.

As regards the practical implementation of mode scramblers,

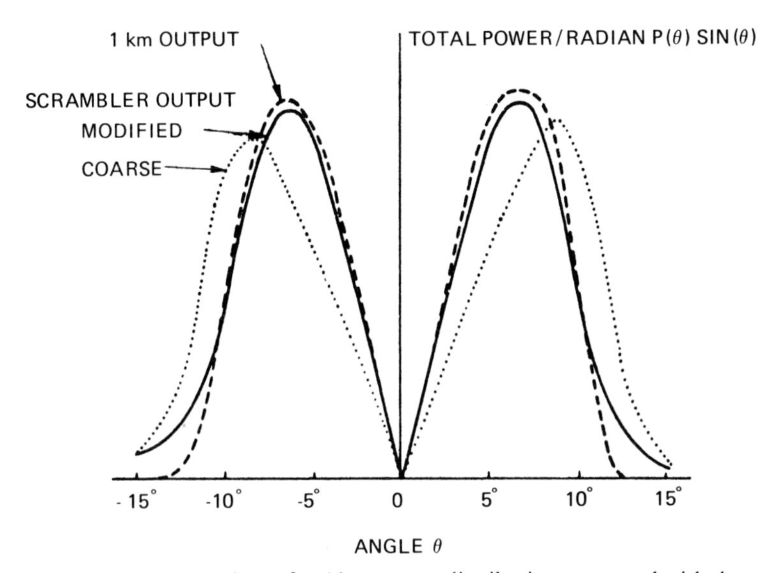

FIG. 10 – Total power plots of a 1 km output distribution compared with the total power produced by a mode scrambler consisting of two sheets of abrasive paper between which the fibre is sandwiched. Use of a modified (shorter correlation length ~ 100 μm) scrambler gives a better result. (After [19]).

several designs have been proposed: *a*) sandwiching the fibre between two sheets of abrasive paper under a certain pressure [19]; *b*) winding a plastic-coated fibre around sharp edges; *c*) shrinking the fibre among metallic wires by means of a heat shrinkable tube [20] (Fig. 11*a*); *d*) etching the fibre end [21] (Fig. 11*b*); *e*) subjecting the fibre to sinusoidal bends by inserting it between and around a straight row of equally spaced pins [22] (Fig. 11*c*). Although all authors claim to have obtained satisfactory results, one obvious shortcoming of these devices, with the exception of the last one, is apparent, i.e. no single parameter can be adjusted to obtain repeatable results. For the last mode scrambler (Fig. 11*c*) three such parameters exist, i.e. rod diameter d, rod spacing s, and the number of periods N. The possibility of specifying the structure of a mode scrambler is very important if any standard method of measurement is required for the purpose of comparison among results obtained in different laboratories.

The third method consists in using an optical source to match the steady state properties of the fibre under test. This can be accomplished if the input light beam has an angular width equal

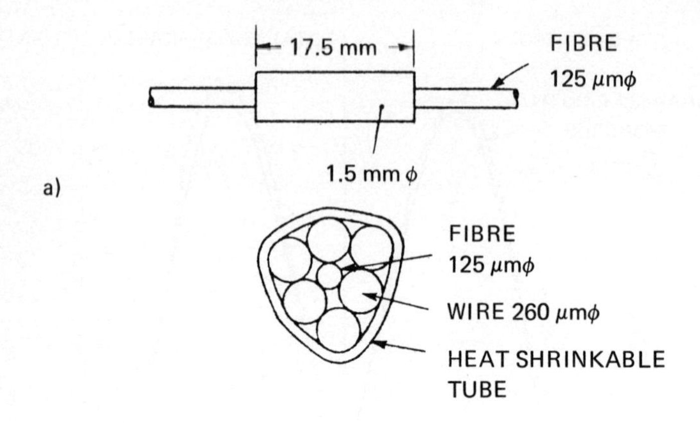

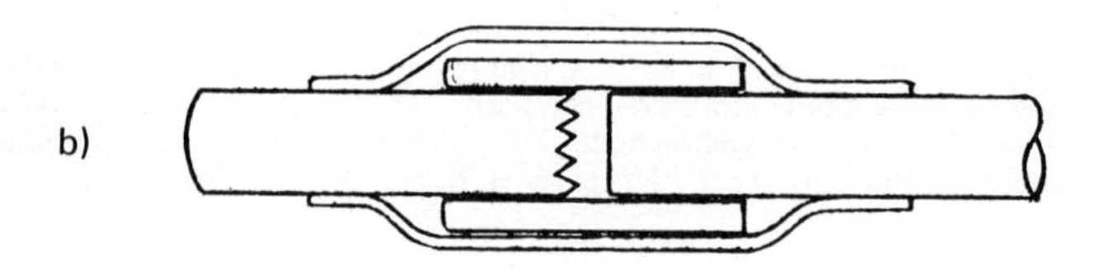

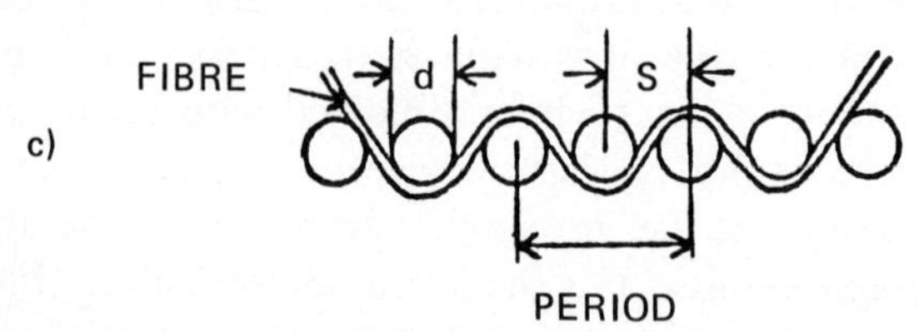

FIG. 11 – Examples of mode scramblers: *a*) shrinking technique (After [20]), *b*) etching technique (After [21]); *c*) bending technique. (After [22]).

to the equilibrium NA, in which case the launching NA remains unchanged during propagation; this is clear from Fig. 12 taken from [23] where the quantity $[\alpha_2(L)/\alpha_s]^4$ is plotted against the length L: α_s is the angular far field distribution at equilibrium and α_2 is the angular distribution measured at a distance L from

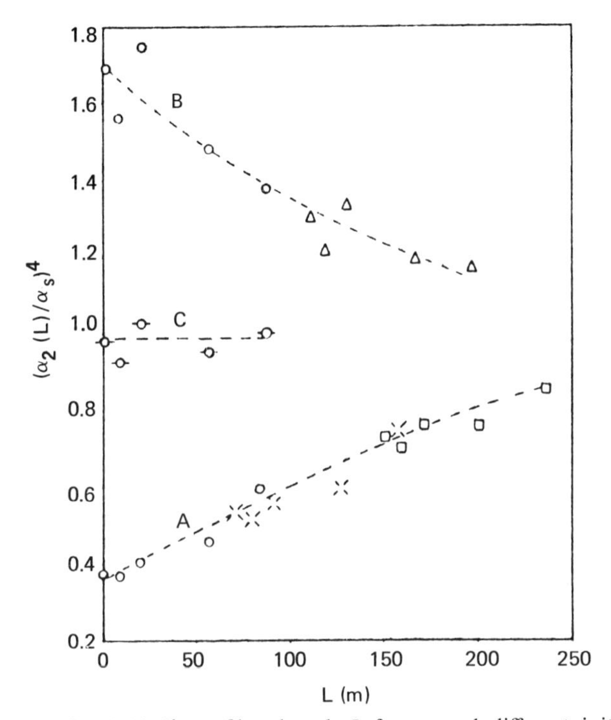

FIG. 12 – Plot of $[\alpha_2(L)/\alpha_s]^4$ vs. fibre length L for several different initial distributions, where $\alpha_s = 15.5°$. (After [23]): *A*) Input NA < equilibrium NA. *B*) Input NA > equilibrium NA. *C*) Input NA = equilibrium NA.

the input end. Input NA larger or smaller than the equilibrium NA tends toward the latter. An input distribution already matched to the equilibrium distribution remains unchanged.

In addition to angular matching, the spot size on the fibre input face should be equal to light power distribution in the fibre cross-section at steady state. This point is discussed in Section 3.9. From a experimental point of view the above conditions may be realized by using a variable NA optical system and variable size optical sources. NA may be varied by using a technique normally applied in optical systems, that of placing behind the last lens of the system a diaphragm of suitable diameter, so that the effective *f*-number of the lens is changed. The *f*-number is defined as the ratio of the focal length to the diameter of the aperture stop. If we consider a collimated light beam impinging on a lens, the aperture stop corresponds to the lens diameter D, and the f-number is given by $f/\# = F/D$ where F is the lens focal length. Placing a diaphragm of diameter D' behind the lens clearly changes the f-number, because the aperture is determined by the diaphragm itself. This is illustrated in Fig. 13*a*; it is clear that changing D' also changes the half-angle of the cone of rays subtended by the aperture with $\theta = \mathrm{tg}^{-1} D'/f$. Thus, a wheel with diaphragms of

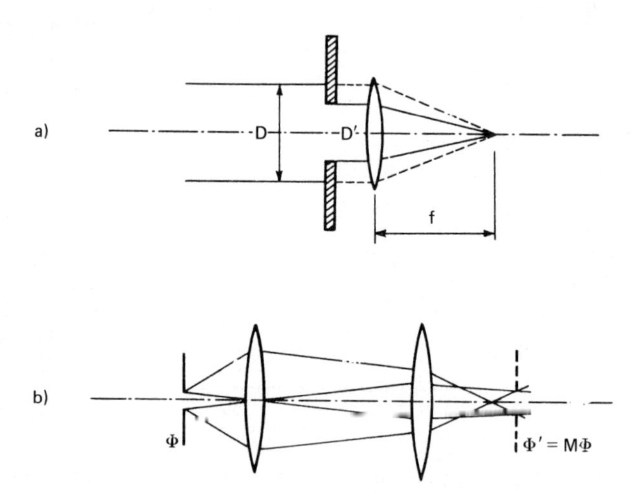

FIG. 13 – Techniques for obtaining controlled launching conditions: *a*) control of lens *f*-number; *b*) control of source size.

different diameter or a variable iris can provide a set of possible angular apertures of the input beam.

Spot size can be similarly varied by placing in the object plane of the optical system a pinhole of given diameter ϕ; the spot size on the fibre ϕ' is determined by the magnification of the optical system M (see Fig. 13*b*), so that $\phi' = M\phi$.

3.3.3 *Total loss*

a) Differential technique

A possible method, based on definition given in Eq. (5), for making an attenuation measurement is to measure $P(z, \lambda)$ at different lengths z_i, then plotting the experimental points on semilog paper.

The various $P(z_i, \lambda)$ should lie on a straight line whose slope has the value $\bar{\alpha}/10$. Taking a fairly large number of experimental data allows a mean square fit to be carried out, resulting in high measurement accuracy. This procedure, however, has the obvious disadvantage of destroying the sample. To avoid this, it has become common practice to perform only two measurements, one at the output end of the full length of the fibre, the other near the input end, cutting the fibre after a short distance z_1 ($\sim$1-3 m from end). The ratio of the two signals is used to obtain the fibre loss in the remaining longer length. In this way the technique can be considered non-destructive. Its main drawback consists in the fact that, using such a short length, the measurement becomes heavily dependent on the launching conditions. The considerations developed in Section 3.3.1 apply completely. An example of the influence of input coupling is shown in Fig. 14 [18], which refers to a loss measurement performed on a 4 km long step index fibre with two different launching NA's, one larger and the other

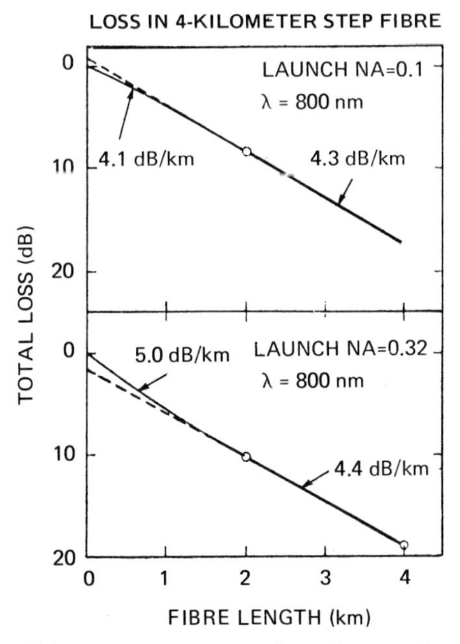

FIG. 14 – Results of loss measurements performed on a 4 km long step index fibre with two different launching NA's. (After [18]).

smaller than the fibre NA. When the launch NA overfills the fibre NA the attenuation measured with a two-metre reference length is 0.7 dB/km greater than the steady state value. This is due to the fact that lossy modes are excited by the said launching conditions, and they are still present after the very short length of fibre used as a reference. Lossy modes contribution is thus fully taken into account when determining the total loss.

The opposite occurs with the small launch NA: in this case low order modes are excited and an attenuation 0.2 dB/km less than the steady state value is measured.

Another experimental confirmation of such effects has already been shown in Fig. 9, where the initial attenuation value in a graded index fibre is 1.2 dB/km higher than the value at longer lengths.

Among the factors responsible for the transient loss, leaky modes play an important role. Leaky modes (see Chapter 3) are modes which are not bound, and radiate through a tunnelling mechanism which leads to a very wide spread of loss values according to the particular leaky mode in question. Loss values range from a few tenths to several thousands dB/km. The influence of leaky modes may be therefore evaluated to try a numerical estimate of the excess transient loss.

For a parabolic index fibre it has been calculated that, when exciting all modes equally, $\sim 25\%$ of the light is launched into leaky modes [24, 25]. The attenuation of leaky modes has been calculated as well [25]; it turns out, for instance, that in a parabolic index fibre (40 μm core radius, NA = 0.2) the leaky modes power at $\lambda = 0.9$ μm is 11.4 % at 1 m, and 7.2 % at 1 km. This means that measuring for such a fibre between 1 m and 1 km the attenuation value would be pessimistic by 0.2 dB. These figures must be taken judiciously, however, as it has been shown experimentally [26] and confirmed by theoretical calculations [27] that the leaky mode loss can increase drastically. This is due, for instance, to small deviations from perfect circular symmetry of the fibre, which cause a loss increase such that an undetectable amount of leaky power is found even after the very short length of 1 m. In any case it appears that a strong contribution to transient loss also comes from lossy guided modes.

The differential technique (also called « two point » or « cut back » technique) lends itself very well to spectral measurement, i.e. measurements carried out versus wavelength. Such measurements are useful for showing up the absorption peaks of metallic impurities, water peaks and scattering peaks. A typical spectral loss measurement apparatus is shown schematically in Fig. 15. It includes a white light source (tungsten halogen lamp

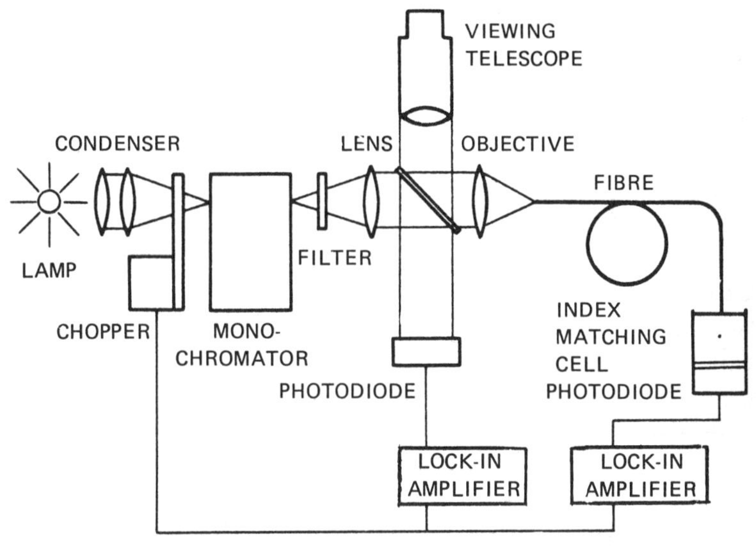

FIG. 15 – Schematic of a typical spectral loss measurement apparatus.

or arc lamp); a condenser to collect and focus the light; a mechanical chopper that provides a modulated signal for a.c. amplification and detection for better sensitivity and stability; a monochromator; a filter used to suppress unwanted diffraction orders from the monocromator grating; a collimating lens; a beam splitter that provides a reference signal used to take into account source power fluctuations; removable viewing optics to observe the fibre input; and an objective to focus the light onto the fibre end. The fibre itself is held in a micromanipulator for precise positioning and signal maximization (see Section 3.3.1); it may be preceded by a short piece of fibre acting as a mode scrambler, as discussed previously (Section 3.3.2). A normal precaution consist in applying near both the input and the output ends of the fibre a « mode stripper » i.e. a device capable of extracting the light travelling in the cladding, which can affect the measurement by adding a spurious signal to the guided optical power. Mode stripping is usually performed by slightly bending the fibre and immerging it in a fluid with a refractive index higher than that of the cladding. If a protective coating is present, it must be removed to make the mode stripper work. The output end of the fibre must be placed in front a of detector: this end may be immersed in a core matching liquid in order to minimize measurement uncertainties deriving from surface irregularities. Signals from the reference detector and the receiving detector are sent to lock-in amplifiers for the measurement. This apparatus allows measurements in the visible and near i.r. region of electromagnetic spectrum to be performed; in the 500-1100 nm region silicon photodiodes or solar cells are usually employed as detectors; in the 1100-3000 nm region, PbS, PbSe (usually cooled) detectors may be used; a germanium photodiode can cover the whole spectrum between 500 and 1700 nm. The dynamic range of the apparatus is in the range of 40 dB at 850 nm for 60 μm core fibres. As an alternative to the monochromator approach, a wheel of interference filters for the wavelength selection can be used.

An optical arrangement using filters and having the additional possibility of changing the source size and beam angular width by means of pinholes and diaphragms (see. Section 3.3.2) is shown in Fig. 16.

An advantage of using filters is the availability of a higher optical power, and hence a higher dynamic range (10-15 dB more), with respect to the monochromator apparatus. A clear disadvantage is the limited selection of wavelengths and spectral widths.

Single wavelength measurements can be performed instead of spectral measurements by using lasers as optical sources. In particular, the use of semiconductor lasers (GaAs or GaAlAs) allows the measurement to be performed in conditions very similar to those found in communication systems usually employing such sources. Using lasers, dynamic ranges of 60 dB or more are possible; on the minus side there is the fact that they are less stable than tungsten-halogen lamps unless a feedback loop is used to control the light output.

The accuracy of the attenuation measurement is limited mainly

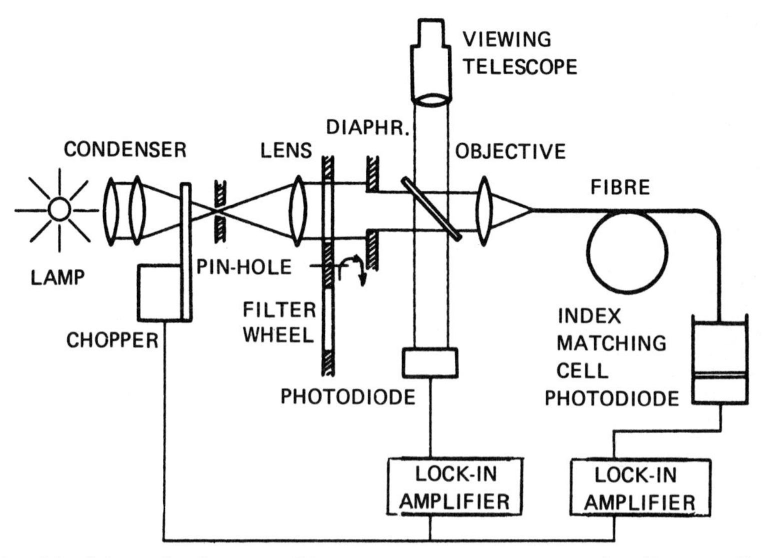

FIG. 16 – Schematic of a spectral loss measurement apparatus using filters to select wavelengths and having the possibility of changing the source size and angular width.

by the repeatability in both input and output coupling conditions. Differences in output signal coupling to the detector are length-independent, so that they may have largely different influence according to the total loss measured; careful end preparation and positioning can limit the fluctuations to ± 0.1 dB or less.

Differences in input coupling are length-dependent through different mode excitation. The repeatability is strictly related to the kind of fibre illumination (see Section 3.3.2). It has been found that, with small spot and small NA excitation it can be well below ± 0.1 dB/km.

b) Insertion loss technique

An alternative approach, which may use the same experimental arrangements as the differential technique, is the one usually called insertion loss technique. Either a fixed source-fibre assembly, or an optical source with definite and suitable size and angular width are used as optical signal sources.

In the first case the fibre under test is spliced or connected by means of a detachable connector to the fixed fibre whose optical output is known; assuming that the optical power is completely coupled to the second fibre (or accounting for connector losses if they are known) a simple measurement of the optical power from the fibre output end gives the loss caused by the insertion of the fibre in the network (« insertion loss »), or, in other words, its attenuation.

The same procedure is followed for the second arrangement, provided the amount of light coupled to the fibre under test is accurately known.

This method has the distinct advantage of being completely non-destructive.

On the other hand, the stability requirements of the optical arrangement are very severe and the accuracy of the measurement is rather poor, being determined by the repeatability of the coupling conditions. It may find application in cases when a non-destructive procedure is required and a low accuracy, quick measurement is satisfactory, such as field tests, optical cable checks and so on.

c) Back-scattering method

A method that overcomes the main drawbacks of the differential technique, being completely non-destructive and requiring access only to one end of the fibre, is the back-scattering method (or optical time domain reflectometry). It was developed by Barnoski *et al.* [28] and implemented by other authors [29-32]. In this approach, a short optical pulse is sent into a fibre; a fraction of the power scattered by the fibre is guided backwards, and the sequence of echo pulses forms an envelope pulse which is received and analyzed at the same input end.

In fact, light travelling in a fibre undergoes a distributed scattering, i.e., approximately isotropic Rayleigh scattering. Considering only this scattering the amount of power scattered at a distance z from the input end in a section of length dz, is

$$P_s(z) = \gamma_s P(z)\, dz \tag{6}$$

where γ_s is the Rayleigh scattering loss coefficient in m^{-1}; it is assumed constant although in general it may be position-dependent, due e.g. to non-homogeneities in the composition of the fibre material.

In an optical fibre $P(z)$ varies according to Eqs. (1), (3).

Assuming an approximately isotropic angular distribution for the scattered power, the fraction of captured power S is given by the ratio of the solid angle of acceptance of the fibre to the total solid angle (this holds for step-index fibres, but is only approximately true for graded index fibres)

$$S \cong \frac{\pi \Delta^2}{4\pi n_0^2} = \frac{\Delta^2}{4n_0^2} \tag{7}$$

where Δ is the value of fibre numerical aperture, $\Delta = \sqrt{n_0^2 - n_1^2}$, n_0 is the fibre refractive index, and n_1 the cladding index. Therefore the power backscattered between z and $z + dz$ is:

$$P_{bs}(z) = \gamma_s SP(z)\, dz\,. \tag{8}$$

The power scattered between z and $z + dz$ reaching the detector, assuming a coupling efficiency to the same equal to η, will be given by

$$P_{bsd}(z) = \eta P_{bs}(z) \exp\left[-\int_0^z \gamma'(z)\, dz\right] =$$

$$= \gamma_s SP(0) \exp\left[-\int_0^z [\gamma(z) + \gamma'(z)]\, dz\right] dz \tag{9}$$

γ' is the attenuation coefficient for the backward propagating light; strictly speaking it is equal to γ only if the same modes with equal power distribution propagate in the two directions; however, for practical use we can consider the two coefficients equal and write

$$P_{bsd}(z) = \eta\gamma_s SP(0) \exp\left[-2\bar{\gamma}z\right] dz\,. \tag{10}$$

The power generated at z is detected after a time $t = 2z/v_g$, v_g being the group velocity of light in the fibre. If the probe pulse has a width ΔT, the total power $p(t)$ falling on the detector at the time t is obtained by summing (10) in an interval $\Delta z = v_g \Delta T/2$. If we can consider $\exp[-\bar{\gamma}z]$ constant over this length, substituting $z \to v_g t/2$ we have

$$p(t) = \eta \frac{c\gamma_s}{2n} SP(0)\, \Delta T \exp\left[-2\bar{\gamma}\frac{v_g t}{2}\right] \tag{11}$$

provided $P(0)$ is constant during ΔT; otherwise its average value must be considered. Therefore the return waveform has an exponential shape from which the total loss coefficient may be evaluated

$$\frac{p(t_1)}{p(t_2)} = \exp\left[-\bar{\gamma}\frac{c}{n}[t_2 - t_1]\right] \Rightarrow \bar{\gamma} = -\frac{n[\ln p(t_1) - \ln p(t_2)]}{c(t_2 - t_1)}. \quad (12)$$

From the complete waveform many points may be taken and a least square fit may be performed to obtain a higher accuracy.

An experimental arrangement used by the authors for carrying out the measurement is shown in Fig. 17. The source is a

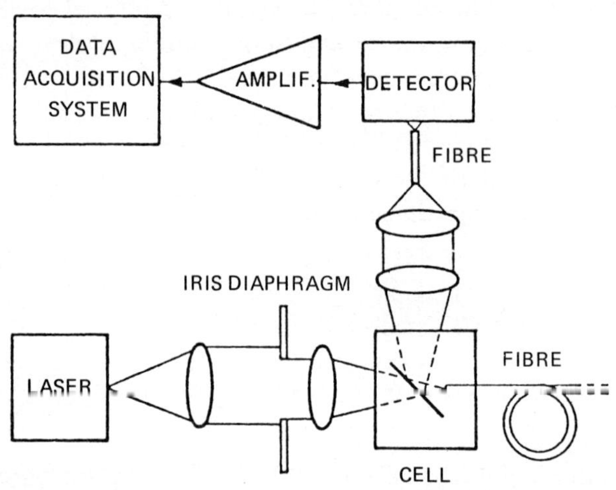

FIG. 17 – Experimental arrangement of backscattering measurement apparatus. (After [31]).

semiconductor GaAs or GaAlAs laser, whose emission wavelength can be chosen in the range 800-900 nm; it is driven by a suitable electronic circuit which includes avalanche transistors providing high current pulses, generating optical pulses in the 4-10 W peak power range. A pulsewidth of 1-40 ns can be easily obtained, the choice depending on the best compromise between resolution (which is obviously better for short pulses) and signal level (which according to (11) depends linearly on the pulsewidth). The laser beam is collimated and focused onto the fibre by lenses; the fibre is inserted in a cell, whose function will be discussed later on, comprising a half-reflecting beam splitter at 45° to collect the backscattered signal; this is collimated and focused onto a detector. In order to reduce background spurious signals to a minimum, angular and spatial stops are placed in front of the detector in such a way as to accept only the light coming from the fibre under measurement; they may be implemented by a proper combination of irises and pinholes or, as in the figure, by means of a short piece of fibre with NA and core size matching those of the measured fibres. The detector is an avalanche photodiode (it could be a photomultiplier [30-33]) followed by a low noise transimpedance amplifier with adequate bandwidth (34 MHz in our experiment). The signal is fed to a boxcar integrator which, operating in a scanning mode, performs an average over a preselected number of pulses to improve the signal to noise ratio (other averaging techniques can of course be employed). The signal from the boxcar can be displayed on an oscilloscope, and recorded by an *x-y* plotter, or by some other means (punched paper tape, cassette tape, etc) for subsequent processing.

From an experimental point of view, a main problem is to prevent the reflection from the fibre input end from falling onto the detector. In fact, if we consider a typical fibre with $\Delta = 0.15$, $\gamma_s = 0.69\ \text{km}^{-1}$, $v_g \simeq c/n_0 = 2.04 \times 10^8$ m/s, and a pulsewidth $\Delta T = 50$ ns, it can be easily calculated from (11) that the backscattered power level is 36 dB below the 4% reflection from a good break; if the sensitivity of the detector-amplifier combination is

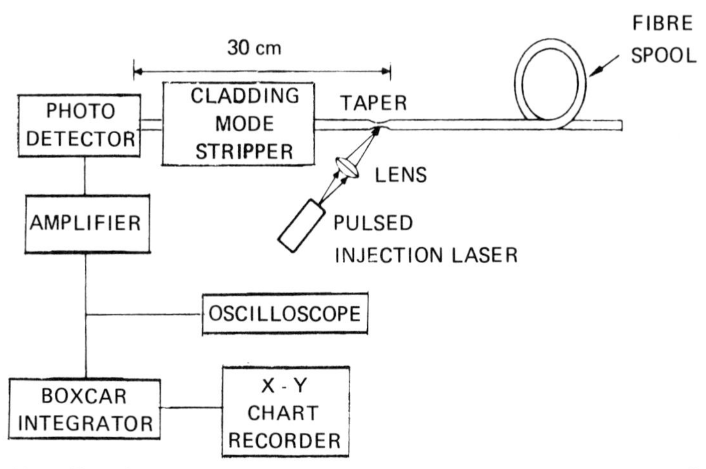

FIG. 18 – Experimental arrangement used to record return optical waveform, using lateral injection of light. (After [28]).

such as to reveal the weak useful signal complete saturation is caused by the reflected pulse, resulting in severe distortion of the exponential waveform and deterioration of data.

Similar reflections from the optics used for collection and focusing of the light beam must also be considered. Different techniques have been proposed to solve this problem: launching into the fibre through a tapered section of it [28] (Fig. 18) which has the disadvantage of difficult fibre preparation, poor coupling efficiency, and excitation of high order modes; electrical gating of the detector to render is not operative while the large signal is present [30, 32] (Fig. 19); use of polarization sensitive beam splitters which reflect light or let if pass according to the state of polarization (they are usable in the assumption that the fibre quickly depolarizes light) [34]; special coupling devices [35] (Fig. 20). The authors used a special cell (Fig. 21) filled with index matching

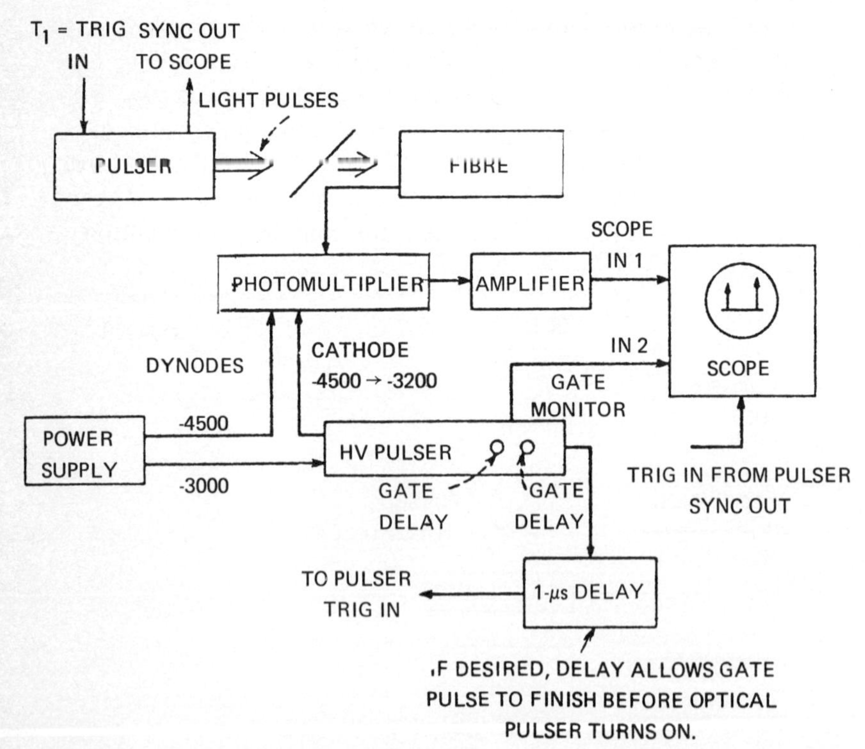

FIG. 19 – Optical time-domain reflectometer, using a gated photomultiplier for suppressing the spurious reflection. (After [30]).

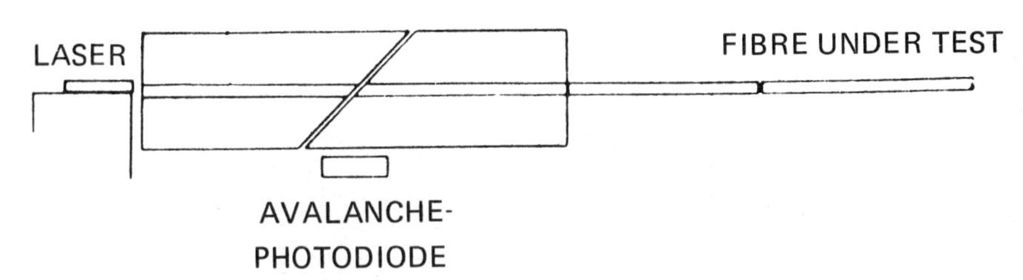

FIG. 20 – Directional coupler making use of a short piece of fibre inserted in a glass capillary and cut at a 45° angle. (After [35]).

liquid to reduce the reflection from the fibre input face; it includes a half-reflecting beam splitter, and the entrance and exit windows are covered by spherical lenses that match the incoming and outgoing approximately spherical wavefronts of the light beam; internal walls are suitably shaped and covered with absorbing material (e.g. black velvet) in order to keep the spurious signal at a minimum.

The fibre is inserted into the cell through a small hole by means of micromanipulators. The construction and use of the device is very simple. The performance is very good and the amount of

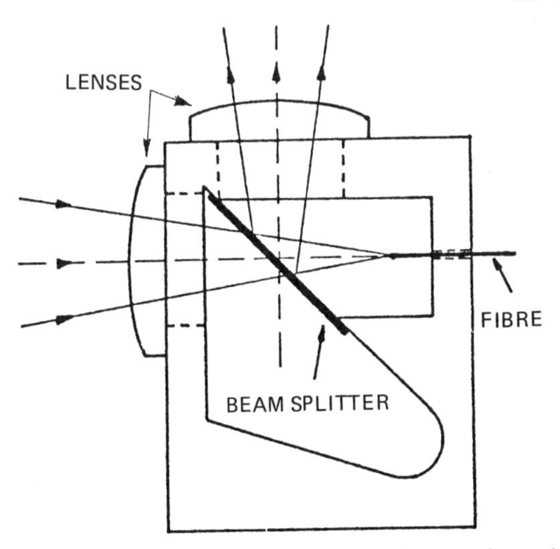

FIG. 21 – Index matching cell used to reduce the reflection from the fibre input face. (After [31]).

possible spurious signal reduction can be judged by the photograph in Fig. 22 which represents the first portion of the backscattered signal. An oscillogram of the whole waveform is shown

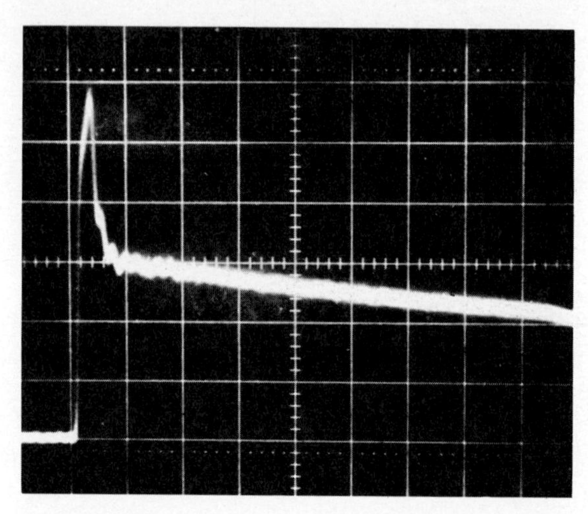

FIG. 22 – Oscilloscope trace of the initial part of a backscattered signal. Horiz. scale: 100 ns/div. Vert. scale: 20 mV/div.

in Fig. 23 before boxcar averaging. A boxcar averaged output is shown in the plot of Fig. 24. We can distinguish the small spurious signal at the input end, the exponential curve, and the Fresnel reflection from the output end (reduced by immersion in glycerine).

As already mentioned, typical advantages of this method are the fact that it is non-destructive and requires access to only one end. These are very desirable features, especially for measuring installed optical cables. Another important feature of the method is that it provides detailed information on the loss characteristics of the fibre along its length so that lossy sections and non-uniform behaviour may be easily identified; in particular, faults occurring in the fibre and giving rise to localized scattering may be visualized and located with an accuracy that, depending on the pulsewidth, can be as high as 1 m. It must be remembered that a 1 ns pulse corresponds to a ~ 0.20 metre path in a glass fibre. Making use of the Fresnel reflection from the output end, the fibre length

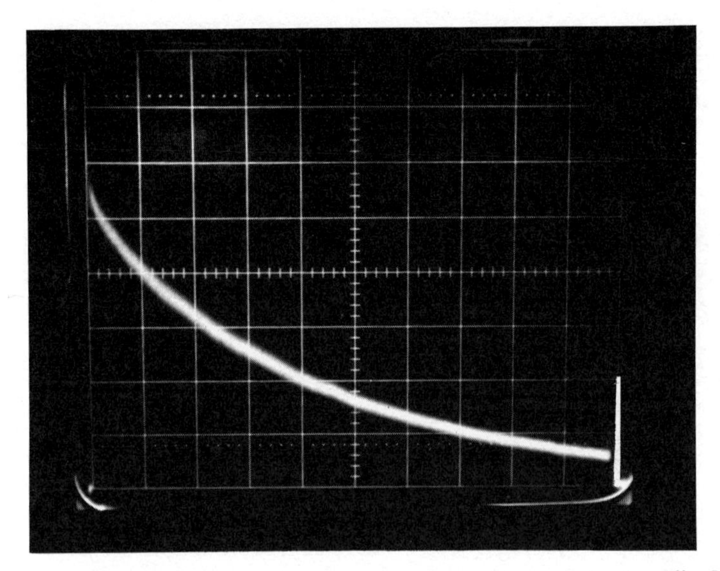

Fig. 23 – Oscilloscope trace of the return waveform Horiz. scale: 1 μs/div. Vert. scale: 10 mV/div.

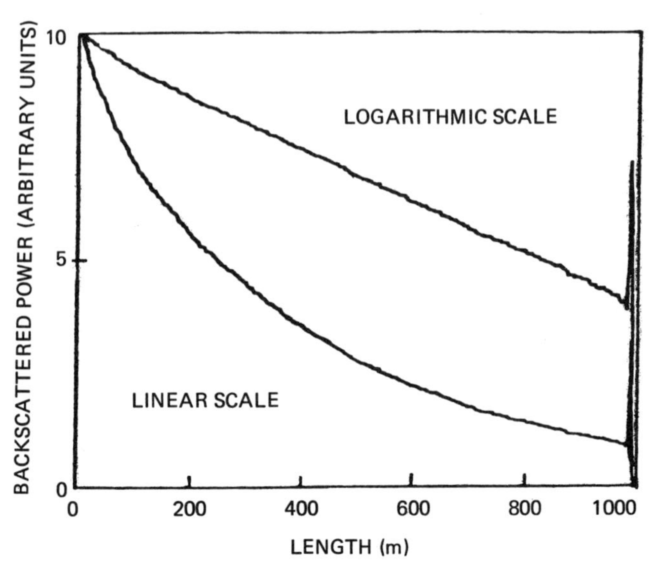

Fig. 24 – Return waveform after boxcar averaging. (After [31])

can also be measured as it is simply $L = v_g T/2$, T being the time employed by the pulse to make the complete round trip. T can be measured directly on the oscilloscope screen, or on a plot recorded on calibrated squared paper, or, for a more precise determination, using a digital time delay generator to make the input and reflected pulse coincide on the oscilloscope. Moreover in long fibres the achievement of steady state propagation may be observed looking for the length after which the logarithmic plot of the backscattered power follows a straight line. In that region the steady state attenuation value may be calculated.

This technique has also some drawbacks. Among the chief ones are: 1) the measurement can typically be performed at a single wavelength only unless tunable lasers are available; 2) a high sensitivity, high bandwidth detection apparatus in required; 3) somewhat sophisticated signal and data processing is necessary; 4) the dynamic range, when semiconductor sources are used, is at present rather limited, being $\sim$40dB when a 60μm core diameter, 0.2 NA graded index fibre is used; this means that fibre lengths with maximum total loss $\sim$20 dB can be measured.

Finally, let us consider the accuracy of the method in determining the attenuation value of an optical fibre. As regards the repeatability it has been shown, by means of a series of measurements on the same fibre, that very high repeatability can be obtained [31]: the measured root mean square deviation was better than 0.04dB for a fibre with attenuation 5.09dB/km at 0.904μm. As regards the « absolute » accuracy of the attenuation value obtained the question is more difficult to answer; a comparison between the loss values obtained by the backscattering technique and those obtained by the two-point technique using similar optical sources in the two cases has been carried out by the authors on a number of good quality graded index fibres [162]. The results are summarized in Table I. The difference is below 0.2 dB in the great majority of cases; from these data it can be concluded that the backscattering technique gives reliable attenuation values.

d) Lateral scattering technique

This technique, like the preceding one, utilizes the light scattered in the fibre core, but in this case the measurement is performed around the fibre, gathering the laterally scattered light power.

TABLE I

Attenuation (dB/km)			
Fiber No.	Backscattering technique	Differential technique	Difference
1NF1	3.9	3.86	+0.04
2NF1	3.5	3.37	+0.13
3NF1	4.22	4.15	+0.07
1NF2	4.1	4.1	0.0
2NF2	4.16	4.3	−0.14
3NF2	3.81	4.0	−0.19
1NF3	3.84	3.88	−0.04
2NF3	3.48	3.44	+0.04
3NF3	3.83	3.52	+0.31
1NF4	3.96	3.75	+0.21
2NF4	3.53	3.5	+0.03
3NF4	3.65	3.61	+0.04
1NF5	3.8	3.6	+0.2
2NF5	3.45	3.08	+0.37
3NF5	3.73	3.67	+0.06
4NF5	3.29	3.11	+0.18

The method is based on the already mentioned assumption that the amount of scattered power in a section of length dz is proportional to the amount of power flowing through that section

$$dP_s(z) = -\gamma_s P(z)\,dz \,. \tag{13}$$

If the light power ΔP_s scattered in two sections of length Δz at distances z_1 and z_2 respectively is measured we have (from (3)):

$$\frac{\Delta P_s(z_1)}{\Delta P_s(z_2)} = \frac{P(z_1)}{P(z_2)} = \exp\left[\bar{\gamma}(z_2 - z_1)\right] \tag{14}$$

and therefore the attenuation $\bar{\alpha}$ in dB/km is given by

$$\bar{\alpha} = 10 \cdot \log e \cdot \bar{\gamma} = 10 \cdot \frac{\log \Delta P_s(z_1) - \log \Delta P_s(z_2)}{z_2 - z_1} \,. \tag{15}$$

If the measurement is repeated at many different positions a least square fit can be applied. The experimental apparatus (see Fig. 25) requires only an appropriately intense light source, and a suitable

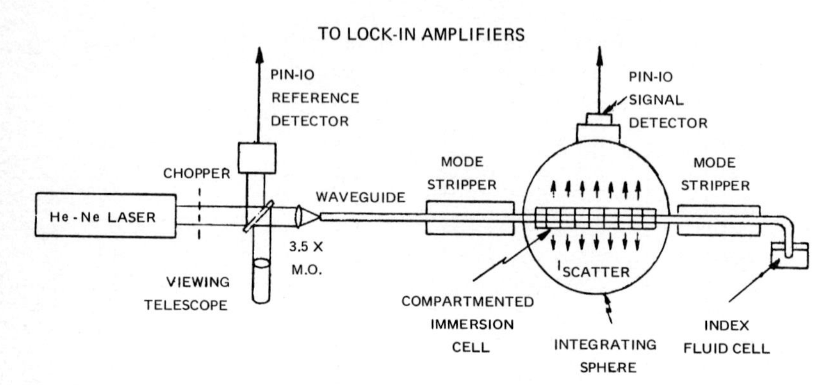

FIG. 25 – Experimental arrangement used in obtaining the total loss and radia. tion loss measurements. (After [37]).

detector with a measuring instrument. Different arrangements have been proposed: 1) a cube detector made of six square solar cells [36] which permits the collection and measurement of all the scattered light, but requires careful selection of components and impedance matching in order to obtain uniform response from the six detectors; 2) integrating spheres [37, 38] (see Fig. 26)

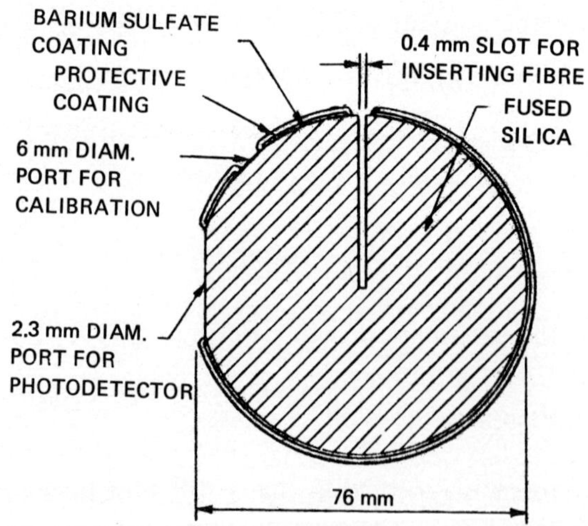

FIG. 26 – Cross sectional diagram of the integrating sphere scattering detector. (After [38]).

which give uniform response independently of the particular scattering distribution but suffer from high loss due to internal reflections; 3) internally reflecting cells [39] (see Fig. 27) which again

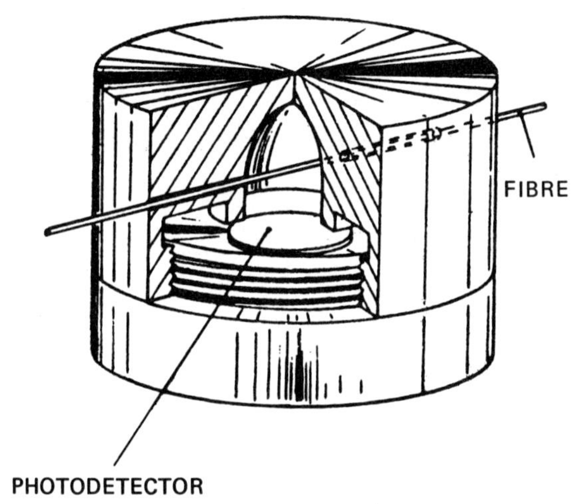

FIG. 27 – Integrating cell. (After [39]).

collect all the light, but may have non uniform response for light scattered at certain angles; 4) two solar cells between which the fibre is sandwiched (see Fig. 28) [163]. This arrangement, at the expense of the small amount of light lost through the small gap between the two detectors, allows an easier selection of components than the cube arrangement.

An important point is that if the measurement is performed with the fibre in air, a portion of scattered light is trapped by the cladding. Provided the fraction of trapped light is constant along the fibre this is not a serious problem for total loss measurements because according to (14) only power ratios are required to calculate the attenuation coefficient. If, however, one is interested in getting all the light out from the fibre a typical procedure is to surround the fibre with a liquid having a refractive index higher than that of the cladding so that no scattered light is trapped, except for the small amount guided by the fibre core. Two other precautions must be taken. One consists in placing immediately before and after the detector two mode strippers, to avoid the detection of light guided by the cladding. Another is to immerse

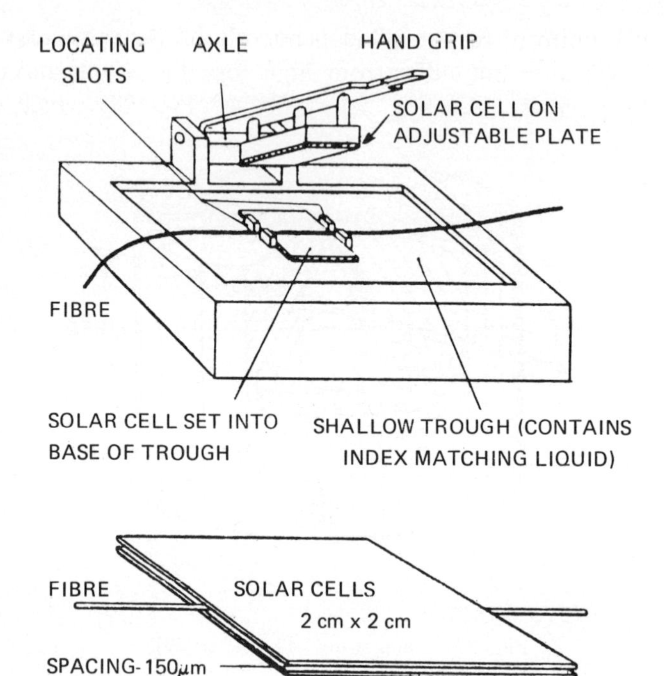

FIG. 28 – Arrangement of scattering photometer. (After [163]).

the output end of the fibre in index matching liquid, so that there is no reflected light contributing to the signal. A result of a measurement performed on a glass fibre is shown in Fig. 29 [39].

In comparison with the back-scattering method this technique has an advantage: the signal level can be higher, as nearly all the scattered light may be collected and measured (except the above-mentioned amount that is trapped in the fibre core, half of which is utilized for the back-scattering measurement) and there is not the effect of double loss caused by the round trip made by the backscattered signal. Moreover, probing the different sections of fibre is performed spatially rather than temporally as in the case of backscattering. Therefore, a modulated signal can be employed, allowing the use of very sensitive detection apparatus.

In contrast, the lateral scattering method shows severe shortcomings: all the fibre must be accessible (a measurement performed on only two sections leads, in general, to rather inaccurate results) and only transparent protective coatings may be tolerated which

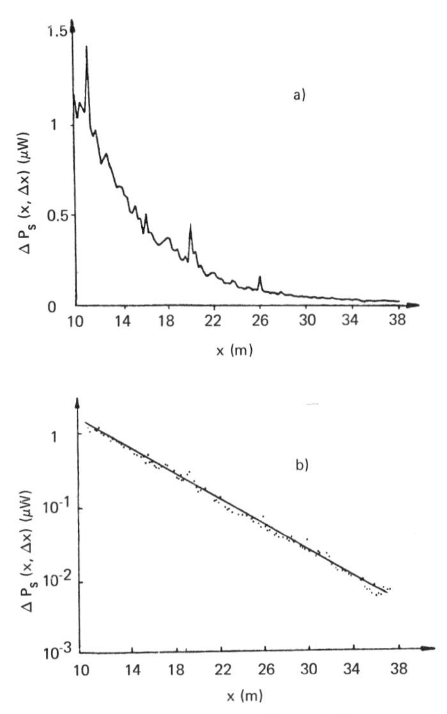

FIG. 29 – Diffusion measurements along the fibre: *a*) linear scale, *b*) semilogaritmic scale. (After [39]).

makes the measurement unsuitable for cabled fibres, or installed optical cables. Moreover, the measurement is long and tedious, as the fibre must be inserted through small holes or slots inside the detector, and moved in steps of precisely known lengths.

e) Pulse reflection method

The method [40] consists in sending an optical pulse into a fibre, allowing for reflection at the output end, and measuring the reflected pulse. A possible experimental apparatus is shown in Fig. 30. A laser and suitable optics provide a pulse to be injected into the fibre; a beam splitter placed between the laser and the fibre provides a reference signal from the laser and allows the reflected pulse to be detected and measured. To increase the reflected power and obtain more controlled conditions than with

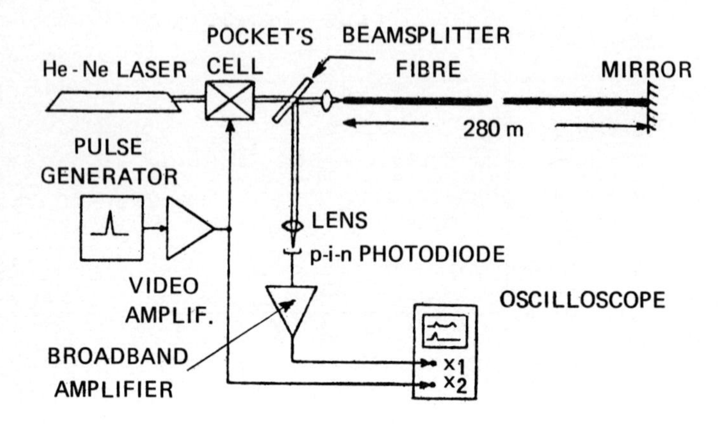

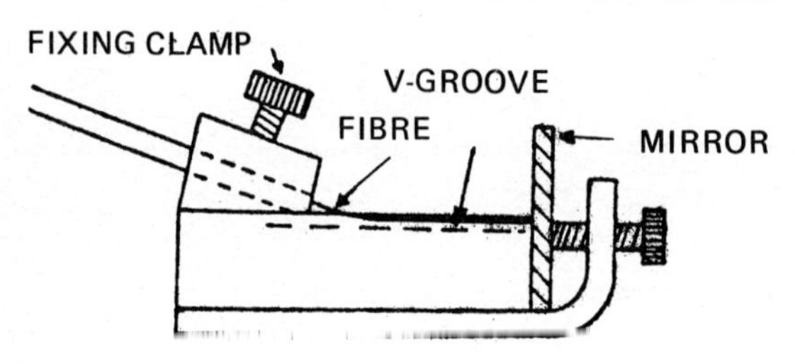

FIG. 30 – Experimental arrangement and mirror mount for attenuation measurement by the pulse reflection method. (After [40]).

a pure Fresnel reflection from the fibre far end, a mirror is fixed perpendicular to the axis of the fibre in a special mounting. To calculate the attenuation, either the signal level after a short length of fibre must be measured for comparison (in this case a correction for the reflection from the fibre input face should be applied, as the two pulses overlap) or reproducible launching conditions must be obtained so that the coupled signal is known and the system can be calibrated.

The main advantage of this method is that a measuring apparatus at the far end of the fibre is not necessary. The end, however, must be available for mirror mounting.

A critical point is also the fibre-mirror alignment, which affects the reproducibility of results.

3.3.4 *Scatter loss*

As already discussed in Chapter 1 the total loss of a fibre consists of a scattering and an absorptive component. It may be useful to measure the relative contribution of the two loss mecanisms for investigating fibre structural characteristics and material composition effects which, in turn, help in identifying the manufacturing parameters to be controlled and material to be selected in order to obtain low loss fibres.

As regards the measurement of the scatter loss there is not much to be added to what was said in Section 3.3.3(*d*), as the experimental apparatus and measuring techniques are similar in the two cases. The only difference is that, while for measuring the total attenuation from a scattering measurement it is necessary that the signal be proportional to the power flowing inside the fibre, but the proportionality constant is irrelevant as only signal ratios are used (see Eq. (14)), as regards the total scatter loss all the scattered light, or an accurately known fraction of it, must be measured in order to evaluate γ_s from (14). Immersion of the fibre in an index matching liquid becomes essential to obtain an accurate measurement; failing this a correction factor may be applied, taking into account the geometrical and physical characteristics [39] of the fibre. Once these requirements are fulfilled the scattering coefficient is given by (from (13)):

$$\gamma_s = \frac{dP_s(z)}{P(z)\,dz}\,. \tag{16}$$

Besides measuring, dP_s and P, dz must also be determined. dz is simply the length of fibre viewed by the detector (i.e. the length inside the various cells or integrating spheres used for measurement). $P(z)$ can be evaluated in two ways: one is to perform the measurement of scattered light close to the output end, so that the optical power coming out from the fibre, which can be easily measured, is essentially equal to $P(z)$ (ignoring the small additional attenuation). The other consists in measuring the output power, but correcting it by the known fibre attenuation coefficient to reduce it to the value $P(z)$ at the measuring point.

Besides total scatter loss, the angular distribution of scattered light is also of interest. From such a measurement information about various scattering mechanisms (Rayleigh, Mie, radiative losses, large-scale imperfections) can be obtained. An experimental apparatus for this measurement is shown in Fig. 31 [41].

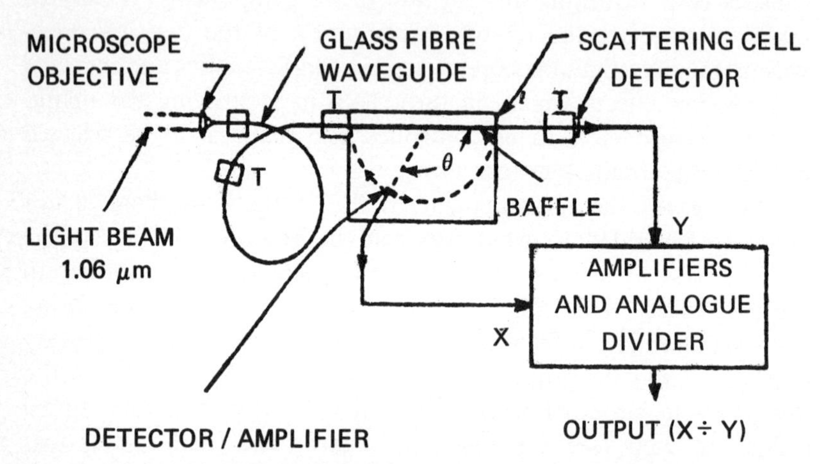

FIG. 31 – Schematic diagram of the apparatus for angular distribution of scattered light measurement. (After [41]).

The fibre is inside a cell filled with liquid with refractive index slightly higher than that of the cladding. The detector scans angularly in an arc corresponding to light scattered from 0° to 180°. Optical power is also detected at the end of the fibre and the ratio of the two signals is recorded in order to compensate for fluctuations and long-term drift. In slightly different arrangements, a light pipe is used to transmit the scattered light to internal detectors (typically photomultipliers) [42, 43]. Due to the very small fraction of light availabe for such measurements, high power sources such as He-Ne laser, krypton ion laser or Nd:YAG lasers are typically used. Wavelength dependent measurements are therefore limited to available laser lines. A typical result is shown in Fig. 32. The weak signal is the Rayleigh component, symmetric around 90°, while the large forward peak is interpreted as energy tunnelling from the highest order guided modes through the cladding; it also contains a smaller component, presumably arising from large scale imperfections.

Of course, all of these measurements can be performed, and, indeed, have been performed on bulk material samples.

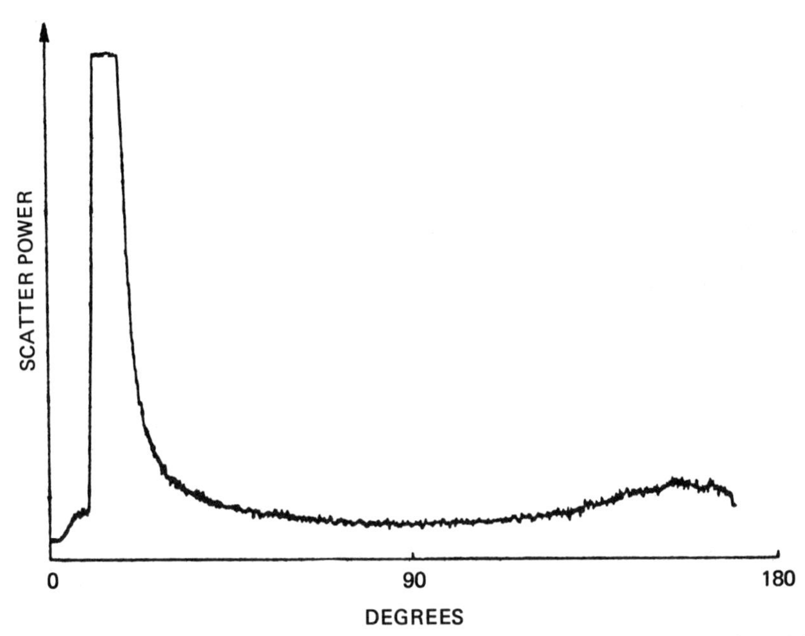

FIG. 32 – Angular scatter spectrum of a multimode fibre showing a large forward scatter peak in the absence of mode selection. (After [43]).

3.3.5 *Absorption loss*

As already shown, absorption losses in optical fibres arise from the presence of transition metal ions and OH-radicals. Measurement of the absorptive component of attenuation is therefore important in order to check the level of impurities control in the starting material and manufacturing process. An indirect way of obtaining this is simply to subtract the scattering loss, measured by some of the above-mentioned techniques, from the total loss. Direct measurements are based on calorimetric methods. They rely on the fact that the absorbed energy is transformed into heat inside the fibre, which causes an increase in temperature; this temperature change can be measured and provides information on the absorption coefficient of the material under study. A general theory giving an expression for the temperature distribution as a function of space and time in a cylindrical rod traversed by a light beam is presented in [44]. Here we present a simpler theory, which can be found in [45, 46], based on the assumption of uniform temperature of the sample and validity of Newton's law predicting a heat leakage proportional to the surface of the sample and to the temperature difference $\theta - \theta_0$ between the sample and the surrounding medium. The energy balance can be written as:

$$\frac{mc}{Sh}\frac{d\theta}{dt} + \theta = \frac{\alpha_a l P_0}{Sh} \tag{17}$$

where: P_0 = optical power flowing through the sample;
l = sample length;
α_a = absorption coefficient;
m = sample mass;
c = specific heat;
h = surface heat transfer coefficient;
S = sample surface.

Moreover, the above equation holds if P_0 can be considered constant along the sample length (in the cm range) or, in other words, if $(\alpha_a + \alpha_s)\, l \ll 1$. If the initial condition is $\theta = \theta_0$ for $t = 0$, we have the following solution:

$$\theta = \left(\theta_0 - \frac{\alpha_a l P_0}{Sh}\right) \exp\left[-\frac{t}{\tau}\right] + \frac{\alpha_a l P_0}{Sh} \tag{18}$$

having posed $\tau = mc/Sh$, τ being the time constant of the heating curve. If, once the equilibrium is reached, the light beam is turned off, we have the following law of temperature change

$$\theta = \theta_1 \exp\left[-\frac{t}{\tau}\right] \tag{19}$$

θ_1 being the equilibrium temperature reached at $t = \infty$. The two temperature functions are shown in Fig. 33. From the derived expressions α_a can be calculated in two independent ways. One consists in measuring the slope of temperature increase, as from (18)

$$\alpha_a = \frac{mc}{lP_0}\left[\left(\frac{d\theta}{dt}\right)_0 + \frac{\theta_0}{\tau}\right]. \tag{20}$$

This approach has been followed by the authors of ref. [44] for

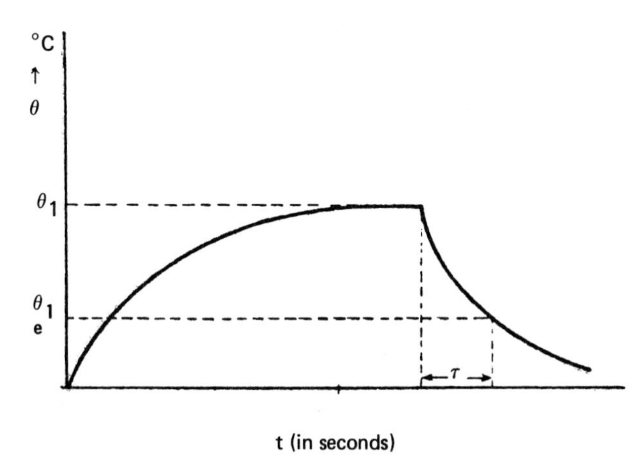

FIG. 33 – Heating and cooling curve for a glass sample. (After [46]).

the measurements of the absorption coefficient of glass rods. The experimental cell is shown in Fig. 34. The glass rod has its ends immersed in index matching liquid, so that polishing is not necessary. The measurement is performed by means of a very thin thermocouple glued to the glass rod for thermal contact. The thermocouple is connected to a similar non-illuminated glass sample, so that the output voltage is proportional to the temperature difference of the two rods. In this way a precise and stable zero is provided, eliminating thermal shift. The signal from the thermocouple is measured by means of a nanovoltmeter.

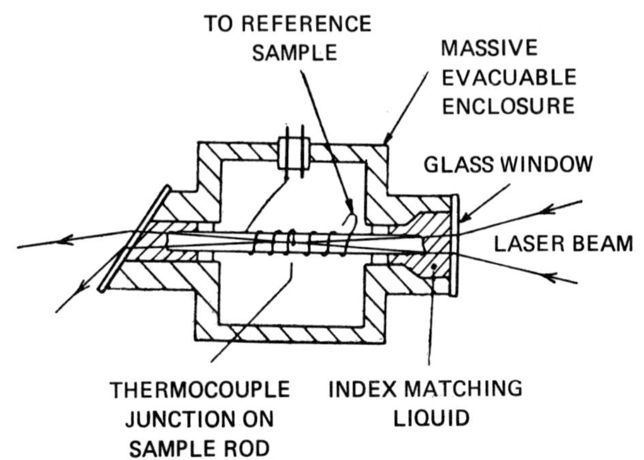

FIG 34 – Schematic layout of apparatus for absorption measurement. (After [44]).

A Brewster angle window prevents any reflected light from reaching the thermocouple. A typical result is shown in Fig. 35, from which a 2.9 ± 0.2 dB/km value of absorption loss can be calculated at the 1060 nm wavelength of a Nd:YAG laser. The accuracy of the measurement depends directly on P_0, as can be seen from (20),

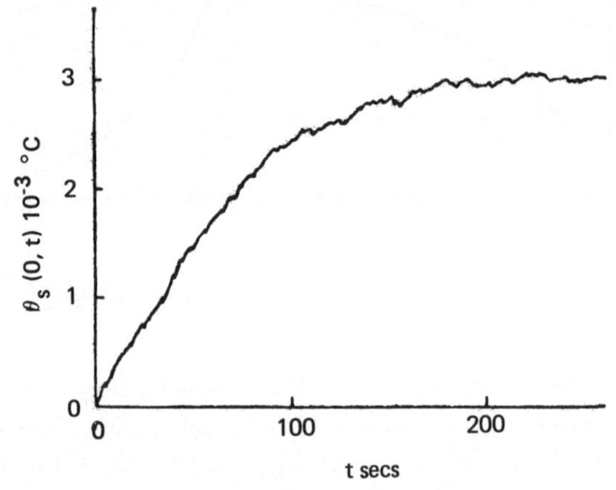

FIG. 35 – The surface temperature at the mid-point of a 2 mm diameter Suprasil-W1 rod as a function of time. (After [44]).

so that laser sources are generally required to obtain a precision less than 1 dB/km with this technique.

A second expression from which α_a can be derived is obtained from the steady state condition, in which

$$\alpha_a = \frac{\theta_1 mc}{lP_0\tau} . \tag{21}$$

This approach has been adopted in ref. [47] for measuring absorption losses of optical fibres. The apparatus (Fig. 36) is similar to the preceding one, except that the rod is replaced by a silica tube filled with a suitable liquid, inside which the fibre is inserted. Calibration of the apparatus can be performed by substituting the fibre with a metallic wire in which a known amount of electrical power is dissipated.

An alternative approach has been followed in [45]: the authors measured the decay time constant after turning off the laser; an experimental curve is shown in Fig. 37. To obtain a high sensitivity the glass rod was placed inside the cavity of a Nd:YAG laser,

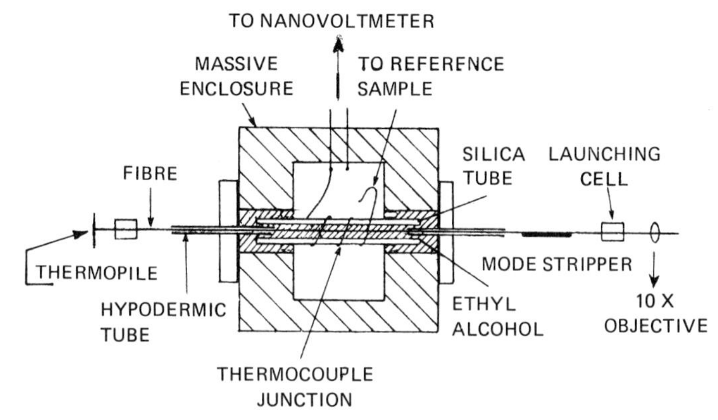

FIG. 36 – Schematic diagram of apparatus for absorption measurement, used in ref. [47]. (After [47]).

thus taking advantage of the very large optical power circulating.

An experimental problem in the described measurements derives from spurious heating effects caused by impurities at the sample ends. To minimize their influence either an extremely careful cleaning is required [45], or a thermal short circuit can be provided by clamping the rod near its ends with metal (brass) collars. Rods with high length to diameter ratio can also be used; in this case the measurement can be performed before the heat being conducted from the ends reaches the measured control zone, affecting the temperature rise due to absorption in the material.

Some techniques based on calorimetric methods have been developed which allow not only the absorption loss to be evaluated,

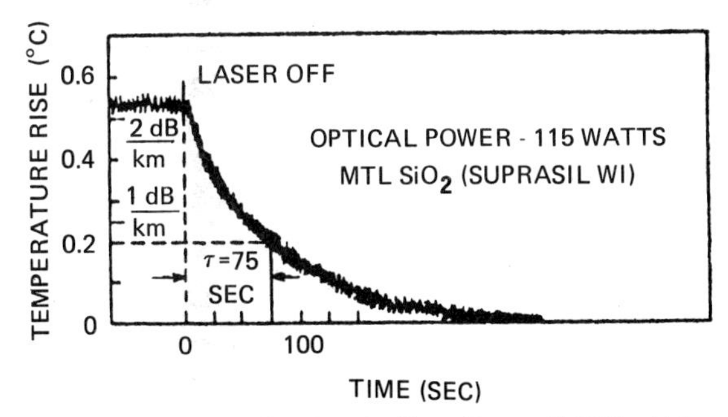

FIG. 37 – Cooling curve for fused silica (Suprasil W1). (After [45]).

but also scattering and total loss. One of these, called the bolometric technique [48, 49], consists in measuring the resistance change, due to temperature rise, of a platinum wire heated by the fibre. Two arrangements have been proposed. The first (Fig. 38*a*) consists in placing the platinum wire parallel to the fibre, bringing the two into thermal contact by a drop of lacquer; the assembly is mounted in an evacuated envelope. The other

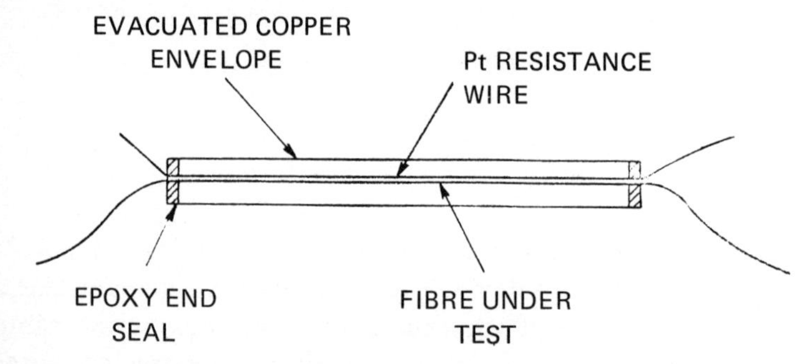

Fig. 38 *a*) – Sketch of parallel wire loss-measuring cell made by attaching a single very fine wire to the fibre under test using a thin lacquer as cement. The cell is about 17 cm long. The thermal time constant is about 1 sec for a 110 μm fibre. (After [48, 49]).

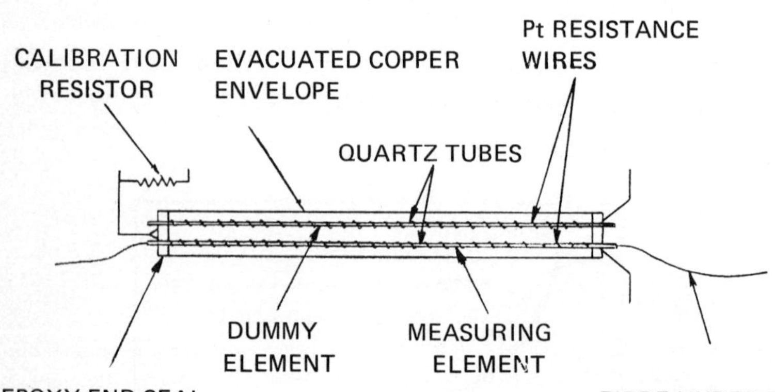

Fig. 38 *b*) – Sketch of tubular loss measuring cell made with two identical tubular elements, one of which is used as the measuring arm and the other as a reference arm to compensate for thermal drifts. The quartz tubes are ~ 15 cm long and 0.4 mm in diameter. The thermal tube constant is about 20 sec for the arrangement shown. (After [48, 49]).

arrangement, shown in Fig. 38*b*, uses two identical thin walled quartz tubes wrapped with a low-pitch spiral of platinum wire, again contacted by means of lacquer. The fibre is inserted inside one of the quartz tubes. End effects are minimized by the large length to diameter ratio, as already explained. The first arrangement has higher sensitivity, but the second is much easier to use, as it is only necessary to insert the fibre in one tube; the other is used as a reference, dummy element.

The flexibility of the method resides in the fact that if transparent lacquer is used the resistance is heated only by the fibre temperature rise, which is due to absorption alone. If an opaque lacquer is also used the scattered light is absorbed and contributes to the temperature change and the total loss can be estimated. The measurement of the very small resistance change can be performed by means of a very high sensitivity a.c. Wheatstone bridge (see Fig. 39).

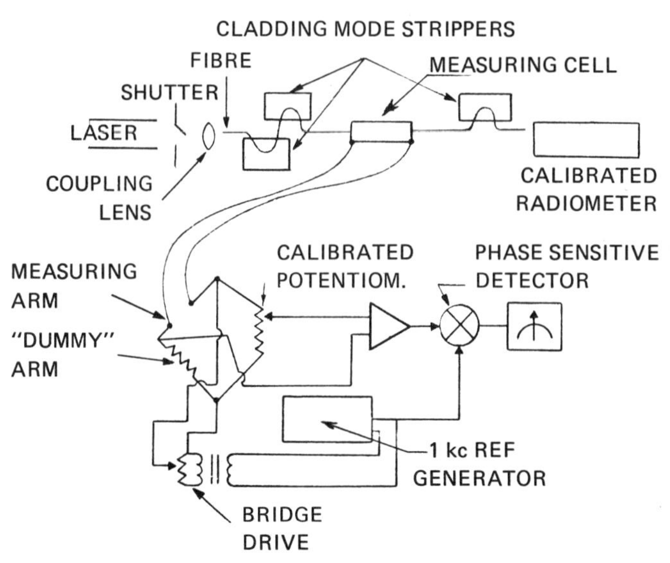

FIG. 39 – Optical connections and ac Wheatstone bridge arrangement for making measurements of fibre transmission loss. (After [49]).

The sensitivity of the apparatus is very good, as losses as low as 0.1 dB/km can be measured with 50 mW of laser power.

A shortcoming of the technique in the tubular arrangement

consists in the fact that, when using the tubular arrangement to measure only absorption loss, low angle scattered radiation gets trapped in the tube and undergoes a large number of reflections. Even if the absorption per each reflection is small an appreciable amount of scattered light is converted into heat by this process, so that the measurement is distorted. A modification getting rid of this problems consists in applying ground glass finish to the tube, thus causing the trapped light to be scattered after few reflections [50]. In addition, the gap between the fibre and the tube is filled by index matching fluid, causing the whole scattered light to escape from the fibre. In [50] a very good agreement was found between the scattering loss measured by this technique and that obtained by more conventional methods (cube detector, see Sections 3.3.3(*d*) and 3.3.4).

Another microcalorimetric technique for measuring both absorption and scattering losses with the capability of separating the two effects has also been proposed [51]: it consists in surrounding the fibre, held inside a transparent cylinder, with an external thin metal cylinder having its inner surface blackened (Fig. 40). The basis of the method is the different thermokinetics of the absorbed and scattered energies; scattered light directly heats the blackened surface immediately after its emission (as the extremely short time required for the light to reach the wall can be completely

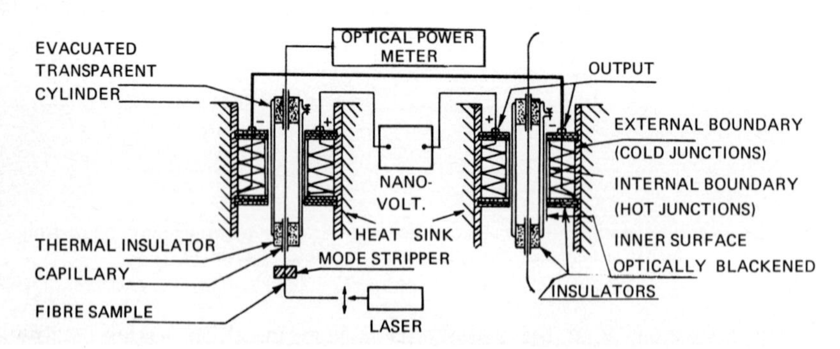

FIG. 40 – Schematic of the experimental apparatus, here shown with fibre samples. The two identical microcalorimetric elements are symmetrically imbedded in a large heat sink. (After [51]).

neglected), while absorbed energy reaches the detector by slow thermal phenomena. The delay time can be controlled by varying a gas pressure inside the transparent cylinder; evacuation of the latter allows suppression of the absorptive component. Thermocouples connected in series are used for measurement and their output is proportional to the total heat flux crossing the internal wall of the metal cylinder. Using light pulses of convenient length, an output pulse is obtained in which two peaks, corresponding to scattering and absorption components, can be easily discerned (Fig. 41*a*). Total attenuation is proportional to the area under

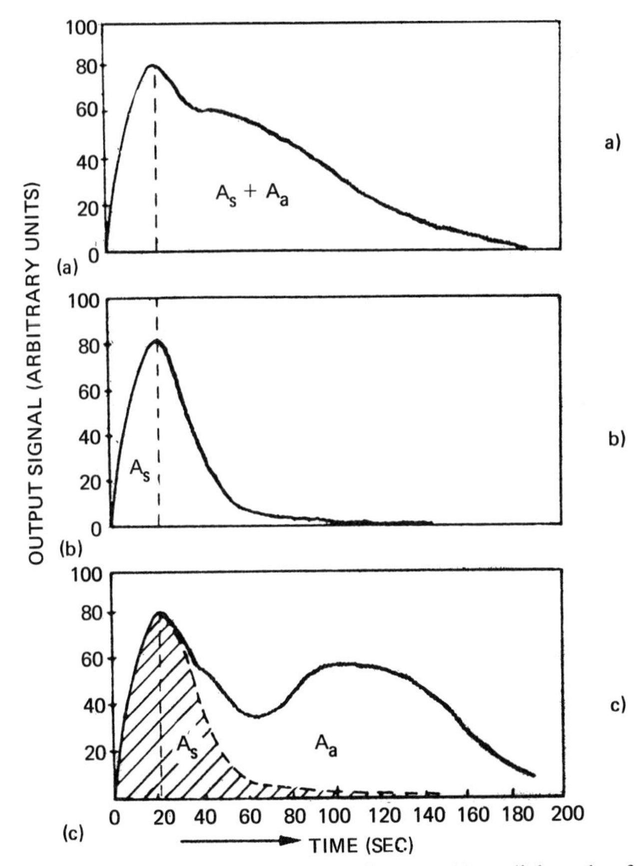

FIG. 41 – Recorded output signals corresponding to a 20 sec light pulse, for several values of the pressure in the hollow cylinder. (*a*) 760 Torr, superposition of absorption and scattering; (*b*) 10^{-6} Torr: Scattered energy is alone detected; (*c*) 100 Torr: the two peaks are far apart. (After [51]).

the waveform. If the absorptive contribution to the signal is suppressed in the way previously indicated, a pulse due to scattering alone is obtained, whose area corresponds to the scattering loss (Fig. 41*b*). Absorption loss can be calculated from the difference between the two measured areas (Fig. 41*c*). The disadvantage of this method is that the cylinder radius necessary to achieve a reasonable time difference results in a rather large wall area; this requires $\simeq 1000$ thermocouples to be connected in series to measure the extremely small temperature rise of the wall.

3.3.6 *Differential mode attenuation*

Fibre material composition and waveguide structure contribute to different losses of the various modes in a multimode fibre. Analysis of the mode dependence of attenuation provides information about the dopant effects on loss, presence of scattering centres, microbending, geometrical variations of the optical waveguide, and the effect of finite cladding thickness. Such information is of great value to the fibre manufacturer for fibre design, choice of best material composition, and for deciding which manufacturing parameters are to be more tightly controlled.

Knowledge of differential mode attenuation is also of use to the experimenter: as attenuation, bandwidth and splice loss measurements are all dependent on launching conditions, i.e. on the particular modes excited, such a knowledge can, in principle, allow extrapolation of measured data to different operative conditions or, at least, give a quantitative basis for interpreting results obtained in different experimental conditions.

For step index fibres the measurement of differential mode attenuation is relatively straightforward. As the modes can be identified by a single parameter θ, i.e. the angle the ray forms with the optical axis (in a geometric optics model), different modes can be excited simply by illuminating the fibre by a plane wave at different angles. Output power after the full fibre length and after a short section, as usual, are compared and the corresponding attenuation is measured. This approach was followed in [52, 53]. The result from [53] is shown in Fig. 42, which confirms that high order modes are more strongly attenuated.

An alternative approach is to excite all modes and to perform

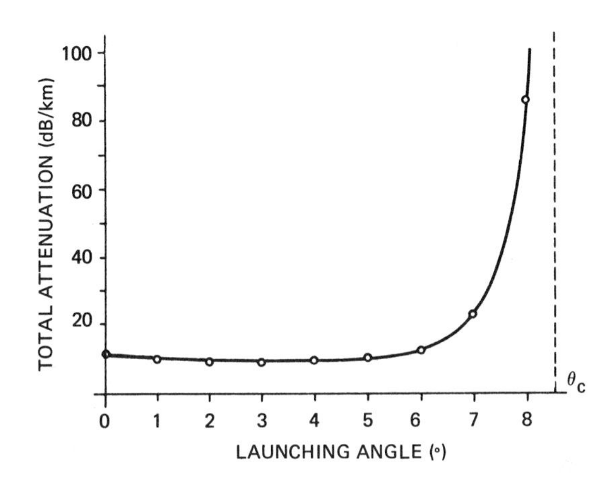

FIG. 42 – Measured angular attenuation of waveguide as a function of input plane wave angle at 632.8 nm. (After [53]).

a measurement of transmitted power as a function of the angle in the far-field.

Basic to all these measurements is, of course, the assumption that mode conversion inside the fibre is negligible, so as to ensure that individual mode properties are actually investigated.

For graded index fibres the situation is more complex [54]: both the angle and the radial position of the incident (or outcoming) ray must be considered in order to specify a mode. This can be seen from the following theoretical analysis, taken from [55]. If we consider a fibre with refractive index distribution given by

$$n^2(r) = n^2(0)\left[1 - \varDelta_1 \cdot \left(\frac{r}{a}\right)^\alpha\right] \tag{22}$$

(where $n(0)$ = refractive index at the core center,

$$\varDelta_1 = \frac{n^2(0) - n^2(a)}{n^2(0)}$$

and a is the core radius), we have for the propagation constant of a ray entering (or emitted from) the fibre end at spatial coordinates (r, ψ) in a direction (θ, φ)

$$\beta = n(0)k\left\{1 - \varDelta_1\left[\left(\frac{r}{a}\right)^\alpha + \left(\frac{\sin\theta}{\sin\theta_c}\right)^2\right]\right\}^{\frac{1}{2}} \tag{23}$$

where $\sin \theta_c$ = maximum numerical aperture. The angular momentum (or azimuthal quantum number) ν is (see Chapter 2)

$$\nu = n(0)\,kr \sin(\psi - \varphi) \qquad \left(k = \frac{2\pi}{\lambda}\right). \tag{24}$$

Since from [55]

$$\beta = n(0)\,k \left[1 - \Delta_1 \cdot \left(\frac{m}{M}\right)^{2(\alpha/(\alpha+2))}\right]^{\frac{1}{2}} \tag{25}$$

where M^2 is the total number of modes, equal to

$$\frac{\alpha}{\alpha + 2}(ka)^2 \sin^2\theta_c,$$

and m the mode group number, one has

$$m = 2M\left[\left(\frac{r}{a}\right)^\alpha + \left(\frac{\sin\theta}{\sin\theta_c}\right)^2\right]^{(\alpha+2)/2\alpha}. \tag{26}$$

From (26) it is clear that a particular mode group can be excited either by varying r keeping θ fixed, or varying θ at fixed r, or varying both θ and r in an appropriate way. This is the basis for the experimental technique proposed in [54]; the corresponding experimental apparatus is shown in Fig. 43. The fibre is excited by an incoherent source (LED), and the fibre output is magnified and imaged onto a pinhole mounted on a 3*D* micropositioner by means of which various portions of the fibre near field can be selected. A photodetector, rotatable around an axis

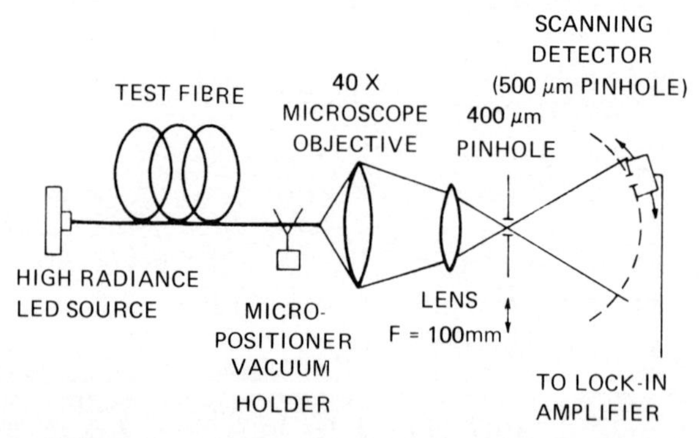

FIG. 43 – Differential attenuation measurement apparatus. (After [54]).

passing through the pinhole, can measure the output power from the pinhole as a function of the angle. Pinhole and detector sizes are chosen so as to have equivalent resolution for spatial and angular coordinates. As the critical point in this kind of measurement is to maintain reproducible conditions at the far end, this is kept fixed and the cut is made near the output end, after which the fibre is re-positioned in front of the light source. Experimental results given in Fig. 44 show the mode loss different behaviour that can be encountered in different fibres.

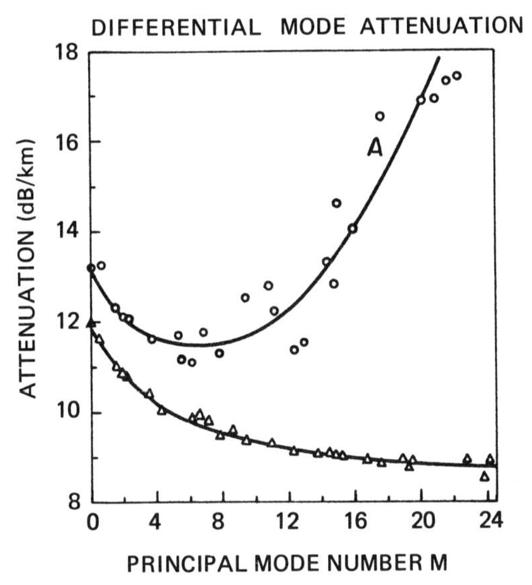

FIG. 44 – Differential mode attenuation in dB/km is plotted vs. the principal mode number m for $\nu = 0$ modes. (After [54]).

According to (26), other experimental approaches can be adopted for selective mode excitation. The one employed in [56] consists in keeping θ fixed (e.g. $\theta = 0$) and moving a diffraction limited spot from a laser (He-Ne or Kr) on the fibre input end. In this case

$$m = M \cdot \left(\frac{r}{a}\right)^{(\alpha+2)/2\alpha} \tag{27}$$

The experimental apparatus is shown in Fig. 45. For precise positioning the fibre is moved by computer controlled translation

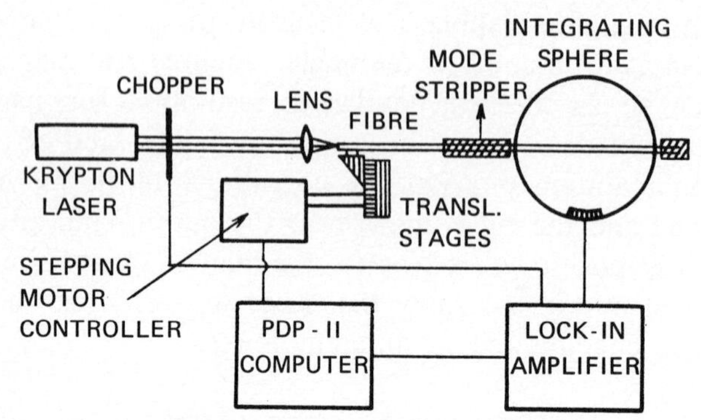

FIG. 45 – Experimental apparatus for automated measurement of mode dependent attenuation and scattering. (After [56]).

stages. The output signal is measured by means of an integrating sphere and recorded as a function of position. This apparatus also allows a measurement of scattering loss as a function of mode number with the procedure described in Section 3.3.4.

The alternative technique consists in keeping r fixed and varying θ [57]; if the laser light is focused onto the fibre core centre, only the $\nu = 0$ modes are excited. This should not be a serious limitation, because experimental evidence seems to indicate that the ν-dependence of mode loss is undetectable. Results of this measurement are given in Fig. 46. It is interesting to note that, as

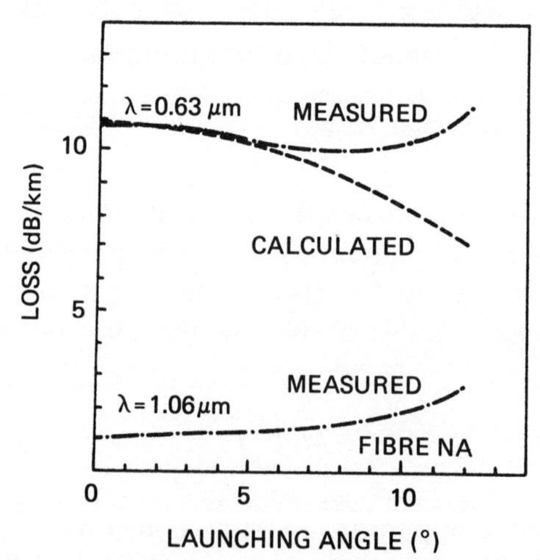

FIG. 46 – Angular attenuations of a graded index fibre. (After [57]).

the optical fibre investigated had very low loss (~1.5 dB/km at 1.06 μm), the major component of optical loss is due to Rayleigh scattering. Therefore, in the measurement at 0.63 μm (He-Ne laser) medium order modes have lower attenuation than low order modes, as the power carried by them is concentrated in a core region with a lower dopant level than that encountered by the low order modes near the core centre. The fact that Rayleigh scattering increases with increasing dopant concentration (at least for GeO_2 doped fibres) explains the result. This interpretation is confirmed by the measurement at 1.06 μm, where the scattering level is so low that differences associated with different composition can no longer be detected. These considerations demonstrate the value of the differential mode attenuation technique as a tool for fibre characteristics investigation.

3.4 Fault location

It is important, especially for installed optical cables, to be able to check whether fibres are fault-free and, if they are not, to locate the fault precisely. The possibility of determining the position of a fault is given by the fact that a certain amount of light is reflected at the discontinuous surface of a fault, and is guided back along the fibre. The fault may be localized if the total time employed by an optical pulse to reach the fault and to come back to the input end is measured. The optical power falling on the detector, reflected from a fault at distance l in a fibre with loss constant α (dB/km), is given by

$$P_r = RP_0 k \cdot 10^{-2\alpha l/10} \tag{28}$$

where P_0 is the output power of the source, k the equipment transmittance, R the reflection coefficient of the fault. For a perfect break perpendicular to exit $R \simeq 4\%$; for irregular breaks R has been measured [58] as $\sim 0.5\%$. It must be remarked that in particular cases no reflected signal is received. This occurs, for instance, if a flat surface is obtained tilted with respect to the axis in such a way that no reflected ray is within the fibre acceptance angle. For measurement purposes, either the described back-scattering method (Section 3.3.3(*c*)) or a pulse reflection method [58, 59] can be employed, the main difference consisting in the sensitivity of the apparatus, which is much higher in a back-scattering arrangement. The time delay of the reflected pulse may be measured with a precise time delay generator; conversion to distance being performed making use of the known pulse velocity in the fibre ($v = c/n(0)$ to a first approximation). The resolution is limited in practice by the laser pulse width for graded index fibres, and by pulse broadening for step index fibres; a 1 m resolution can be achieved in graded index fibres if a few nanosecond wide pulses are used (see Section 3.3.3 (*c*)).

3.5 Impulse response and baseband frequency response

Together with attenuation, bandwidth is an essential parameter for characterizing an optical fibre intended for use in telecommunications, as it determines the information capacity of the fibre. For a complete specification of the fibre information-carrying capacity either the impulse response in the time domain or the baseband frequency response in the frequency domain are to be measured. In the first case, a function $g(t)$ is to be found which, when convolved with fibre optical power input, gives fibre optical power output

$$p_{\text{out}}(t) = g(t) * p_{\text{in}}(t) \tag{29}$$

where $*$ denotes convolution and t is time.

In the frequency domain, a complex function $G(\omega)$ is to be found such that

$$P_{\text{out}}(\omega) = G(\omega)\, P_{\text{in}}(\omega) \tag{30}$$

where ω is the baseband frequency, $P(\omega)$ is the Fourier transform of $p(t)$, and $G(\omega)$ is the Fourier transform of the impulse response $g(t)$

$$G(\omega) = \int_{-\infty}^{+\infty} g(t) \exp\,[-\,i\omega t]\, dt\,. \tag{31}$$

It is possible, in principle, to pass from one type of representation to the other by mathematical means, and this is in fact usually done; moreover, independent measurement of $G(\omega)$ and

$g(t)$ should lead to the same results. The possibility of utilizing the above described concepts rests on the observation that for an incoherent optical carrier the fibre response behaves quasi-linearly in power [60]. It has also been shown that for coherent sources usually employed for such measurements the quasi-linearity holds if their spectral bandwidth is $\simeq 1$ nm [61].

From the experimental point of view there is one basic problem associated with the measurement of the bandwidth of multimode fibres. As the impulse response in such fibres is determined by modal dispersion, i.e. group delay differences among the various modes, besides material and waveguide dispersion associated with each mode (see Chapter 1 and 2), it is vital to specify the launching conditions exactly in order to give full meaning to a measured bandwidth. The question is very similar to that discussed in the case of attenuation measurement, in that in this case as well the use of mode scramblers can prove useful in obtaining reproducible launching conditions and specific bandwidth values independent of the fibre length. Of course, the situation is much simpler for monomode fibres where only material and waveguide dispersion cause pulse broadening. In any case, the source spectral width has to be specified: this is essential for monomode fibres, as just said, but also for multimode fibres with a nearly optimized refractive index profile, in which the relative contribution of modal and material dispersion can be quite comparable. Here as well, measurement result can be strongly influenced by the source spectral width. Moreover, as the material dispersion for a particular material depends on the wavelength, the emission wavelength of the source must also be specified. This is also made necessary by a more subtle effect, called « profile dispersion », which consists in the fact that the refractive index distribution $n(r)$ of an optical fibre changes as a function of wavelength, so that each wavelength « sees » a different index profile.

Time domain and frequency domain measurements will be discussed separately below.

3.5.1 *Impulse response*

The usual method of measuring the impulse response of an optical fibre is to send a very short optical pulse (short in comparison to the total pulse broadening) into the fibre, and to observe the output pulse after it has traversed the fibre length L. A typical experimental apparatus is shown in Fig. 47 (see, for

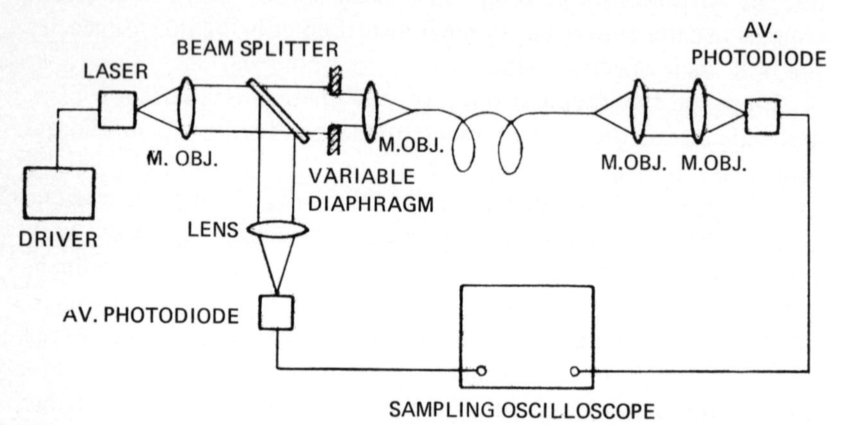

FIG. 47 – Experimental set-up for measuring the impulse response of an optical fibre.

instance ref. [62]). The most widely used optical sources are GaAs or GaAlAs semiconductor lasers which have emission wavelengths at 904 nm or in the range 800-880 nm respectively. Pulsewidths of ~ 100 ps at half height have been obtained either by using a driving circuit consisting of a mercury reed relay, switching a variable delay line [62], or by high speed circuitry with step recovery diodes.

As an alternative mode-locked lasers (krypton [63], Nd:yttrium-aluminum-garnet [63], ND:glass [64], ruby [65]) have been employed with emitted pulse widths as narrow as 7 ps [64]. Such sources have, however, the disadvantage of being expensive, sometimes unreliable, and of providing only a fixed pulsewidth and repetition rate. LED's have also been used, but they provide poor time resolution, typical pulses obtained being ~ 3 ns wide.

For the longer wavelength region, semiconductor laser (InGaAsP/InP) operating at wavelengths around 1.27 μm are becoming commercially available. A successful way of obtaining a tunable laser source consists in generating wavelength shifted pulses in a fibre pumped by high power Nd:YAG laser [66].

The light beam from the laser source is collimated by a microscope objective (when semiconductor lasers are used) and focused onto the fibre held in micromanipulators. A beam splitter provides a reference beam for input pulse measurement. Any of the mode mixers described in Section 3.3.2 can be placed before the fibre to launch an equilibrium mode distribution. The use of a long length of fibre, however, is not practical in this case, because the launched pulse would be distorted and broadened. At the output end of the fibre, microscope objectives collect and focus all the light onto the detector, which must have very high bandwidth and low noise. Photodetectors usually employed are silicon avalanche photodiodes, or *p-i-n* photodiodes which have very fast response. Commercial units are available with total risetime < 75 ps. These detectors cover the range 0.4-1.1 μm; for longer wavelengths germanium photodiodes can be used. High speed photomultipliers with crossed electric and magnetic fields to insure tight control and focusing of electrons, resulting in 0-2.5 GHz bandwidth, have also been employed [61].

For very short optical pulses, in the picosecond range, the detectors mentioned do not provide sufficient time resolution. Techniques have been reported [64, 65] based on a sampling of the repetitive sequence of pulses by means of an ultra-fast optical shutter. Such a shutter is constructed utilizing the birefringence induced in a dielectric liquid through the optical Kerr effect, activated by the passage of a very intense optical pulse. The liquid is carbon disulphide and is contained in a cell placed between the crossed polarizer, so that no light can pass through the system, and the shutter is closed. When a light pulse of sufficiently high intensity (100 MW/cm^2 is the power typically required) passes through the cell, birefringence on the liquid is induced, thus opening the shutter, as part of the light traversing the first polarizer can now be transmitted by the second polarizer. The time of shutter opening is determined by the temporal width of the exciting pulse, the limit being imposed by the time response of the cell, which can be as low as 2 ps. If the gating pulse is finely shifted with respect to the pulse to be measured, a sampling of the latter can be performed. An experimental apparatus of pulse dispersion utilizing this arrangement is shown in Fig. 48. The temporal shift is obtained by means of an optical delay line consisting of a movable prism retroreflector.

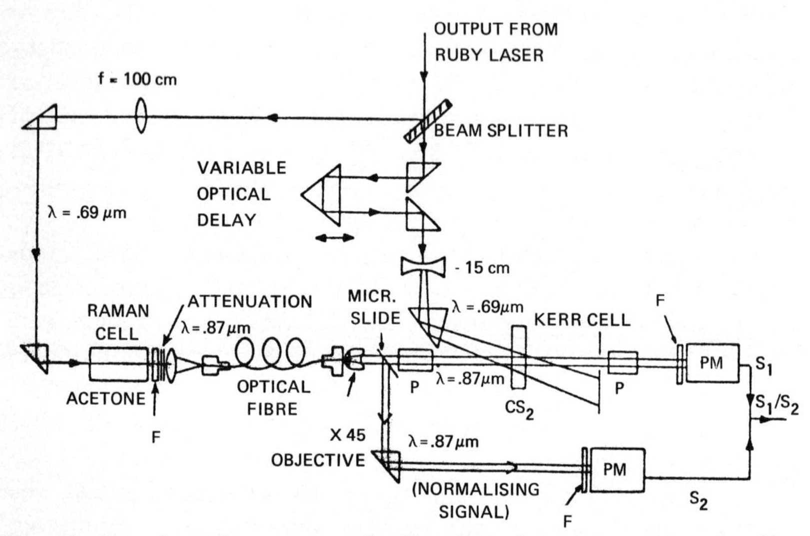

FIG. 48 – Experimental arrangement for fibre dispersion measurements. S = filter (RGN 9) to block radiation at 0.69 μm. P = Glan-Thompson polarizer. PM = photomultiplier. (After [65])

In more usual cases signal from the detector, amplified if necessary, is sent to a sampling oscilloscope of suitably short risetime ($\sim$ 25 ps), on which the output pulse is displayed. Recording by photograph, or *x-y* chart recorder allows further processing. In Fig. 49 the typical input pulse and output pulse from a 1 km long graded index fibre are shown. If the input pulse were a Dirac pulse, the output would directly represent the impulse response of the fibre under test. As it is physically impossible to obtain a Dirac pulse, the input must be deconvolved from the output in order to obtain the correct impulse response [67]. This is simple to do if we can approximate the input pulse and the impulse response with Gaussian functions, i.e.:

$$\begin{aligned} p_{\text{in}}(t) &= p_{\text{in}} \exp\left[-\frac{t^2}{2\sigma_0^2}\right], \\ g(t) &= g_0 \exp\left[-\frac{(t-t_1)^2}{2\sigma^2}\right], \end{aligned} \tag{33}$$

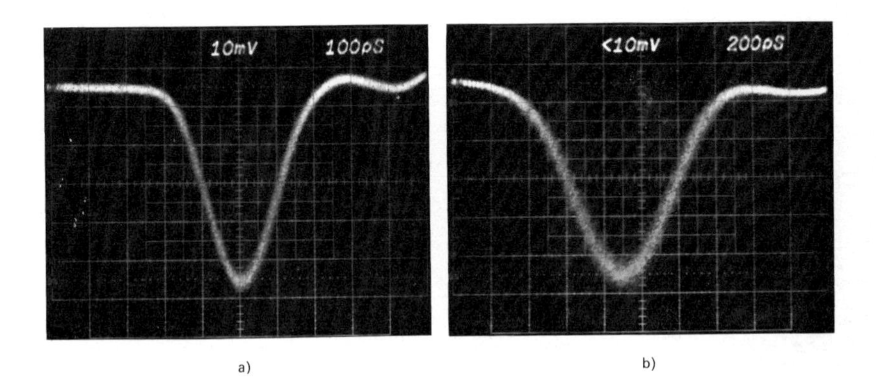

FIG. 49 – Typical input pulse (a) and output pulse from 1 km long graded index fibre (b).

t_1 being the time delay due to the traversing of the fibre, $2\sigma_0$ and 2σ being the full r.m.s. pulsewidths of input pulse and impulse response respectively. The corresponding output pulse is then given by

$$p_{\text{out}}(t) = \frac{g_0 p_{\text{in}} \sqrt{2}\,\sigma\,\sqrt{\pi}}{\sqrt{1 + \sigma^2/\sigma_0^2}} \exp\left[-\frac{(t-t_1)^2}{2(\sigma^2+\sigma_0^2)}\right]. \tag{34}$$

From the last expression the r.m.s. pulsewidth of the output pulse can be easily deduced

$$\sigma_1^2 = \sigma_0^2 + \sigma^2. \tag{35}$$

Conversely the impulse response width can be calculated by the input and output pulse width measurement

$$\sigma^2 = \sigma_1^2 - \sigma_0^2\,. \tag{36}$$

Instead of the 2σ value (which is the Gaussian pulse width at 0.61 of maximum value), the 3 dB of maximum value width is often used. It can be calculated by the same relations and is usually referred to as « pulse broadening », which represents a convenient way of specifying the fibre impulse response by a simple number. It should be noted that the pulse broadening calculated in this way includes all dispersion effects. If the separate contribution of modal and material dispersion are of interest, an approximate calculation may be performed, again assuming Gaussian envelopes, provided one of the two (usually material dispersion) is known; in this case deconvolution yields:

$$\sigma_{\mathrm{tot}}^2 = \sigma_{\mathrm{modal}}^2 + \sigma_{\mathrm{material}}^2$$

where σ_{tot} correspond to the previously defined σ_1 and σ_{modal} and $\sigma_{\mathrm{material}}$ are the standard deviations of the fibre impulse response for modal and material effects respectively. What has been said is valid if the detector response is much faster than the fibre response; if this is not the case, detector response must also be included in the deconvolution process. As it has been pointed out, all the above discussion is valid if pulses and impulse responses can be approximated by Gaussian functions; otherwise, time domain deconvolution is rather difficult [67].

If a certain pulse broadening is measured for a given length of fibre, we can divide it by this length and speak of pulse broadening per unit length (usually ns/km). However, this implies that pulse broadening is considered linear with fibre length, an assumption that may not be valid (see Chapter 2): in fact, it has been demonstrated that if strong mode coupling exists in the fibre the length dependence of pulse broadening tends to follow a square-root law [68]. Experimental determination of length dependence will be discussed later.

3.5.2 *Frequency domain*

For system engineers involved in the design of equalizers and receivers in an optical system, the frequency domain representation is more suitable than the time domain one. One way of obtaining the baseband frequency response $B(\omega)$ in amplitude and phase is simply to calculate it from the measured impulse response by means of Eq. (31). To do this, $g(t)$ is sampled at many points and FFT is used to obtain $G(\omega)$. The calculation may be easily performed by means of a digital-processing oscilloscope (DPO), which can store the waveforms and process them with software facilities so that an « on line » Fourier transform can be performed providing the fibre transfer function.

The problem associated with this approach is that computation errors add to experimental errors, reducing the accuracy of the method. A direct measurement of the frequency transfer function is therefore desirable.

It is usually performed as follows: the source is sinusoidally modulated, and the modulation frequency is varied. The modulated light is focused onto the fibre, whose output is detected by a broadband detector and tuned amplifier (spectrum analyzer, or network analyzer) and recorded. The same measurement is performed either on direct light or after a short reference fibre. The two measurements provide $P_{out}(\omega)$ and $P_{in}(\omega)$, from which

$$G(\omega) = \frac{P_{out}(\omega)}{P_{in}(\omega)} . \tag{37}$$

This technique has been reported by some authors [69-71, 61]. A typical experimental set-up is shown in Fig. 50. The optical

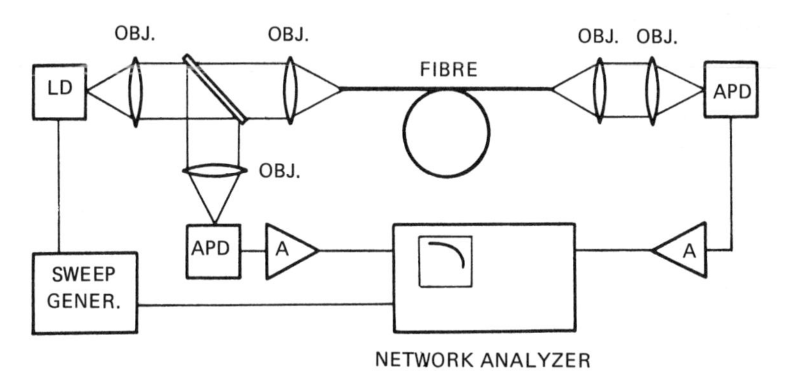

FIG. 50 – Schematic for direct measurement of the frequency transfer function.

source can be an LED (with a limitation in the upper modulation frequency which, at most, can be a few hundred MHz), or a semiconductor GaAs or GaAlAs laser: both can be directly modulated acting on the driving current. Solid state lasers or ion lasers must be modulated by means of external modulators. An incoherent source, a xenon arc lamp, has also been used to perform measurements as a function of wavelength over a wide spectral range (up to 1100 nm) [61]. In this case filters were used for wavelength selection, and an external $LiTaO_3$ electro-optic modulator with 1 GHz bandwidth was employed. Suitable detectors are the same as those indicated from time domain measurement. The measuring instrumentation may consist of:

a) a vector voltmeter

This instrument allows the measurement of both the amplitude and the phase of the received signal with respect to a reference signal; if the light beam from the source is taken as the reference signal, the transfer function can be obtained directly as the ratio between the amplitudes and as the phase difference between the input and output signal. Limitations of such instruments include a limited frequency range ($\sim$ 1 GHz), a not very high dynamic range and the lack of an internal sweep generator, which means that the measurement has, in practice, to be performed at discrete frequencies.

b) a spectrum analyzer

Its main advantage over the vector voltmeter is that an internal generator can provide a continuous display of a frequency swept signal; in addition, the frequency range can be extended to tens of Gigahertz. On the other hand, no phase information is given.

c) a network analyzer

This intrument combines the features of a vector voltmeter with those of a spectrum analyzer, and is therefore the most suitable for such measurements, its use being limited probably due to the high cost. A typical result of a frequency response measurement is shown in Fig. 51. The 3 dB value of the frequency response curve is usually referred to as the « bandwidth » of the fibre, thus providing a convenient single parameter for fibre characterization.

The method is, in principle, more precise than the impulse response and also more suitable from a mathematical point of view, as a simple arithmetic division is sufficient to obtain $G(\omega)$ (see Eq. (31)) while time domain requires a deconvolution integral. The disadvantage of this method is, however, that it is very difficult to measure directly the input to output phase change of the sinusoidal modulation. In fact, the fibre length normally used in these

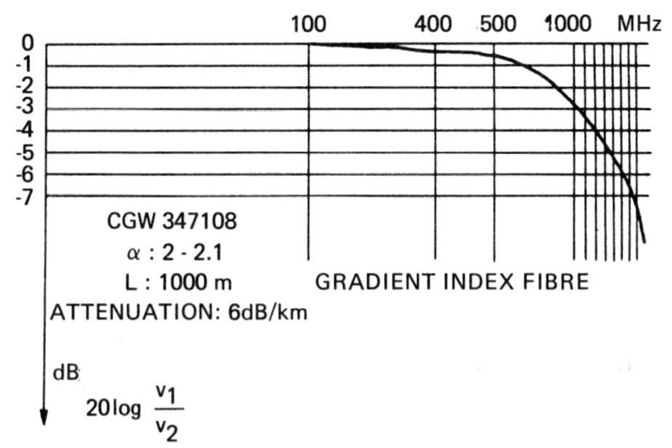

FIG. 51 – Amplitude of a graded index fibre transfer function. (After [70]).

measurements (~ 1 km) contains many modulation wavelengths, and what one is really interested in is the deviation from linearity of the frequency-dependent phase shift. This means that a very small effect must be measured, which is superimposed on a much larger effect. Work is in progress in some laboratories to make this measurement possible. The difficulty can be overcome if the fibre is assumed to behave like a minimum phase network, i.e. $G(j\omega)$ has no zeros in the right half of the complex plane. In this case, $\theta(\omega)$ can be mathematically derived by $|G(\omega)|$. However, serious doubts exist about the validity of this assumption.

Another disadvantage of the frequency domain method with respect to time domain is that currently available techniques do not allow modulation frequencies much larger than 1 GHz, while pulses with very short risetime (~ 40 ps) can be obtained from laser sources allowing the investigation of very high frequency behaviour. In addition short pulses with low duty cycle may be produced with high peak powers, resulting in a high dynamic range of the measurement apparatus. On the other hand, as modulation of a cw wave has rather limited peak power capability, the dynamic range is lower in a swept frequency measurement.

Since the problem of length dependence of bandwidth is completely analogous to that discussed for pulse broadening, the same considerations apply.

At the present stage both methods are useful, and to a certain extent complementary, in analyzing bandwidth properties of optical fibres.

3.5.3 *Differential mode delay*

As already mentioned, in a multimode fibre each mode propagates generally with a different group velocity. The measurement of group delay difference (i.e. the difference in the arrival time among the various modes) is very interesting, as it lends itself very well to verification of theoretical models of pulse dispersion. Moreover, this technique provides a very powerful tool for obtaining information about the refractive index profile and its optimization with respect to group delay equalization; in fact, bumps or small irregularities in the profile shape give rise to separation of modes in groups around certain group delays. Moreover, if we consider a typically studied class of refractive index profile, called α type, already explained in Section 3.3.6, i.e.

$$n^2(r) = n^2(0)\left[1 - \Delta_1 \cdot \left(\frac{r}{a}\right)^{\alpha}\right]^{\frac{1}{2}} \tag{22}$$

one can find, for a particular fibre and wavelength, an optimum α, α_{opt}, which minimizes group velocity differences among the modes. With a group delay difference measurement it is simple to check if the actual fibre refractive index profile is undercompensated or over-compensated. In the first case, $\alpha > \alpha_{opt}$, off-axis rays have a longer travelling time than axial rays, because they travel over a longer geometrical path not compensated by a sufficient decrease of refractive index toward the core-cladding interface. For profiles with $\alpha < \alpha_{opt}$ this decrease is so rapid that the longer propagation length of the off-axis ray is overcompensated by their faster speeds, and in this case axial rays arrive last.

The measurement can be performed in two ways. Either the launching conditions are adjusted so that only one mode, or a small group of modes is excited, or full excitation of all modes is produced at the input and different group of modes are selected at the output end for the measurement. A basic assumption in order to be sure that individual mode properties are measured is that no, or very small, mode conversion occurs. Conversely, if insensitivity to launching conditions is observed, the method can provide information on the strength of mode coupling in the fibre, on the coupling length and so on. The first approach followed in [72-75] can be performed in different ways, as shown in

Fig. 52 [76]. It must be recalled that while for step index fibres the angle of incidence completely specifies a mode, so that plane wave excitation can be used, for parabolic fibres both the angle

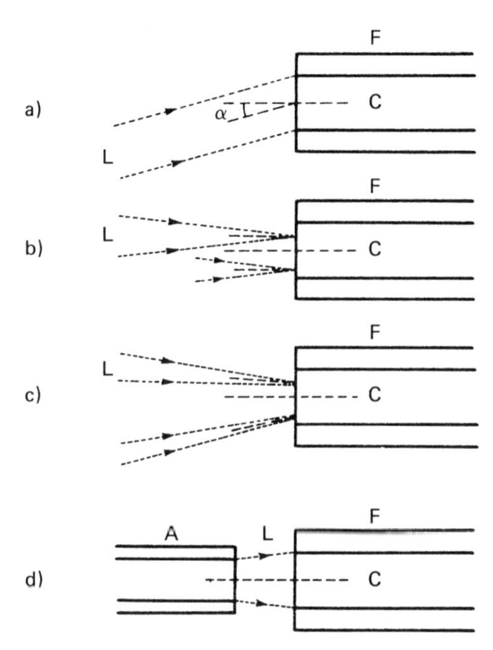

FIG. 52 – Four possible ways of injecting light into an optical fibre. *F* optical fibre with core *C*. *L* light beam. *a*) Parallel beam. The angle of incidence α can be varied. *b*) Strongly focused beam. The acceptance angle is completely filled; the size of the beam spot on the core surface is very small in relation to the core diameter. The spot can be displaced radially. *c*) «Soft focus». The position of the spot and the angle of incidence can both be varied. *d*) «Ray scrambling». A large number of rays are continuously mixed in the auxiliary fibre *A*, and the effect of this is that the end surface behaves as a homogeneous light source with a large radiating surface and a large angle of emission. (After [76]).

of incidence and the position on the fibre core (indicated for instance by the radial distance from the core centre) should be considered (see Section 3.3). For this kind of experimental approach, means for tilting and displacing the light beam (a laser beam) with respect to the fibre must be provided. Usually the fibre is held in precision micromanipulators assuring accurate linear and angular displacement. Results of illumination of a step-index [75] and a graded index [76] fibre at different angles with a plane wave are shown in Fig. 53.

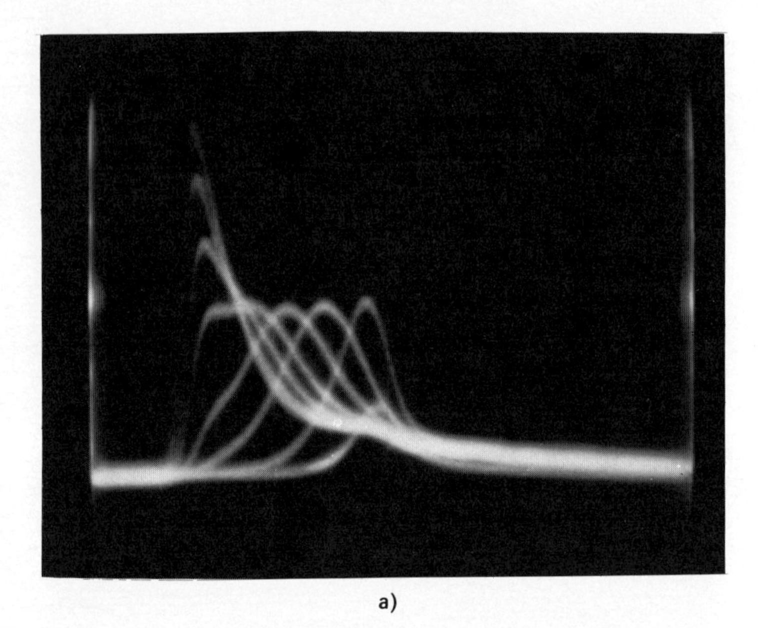

a)

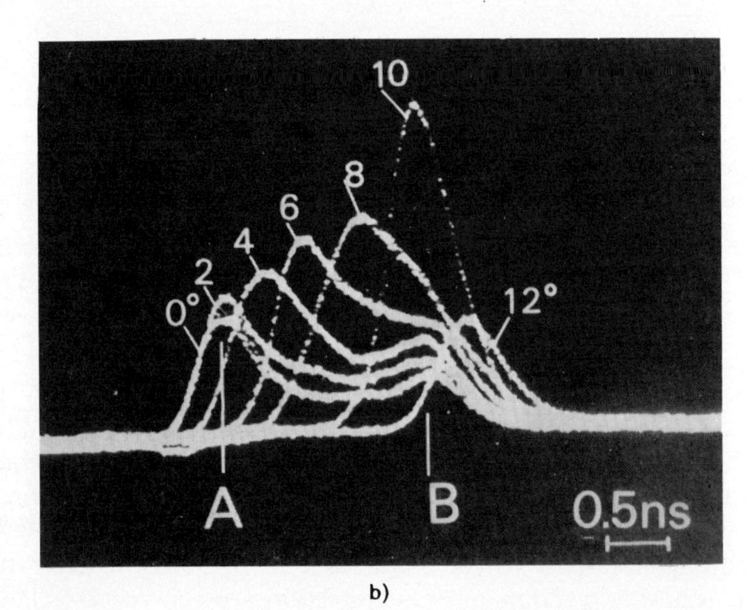

b)

FIG. 53 – Set of typical oscillograms taken at different angles, showing the pulse behaviour and the difference in propagation time as a function of angle: *a*) step index fibres, angles = 0°, 3°, 4°, 5°, 6°, 7°, 8°, 9°; hor. scale 5 ns/div. (After [75]); *b*) graded index fibre. (After [76]).

The second approach, followed in [77], consists in using spatial filters in the form of stops or annular rings to select a group of modes at the output end after having produced full excitation of modes at the input. Figure 54 shows an experimental arrangement

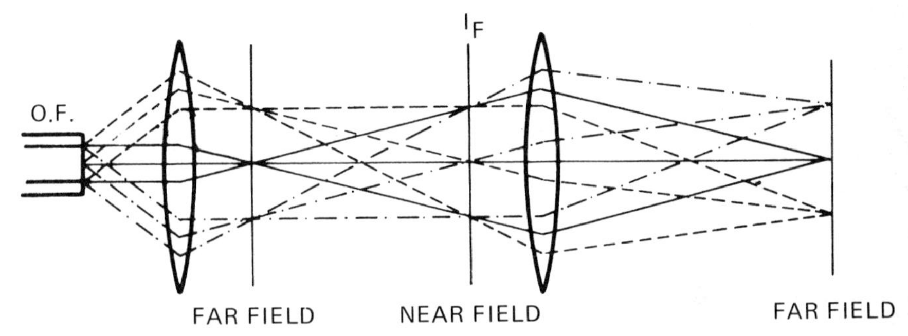

FIG. 54 – Optical system employed for the selection of different modes: Rays corresponding to different emission angles are traced, illustrating the formation of near field and far field images. (After [78]).

taken from ref. [78]. A first lens forms a real image of the fibre cross-section, I_F, providing the near field pattern at the output end. On the other hand, in the back focal plane of the first lens the angular distribution of the light beam emitted by the fibre is transformed into a radial distribution. A real image of this distribution is obtained by the second lens. A detector of suitable size, placed in the center of this image, permits selection of a small range of angles around $\theta = 0°$ $(0° \pm 1.5°)$. Radial scanning is performed by successively placing a series of concentric annular rings of different diameter on the fibre near-field image, I_F.

The complete apparatus is shown in Fig. 55. The optical source is a semiconductor laser focused, by means of microscope objectives, on a short section of fibre in which mode scrambling is achieved through a succession of splices. The fibre under test is connected, by means of a splice, to the short fibre. A beam splitter inserted in the optical path of the output beam provides a signal that, detected by an APD and suitably amplified, acts as a trigger for the sampling oscilloscope used for measurements. Output pulses are detected by an APD with 150 psec risetime displayed on the sampling oscilloscope and recorded on an X-Y plotter. A resolution of 10 psec is obtained, thanks to the very

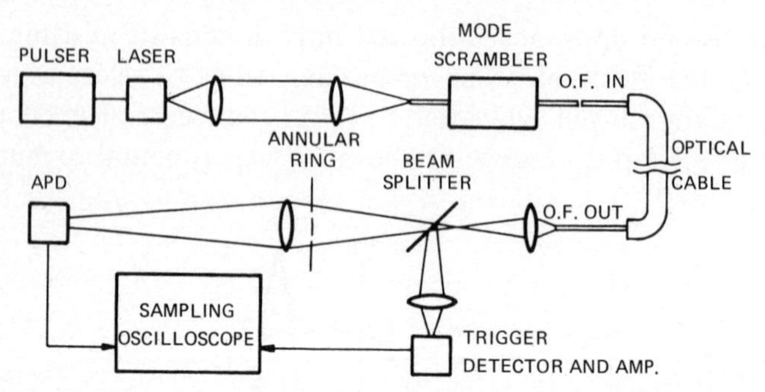

FIG. 55 – Schematic of experimental set up for DMD measurements. (After [78]).

high trigger stability obtained with the said arrangement. It should be stressed that, as discussed in Section 3.3.6, a combination of near field and far field (spatial and angular) selection should be performed in order to achieve true mode separation for graded index fibres, α profile type. Results of measurements at different wavelengths on a very well optimized fibre are shown in Fig. 56.

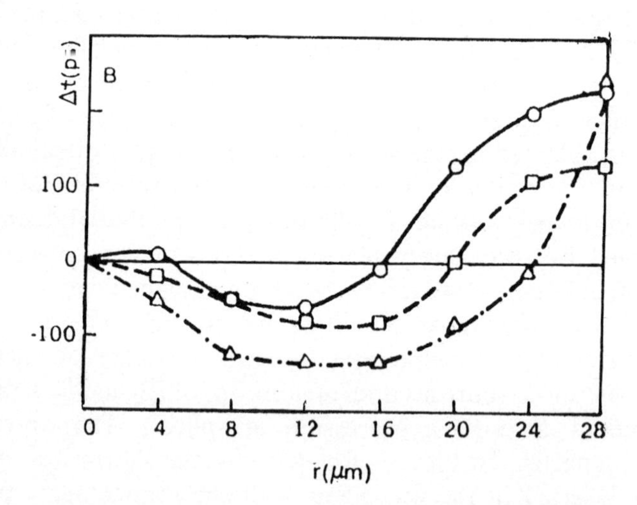

FIG. 56 – Differential mode delay of fibre at different wavelengths. △: λ = = 837 mm; □: λ = 856 mm; ○: λ = 900 mm. (After [80]).

A detailed study employing both techniques has been reported in [79].

3.5.4 *Length dependence of bandwidth*

The length dependence of the impulse response (or frequency response) of an optical fibre is an essential parameter to be specified in order to extrapolate the results obtained for a certain measured length to longer lengths. This, in turn, is of great importance, as the ultimate length of a link operating at a certain bit rate is considerably affected by the length evolution law of the bandwidth. As already mentioned, the pulse broadening increase may follow a law which can range from a linear dependence on length (for mode coupling free, equal mode loss fibres), to a square root law for fibres with strong mode coupling. This is valid for modal dispersion, as material dispersion follows a linear law of addition [80].

The most obvious technique for performing the measurement is to successively shorten the fibre by a known amount and measure the pulse-width, which is then plotted on logarithmic scale as a function of length. This method, despite its evident shortcoming of being fibre-destructive, has been used [81].

A more sophisticated approach called the « shuttle pulse » technique is described in [82, 83]. This consists in pressing partially transparent mirrors against the fibre ends by means of a precisely machined device (see Fig. 57 for the experimental apparatus and holder) with a groove in which the fibre is held, and a mirror at right angles to the fibre axis. Index matching liquid is also used to obtain good contact; the transmittance of the mirrors can be chosen in such a way as to maximize power output after a certain number of round trips.

Results obtained by the described technique are shown in Fig. 58 where it is evident that for the particular fibre under study a square-root dependence is reached after a few hundred metres. The major problem associated with this method is that a small amount of mode conversion may occur at each reflection. As this effect will be included in the data, it is no longer correct to attribute the kind of pulse evolution entirely to the fibre.

Another disadvantage is that power is lost at each reflection. A second sophisticated technique has been proposed [84]: this consists in looping the fibre under test and inserting an ultrasonic deflector between the fibre ends. Figure 59 shows the experimental

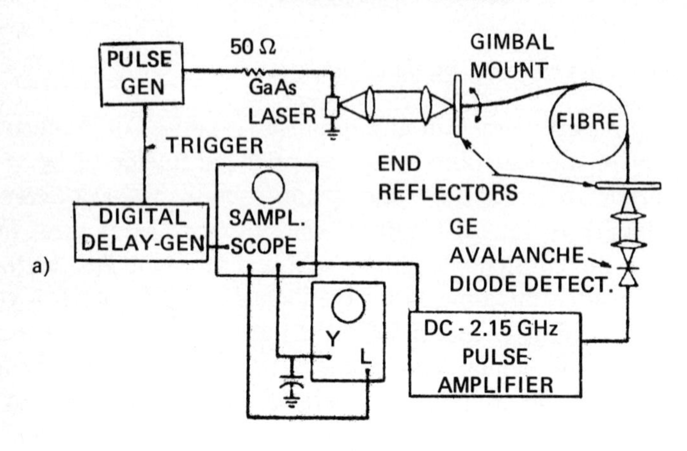

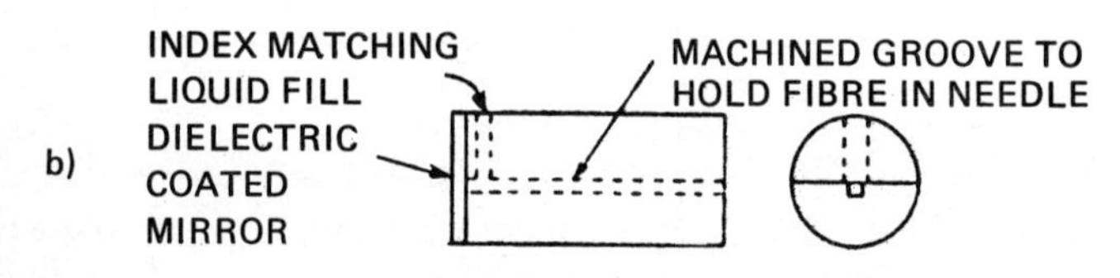

END REFLECTOR SCHEMATIC

FIG. 57 – (*a*) Experimental arrangement for making shuttle pulse measurements with a pulsed GaAs injection laser ($\lambda = 0.9\ \mu m$). It includes the most sensitive detection arrangement, which is a germanium avalanche diode followed by a wideband pulse amplifier. (*b*) Schematic of the holder used to press a fibre against a reflecting mirror. The holder mates with a gimbal mount that can be tilted at an angle relative to the injected laser beam in order to emphasize the launching of high order modes. (After [82, 83]).

apparatus. An acousto-optic deflector is placed between the looped fibre ends and, by means of ultrasonic waves, the laser beam can be deflected to be injected into the fibre. The circulating beam can also be deflected toward a detector. The fibre ends are placed in the focal plane of two lenses, so that the light beam is collimated when crossing the deflector. The efficiency of the deflector, reported in [84], is about 30%.

The laser beam is at first deflected toward the fibre applying rf double pulses to the acousto-optic deflector, with a suitable anticipation to take into account the slower velocity of the ultrasonic pulse with respect to the light pulse. Any circulating pulse can afterwards be extracted by adjusting the time delay for the second driving pulse. Figure 60 shows the results obtained with two step index fibres (*A*, *B*) and two graded index fibres (*C*, *D*).

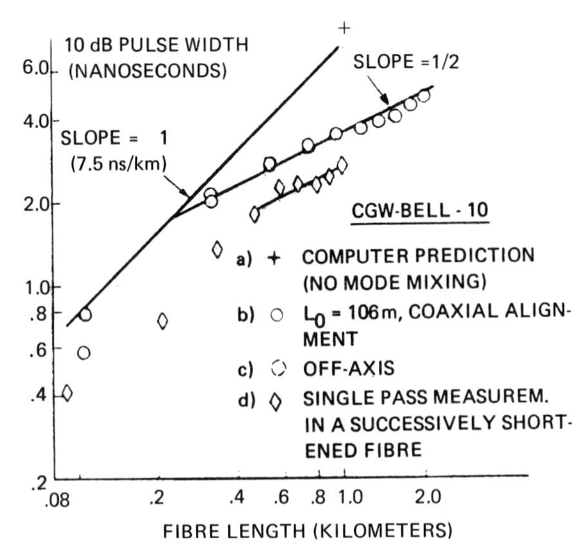

FIG. 58 – Pulse width between 10 dB power points is plotted vs CGW-Bell-10 fibre length on logarithmic scales. (After [82]).

The experimental apparatus introduces a detectable amount of mode mixing, which must be taken into account when interpreting results. With respect to the shuttle pulse method it has the advantage of a smaller overall insertion loss, and the possibility of extracting the pulse after each circulation instead of waiting for a complete round trip.

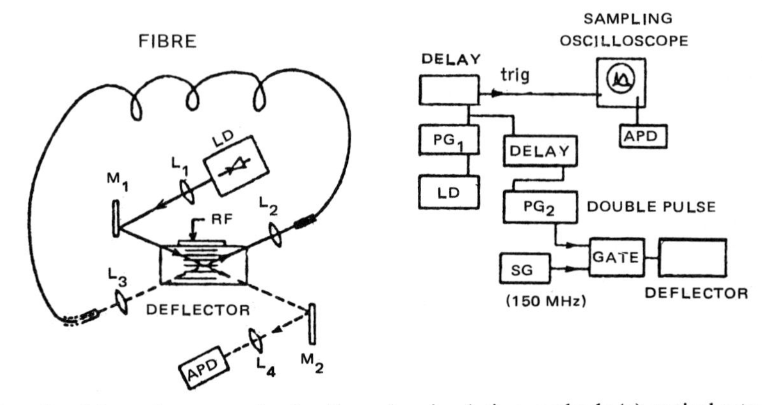

FIG. 59 – Measuring apparatus for the pulse circulation method: (*a*) optical setup; (*b*) deflector driving block diagram. (After [84]).

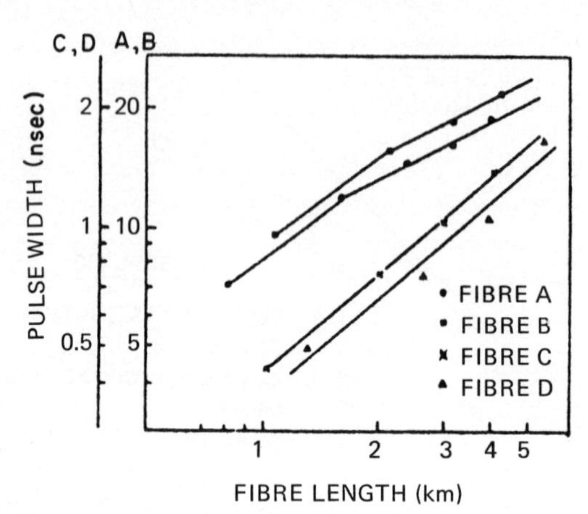

FIG. 60 – Length dependence of 3-dB pulse width for each fibre. Pulse width scales are shown for two types of fibre. (After [84]).

3.6 Material dispersion

The materials forming ordinary optical fibres are dispersive media, i.e., media whose refractive index varies as a function of wavelength. This causes a phenomenon called material dispersion, whereby different wavelengths forming the spectral envelope of an optical carrier travel at different speeds, thus cooperating in determining total pulse broadening. Determination of total pulse broadening is very important, as it may impose a fundamental limit to the maximum information carrying capacity of an optical fibre. For example, in a monomode fibre where modal dispersion, due to multipath propagation, is not present, or in an ideal multimode fibre in which the group velocities of all modes are perfectly equalized, pulse broadening is caused by a combination of material dispersion and waveguide dispersion. The latter is caused by the fact that the group velocity of a mode in a guiding structure, even without the presence of a dispersive medium, depends on frequency. In general, material dispersion and waveguide dispersion interact in a complicated way; however, it has been shown [85] that in certain conditions their effects can be considered « additive ». For modes far from cut-off waveguide dispersion is, in general, negligible [86] compared to material dispersion. Therefore, measurements of material dispersion as a function of wavelength allow the identification of regions of minimum, possibly zero, dispersion, where it is more convenient to operate for maximum bandwidth achievement.

Moreover, when performing measurements of pulse broadening, the spectral width of the optical source used for the measurement contributes to determine the total impulse response. In order to extrapolate the results to other sources it is necessary, in general, to know the contribution of material dispersion and deconvolve it (see Section 3.5.1) from total impulse response to obtain the impulse response due to modal dispersion alone. When another source is used, the effect of material dispersion associated with its own spectral width can be added to the fibre response to obtain the complete expected response. To this purpose, an independent knowledge of material dispersion is obviously required. (What has been said is particularly valid for well-optimized graded fibres, in which modal dispersion and the material dispersion related to the 2-4 nm linewidth of semiconductor lasers are comparable).

In classical optics the term « material dispersion » usually refers to the first derivative of the refractive index with respect to the wavelength, $dn/d\lambda$. For optical fibres it is customary to give it the meaning of delay per unit length per unit wavelength interval. An explicit expression for this quantity may be derived on the assumption that the light pulse is a plane wave, ignoring the waveguide dispersion. In this hypothesis the group delay per unit length is given by

$$\tau(\omega) = \frac{d\beta}{d\omega} \tag{37}$$

β being the phase constant of the plane wave and ω the angular frequency; $\beta = \omega n(\omega)/c$, where $n(\omega)$ is the refractive index dependent on frequency and c the light velocity in vacuum. From the above definitions

$$\tau(\omega) = \frac{1}{c}\left[n(\omega) + \omega\,\frac{dn(\omega)}{d\omega}\right] \tag{38}$$

or, being $\lambda = 2\pi c/\omega$,

$$\tau(\lambda) = \frac{1}{c}\left[n(\lambda) - \lambda\,\frac{dn(\lambda)}{d\lambda}\right] \tag{39}$$

and « material dispersion » is given by

$$\frac{d\tau}{d\lambda} = -\frac{1}{c}\lambda\frac{d^2n}{d\lambda^2}. \qquad (40)$$

If a source of spectral width $\Delta\lambda \ll \lambda$ is used, then the total time spread after a length L is given by

$$\Delta\tau = \frac{d\tau}{d\lambda}\cdot\Delta\lambda\cdot L. \qquad (41)$$

From expression (40) it is clear that material dispersion can be calculated from known dispersion data for the material under study. If these data are not known they can be measured by conventional optical techniques, using refractometric equipment and measuring n as a function of λ on suitably-shaped samples of bulk material. This approach has been followed [87] but has two defects: one is that the drawing process to obtain the fibre can somewhat alter the physical properties of the bulk material; the other consists in the fact that modern material fabrication techniques such as the CVD method (see Chapter 4) result in the production of a complete preform with a core and a cladding from which it is not viable to obtain samples to be measured in a suitable form. A technique reported in [88] overcomes the first objection. The method consists in analyzing the backscattered radiation produced when a collimated light beam impinges transversely upon a fibre. The basis of the method is shown in Fig. 61, which represents the path of a ray of collimated light. It can be shown that as the point of incidence moves, the angle i varies as does φ which passes through a maximum; such a maximum, φ_m, in the geometrical optics limit, depends only on the index of refraction and is given by

$$\varphi_m = 4\sin^{-1}\left[\frac{2}{n\sqrt{3}}\left(1-\frac{n^2}{4}\right)\right]^{\frac{1}{2}} - \sin^{-1}\left[\frac{2}{\sqrt{3}}\left(1-\frac{n^2}{4}\right)\right]^{\frac{1}{2}}. \qquad (42)$$

The experimental apparatus is shown in Fig. 62. The optical source is a Xe-arc lamp, and a set of filters is used for wavelength selection. A scanning photodetector allows the measurement of the backscattered pattern and hence the determination of φ_m.

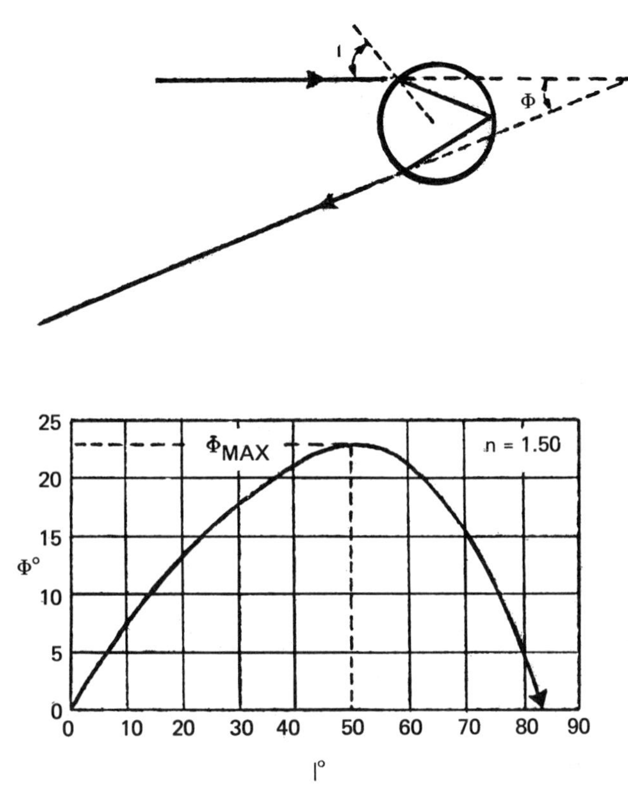

FIG. 61 – Refractive index measurement in uncladded fibres using the backscattered radiation. (After [88]).

The quoted equation applies to unclad step index fibres. The presence of a cladding and grading of refractive index lead to more complicated expressions and to a much greater difficulty in data interpretation. Moreover, the technique gives the dispersion in a localized region of the fibre.

Coming back to Eq. (40), it is evident that if we have two sources emitting at different wavelengths λ_1, λ_2 separated by $\Delta\lambda$, Eq. (41) holds and gives the temporal separation of two pulses simultaneously launched into the fibre. This suggests the most widely used technique, which consists in sending monochromatic pulse at various wavelengths and measuring the arrival time difference. The method was first reported in [89] and the correspond-

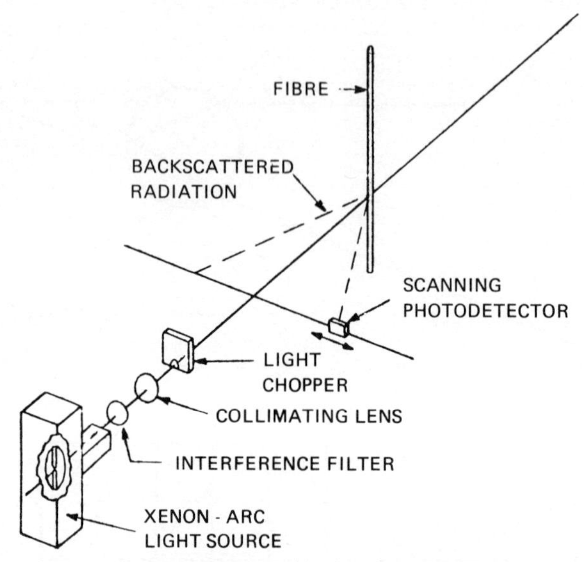

FIG. 62 – Basic optical arrangement for the measurement of material dispersion by backscattering method. (After [88]).

ing apparatus is shown in Fig. 63. The two GaAs lasers have peak emission wavelength of 900 and 860 nm respectively. The pulses before and after traversing the 1 km long graded index fibre are shown in Fig. 64.

Subsequent works are based on this principle, with changes in the optical source used, or in the technique of measuring the delay among different pulses. In [90] a Q-switched, mode-locked

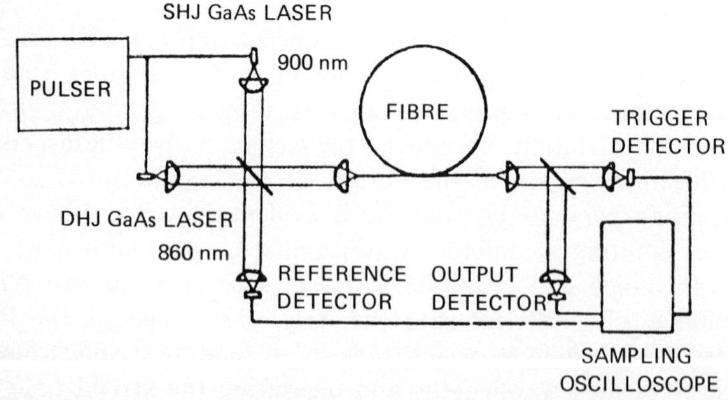

FIG. 63 – Experimental arrangement for measuring transmission time delays vs wavelength using two monochromatic sources at different wavelengths. (After [89]).

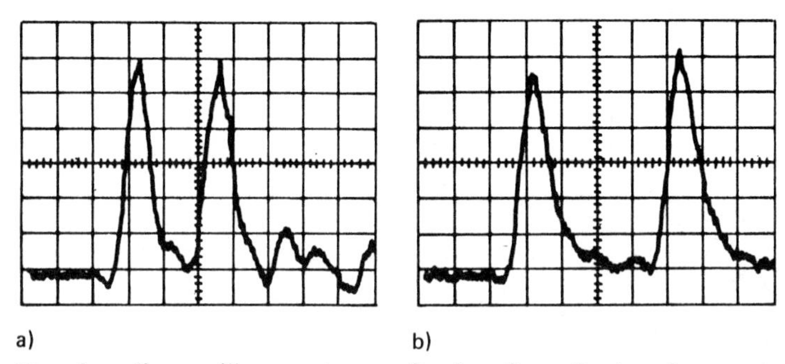

FIG. 64 – Sampling oscilloscope traces of pulses from the two frequency-offset lasers (*a*) before and (*b*) after traversing the 1 km fibre. SHJ laser pulse at 900 nm to the right, DHJ laser pulse at 860 nm to the left. Time scale 2 nsec/km. (After [89]).

ruby laser was used, different wavelengths between 694 and 930 nm being obtained by Raman generation in a liquid filled cell (water or benzene). (See Fig. 65). The radiation directly from the laser

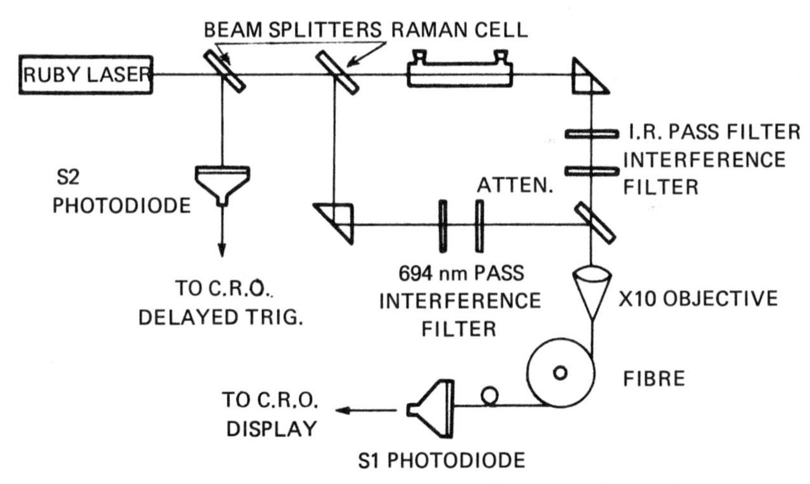

FIG. 65 – Experimental arrangement for measuring transmission time delays vs. wavelength using a ruby laser and a Raman cell. (After [90]).

is also sent into the fibre by means of beam splitters. The corresponding pulse is used as a marker pulse, with respect to which the time delays of pulses at other wavelengths are measured. The results in terms of the defined material dispersion are shown in Fig. 66.

The theoretical prediction of a zero material dispersion point

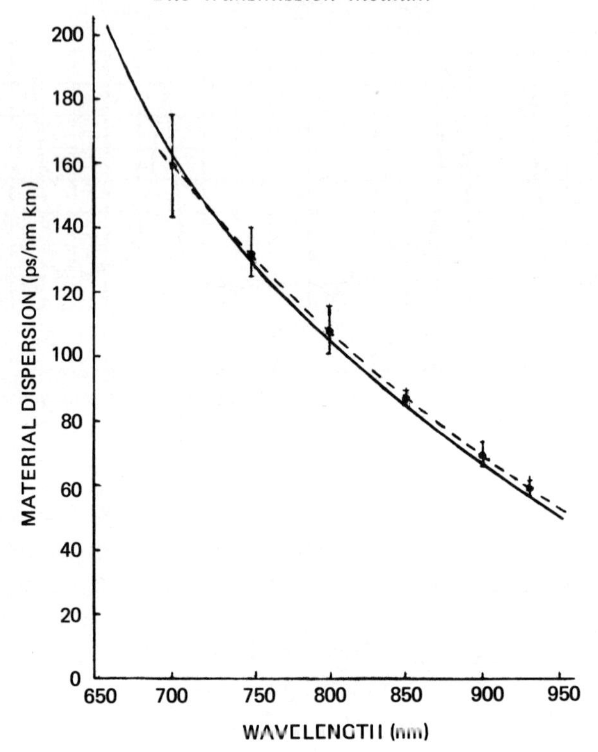

FIG. 66 – Wavelength dependence of the material dispersion of phosphosilicate glass. The dashed curve was obtained using fibres of several different P_2O_5 levels. The solid curve was calculated from the refractive index data of pure silica. (After [90]).

at longer wavelengths stimulated the extension of such measurements to longer wavelengths [91]. Here the optical source is a dye laser followed by a temperature-tuned $LiNbO_3$ parametric oscillator; pulses 0.5 ns wide are obtained by a Pockels cell pulse slicer, in wavelength ranges 580-620 nm and 780-2600 nm. A digital time delay generator provides the time delay necessary to display the pulses on an oscilloscope after which small differences among the various pulses can be read directly on the oscilloscope with a high level of accuracy.

Results of measurements of τ as a function of wavelength for graded and step index fibres with different GeO_2 content are shown in Fig. 67. The corresponding material dispersion is calculated by fitting the experimental points by

$$\tau = a + b\lambda^{-2} + c\lambda^{-4} + d\lambda^{2} + e\lambda^{4} \tag{43}$$

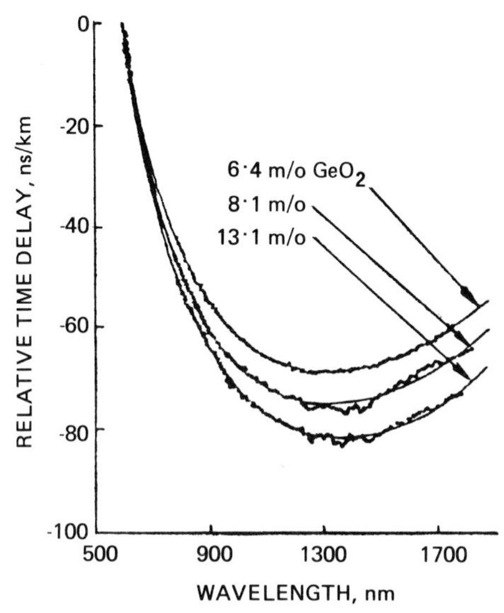

FIG. 67 – Fibre transit time per kilometre relative to that at 580 nm for germania-doped fibres shown. (After [91]).

and deriving $d\tau/d\lambda$. The results in Fig. 68 show that zero material dispersion is effectively reached in a range between 1272 and 1350 nm according to the dopant level.

A different approach to obtain pulses in the region of zero

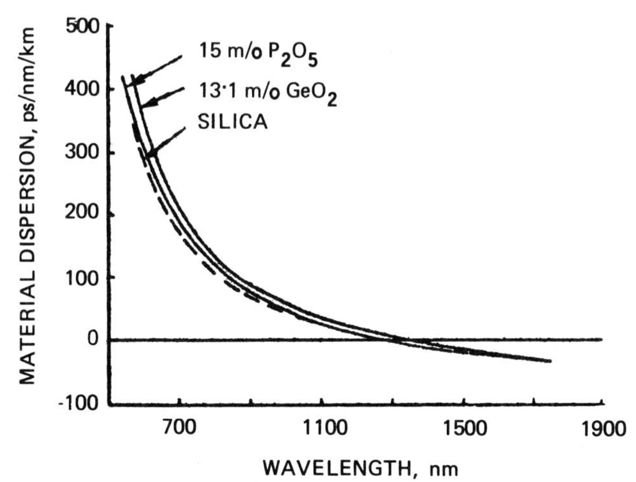

FIG. 68 – Comparison of material dispersion of fibres containing 15 m/o P_2O_5 and 13.1 m/o GeO_2 with that of pure silica. (After [91]).

dispersion has been reported in [66]. A single mode fibre pumped by a Q-switched and mode-locked Nd:YAG laser provides sub-nanosecond pulses in the $\lambda = 1120$-1550 nm range by multiple-order stimulated Raman scattering (see Fig. 69). Results obtained

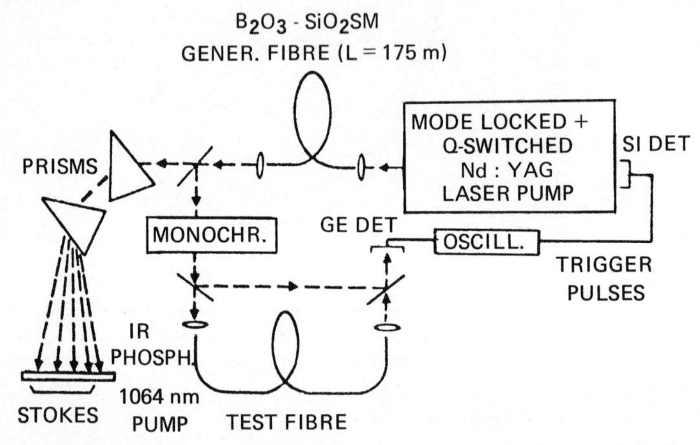

FIG. 69 – Experimental arrangement for measuring transmission time delays vs wavelength using stimulated Raman scattering in a single mode fibre. (After [66]).

from four graded index fibres with different B_2O_3 and GeO_2 content, indicating the achievement of zero material dispersion, are shown in Fig. 70.

Similar measurements were also reported for single mode fibres. Here waveguide dispersion can no longer be neglected; though the two effects of material and waveguide dispersion are not mathematically independent, for an estimate they are computed separately (one in the absence of the other). The total effect is, in this hypothesis, given by

$$\frac{d\tau}{d\lambda} = \frac{d\tau_m}{d\lambda} + \frac{d\tau_\omega}{d\lambda} . \tag{44}$$

If V is the fibre normalized frequency ($V = (2\pi/\lambda)(a \cdot \Delta)$), $d\tau_\omega/d\lambda$ is given by [92]

$$\frac{d\tau_\omega}{d\lambda} = \frac{L}{2\pi c} V^2 \frac{\partial^2 \beta}{\partial V^2} \simeq \frac{L}{c\lambda} \Delta n \, D_\omega(V) . \tag{45}$$

Δn being the core-cladding refractive index difference and $D_\omega(V)$

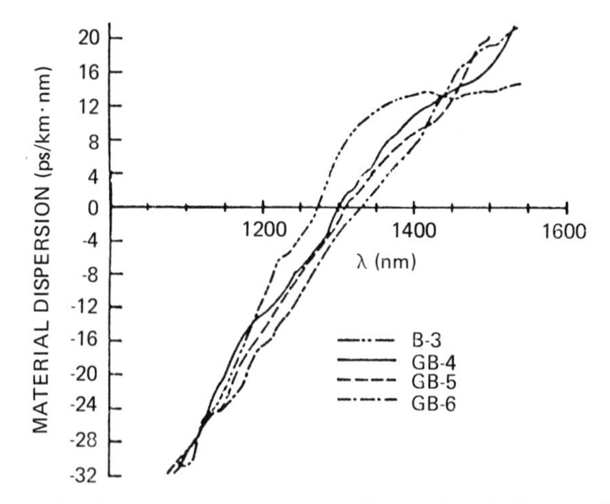

FIG. 70 – Material dispersion vs. wavelength in multimode fibres. Fibre B-3 has a graded borosilicate core. The other fibre cores have a small amount of B_2O_3 and a larger graded GeO_2 concentration. The dopant concentrations were increased from fibre GB-4 to GB-6. (After [66]).

a dimensionless coefficient related to the fundamental mode. When performing similar measurements to those described for multimode fibres on monomode fibres, the contribution of waveguide dispersion must be subtracted from the experimental data in order to get correct results. A result of this analysis is shown in Fig. 71 for two B_2O_3 doped single mode fibres.

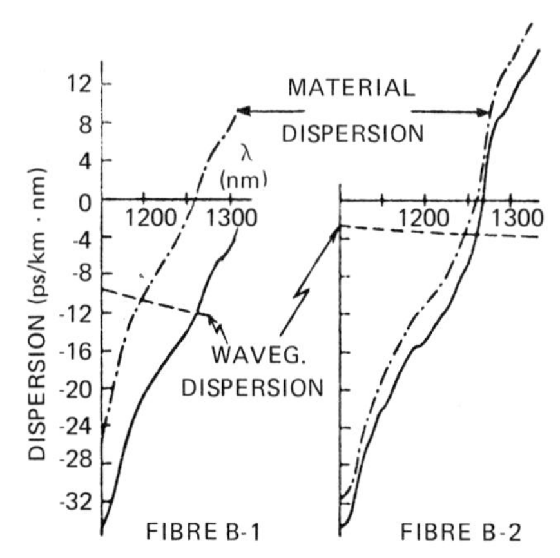

FIG. 71 – Single-mode fibre dispersion (——) vs. wavelength. Waveguide dispersion (– – –) is subtracted to obtain material dispersion (–·–·). Fibre B-1 causes more waveguide dispersion than Fibre B-2 (After [66]).

Measurements on single mode fibres have also been carried out in the 814-905 nm region using pulsed GaAlAs lasers emitting at different wavelengths with temperature controlled wavelength tuning in a small (~ 5 nm) region. Results are shown in Fig. 72.

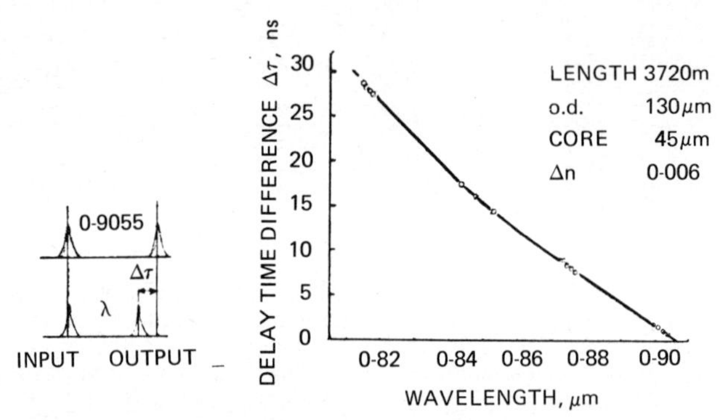

FIG. 72 – Propagation times of pulses transmitted through 3.72 km-length single-mode fibre. (After [66]).

3.7 Refractive index profile

The information carrying capacity of a multimode optical fibre is critically dependent on its refractive index profile, i.e. the refractive index distribution as a function of radius across the core (assuming circular symmetry). Theoretical considerations (see Chapter 2) show that optimized index profiles with a maximum at the core centre can be designed for a fibre, at a specific wavelength, so that mode transit times are nearly completely equalized, thus greatly reducing modal dispersion.

A knowledge of the index profile is therefore important for fibre evaluation, and essential to the manufacturer who has to tailor his production process to the requirements of an exactly shaped index distribution. Very accurate measurement methods must be implemented. To get an idea of the precision required one has to consider that the maximum relative refractive index core-cladding difference for typical fibres is of the order of 1 % or less, and that variations of a few percent within such a difference should be determined to get a reasonably accurate index profile. On the other hand, according to a theoretical study [93] based on numerical computation, in order to realize a truly optimized index profile a 0.01 % relative index variation control is required.

Another requirement for index profile measurement methods is a good spatial resolution. As fine details are to be evidenced in regions corresponding to typical fibre core radii (15-40 μm) a spatial resolution of 1 μm or better is needed. For monomode fibres with fibre core radius of a few microns even better resolution should be necessary, but at this stage exact profile determination seems somewhat less important than for multimode fibres. The refractive index profile also plays a role in determining light acceptance properties and splicing tolerances; in those cases, however, since the dependence on profile shape is much less critical than for dispersion, a fairly good knowledge is more than adequate.

Several methods have been developed for index profile measurement. Those most widely used will be described in the following sections and their relative merits and disadvantages will be discussed.

3.7.1 *Interferometry: parallel to waveguide axis*

This technique consists in viewing a thin slice of optical fibre through an interference microscope. The slice is taken normal to the fibre axis and is polished with flat and parallel faces. Differences in refractive index result in a different path length and a consequent fringe displacement. The fringe pattern thus obtained allows the measurement of the refractive index difference and the determination of the refractive index profile.

Two main approaches may be employed. One is to use a reflected light interferometer [94], the other makes use of a transmitted light interferometer [95-98].

A typical reflected light measurement system is shown in Fig. 73. It is essentially a Michelson interferometer with a microscope to observe fringes. The back surface of the sample is metallized to provide reflection of the light beam and the sample objective is focussed on the back face so that light passes through the slice parallel to the waveguide axis, is reflected and returns to the objective. Photographs of the fringe pattern, taken with high resolution films, allow subsequent analysis.

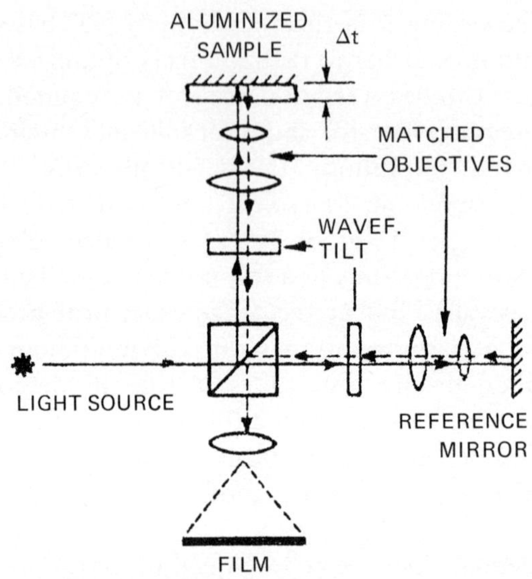

FIG. 73 – Zeiss Model I interference microscope used with transparent samples containing a refractive index gradient. (After [94]).

The transmitted light arrangement makes use of a Mach-Zehnder interferometer, a schematic of which is shown in Fig. 74.

Two methods of fringe visualization are possible with both

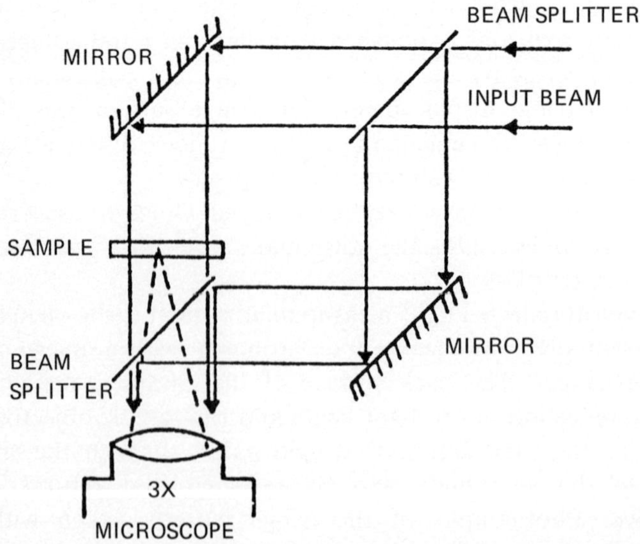

FIG. 74 – A transmitted-light (Mach-Zehnder) interferometer. (After [93]).

techniques. By adjusting the tilt of the reference mirror the sample may be viewed against a field of parallel fringes, so that the fringe displacement is directly proportional to the variation of refractive index. A typical interferogram obtained in this way is shown in Fig. 75. Another possibility is to spread fringes apart until the

FIG. 75 – Photograph of the interference fringe pattern of a clad optical fibre. The region surrounding the fibre (region without fringes) is the mounting cement. A part of the supporting capillary tube is visible in the upper right-hand corner. Note the collinearity of fringes in the fused silica capillary tube and the undoped silica fiber cladding. (After [98]).

entire field of view encompasses a single fringe. In this case the sample is viewed against a uniform flat field, which can be dark or bright. In the sample, fringes in this field connect points of equal refractive index, obtaining maps of iso-refractive index. An example of this mode of operation is shown in Fig. 76.

The quantitative relationship between refractive index difference in two sample points and fringe displacement is:

$$\Delta n = q \frac{\lambda}{t} \tag{46}$$

FIG. 76 – Interferograms of sample with a nearly parabolic profile. (After [96]).

where q is the number of fringe displacements, λ the illuminating wavelength and t the sample thickness.

Errors in the interpretation of the fringe pattern may arise due to several factors: non-parallelism of the two faces, boundary reflections caused by skew rays from the illumination system and waveguide effects caused by multiple reflections at the core-cladding interface. These difficulties are minimized, in general, by using thin sections of highly multi-mode waveguides (0.025-0.5 mm depending on core size and Δn).

Two further problems arise when the interferometric technique is applied to graded index fibres. The first problem is due to the focusing properties of a graded index fibre; rays are bent toward the axis (see Fig. 77) so that a plane wave entering the fibre is focused after a distance that can easily be calculated in the case of the parabolic profile. It can be shown that the focal length is given by [96]:

$$f = \frac{\pi a}{2} \cdot \frac{1}{\sqrt{2\varepsilon}} \tag{47}$$

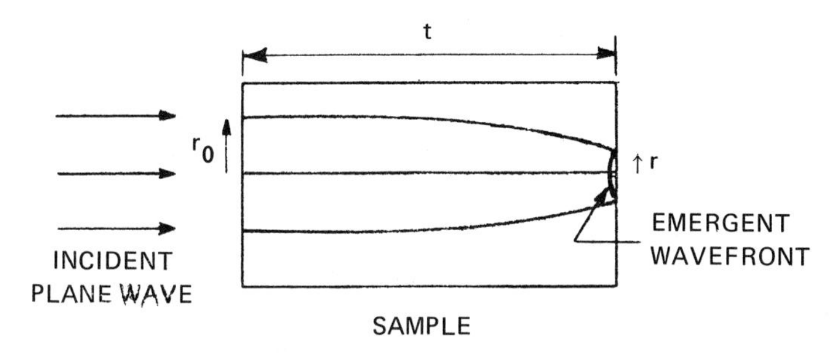

where ε is the maximum relative refractive index difference.

From (47) one can see that f is proportional to the core radius and inversely proportional to the square of the maximum refractive index difference. Due to such an effect, a scale factor should be used in fringe displacement measurement, because the core size as deduced from interference pattern is smaller than the true core size. Moreover, the effective ray path

$$\Delta s = \int_0^t n(r)\, ds \tag{48}$$

should be considered.

The second problem is connected to the first: due to wave front curvature not all the fringes are in focus when viewed through an objective with limited depth of focus. It should be noted that pronounced curvature is related to small diameter fibres that require very high magnification ($1000\times$) for detailed analysis, but high power objectives are just those with smaller depth of focus (0.3 μm nominal for a $100\times$ objective, but the eye can accomodate for a somewhat larger depth).

One way of minimizing the described effects is to keep $t \ll f$ and thus obtain very thin slices. Taking into account the fact that for a typical graded fibre ($a = 30$ μm, $\Delta n = 0.01$) $f = 330$ μm, thicknesses of 100 μm or less may be required. These slices are obtained through a rather long and sophisticated process [97]. A section of a fibre a few centimetres long (or several fibres for time saving) is potted in a capillary tube a few mm in diameter. After curing, several cuts are made by a wafering machine, obtaining samples ~1 mm thick. The sample is brought to the desired final thickness by lapping and polishing. Small core, high N.A. fibres may require thicknesses of 10 μm; handling such samples involves use of vacuum holding techniques. A practical rule for determining slice thickness is that the entire fringe pattern should be in focus simultaneously.

A very thin sample presents another problem: from (46) for a fixed Δn the number of fringe displacements is proportional to t. This means that for very thin slices a single fringe is viewed when operating in the flat field mode. A suggested technique in this case is to insert a known variable path change in one arm of the interferometer and following the single circular fringe as it expands radially outward [95]. This is inconvenient as data must be read step by step from the instrument rather than from the permanent photograph. In the parallel fringe method we have only one fringe displacement between core and cladding, thus rendering the measurement less accurate.

The measurement is generally performed on the interferogram. A fringe passing through the centre of the core is chosen, and is followed at regular steps till the cladding and the displacement with respect to the cladding level is measured.

Due to the large amount of data to be extracted and to difficulties in visual determination of the fringe centre (used for displacement determination), automated analysis techniques have been developed, with digital encoding of fringe pattern by a scanning microdensitometer and computer performed interferogram analysis [98]. A result of such a measurement is shown in Fig. 78.

The accuracy of the method in determining $n(r)$ has been estimated to be ± 2 to 5×10^{-4}. Besides the focusing problem already discussed, differential magnification associated with high power microscope objectives can cause errors. A more subtle effect is due to the polishing process. As the final step of polishing is performed on a soft lap, differential polishing of the fibre depends on fibre composition. Systematic errors of several 10% may occur, due to the differential sample thickness which affects both the absolute value of refractive index and the profile determination. It has been found that the difficulty can be removed by several minutes of polishing on a hard lap.

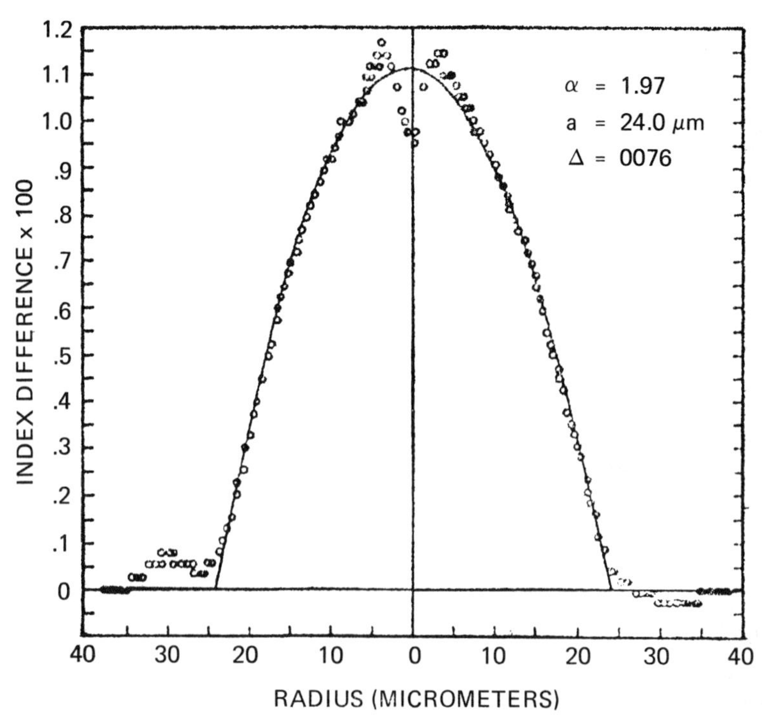

FIG. 78 – Refractive index profile of a fibre as determined by the slab interferometric method. The fitting curve provides the parameters α, a, and Δ which is the maximum refractive index difference. (After [98]).

The spatial resolution is determined by microscope resolution and is approximately 1 μm. In conclusion, the transverse interferometric technique allows the determination of the refractive index absolute value and refractive index profile with high accuracy and good spatial resolution. Moreover, it constitutes a direct measurement of refractive index. Its main drawbacks are that sample preparation is very time consuming and requires high quality processes and that data extraction and analysis involve expensive and sophisticated equipment.

3.7.2 *Interferometry: normal to fibre axis*

Delicate and difficult sample preparation, as required by the previously described technique, have stimulated research into simpler technique. Another kind of interferometric technique, in which the light beam crosses the fibre perpendicular to its axis, has been proposed and developed [99-101]. The method requires no sample preparation. The fringe appearance for the cases in which the refractive index of the surrounding medium is equal to and higher than that of the cladding index is shown in Fig. 79. In [99]

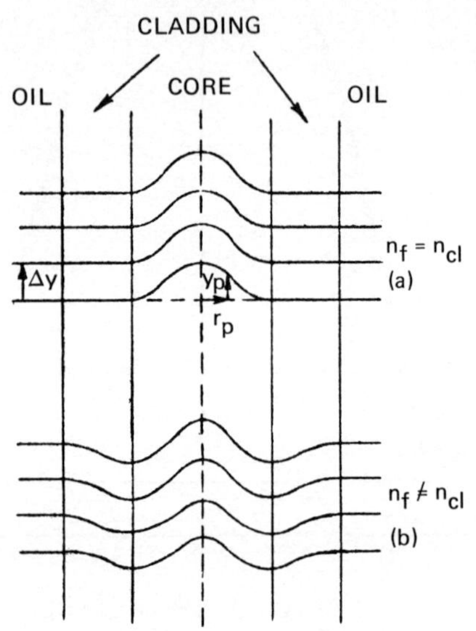

FIG. 79 – Schematic diagram of fringes for a clad fibre: (*a*) ($n_f = n_{cl}$) and (*b*) ($n_f \neq n_{cl}$). (After [101]).

it is demonstrated that the refractive index profile can be obtained from the fringe shifts by an Abel inversion and simple analytical expressions for optical path difference for parabolic profile fibres. In [101] the method is extended to fibres with the index profile already quoted

$$n(r) = n(0)\left[1 - \Delta_1 \cdot \left(\frac{r}{a}\right)^{\alpha}\right]^{\frac{1}{2}}.$$

Assuming that the fibre does not appreciably distort the outgoing waveform, it is calculated that, using the geometry of Fig. 80, the fringe displacement N_p is given by:

$$\frac{N_p \lambda}{4} = \Delta n \sqrt{a^2 - r_p^2} - \frac{\Delta n}{a^\alpha} \int_0^{\sqrt{a^2 - r_p^2}} (x^2 + r_p^2)^{\alpha/2}\, dx \tag{49}$$

where $\Delta n = n(0) - n(a)$.

From this, Δn and α may be calculated by means of numerical

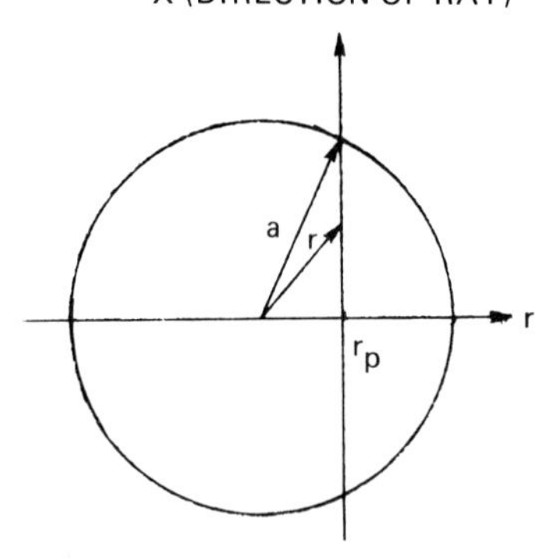

FIG. 80 – Geometry for optical path difference for fringes. (After [101]).

computations based on an iterative process, once N_p is measured as a function of r_p, provided an approximate value of α is known. An accuracy of better than $\pm 10\%$ is possible on Δn and α. A more refined analysis is performed in [100], in which ray bending inside the fibre is taken into account. This allows improved accuracy in index determination. Again an integral equation is found which is solved by an iterative procedure. The experimental apparatus consists of a Mach-Zehnder interferometer. The fibre is surrounded by an oil with a refractive index similar to that of the cladding. It can be temperature tuned (which, conversely, means that thermostatization is necessary for stable conditions). In Fig. 81 the experimental apparatus used in [100]

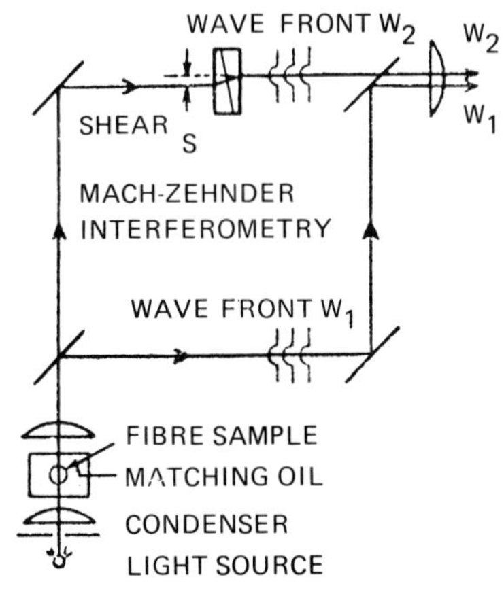

FIG. 81 – Principle of transverse shearing interferometry (After [100]).

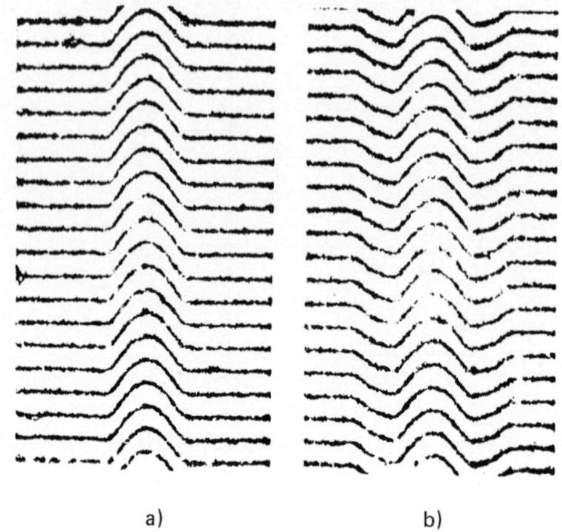

FIG. 82 – Fringes for a clad fibre: (*a*) ($n_f = n_{cl}$) and (*b*) ($n_f \neq n_{cl}$). (After [101]).

is shown. A shearing plate permits either the ordinary interference pattern (for $S < 2a$) or the differentiated optical path difference to be obtained.

In Fig. 82 interferograms obtained by transverse technique are shown [101]. A refractive index profile calculated from experimental data is shown in Fig. 83 [100]. As already stated, the method requires very simple sample preparation, and can be non-destructive. Disadvantages are that it is an indirect method, requiring complicated calculation and a priori knowledge of profile shape

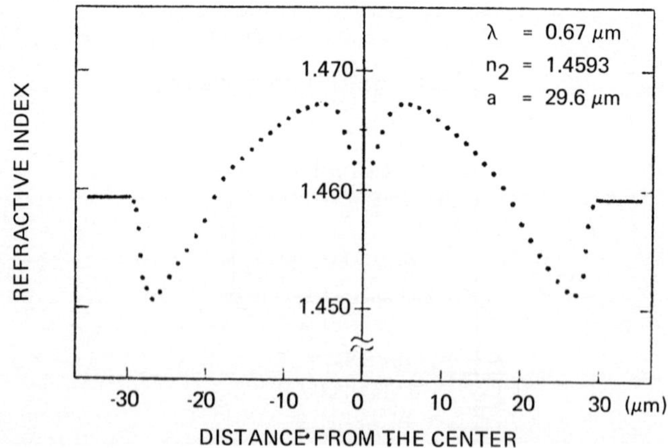

FIG. 83 – Refractive index profile obtained by the technique shown in Fig. 81. (After [100]).

for starting iterations, and that the measured values involve some averaging, as a ray traversing the fibre samples the refractive index over a range of azimuthal angles. Finally, circular symmetry is assumed in the calculation. Slight asymmetries can be taken into account by observing from several directions, and by finding an equivalent index profile.

3.7.3 *Near field method*

This method consists in measuring the near field power distribution across a diameter of an optical fibre and deducing the refractive index profile [102, 103]. The method rests on an observation reported in [55], according to which if we excite equally all modes in a fibre with circularly symmetric profile (e.g. by means of a Lambertian source) the power accepted at each point at a distance r from the core centre is proportional to the corresponding local N.A. If we express it as a fraction of the power accepted at the centre of the core region we have:

$$\frac{P(r)}{P(0)} = \frac{n^2(r) - n^2(a)}{n^2(0) - n^2(a)}. \tag{50}$$

If all modes propagate with equal attenuation and without coupling, the same power distribution is found at the fibre output end, thus enabling a near field measurement of the refractive index to be performed. For small refractive index differences between core and cladding the $P(r)$ distribution is directly proportional to $n(r)$ since $n(r)^2 - n(a)^2 \approx 2n(a)[n(r) - n(a)]$. For the above mentioned conditions—equal mode loss, no mode coupling—to be fulfilled, very short lengths of fibre are to be used for the measurement; otherwise, mode coupling and differential mode attenuation alter the modal power distribution and unreliable data is obtained. Unfortunately, after short length tunnelling leaky modes are still present, and add power to the observed near-field distribution. In the above analysis, however, leaky mode contribution was not taken into account. It can be demonstrated that, for a Lambertian source and excitation of all possible leaky modes one has [104, 105]

$$\frac{P(r)}{P(0)} = \frac{n^2(r) - n^2(a)}{n^2(0) - n^2(a)} \cdot \frac{1}{\sqrt{1 - \left(\dfrac{r}{a}\right)^2}}. \tag{51}$$

Two other factors must be taken into account to obtain an equation applicable to actual experimental conditions. One is that the angle of the Lambertian source is truncated by the launching optics, so that not all possible leaky rays are excited. In this case, if θ_s is the maximum angle at which the source emits with respect to its axis, we have the following expression [26] for the accepted amount of light:

$$\frac{P(r)}{P(0)} = 2\,\frac{n^2(r) - n^2(a)}{\pi[n^2(0) - n^2(a)]}\cdot$$

$$\cdot \mathrm{tg}^{-1}\left[\sqrt{1-\left(\frac{r}{a}\right)^2}\,\frac{\frac{a}{r}\sqrt{1-\frac{n^2(r)-n^2(a)}{\sin^2\theta_s}}}{\sqrt{1-\frac{a^2}{r^2}\left[1-\frac{n^2(r)-n^2(a)}{\sin^2\theta_s}\right]}}\right] +$$

$$+\frac{2}{\pi[n^2(0)-n^2(a)]}\sin^2\theta_s\left[\frac{\pi}{2}-\sin^{-1}\frac{a}{2}\sqrt{\frac{1-n^2(r)-n^2(a)}{\sin^2\theta_s}}\right]. \qquad (52)$$

This expression must be used when θ_s is smaller than the maximum acceptance angle for leaky rays (see Section 2.1.2).

The other factor consists in the fact that leaky rays are very strongly attenuated, even over short distances [106]. To get the correct power distribution at a distance z the following integral must be evaluated:

$$\frac{P(r, z)}{P(0)} = \frac{2}{\pi[n^2(0) - n^2(a)]}\cdot$$

$$\cdot\int_0^{2\pi} d\varphi \int_0^{\theta_M(r,\varphi)} \sin\theta\cos\theta\exp\left[-\gamma_l(r,\theta,\varphi)\cdot z\right]d\theta\,. \qquad (53)$$

Coordinates r, θ, φ, z are defined in Fig. 84. θ_M is the maximum acceptance angle and γ_l is the attenuation coefficient of a ray characterized by certain θ, φ, r; if a short length of fibre is used for the measurement, γ_l can be considered zero for all guided rays. For leaky rays attenuation coefficients have been calculated by several authors (see Section 2.4.3). In [26] the following expression [25] was used:

$$\gamma_l = \frac{e^{-s}}{2k(1+e^{-s}/4)\cdot\int_{r_0}^{r_1}[n^2(r)-k^2-(h^2/r^2])^{-\frac{1}{2}}\,dr} \qquad (54)$$

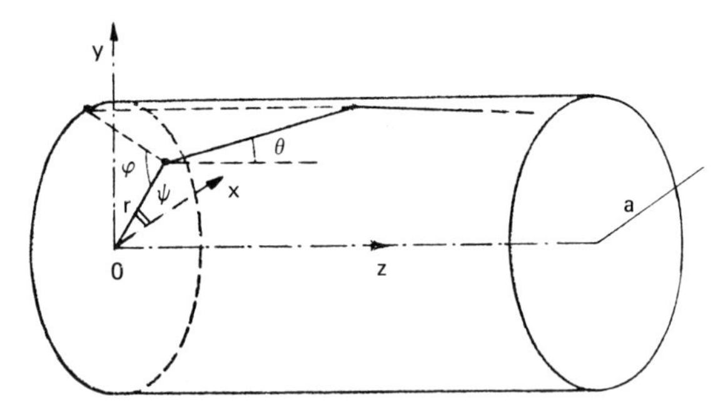

FIG. 84 – Fibre geometry and ray parameters used in the equations of the text.

with

$$s = \frac{4\pi}{\lambda} \int_{r_1}^{r_2} \left[\frac{h^2}{r^2} + k^2 - n^2(r) \right]^{\frac{1}{2}} dr \tag{55}$$

where $k = n(r) \cos \theta$, $h = n(r) r \sin \theta \sin \varphi$, λ is the wavelength in vacuo, r_0, r_1, r_2 are the roots of the equation

$$n^2(r) - k^2 - \frac{h^2}{r^2} = 0 . \tag{56}$$

Assuming a given refractive index profile, numerical calculations can be performed in order to evaluate the integral (55); γ_l can be analytically determined in the important cases of step and parabolic index profile fibres. A set of normalized correction curves, based on the calculation of (54), has been published [106, 107] and is shown in Fig. 85. They are largely independent of refractive index profile, and represent a correction factor $C(r, z)$ by which the measured power distribution $P(r, z)/P(0)$ must be divided in order to get the actual index profile.

A further theoretical problem related to near field measurements consists in the fact that, due to the finite number of guided modes, each with its own power distribution, the near field profile that results from the superposition of individual mode intensities shows a ripple, with a period proportional to $1/V$ where

$$V = \frac{2\pi a}{\lambda} \sqrt{n^2(0) - n^2(a)} .$$

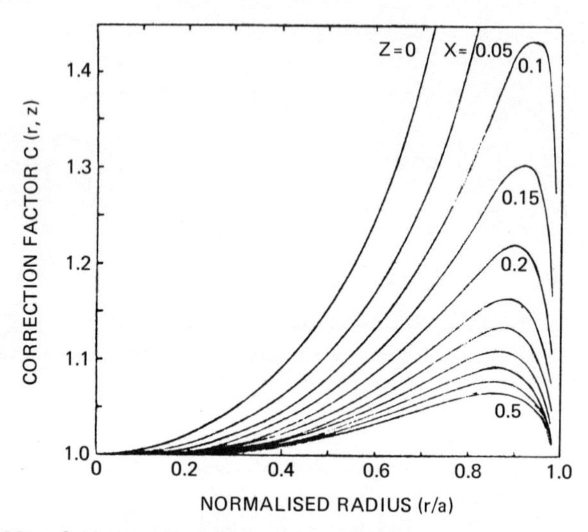

FIG. 85 – Near-field correction factors $C(r, z)$ given as a function of normalised fibre radius for X values from 0.05 to 0.5 in increments of 0.05. The normalisation parameter X describes the fibre core radius, length and numerical aperture, and is given by $X = (1/V) \ln (z/a)$. The curves may be used for fibres having any graded-index profile. Also shown is the result for $z = 0$, when all leaky modes are present. (After [107]).

It is intuitive that for fibres supporting few modes (low V), few ripples with large amplitudes are present. This invalidates the method as an index profile measurement technique [26]. Also, in less extreme cases index profile fidelity and spatial resolution may be degraded by this effect. Numerical calculations and theoretical considerations confirm this statement [108]. In Fig. 86 power density distributions are compared with actual index profiles for step and parabolic profile fibres, and the ripple and its dependence on V are clearly visible. For reasonably large V values, the effect can be minimized by using an optical source with large spectral width. Different adjacent wavelengths cause a continuous shift of the various oscillations, smoothing out the power distribution. It has been shown [109] that a 100 nm linewidth optical source is sufficient to eliminate the effect. Another precaution consists in operating at short wavelength, thus increasing the V value and hence the resolution.

As far as the experimental approach is concerned, two possible techniques can be used. The first consists in illuminating the fibre

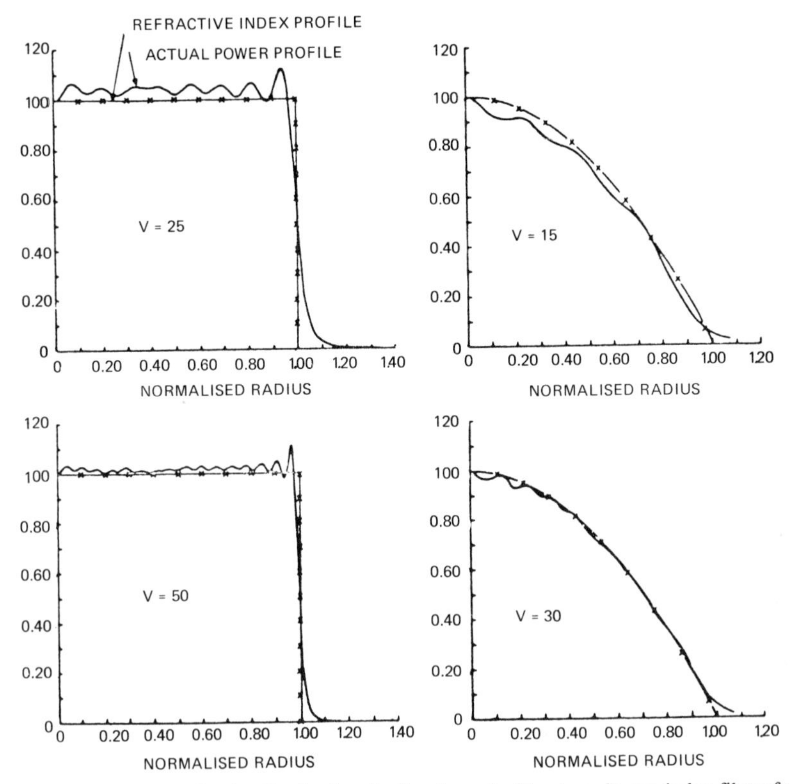

FIG. 86 – *Left:* Power density distributions in (incoherently illuminated) step-index fibres for $V = 25$ and for $V = 50$ *Right:* Power density distribution in (incoherently illuminated) fibres with parabolic profiles. (After [108]).

with a large area Lambertian source and forming an enlarged image of the output end by means of appropriate optics. A schematic of the experimental apparatus is shown in Fig. 87. The optical source can be an LED (which, however, is normally a poor approximation to a Lambertian source), a tungsten-halogen lamp, or an arc lamp.

A pinhole of suitable size can be inserted to act as the effective source. A filter placed in the light beam allows the selection of desired wavelength when white sources are used. The output end of the fibre is magnified by a high resolution optical device that forms an enlarged image of it. Care must be taken that minimal differential magnification occurs (see Section 7.3.2). On the image plane, a small area detector (silicon or germanium photo-

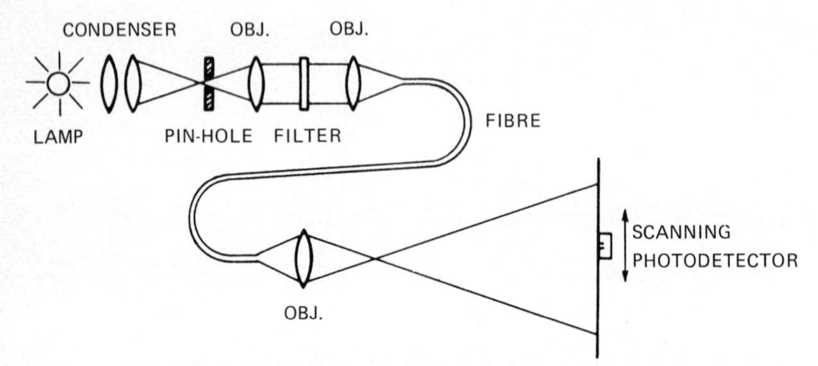

FIG. 87 – Schematic setup of the near-field technique.

diode) is moved on a fibre image diameter, and the signal is recorded as a function of the position. The light beam may be chopped so that lock-in amplifiers may be used for amplification and measurement. Motor-driven translation stages can be used to move the detector. Sharp focusing of the image on the detector may be achieved by visual observation if a definite boundary can be seen (core-cladding boundary in step-index fibres, or index dip in the core centre which appears as a dark hole; the index dip can also help in identifying a fibre diameter).

A lower accuracy version of this approach, allowing quick

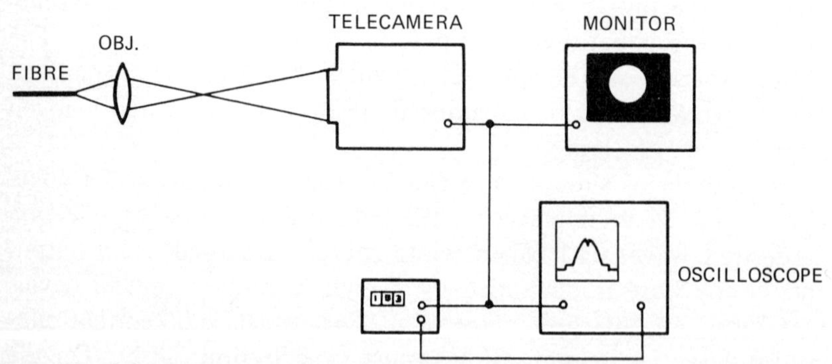

FIG. 88 – Schematic set up of data acquisition system for power distribution at the output face of a fibre in real time.

determination of the index profile, consists in replacing the small detector with a TV camera on which the entire fibre surface is imaged. An electronic device permits the selection of any line of the TV frame, which can be displayed on an oscilloscope giving the power distribution in real time. The experimental set-up is shown in Fig. 88. An oscillogram of a core diameter scanning is shown in Fig. 89.

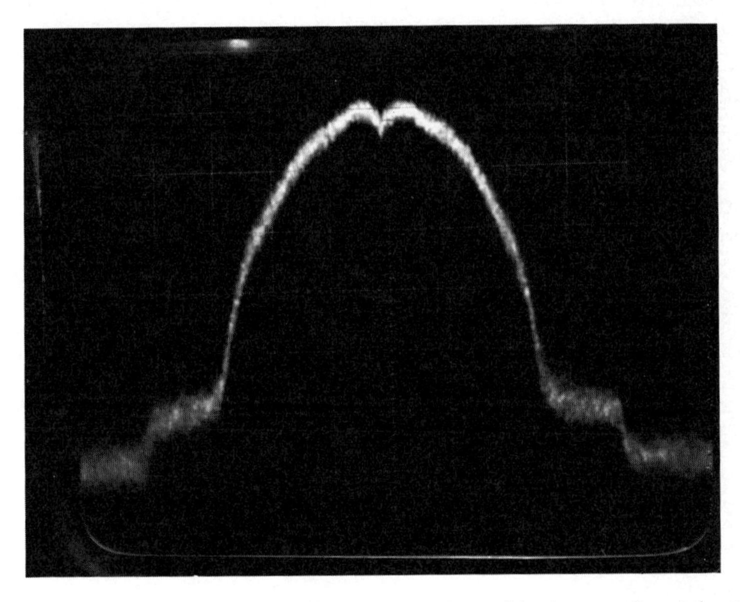

FIG. 89 – Oscillogram of a fibre diameter scanning with the experimental setup of Fig. 88.

The second technique is the reciprocal of that previously described. In this case [110, 111] a small uniform spot is focused onto the fibre end, which is moved by small steps (see Fig. 90). The total power output is monitored as a function of position. Either coherent (He-Ne laser [110]) or incoherent (LED's [111]) sources can be used. In the case of LED's, a high demagnification is needed in order to have a sufficiently small spot ($\sim \lambda$) to ensure high resolution. An experimental apparatus using an LED is shown in Fig. 91. This arrangement permits fibre observation during the measurement.

The main advantage of this technique is that a Lambertian

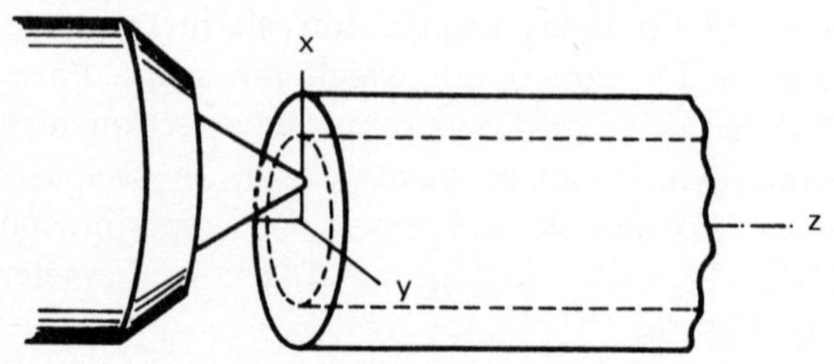

FIG. 90 – In the transmission method, the microscope objective illuminates a small area of the fibre end of the order of λ_0^2. The intensity is assumed uniform. (After [111]).

source uniform over the full cross section of the fibre is not required; in fact, the scanning is always performed with the same spot and hence with a high degree of uniformity. The optical arrangement is also simpler.

A drawback is that very fine motion is required for the input end of the fibre, as typical steps are ~ 1 μm.

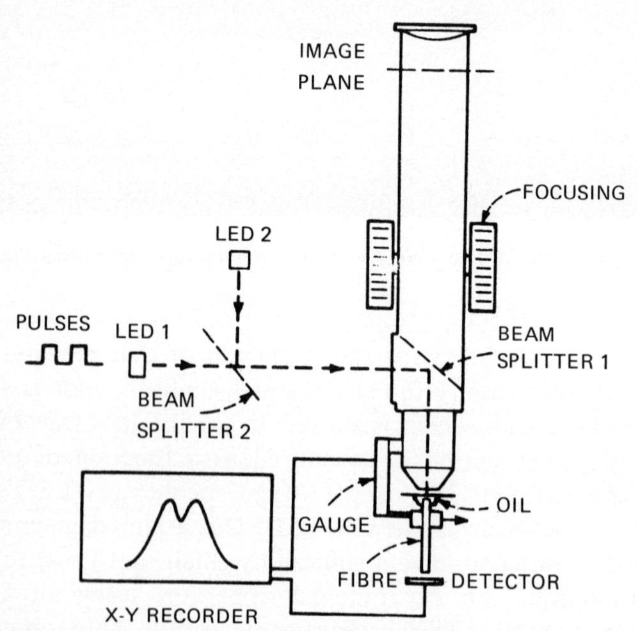

FIG. 91 – Experimental setup of the transmission technique. The fibre is scanned mechanically and its motion is recorded with a gauge. Two LEDs are used for dispersion measurement (After [111]).

Results of near field scanning performed on step index and graded index fibres by the first method are shown in Fig. 92 [26].

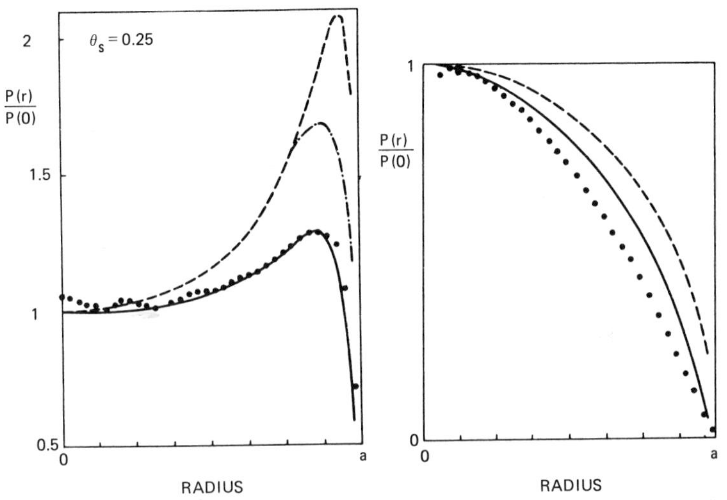

Fig. 92 – Comparison between theory and measured values of P(r)/P(0) for a step and graded index fibre.
--- Eq. (50), -·-·- Eq. (52), —— Eq. (53), ··· Measured values.

Theoretical curves are reported that refer to Eqs. (50), (52), (53) respectively. It can be seen that for the step index fibre good agreement between experimental data and theoretical prediction is found. This means that application of a correction factor to find the true index profile is fully justified. For the graded index fibre no such agreement is found, and it seems that the power distribution reproduces index profile without requiring further correction. This seems contrary to results found by other authors [112]; theoretical studies [27] have pointed out that if near square law fibres deviate from perfect circular symmetry, attenuation coefficients of leaky modes may become much larger. All this confirms the major drawback of the technique, which is the difficulty of controlling the leaky modes effect.

On the other hand, the method is very attractive because it is simple to implement and requires minimum sample preparation. Moreover, it has very high accuracy in the determination of changes of n (a less than 1% $P(r)$ variation can easily be measured, which corresponds to a $< 10^{-4}$ variation in the refractive index in ordinary fibres), and very good spatial resolution. Among the drawbacks, besides the cited problem of leaky modes, there is the general consideration that it is an indirect method, which depends on the source properties and fibre transmission characteristics. It is a method that does not provide an absolute value for the refractive index but only a functional dependence, and that is not applicable to fibres carrying few modes.

3.7.4 *Refracted near-field method*

Another proposed technique, related to the near field technique, has been called by the author RNF (Refracting Near Field) [113]. It consists in using power not trapped by the fibre core, namely

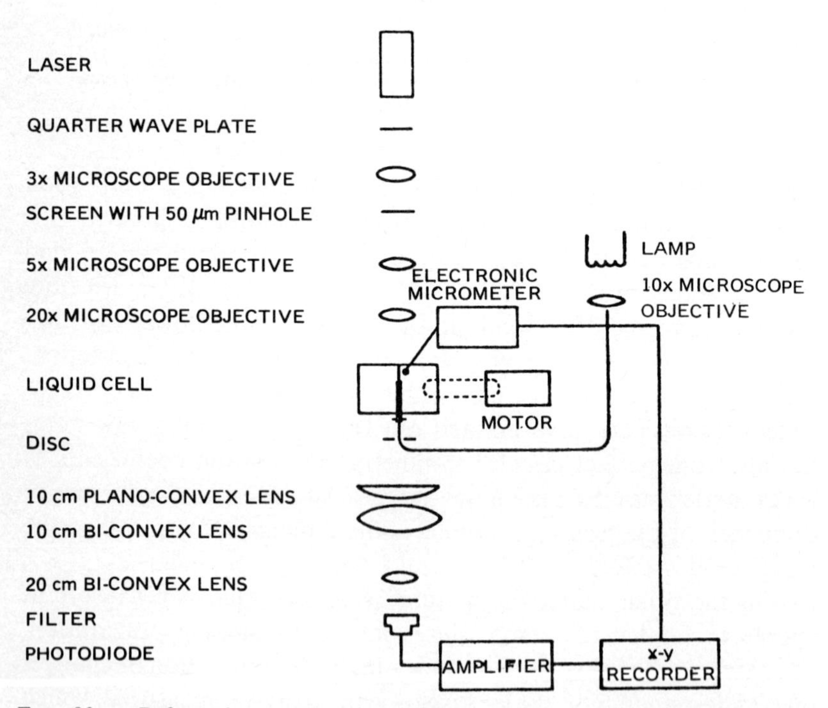

FIG. 93 – Refracted near-field technique; schematic diagram of apparatus. (After [114]).

refracting modes instead of guided (reflected) modes. A qualitative explanation of basic principles underlying this method is given in ref. [113]. Experimental implementation has been described in ref. [114] and is shown in Fig. 93. The light source can be a stable laser. As the reflectivity of light is strongly angle-dependent when linearly polarized light is used, a quarter wave plate is inserted to get circular polarization. The beam is expanded by two microscope objectives and focused on to the fibre end face by a lens with a numerical aperture much larger than that of the fibre. The fibre is held in a cell filled with a liquid whose refractive index is just above that of the cladding.

A disc, suspended by three glass fibres, (see Fig. 94) is placed in such a position to intercept leaky modes.

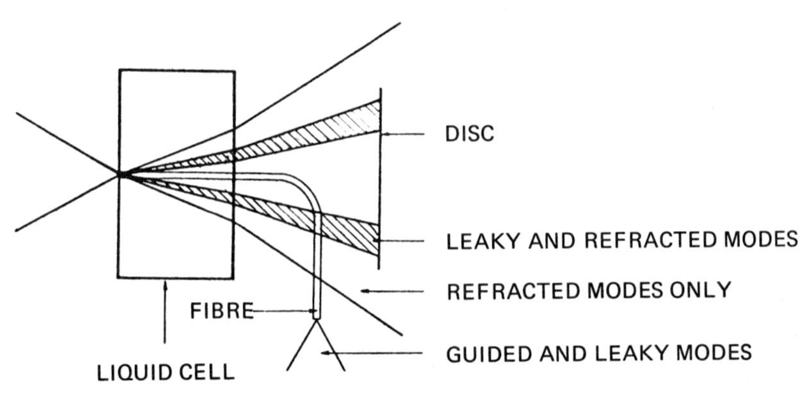

FIG. 94 – Refracted near-field technique; schematic diagram. (After [114]).

The refracted power is collected by a lens systems and imaged onto a photodiode. A tungsten lamp illuminates the fibre from the back to help in alignment. A filter is placed in front of the photodiode to reject ambient light and stray light from the back launching.

The measurement is performed scanning the fibre across the light spot and recording the signal from the photodiode. Two main advantages are offered by this technique over the usual near-field method. One is the strong suppression of leaky rays contribution. This can be seen from Fig. 95, taken from ref. [114], showing the amount of leaky modes rejection in a step index fibre measured with the two methods. The second advantage is that

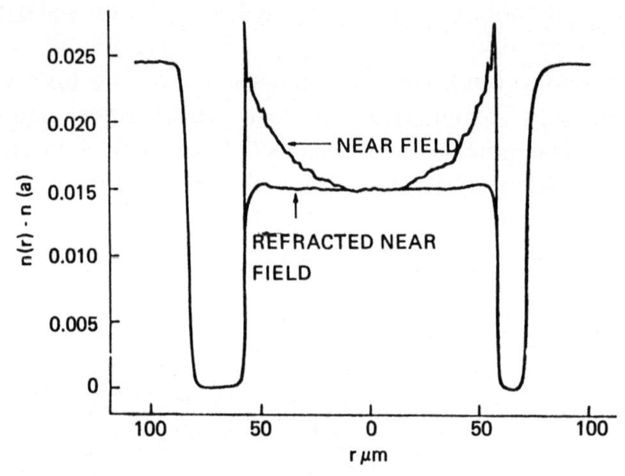

FIG. 95 – The index profile of a step-index fibre obtained by the refracted near-field technique and the power as a function of radius obtained by the near-field technique. (After [114]).

the continuum of irradiated modes is used for the measurement, rather than the discrete spectrum of guided modes. This allows a much greater resolution, as can be seen in Fig. 96, where data of a graded index fibre is shown: fine details are evidenced near the core center. In particular, it is possible to measure a monomode or a two-mode fibre, for which the near-field technique gives a representation of mode power distribution instead of refractive index distribution. An example of the different results obtained for a two-mode fibre is shown in Fig. 97.

3.7.5 *Reflection method*

This method consists in measuring the light power reflected by a mirror-like fibre end face. In order to achieve the required spatial resolution a focused laser beam is used as the light source [115, 116]. The principle of the method is therefore very

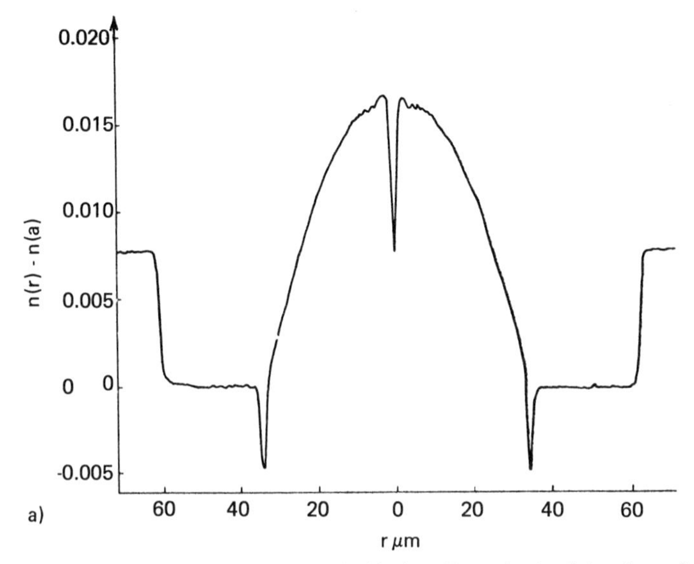

FIG. 96 *a*) – The index profile of a graded-index fibre obtained by the refracted near-field technique. (After [114]).

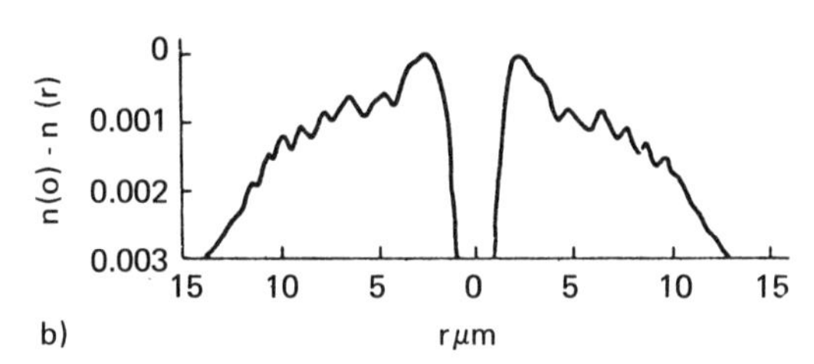

FIG. 96 *b*) – A detail from the fibre whose profile is presented in Fig. 96*a*. (After [114]).

simple: if we consider the light spot as a plane wave impinging on a dielectric surface, the local refractive index may be calculated by the well-known Fresnel reflection formula

$$R = \frac{P_R}{P_I} = \left(\frac{n-1}{n+1}\right)^2 \qquad (57)$$

where P_I and P_R are the incident and reflected light power respectively, n is $n(r)/n_m$ where $n(r)$ is the refractive index of the fibre at a radial distance r from the core centre and n_m the refractive index of the surrounding medium.

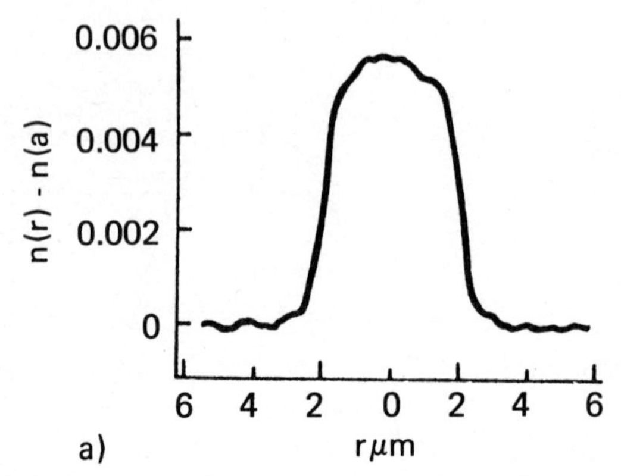

FIG. 97 *a*) – The index profile of a two-mode fibre obtained by the refracted near-field technique. (After [114]).

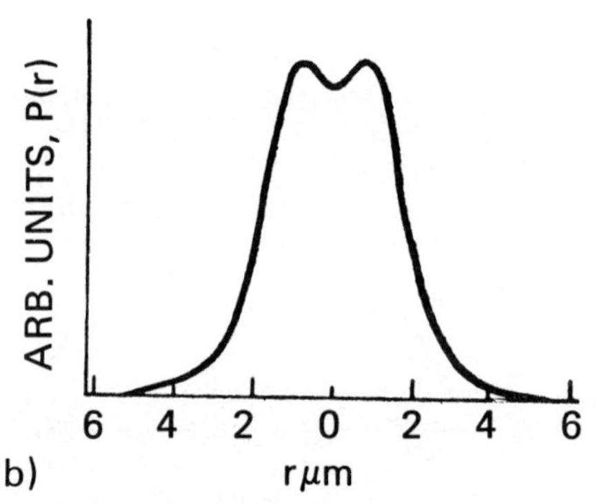

FIG. 97 *b*) – The near-field power distribution of a two-mode fibre obtained by the near-field technique. (After [114]).

An experimental set-up is shown in Fig. 98 [115]. The polarizer and $\lambda/4$ plate are used to avoid feedback of reflected light into the laser resonator. In addition, the $\lambda/4$ plate produces a circularly polarized light beam, useful to avoid polarization dependent reflection from the fibre surface; light is spatially filtered and expanded to obtain a smaller spot. A thick beam splitter provides a reference beam from the ingoing light beam, as well as a double beam from the light reflected at the fibre end face. One beam is used for measurement, the other to check proper focusing onto the fibre surface. A microscope objective is used to focus the

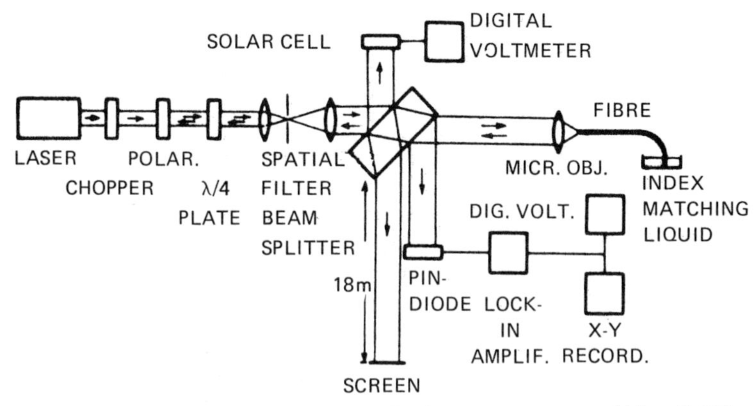

FIG. 98 – Experimental setup for reflection measurement. (After [115]).

light beam onto the fibre. The other end of the fibre is immersed in index matching liquid to avoid troublesome reflections. The fibre itself is moved by means of precision translation stages and the corresponding signal is monitored as a function of position.

From (57), a relationship between the relative variation of reflected light power and refractive index change can be easily obtained; for small changes:

$$\frac{\Delta R}{R} = \frac{4}{n^2 - 1} \Delta n \,. \tag{58}$$

If the measurement is performed with the fibre in air [115, 116] $n_m = 1$ and $\Delta R/R \simeq 3.5\Delta n$. This means that a change $\Delta n = 0.001$ produces only a 0.35% change of R. Considering that for ordinary fibres the total Δn from cladding to core centre is ~ 0.01, it results that the technique requires extremely accurate measurement of R for modest resolution on n. A modification proposed in [26] obviates this drawback. The fibre is surrounded by an oil with a suitable refractive index, so that the factor in front of Δn in (58) may become quite large, leading to very high sensitivity.

The experimental arrangement to perform such a measurement is shown in Fig. 99. It is similar to that previously described, except that the light beam is vertically deflected by a mirror and an oil immersion objective is used for focusing onto the fibre (Leitz 100×, NA = 1.25). The objective is modified by means of a small surrounding cylinder with an *o*-ring allowing a certain

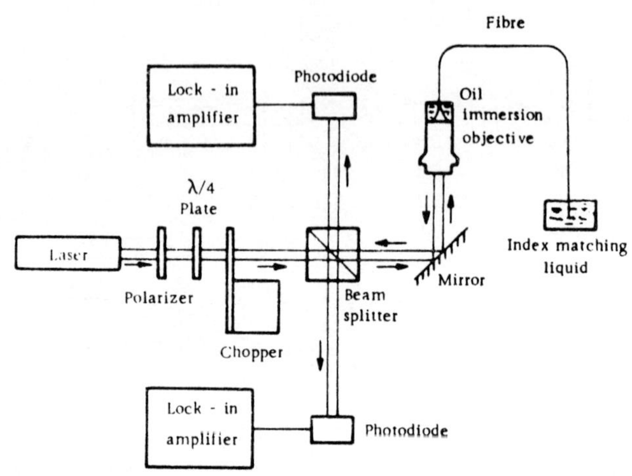

FIG. 99 – Experimental apparatus for reflection measurement. (After [26]).

amount of oil to be contained on top. The fibre is immersed in the liquid and moved through it.

The refractive index of the employed immersion oil is 1.515; accordingly

$$\frac{\Delta R}{R} \simeq 56\,\Delta n$$

so that the sensitivity is comparable to the near field sensitivity. A problem associated with this approach is that the reflected signal is very weak, as $P_R \simeq 3 \cdot 10^{-4} P_I$. Careful and skilled setting up of the apparatus is required in order to avoid spurious reflections from beam splitters (to this purpose thin beam splitters are used) and objectives lenses, which could completely obscure the signal.

This has been very satisfactorily accomplished: in Fig. 100 a comparison of a reflection and a near field measurement on a graded index fibre is shown. The method permits an absolute value of n to be measured through (57). Precise direct measurement is, however, difficult. Hence, the ratio of the reflected to the reference signal is usually measured for a sample of accurately known refractive index, thus calibrating the apparatus, after which other samples' refractive index may be calculated from obtained data. As the cladding index in optical fibres is usually known, it

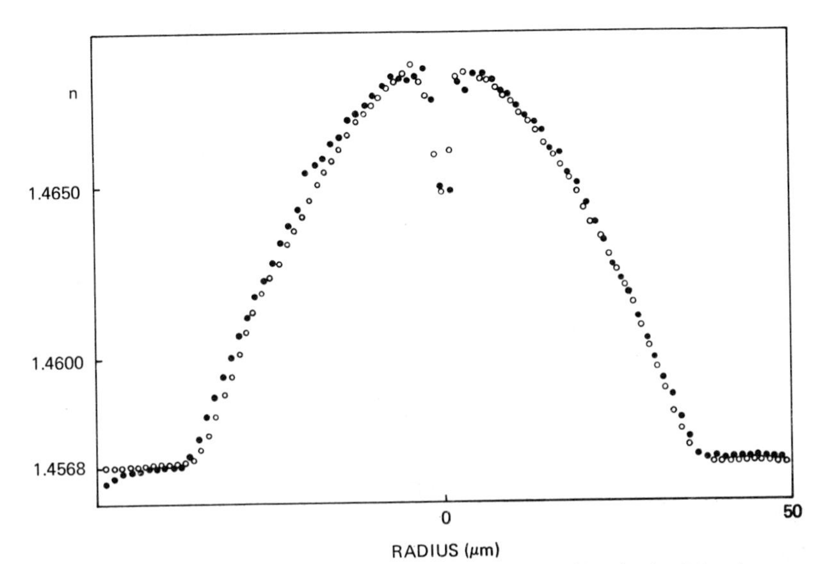

FIG. 100 – Comparison between the corrected index profile obtained by the near-field method [o] and by reflection method [•] for a graded index fibre. (After [26]).

may be conveniently used for calibration for each measurement. This allows a very good accuracy also in the determination of the absolute value of the refractive index, in addition to the inherent good accuracy of differential measurement across the fibre cross section.

The spatial resolution is also very good, being determined by the scanning spot size. For high NA objectives and short wavelength it may be better than 1 μm. This resolution is not completely adequate for monomode fibres where the beam spot size is comparable with the core radius. In addition, the reflected light distribution differs from the refractive index profile according to the index profile itself [117]. However, if the intensity profile or the spot size of the laser beam is known, the refractive index profile may be derived from the reflected power by means of calculation. A particularly simple algorithm is found for laser beams with Gaussian profile. It is shown [117] that spatial resolution of 0.3 μm and accuracies of 5 % in the relative refractive index may be achieved by the corrected reflection method.

For the reflection measurement, it is essential that the fibre end is flat, smooth, and clean: otherwise severe errors arise. To meet these requirements, polishing of the fibre surface has been employed. However, it was found [118, 119] that the usual technique of scoring and pulling, if properly used, produces completely satisfactory surfaces. The fresh cut face is, in addition, very clean, so that no cleaning procedure is required. Sample preparation is therefore minimal.

The technique is attractive, as it has good resolution, both spatial and in Δn, allows an absolute value of n to be measured, and requires little sample preparation. Its main draw-back, apart from the necessity of accurate measurement and careful setting-up of the experimental apparatus, is that it is a surface measurement. Surfaces are, in general, a particular state of matter, not necessarily reproducing the internal composition. It has been pointed out [119] that polishing causes surface alteration due to plastic flow and chemical reactions between the glass and the polishing agent. This can be avoided by fracturing the fibre. However, exposure to the atmosphere may also alter the glass surface; while for Ge and P doped silica fibres reproducible results may be obtained, boron doped fibres show changes in the measured profile as a function of time.

3.7.6 *Forward scattering method*

This method, also called the scattering-pattern method [120], allows the calculation of the fibre index profile by the forward-scattering pattern for a normally incident laser beam. The optical set-up to perform such a measurement is simple, and is shown in Fig. 101. The laser beam illuminates the fibre normally and the forward far-field scattering pattern is detected as a func-

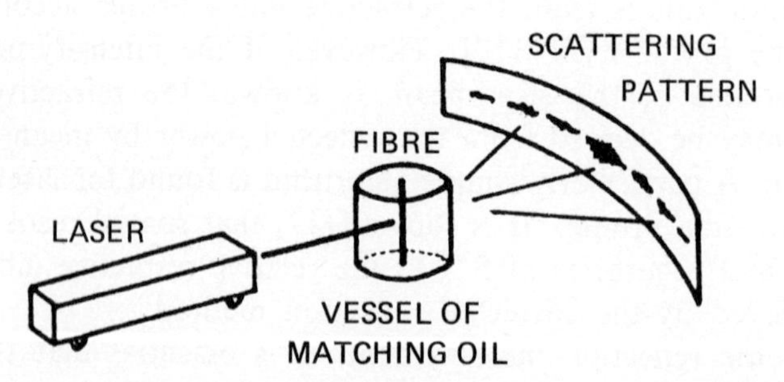

FIG. 101 – Optical setup for measuring the scattering pattern. (After [120]).

tion of the scattering angle θ. The fibre is immersed in matching oil whose refractive index is nearly equal to that of the cladding. In this way the scatter at the fibre surface is almost removed. The vessel of the matching oil has a plane glass window for the incoming beam and a cylindrical window for the scattered light. A wide dynamic range photomultiplier with a slit in front is used as a detector. A cumbersome theoretical analysis [120], by which the scattered field is computed considering distributed equivalent dipoles corresponding to the spatial variation of the refractive index, shows that the following relationship exists between the permittivity $\varepsilon_1(r) = n^2(r) - n_0^2$ ($n_0 =$ oil refractive index) and the scattered field

$$\frac{\varepsilon(r)}{\varepsilon_0} = \frac{k_0^{\frac{1}{2}}}{\pi} \int_0^\infty \text{sign}\,[E_s(\theta)][\sigma(\theta)]^{\frac{1}{2}} J_0\left(2k_0 \sin\frac{\theta_0}{2}\cdot r\right) \sin\theta\, d\theta \tag{59}$$

where $k_0 = 2\pi n_0/\lambda$; $E_s =$ scattered field; $\sigma = 2\pi R(P_s/P_i)$ with R radius of the circle along which the detector moves; P_s and P_i are respectively the scattered and incident power; J_0 is the 0-th order Bessel function of the first kind.

It must be noted that this relationship is valid under the assumption of thin fibres with a very small refractive index difference, or $a\Delta \ll 1$, so that the method gives correct results only if applied to such fibres. An experimental difficulty consists in dermining the polarity of the field required to compute (59). Although in principle a polarity inversion occurs every time E_s becomes zero, in practice noise problems render this approach uncertain. To obtain reliable results an intricate method of polarity determination is employed in [120], which implies that the measurement is to be performed by very small increments, leading to an oversampling by a factor of about 100 with respect to the amount of data needed for index determination. The spatial resolution of the technique is very good; it is given in the theoretical limit, by

$$\Delta r = \frac{\pi}{2k_0 \sin\theta_m/2} \tag{60}$$

where θ_m denotes the maximum scanning angle; the maximum theoretical resolution is therefore $\Delta r = \lambda/n_0$. This implies scanning over a wide angular range which, in conjunction with the preceding requirement, means that a large amount of data is required (> 1000). This leads to the use of an automated measurement system (see Fig. 102).

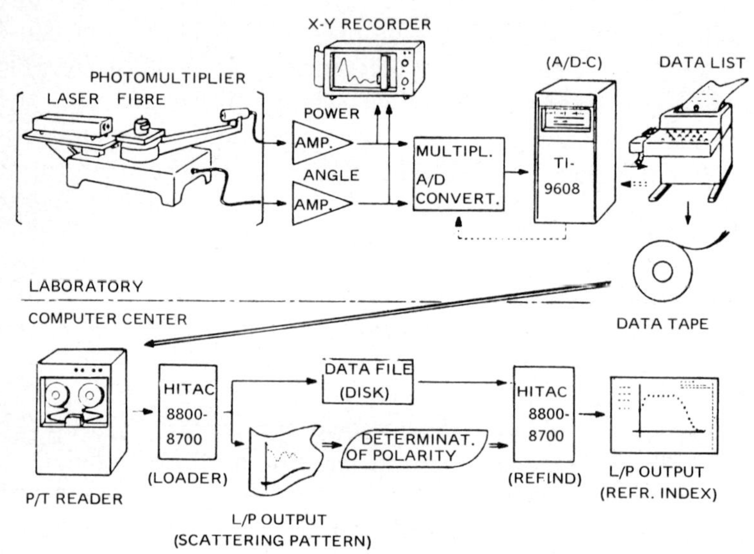

FIG. 102 – Semiautomated system for measuring the refractive-index profile. (After [120]).

An alternative, very sophisticated, approach has been suggested [121]; in this case the problem is treated as an ordinary diffraction problem. The Fraunhofer diffraction has Fourier transform properties so that

$$F\left(\frac{x_f}{\lambda z_0}\right) = kB[\Delta n(r)] \tag{61}$$

where $k = 2\pi/\lambda$ is the wave number, B denotes the Hankel transform also called Fourier-Bessel transform; $F(x_f/\lambda z_0)$ represents the scalar amplitude of one transverse component of the electric field in the diffraction plane; x_f represents the coordinate in an axis normal to the fibre axis in the diffraction plane; and z_0 is the distance of the field point from the fibre. $\Delta n(r)$ is the index difference between core and cladding. Instead of performing the inversion mathematically, it can be performed by the technique of optical data processing, making use of the Fourier transform properties of lenses and of the possibility of spatial filtering in the focal plane [122]. As the expression (61) is formally exact and does not imply any approximation, such as that fibres must be thin, the method can be extended to arbitrary fibres [123]. While the mathematical solution would be too difficult, optical processing obtained by a physical approach is possible.

The procedure for measuring the refractive index profile consists therefore in getting an image of the fibre by means of a two-lens system in which lenses are separated by a distance equal to the sum of their focal lengths, with the fibre placed in the front focal plane of the first lens. A suitable spatial frequency filter is placed in the rear focal plane of the first lens (coinciding with the front focal plane of the second lens), so that measuring the final image intensity distribution the refractive index profile can be obtained by numerical computation. In the case of thin fibres, the refractive index profile can be found directly from the measurement.

3.7.7 *Focusing method*

An accurate, non-destructive, simple to implement method has been proposed called « focusing method » [124]. The fibre is illuminated by a collimated beam perpendicular to its axis, but the measurement of power distribution is carried out in a plane just outside the core. The power distribution in such a plane is determined by the focusing properties of the core region with respect to incident light. The plane must be sufficiently close to the core so that the rays do not cross each other. The geometry of the method is shown in Fig. 103. In Fig. 104 a possible experimental

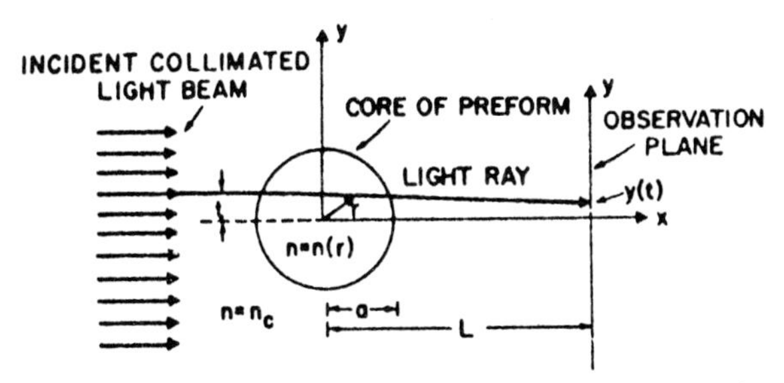

FIG. 103 – Schematic of the geometry of the focusing method. (After [126]).

apparatus is shown [125]. The detector is a vidicon camera and the signal is processed by a video digitizer, which is capable of addressing discrete picture elements in the TV frame and digitally encoding element intensity to 8 bits, or 256 grey levels. Essential to the method is the immersion of the fibre in cladding index matching liquid, so that only the core affects the ray path.

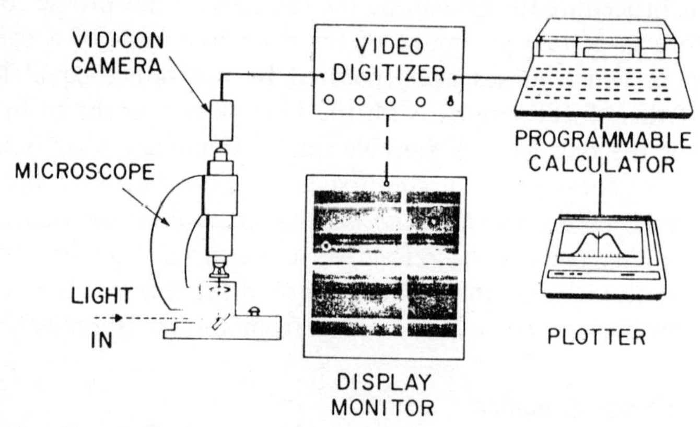

FIG. 104 – Experimental arrangement to measure the index profile of optical fibres by the focusing method. (After [125]).

Once the light distribution $P(y)$ is known (symbols as defined in Fig. 103), the refractive index distribution of the core can be obtained from the integral expression [125]:

$$n(r) - n_c = \frac{1}{\pi L} \int_r^0 \frac{t - y(t)}{\sqrt{t^2 - r^2}} \, dt \qquad (62)$$

In the above expression t has the physical meaning of the distance of an arbitrary ray from the optical axis measured at the input plane and y is the corresponding quantity in the observation plane. The function $t(y)$ is the related to the power distribution by [126]

$$t(y) = \int_0^y P(y') \, dy'$$

from which $y(t)$ can be derived as number pairs y and t.

Examples of refractive index profiles obtained by this method, compared to those obtained by the interferometric slab method are shown in Fig. 105. The agreement is quite good.

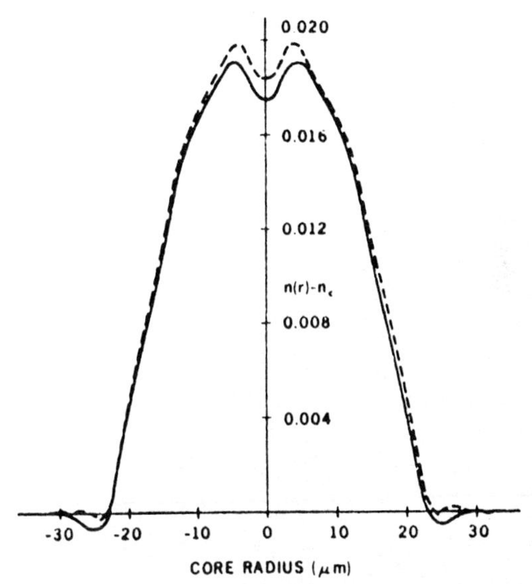

FIG. 105 – Comparison of the refractive-index profile of a fibre obtained by the focusing method (solid curve) and by the interferometric slab method (broken curve). (After [125]).

It may be added that the focusing method, unlike many other methods described so far, lends itself very well to the determination of refractive index profiles of preforms obtained by the CVD method [126].

In general, the method has the advantages of requiring modest sample preparation, of being non destructive and relatively simple to set up. Its major drawback is that very fine details near the core centre are easily lost, due to too strong focusing of very sharp index change.

In addition, the determination of the distance L from core center to the observation plane is difficult to determine, unless an index dip or some way of identifying the core centre is available. Moreover, as with all the other techniques using lateral illumination, circular symmetry must be assumed for the calculations to be valid.

3.7.8 *Electron microprobe and scanning electron beam microscopy*

In modern fibres, the desired index variation in the core is obtained by doping silica with suitable materials. A high resolution electron beam microprobe may be used to quantitatively determine the amount of dopant as a function of radius; an example is shown in Fig. 106. If the relationship between dopant

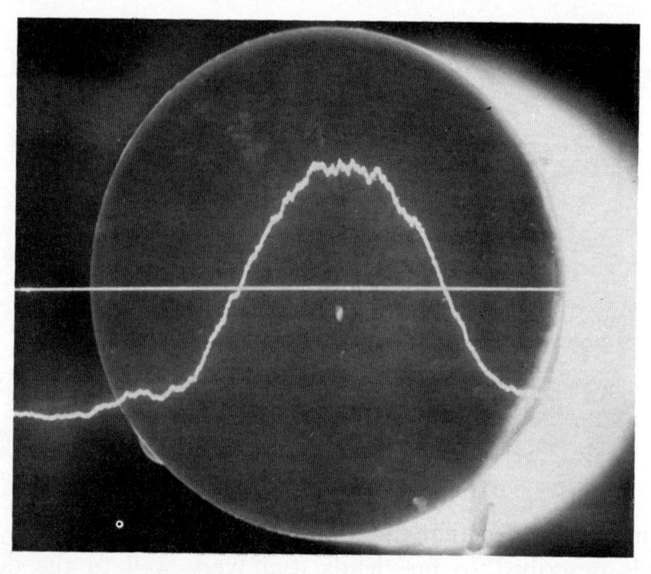

FIG. 106 – Plot of the amount of dopant (Ge) as a function of position along a diameter of the fibre. (Performed at CSELT).

concentration and induced refractive index change is known, the refractive index can be inferred from this measurement. It has been shown that there is a linear relation between dopant concentration and refractive index change (see for instance [127]), thus a direct representation of refractive index profile is obtained. The method has the advantage of very high resolution (a few thousand Å); the accuracy of quantitative analysis is typically ±5%. Accuracy may be improved by averaging over repeated measurements.

Accuracy may be improved by averaging over repeated measurements.

A different technique, also allowing very high resolution, consists in etching the fibre end by a suitable etchant. The etching rate is dependent on the composition, so that a relief pattern reproducing the index profile is obtained at the fibre surface. The resulting sample is viewed at a scanning electron microscope (SEM). Microphotographs may be taken and attractive results

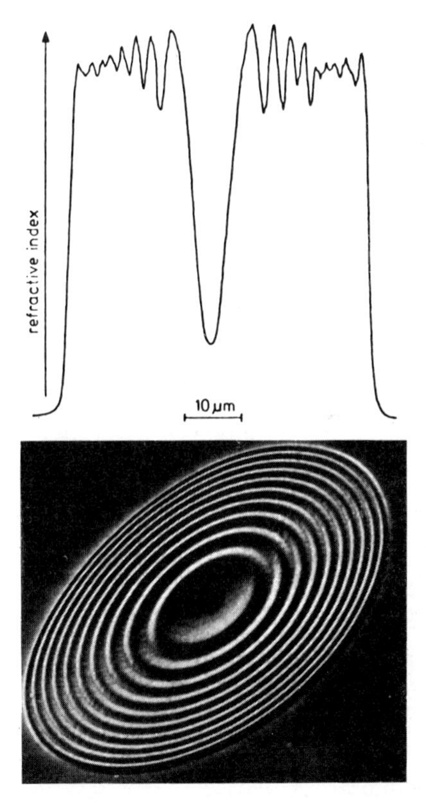

FIG. 107 – Refractive index profile (above) and scanning electron microscope photograph of an etched fibre. (After [128]).

are obtained, such as that shown in Fig. 107 [128]. A quantitative evaluation is in principle made possible by a measurement of the etching profile, once the etching rate is known.

3.7.9 *Profile dispersion*

The index profile of a graded fibre changes as a function of wavelength. This is due to the fact that the material composition varies in the core and in general has different dispersive properties with respect to the cladding material; Δn is thus a function of λ. A profile dispersion parameter P is defined as follows [129]

$$P(\lambda) = \frac{n(0, \lambda)\,\lambda \Delta_1'}{N \Delta_1}$$

where

$$N = n(0, \lambda) - \lambda \frac{dn(0, \lambda)}{d\lambda}; \qquad \Delta_1 = \frac{n^2(0, \lambda) - n^2(a, \lambda)}{n^2(0, \lambda)};$$

$$\Delta_1' = \frac{d\Delta_1}{d\lambda}. \qquad (63)$$

This behaviour has important implications as it means that an index profile optimized for a given wavelength is no longer optimized, in general, at another wavelength. A precise measurement of profile dispersion is therefore important for knowing the amount of this effect and, possibly, in choosing of a composition that minimizes the phenomenon.

For the same reasons considered for material dispersion measurements, it is preferable to perform measurements directly on fibres rather than on bulk material.

An obvious way of performing such a measurement is to employ an accurate method chosen among those described for refractive index profile determination and, using a multi-wavelength source, to repeat the measurement at different wavelengths to get $\Delta n(\lambda)$. This has been effectively done by using transverse interferometry (Section 3.7.1) [130]. The experimental system used in [130] is shown in Fig. 108. Light from a high intensity xenon-arc-lamp is collimated by lens L1 and filtered by 20 nm bandwidth filters in the 500-1100 nm range, in steps of 50 nm. The light is coupled to a Leitz transmission interference microscope by lens L2. Samples are prepared in the manner already described for usual profile measurements. The field of the microscope is detected with a silicon camera tube and viewed on a TV screen. The video signals from two scan lines are displayed on an oscillo-

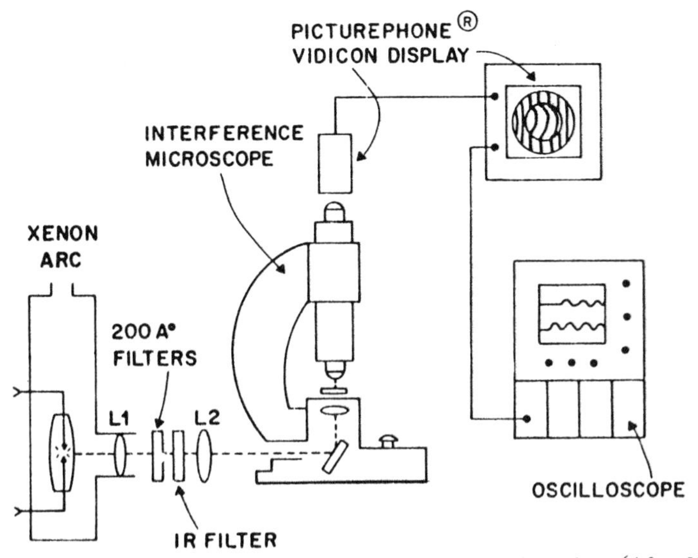

FIG. 108 – Experimental apparatus for measuring profile dispersion. (After [130]).

scope. Two lines are chosen, one passing through the cladding perpendicular to the interference fringes, the other through the core near the fibre axis (see Fig. 109). The displacement h of a fringe between core and cladding may be measured from

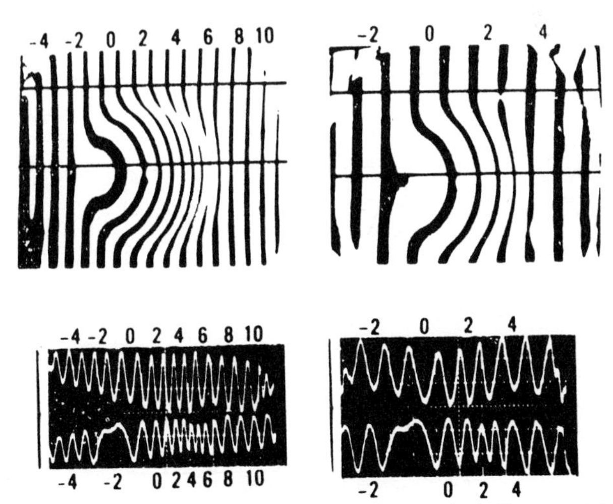

FIG. 109 – Picturephone display of graded index optical fibre at (*a*): 0.5 μm; (*b*): 0.9 μm and respective electronic fringes displays (*c*) and (*d*): corresponding fringes in both displays are labeled. (After [130]).

photographs of the oscilloscope display. The precision in the determination of h is estimated to be 1/100 of fringe. If $\Lambda(\lambda)$ is the spacing between fringes in the uniform cladding and $h(\lambda)$ is the displacement of a fringe at a point in the core, then

$$\Delta n(\lambda) = \frac{\lambda}{t} \frac{h(\lambda)}{\Lambda(\lambda)} . \tag{64}$$

Polynomial representation of the measured $\Delta n(\lambda)$ can be obtained by least squares fit, and hence profile dispersion P and material dispersion can be derived (provided values $n(\lambda)$ for the cladding material are known).

The information provided by such a measurement is extremely valuable as, according to [131], the optimum α value is dependent on λ through

$$\alpha(\lambda) = 2 - 2P \tag{65}$$

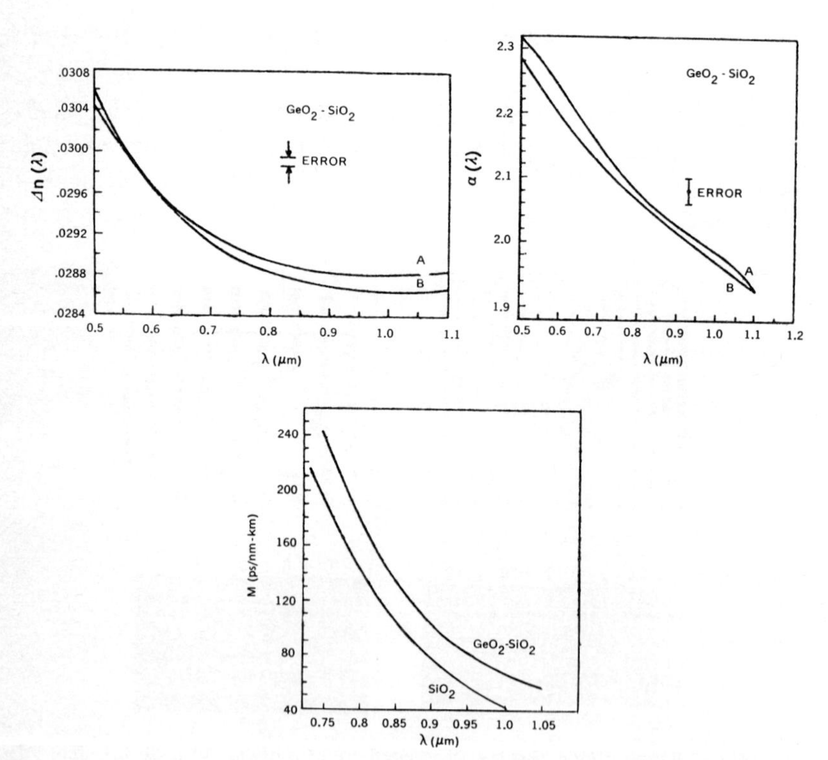

FIG. 110 – Results of measures of $\Delta n(\lambda)$, $\alpha(\lambda)$ and $M(\lambda)$ for GeO_2-SiO_2 fibres. (After [130]).

having ignored a smaller term which is beyond the sensitivity of usual measurements.

Figure 110 shows an example of Δn, $\alpha(\lambda)$, $M(\lambda)$; $M(\lambda)$ is the material dispersion as defined by (40) measured by the described technique for a GeO_2-SiO_2 fibre.

A much simpler technique is described in [132]. In this approach the numerical aperture of the fibre is directly measured as a function of λ. The principle of the method is shown in Fig. 111.

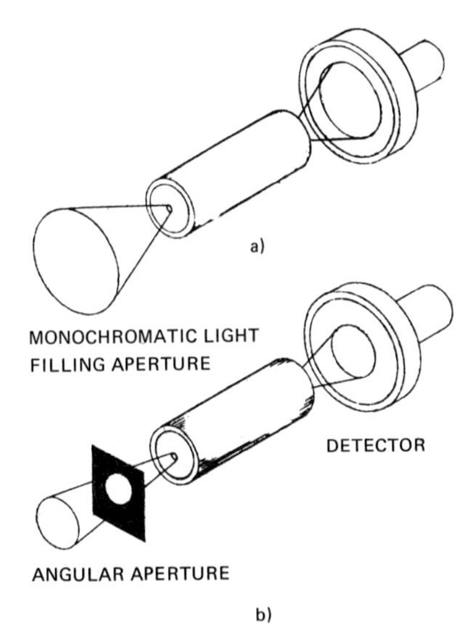

FIG. 111 – Schematic representation of experimental arrangement for measuring variation of NA as a function of wavelength. (After [132]).

A light beam provided by a lamp followed by a monochromator is focused on a small spot on the fibre core centre, so that leaky modes which have strong wavelength dependent loss are not excited. A first measurement is performed with the acceptance angle completely filled by the source. In this case the transmitted power depends on the numerical aperture. A second measurement is performed with a beam angular width smaller by a factor ~ 0.7 than the fibre acceptance angle; in this case no dependence on numerical aperture occurs. Normalizing the first output to the second output the different emission of the source at various wavelengths and the effect of fibre attenuation are compensated.

The spectral range from 400 to 1900 nm can be easily covered, and the parameter P can, according to (62), be deduced from the obtained $\Delta_1(\lambda)$, provided refractive index data of either core or cladding are known. Results of this method agree fairly well with those obtained by the preceding one. An example of measurement on borosilicate and fluorine doped step fibres is shown in Fig. 112.

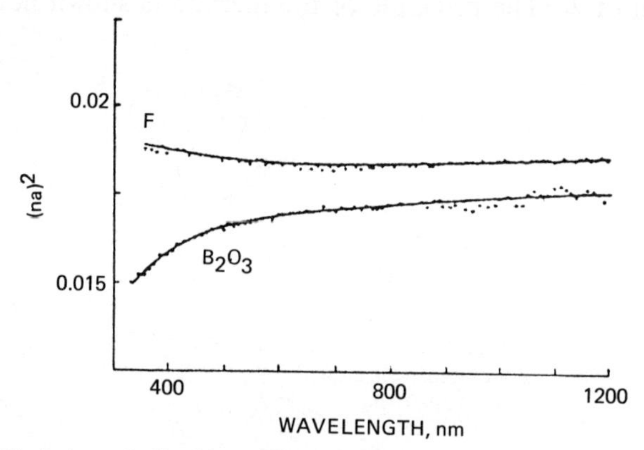

FIG. 112 – Variation of Δ^2 with wavelength for two cladding materials indicated. (After [132]).

3.8 Numerical aperture

The numerical aperture (NA) of a step index fibre has been already defined as $\sqrt{n_0^2-n_1^2}$, and corresponds to the maximum acceptance angle. In the case of graded index fibres only a local NA, $A(r)$, can be defined (see Section 2.1.2), as the refractive index in the core changes along the radius. In this case we have

$$A(r) \equiv \sin \theta_M(r) = \sqrt{n^2(r) - n^2(a)}\ . \tag{66}$$

The NA is an important parameter, as it affects properties such as light gathering efficiency, pulse distortion, microbending loss and curvature loss. Moreover, it enters into the determination of the characteristic parameter V. An obvious way by which it can be determined is to utilize refractive index profile data from those methods that provide absolute values of refractive index (e.g. interferometry, reflectometry).

If such data is not available much simpler techniques may be employed. For step index fibres a far field measurement (see Section 3.9) with an input beam overfilling the fibre NA gives the angular width of the accepted light beam. As a complementary technique an input plane wave may be used. The fibre end is then rotated with respect to the beam axis. Monitoring the output power as a function of the angle permits θ_M to be obtained as the angle at which the power drops to a predetermined amount.

It may be remarked that if this measurement is performed using a very short piece of fibre (~ 1 m), what is obtained is a θ_M as previously defined. If, on the contrary, a long length of fibre is used a θ_M is determined that is in general smaller than the nominal one; this is due to selective attenuation of high order modes (corresponding to high θ values) and mode coupling. The quantity so measured is often referred to as effective numerical aperture. It may be a useful parameter in determining light gathering capability and bandwidth of long fibres, as well as in the achievement of steady state propagation conditions.

As far as graded-index fibres are concerned, the far field method only gives information that is somewhat averaged over the whole range of local NA's; it is, therefore, not a very convenient method if accurate or detailed knowledge is required.

For step index fibres another simple method may be employed [133]. The fibre is put under a microscope on a glass slide and a small drop of index matching liquid is placed on the fibre. A cover glass is placed over the droplet and pressed downward. If the refractive index of the liquid is not matched to that of cladding n, two boundary surfaces appear. The liquid-cladding boundary can be made to disappear by adjusting the fluid index (e.g. by temperature tuning) until it is equal to that of cladding. At this point only a cladding-core boundary in seen; this boundary may be eliminated by further increasing the index of the liquid. This occurs for a refractive index value n_2 of the liquid. Once n_1 and n_2 are known the core index may be determined from the following equation

$$n_0 = \left(\frac{n_1}{r}\right)[\sqrt{1-r^2}-(1-r)] + \frac{n_2}{r}\cdot\frac{1-\sqrt{1-r^2}}{(1-r^2)\sqrt{1-n^2r^2+n^2r^2}} \qquad (67)$$

where $n = n_1/n_2$, $r = r_1/r_2$, $r_1 =$ core radius, $r_2 =$ cladding radius.

An elegant method permitting the measurement of N.A., both for step and graded index fibres of α profile, is an exploitation of interferometry normal to the fibre axis (see Section 3.7.2) [134]. If the refractive index of the oil, n_{oil}, is different from that of the cladding by an amount $\delta n = n_{oil} - n(a)$, a term $\delta n \sqrt{b^2 - r_p^2}$ must be added to equation (49) with the sign changed, b being the cladding radius. If we further choose the particular path through the centre of the fibre, the new equation becomes much simpler:

$$\frac{N_p \lambda}{4} = \delta n\, b - \frac{\alpha}{\alpha + 1}\, a\, \Delta n \,. \tag{68}$$

Δn can easily be calculated from this. The particular expression involving α renders the equation relatively insensitive to variations in α, so that a precise value of this parameter does not have to be known. Experimentally, there are two ways of obtaining Δn: one is to adjust the surrounding oil refractive index until $n_{oil} = n_{cl}$, in which case $\delta n = 0$. The resulting interference pattern is represented in Fig. 113*a*. A measurement of fringe displacement yields

$$\Delta n = \frac{N_p \lambda}{4} \frac{\alpha + 1}{\alpha}\, a \,. \tag{69}$$

The other way is to adjust the oil index until the optical path difference with respect to the fibre core is zero: the situation is like that depicted in Fig. 113*b*. In this case $N_p = 0$ and

$$\Delta n = \delta n \frac{b}{a} \frac{\alpha + 1}{\alpha} \,. \tag{70}$$

Change of liquid refractive index may be accomplished by temperature tuning. If a calibration curve is available, a simple reading of a thermal sensor is enough to give Δn. A possible experimental apparatus utilizing a platinum resistance for temperature measurement and a Watson interference microscope is shown in Fig. 114. The accuracy is estimated to be $\pm 9\%$.

As regards the determination of the NA of single mode fibres, a normal far field measurement is not suitable, as diffraction effects from the small aperture represented by the core predominate so

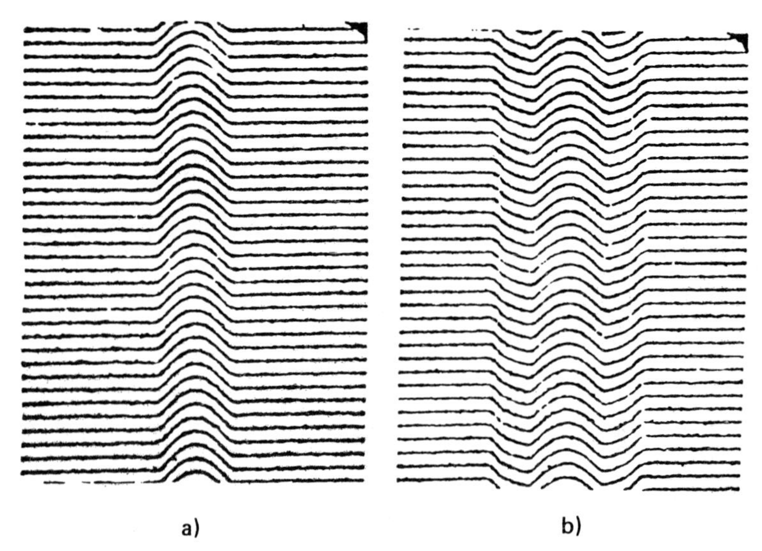

FIG. 113 – Two possible interferograms of a fibre viewed transversely: (*a*) The index of the surrounding fluid equals the index of the fiber cladding. i.e. $\delta n = 0$. (*b*) The optical path through the center of the fibre equals the optical path through the surrounding fluid. i.e. $N = 0$. (After [134]).

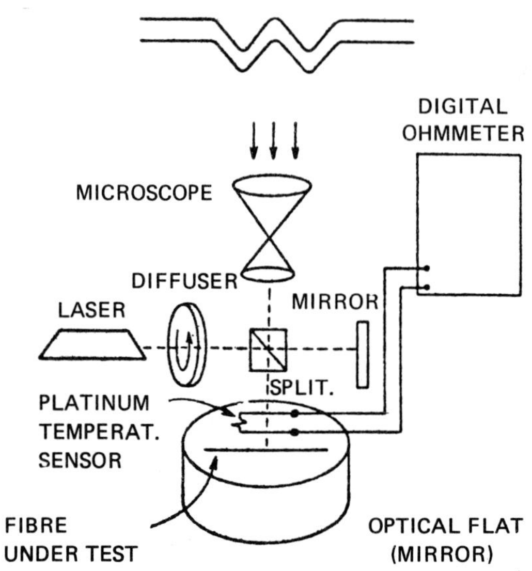

FIG. 114 – Diagram of the apparatus used to measure Δn. The beam splitter and vertical mirror are integral parts of a Watson interference microscope objective. (After [134]).

that simple geometrical optics is not applicable. For the same reason a precise measurement of core diameter is not easy to make.

In [135] a simple and accurate technique enabling the determination of V ($V = (2\pi a/\lambda)\sqrt{n_0^2 - n_1^2}$) and a, and therefore N.A., has been proposed.

The far field of the HE_{11} mode consists of a central lobe with side lobes. It may be shown that, if θ_h is the output angle at which the far field intensity has fallen to one half of that at the central maximum ($\theta = 0$), an unambiguous relationship exists between V and the quantity $\alpha_h = ka \sin \theta_h$ ($k = 2\pi/\lambda$). Such a relationship is represented in Fig. 115. In addition, it may be shown that subsidiary peaks occur at angles and relative intensities which depend on a, n, λ. In particular, the angular width θ_x to the first minimum together with θ_h provides the value of V without any knowledge of a. The ratio $\sin \theta_x/\sin \theta_h$ is also an unambiguous function of V, which is shown in Fig. 115. A simple measurement

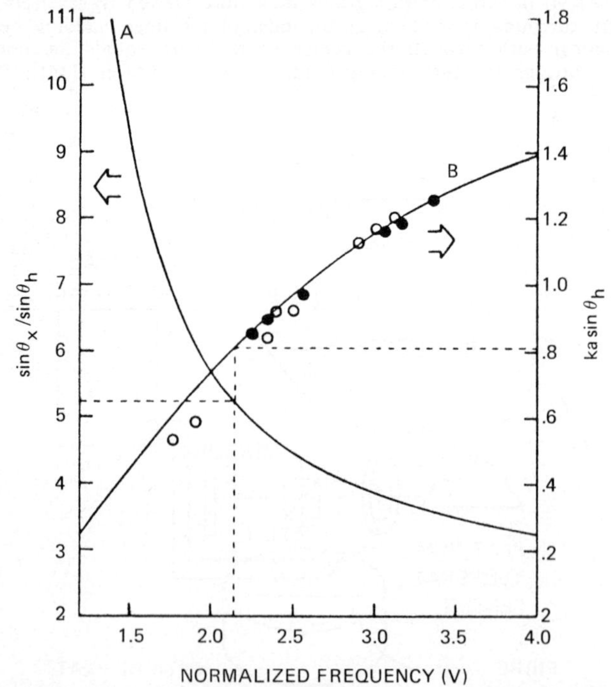

FIG. 115 – Variation of normalised half-intensity angle α_h and the ratio ($\sin \theta_x/\theta_h$) with V. The solid lines are theoretical ones while the points were measured with fibres of core diameter 6.6 μm (o) and 8.1 μm (●) over the wavelength range 0.42 μm to 0.9 μm. (After [135]).

of θ_x and θ_h will provide the V value and a, and therefore the NA. Excellent agreement was found between this technique and more conventional methods.

3.9 Mode power distribution

It is important to know the power distribution among the various modes in an optical fibre, as it provides information on guide structure and mode-carrying characteristics, on source-fibre coupling and, when measured as a function of distance, on mode coupling, coupling length, achievement of steady state distribution, and differential mode attenuation.

The techniques used for mode distribution measurement are essentially near field and far field measurement.

A near field measurement consists in scanning the fibre output end cross-section in order to obtain the power distribution as a function of position. As each mode in a fibre has a characteristic field and, consequently, an intensity distribution in the fibre cross-section (see Section 2.7.2), the aforementioned measurement clearly gives an indication about the kind of modes excited and their relative contribution.

As it is difficult to perform measurements directly in the vicinity of the fibre surface, an enlarged image of the fibre cross-section is obtained through appropriate optics and successively scanned. A schematic of an experimental apparatus is shown in Fig. 87.

The magnifying optics must be carefully designed so that differential magnification of the various regions of the object is minimized and all the light is collected. A small area photodetector (usually silicon or germanium photodiodes), is moved across the image in the desired positions. Motor driven translation stages may be used to avoid eccessive manual effort, and automatic data acquisition by means of punched paper tape, magnetic tape or computer can help in handling large quantities of data.

An alternative approach, generally followed when a complicated near-field pattern is present (as in the case of just one or a few modes propagating in the fibre), consists in replacing the detector by a photographic plate, by which permanent recording of the whole fibre cross-section can be quickly obtained. An example of such a technique is shown in Fig. 116, where the near field pattern corresponding to the $HE_{47,7}$ mode, selectively excited in a multimode fibre, is represented photographically [136]. The mode consists of seven circular rows of 94 bright spots, as predicted by theory (see Section 2.7.3). For a quantitative analysis microdensitometric scanning can be performed.

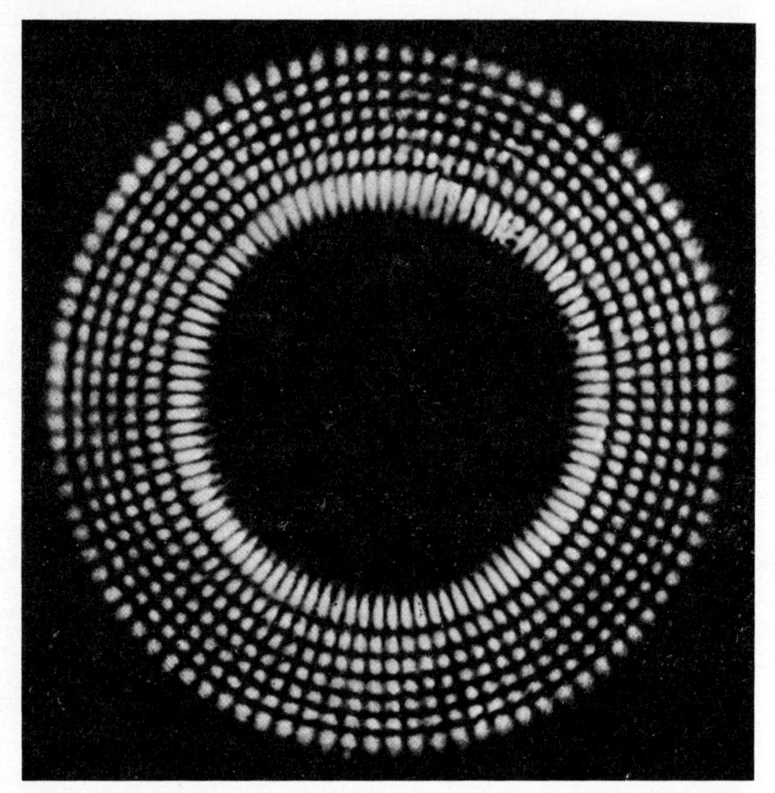

FIG. 116 – Microphotograph of the fibre end face (85 μm diameter) showing the near-field pattern of the $HE_{47,7}$ mode. (After [136]).

At the output end of a fibre, each mode is irradiated with a characteristic angular distribution of its power content. There is thus a relationship between the far-field pattern of a fibre and the mode distribution inside it (see Section 2.3.3). This forms the basis of far-field measurements for characterization of fibres and evaluation of modal distribution. It consists in measuring the power distribution as a function of the angle with respect to the fibre axis (θ) in a meridional plane. Circular symmetry is assumed; otherwise, scanning over a set of meridional planes at different orientations, or photographic recording are required. A possible experimental arrangement is shown in Fig. 117. A small area photodetector moves along a circle with the centre on the fibre

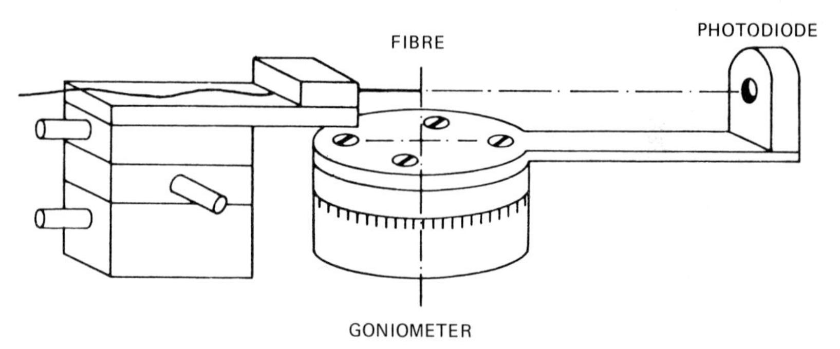

FIG. 117 – An experimental set-up for far field pattern measurement.

end face (typical circle radius ~ 10 cm) and the power falling on the detector area is recorded as a function of the angle. Automatization of the apparatus allows easier data collection.

An alternative approach consists in placing the fibre in the focus of a lens and scanning the other focal plane: fundamental lens properties in the rear lens focal plane produce the Fourier transform (in the variable r, representing the radial distance from the optic axis) of the power distribution as a function of θ. The far-field pattern as a function of θ [20] can be obtained by measuring the power distribution at this focal plane and then Fourier transforming it. Such an approach allows the use of optical data processing techniques: by placing suitable spatial filters in the rear focal plane groups of modes can be selected, or power averaging over the azimuthal angle φ can be performed using annular apertures.

As already discussed in Section 3.3.6, a simple far-field measurement is sufficient to characterize mode distribution in multimode step index fibres, as each mode is identified by a single parameter, e.g. θ. For graded index fibres this is not sufficient, as at a given angle θ contributions arrive from different areas of the core which, according to (26) correspond to different modes. A near-field and far-field measurement are required to characterize mode power distribution in this case, as discussed in Section 3.3.6.

The measurement of mode distribution can give very general information. Correspondingly, the experimental conditions may widely differ. A possible application consists in examining the kind of mode excitation produced by a certain source to be employed, in a telecommunication system, for instance. In this case the source-fibre configuration is determined by the particular situation to be studied.

A different field of application is related to the study of fibre properties. In this case a proper source and coupling conditions must be selected. Light sources for both near and far-field measurements can be standard tungsten-halogen lamps, arc lamps, or LED's. Coherent sources may also be employed, but strong fluctuations arise due to spatial speckle noise generated by interference of coherent beams. This can be avoided, either by averaging techniques (such as the above mentioned azimuthal one, or by time averaging), or by destroying laser coherence by means of moving diffusers.

Launching conditions must be tailored to the particular experiment to be carried out. If all modes with equal power are to be excited (as in the case of NA or refractive index profile measurement) incoherent sources are used, with size and angular radiation distribution such as to completely fill the fibre core and acceptance angle. If only a particular set of modes is to be investigated the techniques described in Section 3.5.3 can be employed; for exciting a pure single mode much more sophisticated techniques, making use of coupling through evanescent waves [136, 137], must be employed.

The analysis of mode distribution is particularly important in the determination of the mode-coupling properties of optical fibres. The steady state length may be determined as that fibre length for which the mode distribution (far-field pattern for step index fibres), remains unchanged, irrespective of variations in the input launching conditions. In turn, the measurement

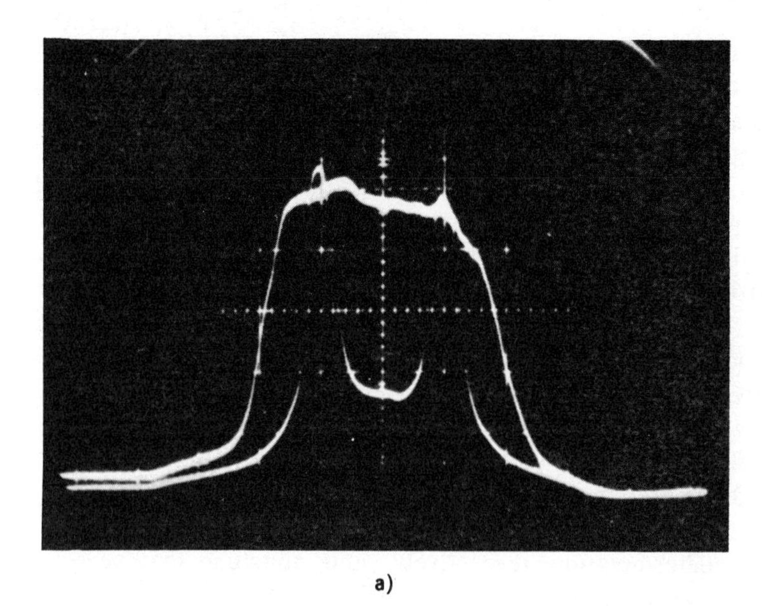

a)

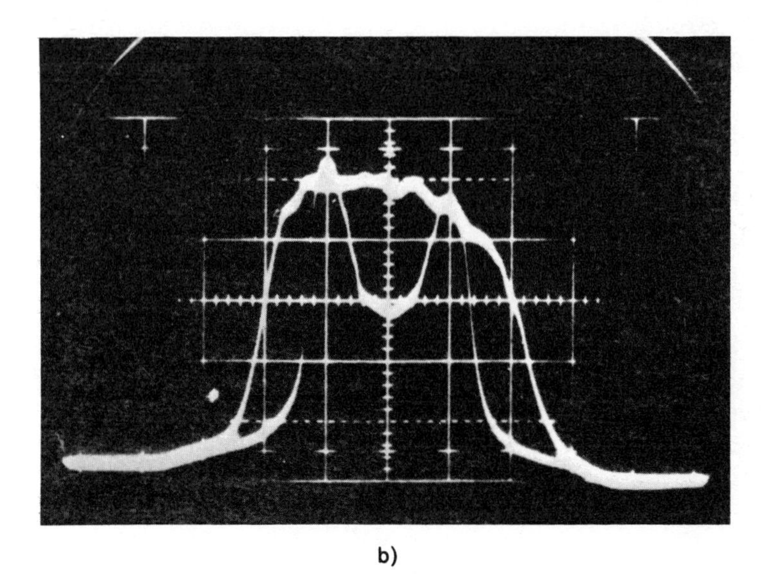

b)

FIG. 118 – Angular intensity distribution at the output of an 82 m length of liquid-core fibre for different transverse pressures. The launching angle of incidence is 6° in air. The double-peaked curve in (*a*) is for a relatively unstressed fibre and that in (*b*) is for a moderate pressure. Steadily increasing pressures result in the flat topped curves in (*a*) and (*b*), respectively. After [139]).

of the steady state radiation pattern allows coupling coefficients to be evaluated [138]. The analysis of radiation pattern changes as a function of the angle of incidence of a plane wave at the fibre input has been suggested for mode conversion coefficient determination [139]. In Fig. 118 the far-field for a 6° launch angle is shown for an unstressed fibre (annular pattern, low mode mixing) and for the same fibre subjected to pressure (flat-topped curve, strong mode mixing).

The hypothesis that coupling occurs only between adjacent modes was experimentally confirmed. It was observed that when low order modes (narrow angle light cone) were launched the angular width of far-field pattern increased slowly and steadily with fibre length, indicating a gradual excitation of higher and higher order modes [140]. At this point we may recall that near-field and far-field measurements are used for index profile and NA determination respectively, and conclude that such techniques, although simple to implement, are very valuable tools for fibre characterization.

3.10 Coupling

This section deals with the problem of optical power coupling between the optical components to be employed in communications systems. Most important are the problems of source-fibre coupling and fibre-fibre coupling.

3.10.1 *Source-fibre coupling*

An important characterization parameter for an optical fibre is the coupling efficiency η, i.e., the amount of optical power gathered from a given source normalized with respect to the total power emitted by the source. It can be defined as:

$$\eta = \frac{P_t}{P_s} \tag{71}$$

(P_t = power accepted by the fibre, P_s = power emitted by the source) and depends on both fibre and source properties. The theoretical problem of light coupling into a fibre is discussed in Chapter 2, Section 2. From an experimental point of view, a very simple arrangement is required to perform the measurement. It is shown in Fig. 119. After the source has been placed in front of the fibre in the desired way, the optical power from the output end of the fibre is measured by a suitable calibrated detector, obtaining P_f.

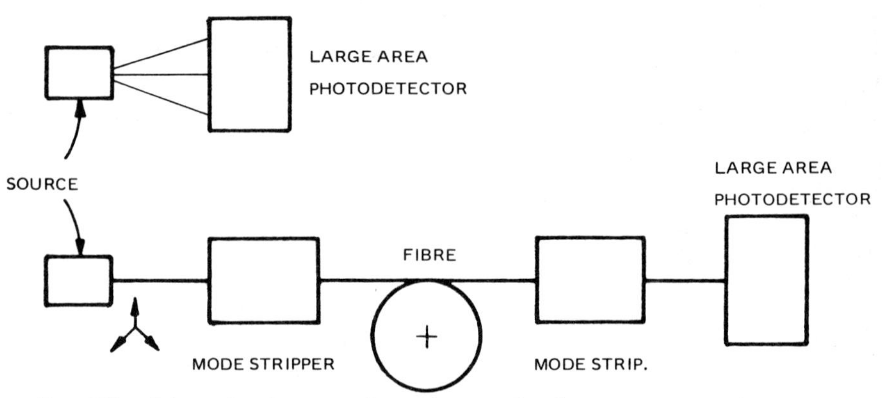

FIG. 119 – Schematic of an experimental set-up for the measurement of coupling efficiency between optical source and fibre.

Total emitted power from the source is then measured with the same detector. Care must be taken to ensure that all the light from the fibre and the source is collected on the detector. This may not be very easy, either because the radiation is emitted in a large solid angle, as in the case of incoherent sources, or because part of the radiation is shielded by the source case or metallic contacts, as often happens for LED's or semi-conductor lasers. The first difficulty may be overcome using integrating spheres.

As far as the interpretation of results is concerned, a problem often encountered in transmission measurements developes (see for instance Section 3.33 or Section 3.7.3): if a short piece of fibre is used, the measured power P_f corresponds to the power coupled into the fibre, assuming negligible fibre attenuation; however, in such a configuration leaky modes and lossy high order modes are included in the measurement, leading to an optimistic figure for η. In order to calculate the power expected after a certain fibre length (needed for system design), the transient fibre loss should be used, rather than the steady state loss.

If, on the other hand, a coupling efficiency value related to steady state conditions is required, the measurement has to be performed using a long section of fibre. The result must be corrected by the known fibre attenuation to obtain the power coupled at the input end.

Often one may be more interested in the relative variation of coupling efficiency as a function of different source-coupling configurations than in the absolute value. It may, for instance, be desirable to measure the improvement of coupling efficiency obtained by means of suitable optical elements such as microlenses, tapered waveguides and the like, in order to compare them and make the best choice. (These topics are discussed in detail in Part III, Section 2).

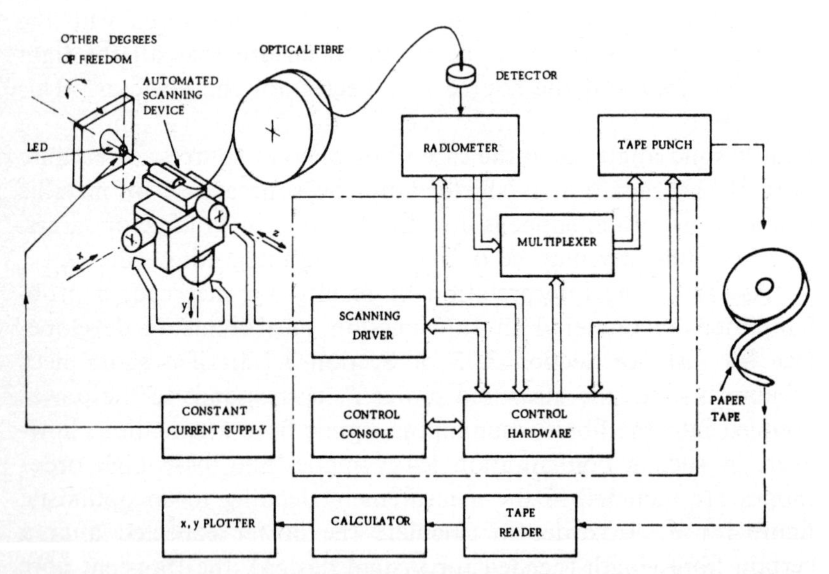

FIG. 120 – Diagram of automated scanning system for source-fibre coupling. (After [141]).

In such cases it is not necessary to measure the total output from the source, because one is mainly concerned with changes in power coupling.

Another point of interest is the change in coupled optical power as a function of the relative position of fibre and source. This data is very valuable for the design of connectors, since it gives an indication of the mechanical tolerances to be respected.

In order to perform this last kind of measurement a precise, calibrated, micromanipulation system is required whereby movements of a known amount may be obtained both in translational and angular coordinates. As a rather large amount of data has to be handled, motor-driven translation and rotation stages are employed, connected with automatic data acquisition system. An example of such an equipment is shown in Fig. 120 [141]. To give an idea of the achievable accuracy of the movement, in Fig. 121 the difference between the predetermined step and the

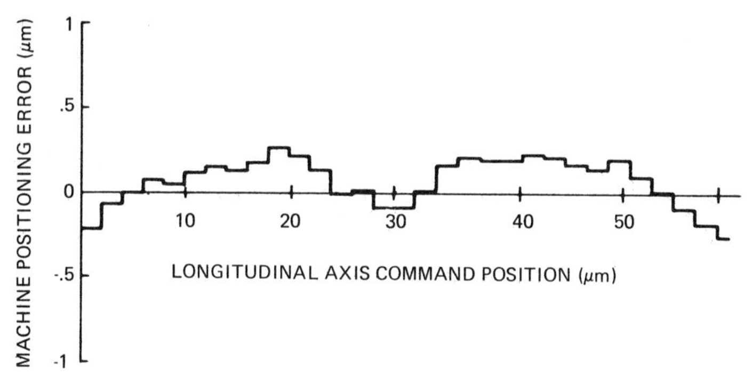

FIG. 121 – Linear positioning error along a longitudinal axis measured with a laser interferometer. (After [143]).

actual displacement (measured with an HP 5526A interferometer) is given for a total displacement ranging from 1 to 50 μm. Less than 0.3 μm difference was found.

The described set-up was used for making LED-fibre coupling measurements (the LED being a Plessey HR 954). As an example of obtained results, a set of curves of equal loss for a 1 km long step index optical fibre, versus lateral displacement v and axial displacement, s, is shown in Fig. 122.

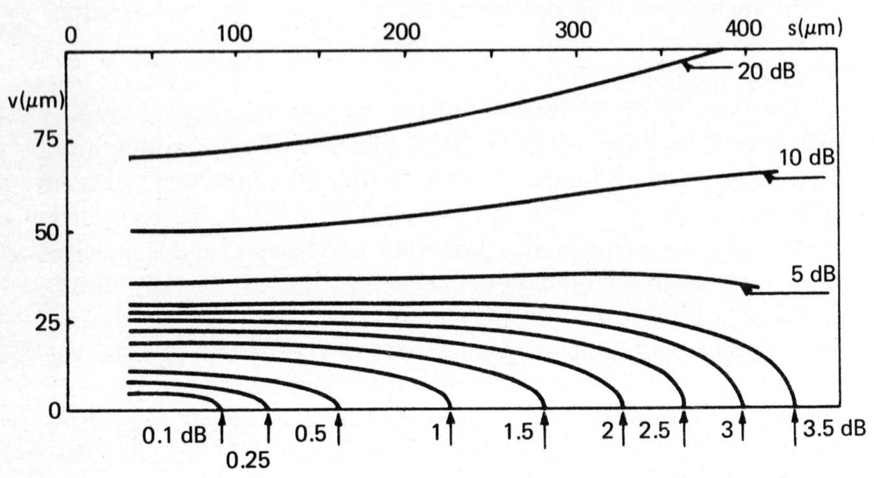

FIG. 122 – Equal loss curves of LED-step index fibre coupling as a function of trans verse and longitudinal displacement. (After [141]).

Very interesting effects may be revealed by such measurements. For instance, in Fig. 123 the amount of power coupled from a GaAs laser into a fibre is shown [142]. The fibre with a plane, mirror-like end face reflected part of the radiation, causing optical feedback into the laser and this, in turn, causes a periodic varia-

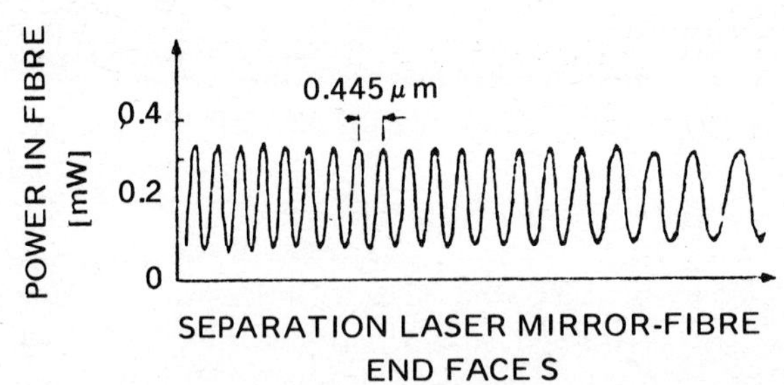

FIG. 123 – Power launched into fibre vs. separation. *s*, between laser mirror and fibre end face. (After [142]).

tion of the coupled power for distances smaller than 50 μm. The measured period was 0.445 μm (the apparent period change was due to the non-linearity of the piezo-electric element used for displacement). This clearly means that proper care must be taken when coupling a fibre to a laser source.

3.10.2 *Fibre-fibre coupling*

The considerations developed in the preceding paragraph for source-fibre coupling measurements apply almost entirely to fibre-fibre coupling, the optical source being replaced by the emitting fibre. The utilization of measurement results is, in this case, related to connectors and splice design and evaluation. Theoretical aspects of the problem are covered in Section 1.2, while the results to be applied to the practical implementation of connectors and splices are discussed in Section 3.2.

The theory and the interpretation of results are simpler for coupling between fibres than for coupling with optical sources, as emission characteristics of fibres can be more easily described. In particular, if the fibres considered are identical a theoretical 100% power transfer efficiency is easily predicted: this is a typical case in which interest is concentrated on losses deriving from deviations from perfect coupling. A test bench very similar to that described in the preceding section is shown in Fig. 124. Two fibres, displaced and tilted by a known amount and seen through a stereo microscope, are shown in Fig. 125. An example

FIG. 124 – Photograph of an automatized set-up for the measurement of coupling losses. (CSELT).

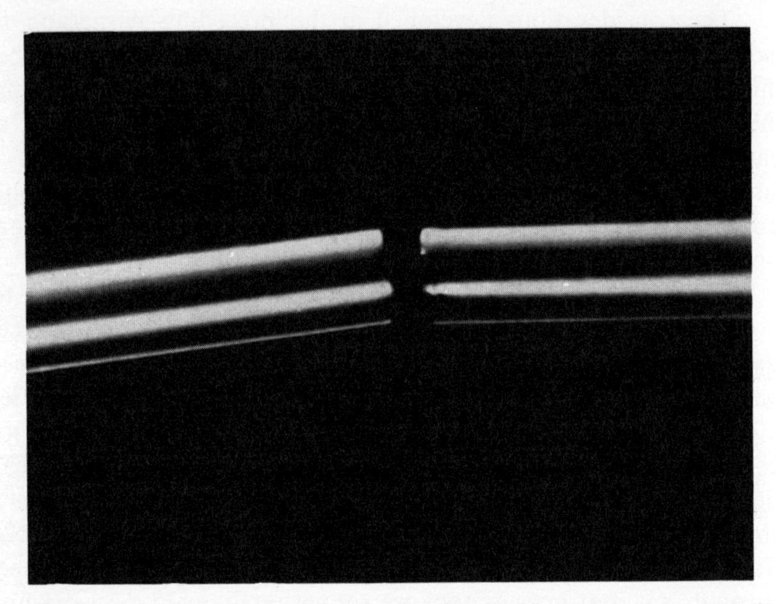

Fig. 125 – Microphotograph of two coupled fibres with linear and angular offsetting. (After [143]).

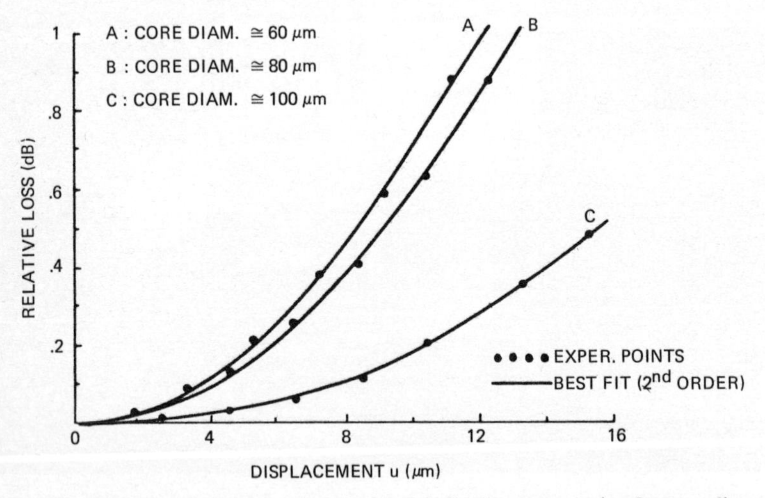

Fig. 126 – Relative power loss versus lateral displacement u in the coupling of graded index fibres with different core diameter. (After [143]).

of lateral displacement losses measured for three graded index fibres of different core diameters is shown in Fig. 126 [143].

The results obtained from these measurements strongly depend on the launching conditions and on fibre length and transmission characteristics.

A detailed discussion of this subject is presented in the next Section, which deals with the measurement of splice losses.

3.11 Splice loss

A basic problem in the implementation of an optical fibre link consists in the splicing of different trunks of optical cables. Splicing is in general required to obtain the desired link length. Several kinds of splices and splicing techniques have been developed and are currently being studied. An essential parameter in determining their quality is the induced loss, i.e. the amount of power lost in the splice due to misalignment of the fibres, and mismatching of both geometrical and physical properties.

The experimental determination of actual splice loss involves some difficulty. In principle, it is a simple matter of measuring

$$L(\mathrm{dB}) = 10 \log \frac{P_{out}}{P_{in}} \tag{72}$$

where P_{out} is the power after the splice and P_{in} the power before the splice. In practice, problems such as steady state propagation before and after the splice, excitation of leaky modes, and coupling of light in the cladding may lead to results that are dependent on launching conditions, fibre length and the kind of fibre used for the measurement. Careful experimental conditions must therefore be ensured in order to obtain unambiguous results and realistic figures. Let us consider in detail the various points.

Ideally a splice measurement simulating the effect of the splice on a long transmission path should be made. This means, first of all, that the experimental arrangement must be such as to ensure the establishment of an equilibrium mode structure at the splice, so that mode dependent effects are removed and isolation from the launching conditions is achieved. The effect of different NA input beams on the loss of a splice, measured with short lengths

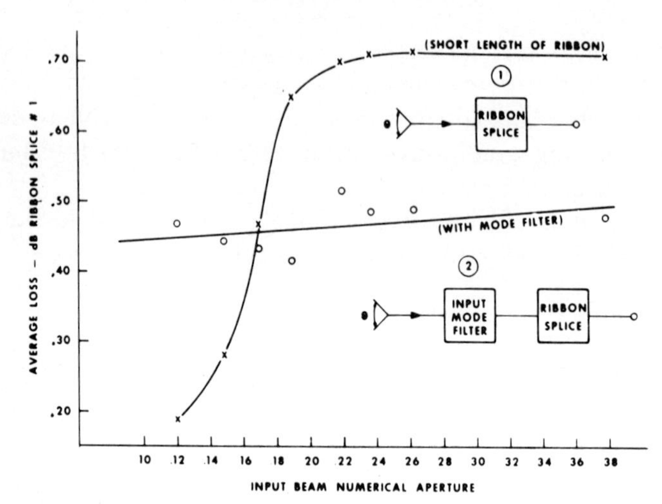

FIG. 127 – Average splice loss with and without input mode filter vs. input beam NA. (After [144]).

of fibre preceding and following the splice itself, is shown in Fig. 127 [144]. The extent of the dependence on input launching conditions when the aforementioned precaution is not taken is clear. Also, after the splice the fibre should be in a condition such that steady state mode distribution is reached before the output end at which the optical power is measured. It has been shown, with both temporal [20] and spatial [144] measurements (see Fig. 128), that even in very good splices a certain amount of mode conversion occurs. This means that the input power distribution before the splice changes after it. If the input were an equilibrium distribution, higher order and leaky lossy modes would be present in the fibre after the splicing. Consequently, the fibre transient loss, which is usually higher than the equilibrium loss, should be used to calculate signal attenuation. As this must be considered an effect of the splice, the resulting loss increase must be assigned to the splice. On the other hand, if a short length of fibre is used after the splice, lossy modes will be present at the output and splice loss will be under-estimated.

Means for obtaining steady state mode distribution in the input fibre have already been discussed in Section 3.3.2: long length of fibre, mode scrambler, and use of input light beam matching the equilibrium light distribution of the fibre. For the output fibre,

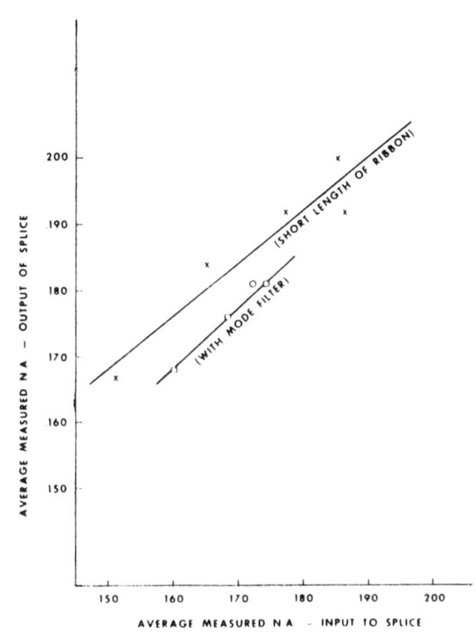

FIG. 128 – Average NA at the output of the splice vs. average NA at the input of the splice. (After [144]).

obviously, only the first two approaches are usable. If necessary, cladding mode strippers should also be employed before and after the splice, to avoid the inclusion of spurious signals.

An experimental arrangement satisfying the above-mentioned requirements is shown in Fig. 129, in which all the possible means

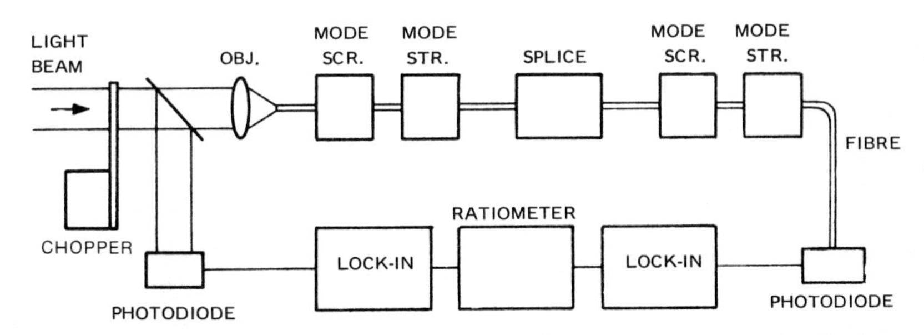

FIG. 129 – Schematic of an experimental set-up for the measurement of splice loss.

of obtaining equilibrium conditions are generically referred to as « mode scrambler ».

The light source can be a laser or an incoherent source (tungsten-halogen lamp). He-Ne lasers are very widely used in splice loss measurements, but tungsten lamps with a properly stabilized power supply have a much better stability ($< 0.1\%$). A reference beam is necessary to eliminate source fluctuations from the measurement. In the apparatus illustrated this is obtained by taking the ratio of the primary signal to the reference signal. An alternative approach is to adjust by means of attenuators the reference beam power level until it equals the signal level. The two signals are sent to a differential lock-in amplifier, and the unbalance is measured as the loss [145].

The measurement procedure must also be carefully established. As optical losses as low as a few 0.01 dB's are to be measured, it is strongly recommended that the input and output ends of the fibre are kept fixed in front of the optical source and the detector respectively. In fact, non-uniformity of response in the photodetector area is, in general, much larger than the desired apparatus sensitivity. For this reason the best procedure for the highest accuracy is to take a fibre, fix its ends, measuring the output signal, and then cut the fibre at the desired length (as emphasized, the length before the detector should be long enough to ensure steady state propagation). Splices are subsequently performed and measured between the two sections so obtained. Any other procedure is very likely to produce much worse results. For instance, a common procedure is to measure the output power from the input fibre, then splice a very short length of fibre and re-measure the output from it, on the assumption that its loss is negligible. On one hand, the fibre has to be re-positioned in front of the detector, which is subject to the described drawback; on the other hand the length of the fibre has to be very short (a couple of metres) for the second assumption to be valid. In this case, however, the arguments against the use of the short fibre after the splice apply. If a long length of fibre is used, the accuracy of the measurement cannot be guaranteed since the fibre section attenuation is not known with sufficient certainty.

The described procedure requires very high stability of the over-all apparatus if a statistical measurement over a large number of splices, implying a long test time, is to be made. The demand for high stability requires additional precautions. For instance, a certain degree of thermostatization is necessary to avoid small mechanical displacements, excess loss induced on the fibre by expansion of the supporting reel and thermal effects on optical components such as filters and so on.

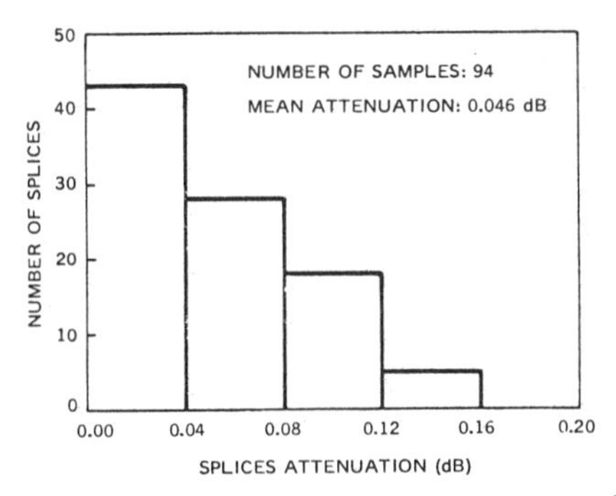

FIG. 130 – Histogram of measured losses for the «springroove®» splice described in ref. [147]. (After [147]).

Accurate control of possible fluctuation sources allows a stability of ± 0.01 dB to be achieved [146], which enables very low loss splices to be tested. An histogram representing results of measurements on such splices [147] is shown in Fig. 130.

3.12 Fibre size and geometrical characteristics

The geometrical characteristics of optical fibres that are worth measuring are, assuming circular cross-section (which is by far the most usual in telecommunications fibres): core diameter, cladding diameter, core-cladding concentricity, and departures from circularities. Core-diameter enters into the characteristic fibre parameter V, and therefore determines properties such as number of modes carried in the waveguide, power confinement, and waveguide dispersion. The cladding-core diameter ratio has a critical effect on microbending loss. The other parameters mentioned may affect splice losses. In addition, good fibre diameter constancy is required to keep the attenuation low. Accordingly, measurement techniques may widely differ to meet different requirements.

3.12.1 *Outer diameter measurement*

To obtain low loss splices it is important to use fibres with the same core and cladding diameter, the latter in particular because most splicing techniques are butt splices and use the fibre outer surface for alignment. In addition, as already said, diameter changes cause additional loss in optical waveguides. A very strict control of fibre diameter is therefore required. The best moment to perform a diameter measurement is at the drawing stage, when unwanted changes may be corrected by a proper feedback action on the drawing apparatus. This problem is dealt with in Section 1.4, where the main measuring methods and equipment are discussed. Here we recall the main characteristics of such methods. They must be non-contacting and high-speed, because the fibre should not be perturbed and the measurement should refer essentially to a specific point, which requires a very high measurement speed in comparison with the drawing speed. In addition, a high accuracy (typically less than 1 μm) is required. Methods satisfying such demands are optical methods, which can be divided into two broad categories. The first consists in casting a shadow of the fibre, either statically or dynamically (e.g. by a moving thin light pencil) over a detector: diameter changes can be determined, and a proper calibration of the apparatus also allows the absolute value of the diameter to be measured. Typical of these methods is the need for very precise fibre positioning to prevent large lateral movements.

The second technique is based on the analysis of the forward or backward scattering pattern produced by the fibre when a plane wave impinges transversely on it. Very good accuracy ($\pm$ 0.25 μm) and a large tolerance on lateral displacement (1 cm) may be attained [148]. More sophisticated data analysis is however necessary; in addition, the measurement is indirect and relies on a theory based on the assumption of circular fibre cross-section. It should be noted that on-line measurements of the kind just described have been implemented for outer diameter control, with the tacit assumption that the core-cladding ratio remains constant.

For off-line measurements much simpler techniques may be employed. A conventional method makes use of a micrometer or a dial gauge, with possible accuracies of $\pm$ 0.5 μm. Another method uses a microscope equipped with a calibrated micrometric eyepiece and suitable magnification. The fibre is placed for visual observation with the cross-section perpendicular to the microscope axis. Care must be taken that the fibre end is flat with well-defined rims. Photographs of the fibre cross-section with a superimposed reference scale allow easier measurement. A clear disadvantage of this last method is that it is destructive; i.e. the fibre must be cut at the point where the measurement is needed.

3.12.2 *Core diameter*

The fibre core corresponds to the inner region of the fibre having a refractive index higher than the surrounding cladding. While this definition corresponds to a well-defined concept for step index fibres, which have a sharp core-cladding boundary, its application may prove difficult in the case of general graded fibres (see for instance Fig. 131 representing a rather common profile

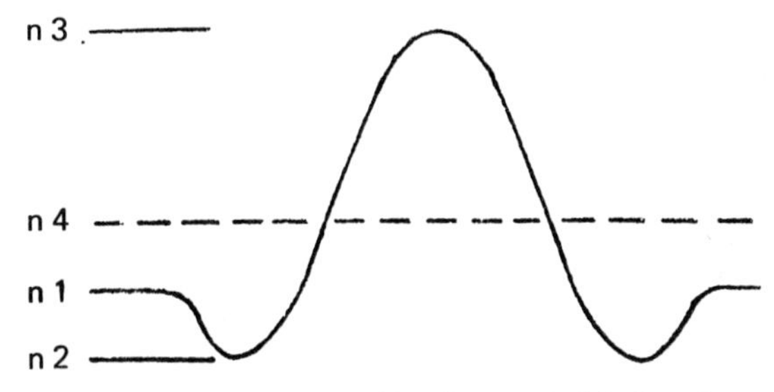

Fig. 131 – Typical refractive index profile of a graded index fibre, showing the difficulty of an unambiguous definition of core size.

in state-of-the-art fibres). Two operative definitions are possible for establishing measurement procedures. The first refers to the refractive index profile of the fibre and assumes the core diameter to be identified by the points where the refractive index falls to a predetermined value. According to this definition all the described refractive index profile measuring methods (interferometry, reflection, near-field, scattering and so on) may be used for core diameter determination. The measurement may also be performed using a microscope, according to the procedure described for cladding diameter measurements, but a certain number of difficulties arise. In fact, of the two possible observation techniques (reflection and transmission), the first is impractical due to the very small difference in reflectivity between the core and cladding material. As regards the second, what one sees is essentially the near field pattern of the fibre. Since the bright region corresponding to the fibre core changes with the illumination system used, the measurement must be undertaken under proper illumination, leading to excitation of all modes, in order to obtain correct results. Even if this condition is fulfilled, a distinction should be made between step and graded index fibres. The former exhibits a definite core-cladding boundary (see e.g. Fig. 4*a*) with a consequent ease of measurement. The latter shows a fading of light distribution toward the cladding, which makes the identification of boundary rather difficult (see Fig. 132, representing a cross-sectional view of a graded index fibre). A photomicrograph can help in performing the measurement, but for a quantitative analysis a densitometric scanning along a diameter should be carried out. This is, however, nothing but a near-field measurement, and therefore a refractive index profile measurement.

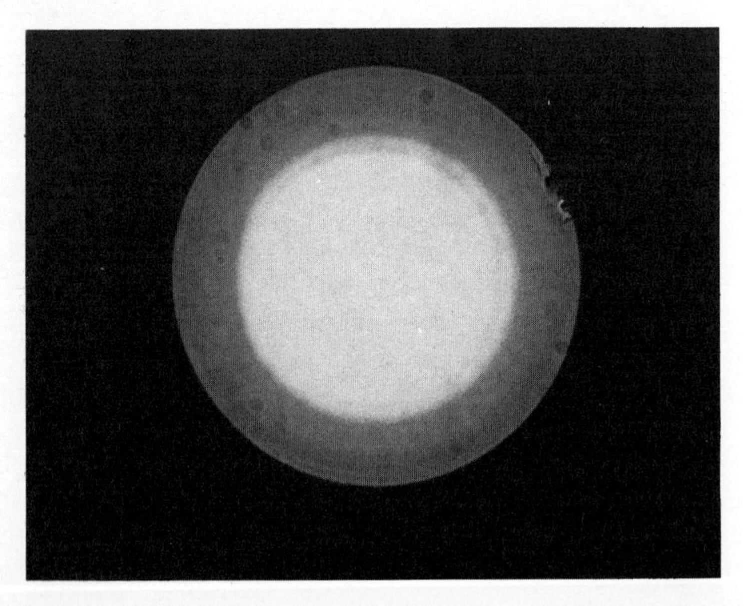

FIG. 132 – Cross sectional view of a graded index fibre with back incoherent illumination.

A different technique for identifying the core region consists in selectively etching the core material according to the procedure discussed in Section 3.7.8. This is particularly useful for monomode fibres because diffraction and waveguide effects render core region identification difficult under transmitted light observation.

The second possible definition concerning the core diameter is related to the propagation characteristics of the fibre, i.e. the core diameter can be considered as corresponding to the diameter of a circular section within which a determinate fraction of the optical power is propagated, e.g. 95 % of the total transmitted power. According to this definition, the measurement can be performed by measuring the light transmitted through a calibrated variable diameter iris placed on an enlarged image of the fibre. A second approach consists in performing a near-field scanning in much the same way as the refractive index profile is measured [149]. Measurements based on this definition are subject to the problem discussed on several occasions, i.e. being a transmission measurement, it is dependent both on fibre length and launching conditions. If performed on a long fibre length it can lead to an « effective core diameter » determination, of the same conceptual relevance as « effective NA » (see Section 3.8).

All these are destructive methods. Non-destructive methods, based again on the analysis of the scattering pattern produced on a normally impinging incoherent plane wave, may also be employed; in fact, cladding and core diameters, as well as refractive index difference can be derived by the scattering pattern, at least in principle. In practice the method can only be applied for step index fibres [150, 151], while for general graded index fibres data extraction is too difficult, unless some of the parameters are known from independent measurements [152].

3.12.3 *Concentricity and ellipticity*

If the outer surface of a fibre is used as a reference for fibre alignement in a splice, it is important that the core be concentric with respect to this surface. The concentricity error, which may be defined as the distance between core and outer surface centres, may be measured by extending the refractive index profile measurement to the rims of the fibre.

A more suitable approach may be to perform the measurement on a photograph taken at a microscope, e.g. with a set of circles of different diameters.

In all the preceding discussion it has been assumed that fibres are of circular cross-section. Due to fabrication tolerances departures from circularity of both cladding and core may occur. This may be an additional loss factor in a splice. Moreover, in monomode fibres core ellipticity causes the fundamental mode to split into two orthogonally polarized modes, which results in the deterioration of the bandwith capacity of such fibres.

Although the non-circularity can be of a very general nature, it often appears in the form of ellipticity. A measure of the ellipticity may be the ratio of the minor axis b to the major axis a, or some other suitable parameter. Both refractive index profile and microphotograph methods can be employed for performing the measurement. A very accurate, non-destructive method for the determination of cladding ellipticity has also been reported [153], based on the same principle explained in Section 3.6 for material dispersion measurement. If the fibre has circular cross-section the location of fringes at extreme sides of the pattern is symmetric around the centre. In the case of an elliptical fibre the fringes are shifted to the right or left in a manner dependent upon the orientation α_i of the ellipse. Experimental set-up is shown in Fig. 133.

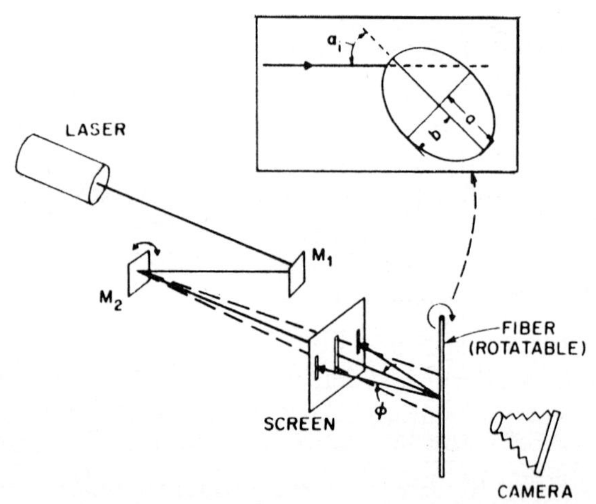

FIG. 133 – Experimental setup to measure fibre ellipticity. (After [153]).

The measurement procedure consists in rotating the fibre by a known amount and observing the corresponding pattern. The fringe appearance at various angles α_i is shown in Fig. 134.

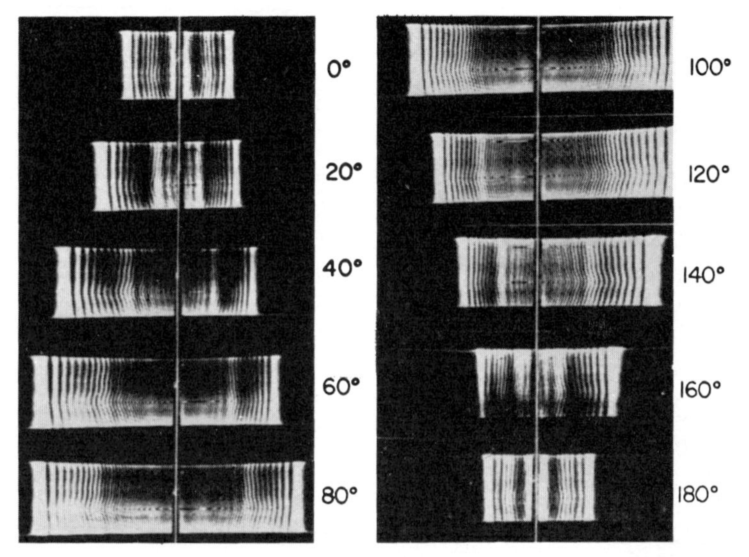

FIG. 134 – Backscattered light patterns at 20° increments from 0° to 180° for an elliptical fibre with $b/a = 0.923$. The white line is drawn at $\varphi = 0°$. (After [153]).

Accuracies of 0.001 in the determination of the b/a value are possible. These are an order of magnitude better than those obtainable by the microscopic method.

3.13 Mechanical properties

Mechanical characteristics of optical fibres, such as Young's modulus, breaking strength, and extensibility, are very relevant for a successful application to telecommunication systems as explained in section 1.1.

From an experimental point of view, testing kilometer lengths of fibre over a period of many years is completely impractical, so that an analytical model is required to extrapolate data obtained on short, numerous samples to long fibre lengths.

To this purpose a Weibull distribution [154-156] is fitted to experimental data in order to extrapolate to longer lengths. The basic concept of Weibull's approach is that the failure of any element in a group of samples must ultimately lead to the failure of the group considered as a whole, or, in other words, the failure of a finite element is a consequence of the failure of the weakest link (as the weakest link of a chain is responsible for the rupture of the total chain). Weibull's distribution defines the breakdown probability of a group of elements where each of them may be seen as a group of smaller elements, or may be collected together in a larger element without changing the distribution law [157].

Weibull's distribution has been applied to optical fibres giving a simple physical interpretation of mathematical parameters. The survival probability for a specimen of length L subjected to a stress smaller than σ is given by [158]

$$s(\sigma, L) = \exp\,[-LN(\sigma)] \tag{73}$$

where $N(\sigma) = \int_0^{\sigma} n(\sigma')\,d\sigma'$ and $n(\sigma')\,d\sigma'$ is the number of flaws per unit length which will fail for a stress between σ' and $\sigma' + d\sigma'$.

According to Weibull, $N(\sigma)$ can be expressed as

$$N(\sigma) = \left(\frac{\sigma}{\sigma_0}\right)^m \tag{74}$$

where m and σ_0 are unknown parameters, obtainable by experimental data from samples of known length, looking for the best fit, or using maximum likelihood estimators [159]. Usually the quantity $\ln\,[(1/L)\ln 1/s(\sigma, L)]$ is plotted against $\ln\sigma$, a linear relationship existing between these two quantities whose slope is given by m; the intercept provides σ_0.

Measurements are normally performed with standard tensile testing machines, with the possibility of measuring breaking stress, elongation, and Young's modulus.

An experimental problem is related to clamping the fibre ends, which may impart localized stresses and alter the failure statistics. This problem has been solved by fastening the fibre to aluminum tabs with epoxy [160]. Another source of induced stress concentration is the non coincidence of the axis of the fibre with the tensile tester pulling axis due, for instance, to eccentricity of gripping jaws, so that care must be taken for fibre alignement.

The test velocity is an important parameter, as the tensile force increases with test velocity [157, 160]. A further problem arises when working with samples of the same gauge length: the majority of samples will fail over a limited range of stress levels and little knowledge of the flaw population outside this range is obtained. A partial solution can be found using several gauge lengths [161].

Even so, it is a delicate matter to extrapolate data to regions of low strength where, of course, a very small amount of data

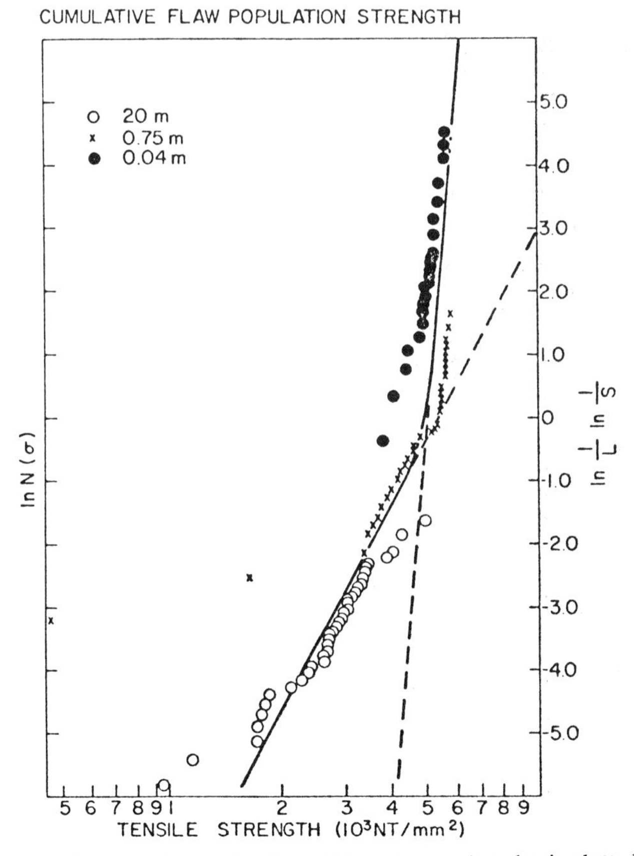

Fig. 135 – Tensile strength data for three different gauge lengths, is plotted to yield a graph of ln $N(\sigma)$ versus ln σ. The fitted lines indicate that the cumulative flaw distributions can be represented as the sum of two Weibull populations. (After [161]).

is available. Extrapolation is valid if only one mode of failure exists.

If the defects that lead to failure in the low strength region are unlike those causing rupture at higher stresses, extrapolation leads to incorrect results. In Fig. 135 an example of measurements [161] of the kind described is shown. The low strength and high strength regions follow two different Weibull distributions.

REFERENCES

[1] T. W. Davies, R. Worthington and K. C. Kao: *The measurement of mode parameters in optical fibre waveguides*, Trunk telecommunications by guided waves, London (September 29-October 2, 1970).

[2] H. H. Witte and R. Khan: *A novel light coupling method for fibers*, Rev. Sci. Inst., Vol. 42, No. 9, pp. 1374-1375.

[3] J. P. Dakin, W. A. Gambling and D. N. Payne: *Launching into glass-fibre waveguide*, Opt. Commun., Vol. 4, No. 5 (1972), pp. 354-357.

[4] D. Gloge, P. W. Smith, D. L. Bisbee and E. L. Chinnock: *Optical fibre end preparation for low loss splices*, Bell Syst. Tech. J., Vol. 52, No. 9 (1973), pp. 1579-1588.

[5] H. Murata, S. Inao, Y. Matsuda and R. Takamashi: *Splicing of optical fiber cable*, First European Conference on Optical Fibre Communication, London (September 16-18, 1975), pp. 93.

[6] P. Hensel: *Simplified optical-fibre breaking machine*, Electron. Lett., Vol. 11, No. 24 (1975), pp. 581-582.

[7] J. E. Fulenwider and M. I. Dakss: *Hand held tool for optical-fibre end preparation*, Electron. Lett., Vol. 13, No. 19 (1977), pp. 578-580.

[8] K. S. Gordon, E. G. Rawson and A. B. Nafarrate: *Fiber-break testing by interferometry: a comparison of two breaking methods*, Appl. Opt., Vol. 16, No. 4 (1977), pp. 818-819.

[9] D. L. Bisbee: *Optical fiber joining technique*, Bell Syst. Tech. J., Vol. 50, No. 10 (1971), pp. 3153-3158.

[10] O. K. Skliarov: *An electric spark method of treating the ends of an optical fiber*, Sov. J. Opt. Technol., Vol. 42, No. 10 (1975), pp. 606-607.

[11] Fr. Caspers and E. G. Neumann: *Optical fibre end preparation by spark erosion,* Electron. Lett., Vol. 12, No. 17 (1976), pp. 443-444.

[12] P. Hensel: *Spark induced fracture of optical fibres,* Electron. Lett., Vol. 13, No. 20 (1977), pp. 603-604.

[13] *Comunicato Siemens AG*, 4. 168i-ZFE, ZI/Presseabteilung Technik (1974).

[14] G. D. Khoe and G. Kuyt: *Cutting optical fibres with a hot wire*, Electron. Lett., Vol. 13, No. 5 (1977), pp. 147-148.

[15] A. Albanese and L. Maggi: *New fiber breaking tool*, Appl. Opt., Vol. 16, No. 10 (1977), pp. 2604-2605.

[16] W. W. Benson and D. R. MacKenzie: *Optical fiber vacuum chuck*, Appl. Opt., Vol. 14, No. 4 (1975), pp. 816-817.

[17] A. Albanese: *Electrostatic fiber holder*, Appl. Opt., Vol. 16, No. 6 (1977), p. 1464.

[18] R. Olshansky, M. G. Blankenship and D. B. Keck: *Length dependent attenuation measurements in graded index fibers*, Second European Conference on Optical Fibre Communication, Paris (September 27-30, 1976), pp. 111-113.

[19] M. Eve, A. M. Hill, D. J. Malyon, J. E. Midwinter, B. P. Nelson, J. R. Stern and J. V. Wright: *Launching independent measurements of multimode fibres*, Second European Conference on Optical Fibre Communication, Paris (September 27-30, 1976), pp. 143-146.

[20] M. Ikeda, Y. Murakami and K. Kitayama: *Mode scrambler for optical fibres*, Appl. Opt., Vol. 16, No. 4 (1977), pp. 1045-1049.

[21] M. Ikeda, A. Sugimura and T. Ikegami: *Multimode optical fibres: steady state mode exciter*, Appl. Opt., Vol. 15, No. 9 (1976), pp. 2116-2120.

[22] S. Seikai, M. Tokuda, K. Yoshida and N. Uchida: *Measurement of baseband frequency response of multimode fibre by using a new type of mode scrambler*, Electron. Lett., Vol. 13, No. 5 (1977), pp. 146-147.

[23] S. Zemon and D. Fellows: *Characterization of the approach to steady state and the steady state properties of multimode optical fibers using LED excitation*, Opt. Commun., Vol. 13, No. 2 (1975), pp. 198-202.

[24] M. J. Adams, D. N. Payne and F. M. E. Sladen: *Mode transit times in near parabolic-index optical fibres,* Electron. Lett., Vol. 11, No. 16 (1975), pp. 389-350.

[25] P. Di Vita and R. Vannucci: *Loss mechanisms of leaky skew rays in optical fibres*, Opt. Quantum Electron., Vol. 9, No. 3 (1977), pp. 177-188.

[26] B. Costa and B. Sordo: *Measurements of the refractive index profile in optical fibres: comparison between different techniques*, Second European Conference on Optical Fibre Communication, Paris, (September 27-30, 1976), pp. 81-86.

[27] K. Petermann: *Uncertainties of the leaky mode correction for near-square law optical fibres,* Electron. Lett., Vol. 13 (1977), No. 17, p. 514.

[28] M. K. Barnoski and S. M. Jensen: *Fiber waveguides: a novel technique for investigating attenuation characteristics*, Appl. Opt., Vol. 15, No. 9 (1976), pp. 2112-2115.

[29] B. Costa and B. Sordo: *Backscattering technique for investigating attenuation characteristics of optical fibres: a new experimental approach,* CSELT Rapporti Tecnici, Vol. V, No. 1 (March 1977), pp. 75-77.

[30] S. D. Personick: *Photon probe - an optical time-domain reflectometer*, Bell Syst. Techn. J., Vol. 50, No. 3 (March 1977), pp. 355-366.

[31] B. Costa and B. Sordo: *Experimental study of optical fibres attenuation by modified backscattering technique*, Third European Conference on Optical Communication, Munich (September 14-16, 1977).

[32] B. Daino and D. Sette: *The measurement of the transmission properties of optical cables*, Eurocon, Venice (May 3-5, 1977).

[33] S. D. Personick: *New result on optical time domain reflectometry*, 1977 International Conference on Integrated Optics and Optical Fiber Communication, Tokyo (July 18-20, 1977), pp. 532-540.

[34] B. Danielson, G. Day and D. Franzen: *Fiber optic metrology at NBS*, Int. Symp. on Meas. in Telecom., Lannion (October 3-7, 1977).

[35] M. Rode and E. Weidel: *Ein Rückstreuverfahren zur Untersuchung von Lichtleitfasern*, NTZ, Vol. 31, No. 2 (1978), pp. 144-146.

[36] A. R. Tynes: *Integrating cube scattering detector*, Appl. Opt., Vol. 9, No. 12 (December 1970), pp. 2706-2710.

[37] D. B. Keck, P. C. Schultz and F. Zimar: *Attenuation of multimode glass waveguides*, Appl. Phys. Lett., Vol. 21, No. 5 (September 1972), pp. 215-217.

[38] F. W. Ostermeyer and W. A. Benson: *Integrating sphere for measuring scattering loss in optical fiber waveguides*, Appl. Opt., Vol. 13, No. 8 (August 1974), pp. 1900-1905.

[39] S. de Vito and B. Sordo: *Misure di attenuazione e diffusione in fibre ottiche multimodo*, LXXV Riunione AEI, Rome (September 15-21, 1974).

[40] B. Hillerich: *Pulse reflection method for transmission loss measurement of optical fibres*, Electron. Lett., Vol. 12, No. 4 (1976), pp. 92-93.

[41] E. G. Rawson: *Measurement of the angular distribution of light scattered from a glass fiber optical waveguide*, Appl. Opt., Vol. 11, No. 11 (1972), pp. 2477-2481.

[42] J. P. Dakin and W. A. Gambling: *Angular distribution of light scattering in bulk glass and fibre waveguides*, Opt. Commun., Vol. 6, No. 3 (1972), pp. 235-238.

[43] M. H. Reeve, M. C. Bierley, J. E. Midwinter and K. I. White: *Studies of radiative losses from multimode optical fibres*, Opt. Quantum Electron., Vol. 8 (1976), pp. 39-42.

[44] K. I. White and J. E. Midwinter: *An improved technique for the measurement of low optical absorption losses in bulk glass*, Opto-Electronics, Vol. 5 (1973), pp. 323-334.

[45] D. A. Pinnow and T. C. Rich: *Development of a calorimetric method for making precision optical absorption measurements*, Appl. Opt., Vol. 12, No. 5 (1973), pp. 984-992.

[46] A. Zaganiaris and G. Bouvy: *L'absorption dans les matériaux pour fibres optiques*, Ann. Télécommun., Vol. 29, No. 5-6 (1974), pp. 189-194.

[47] K. I. White: *A calorimetric method for the measurement of low optical absorption losses in optical communication fibres*, Opt. Quantum Electron., Vol. 8 (1976), pp. 73-75.

[48] R. L. Cohen: *Loss measurements in optical fibers* 1: *Sensitivity limit of bolometric techniques*, Appl. Opt., Vol. 13, No. 11 (1974), pp. 2518-2521.

[49] R. L. Cohen, K. W. West, P. D. Lazay and J. Simpson: *Loss measurements in optical fibers* 2: *Bolometric measuring instrumentation*, Appl. Opt., Vol. 13, No. 11 (1974), pp. 2522-2524.

[50] F. T. Stone, W. B. Gardner and C. R. Lovelace: *Calorimetric measurement of absorption and scattering losses in optical fibers*, Opt. Lett., Vol. 2, No. 2 (1978), pp. 48-50.

[51] A. Zaganiaris: *Simultaneous measurement of absorption and scattering losses in bulk glass and optical fibers by a microcalorimetric method*, Appl. Phys. Lett., Vol. 25, No. 6 (1974), pp. 345-347.

[52] W. A. Gambling, D. N. Payne and H. Matsumura: *Dispersion in low loss liquid-core optical fibres*, Electron. Lett., Vol. 8, No. 23 (1972), pp. 568-569.

[53] D. B. Keck: *Spatial and temporal power transfer measurements on a low-loss optical waveguide*, Appl. Opt., Vol. 13, No. 8 (1974), pp. 1882-1888.

[54] R. Olshansky, S. M. Oaks and D. B. Keck: *Measurement of differential mode attenuation in graded-index fiber optical waveguides*, Optical Fiber Transmission II, Williamsburg (February 22-24, 1977).

[55] D. Gloge and E. A. J. Marcatili: *Multimode theory of graded core fibers*, Bell Syst. Tech. J., Vol. 52 (1973), pp. 1563-1578.

[56] R. Olshansky: *Differential mode attenuation in graded index optical waveguides*, 1977 International Conference on Integrated Optics and Optical Fiber Communication, Tokyo (July 18-20, 1977).

[57] T. Yamada, H. Hashimoto, K. Inada and S. Tanaka: *Launching dependence of transmission losses of graded index optical fiber*, ibid.

[58] Y. Nemo and M. Shimizu: *Optical fiber fault location method*, Appl. Opt., Vol. 13, No. 6 (1976), pp. 1385-1388.

[59] J. Guttmann and O. Krumpholz: *Location of imperfections in optical glass-fibre waveguides*, Electron. Lett., Vol. 11, No. 10 (1975), pp. 216-217.

[60] S. D. Personick: *Baseband linearity and equalization in fiber optic digital communication systems*, Bell Syst. Tech. J., Vol. 52 (September 1973), pp. 1175-1194.

[61] L. G. Cohen, H. W. Astle and I. P. Kaminow: *Wavelength dependence of frequency-response measurements in multimode optical fibers*, Bell Syst. Tech. J., Vol. 55, No. 10 (1976), pp. 1509-1523.

[62] D. Gloge, E. L. Chinnock and T. P. Lee: *Self pulsing* GaAs *laser for fiber dispersion measurement,* IEEE J. Quantum Electron. (November 1972), pp. 844-846.

[63] D. Gloge, E. L. Chinnock, R. D. Standley and W. S. Holden: *Dispersion in a low loss multimode fibre measured at three wavelengths,* Electron. Lett., Vol. 8, No. 21 (1972), pp. 527-529.

[64] D. Gloge, A. R. Tynes, M. A. Dupuay and J. W. Hansen: *Picosecond pulse distortion in optical fibres*, IEEE J. Quantum Electron., Vol. QE-8, No. 2 (1972), pp. 217-221.

[65] H. R. D. Sunak and W. A. Gambling: *Picosecond pulse dispersion in cladded glass fibre,* Opt. Commun., Vol. 11, No. 3 (1974), pp. 277-281.

[66] L. G. Cohen and Chinlon Lin: *Pulse delay measurements in the zero material dispersion wavelength region for optical fibers*, Appl. Opt., Vol. 16, No. 12 (1977), pp. 3136-3139.

[67] Y. G. Goyal, A. Kumar and A. K. Ghatak: *Calculation of bandwidth of optical fibres from experiment on dispersion measurement*, Opt. Quantum Electron., Vol. 8 (1976), pp. 80-82.

[68] S. D. Personick: *Time dispersion in dielectric waveguides*, Bell Syst. Tech. J., Vol. 50 (1971), pp. 843-854.

[69] S. D. Personick, W. M. Hubbard and W. S. Holden: *Measurements of the baseband frequency response of a* 1 km *fiber*, Appl. Opt., Vol. 13, No. 2 (1974), pp. 266-268.

[70] R. Auffret, C. Boisrobert and A. Cozannet: *Vobulation technique applied to optical fibre transfer function measurement*, First European Conference on Optical Fibre Communication, London (September 16-18, 1975), pp. 60-61.

[71] I. Kobayashi and M. Koyama: *Measurement of optical fiber transfer functions based upon the swept-frequency technique for baseband signals,* IECE Trans. Japan, Vol. E-59, No. 4 (1976), pp. 11-12.

[72] G. Rosman: *Variation of pulse delay with launch angle in a liquid-filled fibre*, Electron. Lett., Vol. 8, No. 18 (1972), pp. 455-456.

[73] J. P. Hazan, L. Jacomme and D. Rossier: *Temporal behaviour of a localized index-gradient fibre using variable-angle injection*, Opt. Commun., Vol. 14, No. 3 (1975), pp. 368-373.

[74] D. B. Keck and R. D. Maurer: *Optical pulse broadening in long fibre waveguides*, Optics and Laser Techn. (October 1975), pp. 229-233.

[75] B. Costa and B. Sordo: *Experimental study of pulse distortion in some optical fibres*, XXIII Congresso Internazionale per l'Elettronica, Rome (March 22-24, 1976).

[76] J. P. Hazan and L. Jacomme: *Characterizing optical fibres; a test bench for pulse dispersion,* Philips Tech. Rev., Vol. 36, No. 7 (1976), pp. 211-216.

[77] L. G. Cohen: *Pulse transmission measurements for determining near optimal profile gradings in multimode borosilicate optical fibers*, Appl. Opt., Vol. 5, No. 7 (1976), pp. 1808-1814.

[78] B. Costa, F. Esposto and B. Sordo: *Wavelength dependence of differential mode delay in optical fibres: application to fibre links characterization*, Optical Fiber Communication, Washington D.C. (March 6-8, 1979).

[79] Masahiro Ikeda and Harno Yoshikiyo: *Pulse separating in transmission characteristics of multimode graded index optical fibres*, Appl. Opt., Vol. 15, No. 5 (1976), pp. 1307-1312.

[80] B. Sordo, F. Esposto and B. Costa: *Experimental study of modal and material dispersion in spliced optical fibres*, Fourth European Conference on Optical Communication, Genoa (September 12-15, 1978), pp. 71-79.

[81] E. L. Chinnock, L. G. Cohen, W. S. Holden, R. D. Standley and D. B. Keck: *The length dependence of pulse spreading in the* CGW-Bell-10 *optical fibre*, Proc. IEEE (October 1973), pp. 1499-1550.

[82] L. G. Cohen: *Shuttle pulse measurements of pulse spreading in an optical fiber*, Appl. Opt., Vol. 14, No. 6 (1975), pp. 1351-1356.

[83] L. G. Cohen and S. D. Personick: *Length dependence of pulse dispersion in a long multimode optical fiber*, Appl. Opt., Vol. 14, No. 6 (1975), pp. 1357-1360.

[84] T. Tanifuji and Masahiro Ikeda: *Pulse circulation measurement of transmission characteristics in long optical fibers*, Appl. Opt., Vol. 16, No. 8 (1977), pp. 2175-2179.

[85] D. Gloge: *Dispersion in weakly guiding fibers*, Appl. Opt., Vol. 10, (1971), pp. 2442-2445.

[86] B. Costa and P. Di Vita: *Group velocity of modes and pulse distortion in dielectric optical waveguides*, Opto-Electronics, Vol. 5 (1973), pp. 439-456.

[87] S. Kobayashi, S. Shibata, N. Shibata and T. Izawa: *Refractive index dispersion of doped fused silica*, 1977 International Conference on Integrated Optics and Optical Fiber Communication, Tokyo (July 18-20, 1977).

[88] H. M. Presby: *Variation of refractive index with wavelength in fused silica optical fibres and preforms,* Appl. Phys. Lett., Vol. 24, No. 9 (1974), pp. 422-424.

[89] D. Gloge, E. Chinnock and T. P. Lee: GaAs *twin laser setup to measure mode and material dispersion in optical fibres,* Appl. Opt., Vol. 13, No. 2 (1974), pp. 261-263.

[90] B. Luther-Davies, D. N. Payne and W. A. Gambling: *Evaluation of material dispersion in low loss phosphosilicate core optical fibres*, Opt. Commun., Vol. 13, No. 1 (1975), pp. 84-88.

[91] D. N. Payne and A. H. Hartog: *Determination of the wavelength of zero material dispersion in optical fibres by pulse-delay measurements*, Electron. Lett., Vol. 13, No. 21 (1977), pp. 627-628.

[92] D. Gloge: *Propagation effects in optical fibres*, IEEE Trans. MTT, Vol. MTT-23, No. 1 (1975), pp. 106-120.

[93] T. Okoshi: *Optimum profile design of optical fibers and related requirements for profile measurement and control*, 1977 International Conference on Integrated Optics and Optical Fiber Communication, Tokyo (July 18-20, 1977).

[94] W. E. Martin: *Refractive index profile measurements of diffused optical waveguides*, Appl. Opt., Vol. 13, No. 9 (1974), pp. 2112-2116.

[95] C. A. Burrus and R. D. Standley: *Viewing refractive index profiles and small scale inhomogeneities in glass optical fibers: some techniques*, Appl. Opt., Vol. 13, No. 10 (1974), pp. 2365-2369.

[96] J. Stone and C. A. Burrus: *Focusing effects in interferometric analysis of graded-index optical fibers*, Appl. Opt., Vol. 14, No. 1 (1975), pp. 151-155.

[97] H. M. Presby, W. Mammel and R. M. Derosier: *Refractive index profiling of graded index optical fibers*, Rev. Sci. Instr., Vol. 47, No. 3 (1976), pp. 348-352.

[98] B. C. Wonsiewicz, W. G. French, P. D. Lazay and J. R. Simpson: *Automatic analysis of interferograms: optical waveguide refractive index profile*, Appl. Opt., Vol. 15, No. 4 (1976), pp. 1048-1052.

[99] M. E. Marhic, P. Sotto and M. Epstein: *Nondestructive refractive index profile measurement of clad optical fibers*, Appl. Phys. Lett., Vol. 26, No. 10 (1975), pp. 574-575.

[100] K. Iga and Y. Kokubun: *Precise measurement of the refractive index profile of optical fibres by nondestructive interference method*, 1977 International Conference on Integrated Optics and Optical Fiber Communication, Tokyo (July 18-20, 1977).

[101] M. J. Saunders and W. B. Gardner: *Nondestructive interferometric measurement of the delta and α of clad optical fibers*, Appl. Opt., Vol. 16, No. 9 (1977), pp. 2368-2371.

[102] C. A. Burrus, E. L. Chinnock, D. Gloge, W. S. Holden, Tingye Li, R. D. Standley and D. B. Keck: *Pulse dispersion and refractive-index profiles of some low-noise multimode optical fibers*, Proc. IEEE (October 1973), pp. 1498-1499.

[103] D. N. Payne, F. M. E. Sladen and M. J. Adams: *Index profile determination in graded index fibres*, First European Conference on Optical Fibre Communication, London (September 16-18, 1975), pp. 43-45.

[104] M. J. Adams, D. N. Payne and F. M. E. Sladen: *Leaky rays on optical fibres of arbitrary (circularly symmetric) index profiles*, Electron. Lett., Vol. 11, No. 11 (1975), pp. 238-240.

[105] P. Di Vita and R. Vannucci: *Multimode optical waveguides with graded refractive index: theory of power launching*, Appl. Opt., Vol. 15, No. 11 (1976), pp. 2765-2771.

[106] M. J. Adams, D. N. Payne and F. M. E. Sladen: *Length dependent effects due to leaky modes on multimode graded-index optical fibres*, Opt. Commun., Vol. 17, No. 2 (1976), pp. 204-209.

[107] M. J. Adams, D. N. Payne and F. M. E. Sladen: *Correction factors for the determination of optical fibre refractive-index profiles by the near-field scanning technique*, Electron. Lett., Vol. 12, No. 11 (1976), pp. 281-283.

[108] E. Bianciardi, V. Rizzoli and G. Someda: *Spatial correlation of field intensity in incoherently illuminated multimode fibres*, Electron. Lett., Vol. 13 (1977), No. 1, pp. 25-27.

[109] M. J. Adams, D. N. Payne and F. M. E. Sladen: *Resolution limit of the near field scanning technique*, Third European Conference on Optical Communication, Munich (September 14-16, 1977).

[110] G. R. Newns: *Compound glasses for optical fibres*, Second European Conference on Optical Fibre Communication, Paris (September 27-30, 1976), pp. 21-26.

[111] J. A. Arnaud and R. M. Derosier: *Novel technique for measuring the index profile of optical fibers*, Bell Syst. Tech. J., Vol. 55, No. 10 (1976), pp. 1489-1508.

[112] F. M. E. Sladen, D. N. Payne and M. J. Adams: *Determination of optical fiber refractive index profiles by a near-field scanning technique*, Appl. Phys. Lett., Vol. 22, No. 5 (1976), pp. 255-258.

[113] W. J. Stewart: *A new technique for measuring the refractive index profiles of graded optical fibres*, 1977 International Conference on Integrated Optics and Optical Fiber Communication, Tokyo (July 18-20, 1977), pp. 395-398.

[114] K. I. White: *The measurement of the refractive index profiles of optical fibres by the refracted near-field technique*, Fourth European Conference on Optical Communication, Genoa (September 12-15, 1978), p. 146.

[115] W. Eickhoff and E. Weidel: *Measuring method for the refractive index profile of optical glass fibres*, Opt. Quantum Electron., Vol. 7 (1975), pp. 109-113.

[116] M. Ikeda, M. Tateda and H. Yoshikiyo: *Refractive index profile of a graded index fibre: Measurement by a reflection method*, Appl. Opt., Vol. 14, No. 4 (1975), pp. 814-815.

[117] M. Tateda: *Single-mode fiber refractive index profile measurement by reflection method*, Appl. Opt., Vol. 17, No. 3 (1978), pp. 475-478.

[118] B. Costa and B. Sordo: *Measurement of refractive index profile in optical fibres*, XXIII Congresso Internazionale per l'Elettronica, Rome (March 22-24, 1976).

[119] J. Stone and H. E. Earl: *Surface effects and reflection refractometry of optical fibres*, Opt. Quantum Electron., Vol. 8 (1976), pp. 459-463.

[120] T. Okoshi and K. Hotate: *Refractive index profile of an optical fiber: its measurement by the scattering pattern method*, Appl. Opt., Vol. 15, No. 11 (1976), pp. 2756-2764.

[121] E. Brinkmeyer: *Refractive index profile determination of optical fibres from the diffraction pattern*, Appl. Opt., Vol. 16, No. 11, (1977), pp. 2802-2803.

[122] J. W. Goodman: *Introduction to Fourier Optics*, McGraw-Hill, 1968.

[123] E. Brinkmeyer: *Refractive index profile determination of optical fibers by spatial filtering*, Appl. Opt., Vol. 17, No. 1 (1978), pp. 14-15.

[124] D. Marcuse: *Refractive index determination by the focusing method*, Appl. Opt., Vol. 18, No. 1 (1979), pp. 9-13.

[125] D. Marcuse and H. M. Presby: *Focusing method for non-destructive measurement of optical fibre index profiles*, Appl. Opt., Vol. 18, No. 1, pp. 14-22.

[126] H. M. Presby and D. Marcuse: *Preform index profiling (PIP)*, Appl. Opt., Vol. 18, No. 5 (1979), pp. 671-677.

[127] C. R. Hammond and S. R. Norman: *Silica based binary glass systems-refractive index behaviour and composition in optical fibres*, Opt. Quantum Electron., Vol. 9 (1977), pp. 399-409.

[128] O. Krumpholz: *Measuring parameters of fibres and fibre cables*, Third European Conference on Optical Communication, Munich, (September 14-16, 1977), pp. 469-471.

[129] D. Gloge, I. P. Kaminow and H. M. Presby: *Profile dispersion in multimode fibres: measurement and analysis*, Electron. Lett., Vol. 11, No. 19 (1975), pp. 469-471.

[130] H. M. Presby and I. P. Kaminow: *Binary silica optical fibers: refractive index and profile dispersion measurements*, Appl. Opt., Vol. 15, No. 12 (1976), pp. 3029-3036.

[131] R. Olshansky and D. B. Keck: *Pulse broadening in graded-index optical fibers,* Appl. Opt., Vol. 15, No. 2 (1976), pp. 483-491.

[132] F. M. E. Sladen, D. N. Payne and M. J. Adams: *Measurement of profile dispersion in optical fibres: a direct technique,* Electron. Lett., Vol. 13, No. 7 (1977), pp. 212-213.

[133] Yung S. Lin: *Direct measurement of the refractive indices for a small numerical aperture cladded fiber: a simple method*, Appl. Opt., Vol. 13, No. 6 (1974), pp. 1255-1256.

[134] F. T. Stone: *Rapid optical fiber delta measurement by refractive, index tuning*, Appl. Opt., Vol. 16, No. 10 (1977), pp. 2735-2742.

[135] W. A. Gambling, D. N. Payne and H. Matsumura: *Propagation studies on single mode phosphosilicate fibres*, Second European Conference on Optical Fibre Communication, Paris (September 27-30, 1976), pp. 95-100.

[136] W. J. Stewart: *A new technique for determining the V value and refractive index profiles of optical fibres*, Optical Fiber Transmission, Williamsburg, Virginia (January 7-9, 1975).

[137] J. E. Midwinter: *The prism-taper coupler for the excitation of single modes in optical transmission fibers*, Opt. Quantum Electron., Vol. 7 (1975), pp. 297-303.

[138] L. Jeunhomme and J. P. Pocholle: *Angular dependence of the mode-coupling coefficient in a multimode optical fibre*, Electron. Lett., Vol. 11, No. 18 (1975), pp. 425-426.

[139] W. A. Gambling, D. N. Payne and H. Matsumura: *Mode conversion coefficients in optical fibers*, Appl. Opt., Vol. 14, No. 7 (1975), pp. 1538-1542.

[140] D. Gloge: *Optical power flow in multimode fibers*, Bell Syst. Tech. J., Vol. 51, No. 8 (1972), pp. 1767-1783.

[141] G. Cocito, P. Di Vita, F. Esposto, L. Michetti and E. Thomas: *Theory and experiments on LED/fibre coupling*, Second European Conference on Optical Fibre Communication, Paris (September 27-30, 1976), pp. 281-292.

[142] E. Weidel: *Light coupling problems for* GaAs *laser multimode fibre coupling*, Opt. Quantum Electron., Vol. 8 (1978), pp. 301-307.

[143] G. Cocito, F. Esposto, L. Michetti and E. Vezzoni: *Coupling losses for three different fibre core diameters*, CSELT Rapporti Tecnici, Vol. VI, No. 2 (1978).

[144] A. H. Cherin and P. J. Rich: *Measurement of loss and output numerical aperture of optical fibre splices*, Appl. Opt., Vol. 17, No. 4 (1978), pp. 642-645.

[145] V. Egashira and M. Kobayashi: *Optical fiber splicing with a low power* CO_2 *laser*, Appl. Opt., Vol. 16, No. 6 (1976), pp. 1636-1638.

[146] B. Costa and A. Morelli, unpublished work.

[147] G. Cocito, B. Costa, S. Longoni, L. Michetti, L. Silvestri, D. Tibone and F. Tosco: COS 2 *experiment in Turin: field test on an optical cable in ducts*, IEEE Trans. Commun, Vol. Com. 26, No. 7 (1978), pp. 1028-1036.

[148] D. H. Smithgall, L. S. Watkins and R. E. Frazee: *High-speed, noncontact fiber-diameter measurement using forward light scattering*, Appl. Opt., Vol. 16, No. 9 (1977), pp. 2395-2402.

[149] B. Costa, B. Daino, F. Caviglia, B. Sordo, F. Tosco and M. Zoboli: *Definition of optical fibre characteristics and measurement methods,* CSELT Internal Report. No. 77.07.144 (1977).

[150] L. S. Watkins: *Scattering from side illuminated clad glass fibres, for determination of fibre parameters*, J. Opt. Soc. Am., Vol. 64, No. 2 (1974), pp. 767-772.

[151] P. L. Chu: *Determination of diameters and refractive indices of step-index optical fibres,* Electron. Lett., Vol. 12, No. 7 (1976), pp. 155-157.

[152] D. Marcuse and H. M. Presby: *Light scattering from optical fibres with arbitrary refractive-index distributions*, J. Opt. Soc. Am., Vol. 65, No. 4 (1975), pp. 367-375.

[153] H. M. Presby: *Ellipticity measurement of optical fibers*, Appl. Opt., Vol. 15, No. 2 (1976), pp. 492-494.

[154] W. Weibull: *Proceedings of the Royal Swedish Institute for Engineering Research*, No. 151 (1939).

[155] N. A. Weil, J. S. Islinger and D. W. Levinson: *Investigation of glass fiber strength enhancement through bundle drawing*, Armed Service Tech. Inf. Agency, R.D. 287404.

[156] H. Liertz and U. Oestreich: *Application of Weibull distribution to mechanical reliability of optical waveguides for cables*, Siemens Forsch. U. Entwickl., Vol. 5, No. 3 (1976), pp. 129-135.

[157] U. H. P. Oestreich: *The application of the Weibull distribution to the mechanical reliability of optical fibres for cables*, First European Conference on Optical Fibre Communication, London (September 16-18, 1975), pp. 73-75.

[158] R. Olshansky and R. D. Maurer: *Tensile strength and fatigue of optical fibers*, J. Appl. Phys., Vol. 47, No. 10 (1976), pp. 4497-4499.

[159] E. Helgand and F. R. Wasserman: *Statistics of the strength of optical fibres*, J. Appl. Phys., Vol. 48, No. 8 (1977), pp. 3251-3259.

[160] R. D. Maurer, R. A. Miller, D. D. Smith and J. C. Troudson: *Optimization of optical waveguides. Strength studies*, Technical Report for the Office of Naval Research, No. AD-777118.

[161] C. R. Kurkjian, R. V. Albarino, J. T. Krause, A. N. Vazirani, F. V. Di Marcello, S. Torza and H. Schonkorn: *Strength of* 0.04-50 m *lengths of coated fused silica fibres*, Appl. Phys. Lett., Vol. 28, No. 10 (1976), pp. 588-590.

[162] B. Costa, C. De Bernardi and B. Sordo: *Investigation of scattering characteristics and accuracy of the backscattering technique by wavelength dependent measurements*, Fourth European Conference on Optical Communication, Genoa (September 12-15, 1978), pp. 140-145.

[163] J. P. Dakin: *A simplified photometer for rapid measurement of total scattering attenuation of fibre optical waveguides*, Opt. Commun., Vol. 12, No. 1 (1974), pp. 83-88.

CHAPTER 4

MATERIALS AND TECHNOLOGIES FOR OPTICAL FIBRES

by Giuseppe Cocito, Giacomo Roba and Paolo Vergnano

4.1 General

4.1.1 *Introduction*

The dielectric material to be used for the construction of optical fibres must have the following chief characteristics: excellent transparency at the useful spectral wavelengths, good chemical stability in foreseeable applications and suitability for processing through techniques suitable for all construction phases.

At present, the materials mostly used for telecommunication optical fibres can be subdivided into two fundamental types:

— silica, pure or doped;
— multicomponent glasses.

Attempts at using liquids as the fibre core have not met with much success. Although low attenuation liquids with an appropriate refractive index are in principle available, the chemical inertia of these compounds (generally hydrocarbons) is not sufficient to ensure that the optical characteristics of the fibre will be maintained over a long period. Similarly, although the use of polymers, not only as coating but also as fibre components, is an interesting subject of research and experimentation, it is far from offering optical fibres suitable for long distance connections.

4.1.2 *High silica fibres and glass fibres*

The distinction between these two fundamental types of material is not very clear in principle and is usually correlated with the fibre fabrication processes, which differ considerably depending on whether high silica or conventional glasses are involved.

In any case the materials in question have a vitreous structure, are isotropic and are drawn in fibre form in the fluid state, so as to prevent crystallization. This is despite the fact that the formation of steady and homogeneous glass from a melted mass involves the transition through a temperature range, where nucleation and crystal phase growth can occur. The formation of crystals depends both on the cooling rate and on crystal formation kinetics. For many glasses the growth velocity of crystal nuclei is low and is affected by temperature and stoichiometry, i.e. depending on whether the crystal, which is segregating, has the same bath composition. As far as materials for optical fibres are concerned, those compositions which give vitreous structures prone to devitrification and therefore to aging must generally be avoided.

The basic component is always silica (SiO_2), either natural or synthetic. In the case of high silica fibres, chemical vapour phase deposition (CVD) is usually adopted to obtain high purity silica, with appropriate additives to vary its refractive index. The more the doping is increased, the more the behaviour of the resultant mixture approaches that of a conventional glass, which can therefore be considered a mixture of silica and other vitreous oxides in considerable quantities.

Taking into account that glass properties vary steadily and predictably with the composition, the factor that most clearly distinguishes high silica glass from multicomponent glass is the melting temperature or the temperature at which a fibre can be drawn: if this temperature is within the 800-1200 °C range, we are generally talking about glass fibres, whereas if it approaches or exceeds 2000 °C doped silica fibres are involved.

In the first case, the low softening temperature permits the utilization of traditional furnaces by the « rod in tube » method (Section 4.2.9) or reduced time for melt preparation and fibre production by the double crucible method (Section 4.2.5), which can also be continuous. Fibres having high numerical aperture, suitable for use along medium length sections (data processing, avionics applications etc.) can easily be obtained with these glasses. However, the most recent developments have produced glass fibres with an attenuation of about 5 dB/km, with very accurate purification of raw materials and using clean room techniques during all phases of glass handling, transfer and drawing. A constant characteristic of all the glass fibre fabrication processes is the quite limited range of the refractive index profiles obtained hitherto.

On the other hand, high silica can be obtained with greater purity (fibres show so far a lower attenuation) from a process with self-purifying characteristics, described in Section 4.3.2.

This process eliminates a large part of the contamination sources, depositing through synthesis material layers having the desired composition already in the form and arrangement suitable for being drawn into the fibre. This requires less stringent precautions in raw-materials handling and room cleaning and permits a wide range of refractive index profiles to be easily and accurately determined (in connection with the problem of the distortion of the optical pulse transmitted, see Sections 3.4 and 3.7). The CVD technique for doped silica fibres also allows the attainment of a good core-cladding interface, the reduction of losses due to scattering and an optimum chemical stability of fibres.

The disadvantages of using high silica can be summed up as follows: slowness of the phases of material deposition and drawing, need for high temperature operation in all process phases and consequent difficulties in checking the fibre geometry, and special equipment requirements (very high temperature furnaces, special oxyhydrogen torches etc.).

4.2 Multicomponent glasses for optical fibres

4.2.1 *Introduction*

Glass fibres are manufactured in many different types of material. For the glass composition it is necessary to take into account not only the desired optical properties, but also the requirements that may arise during the drawing process. For instance, in the case of the most common glass fibre drawing technique using the double crucible (Section 4.2.5), it is expedient to utilize a glass couple having a fairly low melting temperature (so as to minimize the phenomenon of corrosion and consequent glass contamination) and similar viscosities within the drawing temperature range, in order to simplify the problems concerning equipment planning and operation stability.

The more typical glass systems are soda-borosilicate. Many variants have been proposed to accentuate different fibre properties. Compositions have been studied to find stable glasses with low crystallization velocity, low melting temperature, high mechanical resistance, low scattering losses etc. Soda-limesilicate glasses [1] or soda-aluminasilicate [2, 3] or alkali-germanosilicate [4] are also used. High numerical aperture fibres (NA = 0.47) with 50 dB/km attenuation at 800 nm have been obtained by means of glasses of the Na_2O - K_2O - PbO - SiO_2 system, in which small variations in the PbO contents give rise to large variations in the refractive indices [5].

4.2.2 *Borosilicate glasses*

The vitreous SiO_2 - B_2O_3 - Na_2O system is widely used for optical fibres. Some of its properties are summed up in Fig. 1 [6]. The isofract lines connect the compositions having a constant refractive index. In addition to the existence of different unusable

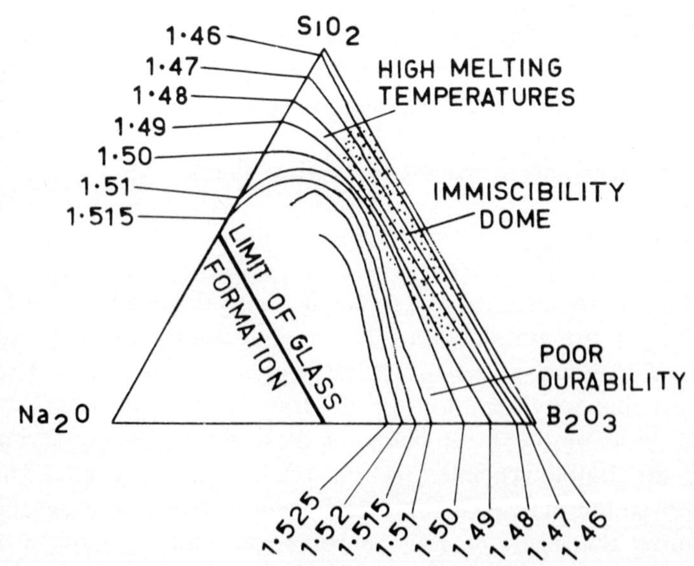

FIG. 1 – Isofracts of the soda-borosilicate system. (After [6]).

regions, it can be noted that the refraction index value is controlled by sodium oxide concentration. The isofract lines are almost parallel to the SiO_2 - B_2O_3 axis, which shows that B_2O_3 has very little effect on the silica refractive index, with the exception of the first region near the pure silica. Yet it must be borne in mind that an excessive Na_2O content can lead to glass crystallization during cooling; this phenomenon is undoubtedly unacceptable. Therefore, if, for the above reasons, both the compositions too near to system edges and those corresponding to the unmiscibility dome are excluded, an intermediate region is left where a couple of glasses suitable for forming the cladding and core of an optical fibre can be chosen. However, the need to ensure compatibility between many physical, chemical and mechanical factors limits the range of appropriate compositions and the attainment of very high numerical aperture fibres becomes difficult. NA values between 0.15 and 0.35 are common [7].

4.2.3 *Absorption losses*

The intrinsic attenuation of glasses at wavelengths about 800-900 nm is very low and can be compared with that of pure silica. Tails of ultra-violet absorption and of infra-red absorption bands are negligible, as are Rayleigh scattering losses.

Thus the fact that glass fibres still show an attenuation higher than that of high silica fibres can only be ascribed to absorption due to impurities. It is mainly a question of transition metal ions, which show considerable absorption bands in the visible and near infra-red even at very low concentrations. Losses due to these impurities can be ascribed to electronic transitions between energy levels associated to the interior incomplete subshell. In practice, these ions are the cause of the glass colouring. The amounts of losses at different wavelengths depend on the impurity concentration, on its oxidation state and on the glass composition where it lies [8].

Table I reports attenuation data for the most frequent glass impurities. From Fig. 2 [6], it can be seen that even inside the same system the glass composition considerably affects the extinction coefficient of an ion (see Section 3.2). In particular, it can be noted that an increase in the refractive index, connected to the Na_2O present, also involves a rapid increment in attenuation due to each Fe^{2+} ppm (part per million). Even the oxidation state of impurities affects the extinction coefficient.

TABLE I (After [8])

Impurity element	Absorption (dB/km) induced at 850 nm for each part per million		
	Na_2O - CaO SiO_2 (1)	Na_2O - B_2O_3 Tl_2O - SiO_2 (2)	SiO_2 (fused silica) (3)
Fe	125	5	130
Cu	600	500	22
Cr	10	25	1300
Co	10	10	24
Ni	260	200	27
Mn	40	11	60
K	—	40	2500

(1) G. R. Newns, P. Pantelis, J. L. Wilson, R. W. J. Uffen and R. Worthington: *Absorption losses in Glasses and Glass Fibre Waveguides*. Opto-Electronics, Vol. 5, No. 4 (July 1973), p. 289.

(2) T. Uchida: *Preparation and properties of Compound Glass Fibres*. URSI General Assembly Commission VI, Lima (August 18, 1975).

(3) P. C. Schultz: *Optical Absorption of the Transition Elements in Vitreous Silica*. American Ceramic Society Journal, Vol. 57, No. 7 (July, 1974), p. 309.

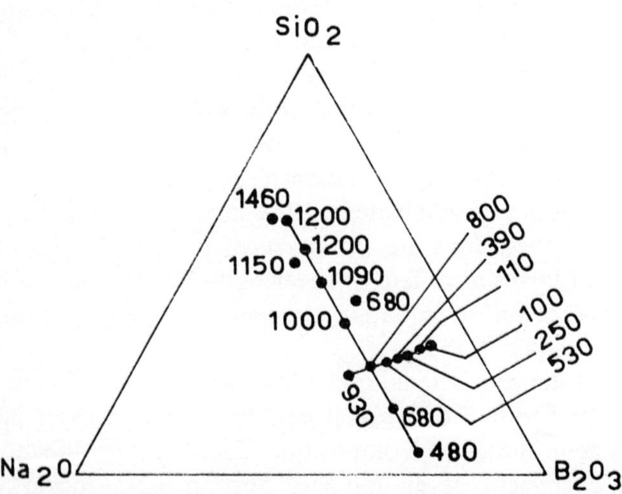

FIG. 2 – Extinction coefficients of Fe^{2+} in soda-borosilicate glasses [(dB/km)/ppm]. (After [7]).

Figure 3 shows that the oxidation state of iron and copper acts in the opposite direction on the attenuation [9]. The ratio between ion concentrations at the different oxidation states is mainly connected to the reaction of melted glass with the atmo-

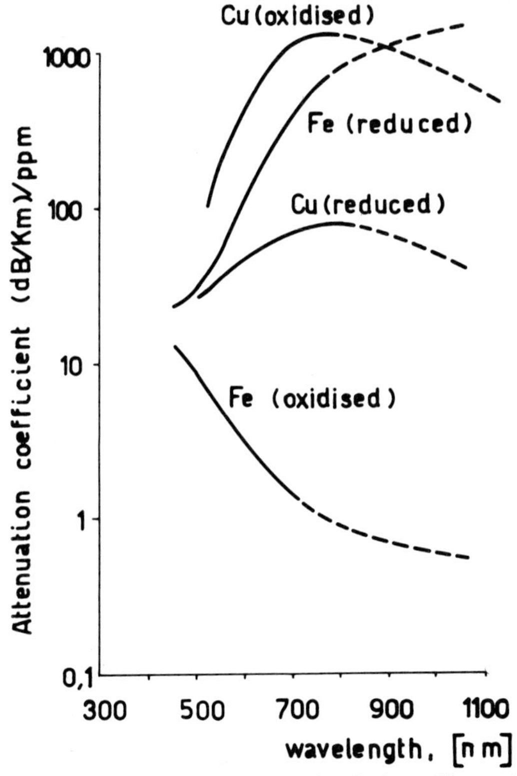

FIG. 3 – Absorption losses of Fe and Cu doped soda-borosilicate glasses showing oxidised and reduced states. (After [9]).

sphere where it is prepared, which can have an oxidative (O_2) or more or less highly reducing ($CO_2 + CO$) effect.

Another kind of impurity consists of the presence of water dissolved in the glass in the form of hydroxyl groups (—OH) causing a vibrational absorption with a main peak at 2.77 μm with overtones at 1.38 and 0.95 μm and various other minor peaks in connection with combination bands. Figure 4 shows the hydroxyl group absorption spectrum for a borosilicate glass.

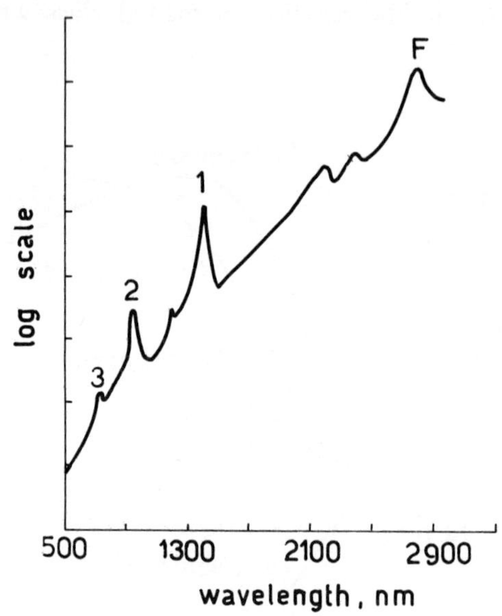

FIG. 4 – Hydroxyl overtone and combination bands in soda-borosilicate glass. (After [9]).

The maximum water portion in a glass is already contained in the raw materials. The bond is so strong that it is quite difficult to reduce the absorption peaks of —OH groups [9] during the manufacturing process.

4.2.4 *Purification and analysis techniques*

Detection and measurement of such small traces of impurities, which are, however, determinant as regards material quality, has necessitated the development of very sophisticated analysis methods in addition to the purification procedures which have allowed the production of materials suitable for optical fibres. The materials are oxides (SiO_2, Na_2O, CaO, B_2O_3, Tl_2O, etc.) prepared from carbonates or acids or obtained through synthesis: SiO_2 from silicon tetrachloride ($SiCl_4$), from sylanes (SiH_4) or from ethyl-orthosilicate ($Si(OC_2H_5)_4$). Silica could also be obtained directly from the natural one, but in this case the difficulties of obtaining the required purification would be almost insurmountable. Liquid or gaseous silica compounds, on the other hand, can be decontaminated through double distillation [10], partial precipitation, active carbon absorption, impurity complexation etc. [11]. After the purification of basic materials, SiO_2 is obtained from sylanes through pyrolisis, from silicon tetrachloride through vapour phase reaction with high temperature oxygen and from ethyl-orthosilicate through hydrolysis followed by drying and calcination.

Carbonates and other soluble compounds can be purified through coprecipitation, extraction by means of solvent, impurity complexation, or ion exchange. The transformation from carbonates into oxides through the elimination of CO_2 occurs generally during an intermediate phase (melting) before the drawing operation.

The analysis techniques most suitable for measuring impurities content are flameless atomic absorption, neutron activation, mass spectrometry, and X-ray fluorescence.

At the end of the purification phase, a mixture of compounds is obtained having less than 10 ppm for each impurity measured separately.

4.2.5 *Glass drawing with double crucible*

At present the double crucible method is the most widely used technique for the production of glass fibres. Two cylindrical containers having a conical bottom, each with a nozzle at the bottom, are kept concentric (Fig. 5).

The inner container holds glass for the core, the outer glass for the cladding. Usually the actual drawing phase is preceded by the preparation of glass with the desired composition in a proper container heated by the furnace or directly by R.F., in that over 800 °C alkaline glasses have a sufficient ion conductivity. The purified components are melted in this container (Fig. 6) and treated to obtain a homogeneous mass free of bubbles and stripings.

Special care must be taken in the degassing, which often follows a phase in which a dry gas is placed in contact with the glass to reduce the concentration of —OH ions.

A large number of the hydroxyl groups can be eliminated

FIG. 5 – Double Crucible (After [18]).

in advance by keeping the raw materials in a dry atmosphere at some hundreds of degrees for a long period (several days). The handling and transfer of ultrapure materials require strict precautions to prevent further contamination. Clean room techniques are necessary; yet, in spite of all precautions, there is always a slight increase in the impurity level inside the melted glass as compared to materials just purified, even if the container is made of vitreous quartz and is isolated from the external environment. Cylindrical rods are drawn from the melted bulk and then kept in special containers; these rods feed the double crucible. Nozzle dimensions for glasses having a soda-borosilicate or soda-lime-silicate composition are about 0.5mm for the core crucible and up to 3 mm for the cladding [12].

The fibre drawing takes place at speeds varying from 0.5 to 5 m/s. Step-index fibres can be obtained by means of the double

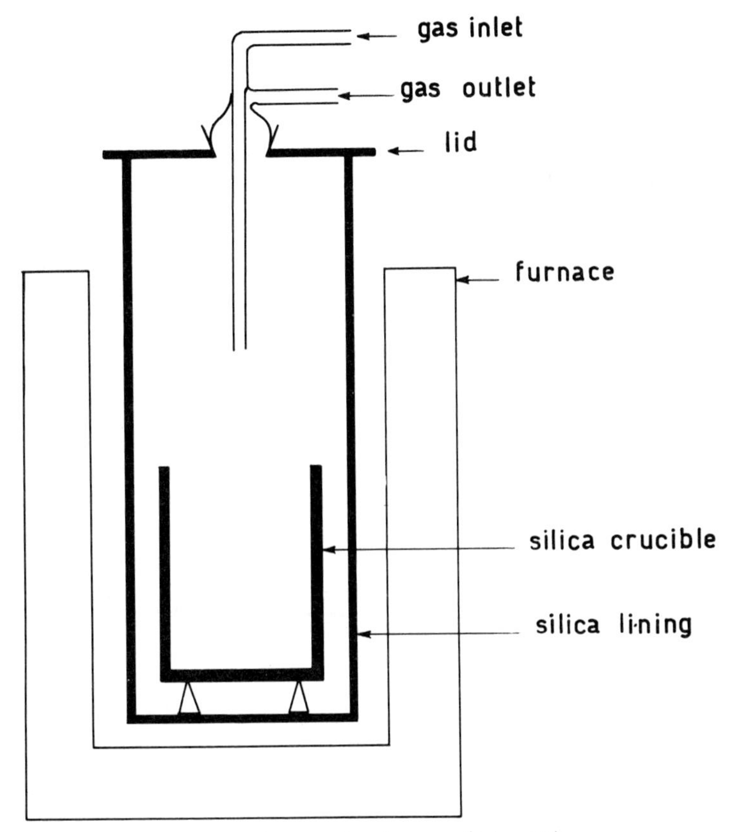

FIG. 6 – Glassmaking furnace (After [9]).

crucible, the shape of which is illustrated in Fig. 5, as the diffusion region between glasses is very narrow. Furthermore, the viscosity at the operation temperature (800-1000 °C) is so high as to prevent considerable glass mixings.

Glasses in the double crucible can also be heated by the R.F. furnace, if possible using the double crucible material as a resistor.

4.2.6 *Materials for double crucible manufacturing*

The double crucible is often manufactured in platinum or platinum alloy enabling operating temperatures up to over 1400 °C.

The homogeneity of the melted mass is ensured by a stirrer, also in platinum. However, this material is not completely inert and a considerable contamination takes place when the platinum dissolves in the glass.

In a non-oxidative environment, the platinum solution in the colloidal state can turn the glass yellow [13]. In an oxidative atmosphere Pt is in the form of Pt^{2+} or Pt^{4+} ions and the glass loses the yellow colour as the absorption peak shifts towards U.V. The use of iridium as a material for the double crucible has also been proposed, but the presence of Ir or IrO_2 in the glass causes unacceptable losses due to scattering.

An alternative may be represented by silica glass, whereby high purity double crucibles can be manufactured. On the other hand, these crucibles at high temperatures are quickly corroded by the melted glass, where stripings, bubbles and nonhomogeneities appear. In this case, R.F. heating is recommended, which maintains the glass at high temperature without directly heating the silica walls. In this way, however, convections producing considerable nonuniformities are generated inside the melt [14]. The use of mixtures with the lowest possible melting point is clearly advantageous.

The fact that glass adheres to silica must not be neglected, as breaks occur very easily during cooling owing to tensions due to different expansions coefficients of the materials [11].

Some attempts to use amorphous carbon or graphite double crucibles have given quite encouraging results, in that these materials are not wet by the melted glass and in a controlled atmosphere (Ar+CO) reduction phenomena of some oxides (particularly Na_2O) by carbon can be eliminated or at least limited [2, 11]. Manufacturing difficulties excepted, alumina also seems to be a promising material [5].

4.2.7 *Further aspects of double crucible drawing*

Problems of glass contamination from the container are sometimes accompanied by other difficulties. For instance, the development of small gaseous bubbles within the feeding region of the glass rods is frequent. On account of the high viscosity of the vitreous system, these bubbles can persist till they appear within the fibre. This disadvantage can be avoided by limiting the feed velocity of the rods to a few cm/min. If the platinum double crucible is used in an oxidative atmosphere, microscopic oxygen bubbles, located at the core-cladding interface and generated by the electrolytic couple formed by the two glasses, frequently develop.

When these small bubbles move towards the fibre interior, losses due to scattering reach excessive levels. Other scattering regions may also be produced owing to the presence of carbon traces deriving from organic residues inside the raw materials. This disadvantage can be eliminated treating materials in advance with oxygen at high temperature.

Other defects introduced into the fibre during drawing may be of the geometric type. The core-cladding concentricity, ellipticity, fibre diameter constancy, constancy of the diameter cladding-diameter core ratio etc. depend on the reliability of the double crucible construction and fibre winding. A drawing meniscus (neck down belt) which is stable and well developed downstream the double crucible must be ensured. This can be obtained by increasing both the melt level and the temperature whilst reducing, as far as possible, the possibility of crucible contamination. If the drawing process is continuous, an accurate check of the melt level inside the crucible is also required [13].

4.2.8 *Light focusing fibre*

An interesting variant of the double crucible drawing technique consists in using equipment with which core and cladding glasses are kept in contact, at high temperature, for a long enough period to cause an ion diffusion between both materials (Fig. 7). If the double crucible is dimensioned in such a way as to regulate the conditions of the ion exchange, a range of refractive index profiles from the step up to a quasi-parabolic curve can be obtained.

Although both glasses usually have a very similar soda-borosilicate composition inside the core glass, thallium oxyde (Tl_2O) is added to increase the refractive index.

Thallium is chosen since the monovalent ion Tl^+ has a small ray and considerable mobility. It is consequently suited for diffusion within the core-cladding contact region, thus determining a graded behaviour of the refractive index. In the primary dis-

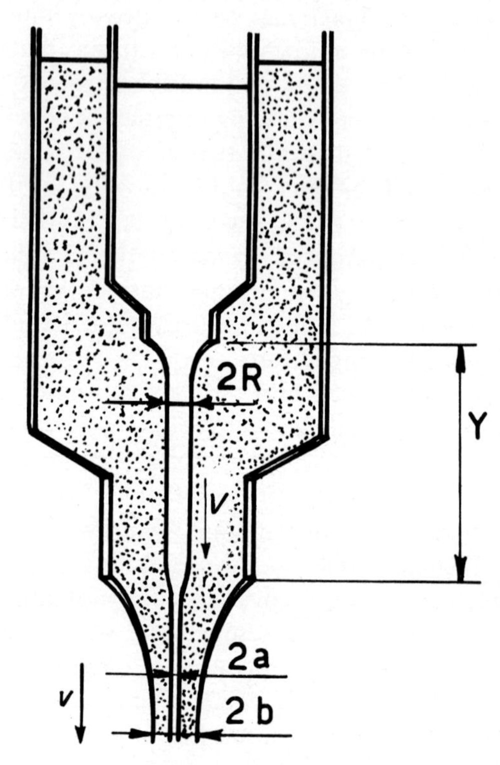

Fig. 7 – Ion exchange process between core and cladding glasses in a double crucible. [After [14]).

continuous process step-index preforms with thallium glass core having a diameter of some millimetres were manufactured.

These were immersed in melted salts (KNO_3 or the like) at 500 °C so as to allow the desired thallium diffusion, and later drawn into the fibre. Afterwards the process became continuous using the above mentioned double crucible [14].

Assuming that a fibre with the core ray equal to a is drawn at a speed v, it can be said that the glass flow rate is (see Fig. 7)

$$Q = \pi R^2 V = \pi a^2 v\,.$$

By defining an exchange parameter K as

$$K = D \cdot t / R^2$$

where D is the diffusion coefficient of Tl^+ ion inside the melted glass and t is the time for which the exchange is possible and considering that

$$Y = t \cdot V$$

we then obtain

$$K = Y \cdot D / a^2 v \, .$$

The value of K is determinant as regards the refractive index profile inside the fibre. If $K = 0$, no diffusion occurs and we get a step profile. If K grows, the profile tends to assume a bell-shaped behaviour (Fig. 8) [15].

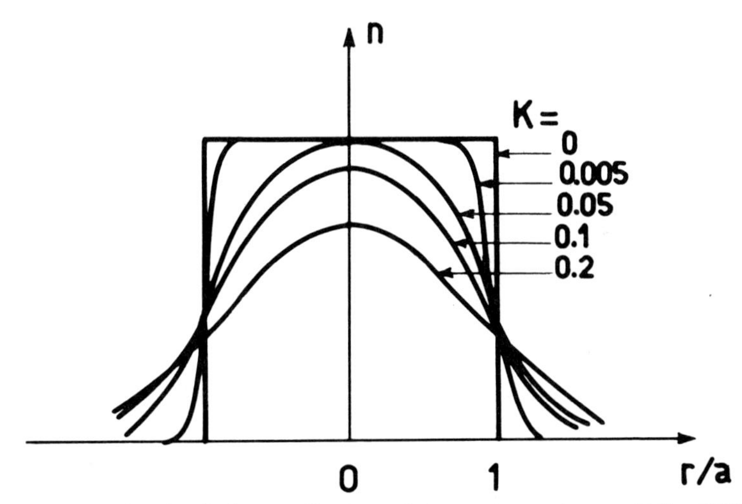

FIG. 8 – Refractive index profile formed by an ion-exchange. (After [15]).

The profile best approximating a parabolic shape (see Section 3.7) is characterized by $K = 0.081$. To reach this objective, the most easily controlled factor is Y. As a matter of fact, for the reasons already illustrated in connection with the conventional crucible, the drawing temperature must not be too high (consequently D is limited) and v cannot be too low (0.3-0.5 m/s). To obtain an adequate exchange Y has therefore to be increased, without however exceeding the limit imposed by the beginning of glass devitrification caused by excessive permanence at high temperature (see Section 4.1.2).

It must also be noted that the region of the Na_2O - Tl_2O - B_2O_3 - SiO_2 system which can be useful for the glass selection is quite limited (Fig. 9) [16]. Fibres with reduced losses at 5 dB/km

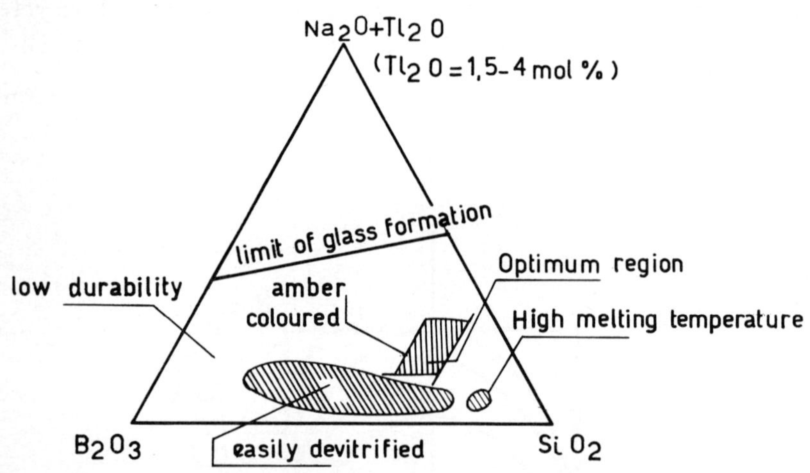

FIG. 9 – Optimum region for low-loss glasses. (After [16]).

at $\lambda = 0.83\ \mu m$ and bandwidth up to 1 GHz·km have been prepared by means of the procedure illustrated above. Step-index fibres produced by the same equipment can have numerical apertures up to 0.3. Attempts to produce graded-index fibres by means of the same technique through the exchange between Na^+ and K^+ ions in borosilicate glasses have led to fibres with insignificant numerical aperture. However, the results obtained with germaniasilicate glasses [17] show that the $Na^+ \leftrightarrows K^+$ exchange has a large effect on the refraction index and occurs at considerable velocity. Thus, refractive index differences between core and cladding up to 1.7% and overall attenuation of about 20 dB/km have been obtained.

4.2.9 *Rod in tube method*

In this technique a glass rod with a high refractive index is inserted in a glass tube with a lower refractive index. The assembly is then drawn into the fibre in a furnace (Fig. 10) [18].

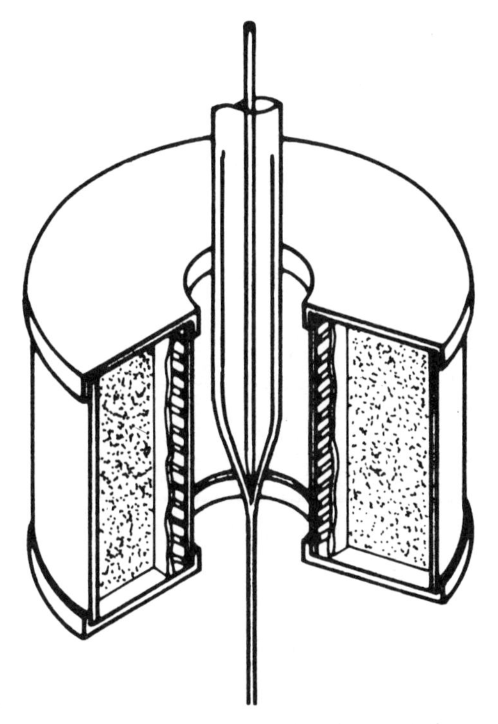

FIG. 10 – Rod in tube. (After [18]).

If the tube and rod feeding occurs at the same velocity, the fibre reproduces, in the diameter core-diameter cladding ratio, the relevant dimensions between rod and tube. The fibre geometrical characteristics can be well checked if the rod diameter is practically equal to the tube inner diameter. Furthermore, it is necessary that the viscosity curves of both glasses are similar in the drawing temperature range and that the heat expansions coefficients are compatible. Before drawing, the surfaces of rod and tube must be thoroughly decontaminated.

This method, despite the advantages of simplicity and absence of contamination from the contained, presents the risk of including gaseous bubbles and foreign particles on the surface separating core and cladding. The drawing furnace can be replaced by lasers, gas burners or R.F. heating.

4.2.10 « *Composite rods* » *drawn from the melt by the* « *floating-crucible* » *method*

A method for producing glass preforms with excellent properties of core-cladding interface consists in drawing a glass rod (core diameter = 2.5 mm, cladding diameter = 5 mm) by means of

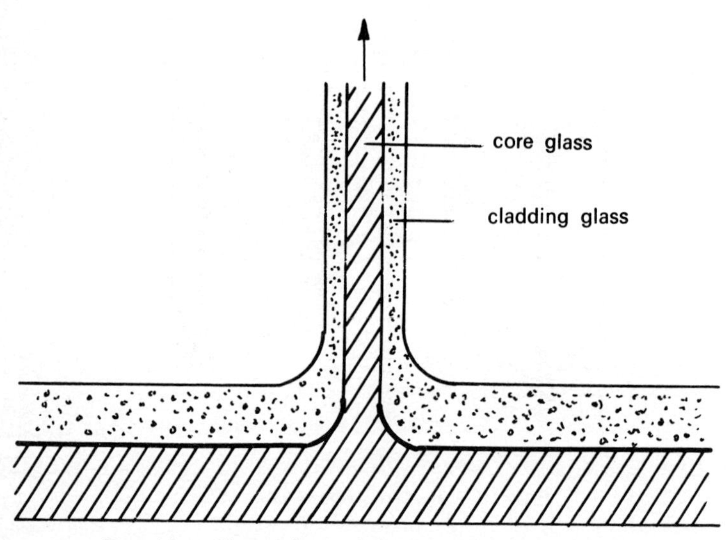

FIG. 11 – Composite rod from « floating crucible ».

the floating crucible technique. An inner crucible containing the cladding glass is placed in contact with the layer of the core melted glass, contained in an external crucible. A column of glass, which is automatically covered by the cladding, can be drawn through a hole on the base of the inner crucible. Any mixing between glasses must be avoided, as this involves some difficulty when glasses with very similar physical properties are worked (Fig. 11) [19]. From this preform the true fibre can be obtained by simple drawing.

4.3 Manufacturing techniques for high-silica optical fibres. Chemical Vapour Deposition (CVD)

4.3.1 *Introduction*

The main processes for manufacturing preforms for high-silica optical fibres generally make use of Chemical Vapour Deposition (CVD), which is already widely employed for semiconductor production.

Silica, and other oxides used as dopants, are synthesized by vapour-phase oxidation at a high temperature. Compounds, which are often halides of the elements desired, are vapourised in a gas stream (oxygen or a neutral gas such as argon). After further oxygen enriching, these compounds flow into a high temperature region, where they react with oxygen to provide a dispersion of fine particulate glass called « soot ». The « soot » may be collected as a powder or may impinge and be deposited on a hot surface to form a porous solid.

Halides used for this process are, at room temperature, in liquid state or in vapour form and can be highly purified through different processes. The most widely used process is distillation, which produces extremely pure glass required for low attenuations [20].

The vapour oxidation, for the preparation of a structure (preform) to be drawn into the optical fibre, can be carried out in a variety of ways, which differ both in the oxides synthesis phase and glass deposition geometry. These processes can be separated into two different categories [21]: Inside Vapour-Phase Oxidation [22-24] (IVPO or « Inside Process ») and Outside Vapour-Phase Oxidation [25] (OVPO or « Outside Process »).

4.3.2 *Inside Vapour-Phase Oxidation (IVPO) process*

The IVPO process consists in depositing high-SiO_2 glass layers on the internal walls of a silica glass tube. This is the support for the material to be deposited. At the same time, it prevents contamination produced by both containers and surrounding environment, which represents an important problem for the above-mentioned simpler processes. The final structure or preform must have the characteristics suitable for optical fibres. Therefore, during the deposition process, other oxides are added to the silica, modifying its refractive index. By varying the dopant concentration, it is possible to control both the numerical aperture and refractive index profile and then to obtain preforms suitable for step-index, graded-index, monomode, etc. optical fibres.

As shown in Fig. 12, reagents are fed into a silica glass tube, which is slowly rotated around its axis by a standard glass lathe.

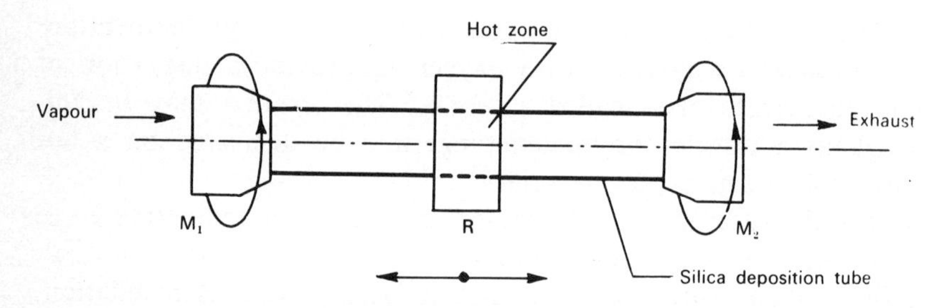

FIG. 12 – Schematic diagram of IVPO deposition system.

Heater *R*, which can be a tungsten or graphite resistance furnace or a multiple oxy-hydrogen burner, moves slowly and axially along the tube in the reagent flow direction. At the end of the active pass, it is driven back quickly and automatically to the back of the tube and the cycle repeats itself.

In this case it is possible to regulate preform diameter by controlling pressure inside the tube [26, 27].

The slow rotation of mandrels M1 and M2 supporting the tube is necessary to ensure that the hot reaction zone affecting the substrate for a short part of its length is free of temperature nonhomogeneities, which would cause a nonuniform layer thickness. The direction of slow traverse of the heater is due to the fact that some of the oxides generated by the hot zone reactions are moved by the gas stream and form as soot on the tube walls, down-stream the hot zone.

This implies the immediate deposition and annealing of a glass layer, which forms a porous solid. If the hot zone temperature, depending on the temperature and speed of translation of the heater, is sufficiently high, the soot actually vitrifies. The final result is a solid glass layer, which is both transparent and bubble-free.

This process is called Modified Chemical Vapour Deposition (MCVD) [28, 29] and is generally preferred to the first (CVD), within the IVPO techniques, as it ensures imperfection-free optical fibres.

Each glass layer, typically about 5-15 μm thick, is uniform along the total length of the supporting tube, because temperature, traverse speed and reagent concentration are kept constant during the deposition. On the other hand, the number of deposited layers is limited during the preparation by the preform stress due to the difference between the thermal expansion coefficients of the silica glass support tube and the binary or ternary glass being deposited (silica + dopant oxides). Typically, it is possible to accumulate 80-100 glass layers before the deposit fractures, corresponding to a total glass thickness of the order of 1 mm. Furthermore, these parameters, as well as the difference in the viscosity curves between core and cladding materials, are very important for the preform drawing process; the choice of dopant materials and their concentrations depends both on optical properties, described below, and on their effects on the physical properties of such a complex structure as the optical fibre.

4.3.3 *Parameters for dopant choice*

Figure 13 shows the effect of some of the oxides most commonly used as dopants on thermal expansion coefficients of binary silicate glasses.

The same oxides also affect the transformation temperature (Tg) of the fused quartz. This temperature, corresponding to a viscosity of about $1.5 \cdot 10^{13}$ Poise [30], represents a value below which glass stresses are not quickly relieved by internal structural relaxation.

The effect of the addition of these oxides is a decrease in *Tg*: the most effective oxides are P_2O_5 and B_2O_3 with respect, for example, to GeO_2 or TiO_2.

The physical properties of the glasses cause mechanical stress

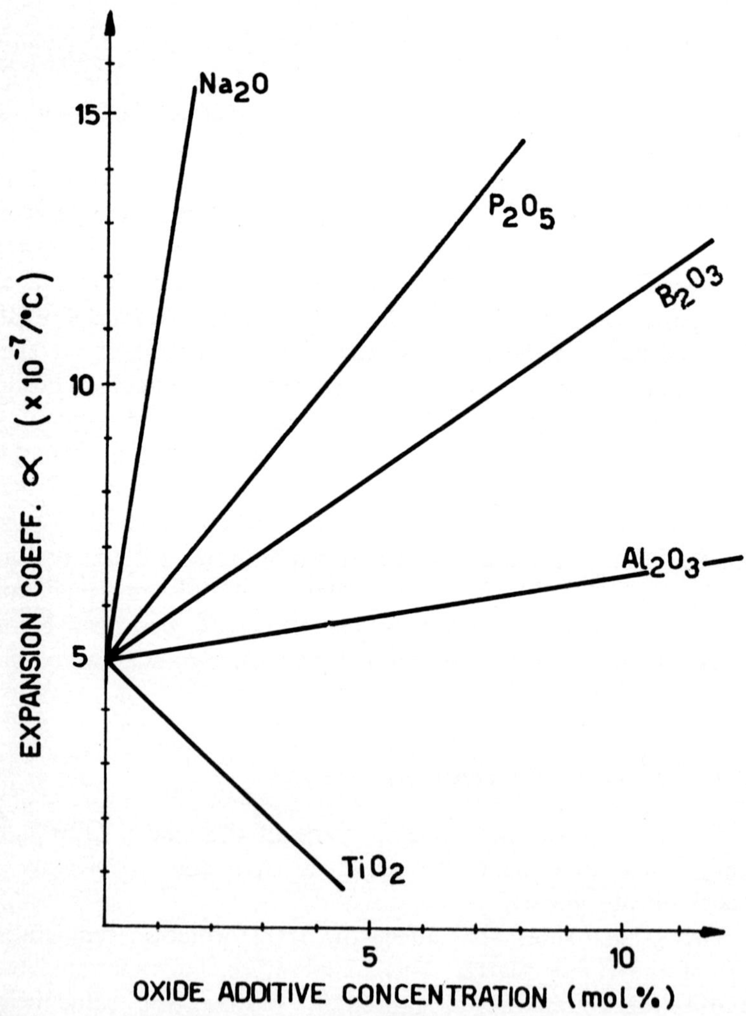

FIG. 13 – Thermal expansion of binary silicate glasses (lines obtained from data). (After F. Franceschini, « Il Vetro », Hoepli, 1955 and F. V. Tooley, « The Handbook of Glass Manufacture - Books for Industry », 1974).

in the fibre. If the core glass has a thermal expansion coefficient higher than the cladding glass and Tg shows the same trend, the core glass undergoes a tensile stress and the cladding glass a compressive stress during the cooling process.

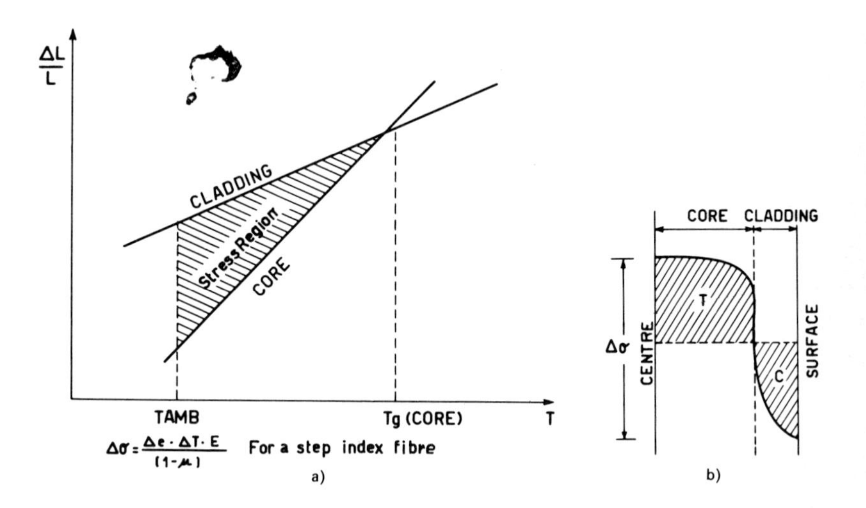

FIG. 14 – Origin of stress in preforms. (After [30]). *a*) Thermal expansion mismatches between *core* and *cladding*. *b*) Stress distribution across the *core* and *cladding* diameters.

These stresses, which for a step-index optical fibre can be represented by following relation, are not constant along the fibre diameter and show the distribution given in Fig. 14 [30]

$$\sigma = \frac{\Delta e \cdot \Delta T \cdot E}{1 - \mu}$$

Δe = difference between the thermal expansion coefficients of the core glass and the cladding glass;

ΔT = difference between the lowest transformation temperature (between core and cladding) and the room temperature;

E = Young's modulus;

μ = Poisson's ratio.

Within reasonable limits, glass fractures occur only under tension and generally at a free surface. Thus, it is important to know the nature of the stress within the preform, and to ensure that free surfaces do not undergo a tensile stress at lower than transformation temperatures.

Let us now examine how oxides, whose property of affecting physical characteristics has already been dealt with, can modify **the refractive index of the pure silica.**

If we dope pure silica with fluorine or B_2O_3, we obtain glasses

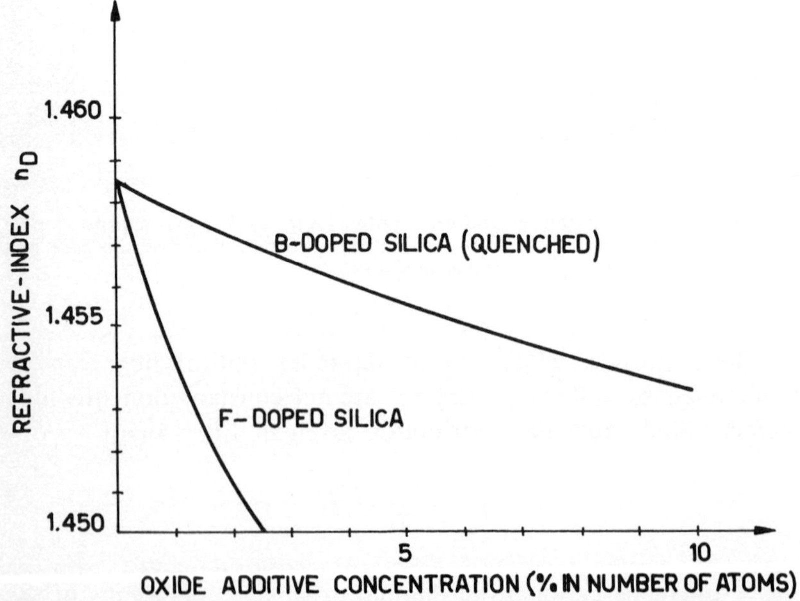

FIG. 15 - Refractive index of binary silicate glasses for fibre cladding. (After K. Abe, « Fluorine doped silica for optical waveguides », 2nd ECOC, Paris, 1976).

suitable for fibre cladding, because they produce a decrease in the refractive index of SiO_2. This effect is shown in Fig. 15.

The SiO_2 - B_2O_3 glasses have a refractive index which depends on the quenching of a metastable phase during fibre drawing. On the contrary, the fluorine causes a refractive index reduction related to its presence in the glassy molecular matrix. Thus, if NA of the fibre is too low owing to long term annealing of B_2O_3 doped silica, it can be convenient to use fluorine instead of boron oxide.

If we dope pure silica with P_2O_5, GeO_2 or TiO_2, we obtain glasses suitable for fibre core, because their presence in the glassy matrix causes an increase in the refractive index of SiO_2, as shown in Fig. 16.

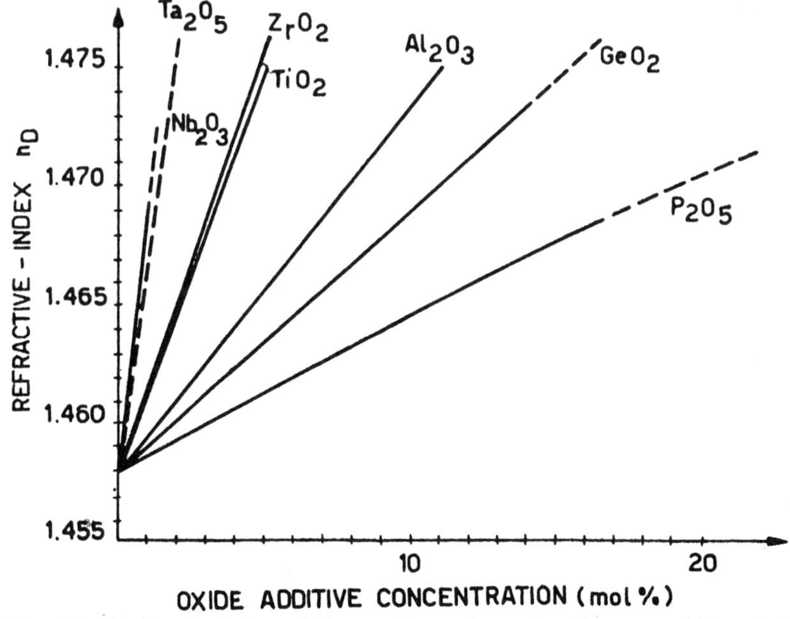

FIG. 16 – Refractive index of binary silicate glasses for fibre core. (After C. K. Hammond and S. R. Norman, « Opt. Quantum Electron. », No. 9 (1977), p. 399 and « Corning Glass Works », (May 1971). 13325/71 Heading CIM).

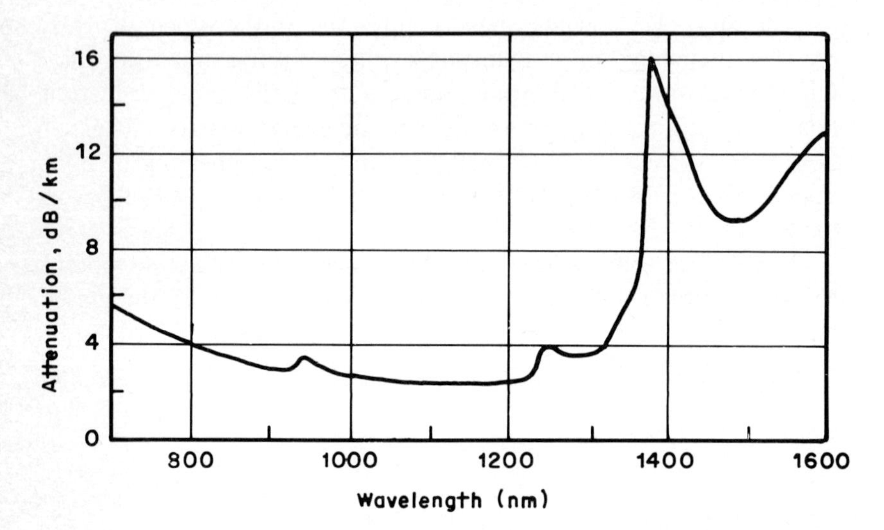

FIG. 17. – Spectral attenuation curve typical of the optical fibres obtained with the IVPO method.

A similar effect is obtained by the oxides: Al_2O_3, Ta_2O_5, ZrO_2, Nb_2O_3. An analysis of Figs. 14, 15 and 16 serves to evaluate the parameters which govern the choice of the materials to be used as dopants. By way of example, a step-index optical fibre with GeO_2 doped silica core and undoped fused silica cladding and numerical aperture NA = 0.2 can present a differential axial stress of about 100 kg/cm^2 assuming $E = 10^7$ and $\mu = 0.2$. The use of B_2O_3 as a cladding glass dopant can reduce stress problems, because this material balances expansions and viscosities between core and cladding. At the same time, it reduces the refractive index of the pure silica and allows greater numerical apertures to be obtained [30].

Boron, phosphorus and germanium cations are the most widely used materials for doping optical fibres with the most suitable characteristics (low attenuation and high numerical aperture), even though other dopant materials have been used for producing optical fibre preforms by the MCVD method. The respective oxides, together with the silica, are synthesized by the following reactions:

$$4BCl_3 + 3O_2 \longrightarrow 2B_2O_3 + 6Cl_2$$
$$4POCl_3 + 3O_2 \longrightarrow 2P_2O_5 + 6Cl_2$$
$$GeCl_4 + O_2 \longrightarrow GeO_2 + 2Cl_2$$
$$SiCl_4 + O_2 \longrightarrow SiO_2 + 2Cl_2$$

Figure 17 shows a spectral attenuation curve typical of the optical fibres obtained with such a manufacturing method.

If GeO_2 is used instead of P_2O_5, optical fibres are produced with a larger numeric aperture, as shown in Fig. 16. Nevertheless, the presence of the germanium oxide in the silica matrix does not sufficiently reduce the transformation temperature of the glass thus obtained (Fig. 13): to prevent the supporting tube undergoing a temperature so high as to cause undesired deformations a small percentage ($\sim 3\%$) of phosphorus oxide is generally added.

4.3.4 *Deposition phase*

Figure 18 shows the full process schematically. The vapours of $SiCl_4$, $GeCl_4$, $POCl_3$ are obtained by bubbling oxygen, whose

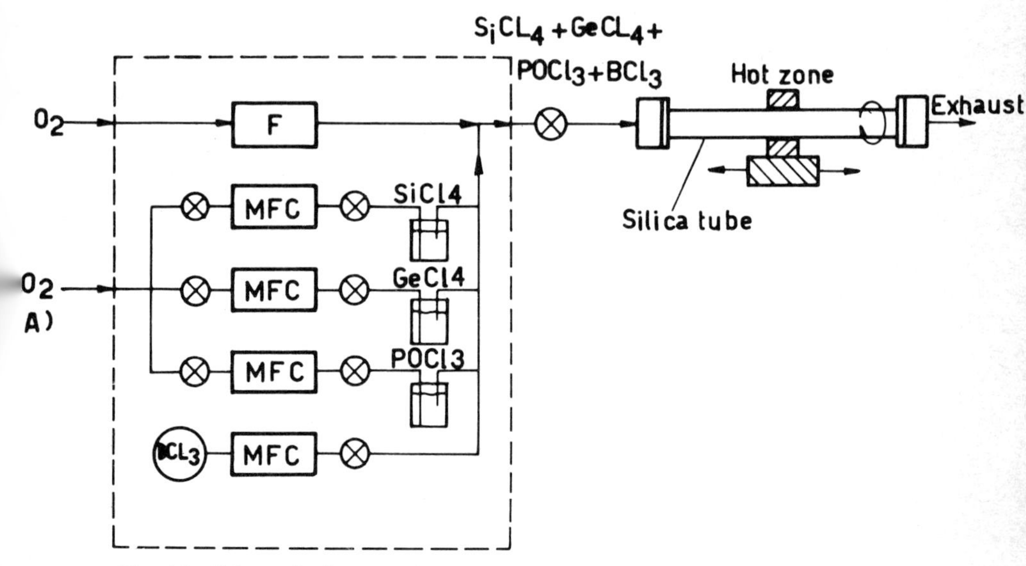

FIG. 18 – Schematic diagram of MCVD process, MFC denotes *Mass Flow Controller*, F denotes *Flow Meter*.

flow is controlled and kept constant by a Mass Flow Controller (MFC), in Drechsel bottles containing the halides in the liquid state. On the other hand, BCl_3 evaporates spontaneously and requires only a flow control.

All the reagents, present in the gaseous phase in percentages established according to the refractive index value required for the layer to be deposited, have further oxygen added; this addition gives rise to a suitable flow speed. They are then fed into the tube where the reactions take place and the glass is deposited as

TABLE II (After [32])

Method	$SiCl_4$ cc/min	$GeCl_4$ cc/min	PCl_3 cc/min	O_2 cc/min	He cc/min	Thikness μm	Layers n
Improved	350	350	40	1000	500	50	50
Conventional	60	60	—	1000	—	10	50

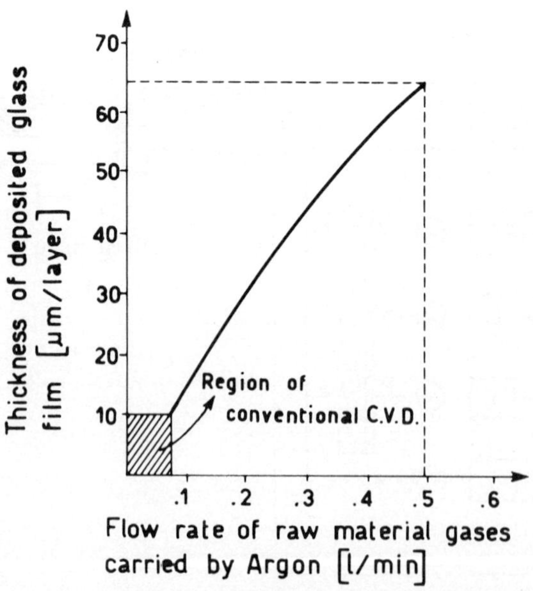

FIG. 19 – Flow rate of raw material gases vs. thickness of deposited glass film. (After [32]).

successive strata [31]. The temperature of the deposition zone is ~1500 °C, for the MCVD process. An inert gas such as Argon is often used instead of oxygen as the carrier gas. The oxygen quantity stoichiometrically necessary for the reactions is then added to the reagent stream.

By a further addition of He gas in the reagent stream, we obtained preforms for step-index optical fibres, with a GeO_2 - P_2O_5 - SiO_2 core and deposited strata 6 times thicker than those formed with the conventional process. With this process a large quantity of soot cannot be melted into an homogeneous glass film, because the excess oxygen and the carrier gas, such as Argon, absorbed by the soot are also absorbed by the glass film, thus producting bubbles during the subsequent fusion process. A gas with a small atomic ray, such as He, is likely to purge other gases and flows easily through the glass network structure during the fusion process, preventing bubble formation. Thus, the time required for preform preparation is greatly reduced. Table II and Fig. 19 show a comparison between the two methods [32].

4.3.5 *Collapsing phase and final preform characteristics*

At the end of the deposition process the preform is collapsed by increasing the traverse heater temperature value to obtain plastic silica. Thus, the surface stresses imply a decrease in the tube section sometimes up to a point where the internal hole is completely eliminated. This prevents both glass fractures during cooling, caused by the free internal surface, and pollution due to external agents during the subsequent drawing phase.

The high temperature (~1900 °C) required for tube collapsing implies the evaporation in the inner strata of a part of the previously deposited dopant oxide. This causes a dip at the core centre in the refractive index profile. This undesired effect, which cannot be fully eliminated, is greatly reduced by depositing a strate rich in dopant before the collapse or by keeping a dopant vapour stream in oxygen under proportions controlled during the collapsing phase [29].

The refractive index profile of the optical fibres obtained with the CVD methods varies discrete quantities. An analogous effect has been noted in the dopant concentration profile at the interface between successive strata of both the P_2O_5 silica doped core and GeO_2 silica doped core fibres. It consists in concentric bright and dark rings due to fluctuations of the P_2O_5 or GeO_2 content in each fused silica strate. In GeO_2 silica doped core fibres, GeO_2 concentration decreases towards the middle, as shown in Fig. 20*a*. On the other hand, in GeO_2 silica doped core fibres with P_2O_5 additive (2-3%) the P_2O_5 content in each deposited strate decreases towards the middle, just as the GeO_2 in the previous case, but the GeO_2 concentration profile presents an opposite slope (Fig. 20*b*).

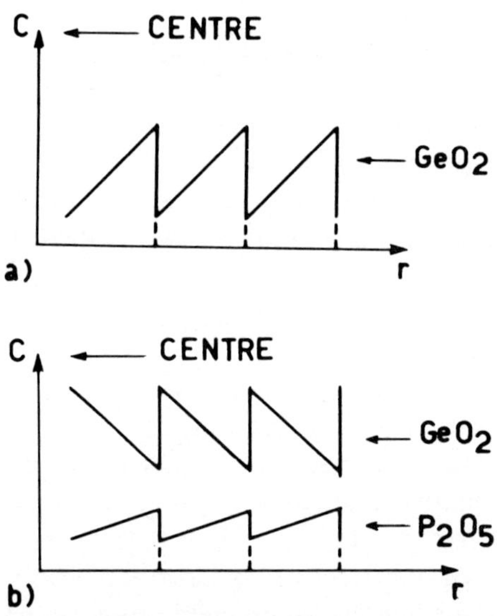

FIG. 20 – Concentration profile anomalies for step-index fibres *a*) GeO_2 doped silica core *b*) GeO_2-P_2O_5 doped silica core: r = preform radius; C = concentration of the dopant material. (After [34]).

The nonhomogeneous GeO_2 distribution can depend on:

1) GeO_2 outwards diffusion during fusion of deposited soot;
2) further chemical reaction between GeO_2 and $SiCl_4$ during deposition and fusion processes;
3) nonhomogeneous GeO_2 distribution in deposited soot due to chemical reaction kinetics influenced by both operating temperature and difference of free energy required for $GeCl_4$ and

$SiCl_4$ oxidation. Halide oxidation takes place after the metal-chlorine bond cracking due to thermal excitation and continues at different rates according to the various elements. There is not much information on the reaction kinetics. Nevertheless, the reaction rate between $SiCl_4$ and O_2 becomes sensitive at temperatures above 900 °C to form silicon oxychlorides and above 1000 °C to obtain SiO_2. This reaction is very rapid above 1300 °C.

The corresponding germanium reactions take place at slightly lower temperatures. This helps to explain this ring structure, as it is likely that when reagent mixture approaches the hot zone, GeO_2 is formed therein before silica deposition starts [33, 34].

All these factors warrant the first case (GeO_2 silica doped core), while in the second case (GeO_2 silica doped core plus P_2O_5) the presence of phosphorus may lead to a great decrease in the operating temperature. Therefore, while P_2O_5 (a relatively volatile component) diffuses towards outside and evaporates acting as did the germanium in the previous case, the latter finds a temperature which is sufficient for diffusion and densification towards the preform centre, but not for evaporation.

The result of these observations is that the refractive index profile slope of each strate can be adjusted by adjusting P_2O_5 and GeO_2 concentrations [34].

4.3.6 *Plasma-activated CVD (PCVD)*

Figure 21 shows the operating principle of this method. The reagent transformation reactions from chlorides to oxides are stimulated by a nonisothermal plasma. As in the previous case and as shown in Fig. 21, reagents are fed into a silica tube, whose internal pressure is kept at values between 1 to 50 torr by a vacuum pump. The tube is not traversed by the heater but by a microwave resonant cavity connected to a generator in the frequency range from 2 to 3 GHz with powers of about 100-500 W [35, 36]. Deposits of uniform thickness are obtained without tube rotation owing to the symmetry of the cavity and thus of the plasma zone.

The reaction takes place even if the tube is at room temperature, but in this case the deposited glass layer is cracked. To obtain a stable, crack-free glass, it is sufficient that the whole tube

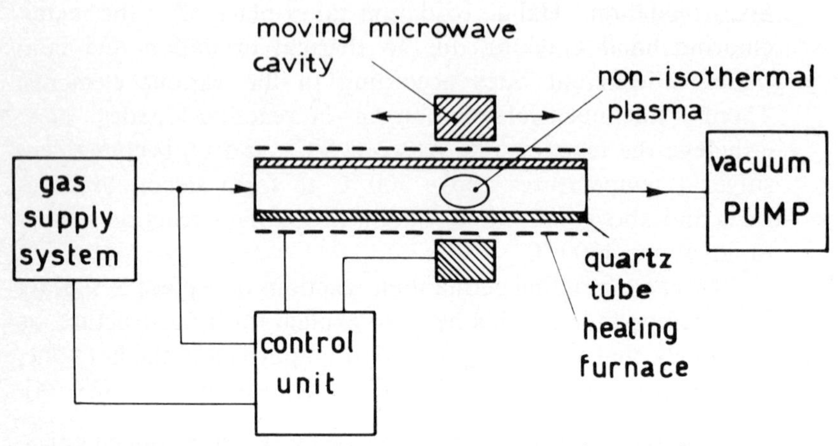

FIG. 21 – Schematic view of the PCVD. (After [35]).

be brought to temperatures between 1000 and 1100 °C by a resistance furnace. The chlorine presence in the deposited SiO_2 layers decreases as the temperature increases. At 1000 °C the chlorine content is still quite high, $\sim 0.1\%$, but the spectral region from 0.6 to 1.1 μm does not have an absorption band due to chlorine.

The plasma-activated CVD process fully prevents soot forming, because the thin glass film formed on the internal walls of the silica tube is obtained by surface-reactions. Furthermore, the deposition efficiency reaches 90-100% and deposition-rates in the range 50-100 μm/min can be obtained in steady-state and with a reaction zone of about 1 cm.

As the plasma instantaneously follows the reactor, the deposition zone can be moved at a high speed.

Therefore, the number of deposited layers is very high (usually it varies from 500 to 3000 without increasing the total deposition time) and a very fine control of the refractive index profile is obtained. The minimum attenuation value reached by GeO_2 doped silica fibres manufactured by the PCVD method is 1.4 dB/km at 1050 nm [17].

4.3.7 *Outside Vapour-Phase Oxidation (OVPO) Process*

The « outside process » consists in depositing the soot glass produced, by the same chemical reactions found in the IVPO processes, on the external surface of a silica or graphite rod.

As shown in Fig. 22, the reagent vapours are fed into a methane, propane or hydrogen burner having its flame directed to-

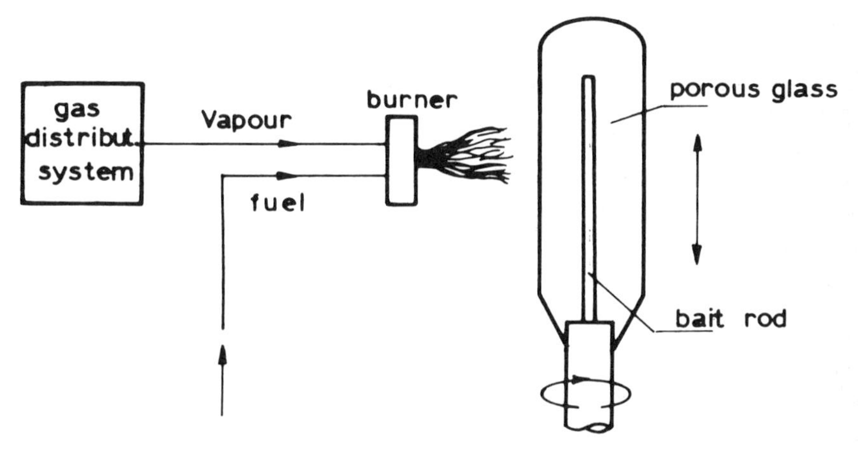

FIG. 22 – Schematic diagram of OVPO process.

wards the silica rod. The latter is rotated and moved along its axis so that the formed soot particles collide, deposit and form a porous glass layer. The concentration of dopants fed into the burner is different for each deposited layer, so as to modify radially the refractive index. Unlike the IVPO processes, the first deposited layers constitute the preform core while the last layers form the cladding. At the end of the deposition, the bait rod, on which the preform has been obtained, is taken away because it would constitute an anomaly in the refractive index profile at the core centre. The deposited material is easily contaminated by external agents and in particular by hydroxyl groups generated by the flame. The porous preform obtained is fed into a furnace where a high temperature dry gas stream penetrates through the glass pores and frees the glass from OH radicals. Notwithstanding this process, the fibres obtained from preforms produced with the OVPO process have an —OH content of 50-200 ppm, which is considerably higher than that of the fibres obtained with the IVPO process. However, by a new system consisting of chlorine drying a porous preform at the same time as consolidation [37], it is possible to reduce OH content. At the end of this purification phase, the preform is thermally consolidated to obtain a transparent glass after which it can be drawn into a fibre.

Using this technique, P_2O_5 doped or GeO_2 doped optical fibres with characteristics similar to those of the IVPO techniques have been obtained; the only exception is the higher attenuation peak in connection with the absorption of the hydroxyl groups [27, 30].

4.3.8 *Other CVD synthesis techniques*

Some processes do not fall within the two previously described categories but use the CVD technique for silica glass production. They are:

a) Silica glass preparation using a CO_2 laser [38].

Figure 23 shows a schematic diagram of this process, which allows the manufacture of silica glass rods free from imperfections and with a low content of OH radicals ($\sim$ 3 ppm). The heat source is a CO_2 laser whose radiation, absorbed by the silica glass, permits high temperatures ($\sim$ 1700 °C) necessary for the process, without the contamination risks involved in other methods.

The laser beam is scanned to obtain a uniform power distribution and to radiate a SiO_2 seed. The latter is placed in front of a nozzle within a glass vessel, with the aim of both isolating the reaction zone from its surroundings and preventing pollution by dust and water.

A pure silicon tetrachloride vapour, obtained with the methods previously described, is transported by oxygen, flows through the nozzle and impinges on the SiO_2 seed, where, because of the high temperature, it reacts with the oxygen and forms a SiO_2 soot.

The latter is melted and vitrified by the laser beam so that a transparent silica glass layer grown on the seed is obtained. To obtain a constant diameter silica glass rod, the seed is rotated and moved towards the glass vessel outside at the same speed at which silica grows.

The system presents a high deposition efficiency, as a 200 W-CO_2 laser allows the production of a silica glass rod 10 mm in diameter and 50 mm in length in a time of about 2.5 hours.

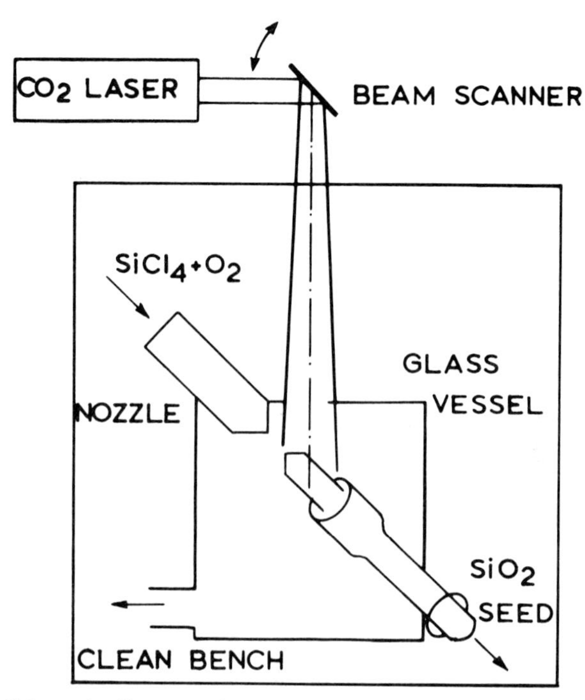

FIG. 23 – Schematic diagram of the preparation system of SiO_2 glasses using a CO_2 laser. (After [38]).

b) Continuous manufacture of optical fibre preforms with high silica content [39].

This method, called by the authors « Vapour phase Verneuil method », allows a continuous preparation of preforms with high silica content. This characteristic and the growing direction of the synthesized glass particles distinguish this method from the more conventional methods.

In fact, the preform grows in the axial direction through a deposition on the end surface rather than in the radial direction (deposition on the side surface) as in the OVPO process.

Figure 24 shows the system schematically. The reagent gases ($SiCl_4$, $GeCl_4$, $POCl_3$) are fed into the oxy-hydrogen burner, whose flames are directed towards a circular target which receives the resultant flow. The flames allow the preparation and deposition of the glass soot on the circular target as porous rod glass.

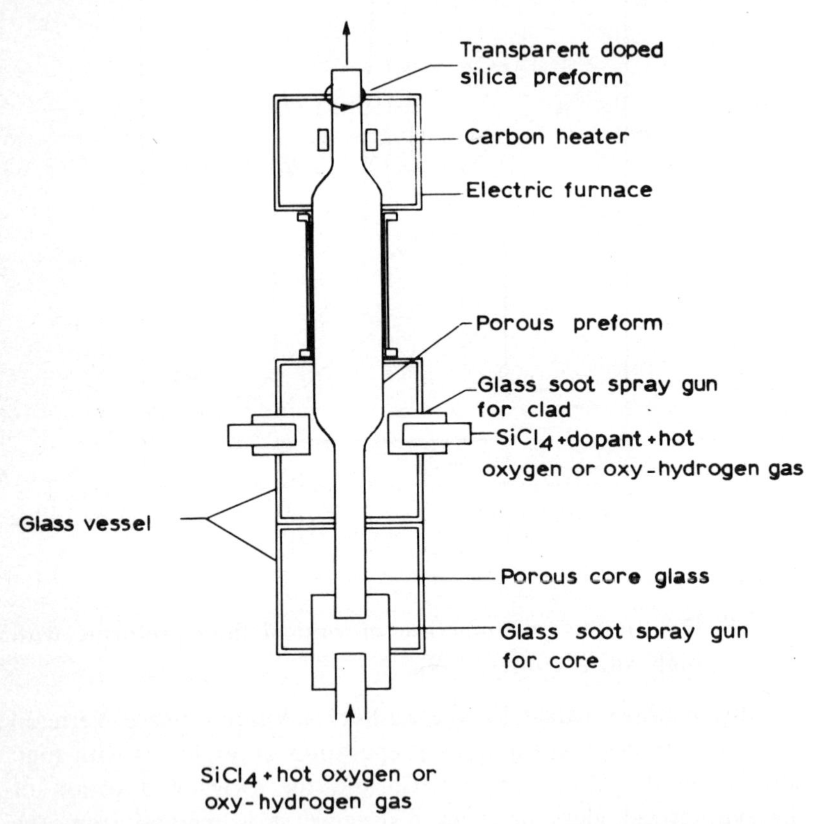

FIG. 24 – Schematic diagram of Vapour Phase Verneuil method. (After [39]).

With the aim of obtaining a diameter uniformity and a continuous preform growth, as in the previous case, the target is rotated around its axis and moved upward at the same speed at which the glass deposit is growing.

In this way the fibre core is formed, while a second burner or a set of burners perpendicular to the rod glass axis deposit on the external surface a glass particle layer, with a lower refractive index, which constitutes the cladding. In the deposition zones, the preform is protected through a glass vessel against atmospheric contamination.

Oxy-hydrogen burners can be substituted by a RF plasma torch, DC plasma torch, CO_2 laser or electric resistance heat torch; in this case the halides react directly with the oxygen and thus the presence of water in the preform is limited.

The refractive index preform profile can be controlled by adjusting the dopant space distribution in the burners.

Graded-index fibres can be obtained by bringing the second burner nearer to the first, so as to ease the diffusion process of the dopants in the hot zone and then to decrease the refraction index gradually and radially [40].

The quasi-parabolic profile obtained is similar to that of the « SELFOC method », because both techniques make use of a diffusion process. A finer control can be achieved by using more burners so that the dopant space distribution in the flame is a parabolic curve. The porous preform obtained is fed into a furnace for the annealing process and then takes the final aspect of a solid, transparent glass free of rough imperfections. Typically, the preforms obtained with this method allow GeO_2 - P_2O_5 doped fused silica optical fibres 130 μm in external diameter and 60 μm or 90 μm in core diameter to be drawn.

The spectral attenuation in the region between 750 and 1100 nm varies about 5 dB/km, with the exception of the peak at 950 nm (9-10 dB/km), corresponding to the absorption of the hydroxyl groups presenting a concentration of the order of 5 ppm.

c) P_2O_5 - GeO_2 - Ga_2O glass fibre for optical communications [41].

The double crucible method is used for fibre manufacturing, but the glass materials are prepared employing the flame hydrolysis method to obtain a lower impurity content.

The vapours of the chlorides of the materials used ($POCl_3$, $GeCl_4$, $GaCl_3$) are fed into an oxy-hydrogen torch and the resulting soot oxides are collected in a high purity quartz crucible.

The compounds obtained are then fused for approximately 1 hour at about 1440 °C in a quartz crucible and made transparent by bubbling with dry gases.

The two crucibles are then joined to form the double crucible and to draw the fibre at temperature of about 900 °C.

The refractive index of these glasses is controlled by acting on the Ga_2O_3 percentage. The presence of Ga_2O_3 also improves the chemical stability of these compounds, whose properties are shown in Table III.

TABLE III (After [41])

	Core glass	Clad glass
Refractive indices	1.6015	1.5908
Expansion coefficients	$6.48 \times 10^{-6}/°C$	$6.49 \times 10^{-6}/°C$
Transition temperatures	632 °C	620 °C

This method uses low softening glasses and thus silica crucibles can be used for both drawing and melting. Furthermore, it employs crucible techniques which permit continuous drawing. Minimum attenuations of 15 dB/km at $\lambda = 0.83$ μm have been obtained, and better results are foreseen.

4.3.9 *Plastic cladding silica core fibres and single-material fibres*

Optical fibres with a structure different from that of the fibres so far described have been prepared with different methods. In the plastic clad, silica core fibre, the core is made of pure silica, while the cladding, made by polymers, can either adhere to the core or be extruded as a loose jacket around the core (Fig. 25). For the latter, it is advisable to use polymers having suitable extrusion characteristics and a low refraction index (FEP teflon, PFA teflon, etc.) [42, 43] and high purity silica (Suprasil, Spectrosil) with a low content of hydroxyl groups.

Low attenuation (below 3 dB/km) and high numerical aperture (above 0.4) fibres are obtained, even if the polymer presents a $5 \cdot 10^5$ dB/km attenuation due to its crystalline structure. For plastic coatings adhering to the fibre, the best results have been obtained with highly transparent (900 dB/km at 0.77 μm) silicone resins, protected by a polymer layer without special optical pro-

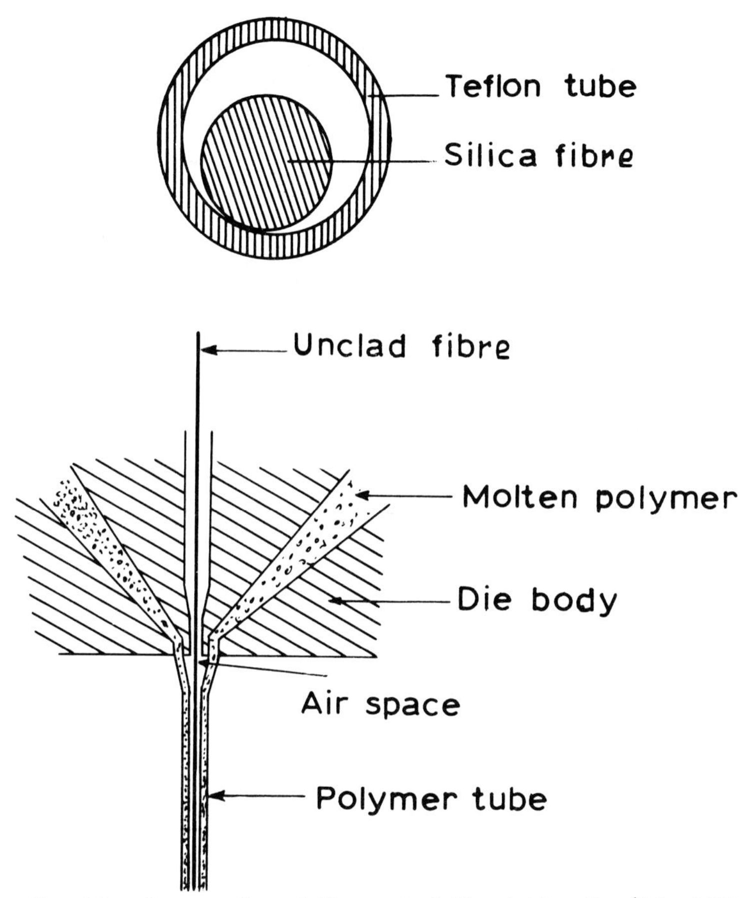

FIG. 25 – Cross section of fibre and of fibre tubing die. (After [42]).

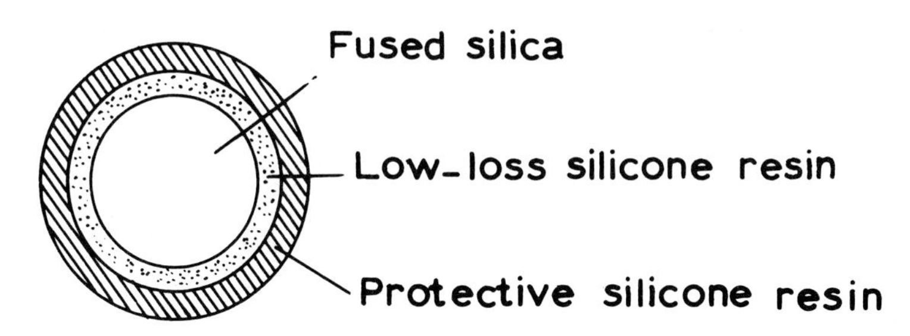

FIG. 26 – Cross section of silicone-clad fused-silica-core fibre. (After [44]).

perties and with a purely mechanical function, as shown in Fig. 26 [44].

Monomode and multimode optical fibres have also been obtained by using silica only. The core is a small-diameter rod centred with respect to the external protection tube by a thin quartz support (Fig. 27).

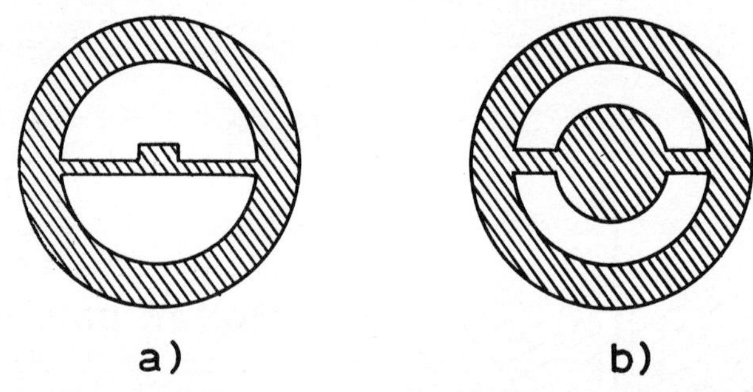

FIG. 27 – Single material fibres. Cross-section of rectangular core in single-mode fibre (*a*) and cylindrical core in multimode fibre (*b*). (After [45]).

The behaviour of such a fibre is similar to an uncovered fibre made of the same material. The use of Spectrosil WF quartz has produced fibres with a minimum attenuation of 3 dB/km [45].

4.4 Drawing of high-silica fibres

4.4.1 *Drawing equipment*

The equipment for pulling optical fibres with a high silica content from preforms obtained with one of the methods discussed above is schematized in Fig. 28. Equipment consists of the following basic component parts:

1) a preform feeding apparatus;

2) a heater;

3)-4) a temperature regulating and measuring apparatus;

5)-6) a fibre diameter regulating and measuring apparatus;

7)-8) a fibre coating and drying system;

9) a fibre pulling and winding drum.

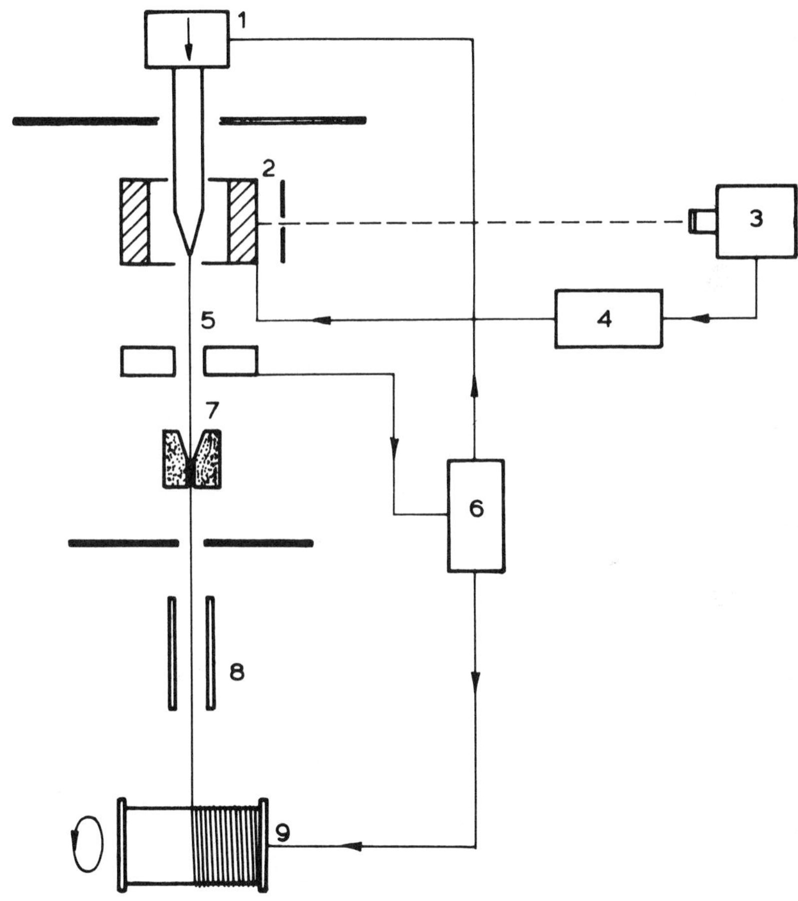

FIG. 28 – Schematic view of the equipment for drawing and coating silica-glass fibres. 1) Preform - 2) Heating furnace - 3) Pyrometer - 4) Temperature regulator - 5) Optical equipment for measuring the diameter of the fibre - 6) Speed regulator - 7) Fibre coating apparatus - 8) Tubolar furnace for coating drying - 9) Drawing and winding machine.

The preform is attached to a feeding system 1 which allows it to be inserted and to move at strictly controlled speed into a high temperature area 2 (above 2000 °C) produced by heaters such as resistance or induction furnaces, oxy-hydrogen burners, CO_2 lasers, which are described later on in more detail.

Temperature in the hot area is normally regulated by an optical pyrometer 3 or by a tungsten-rhenium or graphite-borongraphite thermo-couple and is kept constant by a regulator 5 which controls the power supply to the heater.

In this area, the pulling process is subject to vitreous stresses, surface tension effects and quenching rates and may be regarded as a problem of fluid dynamics.

To meet the requirements for stringent tolerances on the mechanical characteristics of optical fibres used in the telecommunications field, it has been necessary to examine the process from the fluid dynamics standpoint.

In particular, it is interesting to know the response of these processes to different stresses, especially those causing variations in fibre diameter. Subsequently, a model will be described which is based on the theoretical study of the dynamic response to a variety of physical disturbances which may be represented by mechanical vibrations, thermal transients, stream of environment gas and, in an extreme case, also by acoustic noise. This model is suggested by F. T. Geiling [46].

Let us assume that

z = axial coordinate
r = radial coordinate
v = axial component of velocity
u = radial component of velocity
ϱ = fluid density (assumed constant)
σ = surface tension
μ = Newton's viscosity as a function of temperature
p = pressure.

In a system of cylindrical coordinate (r, z, θ) Euler's equations for the conservation of mass and momentum are:

$$v_z + u_r + u/r = 0 \tag{1}$$

$$\varrho[u_t + uu_r + vu_z] = \frac{\partial \tau_r}{\partial r} + \frac{1}{r}\tau_r + \frac{\partial \tau_{rz}}{\partial z} - \frac{1}{r}\tau_\theta \tag{2}$$

$$\varrho[v_t + uv_r + vv_z] = \frac{\partial \tau_z}{\partial z} + \frac{\partial \tau_{rz}}{\partial r} + \frac{1}{r}\tau_{rz} \tag{3}$$

where τ_r, τ_z, τ_{rz} indicate the stress components, whereas indices r, z, t indicate partial derivatives.

For a Newtonian incompressible liquid the stress components are linked to one another through the following relationships:

$$\begin{aligned} \tau_z &= -p + 2\mu \frac{\partial v}{\partial z}; \qquad \tau_\theta = -p + 2\mu \frac{u}{r}, \\ \tau_r &= -p + 2\mu \frac{\partial u}{\partial r}; \qquad \tau_{rz} = \mu \left(\frac{\partial v}{\partial r} + \frac{\partial u}{\partial z} \right). \end{aligned} \tag{4}$$

We now introduce relation $r = a(z, t)$ and boundary conditions for relation (1)-(3):

— at $z = 0$: initial part of the viscous zone at preform side

$$\begin{aligned} v(r, 0, t) &= v_0(r, t) \\ u(r, 0, t) &= u_0(r, t) \\ a(0, t) &= a_0(t) \end{aligned} \tag{5}$$

— at $z = L$: final part of the viscous zone, at fibre side

$$v(r, L, t) = v_L(r, t) \tag{6}$$

— at $r = a(z, t)$: the kinematic condition

$$v = u \frac{\partial a}{\partial z} + \frac{\partial a}{\partial t}. \tag{7}$$

As the fluid flow in the draw down region is essentially unidirectional, it is possible to derive from (1) and (3) the equations for conservation of volume and axial momentum:

$$(a^2 v)_z + (a^2)_t = 0 \tag{8}$$

$$\varrho(a^2 v^2)_z + \varrho(a^2 v)_t - 3(a^2 \mu v_z)_z - \sigma a_z = 0 \tag{9}$$

which estabilish that the draw force is uniform along the fibre and vary with t only. In case of steady state (independent of time) relations, (8) and (9) lead to the following solutions

$$a^2 v = \text{cost.} = Q \tag{10}$$

$$3a^2 \mu v_z - \varrho Q v + a\sigma = C. \tag{11}$$

Leaving out radius a from (10) and (11) we introduce the normalized coordinate $\psi = v/v_0$, $\xi = z/L$, $(\ )' = \partial(\ )/\partial\xi$ to write the equation

$$\psi' - D\psi = - W_e \psi^{\frac{1}{2}} + R_e \psi^2 \tag{12}$$

where $D = LC/3a_0^2 v_0^2 \mu_0$: non-dimension quantity equivalent to C in (11)

$W_e = \sigma L/3a_0 v_0 \mu_0$: Weber's number

$R_e = v_0 L\varrho/3\mu_0$: Reynold's number

where the elementary solution when $W_e = R_e = 0$ is: $\psi(0) = = \exp[\xi \ln E]$

for the following boundary conditions $\begin{cases} \psi = 1 & \text{with} \quad \xi = 0 \\ \psi = E & \text{with} \quad \xi = 1 \end{cases}$

where $E = v_L/v_0 =$ « draw-down ratio »; $D = \ln E$.

Let us write the first order solutions of (7) and (8)

$$\begin{aligned} \bar{a} &= a(z)[1 + \hat{a}(z, t)] \\ \bar{v} &= v(z)[1 + \hat{v}(z, t)] \\ \bar{\nu} &= \nu(z)[1 + \hat{\nu}(z, t)] \end{aligned} \tag{13}$$

where:

$a(z) =$ radius, $\hat{a}$
$v(z) =$ velocity, $\hat{v}$
$\nu(r) =$ kinematic viscosity, $\hat{\nu}$
} first order perturbations

and assume

$$\zeta = \alpha\xi; \qquad \alpha = \ln E/2; \qquad \tau = v_0 \alpha t/L. \tag{14}$$

Substituting (13) in Eqs. (7) and (8) we obtain the following equation

$$\left(\frac{\partial}{\partial\xi} + \frac{1}{\psi}\frac{\partial}{\partial\tau}\right)(\hat{a}' + \alpha\hat{a}) = \frac{\exp[-\alpha\xi]}{2}\hat{\nu}' \tag{15}$$

$$\psi = \exp[2\alpha\xi]$$

$$R_e = 0, \qquad W_e = 0$$

quenched inertialess base state

$$\frac{\partial^2 \hat{a}}{\partial \zeta^2} + \frac{\partial^2 \hat{a}}{\partial \zeta \, \partial \tau} \exp[-2\zeta] = \frac{\partial \hat{v}}{\partial \zeta} \tag{16}$$

$$\psi = \exp[2\alpha\xi]$$

$$R_e \neq 0, \qquad W_e = 0.$$

unquenched inertialess base state.

Equation (16) has the following general solution:

$$\frac{\partial \hat{a}}{\partial \zeta} = f(\tau + \tfrac{1}{2}\exp[-2\zeta]) \Rightarrow \hat{a}_c = \int f(\tau + \tfrac{1}{2}\exp[-2\zeta])\, d\zeta + \varphi(\tau)$$

$\varphi(\tau)$ = arbitrary time function.

If $\hat{a}(0, \tau) = \sin(\omega\tau)$ (particular solution) we can reconstruct a solution in the following form: $\hat{a}(\zeta, \tau) = A_1(\zeta)\sin\omega\tau + A_2(\zeta)\cdot$ $\cdot\cos\omega\tau$; with $A(\omega, \zeta_L) = [A_1^2(\omega, \zeta_L) + A_2^2(\omega, \zeta_L)]^{\frac{1}{2}}$.

In Fig. 29*a*, the amplitude of the radial perturbation at $\zeta = 0$, for $\zeta_L = 2$, is shown, normalized with respect to $\hat{a}_0$ as a function of ω.

Peaks there illustrated are termed « draw resonances », and might be natural frequencies of interaction between the inertial system and some restoring force.

In Fig. 29*b* and *c* the profiles of radial perturbations $\hat{a}(\zeta, \tau)$ are indicated for $\omega = 100$ and $T = 0$ and $T = 0.25$ respectively, which illustrate the space amplification of surface perturbations which occur along the draw path [46].

The temperature when the preform is introduced into the draw-down zone must be such that the preform can reach its softening point where the quartz has sufficient viscosity to be drawn.

The fibre is drawn by the drum 9 at a pulling speed in the range of 0.5-1.5 m/s and wound on the same drum.

The fibre diameter is established considering that the ratio of feeding speed to pulling speed is equal to the ratio of the cross-sections of the fibre and the preform respectively.

Between the heater and the winding drum the fibre is checked by the equipment for the on-line measurement of diameter 5,

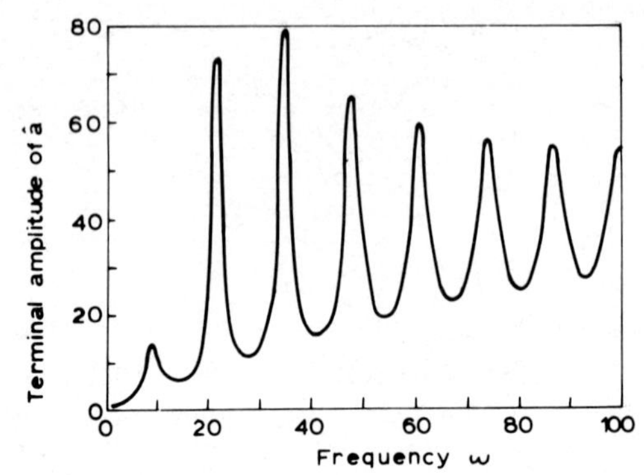

FIG. 29 *a*) – Tensile fibre model: frequency response of surface perturbation at $\zeta_L = 2$ for base state. (After [46]).

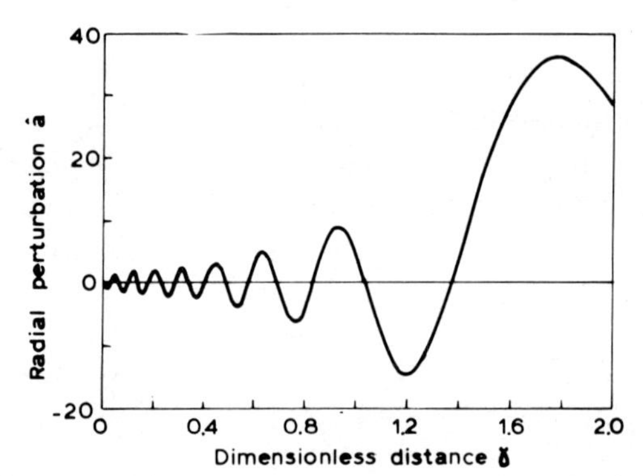

FIG. 29 *b*) – Surface pertubation for base state with $\omega = 100$, $\zeta_L = 2$, at $T = 0$. (After [46]).

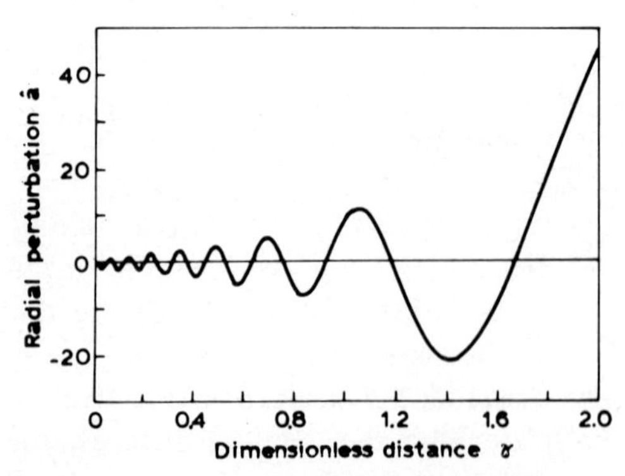

FIG. 29 *c*) – $\hat{a}$ at $T = 0.25$. (After [46]).

which allows a continuous control of the diameter and, if necessary, of the regularity of the cross-section.

Together with the regulator 6 which interlocks the feed rate of the preform and the pulling rate of the fibre, this apparatus keeps constant the fibre diameter during the whole process, with a diameter stability which may reach 0.1 %.

During the drawing in the interface area between the preform and the fibre, a zone is produced which is generally termed « neck down » and which must take on a conical shape such as to allow the regularity of the cross-section and the uniformity of the fibre diameter.

Within this zone the cladding and the core keep their respective geometrical relations and, as a result, a fibre is obtained having a core-to-clad ratio and a refractive index profile which are faithfully coincident with those of the preform. In some cases the diameter reduction ratio is greater than 300:1. Before it is wound on the drum the fibre is covered with a thin polymer layer by an apparatus which is generally formed by a nozzle 7 containing the polymer into which the fibre is dipped for coating and a drier 8. This is required in order to improve the mechanical properties of the fibre and to allow the fibre to be handled and protected from damage to its outside surface which might modify its optical characteristics.

4.4.2 *Heater*

The implementation of low loss high silica optical fibres has required the improvement of high temperature heating sources up to and above 2000 °C for the drawing process.

As already mentioned, heat sources capable of reaching temperatures in the 1800-2500 °C range which are required for drawing silica preforms into optical fibres include oxy-hydrogen burners, electrical resistance or induction furnaces, and CO_2 lasers [47, 48].

a) Oxy-hydrogen burners

Of all burners, the oxy-hydrogen burners in various configurations (for example ring-burners) are the most economical and the easiest to use. However, insufficient control of temperature, due to flame instability, may produce differences in the fibre diameter which makes them unsuitable for their application.

The fibre diameter constancy is a necessary condition to obtain loss attenuations, in particular at joints. In spite of this, it has also been possible to produce optical fibres with these heat sources having minimum transmission loss as low as 2 dB/km. This means that, besides an inevitable increase of some p.p.m. in the content of hydroxyl groups, no further appreciable contamination occurs during the drawing process.

b) Resistance furnace

A resistance furnace offers advantages similar to those of the oxy-hydrogen burners, given that is simple both to construct and use.

It can also, however, provide a stricter control of temperature stability and uniformity in the hot zone, both longitudinally and radially. Thus the preform also keeps its circularity during its conversion to fibre.

Figure 30 shows a cross-section of this apparatus. Additional characteristics of the furnace are:

— low thermal mass and high stability, so that the response time to temperature variations may be very short;

— very good resistance to thermal shocks, which makes it possible to reach the operating temperature from room temperature in about 1 minute. Usually the material used for the resistance furnace is graphite, which fully meets the above requirements, but tungsten can also be employed. If graphite is used, material characteristics permit the resistance to be obtained by processing a single block, whereas with tungsten it is necessary to guarantee a sufficient elasticity and a good temperature uniformity to avoid fractures due to thermal stress. Therefore a group of filaments is required which are adeguately stranded and connected to one another.

A controlled atmosphere of inert gases, such as Argon, inside the furnace prevents heater oxidation at the high operating temperature, as this would adversely effect heater operation and life. Moreover, is possible to regulate fibre diameter by controlling gas flow [49].

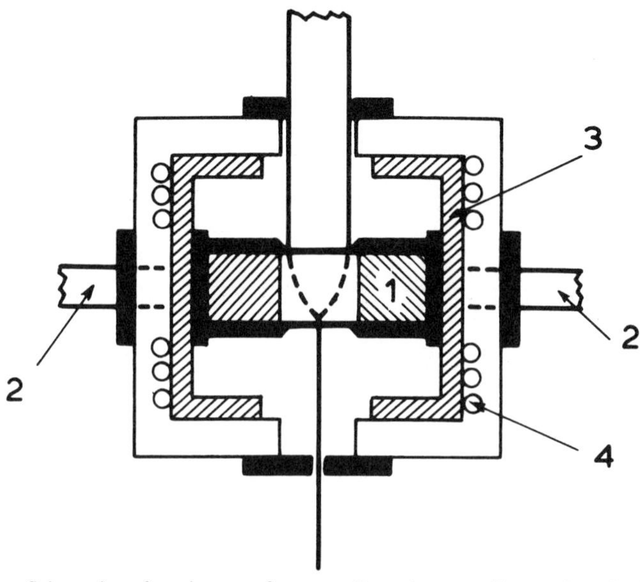

FIG. 30 – Schematic of resistance furnace. 1) resistance; 2) supply of electrical current; 3) thermal insulating material; 4) water cooling.

This type of heater still suffers, even if to a lesser extent than the type described above, from some contamination of the protective atmosphere of inert gas and, as a consequence, of the fibre surface, by the heating unit. Additionally, this unit still does not have a particularly long operational life [48, 50, 51].

c) Induction furnace

The resistance limits are partly overcome by induction furnaces, in which the heater is formed by a susceptor heated by r.f. induction.

The susceptor normally consists of a small, cylindrical graphite element that is easily built because of its small size and offers a sufficiently fast response to variations of input power, thus allowing a good automatic control of temperature.

Recently [51], a furnace has been developed for operations with long-life expectancy and minimal furnace atmosphere contamination in an oxygen-bearing atmosphere, which is featured with the high-frequency (MHz) induction of an oxide susceptor.

The basic material of the susceptor is yttria stabilized zirconia, which offers the lowest resistivity and vapour pressure of all materials suitable for this purpose, thus allowing maximum coupling efficiency and minimum contamination of the furnace atmosphere.

Figure 31 shows a cross-section of the oxide susceptor furnace. The shape of the susceptor is such that maximum resistance to thermal shocks and maximum flexibility for the heating profile are obtained.

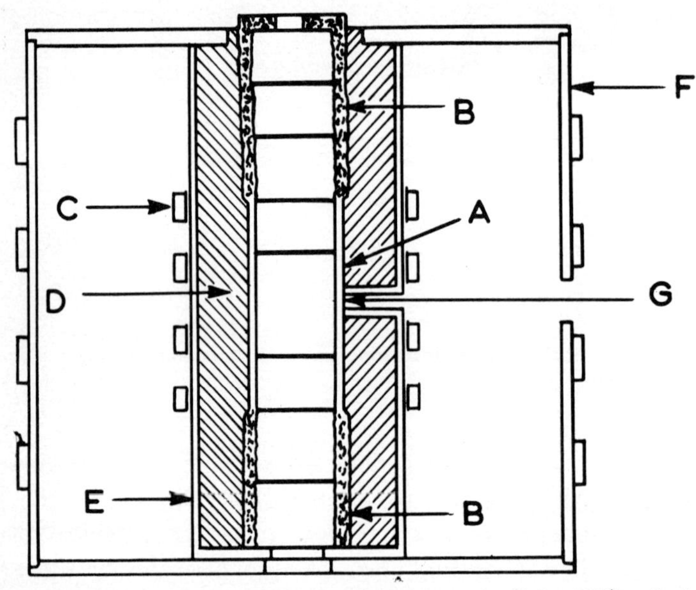

FIG. 31 – Vertical section of zirconia induction furnace. (After [51]). *A)* zirconia susceptor ring; *B)* zirconia support rings; *C)* induction coil; *D)* zirconia insulation; *E)* fused quartz; *F)* copper shell; *G)* temperature monitoring.

To reach the maximum efficiency for the induction heater of a hollow cylinder, it is required that the penetration depth of the e.m. field be nearly equal to the thickness of the cylinder walls.

Walls of a thickness considerably smaller than the penetration distance of the e.m. field would increase coil losses, whilst thicker walls are heated with a good efficiency only on the surface.

For processes at 2000 °C with a yttria stabilized zirconia susceptor and walls of a few millimeters, frequencies in the range of 1-10 MHz are required.

The resistivity at low temperatures of the zirconia susceptor

($>10^4$ ohm/cm) is too high to allow direct coupling to the induction field.

The ZrO_2 susceptor is pre-heated up to 1000 °C through coupling with a carbon rod. At this temperature ZrO_2 presents a sufficiently low resistivity to be directly coupled with the r.f. source. At about 1400 °C the carbon rod can be taken out without causing thermal shocks to the susceptor. In this way the operating temperature can be reached in about 30 minutes.

With different heating profiles it is possible to draw the fibre from very low pulling rates up to 10 m/sec.

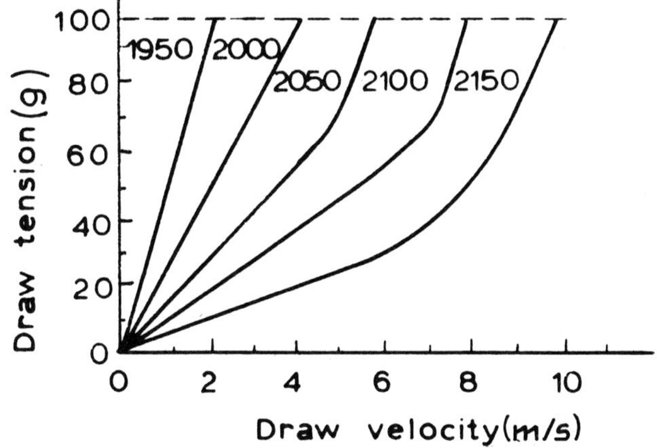

FIG. 32 – Fibre drawing parameters. (After [51]).

Figure 32 shows the measured functional relationship between the drawing parameters of tension, temperature and draw velocity under one set of conditions.

d) CO_2 laser

As has been previously mentioned, with both resistance furnace and of induction furnace, the furnace shape and the methods of use are particularly critical for obtaining the optimal properties of optical fibres.

From this standpoint, the drawing procedure using a CO_2 laser offers better characteristics and a cleaner drawing atmosphere. However, it requires a sophisticated optical system for the distribution of the laser beam over the preform along a given axis.

A laser draw schematic is shown in Fig. 33 [52]. A CO_2 laser beam is reflected by a rotating mirror 1 which is assembled slightly eccentrically so as to produce a radiation ring that in

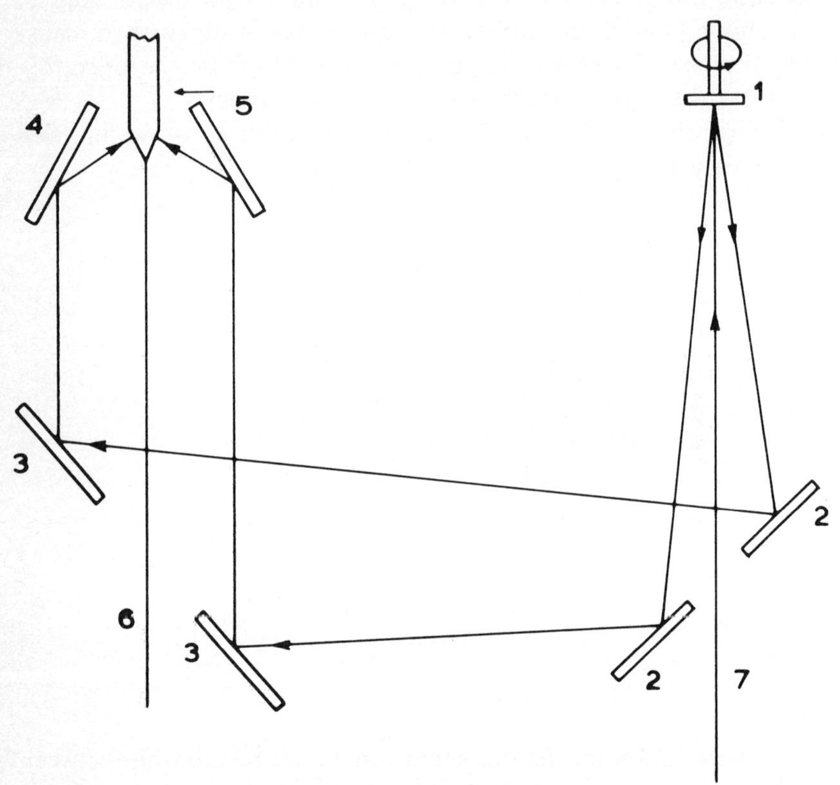

FIG. 33 – Laser draw schematic. (After [52]). 1) rotating disc mirror; 2-3) mirrors; 4) focus mirror; 5) preform; 6) optical fibre; 7) CO_2 laser beam.

turn is reflected twice by mirror 2 and 3 and directed to a conical reflector, 4, which focusses it on the preform. The heated zone can consequently be very small.

By virtue of this fact and of the rapid response to thermal shocks, the system presents a small thermal inertia.

In laser fibre drawing, however, the 10.6 μm CO_2 laser radiation is only absorbed within a thin surface layer (~ 15 μm), which results in an excessive increase in surface temperature [53].

Heat is transferred within the preform via lattice and radiation thermal conductivity. Now, as a high drawing temperature and a low radial temperature gradient are indispensable for ensuring constant drawing conditions and a reasonable drawing tension level, this might mean a reduction in the drawing velocity.

An increase in laser power does not result in an increase in drawing speed because, as the radiation is only absorbed on the surface, a considerable surface vaporization occurs.

4.4.3 *On-line measurement of optical fibre diameter*

The drawing process must allow the fabrication of optical fibres with good geometrical characteristics. The mechanical and optical properties of the resulting fibre are strongly affected in the course of the drawing process by controllable parameters, the most important of which are the constancy of the core and cladding diameter and the appearance of the outside surface. In practice the propagation of modes inside the fibre is affected by variations in core diameter, whereas the mechanical strength is endangered by flaws on the outside surface.

A suitable combination of temperature and drawing speed such as to reduce mechanical stresses leads to the creation of a smooth and regular surface on the outside of the fibre.

Starting assumptions for fibre diameter constancy are: preform regularity and drawing speed uniformity. However the correction of unavoidable defects makes it absolutely necessary to provide an on-line measurement of the fibre diameter and to derive from this measurement an appropriate signal controlling a servo-device so that any variation in fibre diameter can be counterbalanced in the course of the drawing process.

Normally, the correction is carried out on the rotation speed of the drum on which the fibre is wound. If the difference from the required diameter is progressive (for instance, owing to the taper of the preform), a check is also required for the feeding speed of the preform, so that the drawing rate is kept around the optimum speed, also with respect to furnace performance (such as operating temperatures, dimensions of the hot zone, etc.).

The on-line measurement of the fibre diameter can be accomplished in different ways, normally including the adoption of optical measurement systems, since it does not appear suitable to use mechanical tracer points operated by fibre flaws.

Leaving aside the difficult problems arising in the manufacture of a tool for the dynamic detection of diameter variations in the region of one micron, it should be stressed that any contact with the uncoated fibre that is not strictly necessary is to be avoided.

a) Method of fibre profile projection

One of the optical methods for on-line measurement of fibre diameter during the drawing process, with detection of a special signal for adjusting the drum rotation speed is that based on the projection, after magnification, of the fibre profile on a set of

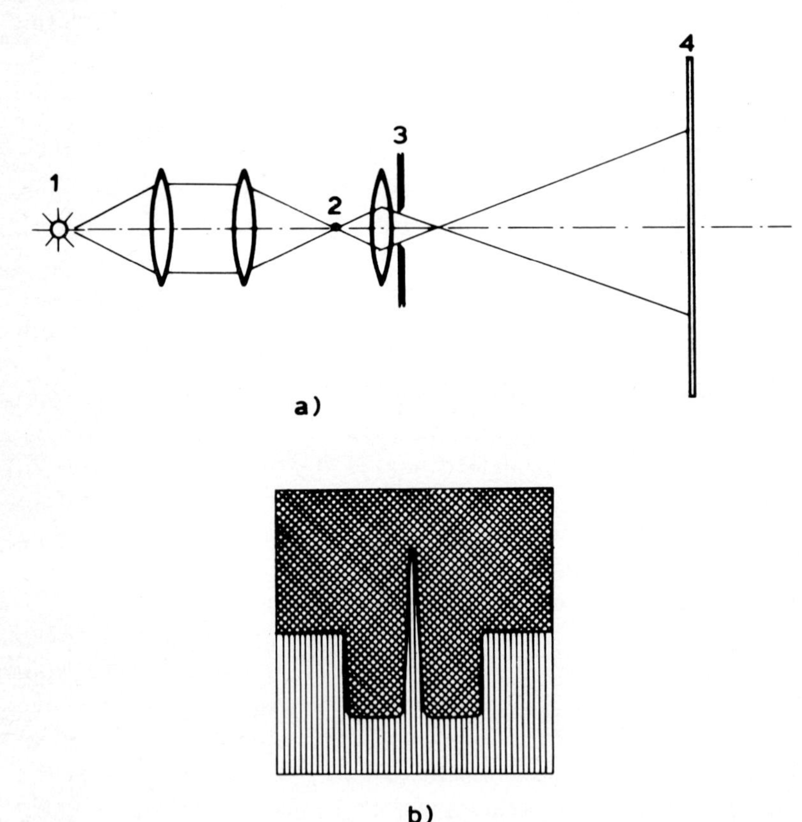

FIG. 34 – *a*) Schematic view of the experimental arrangement of the fibre profile projector: 1) source (LED or incandescent filament); 2) optical fibre (cross section), 3) diaphragm, 4) linear photodiode array (512 or 1024 elements). *b*) Visualization of the array signal.

photodiodes: the number of darkened photodiodes (Fig. 34) is proportional to the diameter dimension.

Owing to the fibre transparence, the central part of the path is lit.

This method is sensitive to displacements of the fibre along the lens axis, which may produce out-of-focus of the track on the sequence of photodiodes.

b) Dynamic method

This procedure of adjustment of optical fibre dimensions is based on a dynamic measurement [54].

A swinging mirror is used to divert an accurately collimated laser beam through a fibre.

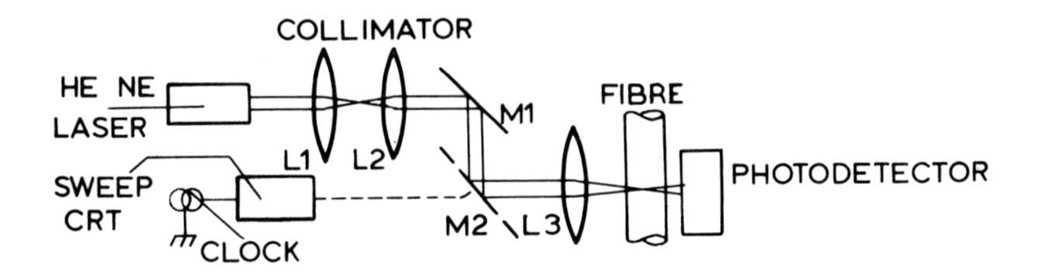

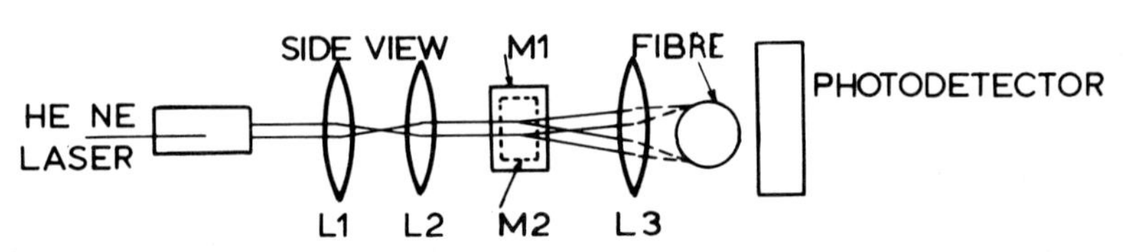

FIG. 35 – Experimental arrangement for measuring fibre diameter. Optical components are emphasized. An oscillating mirror (M_2) is used to deflect a laser beam at a constant rate through a small angle and then returns the beam to its original start position. The photodetector converts the laser beam shadow, cast behind the fibre, into an electrical pulse whose width is directly proportional to the fibre o.d. (After [54]).

If the angular speed of the beam is known, the time during which the beam is intercepted by the fibre can be correlated with the diameter.

It is necessary to record the response of a photodetector behind the fibre (Fig. 35).

Also, it is required that the fibre axis does not considerably change its position during measurements.

c) Interferometric method

This method is based on the comparison between the wave front reflected by the fibre surface and that reflected by a reference surface (bent or flat) [55].

The fibre and the reference surface (which must be accurately machined and, if bent, of known diameter) are to be placed on the arms of a Linnik interferometer (Fig. 36).

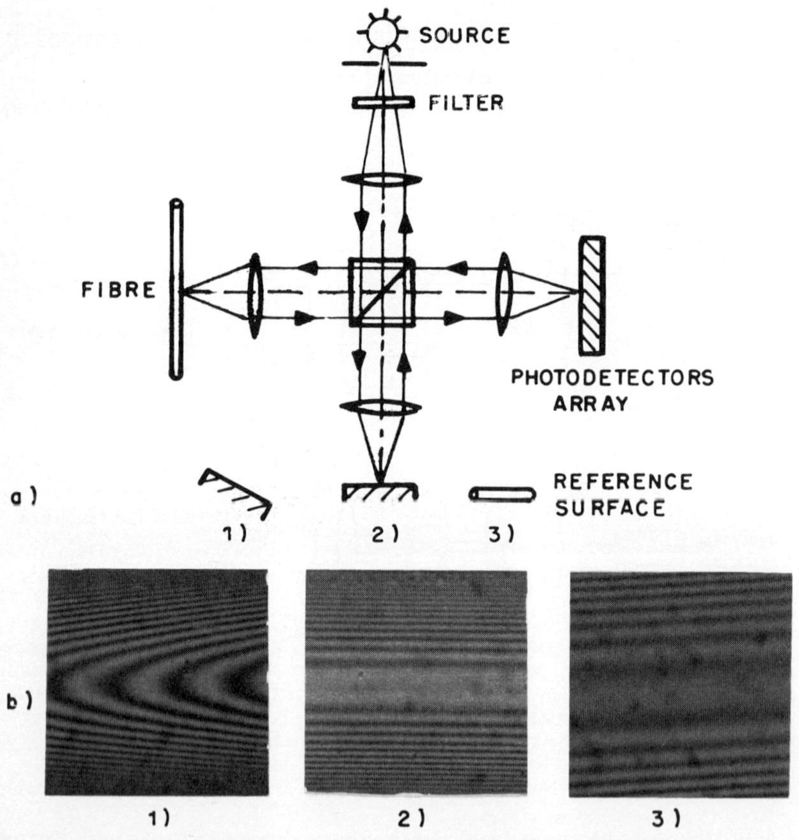

FIG. 36 *a)* Optical arrangement of Linnik-type micro-interferometer for fibre measurement. *b)* Three different forms of wavefronts with expected interference patterns: *1)* flat and tilted. *2)* flat and parallel. *3)* cylindrical and parallel.

The interference fringes are projected on the photodetector, thereby assuming the shape illustrated in the figure.

As may be seen, if the reference surface consists of a cylinder with radius close to that of the fibre, fringes are clearly spaced. This fact may, however, cause a loss of information, since given the high magnifying rates of the system, the photodiode can check only a small number of fringes. This method is also suitable for checking the quality of the exterior surface of the fibre, since any flaws alter the diffraction shape as against the theoretical case reproduced in the drawing.

Also this measurement is particularly sensitive to any displacements of the fibre, which therefore must be guided by a precision pulley system.

It should be noted that the core size can be measured with this method by dipping the fibre into a liquid having the same refractive index as that of the cladding.

d) Opto-photoelectric sensor

This method takes advantage of the fibre properties as a cylindrical lens.

By lighting the fibre with a beam perpendicular to its axis and by observing it through a microscope located at a certain angle with respect to the beam, one can note two parallel bit tracks, due to reflection and refraction.

Now, by lighting with another beam symmetrical to the first one with respect to the microscope axis, we obtain another two tracks symmetrical to the previous ones (Fig. 37*a*) [56]. The location of the tracks is a function of the geometrical and optical parameters of the system (Fig. 37*b*):

$$h = r \sin\left[\frac{\alpha_1 - \alpha_2}{2} + \arcsin\left(\frac{\sin^2(\alpha_1 + \alpha_2)}{\sin^2(\alpha_1 + \alpha_2) + [n - \cos(\alpha_1 + \alpha_2)]^2}\right)^{\frac{1}{2}}\right]$$

where: r = fibre radius
n = refractive index.

By measuring h and knowing all the other values, fibre radius r can be derived. A schematic view of the measurement system is shown in Fig. 37*a*.

The objective focusses the beams coming from the fibre on a modulator formed by a rotating disk with short radial grooves, placed at a distance slightly greater than that between the light

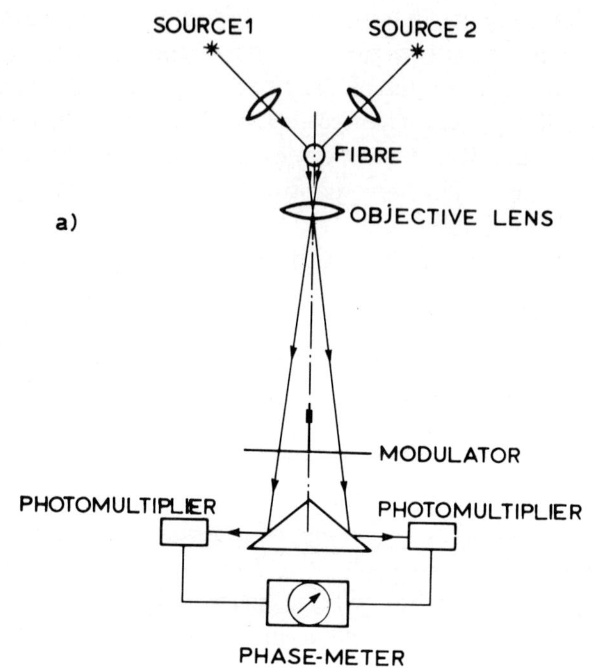

FIG. 37 *a*) – Scheme of the opto-photoelectric sensor.

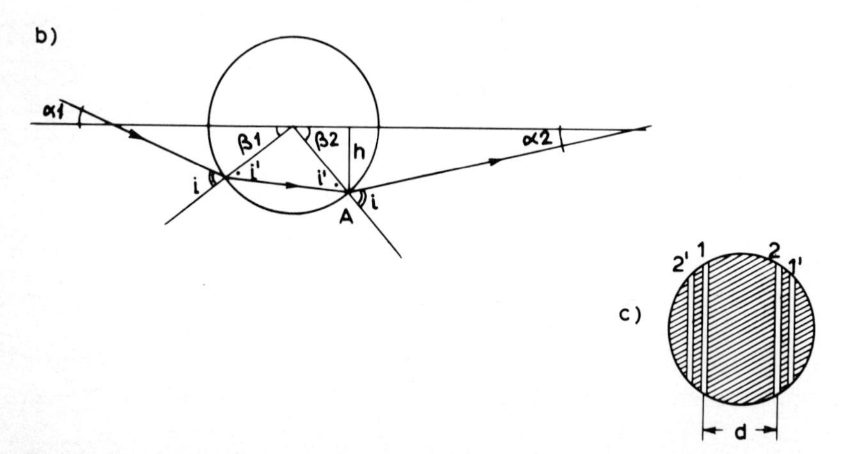

FIG. 37 *b*) – Path of a ray in glass fibre (cross section). *c*) The microscope field when viewing a glass fibre illuminated by the two beams: 1 and 2 are fairly bright bands due to refraction, 1′ and 2′ are low-brighteness bands due to reflection on the cylindrical surface of the fibre. (After [56]).

beams. In this way the two photomultipliers produce two sets of pulses with a phase difference proportional to the fibre diameter and detectable by means of a phase-meter.

This measurement is not affected by small vibrations of the fibre along the objective axis.

e) Interference fringe method

This measurement requires the use of a monochromatic and coherent light beam which, arriving perpendicularly at the fibre, is partly reflected by the surface and partly refracted [57] (Fig. 38). Observing from a given angle the light coming out from the fibre, interference fringes having a spacing proportional to the fibre diameter can be seen.

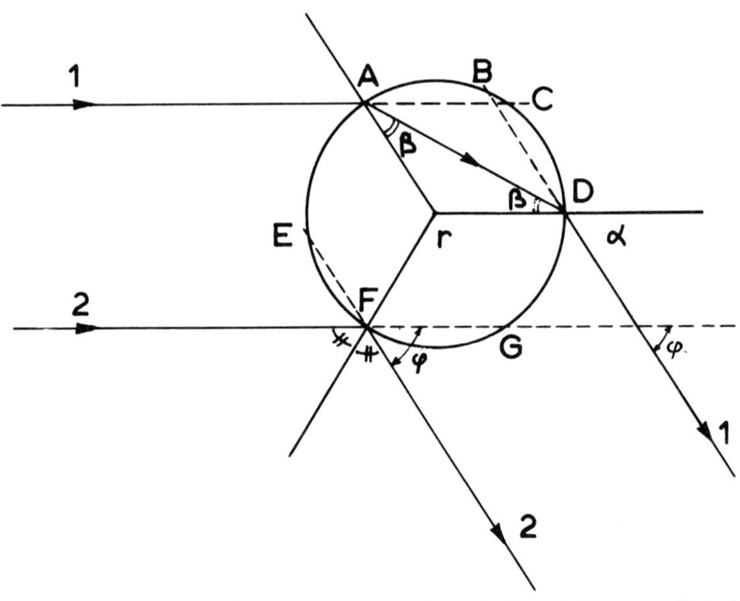

FIG. 38 – Path difference between the ray refracted (1) and the ray reflected (2) (After [57]).

These fringes are caused by the interference between the light reflected and that refracted by the fibre, owing to the different beam paths.

The most suitable angle φ is 60°, as in this case spacing between fringes is proportional to diameter.

Since, for this measurement, variations of 1% in the refractive index are negligible, the diameter of the fibre with or without cladding can be measured.

This method is affected by small vibrations of the fibre, as these may cause fringe displacement. However, no variation would occur in fringe spacing, which is the parameter correlated with fibre diameter.

Other methods for the assessment of fibre diameter are based on the distribution of light fringes produced by the scattering of a laser beam impinging perpendicularly on the fibre axis [58-60].

4.4.4 *Fibre coating*

To preserve the optical and mechanical characteristics of the fibre its surface must be protected by a polymer layer to prevent physical damage and deterioration of properties by the environment.

Resin must be applied before the fibre is wound on the drawing apparatus drum so that it hardens before coming into contact with the drum.

The suitable thickness for the resin layer is of the order of some microns. The layer must not give rise to stresses, either during its application on the fibre surface, or during the drying process.

Materials must be dissolved in high-volatility solvents or be in the prepolymer liquid state with suitable viscosity. In the latter case polymerization occurs on the fibre immediately after application.

Numerous types of resins have been suggested to obtain the best combination of the required properties.

A suitable protective layer for the fibre must meet the following requirements:

— concentricity with respect for the fibre, i.e. uniform thickness of the film. Otherwise, tensions tending to bend the fibre may arise during hardening;

— longitudinal uniformity and complete coating of the fibre surface;

— film smoothness;

— good abrasion resistance;

— short hardening time;
— high application speed;
— softening temperature range high enough to permit subsequent buffering to be easily accomplished;
— chemical stability of the layer;
— easy « strippability » with appropriate solvent such as to allow fibre cutting and jointing operations;
— minimum deterioration in fibre performance during temperature variations (owing to expansion coefficient difference).

The suggested polymers have to some extent these properties. In general, all require the presence of additives to improve the adhesion to the fibre surface.

These additives are polyfunctional silanes which represent, by virtue of their intrinsic properties, a very good intermediate binder between polymer and glass [61].

The materials most used for coating are fluorinated resins and acrylics in acetone solution, vinyl-acetate dissolved in trichloromethane, cellulosic acetate dissolved in acetone, and silicones and epoxies applied without solvents.

Good results are also obtained with epoxy-acrylate resins without solvent, with a U.V. sensitizer added which starts the polymerization [62].

4.4.5 *Coating apparatus*

The choice of the coating method depends to a large extent on the protective film quality. It is necessary to adopt a technique allowing stresses on the fibres to be kept to a minimum. Various experimented methods include coating through dipping, with wick, spray, with roller, electrostate powder process, by extrusion, from « hot melt » and sizing die arrangement. Only the last method is widely used. Different configurations have been suggested for the coating crucible method.

An example of a nozzle suitable for the application of fluorinated resins in a acetone solution with viscosity equal to 3.7 poises is given in Fig. 39 [63].

The operation is correct if the solution viscosity, the fibre speed and the fibre diameter keep within the specified limits during the whole process.

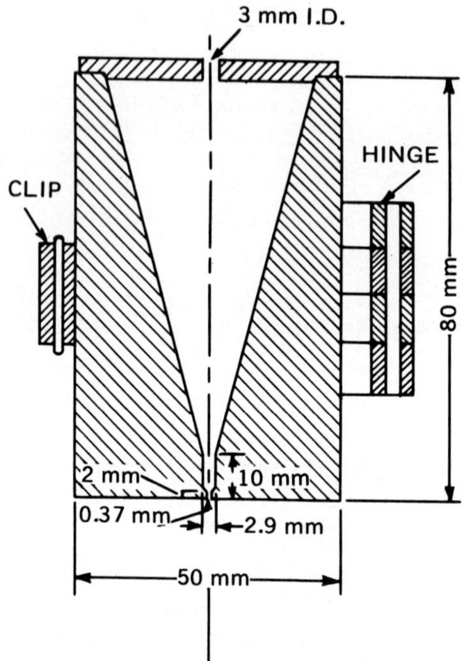

FIG. 39 – Coating crucible (After [63]).

In the operating conditions the thickness of the resulting coating is a function of the drawing parameter according to the following relationship:

$$t = \left\{a^2 - \frac{L}{\ln b/a}\left[\frac{(b^2 - a^2)}{2} - a^2 \ln b/a\right]\right\}^{\frac{1}{2}} - a$$

where: t = film thickness

$2a$ = fibre diameter

$2b$ = nozzle diameter

L = volume loading factor (dried polymer volume per solution volume unit).

It appears that thickness does not depend upon fibre speed in laminar flow conditions, and varies slightly with fibre diameter.

Other types of coating crucibles based on the same principle make use of various apparatus for centring and guiding the fibre.

The nozzle diameter is 2-3 times that of the fibre. A smooth working of the nozzle is helped by keeping the temperature constant and the fluid level and resin viscosity steady by means of suitable devices.

To assist the formation of an even film, which is above all concentric with respect to the fibre, a flexible coating equipment in silicone rubber has been developed which is not liable to attacks by solutions commonly used for coating [64].

The advantage of this device (Fig. 40) is that it allows an automatic centring of the fibre with respect to the nozzle as a consequence of the action of hydrodynamic forces generated by the convergent resin flow.

This makes it possible to counterbalance alignment errors by making the resin distribution uniform over the optical fibre surface.

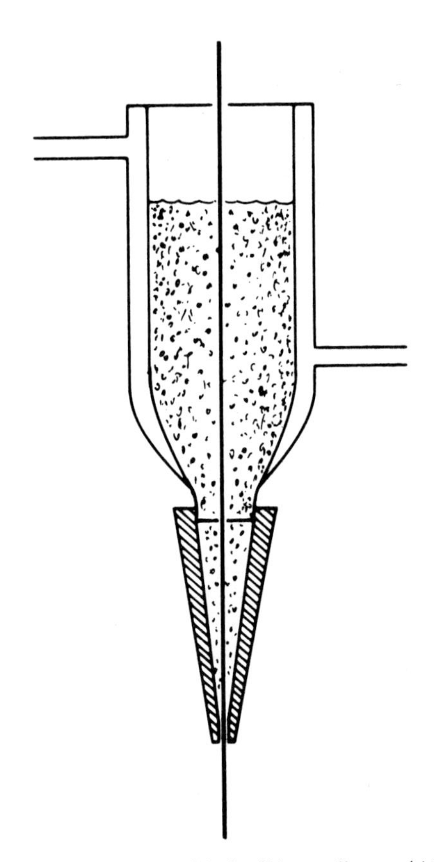

FIG. 40 – Coating crucible with flexible applicator (After [64]).

4.4.6 *Drying of the film*

The final operation in the fibre coating process is the drying of the resin immediately after its application.

Generally a tube-shaped furnace is used, which, if necessary, is kept in a controlled environment. It must be heated to a temperature sufficient to bring about the evaporation of the resin solvent (or the polymerization) in the time taken by the fibre to pass through.

It is important that the solvent evaporate without reaching the boiling temperature, so as to prevent the formation of bubbles inside the resin layer.

In some processes [63], a second heating stage follows the first furnace in order to produce a temporary melting of the resin and to obtain as a result a shiny and clear layer.

The aforementioned protection does not yield adequate mechanical characteristics for direct cabling. Therefore, it is necessary to protect the optical fibre with a *secondary coating* consisting of an extrusion of plastic material. This is considered in detail in Chapter 1 of Part III (Section 1.2).

REFERENCES

[1] M. Zief, J. Horvath and N. Theodorou: *Chemicals for low loss fibers*, Optical Fiber Transmission, Williamsburg (January 7-9, 1975), pp. TuB3.

[2] R. G. Gossink and H. M. J. M. Van Ass.: *Preparation of ultrapure raw materials and the melting of low-loss compound glasses*, Optical Fiber Transmission, Williamsburg (January 7-9, 1975), pp. TuB4.

[3] D. A. Pinnow, L. G. van Uitert, T. C. Rich, F. W. Ostermayer and W. H. Grodkiewicz: *Investigation of the soda aluminosilicate glass system for application to fiber optical waveguides*, Optical Fiber Transmission, Williamsburg (January 7-9, 1975), pp. TuA3.

[4] C. M. G. Jochem, T. P. M. Meevwsen, F. Meyer, P. J. W. Severin and G. A. C. M. Spierings: *Technology of alkali germanosilicate graded-index fibres*, Fourth European Conference on Optical Communication, Genoa (September 12-15, 1978), pp. 2-8.

[5] H. Aulich, J. Grabmaier, K. H. Eisenrich and G. Kinshofer: *High-aperture, medium-loss alkali-leadsilicate fibers prepared by the double crucible technique*, Optical Fiber Transmission II, Williamsburg (February 22-24, 1977), pp. TuC5-1-4.

[6] K. J. Beales, C. R. Day, W. J. Duncan and G. R. Newns: *Low-loss optical fibres prepared by the double crucible technique*, Third European Conference on Optical Communication, Munich (September 14-16, 1977), pp. 18-20.

[7] G. R. Newns: *Compound glasses for optical fibres*, Second European Conference on Optical Fibre Communication, Paris (September 27-30, 1976), pp. 21-26.

[8] P. W. Black: *Fabrication of optical fiber waveguides*, Electrical Communication, Vol. 51, No. 1 (1976), pp. 4-11.

[9] K. J. Beales, C. R. Day, W. J. Duncan and G. R. Newns: *Preparation of sodium-borosilicate glass fibre for optical communication*, Proc. IEE, Vol. 213, No. 6 (June, 1976), pp. 591-596.

[10] R. G. Gossink *et al.*: Materials Research Bulletin, Vol. 10, No. 1 (January 1975).

[11] J. P. Parant, Ch. Le Sergent, P. Lerner, S. Galaj and Cl. Brehm: *Optimisation et caractérisation des matériaux pour fibres optiques sodocalciques*, Second European Conference on Optical Fibre Communication, Paris (September 27-30, 1976), pp. 33-36.

[12] C. R. Day, J. E. Midwinter, G. R. Newns, M. H. Reeve, K. I. White and M. C. Brierley: *Loss mechanisms in glass fibre produced by the double crucible technique*, Optical Fiber Transmission, Williamsburg (January 7-9, 1975), pp. TuC2.

[13] J. G. Tithmarsh: *Fibre geometry control with the double crucible technique*, Second European Conference on Optical Fibre Communication, Paris (September 27-30, 1976), pp. 41-45.

[14] K. Koizumi, Y. Ikeda, I. Kitano, M. Furukawa and T. Sumimoto: *New light-focusing fibers made by a continuous process*, Appl. Opt., Vol. 13, No. 2 (February, 1974), pp. 255-260.

[15] R. Ishikawa, M. Seki, K. Kaede, K. Koizumi and T. Yamazaki: *Transmission characteristics of graded-index and pseudo step-index borosilicate compound glass fibers*, 1977 International Conference on Integrated Optics and Optical Fiber Communication, Tokyo (July 18-20, 1977), pp. 301-304.

[16] T. Inoue, K. Koizumi and Y. Ikeda: *Low-loss light-focusing fibres manufactured by a continuous process*, Proc. IEE, Vol. 123, No.6 (June, 1976), pp. 577-580.

[17] D. Kuppers, H. Lydtin and F. Meijer: *Preparation methods for optical fibres applied in Philips research*, 1977 International Conference on Integrated Optics and Optical Fiber Communication, Tokyo (July 18-20, 1977), pp. 319-322.

[18] A. D. Pearson and W. G. French: *Low-loss glass fiber for optical transmission*, Bell Laboratories Record (April, 1972), pp. 103-109.

[19] C. E. E. Stewart and P. W. Black: *Optical losses in soda lime/silica cladded fibres produced from composite rods*, Electron. Lett., Vol. 10, No. 5 (1974), pp. 53-54.

[20] D. N. Payne and W. A. Gambling: *Preparation of water-free optical-fibre waveguide*, Electron. Lett., Vol. 10, No. 16 (1974).

[21] R. D. Maurer: *Doped-deposited-silica fibres for communications*, Proc. IEE, Vol. 123, No. 6 (June, 1976), pp. 581-585.

[22] D. B. Keck and P.C. Schultz: U. S. Patent 3711262.

[23] J. B. Mac Chesney, P. B. O'Connor, F. U. Di Marcello, J. R. Simpson and P. D. Lazay: *Preparation of low loss optical fibers using simultaneous vapor phase deposition and fusion*, 10th International Congress on Glass, Kyoto-Japan (July, 1974), pp. 6-40.

[24] P. Geittner, D. Kuppers and H. Lydtin: *Low-loss optical fibers prepared by plasma-activated chemical vapour deposition (CVD)*' Appl. Phys. Lett., Vol. 28 (1976), pp. 645-646.

[25] D. B. Keck, P. C. Schultz and F. Zimar: U. S. Patent 3737393.

[26] M. Okada, M. Kawachi and A. Kawana: *Improved chemical vapour deposition method for long-length optical fibre*, Electron. Lett., Vol. 14, No. 4 (1978), pp. 89-90.

[27] P. D. Lazay and W. G. French: *Control of substrate tube diameter during MCVD preform preparation*, Optical Fiber Communication, Washington D.C. (March 6-8, 1979), pp. 50-52.

[28] G. W. Tasker and W. G. French: *Low-loss optical waveguides with pure fused* SiO_2 *cores*, Proc. IEEE, Vol. 62 (1974), p. 1281.

[29] W. G. French, G. W. Tasker and J. R. Simpson: *Graded index fiber waveguides with borosilicate composition: fabrication techniques*, Appl. Opt., Vol. 15, No. 7 (July, 1976), pp. 1803-1807.

[30] Technical Information Exchange - C.G.W.

[31] W. A. Gambling, D. N. Payne, C. R. Hammond and S. R. Norman: *Optical fibres based on phosphosilicate glass*, Proc. IEE, Vol. 123, No. 6 (June, 1976), pp. 570-576.

[32] T. Akamatsu, K. Okamura and Y. Ueda: *Fabrication of long length fibers by improved C.V.D. method*, Optical Fiber Transmission II, Williamsburg (February 22-24, 1977), pp. TuC3-1-4.

[33] W. G. French and L. J. Pace: *Chemical kinetics of the modified chemical vapour deposition process*, 1977 International Conference on Integrated Optics and Optical Fiber Communication, Tokyo (July 18-20, 1977), pp. 379-382.

[34] T. Akamatsu, K. Okamura and M. Tsukamoto: *The anomalous concentration profile of* P_2O_5 *and* GeO_2 *in silica fiber*, 1977 International Conference on Integrated Optics and Optical Fiber Communication, Tokyo (July 18-20, 1977).

[35] P. Geittner, D. Kuppers and H. Lydtin: *Low-loss optical fibers*

prepared by plasma-activated chemical vapour deposition (PCVD), Appl. Phys. Lett., Vol. 28, No. 11 (June, 1976), pp. 645-646.

[36] J. Koenings, D. Kuppers, H. Lydtin and H. Wilson: *Deposition of* SiO_2 *with low impurity content by oxidation of* $SiCl_4$ *in a non-isothermal plasma*, Fifth International Conference on Chemical Vapour Deposition (1975).

[37] B. S. Aronson, D. R. Powers and R. G. Sommer: *Chlorine drying of a doped deposited silica preform simultaneous to consolidation*, Optical Fiber Communication, Washington D.C. (March 6-8, 1979).

[38] S. Kobayashi, S. Sudo, T. Miyashita and T. Izawa: *Preparation of silica glass using a* CO_2 *laser*, Appl. Opt. Vol. 14, No. 12 (1975).

[39] T. Izawa, S. Kobayashi, S. Sudo and F. Hanawa: *Continuous fabrication of high silica fiber preform*, 1977 International Conference on Integrated Optics and Optical Fiber Communication, Tokyo (July 18-20, 1977), pp. 375-378.

[40] T. Izawa, S. Sudo, F. Manawa and T. Edahiro: *Progress in continuous fabrication process of high-silica fiber preform*, Fourth European Conference on Optical Communication, Genoa (September 12-15, 1978), pp. 30-36.

[41] K. Inoue, J. Goto, T. Arima, O. Wakamura and T. Akamatsu: P_2O_5 - GeO_2 - Ga_2O_3 *glass fibers for optical communication*, 1977 International Conference on Integrated Optics and Optical Fiber Communication, Tokyo (July 18-20, 1977), pp. 387-390.

[42] P. Kaiser, A. C. Hart Jr. and L. L. Blyler Jr.: *Low-loss FEP-clad silica fibers*, Applied Optics, Vol. 14, No. 1 (January, 1975), pp. 156-162.

[43] L. L. Blyler Jr., A. C. Hart Jr., R. E. Jaeger, P. Kaiser and T. J. Miller: *Low-loss, polymer-clad silica fibers produced by laser drawing*, Optical Fiber Transmission, Williamsburg (January 7-9, 1975), pp. TuA5.

[44] S. Tanaka, K. Inada, T. Akimoto and M. Kozima: *Silicone-clad fused-silica-core fibre*, Electron. Lett., Vol. 11, No. 7 (April, 1975), pp. 153-154.

[45] P. Kaiser and H. W. Astle: *Low-loss single-material fibers made from pure fused silica*, Bell Syst. Tech. J., Vol. 53, No. 6 (1974), pp. 1021-1039.

[46] F. T. Geyling: *Basic fluid dynamic considerations in the drawing of optical fibers*, Bell Syst. Tech. J., Vol. 55, No. 8 (1976), pp. 1011-1056.

[47] M. I. Cohen and R. J. Klaiber: *Drawing of smooth optical fibers*, Optical Fiber Transmission II, Williamsburg (February 22-24, 1977), pp. TuB4-1-3.

[48] P. Kaiser: *Contamination of furnace-drawn silica fibers*, Appl. Opt., Vol. 16, No. 3 (1977), pp. 701-704.

[49] K. Imoto, S. Aoki and M. Sumi: *Novel method of diameter control in optical-fibre drawing process*, Electron. Lett., Vol. 13, No. 24 (1977), pp. 726-727.

[50] V. C. Paek and R. B. Runk: *Physical behavior of the neck-down region during the furnace drawing of silica fibers*, Optical Fiber Transmission II, Williamsburg (February 22-24, 1977), pp. TuC1-1-4.

[51] R. B. Runk: *A zirconia induction furnace for drawing precision silica waveguides*, Optical Fiber Transmission II, Williamsburg (February 22-24, 1977), pp. TuB5-1-4.

[52] M. R. Montierth: *Recent advances in the production of multimode optical fibres*, Journal of Electronics Materials, Vol. 6, No. 3 (1977), pp. 349-372.

[53] M. A. Saifi: *Limitation in CO_2 laser fiber drawing*, Optical Fiber Communication II, Williamsburg (February 22-24, 1977), pp. TuC2-1-4.

[54] L. G. Cohen and P. Glynn: *Dynamic measurement of optical fiber diameter*, Rev. Sci. Instrum. Vol. 44, No. 12 (December, 1973).

[55] W. S. Kapany: *Fiber Optics*, Academic Press., (1967).

[56] A. I. Inyushin and L. A. Shiffers: *Opto-photoelectric sensor for measuring diameters of glass fibers during drawing*, Optical Technology, Vol. 38, No. 8 (August, 1971).

[57] K. C. Kao, T. W. Davies and R. Worthington: *Coherent light scattering measurements on single and cladded optical glass fibres*, Radio and Electronic Engineer, Vol. 39, No. 2 (February, 1970).

[58] D. Marcuse and H. M. Presby: *Light scattering from optical fibres with arbitrary refractive-index distribution*, J. Opt. Soc. Am., Vol. 65, No. 4 (April, 1975), pp. 367-375.

[59] P. L. Chu: *Determination of the diameter of unclad optical fibre*, Electron. Lett., Vol. 12, No. 1 (January, 1976), pp. 14-16.

[60] W. Fink and W. Schneider: *A coherent-optical method for measuring fibre diameters*, Optica Acta, Vol. 21, No. 2 (1974), pp. 151-155.

[61] R. Muto, N. Akiyama, H. Sakata and S. Furuuchi: *Adhesion of ethylene-tetrafuoroethylene copolymer with silicate glasses*, Journal of Non-Crystalline Solids, Vol. 19 (1975), pp. 369-376.

[62] H. N. Vazirani, H. Schonborn, and T. T. Wang: *U. V. cured epoxy-acrylate coatings on optical fibers. Chemistry and application*, Optical Fiber Transmission II, Williamsburg (February 22-24, 1977), pp. TuB3-1-4.

[63] P. W. France and P. L. Dunn: *Protection des fibres optiques par une solution plastique*, Second European Conference on Optical Fibre Communication, Paris (September 27-30, 1976), pp. 177-180.

[64] A. C. Hart and R. V. Albarino: *An improved fabrication technique for applying coatings to optical fiber waveguides*, Optical Fiber Transmission II, Williamsburg (February 22-24, 1977), pp. TuB2-1-4.

PART II
SOURCES AND DETECTORS

INTRODUCTION

by Giorgio Randone

In an optical fibre telecommunication system the selection and design of the transmitter and receiver depend chiefly on the properties of the transmission medium. For example, the wavelength of the optical carrier must be that of minimum cable attenuation and optimization of coupling between fibre and photoemitter or photodetector requires devices with special geometries. More generally, radiation sources and detectors must be compact, efficient, reliable and, especially, safeguard the cost competitiveness of an optical fibre telecommunication system with regard to alternative systems currently in use or under development. These requirements can be met fully by semiconductor devices. In the photoemitter field, existing technology is sufficiently developed to allow the preparation of materials suitable for light emission at the minimum attenuation wavelength of optical fibres. As regards detectors, on the other hand, the utilization of both PIN and avalanche silicon photodiodes has proved to be most satisfactory for operation up to 1 μm wavelength. Further improvement is possible using special geometries or integrated or hybrid assembly techniques for the detector or input preamplifier.

The working principles of these devices, their present limitations and the development forecasts, mostly as regards light sources reliability, will be described in the following. The neodynium pentaphosphate laser will be briefly illustrated as an alternative to semiconductor sources. Although this laser is currently only at the preliminary development stage, it offers very interesting prospects for use with monomode fibres in high capacity and long-distance advanced systems.

Taking into account the evolution of the transmitting medium, it is also necessary to point out that very low attenuation and fibre dispersion can be reached at higher wavelengths than those (0.8-0.9 μm) used by more conventional devices. This justifies the intensive work now in progress towards the production of components and materials suitable for the wavelength interval between 1 and 1.3 μm. At these wavelengths, selection of materials and associated preparation methods are still the main source of difficulty. Thus, a description of these methods and the realistic medium-term development prospects of devices suitable for use in this field will be presented in some detail.

CHAPTER 1

ELECTRO-OPTICAL SOURCES

by Giorgio Randone

1.1 Material choice and preparation technologies

1.1.1. *General considerations*

As mentioned above, the choice of a radiation source suitable for optical cable telecommunication systems is connected with two typical parameters of the transmission medium: attenuation as a function of wavelength and material dispersion always associated with the more or less imperfect source monochromaticity. A further contribution to dispersion is brought about by the different propagation velocity of the different modes in a multimode fibre; as this effect is to a large extent independent of the source type adopted, but on the other hand dependent on the waveguide characteristics (e.g. whether of the « step index » or « graded index » type), this factor does not a priori condition the selection of the different sources, which will be examined in the following. It is, however, important to note that this observation ceases to be valid in the case of monomode fibres which permit, or at most necessitate, the use of high spatial coherence sources.

As far as attenuation is concerned, two wavelength intervals have been taken into consideration to date, where losses can reach a minimum value; for the early fibres (Fig. 1) a first low attenuation window was found at wavelengths around $\lambda = 0.82$ μm, while, more recently (see Fig. 16, Part I, Chapter 1) a second window with even lower attenuation was obtained at wavelengths from 1 to 1.3 μm.

A semiconductor material for generating the optical carrier at the required wavelength can be prepared for each of the two low attenuation windows: for example, for the first region such a

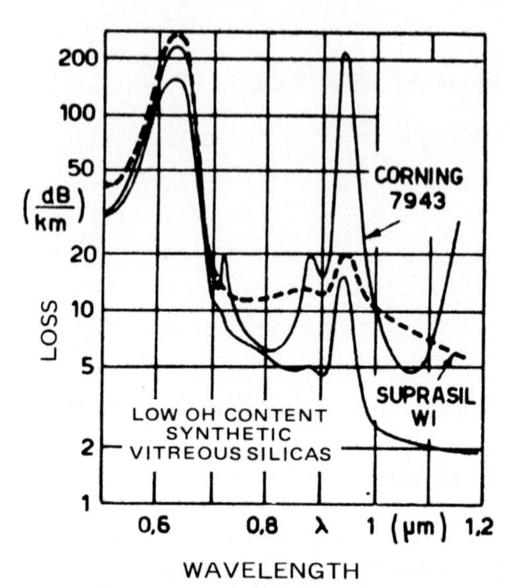

FIG.1 – Wavelength dependence of the attenuation in an early quartz optical fibre; attenuation minima are clearly shown at 0.82. (After S. E. Miller *et al.*, Proc. IEEE, Vol. 61, No. 12, p. 1703).

material can be formed by the $Ga_{1-x}Al_xAs$ mixed ternary system, while a material suitable for the second region is made up, for instance, by the similar $Ga_{1-x}In_xAs$ ternary system. In particular, the emission wavelength of the simplest GaAs ($\lambda = 0.9$ μm) binary compound is located at the edge of the first region with non-prohibitive attenuations in terms of application. Therefore, GaAs emitters can be used in systems in which it is not essential to attain minimum attenuation values. This avoids the need to employ sources derived from ternary compounds and offers the advantage of notable construction simplicity.

As far as dispersion is concerned, Fig. 2 indicates how it decreases, when the wavelength increases, until it vanishes (at first order) for $\lambda = 1.27$ μm [1]; at such a wavelength, in other words, the length of the section usable with the same power injected into the fibre and for a given transmission capacity is no longer conditioned by the spectral width of the source. Apart from this special case, however, the dispersion for incoherent sources (LED) operating

[1] The zero-dispersion wavelength depends of course on the fibre core doping; the value 1.27 μm refers to the undoped silica.

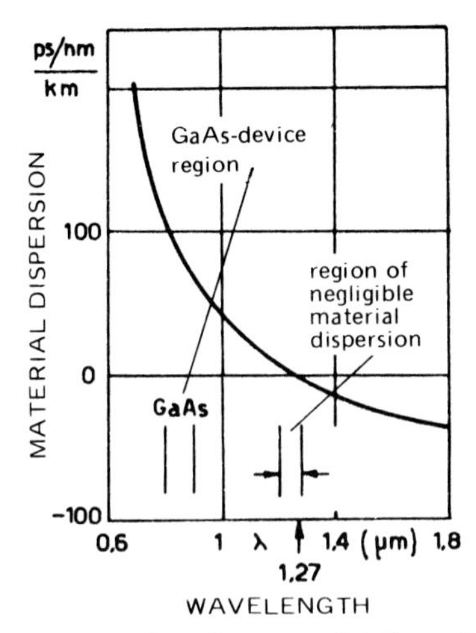

FIG. 2 — Wavelength dependence of the fibre material dispersion. (After D. N. Payne, W. A. Gambling, Electron. Lett., Vol. 11, No. 8, p. 176).

in the first region is almost one order of magnitude greater than for those operating within the second region with the same spectral width.

This is shown in Fig. 3, where different transmission capacities are compared, which can be obtained with sources having the

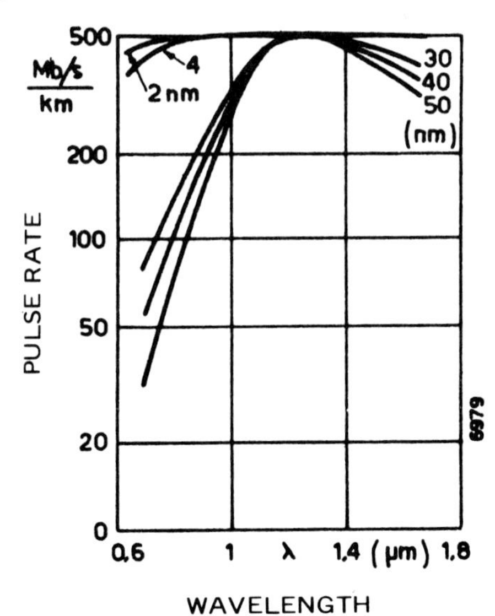

FIG. 3 — Transmission rate as a function of the optical carrier wavelength, and for different source spectral widths (Modal dispersion — 1 ns/km). (After D. N. Payne, W. A. Gambling, Electron. Lett. Vol. 11, No. 8, p. 176).

typical spectral width of semiconductor lasers (2-4 nm) and LED's (30-50 nm) operating at wavelengths between 0.7 and 1.6 μm.

Hence, it can be concluded that while dispersion does not constitute a serious limiting factor for coherent sources, such as lasers, the use of the second region is extremely interesting and desirable for incoherent sources such as LED's. However, those materials which are the most promising for obtaining devices suitable for emission at $\lambda \geqslant 1$ μm to match the transmission medium characteristics unfortunately require a more complex, and at present, less-developed technology.

This aspect calls for discussion of the main properties of binary and ternary semiconductor compounds of the III-V group and more generally, of the reasons why for light generation purposes these materials are preferred to more conventional germanium and silicon semiconductors. The latter point is connected with a fundamental difference [1] in the nature of the recombination transition (Fig. 4) between conduction band and valence band, which is indirect for Si and Ge and direct for many of the III-V compounds (see Table I, Part V, Chapter 3).

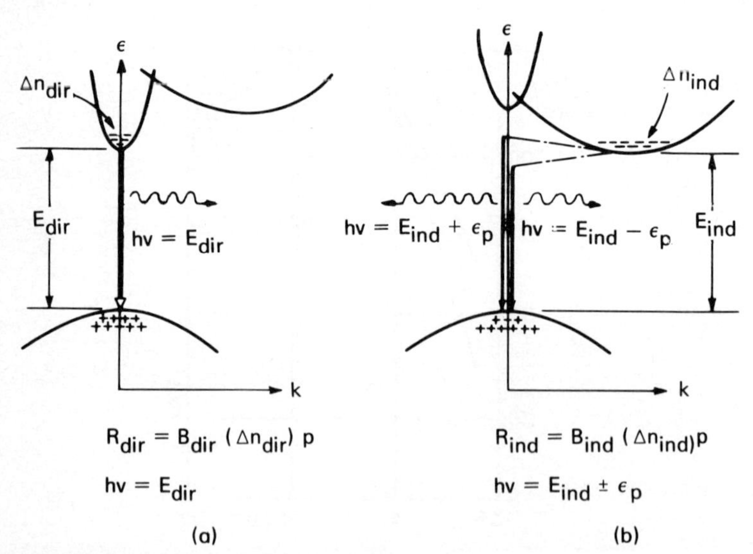

FIG. 4 – Recombination transitions in a direct and an indirect gap semiconductor. Momentum conservation is allowed in the second case by the emission or absorption of a lattice vibration (phonon). (After C. J. Nuese *et. al.*, IEEE Spectrum, Vol. 9 (May 1972), pp. 28–38).

Maximum recombination efficiency with radiation emission and very fast decay times ($\leqslant 1$ ns) characterize direct transition; this assures modulation rates equally fast in principle. The wavelength of the radiation emitted during this recombination process is entirely determined by the energy associated to the forbidden gap between the two semiconductor bands. For instance, such an energy at 300 °K is about equal to 1.4 eV for the GaAs and the respective wavelength yields $\lambda = 0.9$ μm. As was discussed earlier, such a wavelength does not correspond to any of the minimum attenuation regions of the optical fibre, even if it drops only at the edges of the high attenuation regions between the first and second low attenuation windows. The above mentioned Table I contains the transition nature, the relevant energy and the associated wavelength for some III-V compounds. It can be noted that no binary compound can be used directly at the wavelengths of interest of 0.82 μm and 1.30 μm respectively. However the crystal structure of the binary III-V compounds allows, within certain limits, the creation of ternary solid solutions $III_{Ax}III_{B1-x}V$, in which a certain fraction x of an element is replaced by a different element in the same column of the periodic table. Actually the two $Ga_{1-x}Al_xAs$ and $Ga_{1-x}In_xAs$ ternary compounds mentioned above are derived from the solid solution of the GaAs, AlAs and GaAs, InAs binary compounds respectively.

The main characteristic of the ternary compounds of this species are « intermediate » between those of the two binary starting compounds. For instance Fig. 5 gives, as a function of the atomic percentage x of Al or In as compared to Ga, the evolution of the energy gap for the $Ga_{1-x}Al_xAs$ and $Ga_{1-x}In_xAs$ ternary compounds and for the $GaAs_{1-x}P_x$ ternary compound, where phosphorus can replace arsenic.

Hence, by the appropriate choice of x, a material can be obtained the characteristic emission wavelength of which is at 0.82 μm ($E_g = 1.51$ eV) and at 1.06 μm ($E_g = 1.18$ eV). An interesting difference between the two ternary materials examined concerns the nature of the recombination transition. Whereas for the $Ga_{1-x}In_xAs$ system, both GaAs and InAs have a direct transition, which is thus maintained by the ternary compound for all values of x, for the $Ga_{1-x}Al_xAs$ system, the AlAs compound has an indirect transition (Fig. 4*b*). This signifies that the re-

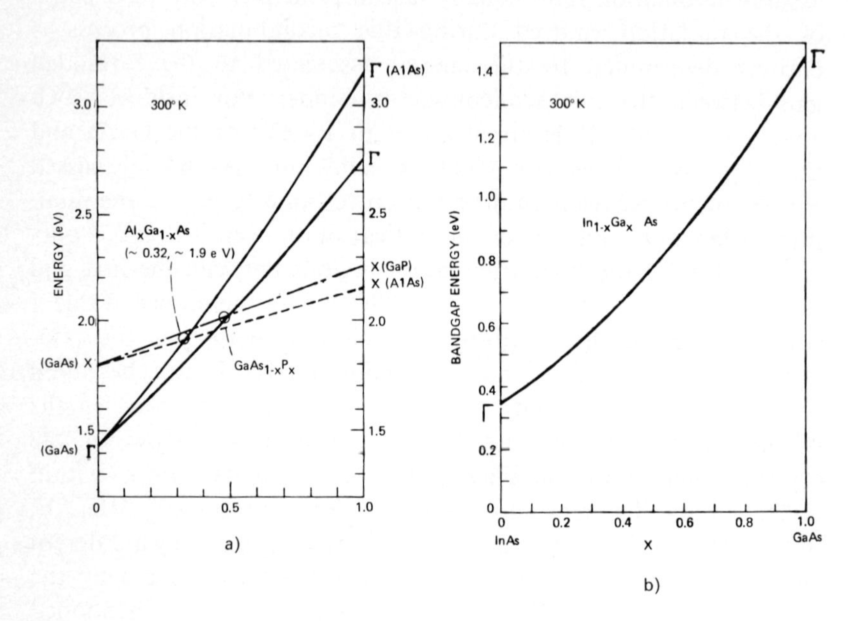

FIG. 5 – Band gap energy dependence on alloy composition (x), for a few ternary III-V mixed compounds. The Γ and the X lines show the shift of the direct and indirect gap respectively.

combination transition of the ternary compound is direct for x between 0 and 0.35 and indirect for $x > 0.35$. This is shown in Fig. 6, which gives the evolution of the radiative efficiency, as the ternary compound composition varies. It can be seen that this efficiency drops rapidly on coming into the region characterized by indirect transition. The extensive utilization of semiconductor materials, for which the energy gap value and the direct transition characteristic can be determined within the limits reported for instance in Fig. 1 (Part V, Chapter 3), allows, in principle, the development of devices suited to the optimal exploitation of the transmission medium characteristics. Since the limits and difficulties relevant to such an utilization depend on the synthesis methods of such materials, let us go a little deeper into the question.

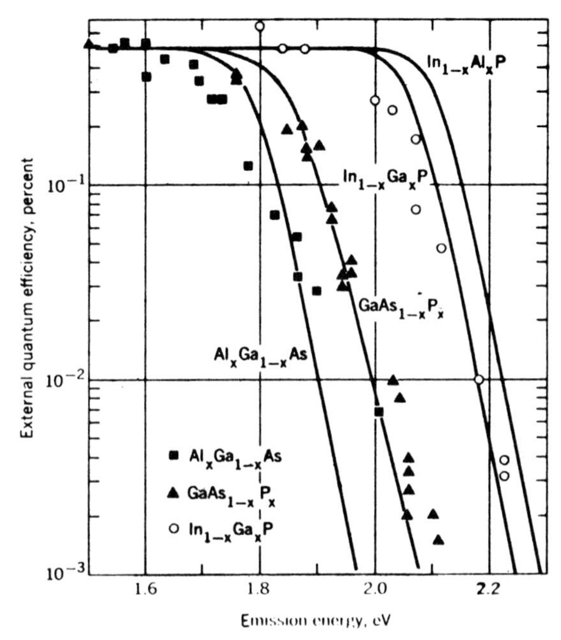

FIG. 6 – Relative quantum efficiency for several ternary alloys. The efficiency drop indicates the crossover between the direct-indirect recombination transition (After C. J. Nuese *et al.*, IEEE Spectrum, Vol. 9, (May 1972), pp. 28–38).

1.1.2. *Material preparation techniques through epitaxy*

The continuous improvement in opto-electronic device performance in recent years is chiefly due to the availability of very sophisticated techniques for preparation of materials. These are generally based on the epitaxial deposition, on the substrate, of a thin layer of material with well-defined electrical and optical characteristics. The grown layers have the same crystal structure but the stoichiometric composition and the electrical conductivity may be different from those of the substrate.

The deposition on the substrate can be carried out using two methods: liquid phase epitaxy [2] or vapour phase epitaxy [3]. In these cases the process is controlled by essentially thermodynamic parameters, such as temperature and system free energy. A third method [4] uses molecular beams of the elements forming the material to be grown, which impinge on the deposition substrate according to well-defined intensity ratios. Since this last method does not have to comply with very restrictive thermodynamic conditions, it appears, in principle, very flexible.

In particular, the three delineated methods have been applied very successfully for the growth of compound materials formed by binary, ternary and sometimes also quaternary alloys of elements of columns III and V of the periodic system. The materials in question are semiconductors prevalently with direct gap, having extremely attractive characteristics from the standpoint of optoelectronic devices.

The epitaxial growth process involves the deposition of a thin layer (from 0.1 to some tens of μm) of material, which maintains the structure and the crystal orientation of the substrate. For the LPE or VPE methods the deposition is controlled by the fact that the substrate is brought in contact with its liquid or vapour phase under appropriate oversaturation conditions. Therefore, the precipitation of the liquid or vapour phase helps in reducing the free energy of the solid-liquid system or solid-vapour system respectively.

Even if the thermodynamic principles governing both processes are similar, technologies and results associated to both growth types from the liquid phase to the vapour phase are very different: generally although the vapour phase method gives the best results from the standpoint of material doping control and epitaxial layer crystal perfection, it neither permits the simple preparation of multiple structures nor it is easily adapted to ternary compound deposition.

The liquid phase method, on the other hand, is particularly suited to the preparation of highly-doped multiple structures, which are, as will be seen in the following, suitable for photoemitting devices. The basic diagram of a LPE system is shown in Fig. 3 (Part V, Chapter 3); a practical LPE reactor is shown in Fig. 7. The critical points in adopting this technique concern the uniformity in the temperature profile of the reaction furnace and the use of multi-bin crucibles (Fig. 8). These allow the sequential deposition of a multiple structure, the various layers of which can be wholly different as regards their composition and thickness.

Without going into detail, for which we refer to the literature, it is interesting to illustrate the main aspects of the process which

FIG. 7 – Liquid phase epitaxy system. (Courtesy of SGS-ATES-CSELT).

enable the growth of the multistructure suitable, for instance, for obtaining a double heterojunction laser in the $Ga_{1-x}Al_xAs$ system.

This structure in its simplest version involves the deposition of at least three layers in the sequence $Ga_{0.7}Al_{0.3}As$ doped n ($5 \cdot 10^{17}/cm^3$, Sn), GaAs (not deliberately doped), $Ga_{0.7}Al_{0.3}As$ doped p ($10^{18}/cm^3$, Ge) from a substrate of GaAs (n^+, $2 \cdot 10^{18}$, Si) (2).

The drawing of a type of graphite crucible used for depositing the sequence of layers is shown in Fig. 8*b*). It is made up of a slider, in an appropriate recess containing the substrate, and several melt containers, which can be moved to sequentially overlap the substrate. Each bin is loaded with the exact quantities of Ga, GaAs and, where required, of Al Ge and Sn to form the saturated liquid phase, at a contact temperature between the melts and the substrate. The required material quantities can be obtained from

(2) A fourth layer of p^+ GaAs is usually grown on top of the double heterounction in order to facilitate the application of the electrical contacts.

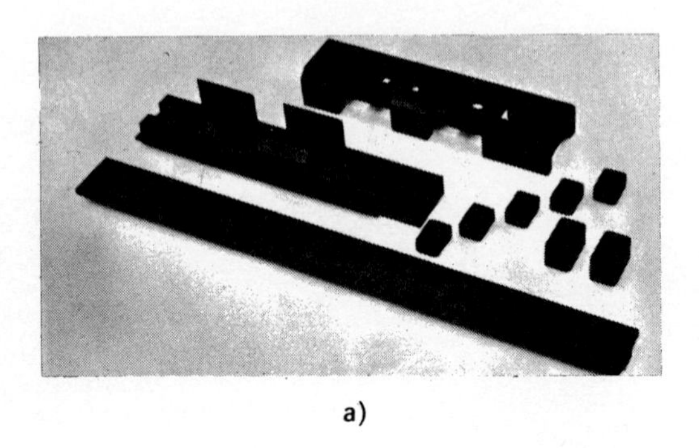

a)

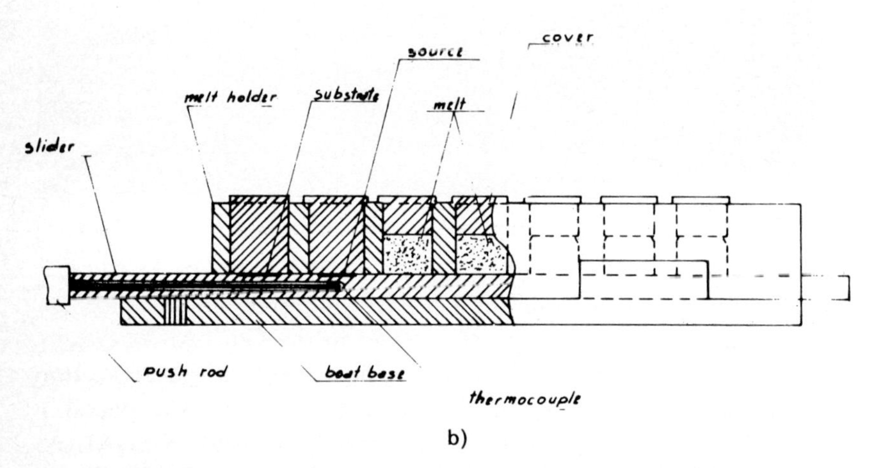

b)

Fig. 8 – *a*) Disassembled view of a LPE graphite crucible (Courtesy of SGS-ATES-CSELT). *b*) Schematic diagram of the multi-bin crucible.

the phase diagrams of the pertinent systems [5]. The crucible, containing the substrate and the different growth melts so prepared, is then placed inside a furnace under hydrogen flow. The temperature is then increased to just above the growth temperature. Afterwards the furnace temperature is reduced at a constant rate and, at the pre-established temperatures, the substrate is sequentially shifted under the growth melts where it is left for the time required to achieve the desired thickness.

This procedure is, in principle, suitable for the epitaxy growth of any ternary or quaternary compound of the III-V group, provided that the phase diagram is known and that the lattice constant of the substrate material does not differ appreciably from that of the material to be deposited. The latter constraint is unfortunately very restrictive, as can be seen from Fig. 9, which shows the dependence of the lattice constant for some mixed ternary compounds as a function of the *x* concentration of the replaced element.

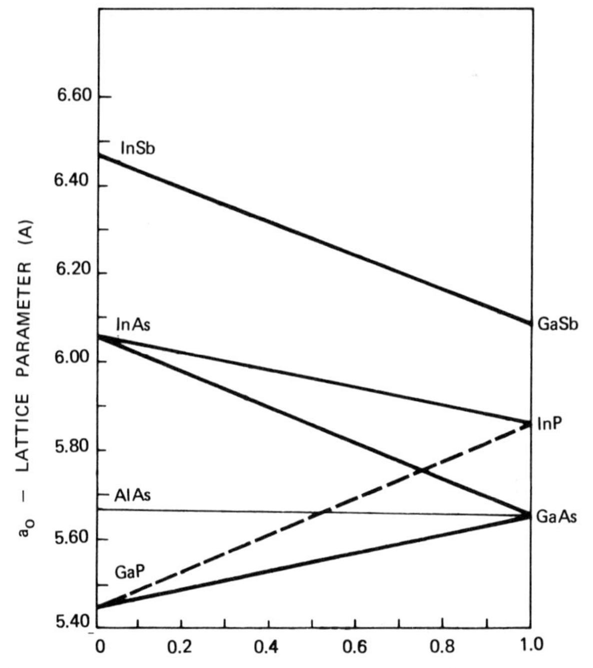

FIG. 9 – Lattice constant variation for several ternary alloys as a function of *X* composition.

As a matter of fact, the selection of the $Ga_xAl_{1-x}As$ ternary alloy was not random, but governed to the fact that among all possible III-V ternary systems, only the GaAs-AlAs one keeps practically the same lattice constant for all possible compositions. As the LPE method allows only the deposition of a material having a definite composition with respect to the substrate, the possible lattice constant difference between the layer and substrate is absorbed through the interface by producing lattice defects, typically dislocations. As these introduce non-radiative recombination centres, the material ceases to be suited to opto-electronic applications.

For this reason it is, for instance, not possible to deposit directly, through LPE, on a GaAs substrate a ternary material of the $Ga_{1-x}In_xAs$ species, suitable for emission at $\lambda > 1.06$ μm. Criteria for overcoming this difficulty will be considered in the following.

The Vapour Phase Method (VPE) has been utilized hitherto mainly for the preparation of high-purity GaAs material to be applied to microwave devices. Work is however in progress to obtain by this method material suitable for optoelectronic devices.

Although this method is technologically more complex than the LPE method, the results can be reproduced and controlled with greater ease. Furthermore, since the composition of the material deposited can be controlled by varying the flow of the gases feeding the reactor during growth, it is possible to have gradual variations in composition by eliminating or reducing to a minimum the formation of crystal defects ensuing from lattice constant differences.

The principles of the VPE method are relatively simple: after the saturation of a carrier gas, the elements forming its saturated vapour phase are made to flow over the substrate at the growth temperature. By operating appropriately on the saturation and growth temperatures the controlled deposition of the vapour phase on the substrate can be obtained. A schematic example of this procedure is reported in Fig. 10 which shows a system suited to the deposition of the $Ga_{1-x}In_xAs$ ternary material on a GaAs substrate; in the particular case illustrated, the carrier gas is formed by HCl, which flows over the Ga and In sources; As is directly carried into the reaction region through the AsH_3 compound: a dopant (Zn) can be introduced separately. Thus the problem of lattice constant differences can be avoided, in that the deposition can take place letting the In concentration to be deposited on the substrate grow gradually in the vapour phase. This is the same as reducing within tolerable limits the concentration of defects arising from the difference between substrate and epitaxial layer lattice constants.

The third epitaxial deposition method considered concerns

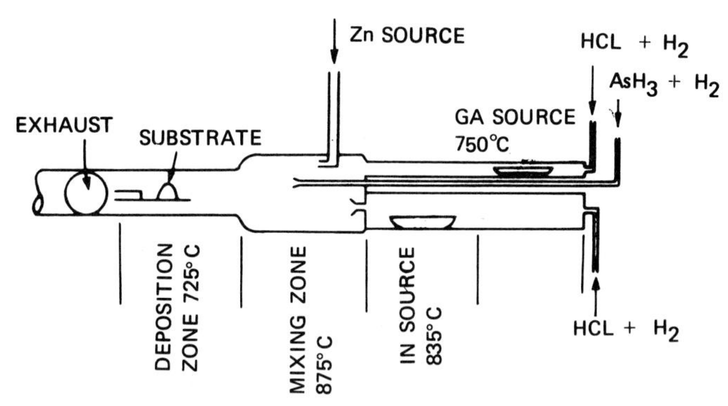

FIG. 10 – Schematic diagram of vapour phase epitaxy system, suitable for the growth of the ternary alloy $Ga_{1-x}In_xAs$.

Molecular Beam Epitaxy (MBE). In this case the deposition occurs by the controlled evaporation on the substrate of the molecular beams formed by the elements to be deposited in not particularly restrictive thermodynamic conditions. At present the method appears rather complex owing to the necessity of operating under ultravacuum conditions and of using sophisticated analysis techniques for the process control. On the other hand, since the MBE is a material preparation system of general application, it is the most suitable method for carrying out advanced research activities on materials in the opto-electronic sector and for implementating integrated optics structures. Examples of this kind will be given in the following with reference to distributed feedback lasers. A more detailed description of the MBE system and associated analysis and control procedures will be given elsewhere in this book, in Part V, Chapter 3.

1.1.3. *Materials suited to emission at* $\lambda > 1$ μm

As will be described later on the ground of the data given in Part V, Chapter 3 (see for instance Table I and Fig. 1, pp. 848-849), it would not be difficult to select ternary compounds of the $III_{A_{1-x}}III_{B_x}V$ species or of the $IIIV_{A_x}V_{B_{1-x}}$ species able to cover, with an appropriate selection of x, the wavelengths interval between 1 μm and 1.5 μm, where the optical fibres can achieve optimum performance in terms of attenuation and dispersion. For instance, the above mentioned $Ga_{1-x}In_xAs$ [6] or $GaAs_{1-x}Sb_x$ [7] ternary systems could be deposited on the GaAs substrate or the $InAs_xP_{1-x}$ [7] compound could be deposited starting from a substrate of InAs. Unfortunately, such a solution, so simple in principle, cannot be directly adopted, in that, as shown in Fig. 9, the lattice constants of the binary compound pairs examined (GaAs and InAs, GaAs and GaSb, InAs and InP respectively) are too different to ensure that the epitaxial layer with ternary composition is free from crystal defects due to this difference on the growth interface. In this section, without going into detail about the structures of the LED or laser devices suited to emission at these wavelengths, which will be discussed later, some general criteria for eliminating or reducing these difficulties will be illustrated. A method, which is simple in principle, consists in depositing a sequence of layers, the composition of which varies gradually [8] till it attains the one determining the required wavelengths. Thus the total stress due to the lattice mismatch can be minimized and also distributed over a greater thickness of epitaxial material.

This procedure is particularly suited to the VPE growth method [9], as mentioned earlier, in the case of a $Ga_{1-x}In_xAs$ system whereas in the case of an LPE system its implementation becomes difficult due to the necessity of using crucibles with a large number of melt containers. A second, less simple but more viable, method consists in the deposition of a material having a quaternary [10, 11] composition (the $III_{Ax}III_{B1-x}V_{A'y}V_{B'1-y}$ species). Here the addition of a first element to the starting composition substrate (for instance the $III_AV_{A'}$ compositions) allows the required emission wavelength to be obtained, whilst the addition of the second element compensates the lattice constant variation with respect to the substrate. The application of such a method can be better understood by referring to the diagram of Fig. 11, where the variations related both to the emission wavelength (on the abscissa axis) or to the lattice constant (on the ordinate axis) are reported for a certain number of ternary compounds of the III-V group. Considering, for instance, an InP substrate, the diagram of Fig. 11 indicates that moving along the line GaAs → → InAs, material can be obtained emitting radiations at the wavelengths between 1 μm and 1.5 μm, while the lattice constant of the $Ga_{1-x}In_xAs$ compound can be maintained identical to that of InP by replacing a fraction of the As atoms by the P atoms.

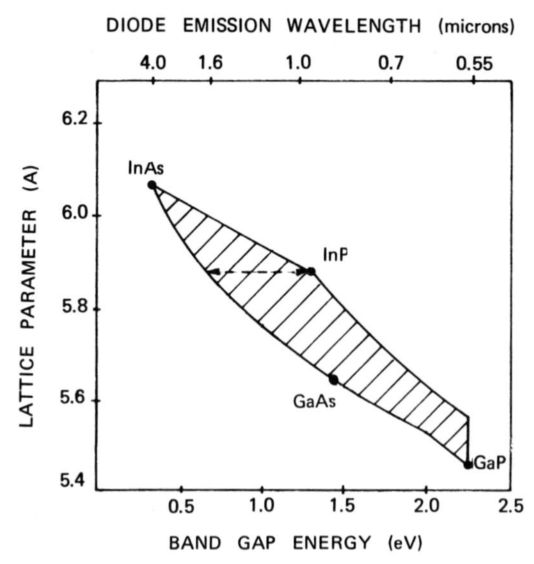

FIG. 11 – Composition range where a quaternary compound $Ga_{1-x}In_xAs_{1-y}P_y$ can be grown lattice-matched on a InP substrate.

Practically this means the deposition of the $Ga_xIn_{1-x}As_yP_{1-y}$ quaternary composition material on an InP substrate and it can be demonstrated, on the ground of the Vegard law, that such a quaternary compound keeps the same lattice constant as the InP substrate for $y=2x$ with $0 \leqslant x \leqslant 0.5$. The deposition of a quaternary material can be carried out through LPE without particular difficulties and very encouraging results have already been obtained in this direction. Further details of these results will be given with reference to the emitting devices operating at these wavelengths.

1.2 Radiative recombination process

1.2.1. *Properties of p-n homojunctions*

Radiation emission from a semiconductor is due to the recombination transition between electrons and holes, which takes place when the material has been excited outside the thermodynamic equilibrium state [12] (Fig. 4, Section 1.1.1).

Under equilibrium conditions the number of recombination transitions between the conduction band and the valence band $R_{c\to v}$, must be equal to the number of inverse transitions due to the thermal excitation $R_{v\to c}$ and, if n_0 and p_0 are the equilibrium concentrations of electrons and holes respectively, we get

$$R_{c\to v} = R_{v\to c} = Bn_0p_0 \tag{1}$$

where B is a constant, characteristic of the material and type of transition assuming that no selection rules exist.

If the equilibrium condition is disturbed, for instance by the injection of g electrons per second in a p-type material, the electron density (minority carriers) produced in excess in the conduction band becomes

$$\Delta n = \frac{g}{B(n_0 + p_0)} = g\tau \tag{2}$$

where τ is defined as the lifetime of the minority carriers. If the electron generation ceases, the excess concentration decreases towards the equilibrium condition according to the law

$$\Delta n = g\tau \exp\left[-t/\tau\right]. \tag{3}$$

In a direct gap material the B constant can be evaluated in the order of 10^{-9} cm^3 s^{-1}; consequently, in a semiconductor of this type, in which the majority carrier concentration is in the order of 10^{18}/cm^3, the average life of the minority carriers is $\sim 10^{-9}$ s.

In an indirect gap semiconductor this value is some orders of magnitude greater. This is one of the reasons whereby the radiative recombination efficiency is proportionately reduced in the case of band to band transition. As a matter of fact, other recombination processes are possible [13], which take place without radiation emission and where the energy associated to the transition is generally transferred, in the form of heat, to the crystal lattice. Since these undesired processes are also characterized by a lifetime as illustrated earlier, it stands to reason that in terms of competition the radiative recombination will occur more efficiently in those materials in which this process has a smaller time constant τ. Different excitation mechanisms can accomplish the excess concentration Δn of minority carriers; in practice, the most suitable mechanism from the application standpoint is that in which the minority carriers injection occurs through a forward biased *p-n* junction [14]. A *p-n* junction can be obtained, for instance, through the diffusion of a *p*-dopant in a material of *n*-conductivity or by epitaxial growth of a *p*- or *n*-layer on a substrate of opposite conductivity.

The band structure in a *p-n* junction is illustrated in Fig. 12:

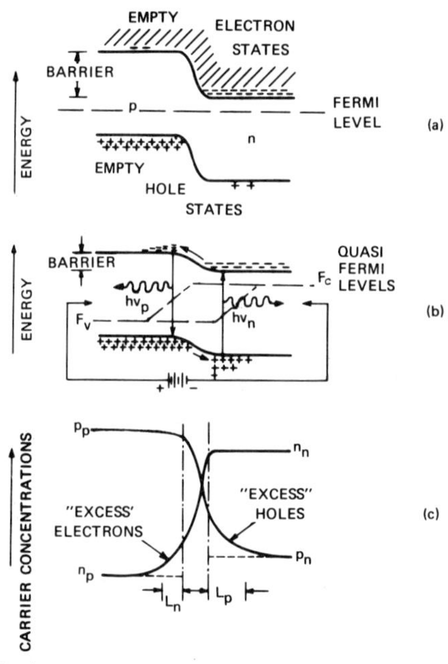

FIG. 12 – Basic principles of a *p-n* junction operation: *a*) Without external bias; the constant Fermi level represents the equilibrium condition. *b*) With forward bias; the non equilibrium carrier distribution across the junction is described now by the « quasi Fermi » levels F_c and F_v for the conduction and valence band respectively. *c*) The minority (excess) carrier diffusion across the junction (electrons on the *p*-side and holes on the on-side) is controlled by the diffusion length L_n and L_p.

without external bias, the system is in equilibrium and therefore the Fermi-level remains constant through the junction. Consequently a potential barrier, the amplitude of which is nearly equal to the forbidden energy gap of the material, prevents the carrier diffusion through the junction. When this potential barrier is lowered by applying an external bias, a current can flow by diffusion through the junction, by producing a certain excess concentration of minority carriers (electrons on the *p*-side of the junction and holes on the *n*-side), which in general recombine with radiation emission.

The current density flowing into the diode can be written in the form

$$J = e\left(\frac{D_e n_p}{L_e} + \frac{D_h p_e}{L_h}\right) \exp\left[\frac{eV}{KT} - 1\right] \tag{4}$$

where $D = \mu(KT/e)$ and $L = \sqrt{\tau D}$ indicate the diffusion constant and the diffusion length respectively (the *e* or *h*-index refers to electrons and holes respectively), τ the average life of minority carriers, k the Boltzmann constant, e the electron charge, V the external voltage applied to the junction and n_p, p_e the concentration of minority electrons in the *p*-material and holes in the *n*-material respectively. The relationship (4) takes into account symmetrically the contribution to the current of electrons and holes; however, in practice, the relevant μ_e and μ_h mobilities are very different and, in general, in the III-V compounds, $\mu_e \gg \mu_h$ (for instance, in GaAs at 300 °C, for material doping of the order of $10^{18}/\text{cm}^3$, $\mu_e \simeq 20\mu_h$). Consequently, the electron injection in the *p*-region is dominant and the radiation is therefore generated in the *p*-side of the junction. This involves an appropriate choice of the geometry and assembly of the photoemitter device for optimizing the transfer outside of the optical power. Furthermore, as the bias voltage applied to the device is, in practice, such that $eV \gg KT$, the relationship can be written more simply

$$J = A \exp\left[\frac{eV}{KT}\right]. \tag{5}$$

The electric field in the *p-n* junction is associated to a distribution of fixed electric charges, and the region, where this distribution, is located, is called the spatial charge region of the junction. The simple relationship (4) is based on the assumption that in this region minority carrier recombination does not occur. Frequently this is not the case and a portion of the current flowing through the junction is actually due to the non-radiative recombination through deep energy levels inside the forbidden band (Fig. 13). These are caused by impurities accidentally present inside the material, or by metallurgical defects near the *p-n* junction.

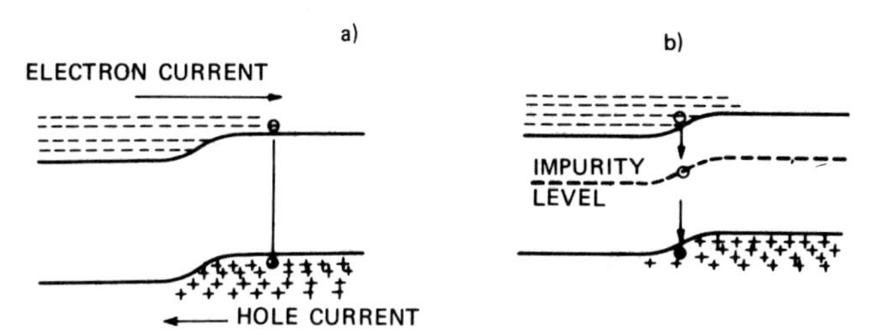

FIG. 13 – Radiative (*a*) and non radiative *b*) recombination processes. In the *b*) case recombination occurs through a deep impurity level in the space charge region.

The dependence of the current density, associated to the recombination in the spatial charge region, is of the type

$$J \equiv \exp\left[\frac{eV}{nKT}\right] \tag{6}$$

where n has generally a value between 1 and 2. This current, which does not contribute to the radiation generation, must be reduced as far as possible, by trying to minimize the number of deep levels in the forbidden band; in particular, this justifies the cautious use of epitaxial *p-n* junctions between materials having a different lattice constant, in that the defects, which necessarily arise on the growth interface, are very likely responsible for a high concentration of non-radiative recombination centers. In practice, the influence of the non-radiative recombination in the spatial charge region prevails at low injection levels. Therefore, the current-voltage characteristic of a forward biased device can be in many cases divided into three regions, according to the increasing voltage levels: the first, where $I \equiv \exp[eV/2KT]$; the second, where the diffusion current prevails, in which $I \equiv \exp[eV/KT]$ and finally, at high current, where the characteristic is controlled by the series resistance of the device and hence $I \equiv V/R_s$ (Fig. 14).

The radiative internal quantum efficiency in the condition of prevailing diffusion current can reach very high values in excess of 50%. However, since this radiation is generated inside the device, owing to the internal absorption and reflection losses on the semiconductor surface, the radiation fraction actually available outside is very reduced (1-10%). Therefore a very important aspect, when designing an emitter device, is the choice of a geo-

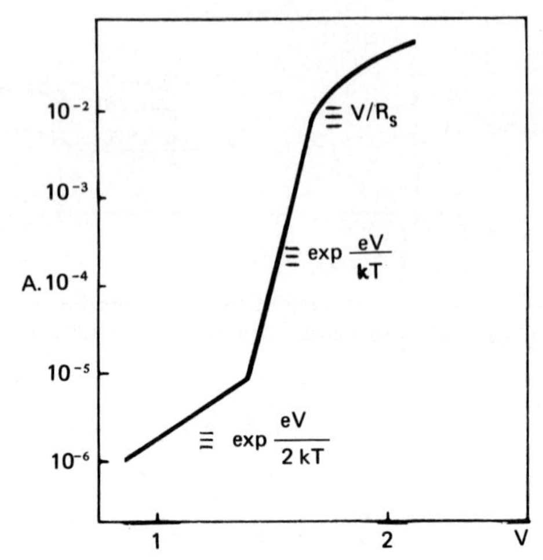

FIG. 14 – Typical current-voltage characteristic of a forward-biased GaAs *p-n* junction.

metry suited to the optimization of its external quantum efficiency. This is particularly relevant when it is necessary to provide the generator and an optical fibre for the coupling. This topic will be discussed in detail later on.

1.2.2. *Spontaneous emission and stimulated emission*

Formally in a *p-n* junction the same excitation mechanism is used for the attainment both of the spontaneous and stimulated emission. In general [15] (Fig. 15), three fundamental processes control the interaction between radiation and matter: an absorption process occuring when a photon is absorbed by an atomic system which steps from its fundamental level to an excited level;

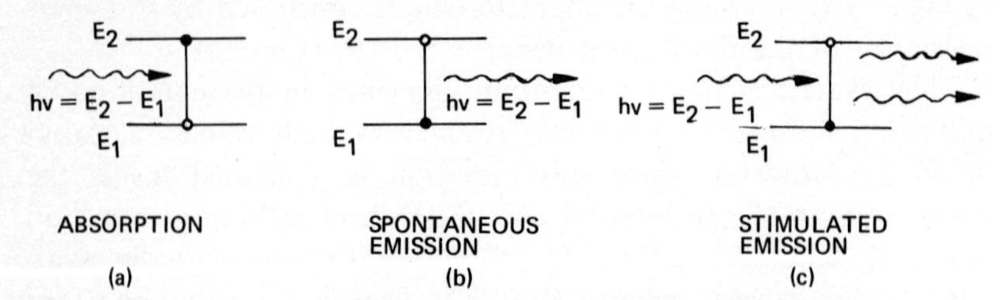

FIG. 15 – Fundamental interaction processes between radiation and matter, in the case of a simple two-level atomic system.

a process of spontaneous emission, taking place when an atomic system starting from an excited level returns to its fundamental level by emitting a photon, the $h\nu$ energy of which corresponds to the energy difference between the two levels; a process of stimulated emission occurring when a photon, the energy of which is equal to the energy difference between excited level and fundamental level, impinges on an atomic system already in the excited level. In this last case, the incident photon is not absorbed, but causes the stimulated de-excitation of the atomic system through the emission of a second photon having the same direction and phase as the incident photon.

To obtain the stimulated emission, most atoms forming the material must therefore lie in the excited level rather than in the fundamental level. An atomic system complying with this condition is in a population inversion state.

In a semiconductor, in a state of thermodynamic equilibrium, the occupation probability of an energy level E is given by the Fermi-Dirac distribution

$$f(E) = \left(1 + \exp\left[\frac{E - F_0}{KT}\right]\right)^{-1} \tag{7}$$

where F_0 denotes the Fermi level, K the Boltzmann constant and T the absolute temperature. Once the density of state distribution function $n(E)$ is known, the distribution (7) allows the number of actually occupied states to be calculated. For instance, for the conduction band, this number can be written

$$n = \int_{E_c}^{\infty} n(E) f(E) dE \tag{8}$$

where E_c is the energy of the conduction band bottom. Similar considerations can be extended to the valence band. The equilibrium state (for $T = 0$ °K) is illustrated in Fig. 16*a*) for a direct gap semiconductor.

In the case of a semiconductor which is in an excited state as a consequence of the injection of minority carriers, a similar procedure can be used, provided that in the distribution function (7) the so-called F_c and F_v quasi-Fermi levels, for the conduction and valence bands respectively, replace the F_0 energy. In this case,

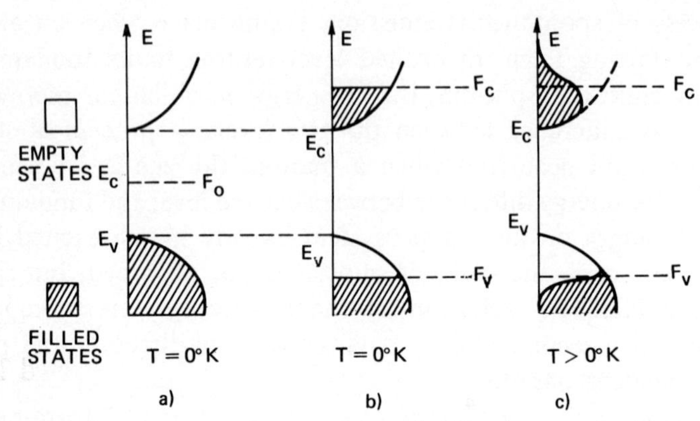

FIG. 16 – Density of states and occupation probability for a semiconductor: *a*) in the equilibrium state at 0 °K; *b*) in the inverted population state at 0 °K; *c*) in the inverted population state at $T > 0$ °K.

if an appropriate excitation mechanism makes possible the transfer of a large number of electrons from the valence band to the conduction band (Fig. 16*b*, *c*) the distribution of the occupied levels in the semiconductor is inverted as regards the equilibrium condition. By assuming the bottom of the conduction band E_c as energy reference level, we can calculate the emission probability for the photons of energy $h\nu$ nearly equal to the transition energy between the occupied states in energy conduction band E and the empty states of energy $(E - h\nu)$ in the valence band. As a matter of fact, the emission rate is proportional to the product between the density of the filled upper levels $n_c(E) f_c(E)$, where

$$f_c(E) = \left(1 + \exp\left[\frac{E - F_c}{KT}\right]\right)^{-1} \tag{9}$$

and the density of the empty lower levels $n_v(E - h\nu)\,[1 - f_v(E - h\nu)]$ where

$$f_v(E) = \left(1 + \exp\left[\frac{E - F_v}{KT}\right]\right)^{-1}. \tag{10}$$

The total probability is obtained by integrating on all energies and is proportional to

$$W_{\text{emission}} \equiv \int_E n_c(E) f_c(E) n_v(E - h\nu)\,[1 - f_v(E - h\nu)]\,dE \tag{11}$$

The absorption probability for the photons of the same $h\nu$ energy can be likewise calculated; it is proportional to

$$W_{\text{absorption}} \equiv \int_E n_c(E)[1 - f_c(E)]\, n_v(E - h\nu)\, f_v(E - h\nu)\, dE\,. \quad (12)$$

Since in the two previous expressions we have the same proportionality constant for both transition types, the stimulated emission corresponds to the case in which $W_{\text{emission}} > W_{\text{absorption}}$. By substituting the relations (9) and (10) in the expressions (11) and (12), it is easy to verify that the foregoing inequality involves a similar condition for the quasi-Fermi levels, which can be written

$$F_c - F_v > h\nu\,. \quad (13)$$

The relationship (13) indicates the condition [16] necessary to obtain the population inversion inside the semiconductor, where the occupation probability is greater for the higher energy state. In a *p-n* junction, a simple method for complying a priori with the condition (13) is to heavily dope the *n*- and *p*-material, until the relevant Fermi level enters into the conduction band and valence band respectively (degenerate doping). The band diagram of the biased *p-n* junction in direct conduction is shown in Fig. 17. In a high injection condition a region exists which extends towards the *p*-side of the junction, owing to the greater electron mobility,

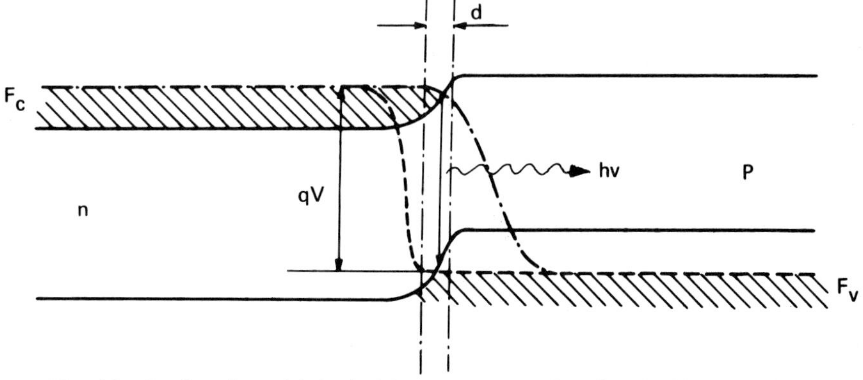

FIG. 17 – *P-n* junction with both sides degenerately doped and with strong forward bias. The quasi Fermi levels separation is higher than the recombination energy $h\nu$ in the narrow region *d*. Note that carriers are also injected outside the region *d*, where the condition $F_c - F_v > h$ does not hold. These carriers do not contribute to the stimulated emission.

where the separation between the quasi-Fermi levels is greater than the energy associated to the band-to-band recombination, thus verifying the condition (13).

In general, this doping distinguishes the material suitable for lasers from the material suitable for LED's, since the maximum quantum efficiency for incoherent emitters is reached at doping levels lower than those required for degenerate bands.

In this way, the first condition required for obtaining laser oscillations (i.e. the availability of material suitable for radiation amplification) can be met; the resonant cavity for establishing the optical feedback can be obtained through a resonant structure of the Fabry-Perot type formed by two cleaved surfaces of the same semiconductor crystal (Fig. 18).

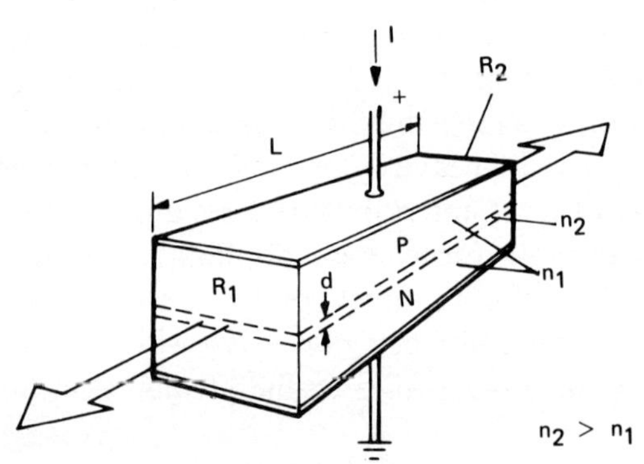

FIG. 18 – Idealized structure of a homojunction semiconductor laser with a Fabry-Perot resonant cavity.

In the particular case of GaAs (index of refraction $n = 3.6$), the reflection coefficient at the GaAs—air interface, for instance, is equal to $R = 0.32$; the structure of Fig. 18 can be easily obtained by orientating the crystal with the *p-n* junction plane parallel to the crystal plane $\langle 100 \rangle$, thus using the natural cleavage planes $\langle 110 \rangle$ as mirrors.

The threshold condition for the laser oscillation can be written for the cavity of Fig. 18 as a function of the gain necessary for compensating the losses due to reflection and absorption for a light round-trip along the cavity in the form

$$R_1 R_2 \exp \{(g_{th} - \alpha) L\} = 1 , \tag{14}$$

hence

$$g_{th} = \alpha + \frac{1}{L} \log \frac{1}{\sqrt{R_1 R_2}} \tag{15}$$

where L denotes the cavity length, R_1 and R_2 the reflectivity of the two terminal mirrors, α the cavity losses. For the idealized uniformly excited slab in Fig. 18 a nominal threshold current density can be defined for a unit radiative quantum efficiency as

$$J_{\text{nom}}(\text{Amp/cm}^2) = I/A = ed\overline{W} \tag{16}$$

where I is the current trough the slab, A its cross sectional area, d the slab thickness, e the electronic charge and

$$\overline{W} = \int_0^\infty W_{\text{emission}}(E)\,dE$$

is the total radiative recombination rate per unit volume.

By using the Einstein relationships, relating the spontaneous emission probability to the stimulated emission and absorption probabilities, Stern [17] has calculated the J_{nom} dependence on the gain $g(E)$ in the form

$$J_{\text{nom}} = \frac{8\pi de\bar{n}^2 E_{g_{\max}}^2 \gamma \Delta E}{c^2 h^3} g(E_{g_{\max}}) \tag{17}$$

where: $\bar{n}$ is the refractive index of the semiconductor;
$E_{g_{\max}}$ is the energy of the maximum gain;
ΔE is the spontaneous emission spectral half-width;
γ is a complicated function which takes into account the temperature dependence and non-linear dependence of gain on current.

For a real semiconductor the threshold current density can be written now

$$J_{th} = \frac{J_{\text{nom}}}{\eta_i \Gamma} \tag{18}$$

where η_i is the internal quantum efficiency and Γ is the fraction of the propagating optical mode that is within the slab. By substituting (18) and (17) into (15), the threshold condition can be written

$$J_{th} = \frac{d}{\eta_i \Gamma} \left\{ \frac{8\pi e \bar{n}^2 E_{g_{\max}} \gamma \Delta E}{c^2 h^3} \right\} \left(\alpha_L + \frac{1}{L} \log \frac{1}{\sqrt{R_1 R_2}} \right). \tag{19}$$

The linear dependence of J_{th} on the active region thickness d is experimentally well observed in the double heterostructure injection lasers, where the uniform excitation of the active region is made possible by the electrical confinement. When d becomes smaller than 0.1 μm, due to the break in the optical confinement, Γ decreases and no further reduction of J_{th} can be obtained. A more detailed account on these aspects will be given in Section 1.4. The cavity losses are mainly due to two factors, namely free carrier absorption, responsible for the attenuation in the active (inverted) region of the slab, and electromagnetic field penetration in the passive regions near the central region.

The external quantum efficiency η_{ext} of a semiconductor laser with a Fabry-Perot cavity can be calculated [18] by the relationship

$$\frac{1}{\eta_{\text{ext}}} = \frac{1}{\eta_i} \left[1 + \frac{\alpha L}{\log (1/R)} \right] \tag{20}$$

where we assume $R_1 = R_2 = R$.

The electromagnetic field distribution inside the Fabry-Perot resonant cavity of Fig. 18 can be determined on the ground of the Maxwell equations, if it is possible to assign the refraction index distribution $n(xy)$ in the plane perpendicular to the radiation propagation direction. This depends on the device structure and therefore the problem must be dealt with referring to the particular cases, considered in the following. On the other hand, as the distribution in the plane xy is correctly assumed constant along the z-axis (propagation direction), in general, the frequency spacing of the so-called longitudinal modes can be simply calculated. As a matter of fact, denoting with (integer) N the order of the longitudinal mode, the resonance condition requires the arrangement of $N(\lambda/2\bar{n})$ half-wavelengths between the cavity mirrors, separated by L distance. In other terms the following result must be obtained

$$N\lambda = 2L\bar{n} \,. \tag{22}$$

As the $\Delta\lambda$ spacing between two adjacent modes can be achieved by differentiating (22) with respect to λ, we find for large N

$$\frac{L\Delta\lambda}{\lambda^2\Delta N} \simeq \frac{-1}{2\bar{n}[1-(\lambda/\bar{n})(d\bar{n}/d\lambda)]}. \tag{23}$$

The term between brackets in the relation takes into account dispersion, which in the case of a semiconductor material is not at all negligible as in the case with a gas laser. In the case of GaAs and for a cavity having a length $L = 0.5$ mm, the relation (23), by assuming $\lambda \simeq 900$ nm, $\bar{n} = 3.6$ and $dn/d\lambda \simeq 10^{-4}$, allows $\Delta\lambda \simeq 0.3$ nm to be evaluated, in good agreement with the experimental data, considering the uncertainty in the evaluation of $\bar{n}$ inside the active region.

1.2.3 *Heterojunctions*

In general, a heterojunction is formed by bringing into contact two semiconductor materials having different energy gaps. It can be implemented in the semiconductors of the III-V groups through the methods described in Section 1.1. It can be said that the progress registered during the past ten years by the semiconductor laser devices is mainly connected to the extensive use of the heterojunction structure. Different heterojunction types can be used according to the type of conductivity (n or p) of its two component semiconductor materials. The calculation of the relevant band structure can be carried out following the same criteria adopted for determining, for instance, the p-n homojunction structure, shown in Fig. 12 even if in this case an explicit evaluation is much more complicated [19]. However, in our case, we can limit our discussion on heterojunction properties to the effects of electrical and optical confinements. This can be done in purely qualitative terms, referring to some cases of particular interest.

The two heterojunction structures which have led in particular to the improvement of the performance of semiconductor laser photoemitters are illustrated in Fig. 19. For comparison purposes, a simple homojunction structure identical to that illustrated in Fig. 17 is also shown. The same figure shows the distribution of the minority carriers injected through the p-n junction, the evolution of the refraction index and the near-field distribution of the

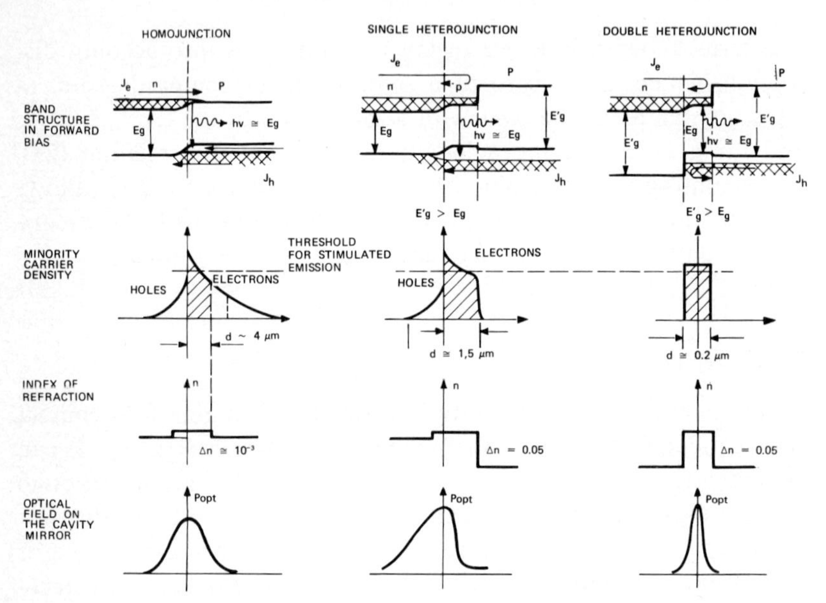

FIG. 19 – Comparison between the basic properties of a *p-n* homojunction, a single heterojunction and a double heterojunction.

optical power on the device emission surface, assumed to have the same laser structure illustrated in Fig. 18; the potential energy in the three diagrams is schematically shown under strong forward bias. As can be noted, the heterojunction between the materials having gaps of forbidden energy E_g and E_g' with $E_g' > E_g$, has the effect of producing a potential barrier preventing the minority carrier diffusion outside the radiative recombination region. In the case of the homojunction, where there are not such barriers, the minority carrier concentration decreases according to an exponential law governed by the diffusion length L_e; in this case the device region, where the relation (13) is verified, is very large ($d \simeq$ 3-4 μm); furhermore, a considerable fraction of the injected carriers does not contribute to the stimulated emission. Practically, for a homojunction laser, this means a very high threshold current density ($J_{th} \simeq 10^5$ A/cm^2 at 300 °K). In the case of the single heterojunction, the *p-n* junction, through which the radiative recombination takes place, is still made up of materials having identical gap E_g, whereas the heterojunction is located in such a way as to block the carrier injection having the greatest diffusion length (in this case the electrons injected into the *p*-material). Thus, the thickness of the recombination region and, consequently, the excitation current density required for reaching the stimulated emission threshold can be reduced (compare the relationship (19)).

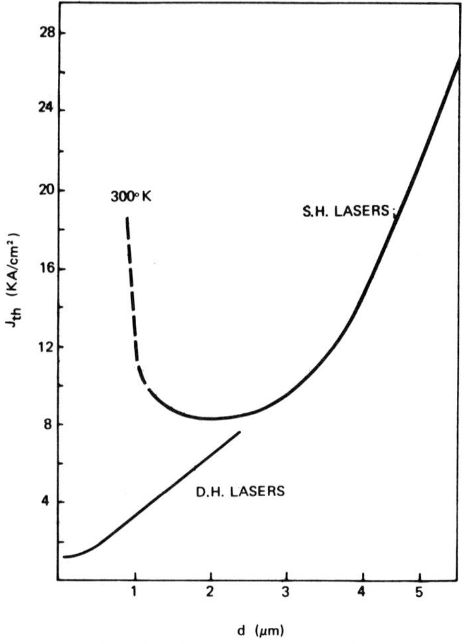

FIG. 20 – Threshold current density dependence on the active region thickness *d*, for a single heterojunction and double heterojunction structure.

Figure 20 shows the evolution of the threshold current as a function of the *d* thickness of the active region for a single heterojunction laser of the $GaAs/Ga_{1-x}Al_xAs$ type; J_{th} is minimum ($\sim 8 \cdot 10^3$ A/cm^2) for *d* equal to ~ 1.5 μm. Moving towards a lower thickness, J_{th} rises suddenly. This is due to the injection of holes into the *n*-side of the junction, which ceases to be negligible when *d* can be compared to the relevant diffusion length L_h and which permits the radiative recombination outside the region, where the population inversion exists. The active region, having an E_g gap, inside the so-called double heterojunction structure appears delimited on both sides by material having a gap $E'_g > E_g$; thus both electrons and holes are confined inside the same region, the thickness of which can therefore be reduced by about an order of magnitude as regards the value mentioned above. However, a second substantial advantage is connected to the double heterojunction symmetrical structure and concerns the fact that the refraction index of the central layer of energy gap E_g is greater than that of the adjacent layers with a gap E'_g. Consequently, the structure forms a planar dielectric guide, where the radiation is confined inside the central layer, in which the stimulated emission occurs, and does not appreciably penetrate into the external layers, where it would undergo a high attenuation. The combined effect of both electric and optical confinement operating in a double heterojunction allows the threshold current density to be reduced to about 1000 A/cm^2; at this current level laser devices, operating in direct current at room temperature, can be implemented. The earlier considerations are not connected to particular materials, the only constraint being that the heterojunctions must be free from non-radiative recombination centres; since such centres are mainly due to crystal defects, the imposed condition is met if the materials forming the heterojunction have the same lattice constant. This leads us back to the considerations discussed in the first section. In particular, referring to Fig. 5*a*) it can be observed that, to get a laser emission at wavelengths between 0.8 and 0.9 μm, the $Ga_{0.6}Al_{0.4}As/Ga_{1-x}Al_xAs/Ga_{0,6}Al_{0.4}As$ double heterojunction structure with x between 0 and 0.1 can be used and deposited on a GaAs substrate. At wavelengths between 1 and 1.6 μm the $InP/Ga_{1-x}In_xAs_{1-y}P_y/InP$ structure can be adopted, the substrate of which is made up of the same InP, which is on the whole lattice matched, when $y \simeq 2x$ and x is between 0.1 and 0.4. This allows the interval at the indicated wavelengths to be covered. Many other structures are, in principle, possible, although the two explicitly mentioned have at present delivered the most satisfactory results. While double heterojunction structures are essential for the practical utilization of laser devices, in the case of LED emitters the considerable advantages offered by these structures are not decisive. Since the preparation of double heterojunction laser structures still involves considerable difficulties as regards the control or the reproducibility of the material growth process, regardless of the methods adopted, it is evident that LED devices, which can be implemented more simply and more reliably also without resorting to the structure recalled in this section, are likely to remain at the center of interest in the short and medium term from the point of view of applications.

1.3 Incoherent emitters

1.3.1 *General remarks on LED structures*

Light-Emitting Diodes (LED), for use in optical fibre communication systems, must necessarily have high efficiency, high frequency response, narrow spectral linewidth and high reliability characteristics. To meet these requirements, various LED structures have been developed, roughly divisible into two categories: surface emitting LED and edge emitting LED. In both cases, the main object in the device design is the optimization of best optical coupling from the LED to the fibre. The incoherent optical power that can be coupled into an optical fibre depends directly upon the source radiance B. It can be evaluated in the case of direct coupling from the simple relation

$$P_f = \pi T B \Delta^2 A_f \tag{24}$$

where T indicates the reflection losses, Δ the fibre numerical aperture, A_f the fibre core area.

For low dispersion optical fibres, for use in long-distance communication links, a typical value for NA is 0.15, and the core diameter ranges from 60 to 80 μm. Therefore, since the fibre parameters are set by overall considerations, a very high source radiance is required to increase P_f. To illustrate how this can be obtained from an incoherent emitter, we shall refer to Fig. 21, which shows a cross-section of a typical diffused *p-n* junction light emitting device. Light is generated by the recombination process uniformly on the *p*-side of the junction; this region is also the main source of device heating and because thermal satu-

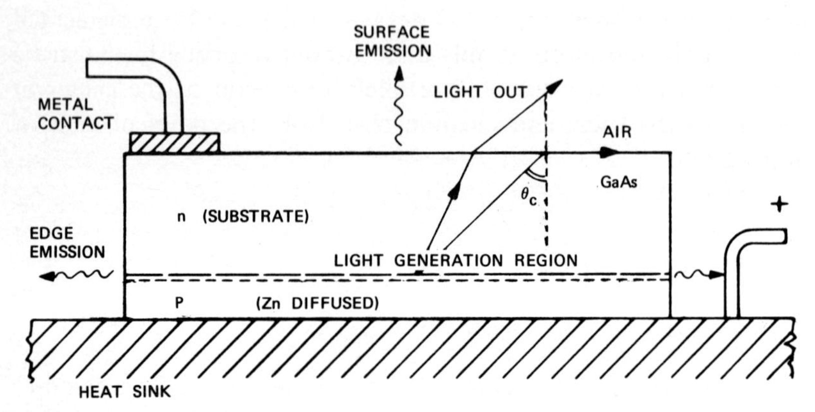

FIG. 21 – Cross-section of a typical *p-n* junction GaAs LED device. θ_c is the total reflection angle at the GaAs-air interface.

ration is the power limiting mechanism in the LED operation the device is mounted *p*-side down on the heat sink; radiation is only weakly absorbed in the *n*-type material because the heavily doped *p*-type layer has a slightly smaller energy gap than the *n*-type. Nevertheless, only a small fraction of the optical power generated at the junction can be transmitted outside the device; this is due to the high refraction index of the semiconductor which causes a severe total reflection loss. If we call θ_c the critical angle for the total reflection, the loss factor can be calculated for the structure in Fig. 21 as follows

$$F = \frac{\int_0^{\theta_c} \sin\theta \, d\theta}{2\int_0^{\pi/2} \sin\theta \, d\theta} = \frac{1 - \cos\theta_c}{2} \simeq \frac{1}{4\bar{n}^2} \tag{25}$$

where $\bar{n}$ is the material refraction index. Besides the F loss factor, a transmission loss must be accounted for, which, at normal incidence, can be written

$$T = 1 - \frac{(n-1)^2}{(n+1)^2} = \frac{4n}{(n+1)^2} \, .$$

Even in the absence of absorption loss in the *n*-type material and 100% internal radiative quantum efficiency the overall attenuation factor $T \times F$, due to the refraction index is very high. Assuming, for instance, $\bar{n} = 3.6$, for the GaAs material only 1% of the generated optical power is available outside the simple LED structure shown in Fig. 21. The outgoing radiation is uniformly distributed on the exit surface which acts as a Lambertian source. For display applications, the external efficiency can be greatly improved by covering the device surface with transparent epoxy lenses (refraction index $n \simeq 1.6$) or by shaping conveniently the semiconductor diode chip; a number of such optical systems are shown in Fig. 22. In general, although the epoxy lens reduces the loss due to the total reflection, enlarges the source size and correspondingly narrows the source radiation pattern, it cannot however increase the device radiance. Therefore, this approach is not suitable for improving the coupling of the LED with the optical fibre, at least as long as the source area A_s, (total chip area in the structure of Fig. 21) is very large in comparison with the fibre area A_f. Since linear dimension of the LED chip area cannot be much less than 500 μm for technological reasons a reduction in the radiating source can be obtained by limiting

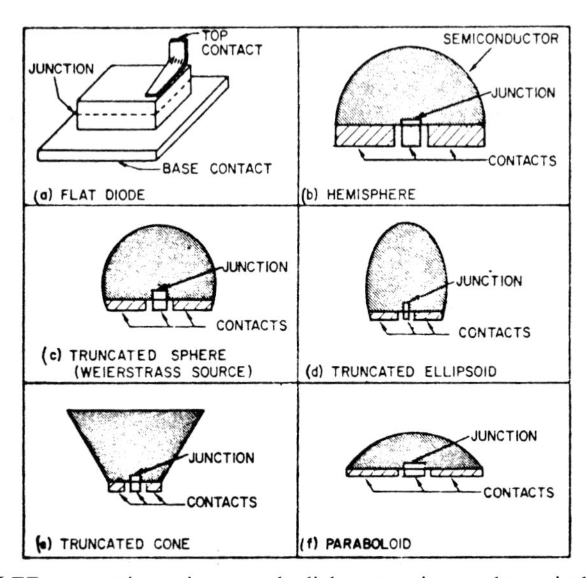

FIG. 22 – LED geometries to increase the light extraction or the optical efficiency.

the current injection to a restricted area on the LED chip. By operating, in this case, the LED at the same current, the power dissipation over the whole chip will remain unchanged, but the current density injected through the restricted area will increase and, as a consequence, a high source radiance can be obtained.

These are the basic considerations that lead to the surface emitting high radiance LED whose properties will be described in detail in the next section. Radiation which is emitted from the device edge, in the plane of the *p-n* junction (see Fig. 21), could also be efficiently coupled to an optical fibre; in this case, the source radiance can be increased by making the emitted light radiation pattern more directional than a Lambertian distribution would allow. This effect cannot be obtained in a simple *p-n* junction but requires a narrow double heterostructure active region where the dielectric waveguide works as a spatial filter for the incoherent optical power emitted from the device edge. In other words, an efficient edge emitting LED is very similar to a laser structure operating below the threshold for stimulated emission. In these conditions it is still uncertain whether it would be more advantageous to use an edge emitting LED rather than the laser itself.

1.3.2 *High radiance surface LED*

A cross section of a surface emitter constructed in accordance with the design criteria mentioned above is shown in Fig. 23. This structure, constructed using a GaAs diffused-junction homostructure material, was first described by Burrus *et al.* [20] in 1970 and has since been implemented in several other laboratories [21, 22]. The device technology consists first of a Zn diffusion process where the *p-n* junction is formed. This process usually takes place in a closed quartz ampoule using a zinc-gallium-arsenic mixed source. Temperature and diffusion time can be chosen in such a way as to obtain the best compromise between device speed and radiance. Actually the device rise time is a monotonically decreasing function of the *p*-layer doping, while the radiative recombination efficiency increases with the *p*-layer doping up to a maximum and then drops because of the increasing concentration of the non-radiative recombination centres in the

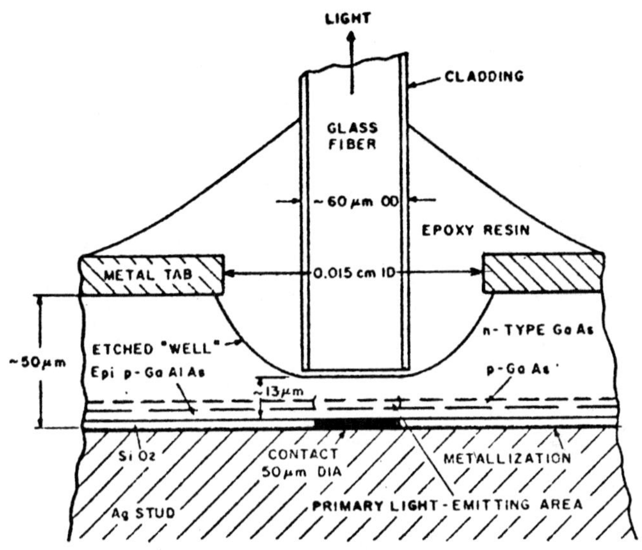

FIG. 23 – Cross-section of high radiance surface emitters, coupled to an optical fibre. (After [20]).

heavily doped material. The diffused wafer is then covered on the *p*-side with an SiO_2 layer and 50 μm in diameter windows are opened by standard photolithographic techniques in this insulation layer. The ohmic contacts are metallized and wells are etched on the *n*-side up to 20 μm from the *p-n* metallurgical junction. This is to reduce internal absorption of the forward emitted radiation and to allow the direct coupling of an optical fibre in close proximity to the emitting surface. A gold pad is then electroplated onto the *p*-side through a photoresist mask. This gold pad gives mechanical strength to the very thin active region of the LED chip and also serves as an integral heat sink in close contact with the primary heat source region of the diode. The diode chip, before mounting on its microwave pill-package, is shown in Fig. 24. Typical device performance is given in Fig. 25 which shows peak radiance versus input current, in Fig. 26 which shows near field optical distribution over the emitting area, and in Fig. 27 which shows the spectral distribution of the emitted radiation. The same technology can be applied if a double heterostructure wafer growth by liquid phase epitaxy is used as a starting material instead of a simple diffused homo-

FIG. 24 – Surface high radiance LED chip. It is shown the etched well and *n*-side metal contact pad. (Courtesy of SGS-ATES/CSELT).

junction: in this case the peak emission wavelength can be shifted, by choosing the active layer composition. For instance, the system $Ga_{1-x}Al_xAs$, for $0 < x < 0.1$ has been used to cover the wavelength range from 0.8 μm to 0.9 μm, and the system $In_xGa_{1-x}As_yP_{1-y}$ can be used in the wavelength range from 1.1 to 1.5 μm. A homojunction high radiance LED has been developed using vapour phase epitaxy with an $In_xGa_{1-x}As$ *p*-layer deposited on a GaAs substrate, and a grading layer of increasing In concentration to obtain a lattice matched *p-n* junction [23]. The peak emission wavelength in this case is 1.06 μm. Although some improvements particularly in the external quantum efficiency are generally observed in the performance of double heterostructure high radiance LED's due to the superior quality of the epitaxial material, the basic device properties remain the same as those of the diffused homojunction structure.

Different technologies have been applied to obtain a better

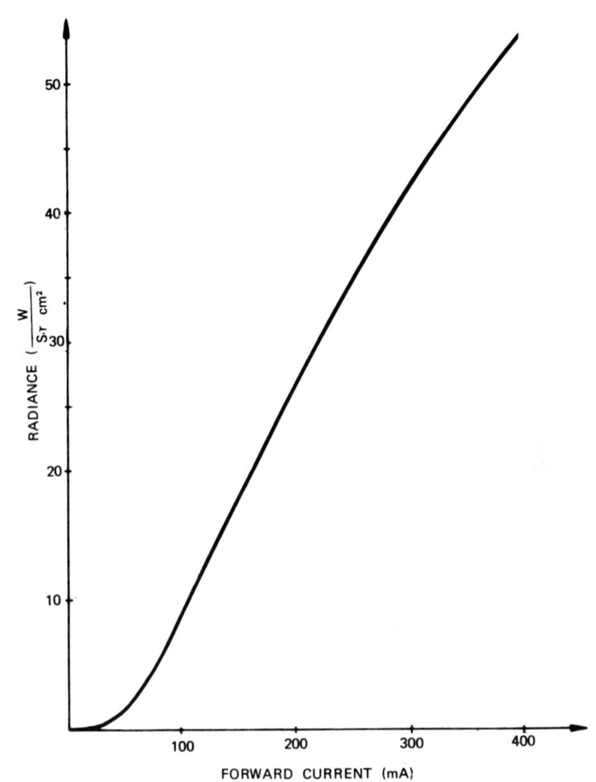

FIG. 25 – Radiance vs. input current for a diffused Burrus type LED.

current confinement than that allowed by the simple SiO_2 insulating mask; particularly it should be noted that in the structure of Fig. 23 the SiO_2 mask restricts the current flux to a small area in the junction, while the junction itself extends over the whole chip area. Therefore, the depletion layer parasitic capacitance is much larger than necessary, and this can substantially reduce the useful LED bandwidth. Proton bombardment techniques have been used to achieve both a tight current confinement and a reduction of the depletion layer capacitance [24]. Consequently, an almost flat near field distribution profile over the emitting area can be obtained and the high frequency device performance can be improved. Although less effective than proton bombardment, a planar diffusion [25] through a suitable diffusion mask can give similar results.

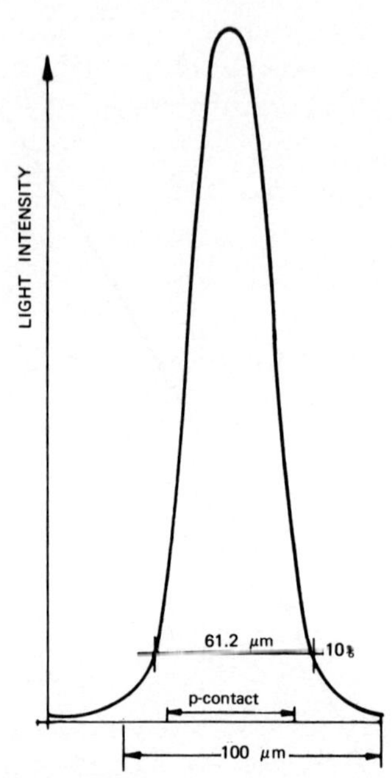

FIG. 26 – Near field optical distribution for a high radiance LED. Nominal injection contact diameter 50 μm; driving *dc* current 300 mA; peak radiance 40 w/sr cm².

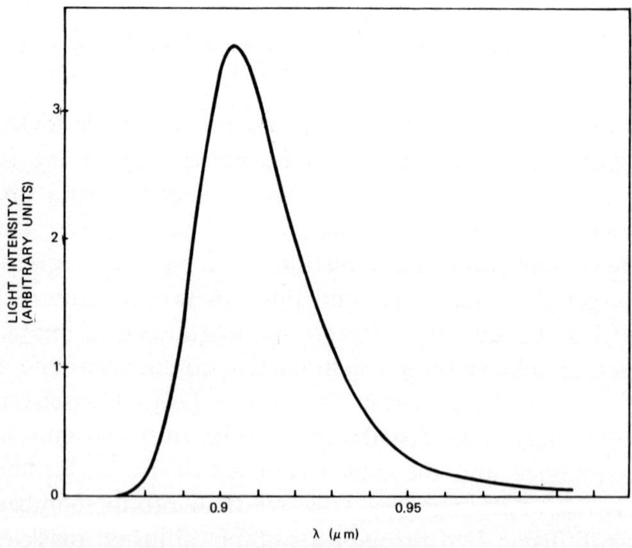

FIG. 27 – Spectral distribution of the light emitted from an high radiance LED. Driving current 250 mA.

Radiances available from the surface emitters just described usually range from 30 to 60 Watt/Ster·cm^2; these values correspond to the peak of the near field distribution, and therefore they cannot be directly used to calculate the optical power coupled to a fibre by (24). In fact, relation (24) actually holds strictly in the hypothesis that the source area $A_s \geqslant A_f$ and B remains constant over that area. The near field analysis allows the definition of an effective radiance (referred to the nominal area of the injection contact) which is generally less than the peak value by a factor of two, in the case of an oxide insulated LED. In this case, taking into account the fibre parameters, relation (24) gives an optical coupled power P_f from $-$ 15dBm to $-$ 17dBm in agreement with the experimentally measured values.

1.3.3 *Edge emitting high radiance LED*

Several attempts have been made to develop edge emitting LED structures with better radiance and efficiency characteristics than surface emitters. These early edge emitters were generally based on the assumption that a Q-spoiled resonant cavity laser structure, operating in a superradiant mode below the lasing threshold would retain high radiance near field laser pattern. At the same time, due to the absence of effective laser operation, improved degradation characteristics are to be expected. High radiances, in excess of 100 Watt/St·cm^2, have been satisfactorily reported for these double heterostructure edge emitters [26]. However, at that time, laser reliability was also undergoing a significant improvement and thus the real advantages of this approach remain somewhat questionable. Consequently, these devices have not been developed beyond the laboratory prototype stage. Nevertheless, as an incoherent emitter should allow a more simplified operation from the system viewpoint than a laser source, work has continued on the edge emitting LED to fully explore the potential of such sources [27]. A recent (1976) interesting result has been obtained by Ettenberg *et al.* [28], with the device structure presented in Fig. 28. The key feature of this double heterostructure edge emitting LED is the extremely thin (0.05 μm) active region, sandwiched between two relatively thick confining layers with high aluminium concentration. The laser operation in this dielectric

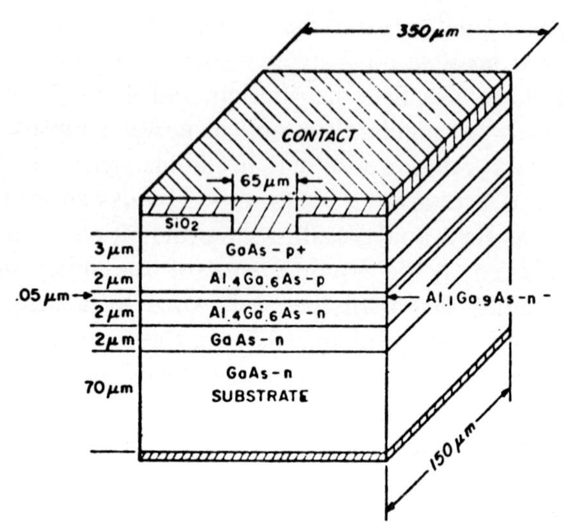

FIG. 28 – Edge emitting LED with very thin active region. (After [28]).

waveguide is in fact impaired, because the fundamental oscillating mode cannot be confined to the active region but spreads over the unpumped confining layers where it suffers from high losses. This fact will not, however, adversely affect the spontaneous emission which is strongly coupled to the fundamental guided transverse mode. The far-field beam width in the direction perpendicular to the junction can be calculated by the approximate relation [29]

$$\theta_{\perp} \cong \frac{20 \Delta x d}{\lambda_0} \tag{26}$$

where Δx is the fractional Al concentration difference across the heterojunctions, d is the active layer thickness, and λ_0 is the emission wavelength. The experimentally measured values of $\theta_{\perp}$ are in good agreement with those calculated by (26) while the far-field beam width in the direction parallel to the junction can be fitted by a regular Lambertian distribution. These results are convincing evidence that a strong coupling exists between spontaneous emission and the dielectric waveguide, and that the device is not operating in the superradiant mode. As a consequence of the narrowing in the solid angle of the radiation emission, a radiance in excess of 1000 Watt/St·cm^2 has been measured for these LED's and an optical power as high as 800 μwatt has been coupled in a low aperture (NA = 0.14) low loss, step-index optical fibre with 90 μm core diameter. Therefore, a significant increase in the power coupling into an optical fibre can be expected from these devices with respect to other LED emitters. Furthermore, the coupled power is quite comparable to that available from a laser source. Another interesting consequence of the tight carrier confinement in a thin active region is that a very high modulation rate can be expected from these edge emitters. In fact, the average carrier density in the recombination active region depends on the injection current density J according to the relation

$$\Delta n = \frac{J\tau}{qd} \tag{27}$$

where τ is the minority carrier lifetime, q the electronic charge and d the active region width. For $d \simeq 0.05$ μm, at 400 mA drive current ($J = 3500$ A/cm^2), the injected carrier density greatly exceeds the background doping carrier density and a fast recombination time is obtained. Accordingly, a 250 Mhz bandwidth has been measured for these LED's at 400 mA drive current. In conclusion, the performance of this edge emitting LED, which still requires a very sophisticated LPE growth technology in comparison with the more conventional structures of the surface high radiance LED, clearly indicates that these devices cannot be ignored as potentially useful incoherent optical sources for fibre data transmission systems, even if at present they are only in an early stage of development.

1.3.4 *Coupling problems and system considerations*

There are various system requirements which can be satisfied by the incoherent emitters decribed above. Even the most simple GaAs diffused Burrus type LED will be more than a suitable light source for short-distance optical data links for ships, aircraft or internal plant. For long-distance communication systems, on the other hand, incoherent emitters operating in the 0.8-0.9 μm wavelength range suffer a serious drawback in that their spectral width, due to the fibre material dispersion, reduces the useful data rate × distance product.

Figure 29 shows how pulse delay spread and data rate × distance product vary with source spectral width for typical silica doped multimode fibres operating in the 0.8-1.3 μm wavelength range.

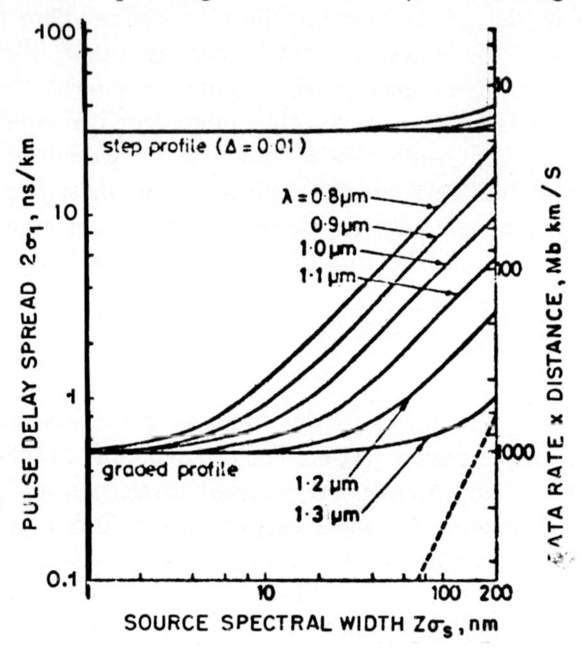

FIG. 29 – Pulse delay spread $2\alpha_f$ and transmission capacity, against source spectral width $2\sigma_s$, for germanium borosilicate multimode fibres. (After: W. M. Muska, Tingye Li, T. P. Lee and A. G. Dentai, Electron. Lett., Vol. 13, p. 605).

If a 30 nm spectral width is assumed, a 150 Mbit/s km data rate × distance product can be estimated for LED sources operating at 0.8-0.9 μm wavelengths. An improvement of more than one order of magnitude in this parameter is to expected for an LED source operating at 1.3 μm wavelength, where the fibre exhibits its minimum material dispersion. Pulse spreading at this wavelength is governed by higher order effects and is proportional to the square of the source spectral width; this ultimate limit is shown by the broken line in Fig. 29, and therefore the potential bandwidth of an optical link, using a 1.3 μm LED source with 100 nm spectral width, can be as large as 250 Mbit/s over a 10 km repeater spacing [30]. Recent experimental results, obtained by using a Burrus type high radiance LED, with a quaternary InGaAsP active layer, are in good agreement with the theoretical predictions [31]. Therefore, a bright future is to be expected for LED sources operating at 1.3 μm wavelength where there is the additional advantage of a reduced fibre absorption loss.

Even if the total optical power available from an incoherent high radiance source and from a laser source are fairly similar, the LED suffers a considerable disadvantage due to the very low LED-fibre coupling efficiency; from rel. (24) and by writing the source emitted power as

$$P_s = \pi B A_s \tag{28}$$

where A_s is the source effective area, the coupling efficiency can be given as

$$\eta_f = \frac{P_f}{P_s} = T\Delta^2 \frac{A_f}{A_s}. \tag{29}$$

For a flat ended fibre, (29) merely states that a maximum coupling efficiency can be achieved if $A_f = A_s$. In this case no gain would result by making $A_s < A_f$. However, when $A_s < A_f$, η_f can be increased if the source emitting area is imaged through a suitable optical system on the fibre core area. This effect can be explained, since the optical system focuses on the area A_f an enlarged (virtual) image of the LED area A_s whose radiance increases with the area ratio A_f/A_s; this is true only if the same optical power is available from the reduced size emitting area. A suitable optical system can be obtained by shaping the fibre in the form of a spherical lens. Spherical lenses with different curvature radii can be produced simply by controlling the thermal melting conditions of the fibre end; some examples of such spherically ended fibres are shown in Fig. 30. Figure 31 gives typical experimental results of the coupling efficiency for a few LED-fibre pairs; a power gain from 2 to 4 with respect to a flat ended fibre has been already reported [32, 33]. Better coupling conditions can be expected as both the spherical radius and the source diameter become smaller, but limitations arise since with a source diameter below approximately 25 μm, the LED is power limited by a thermal saturation effect. Furthermore, a simple melting process will not always produce with the required accuracy a spherical radius less than the fibre cladding radius. A careful mechanical position-

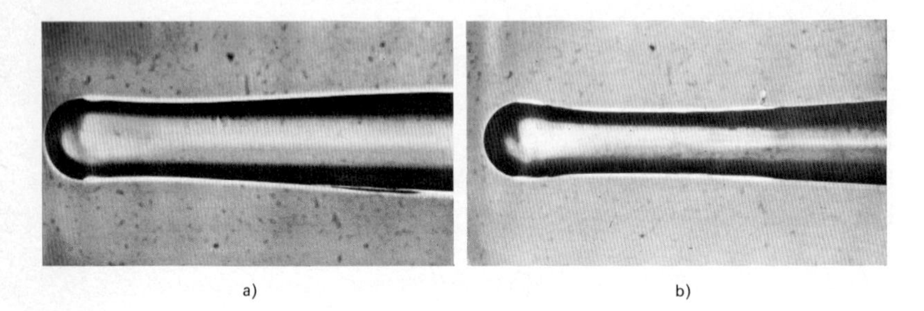

FIG. 30 – Examples of bulb-ended pigtails: *a*) radius of curvature 52 μm; *b*) radius of curvature 43 μm.

ing is required to get the best advantages from this coupling procedure. The LED near-field pattern, which determines the effective source size, and the rounded fibre curvature must also be accurately known if a proper LED-fibre power coupling is to be obtained. These operations could be more suitably performed by

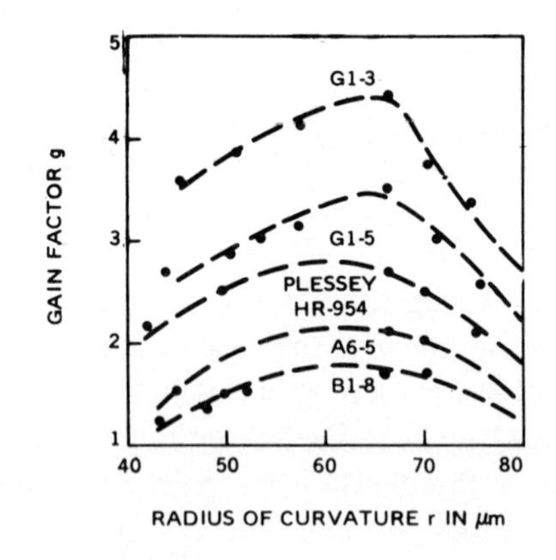

LED	RADIANCE (B) (*) WATT/CM² · STER	WIDTH AT 0.1 B (*) μm	WIDTH AT 0.5 B (*) μm
G1-3	79.0	35.7	18.6
G1-5	52.9	41.4	22.9
PLESSEY HR-954	35.7	59	35
A6-5	26.7	80	48
B1-8	21	100	62

(*) THESE VALUES REFER TO 300 mA d.c. BIAS CURRENT

FIG. 31 – Gain factor against fibre bulb radius of curvature for several high radiance LED's with increasing near-field distribution width. (After [33]).

using an integral LED-fibre mounting package where a short length of rounded fibre (the so-called « pig tail ») is firstly adjusted to the LED for maximum optical coupling, and then held firmly in place. An example of such an integral package is shown in Fig. 105 (Part III, Chapter 3) An exploded view of a practical example is shown in Fig. 32. A fibre-fibre joint, which in general does not require a critical adjustment, can be used for coupling the « pig tail » to the transmitting fibre, as will be discussed in detail in Part III.

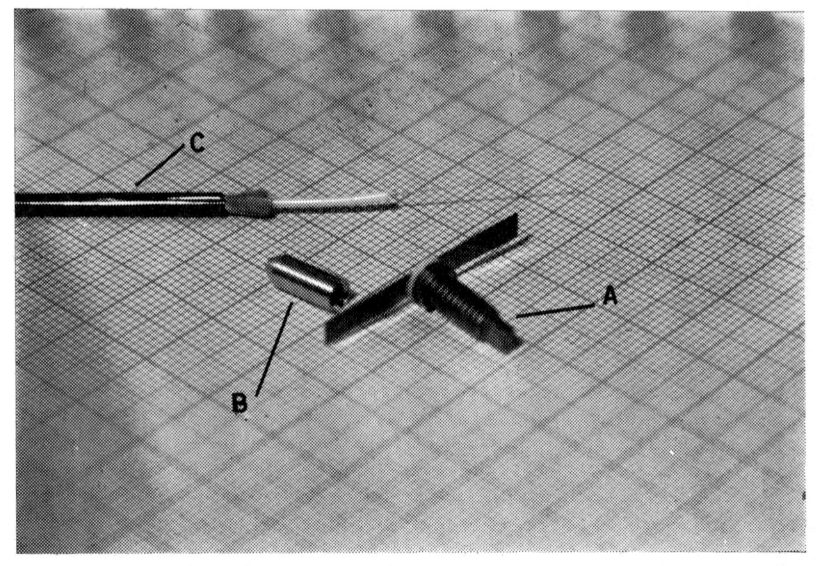

FIG. 32 – LED-pigtail integral assembly: *a*) LED package; *b*) fibre positioning holder; *c*) tight single fibre cable.

1.4 Coherent sources

1.4.1 *Introductory remarks on laser devices*

Stimulated emission in semiconductors and the basic properties of the Double Heterojunction (DH) multilayer structures have already been described in a previous section. In the following, it will be shown how these basic properties can be tailored to obtain laser devices suitable for specific applications.

For instance, the DH layer thickness, composition and doping can be adjusted either to obtain the best peak optical power output in pulsed operation (as in the case of the Large Optical Cavity, LOC device [34]), or to reduce the threshold current density down to the minimum value required for CW laser devices [35]. Although a simple laser structure (the so-called Broad-Area configuration) can be made directly by cleaving and cutting the DH wafer in the form of a Fabry-Perot resonant cavity (see Fig. 18), a practical device requires also a lateral confinement, which is obtained by restricting the current injection through a narrow stripe perpendicular to the Fabry-Perot cavity mirrors. Lasers with this current restriction are called stripe geometry lasers [36]. They require less current because the threshold current is proportional to the active area, and therefore thermal dissipation problems can be greatly reduced. Moreover, the stripe width is small enough to avoid multifilament instability usually found in broad area devices. Several stripe geometry structures have been implemented and a few examples will be described in this section. Stripe geometry lasers, with a 3-layer DH structure, made with $Ga_{1-x}Al_xAs$ ternary material can now be regarded as well-known conventional devices. They are able to launch more than 1 mwatt of optical power into a multimode fibre, in the wavelength range from 0.8 to 0.9 μm. The laser emission spectral width is at least one order of magnitude less than in the case of a LED source, so that the fibre material dispersion for these wavelengths is not a limiting factor for the system bandwidths under consideration at present. Even if these relatively simple laser devices can already operate as efficient transmitters in the optical fibre data communication systems currently under development, a strong research effort is still in progress in order to improve semiconductor laser performance. Advanced laser structures have been developed by using a five-layer heterostructure with a separate confinement for the current carriers and the optical field [37]. In other cases, the Fabry-Perot feedback mechanism is replaced by Bragg scattering from a periodic spatial variation of the refractive index within the active medium of the laser. This distributed optical feedback mechanism can provide a single longitudinal mode spectral selectivity that is not easily available from the Fabry-Perot cavity [38]. Both these advanced structures will be described in some detail in this sections. Semiconductor laser emission at wavelengths in the 1-1.3 μm range, requires III-V ternary or quaternary DH structures which are different from the well-known $GaAs/Ga_{1-x}Al_xAs$ system. General criteria for the choice and preparation of these materials have already been discussed in Section 1.1. A few examples of DH laser devices operating at $\lambda > 1$ μm wavelength will be illustrated here. An interesting indication arises from the results reported to date. In spite of the increasing difficulties related to material growth technology, these laser sources seem to be more degradation-resistant [39] in comparison with devices based on DH GaAs/GaAlAs.

At the end of this section, the basic properties of Neodimium laser sources will be briefly described. These light sources are not of practical advantage for multimode systems with respect to semiconductor lasers. They are severely limited by the complexity of their operation. If monomode fibres and integrated optics are to play a significant role in long term system development, the very high spatial coherence of the Neodimium laser will become very attractive and probably essential in order to meet the single mode system requirements.

1.4.2 *Stripe-geometry structures*

A schematic of a stripe geometry semiconductor laser is shown in Fig. 33. In the x-direction carrier confinement in the active region $d_\perp$ is provided by the double heterostructure which forms also the dielectric step waveguide (Fig. 19). In the transverse y-direction, due to the fact that the injection current can flow only through the narrow stripe w, the optical field builds up in the central part of the active region, which is beneath the stripe. In general, because of the lateral current spreading under the injection stripe the optical field distribution in the y-direction is larger than the nominal stripe width w. Due to this effect, the threshold current for a stripe geometry device is somewhat higher than that calculated by multiplying the threshold current density of a broad area device by the nominal injection contact area. Since the lateral carrier spreading increases as the stripe width

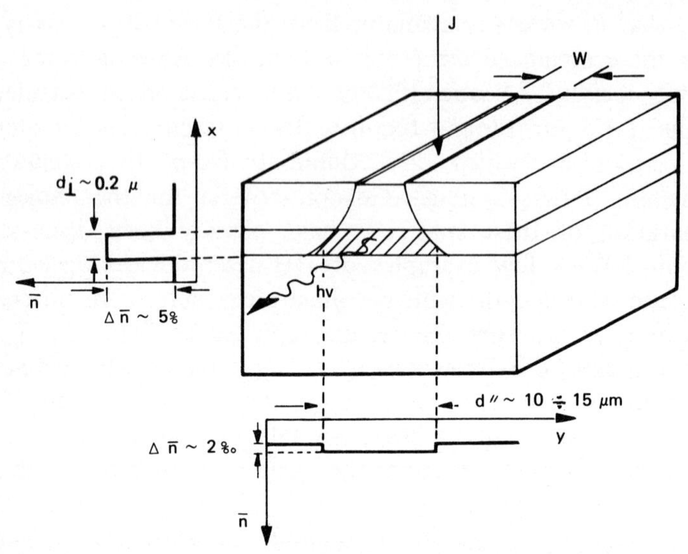

FIG. 33 – Schematic of a stripe-geometry semiconductor laser. Guiding in x and y directions is provided by a $\Delta\bar{n}$ dielectric step whose order of magnitude is indicated only for illustration purposes.

is reduced, a minimum useful value of w is in the range from 10 to 15 μm. Attempts [40] to reduce the stripe width to values under 10 μm cause a rapid increase in threshold current density and a decrease in the differential quantum efficiency of the device. Therefore, at least for stripe geometry devices where the y confinement is provided by some kind of surface insulation, a large asymmetry exists between the optical field distributions in the x- and in the y-directions, respectively.

In particular, in the x direction, the radiating aperture width on the cavity mirror will be of the order of 0.2-0.3 μm while in the y-direction it will be larger by more than an order of magnitude. By diffraction, the far field radiation angular divergence would be of the order of 40-50° in the xz plane, and of 5-10° in the yz plane. These numbers can be roughly estimated from the Airy diffraction relation

$$\Delta\theta = 1.22\frac{\lambda_0}{d} \tag{30}$$

(which expresses the $\Delta\theta$ angular divergence (in radians) for the light of wavelength λ_0, diffracted through a slit of width d). The values of $\Delta\theta$, given by (30), are in good agreement with the experimentally measured values. However, it should be pointed out that a standard diffraction theory is not applicable to the case of light emanating from a dielectric active waveguide since the radiating aperture is itself a source of electromagnetic field.

Therefore, a rigorous analysis of the beam divergence from a stripe geometry DH laser, which governs the coupling efficiency of the laser with an optical fibre, requires the complete solution of the guided modes of the dielectric waveguide shown in Fig. 33.

If the transverse field distribution in the resonant cavity is known, the laser beam divergence can be calculated as the Fourier transform of the electric field on the radiating cavity mirror. The two-dimensional dielectric waveguide, which characterizes the stripe geometry DH laser, has an index of refraction profile on (x, y) which is considerably asymmetric in the x- and y-directions respectively. In particular, the abrupt aluminium concentration change across the active layer and the confining layers in the x-direction results in a 5% step variation for the refractive index. In the y-direction, even if the guiding mechanism has yet to be fully understood, the experimentally measured far-field can be accounted for by a small parabolic decrease of the refractive index from the centre to the stripe edges. The absolute amount of this change is at least an order of magnitude lower than the step in the x-direction. Therefore, as a first approximation, we can assume that guiding can be considered separately in the orthogonal directions. In the x-direction (Fig. 34) the waveguide is identical to a symmetric planar (slab) waveguide which has been dealt with in detail by Marcuse [41]. For a TE mode configuration ($E_z = 0$), which is experimentally observed in the laser emission, the field component E_y is the solution of the scalar wave equation

$$\frac{\partial^2 E_y}{\partial x^2} + \frac{\partial^2 E_y}{\partial z^2} + \bar{n}^2(x)\,k_0^2 E_y = 0 \tag{31}$$

where

$$\bar{n}(x) = \begin{cases} \bar{n}_1 & \text{for } |x| \leqslant \dfrac{d}{2} \\[2ex] \bar{n}_2 & \text{for } |x| > \dfrac{d}{2} \end{cases}$$

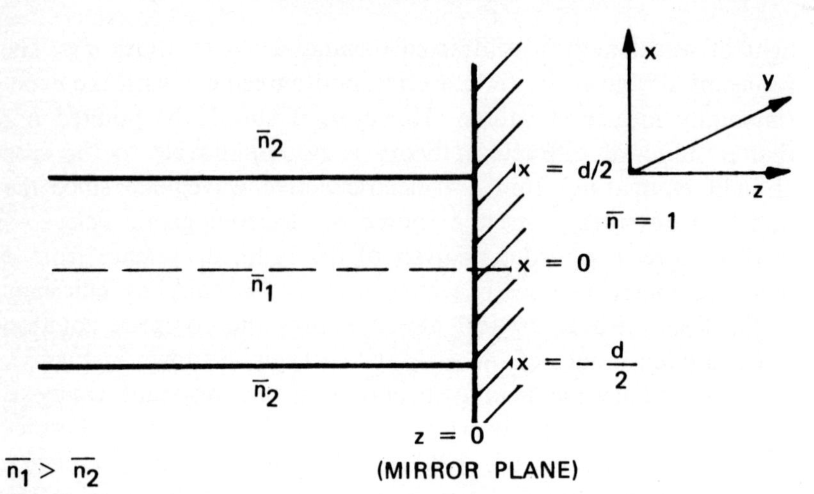

FIG. 34 – Slab dielectric waveguide, for calculating the near field pattern in the x direction.

and $k_0 = 2\pi/\lambda_0$ is the free space propagation constant. Solutions of Eq. (31) are periodic functions for $|x| < d/2$, with an exponential tail for $|x| > d/2$. For example, for even modes, they can be written as

$$\begin{aligned} E_y &= A_e \cos kx \exp[i\beta z] \\ H_z &= -(ik/\omega\mu_0) A_e \sin kx \exp[i\beta z] \qquad \left(|x| < \frac{d}{2}\right) \end{aligned} \tag{32}$$

and

$$E_y = A_e \cos\left(k\frac{d}{2}\right) \exp\left[-\gamma\left(|x| - \frac{d}{2}\right)\right] \exp[i\beta z]$$

$$H_z = \left(-\frac{x}{|x|}\right)\left(\frac{i\gamma}{\omega\mu_0}\right) A_e \cos\left(\frac{kd}{2}\right) \exp\left[-\gamma\left(|x| - \frac{d}{2}\right)\right] \exp[i\beta z] \qquad \left(|x| > \frac{d}{2}\right) \tag{33}$$

where the time dependence $\exp[-i\omega t]$ has been omitted and the coefficient A_e is given in [41]. The eigenvalue equation is obtained applying the boundary condition that H_z is continuous at the dielectric interface. For even modes:

$$\tan k\frac{d}{2} = \frac{\gamma}{k}. \tag{34}$$

Taking into account that for the field components given in Eqs. (32) and (33)

$$\bar{n}^2 = \bar{n}_1^2 k_0^2 - \beta^2 \quad \text{and} \quad \gamma^2 = \beta^2 - \bar{n}_2 k_0^2 \tag{35}$$

where β is the z-component of the wavevector, Eq. (34) can be solved numerically to give k. Hence γ and β can be evaluated from Eqs. (35).

In particular, for even modes, the mode order can be determined from Eq. (34) under the condition

$$2N\pi < \sqrt{\bar{n}_1^2 - \bar{n}_2^2}\, dk_0 \qquad N = 0, 1, 2, \ldots \tag{36}$$

for given $\bar{n}_1$, $\bar{n}_2$ and d. For a 5% step in the dielectric waveguide fundamental ($N = 0$) mode operation is maintained up to $d \leqslant 0.7$ μm If odd modes are considered, relation (36) is still valid by replacing N with $N/2$. Therefore, the first odd mode ($N = 1$) could be excited if $d > 0.35$ μm. By reducing d below the value required for fundamental mode operation, with a constant dielectric step, the electric field spreads over the active region boundaries, and a reduced beam divergence is obtained as a consequence of a wider near field. A higher gain is required to sustain a mode which propagates outside the active region, and a penalty is paid in this case by an increase in the threshold current.

We consider now confinement in the y-direction. The same Eq. (30) can be written also in this case, by replacing $\partial^2/\partial x^2$ with $\partial^2/\partial y^2$ and by using a parabolic refractive index profile $n^2(y) = n_s^2(1 - y^2/y_0^2)$. For small y, we can write $n(y) \simeq n_s(1 - \frac{1}{2}(y^2/y_0^2))$, and n_s (the index of refractions along the stripe axis) and y_0 (which is related to the stripe width) are both adjustable parameters. If we assume for $E_y(y, z)$ a z-dependence in the form

$$E_y(y, z) = E_y(y) \exp[-i\beta_y^z) \tag{37}$$

Eq. (31) becomes

$$\frac{\partial^2 E_y}{\partial y^2} = \left[\beta_y^2 - k_0^2 (n_s^2)\left(1 - \frac{y^2}{y_0^2}\right)\right] E_y \tag{38}$$

which is identical to the well-known Schrödinger equation for the harmonic oscillator whose solutions are the Hermite-Gauss [36] functions defined as

$$E_y^\nu(y) = H_\nu\left(\sqrt{2}\,\frac{y}{W_0}\right)\exp\left(-\frac{y^2}{W_0^2}\right) \tag{39}$$

$W_0^2 = \lambda_0 y_0/\pi n_s$ is known as the « beam radius » [42] which indicates the degree of confinement of the fundamental mode, and H_ν is the Hermite polynomial of order ν defined as

$$H_\nu(x) = (-1)^\nu \exp[x^2]\,\frac{d}{dx^\nu}\exp[-x^2]\,. \tag{40}$$

The function $|E_y^N|^2$ (relation (32), (33)) is plotted in Fig. 35 for $N = 0, 1, 2$. An experimentally measured near-field intensity and

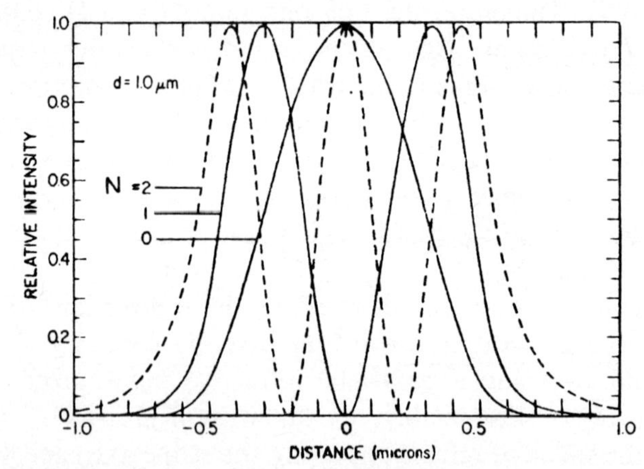

FIG. 35 – Calculated $[E_y^N]^2$ for $N=0$, 1, 2. Dielectric step of the waveguide edge $\Delta n=0.05$. (After [104], p. 469).

far-field patterns along the y-direction for different stripe widths are shown in Fig. 36. For the propagation constant β_y^ν of a mode of order ν we can then write

$$\beta_y^\nu = \sqrt{n_s^2 k_0^2 - (2\nu + 1)\,\frac{n_s k_0}{y_0}} \tag{41}$$

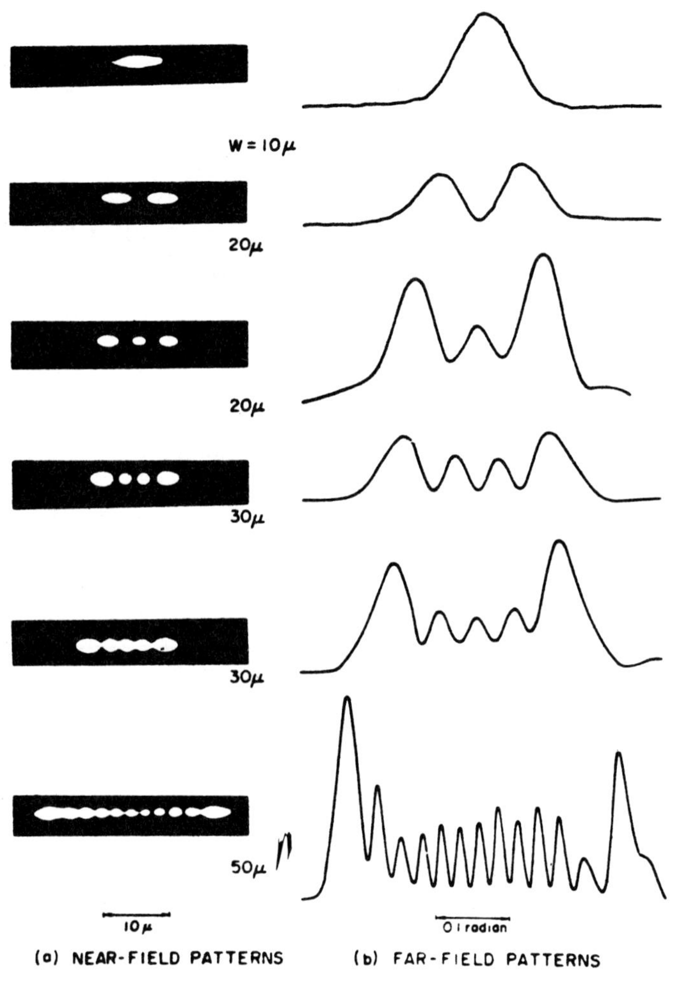

FIG. 36 – (*a*) near-field (*b*) far-field pattern. Different parallel modes for different stripe width (W) (After Yonezu *et al.*, Japan. J. Appl. Phys. Vol. 12 (1973), p. 1585).

By introducing the effective index $\bar{N}^\nu$ of the waveguide ($\bar{N}^\nu = \beta^\nu/k_0$), the condition for the mode order ν as a function of the waveguide parameters y_0 and n_s can be estimated as follows. For a guided mode, $\bar{N}^\nu$ is bounded by

$$n_1 < \bar{N}^\nu < n_s \tag{42}$$

where n_1 is the refractive index far from the stripe edges in the active region. From (41) (42) and by the definition of $\bar{N}^\nu$ we obtain the condition

$$\bar{N}^\nu = \sqrt{n_s^2 - \left(\nu + \frac{1}{2}\right)\frac{2n_s}{k_0 y_0}} > n_1 \, . \tag{43}$$

This can be written as

$$k_0 y_0 \Delta n > \nu + \tfrac{1}{2} \tag{44}$$

where $\Delta n = n_s - n_1$, and $n_s^2 - n_1^2$ has been approximated by $2n_s \cdot \Delta n$. A comparison can be made now with the experimental results shown in Fig. 36; Δn may be estimated to be of the order of 10^{-2}. (By assuming $w \simeq 2y_0$ for the stripe width). This value is very small with respect to the dielectric step in the x-direction. (In this case $n_1 - n_2 \simeq 0.2$). Therefore, the hypothesis on which this simple calculation has been carried out may be considered realistic. The physical nature of the mode guiding mechanism due to the small parabolic variation of the refractive index along the y-direction is still uncertain. For instance, observations [43] of the astigmatism in the emitted beam are not consistent with a real dielectric constant profile alone. A gain-guiding mechanism, related to the imaginary part of the dielectric constant, has been recently [44, 45] reported to be very important in determining the waveguide characteristics in narrow stripes ($\sim$ 10 μm width). On the contrary, in wider stripes a decrease in carrier density at the centre of the stripe with respect to the edges is caused by stimulated recombination. As the dielectric constant profile is inversely related to the carrier density, the dielectric constant decreases with the distance from the stripe axis. Experimental observations are in good agreement with the parabolic dielectric waveguide we have already described. If the lateral current spreading beneath the stripe causes a temperature distribution (the stripe axis is hotter than the edges), the index of refraction should vary accordingly [46]. This effect can also provide a self-focusing guiding mechanism. Even if discrimination between the above-mentioned effects still remains a difficult task, the overall understanding of the transverse modal behaviour in a stripe geometry DH laser can be considered good enough from the application point of view [47]. Quite a large number of device structures and technologies have been used to construct stripe geometry lasers: a few examples are shown in Fig. 37 in order of increasing complexity. In the most popular device structure (Fig. 37*b*)) [48], the stripe contact is defined by an oxide (SiO_2) dielectric insulator on which the stripe window is opened by a photolithographic technique. Proton bombardment (Fig. 37 *c*)) [49] or oxygen implantation [50] have been used to increase the resistivity of bulk material outside the stripe contact. In this case a more efficient current confinement is obtained since the insulating regions are shifted very close to the active layer. By avoiding a dielectric surface insulation, better thermal dissipation properties are also a distinct advantage in these structures. In Fig. 37*d*) a planar Zn diffusion [51], creating two reverse biased *p-n* junctions near the stripe edges perpendicular to main current flow, is used as a con-

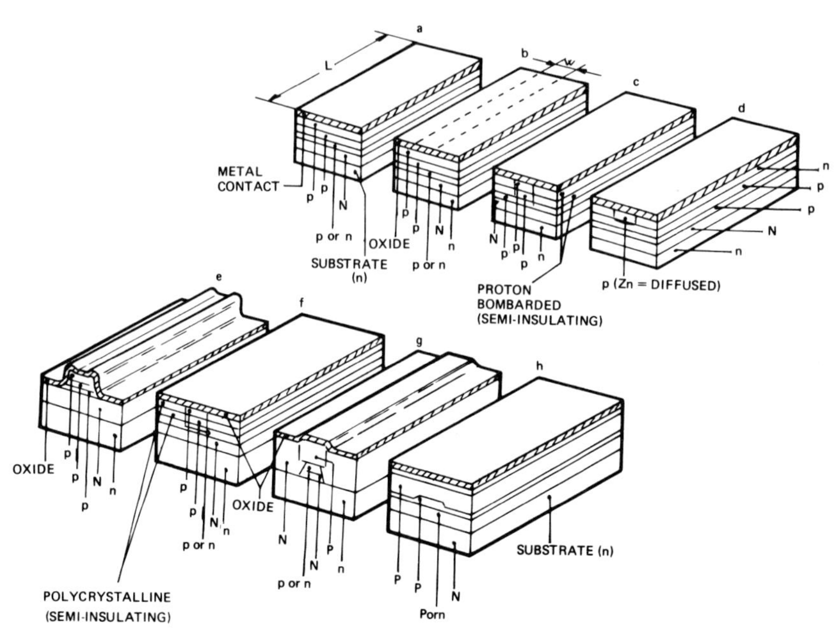

FIG. 37 – DH laser configurations. *(a)* Broad. *(b)* Stripe contact. *(c)* Proton bombardment (semi-insulating) stripe isolation. *(d)* Doping profile stripe. *(e)* Stripe mesa. *(f)* Embedded stripe. *(g)* Buried heterostructure. *(h)* Rib-guide stripe. The top p-GaAs layer shown in most cases, provides an easier contact to the metal than $Al^xGa_{1-x}As$. (After [104], p. 469).

finement mechanism. In all these structures, where only current confinement and spatially varying carrier distribution is responsible for the light guiding in the y-direction, the above considerations describe adequately the properties of transverse modes. Other stripe configurations shown in Fig. 37 provide an optical confinement in the lateral dimension which is based on a controlled and relatively large dielectric step at the stripe edges. For the stripe mesa [52], the embedded stripe [53] and the rib-guide stripe [54] we shall refer the readers to the original papers. We will give instead a more detailed description of the buried heterostructure (Fig. 37*g*)) (BH) stripe geometry with both transverse (y) optical and carrier confinement [55], which is the most successful among recent devices. Device technology is far more complicated than for the previously described geometries. It requires a two step liquid phase epitaxy process: first, the conventional DH structure is grown, then, the stripe is defined by a very narrow (~ 1 μm) mesa structure made by a chemical etching. Finally, the etched regions are replaced by a lightly doped n type $Ga_{1-x}Al_xAs$ layer. grown by LPE. Al concentration x in this layer can be selected in order to allow the fundamental mode operation in the lateral dimension. (For $w = 1$ μ, for instance, Eq. (36) predicts a value of $\Delta n \simeq 0.025$ for $N = 0$, which can be achieved by an Al concentration of 5% in the regrown layer). A fairly symmetrical active region is obtained which is completely surrounded by a higher band gap material that prevents the out diffusion of electrons. An extremely low threshold current of 10 mA has been achieved in these devices together with a stable, kink-free zero order mode operation in the light output characteristics up to current many times higher than threshold.

In conclusion, a large class of stripe geometry DH lasers has been developed, starting from the first SiO_2 insulated device. A fairly good understanding of light guiding properties in these structures has been reached. At present these devices can be efficiently coupled to multimode optical fibres and device design can be tailored to meet specific coupling conditions. Despite the increasing technological difficulties, devices like the BH laser or the rib-guide stripe laser, which are still at the development stage, hold the promise of large stability and reliability improvements over the currently available DH stripe geometry semiconductor lasers.

1.4.3 *Advanced laser devices*

Laser devices described so far consist basically of a four-layer DH, where the last layer (the « cap » layer) is grown for electrical contacting purposes only and has little or no influence at all on laser operation. The optical feedback sustaining laser oscillations is due to partial light reflection from the two opposite cleaved planes which form the mirrors of the Fabry-Perot resonant cavity. In this section we shall describe semiconductor laser structures based on a more complex multilayer DH and on a different optical feedback mechanism. In the first case, two epitaxial layers are added to the conventional DH structure, in order to separate the optical waveguide from the carrier recombination active region.

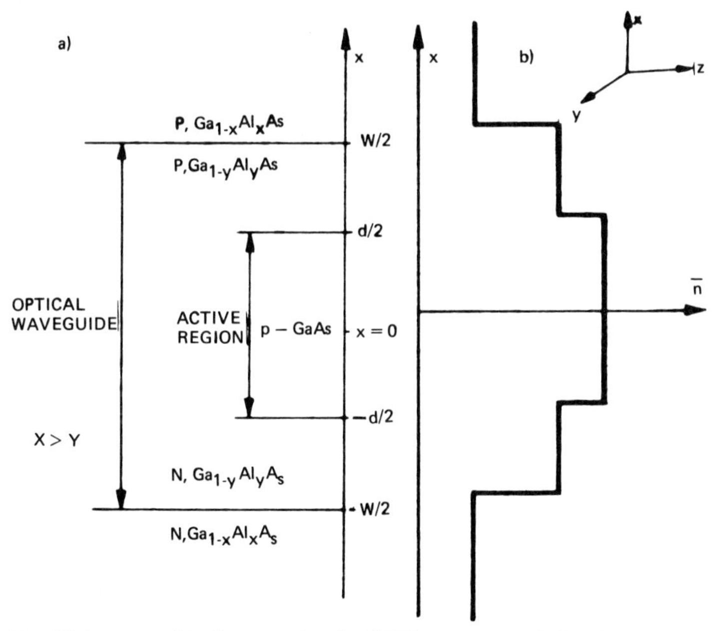

FIG. 38 – Slab waveguide diagram for the SCH structure: *a)* five layers structure; *b)* refractive index profile.

In the second case, Bragg diffraction from a periodic grating rather than a mirror reflection is used for the optical feedback. The Separate Confinement Heterojunction (SCH) layers structure is shown in Fig. 38 together with the refractive index profile which forms the two step optical wave-guide. Considerations which lead to the SCH structure [56, 57] are based on the fact that:

1) carriers are more easily confined than the optical field by changes in the band-gap at the heterojunction,

2) the gain in the heterostructure laser increases superlinearly with injection current density [57]. Therefore, a small Al concentration step is sufficient to confine carriers in a very thin active region. At the same time the optical fleld actually spreads over the three inner layers which form the optical waveguide, and is confined by the large refractive index step at the boundaries with the external layers. In particular, the Al concentration variation at the heterojunctions between the active region and the adjoining layers is of the order of 0.1 At%, and rises to about 0.2 At% at the heterojunction between these layers and the external layers. Therefore, the total Al concentration variations across the SCH and the simple DH structure are very similar.

Transverse field distribution $E_y(x)$ for the SCH structure can be determined by means of Eq. (31), by imposing the appropriate boundary conditions [58] on all dielectric interfaces. A comparison of the calculated intensity distribution $|E_y(x)|^2$ for several SCH lasers for varying optical cavity widths w and a fixed active region thickness d, with a DH laser having the same value of d is shown in Fig. 39. It can be shown by inspection that the SCH structure

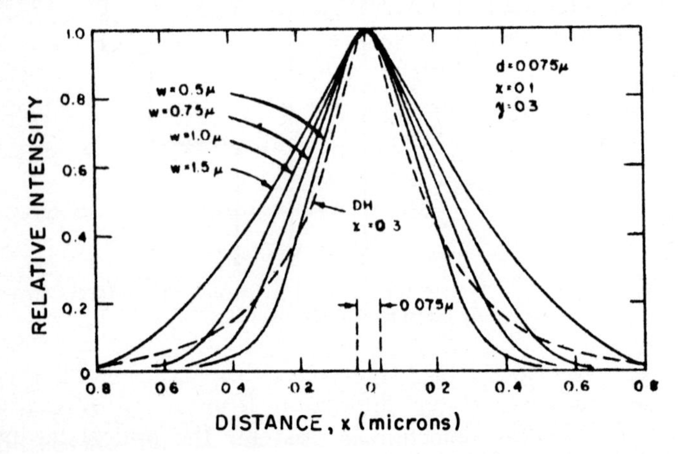

FIG. 39 – Calculated near field distribution for a SCH structure with fixed d and and varying w; a similar distribution for a DH structure with the same d is also shown for comparison. (After [58]).

provides a stronger optical confinement, at least for a relatively small w ($w \leqslant 0.5$ μm). In these conditions the confinement factor Γ, see (18), is larger for the SCH structure than for the DH. This is the essential reason why a very low threshold current density has been obtained in these devices. In particular J_{th} of 575 A/cm^2 has been reported for a SCH laser with $w = 0.4$ μm and $d = 0.04$ μm [59]. The successful growth by liquid phase epitaxy of such an extremely thin layer in a rather complex multilayer structure is a good demonstration of the high accuracy in the process control achieved by this technology [60].

Distinct advantages are obtained with a SCH structure for $w > 0.5$ μm if d is maintained below 0.1 μm. The Γ factor increases slightly faster for the DH than for the SCH structure with increasing d. The threshold current density is actually higher for a SCH laser with $d > 0.05$ μm but the relative increase in J_{th} is very small because of the gain current superlinearity. Therefore, a large optical cavity ($w \simeq 1$ μm) laser can be obtained with a relatively low threshold current density (~ 1.5 KA/cm^2). These devices should permit high peak optical power operation, with a small beam divergence, without catastrophic mirror damage. In particular, since the gain region is located at the maximum of the electric field distribution for the fundamental transverse mode, excitation of higher order modes has not been observed in these devices even at high pumping levels. Apart from the specific properties of the SCH laser, which have been briefly described, the separate confinement concept may be very important in the future development of monolithic integrated optical circuits. The distributed feedback injection laser, which we shall discuss in the following, is a good example for this application.

Single longitudinal mode selection is difficult to obtain in a semiconductor laser with a Fabry-Perot optical cavity. Owing to the broad gain spectrum, even at moderate current drive above threshold, several longitudinal modes are excited: a typical lasing spectrum with reference to this case is shown in Fig. 40. Spectral selectivity could be provided by operating the laser in an external cavity [61]. This approach does not seem particularly promising from the application point of view. Bragg diffraction from a periodic spatial variation of the refractive index within the gain medium of the laser, as suggested by Kogelnik and Shank [62],

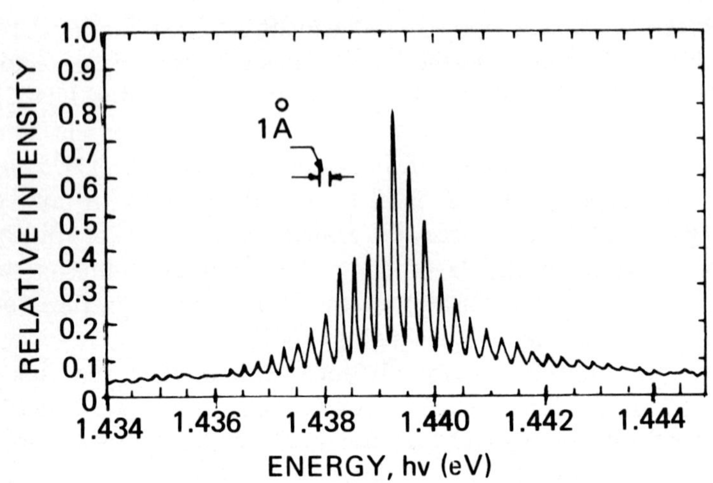

FIG. 40 — Spectral distribution of radiation from a semiconductor laser with a Fabry-Perot cavity. (After [104], p. 469).

may provide a feedback mechanism to replace the Fabry-Perot cavity mirrors. This occurs if the variation period Λ is fixed according to the Bragg condition for a back (180°) reflection as

$$\Lambda = \frac{m\lambda_0}{2\bar{n}} \qquad m = 1, 2, \ldots \tag{45}$$

where m is the grating order, λ_0 the free space wavelength and $\bar{n}$ the mean refractive index of the medium. In this case, possible resonance frequencies have approximately the same spacing in wavelength as the Fabry-Perot cavity, but the threshold gain increases sharply with deviations from the wavelength selected from the Bragg condition (45). A single longitudinal mode operation can be obtained by means of this feedback mechanism with a spectral bandwidth $\Delta\lambda \leqslant 0.1$ nm.

For a semiconductor laser, a periodic variation of the refractive index may be caused by a corrugation embossed on a dielectric interface near the active region of the device. Such corrugation may be obtained by ion milling or chemical etching through a suitable mask made by photolithographic techniques. A sophisticated technology is required in this process [63] as the period Λ of the corrugation (0.11 μm for a first-order grating) and the corrugation height (< 20 nm) must be highly uniform over the whole laser length (~ 0.1 cm). Schematic representations of several Distributed Feedback Bragg (DFB) laser configurations are shown in Fig. 41. In all cases, the radiation field, generated in the active

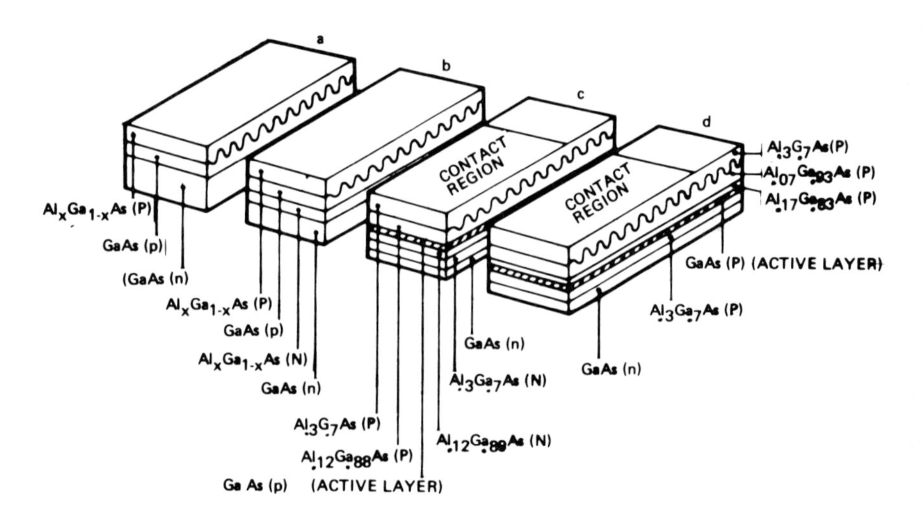

FIG. 41 – Schematic representation of several DFB lasers. By using a stripe-geometry configuration (not shown) structures *c) d)* are suitable for c.w. operation. (After [104], p. 469).

region and propagating in the heterojunction plane, interacts with the corrugation pattern so that the light intensity is continuously diffracted back and forth. A crucial point in the operation of the DFB semiconductor laser is noteworthy. In order to have a sizeable diffraction efficiency, the corrugation grating must be very close to the radiative recombination region. Since the corrugation fabrication methods always introduce a large density of non-radiative recombination centres, a low quantum efficiency and a high threshold may result if the electrically active region of the laser is directly in contact with the corrugations. For this reason, only low temperature (77 °K operation has been achieved with structures of the type shown in Fig. 41 *a*, *b*). In fact, at 77 °K, the minority carrier diffusion length is very small [64]: therefore, carrier recombination is maintained far from the corrugated interface. This difficulty has been overcome in the structures in Fig. 41 *c, d*) where the separate confinement principle has been applied. In this case the active region is physically insulated from the feedback grating which is placed on the boundary of the dielectric optical cavity. Both DFB-SCH structures have been made by a two-step epitaxial growth process. For the Fig. 41 *c*) device the first four layers are grown by liquid phase epitaxy. The grating pattern is then etched by ion milling while the last two layers are grown by Molecular Beam Epitaxy (MBE) [65]. (The « cap » GaAs layer is not shown in Fig. 41 for the sake of clarity). The structure in Fig. 41 *d*) is almost identical, but a further layer with low aluminium concentration is added during the first LPE growth step. Starting from this layer after the grating application, the structure was completed again by LPE [66]. In spite of the fact that LPE growth has been reported to be very difficult on a $Ga_xAl_{1-x}As$ layer, after exposure to oxidizing ambient gases, several authors found that it was still possible by keeping Al concentration $\leqslant 0.07$. Room temperature CW operation was achieved for the last structure when a stripe geometry contact and proper heat sink were added.

Perhaps one of the most important consequences of the elimination of the cleaved end mirrors allowed by the distributed feedback mechanism might be the construction of active laser sources directly integrated in a dielectric passive waveguide. An example of a truly integrated optical circuit based on these concepts has already been presented [67]. It consists of a six-channel optical multiplexer where six DFB-SCH lasers are grown on the same substrate chip. Each laser is tuned to a single optical wavelength by a small change in the corrugation period, and is then taper-coupled to a dielectric waveguide. The six individual waveguides, which are separated by chemical etching, combine near the chip edge and a single multiplexed optical output is then available for further processing. More details on these integrated optic devices will be given in Chapter 3 of Part V.

1.4.4 *Laser devices for operation at* $\lambda > 1$ μm.

Several potential advantages of optical fibre long distance communication systems operating in the wavelength range between 1.0 and 1.3 μm have been illustrated in this chapter.

A rather large choice currently exists, at least in principle, among several ternary and quaternary III-V alloys, for obtaining a lattice matched double heterostructure suitable for laser operation at the required wavelengths [68].

Problems related to the growth of these materials by Liquid phase or Vapor phase epitaxy have already been reviewed in Section 1.1.3; here we shall briefly describe a few laser structures by which, to date, CW operation has been achieved at these wavelengths.

It should be pointed out that the fundamental principles of laser operation, which have been described in this section mainly with reference to the GaAs/GaAlAs double heterostructure, are largely independent of the material system taken into consideration, provided that the band gap step at the heterojunctions is adequate for carrier and optical confinement.

CW laser operation up to 1.12 μm wavelength was achieved by a structure where the active region consisted of the $In_xGa_{1-x}As$ alloy, with lattice matched confining layers of $In_yGa_{1-y}P$ [9]. The structure was grown by Vapor Phase Epitaxy, on a GaAs substrate, and a compositional grading layer of $In_xGa_{1-x}As$, with x increasing in several small steps, was first grown on the substrate in order to minimize the lattice mismatch effects in the device active layer.

Very low threshold current density (1000 A/cm^2) and high differential quantum efficiency (55%) have been reported for these laser devices.

CW laser operation at the same wavelength was also achieved with an active region of $GaAs_{1-x}Sb_x$ and confining layers of the quaternary alloy $Ga_yAl_{1-y}As_{1-x}Sb_x$ [69]; in this case the structure was grown by liquid phase epitaxy starting from a GaAs substrate and a step compositional grading was achieved by growing three layers of $GaAs_{1-x}Sb_x$ with increasing x up to the value required to lattice-match the device active layers. The successful epitaxial growth of this very complicated seven layer structure is a clear demonstration of the feasibility of the general principles already discussed in Section 1.1 even if it still remains for the time being at the level of a difficult laboratory experiment.

The most promising laser structure presented to date for operation at $x = 1.3$ μm is shown in Fig. 42; the active layer is made

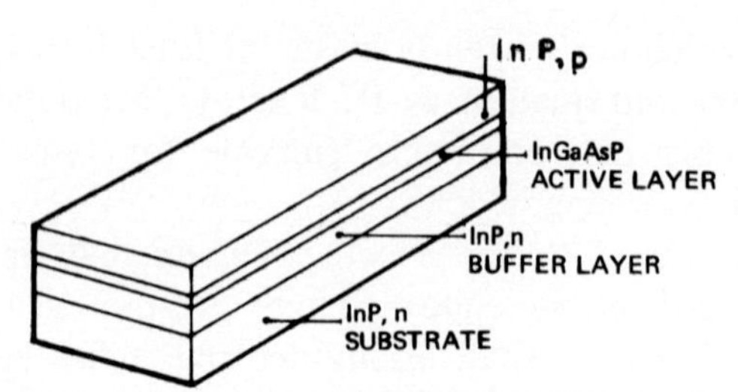

FIG. 42 – Schematic structure of a InP/InGaAsP/InP double heterostructure, tunable from 0.92 to 1.7 μm.

of the quaternary alloy $In_{1-x}Ga_xAs_yP_{1-y}$, which is lattice matched to the simple binary compound InP, when $y = 2.2x$; by an appro- appropriate choice of x, the emission wavelength could be tuned from 0.92 μm to 1.5 μm. The main advantage of this system over the others already described is that it requires a close compositional control of only thc single quaternary active layer, without any need for a compositional grade with respect to the substrate. Both LPE [11, 70] and VPE [71] growth technologies have been applied to this system to get double heterostructure laser devices with performances quite comparable with those of devices based on the GaAs/GaAlAs system.

By using this laser source, a dramatic increase in the repeater spacing, up to more than 50 km, in a 32 Mbit/s experimental optical fibre link, has been already achieved [72].

1.4.5 *Laser emission from the Neodimium compounds*

A conventional Nd: YAG laser source is basically made of a Nd-doped Yttrium Aluminum Garnet rod, which is optically pumped by an external light source. Several watts of CW optical power can be obtained from a Nd: YAG laser, with very high spatial and temporal coherence, at the wavelengths $\lambda = 1.06$ μm and 1.32 μm, which are well matched with the optical fibre low attenuation and dispersion spectral window. A source of this kind does not represent, however, a suitable solution for a fibre optic system for reasons of size and efficiency. Furthermore it suffers from two serious drawbacks: the limited reliability of the external optical pumping system (the intrinsic reliability of the active Nd-doped material, on the contrary, is very good), and the need for the laser beam to be externally modulated due to the long lifetime of the excited state responsible for the laser transition.

Recently, however, new stoichiometric Nd laser materials have been developed [73-75] (the NdP_5O_{14}, Neodimium-Pentaphosphate and the $LiNdP_4O_{12}$, Lithium-Neodimium-Tetraphosphate for instance, and many others), in which the Nd atoms mass fraction has increased by about a factor of 30 with respect to the conventional Nd doped YAG material.

In this case, the large Nd mass fraction results in high optical gain, and both the active material volume and the external optical pump power to reach the threshold for stimulated emission may be drastically reduced. Therefore, the compatibility of the Nd laser source with a fibre optic system in terms of size and efficiency would be ensured. Its advantages over the direct use of semiconductor sources include the width of the spectral line ($\leqslant$0.1 Å), compared with about 20 Å for semiconductor lasers and 500 Å for LEDs, and the possibility of obtaining a resonant laser cavity able to inject a single fundamental spatial mode in a monomode optical fibre. In particular, the required external optical pump power could be supplied in this case by one or more LEDs or semiconductor lasers operating at 0.84 μm wavelength [76, 77] and therefore semiconductor devices made with the GaAlAs material might be used.

A typical configuration of a miniaturized Nd laser source is schematically shown in Fig. 43.

Besides the pumping scheme shown in Fig. 43, where the excitation source is placed behind the Nd rod, it is possible and pro-

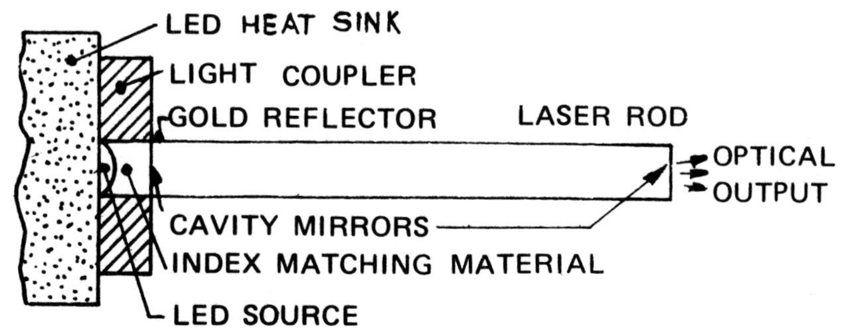

FIG. 43 – Schematic arrangement for a Nd laser source longitudinally pumped with a single LED.

bably advantageous to have a transversal configuration in which several LED or laser diodes can be used simultaneously.

In this case, the available optical power injected into the fibre, could be increased, at least in principle, up to the level required for a specified optical link length. The transversal pumping configuration might be also suitable for working as a true optical amplifier, thus avoiding the need for the complex electro-optical circuitry required in a conventional repeater.

A few experimental results are available to date [78], which allow the rough estimate that more than 30 mWatt of optical power could be obtained from a Nd laser source in an optimized transversal pumping configuration. Several difficulties concerning the material growth and device technology must still be overcome [79] before these potentially outstanding performances could find practical application; a feasibility test has, however, already been carried out in a Japanese laboratory [80]; a transmission rate of 800 Mbit/s over 800 m long monomode optical fibre was achieved using a $LiNdP_4O_{12}$ laser source, longitudinally pumped with a semiconductor laser, and intensity-modulated with an $LiNbO_3$ electro-optical modulator (see Part V, Chapter 3).

In conclusion, it seems clear that the Nd laser sources, described in this section, bearing in mind the complications arising from the use of an external electrooptical modulator, cannot compete with semiconductor sources in the optical fibre system currently under development. Nevertheless, the future demand for very high capacity long distance systems, where integrated optics and monomode fibres are likely to find widespread application, could completely change this point of view, and the less simple Nd laser transmitter operation will be largely compensated by the full exploitation of its rather unique properties.

1.5 Transient properties and modulation techniques

1.5.1 *LED emitters*

One of the most significant advantages in the use of semiconductor diodes, with coherent and incoherent emission, in optical fibre transmission systems, concerns the simplicity by which these sources can produce a modulated optical signal through the direct modulation of the injection current. In principle, the possible modulation bandwidth is connected to the lifetime τ_s of minority carriers (2) by the relation

$$\frac{P(\omega)}{P_0} = \left[\frac{1}{1 + (\omega\tau_s)^2}\right]^{\frac{1}{2}} \tag{46}$$

where $P(\omega)$ is the modulated optical power and P_0 the available power in the steady excitation.

In an LED doped near the optimum level to get maximum quantum efficiency, τ_s is in the order of $\sim 2 \cdot 10^{-9}$ s. Therefore the available bandwidth (3 dB) can be much higher than 100 MHz [23]; in a laser, due to the stimulated emission, τ_s reduces by at least one order of magnitude and, consequently, modulation rates at some GHz are possible.

Practically, numerous difficulties must be overcome to develop devices operating near the intrinsic limits.

In the case of high radiance LED's, discussed earlier, the greatest limitation is connected to the junction capacity due to the spatial charge region, which can assume considerable values when the diode is biased in direct conduction [81].

As a matter of fact, the charge time of the overall diode capacity, in the case of a surface emission structure can be expressed as a function of the recombination time τ_s of minority carriers by the relationship

$$\tau = \frac{C_d}{g_d}\left(1 + \frac{C_J}{C_d}\right) = \tau_s\left(1 + \frac{C_J}{C_d}\right) \tag{47}$$

where $C_d = eI_0\tau_s/nkT$ denotes the diffusion capacity and $g_d = eI_0/nkT$ the diode conductance, while C_J indicates the spatial charge capacitance, I_0 the polarization current in direct conduction and nKT is the same factor appearing in the relationship (6). On the other hand, C_J can be approximated by a relation of the type

$$C_J \cong \frac{Ac_0}{(1 - V_d/\varphi)^{\frac{1}{2}}} \tag{48}$$

where c_0 denotes the junction capacity per surface units at zero bias, A the actual junction area, φ the barrier potential and V_d the bias voltage.

To reduce this capacity the actual junction area should be diminished through planar techniques or proton bombardment; for instance, in the device illustrated in Fig. 23, only the injection current is delimited by the SiO_2 mask, but the *p-n* junction extends over the whole device area.

Consequently, A is nearly 100 times larger than the A_e area relevant to the emitting surface. An experimental confirmation of the above considerations is given by Fig. 44, which shows the optical pulse emitted by a LED having a large area and by a LED

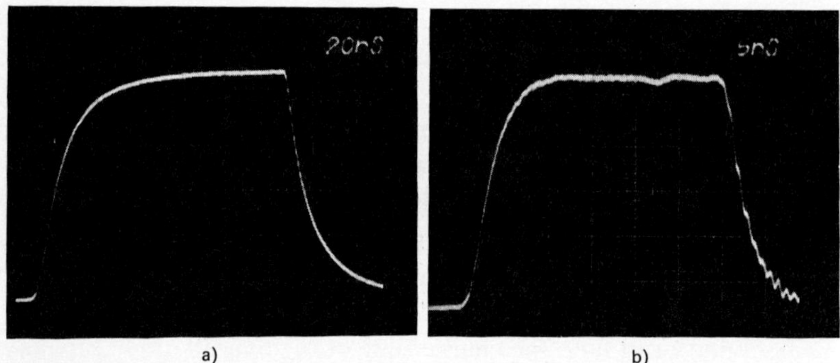

a) b)

FIG. 44 – Optical pulse from: *a)* Burrus type LED. *b)* Planar diffused LED. (After: C. P. Basola *et al.*: Alta Frequenza, Vol. 47 (1978), p. 208).

in which the junction area coincides with the emitting surface area through planar diffusion. The rise times measured in both cases amount to 25 and 5 ns respectively; in both cases the recombination time must be the same, since the *p-n* junction doping is identical. Therefore the considerable decrease in the rise time can be ascribed to C_J reduction.

However, in general, the utilization of the relatively complex technologies, necessary for the attainment of these results, is at present not justified for LED's, in that the dispersion of the transmitting medium associated to the LED spectral width, at least in the interval 0.8-0.9 μm, would not allow the advantages obtained enlarging the trasmitter modulation bandwidth to be exploited (compare Fig. 3). For LED's operating at $\lambda > 1$, on the other hand, a rise in the modulation bandwidth would directly involve an increase in the system transmission capacity. This justifies the research and experimentation carried out on complex structures, even when immediate application is not possible.

However, at present, modulation bands between 50 and, 100 MHz fall certainly within conventional LED device performance.

1.5.2 *Laser emitters*

Wide modulation rates can be attained by double heterojunction lasers; many authors [82, 83] have indicated, by using different configurations, 3dB cut-off frequencies higher than 1 GHz. Therefore, the laser appears henceforth suited to all applications also in advanced systems, which do not require at present transmission rates higher than some hundreds of Mbit/s. Figure 45*a*) gives the typical output characteristic of a double heterojunction laser with a strip geometry. The curve discontinuity denotes approximately the device threshold current I_{th}. Since the laser excitation due to a current step occurs after a delay constituted by the time ($2 \sim 3$ ns) necessary to establish the population inversion, usually the device is forward biased near the threshold I_{th} and the signal current overlaps such a bias.

By comparing Fig. 45*a*) to the typical output characteristic

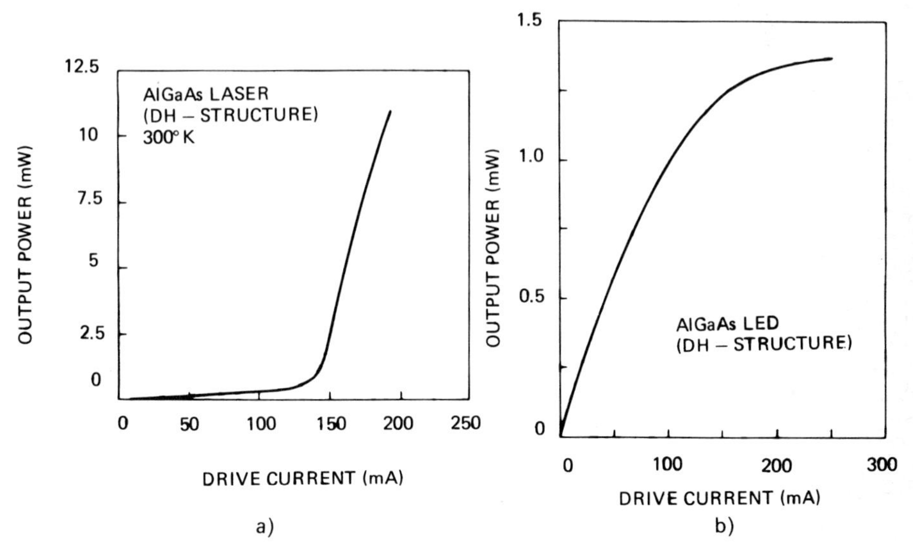

FIG. 45 – Light output characteristics for: *a)* DH laser diode; *b)* LED emitter. (After S. E. Miller *et al.*, Proc. IEEE, Vol. 61, No. 12, p. 1703).

of a LED (Fig. 45*b*)), it is immediately apparent that, to obtain the same modulation efficiency, the required signal electric power is inversely proportional to both curves' slope. This slope, which is directly connected to the device differential quantum efficiency, is more than an order of magnitude greater for the laser than for the LED. This involves a considerable simplification in planning the transmitter power stage, using the laser instead of the LED.

The laser maximum modulation frequency is limited not so much by the relationship (46), representing the characteristic cut-off of a low-pass filter, as by the interaction between the electromagnetic field inside the resonant cavity and the concentration of the electron injected into the device active region.

Equations determining the time dependence of the N_{ph} photons density and of the N_e electrons concentration inside the resonant cavity (rate equations) can be written in the form [84]

$$\begin{aligned} \frac{dN_{ph}}{dt} &= N_{ph}G - \frac{N_{ph}}{\tau_{ph}} + \alpha\frac{N_e}{\tau_s} \\ \frac{dN_e}{dt} &= \frac{i}{ed} - N_{ph}G - \frac{N_e}{\tau_s} \end{aligned} \qquad (49)$$

where τ_{ph} and τ_s denote the photons lifetime in the cavity and the lifetime for the electrons spontaneous recombination ($\tau_s \gg \tau_{ph}$) respectively, G is the active medium gain, which is assumed to be proportional to the N_e electronic density in the form $G = (1/\tau_{ph})(N_e/N_e^{th})$, where N_e^{th} is the electronic concentration at the stimulated emission threshold, i the current density feeding the laser, e the electron charge, and d the active region thickness. Lastly the term $\alpha(N_e/\tau_s)$ denotes the contribution of spontaneous emission into the stimulated emission modes. Therefore, as a first approximation $\alpha = m/M$, where m is equal to the number of oscillating modes and M is the total number of possible modes. According to experimental measurements [85], the value of α can be evaluated between 10^{-3} and 10^{-5}. Equations (49) are assumed to be valid for all cavity oscillating modes, even if each mode can possess a slightly different threshold and G gain. Also in this simplified form the non-linear system (49) cannot be simply solved in the analytic form except in the approximation for small signals. In this case Eqs. (49) can be linearized and therefore used to obtain-the time dependence of the small variations n_{ph} and n_e as regards the stationary values $\bar{N}_{ph}$ and $\bar{N}_e$ calculated by setting $dN_{ph}/dt = dN_e/dt = 0$ in (49). Thus the linear differential equation of the II order can be written for the n_{ph} variation in the form [86]

$$\ddot{n}_{ph} + \gamma\dot{n}_{ph} + \omega_0^2 n_{ph} = i_{\text{small signal}}(t)\left(\bar{N}_{ph}\frac{\bar{N}_e}{N_e^{th}\tau_{ph}} + \frac{\alpha}{\tau_s}\right)\frac{1}{ed} \qquad [50]$$

where

$$\gamma = \frac{J_0 + 1 + \omega_0^2\tau_s\tau_{ph}}{2\tau_s} + \frac{2\alpha}{\tau_{ph}(J_0 - 1 + 2\alpha + \omega_0^2\,\tau_s\tau_{ph})},$$

$$\omega_0^2 = \frac{1}{\tau_s\tau_{ph}}\sqrt{(J_0 - 1)^2 + 4\alpha J_0} \quad \text{and} \quad J_0 = \frac{\bar{i}}{i_{th}}$$

is the direct bias current through the diode normalized with respect to the threshold current. The results which can be obtained from this simplified analysis are in a good qualitative agreement both with the experimental data and with the most comprehensive theoretical results derived from computer solutions and can be summed up as follows.

1) In the case of a step excitation in which the current varies rapidly from zero to J_0, the radiation emitted by the device has a damped oscillating behaviour the period of which is given by the relationship

$$\omega_R = \sqrt{\omega_0^2 - \left(\frac{\gamma}{2}\right)^2} \qquad (51)$$

At threshold $J_0 = 1$ and therefore $\gamma_{th} = 1/\tau_s$, then assuming $\alpha = 0$, the order of magnitude of the damping time constant is identical to the spontaneous emission recombination time. For $\alpha > 0$, γ increases, therefore the oscillating transient time reduces considerably up to the point where, if $\gamma/2 \geqslant \omega_0$, the excitation response ceases to be oscillating. Figure 46 shows this situation. Physically it can be observed that α has a considerable influence on the transient behaviour of the laser device; as a matter of fact, the contribution of the term $\alpha(N_e|\tau_s)$ in the Eqs. (49) reduces the rate of increase of the electronic concentration N_e, to which the G gain of the active medium is proportional. Unfortunately, α is not a parameter which can be controlled on the ground of the device design and, consequently, the only possibility of control-

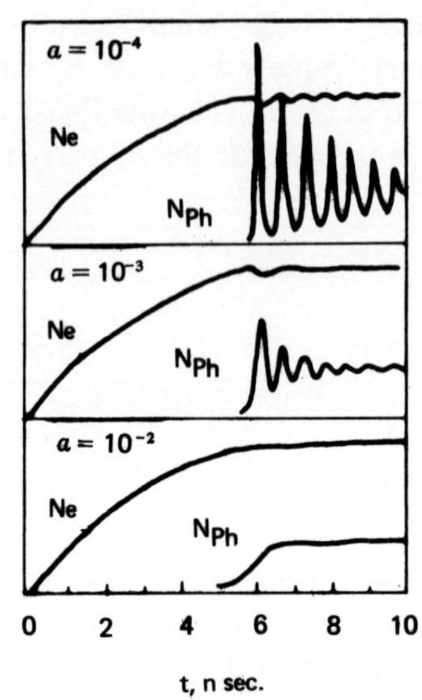

FIG. 46 – Damped oscillations in a semiconductor laser light output, for a step excitation current, and for different values of the α parameter.

ling the transient behaviour of the laser device consists in choosing accurately the bias current J_0 as regards the signal current, so as to limit as far as possible the transient variation in the G gain of the active medium. In the case of PCM pulse modulation with repetition frequencies near the laser resonance frequency (51), the pulse duration and its repetition frequency can be furthermore determined on the ground of Eqs. (49) conditions [87] in such a way as to reset at each pulse the same starting level for the electronic concentration N_e. Thus the resonance frequency in the order of some GHz can be exploited to extend the modulation bandwidth.

2) By assuming in Eq. (50) $i(t) = \exp[i\omega t]$ the modulation efficiency $F(\omega)$ can be evaluated in the form

$$F(\omega) = \frac{|n_{ph}(\omega)|}{|n_{ph}(o)|} = \frac{\omega_0^2}{\sqrt{(\omega_0^2 - \omega^2)^2 + \gamma^2\omega^2}}\,. \tag{52}$$

Figure 47 reports the course of $F(\omega)$ and confirms qualitatively the above considerations. In particular, it can be observed that the frequency response is relatively flat up to frequencies in the order

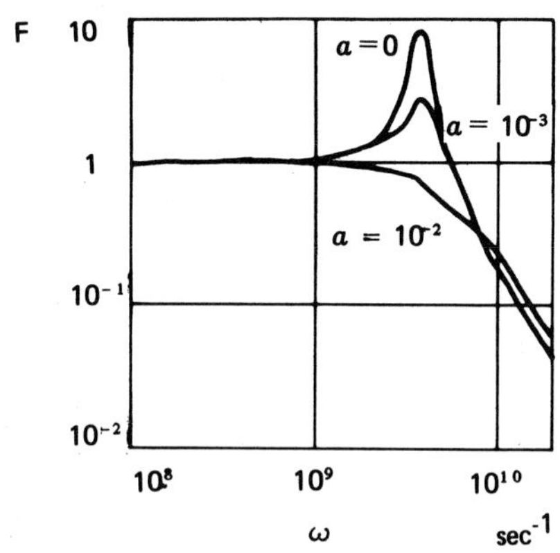

FIG. 47 – Modulation characteristic for a semiconductor laser in the small signal regime.

of 500 MHz. This brings us to the conclusion that the previously enumerated difficulties should not constitute an important obstacle to medium term applications. Moreover, numerous experimental [88] results relevant to high speed modulation experiments (2-3 Gbit/s) confirm that the theoretical forecasts, based on the rate equations analysis, are sufficiently accurate to enable the attainment of results largely in excess of expected future requirements.

On the other hand, a difficulty connected to the transmission medium dispersion depends on the laser spectral line broadening during the switching transient [89]; this broadening is due to the same physical cause responsible for the damped oscillations in the radiation emission, i.e. to the fact that the electronic density N_e increases during the transient to a value higher than the stationary value (Fig. 48). Thus, besides the longitudinal modes, which are excited under stationary conditions, corresponding to the $N_e(\lambda)$ distribution, also the longitudinal modes within the $\Delta\lambda$ interval are excited, in which, during the transient $N_e'(\lambda)$ is higher than the level required to reach the G_{th} gain in order to obtain stimulated emission. This larger spectral width can in fact severely limit the repetition section length of a communication system owing to the material dispersion. Practically $\Delta\lambda$ is proportional to the damped oscillation amplitude by which the total optical power is emitted during the transient and, consequently, the same

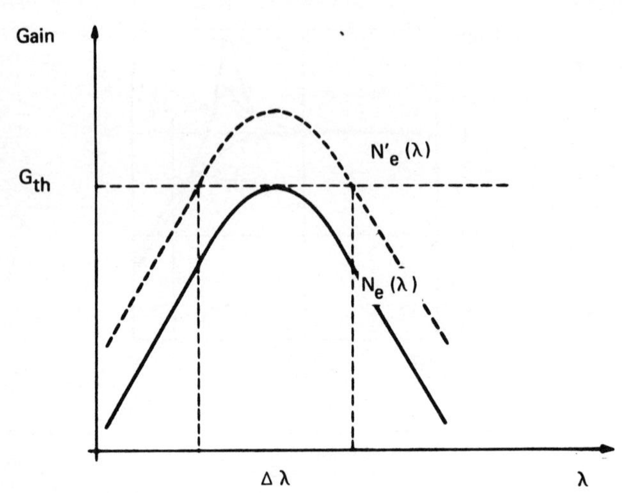

FIG. 48 – Gain increase for a step variation in the electronic population, during a pulse excitation of a semiconductor laser.

methods used for reducing the oscillations give rise to a reduction of $\Delta\lambda$. In particular [90], if the laser bias current is maintained slightly above the threshold for CW operation, so that a single longitudinal mode is constantly excited, the excess number of electrons injected into the device during modulation pulses, owing to the stimulated emission, does not contribute to the excitation of the other modes, in addition to the already excited one. This operational condition for the bias current does not contrast with previous indications even if the device operates, by emitting a non-negligible optical power level, also when modulation is missing.

From the foregoing considerations, it becomes clear that the operating conditions of a laser device must be kept stable through an appropriate optical feedback circuit [91]. This avoids the drift of the device operating point when temperature variations occur and the threshold current increase owing to gradual degradation. In conclusion, the analysis of the operation of a semiconductor laser in the transient condition, carried out through the rate equations under small signals condition, can be considered wholly adequate as regards the use of such devices in optical fibre communication systems, the implementation of which can be at present expected. In any case, it must be recalled that a close investigation about the transient behaviour of a semiconductor laser is still very far from being exhausted at the level of basic research, as far as the analysis under large signals regime is concerned, in which the non-linearity intrinsic to the interaction between radiation and matter plays a determinant role.

For instance, experimental results have been reported for devices capable of self-oscillation when emitting light-pulses shorter than 150 ps, at repetition frequencies in the order of several GHz; the possibility of controlling such effects, for instance, by coupled structures [92, 93] indicates clearly that current and future requirements in the field of data transmission via optical media can be satisfied to a much greater extent by the semiconductor laser.

1.6 Reliability problems

1.6.1 *Analysis methods*

A severe limitation to the applicability of electroluminescent devices to optical communication systems may lie with the unsatisfactory reliability of these components. As a matter of fact, available data on the operational average life of photoemitter devices at 0.82 μm reveals that lifetimes longer than 10^4 hours for lasers and 10^5 hours for high radiance LED's cannot realistically be expected to date.

Therefore, the present MTBF of photoemitter devices of the laser type is still at least one order of magnitude lower than that required for telephone applications. However, the analysis of the different degradation mechanisms has not to date revealed any process that limits intrinsically the reliability of these components or that cannot be reduced by an overall improvement of material growth technology and device processing. This prospect appears to support the steady progress observed experimentally during the past ten years, in the average life of photoemitter devices. In fact, only in 1971 [20, 94] the average life of high radiance LED devices could be evaluated in the order of 10^3 hours, whilst the average life of lasers varied from 10 to 10^2 hours.

From the experimental standpoint, the reliability of a photoemitter component can be statistically determined by means of accelerated aging tests, in which the stress factor, responsible for accelerated degradation, is the temperature at which the device operates during the test. If the mechanism, to which the degradation is connected, possesses a determinate activation energy, the so-called Arrhenius diagram allows the extrapolation of the average life at the device operating temperature from aging tests carried out at high temperatures.

Figure 49 gives an example of the results to be achieved by

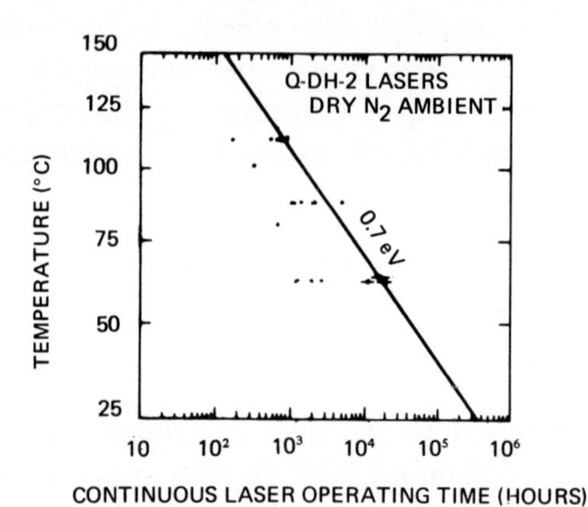

FIG. 49 – Arrhenius plot for a high temperature aging test for semiconductor lasers. Note the room-temperature extropolated intercept. (After [95]).

this procedure in the case of laser devices [95] and shows that at room temperature an average life in the order of 10^5 hours can be extrapolated. For LED devices similar evaluations [96] lead to operating lives of more than 10^6 hours. Unfortunately these results, particularly in the case of the laser, must be considered very cautiously, as statistically correct results can be extrapolated from the Arrhenius diagram only under the assumption that the temperature acts solely on the degradation mechanism, without affecting appreciably the device operation. This, for instance, is not true in the case of the laser, in which the threshold current increase rapidly with temperature (Fig. 50). Therefore both the bias current and the optical power emitted by the device cannot be kept constant at different temperatures. Furthermore it has not been demonstrated whether the temperature affects directly the degradation mechanism or whether the latter depends on the current intensity applied to the device or to the optical flow passing through it. Therefore, the correct application of the

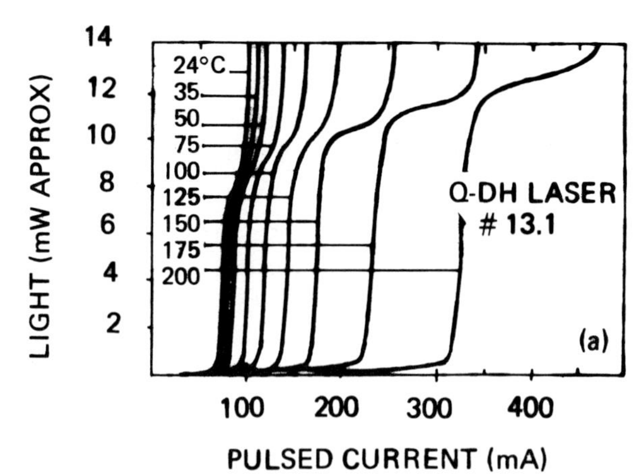

FIG. 50. – Temperature dependence of pulsed threshold current for a laser diode. (After [95]).

method connected to the Arrhenius diagram requires a further in-depth analysis of the physical effects responsible for the degradation. For instance, a very fast degradation mechanism has been found in the residual strain [97] due to the soldering of the device to the heat sink. Generally a good yield from devices free from fast degradation can be obtained through a careful selection of devices without mechanical faults due to the processing steps. The improvement in device reliability, which in the case of LED's can now be considered quite satisfactory, is mainly due to the elimination of these degradation effects. On the other hand, very slow degradation effects must still be singled out in the case of lasers, which take place within operation times between 10^4 and 10^5 hours.

The degradation mechanisms identified and partly eliminated concern both bulk and surface effects. While in the first case they can be considered as common to LED's and lasers, in the second case the phenomena concern specifically the laser, as they are mainly connected to the failure of the resonant cavity mirrors. Gradual degradation mechanisms for bulk effects can be generally ascribed to the presence and development of non-radiative recombination centres within the device active region. These centres can be formed either by dislocation networks or by impurity precipitates and it has not yet been ascertained definitely whether these centres are connected to small defects originally present inside the material or are produced directly during the device operation.

Obviously, in the first case the failure will be eliminated by the improvement of the material growth methods, while in the second case an upper limit to the device operative life would anyhow exist.

Convincing experimental evidence has verified the presence of the first gradual degradation mechanism. As a matter of fact, during the device operaton the growth of dark regions (the so-called dark-lines), (Fig. 51), produced by already existing dislocations

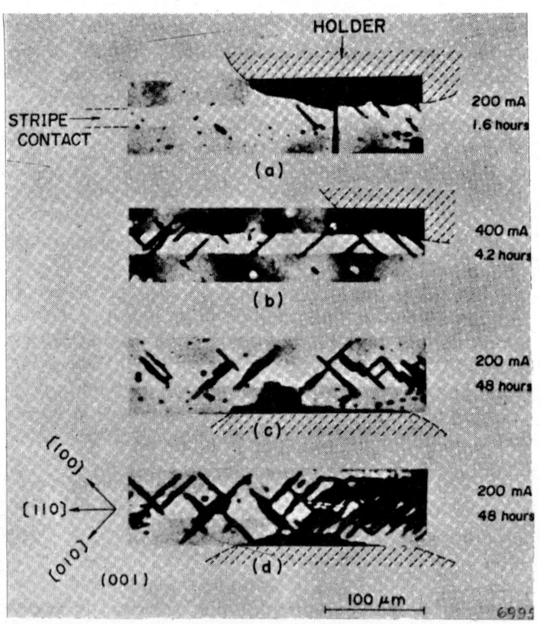

FIG. 51 – Dark lines development during operation of a stripe-geometry semiconductor laser. (After Y. Nannichi and I. Hayashi, J. of Crystal Growth, Vol. 27 (1974), p. 128).

within the active region of lasers and LED's, has been observed in some detail [98]. Therefore, an improvement in the laser average life has been achieved minimizing the initial concentration of such dislocations, which is at least partly connected to the different lattice constant of double heterostructure materials: i.e., in the $Ga_xAl_{1-x}As$ system a substantial improvement has been achieved [99] by compensating the small difference (see Fig. 3 Part V, Chapter 3) adding phosphorus to the confinement layers. Moreover, the average life of the devices tuned for a 0.82 μm emission, containing about 5% of Al within the active region, is systematically longer than the average life of those in which the active region is simply constituted by GaAs. Considering the second degradation mechanism, since the spontaneous generation of lattice defects can be, with reasonable certainty, considered dependent on activation energy and hence on temperature, the results obtained by means of the Arrhenius diagram previously described may be suitable for this case. In other words, such an analysis would allow the conclusion that the limit of 10^5 operation hours at room temperature is a correct estimate, when assuming that degradation is solely due to the thermal activation of the defects in question.

It is well-known that laser degradation, caused by the failure of the resonant cavity mirrors, is rapid and irreversible when the optical power transmitted through the mirror exceeds the density of about $5 \cdot 10^6$ Watt/cm^2. In a device with stripe geometry of 13 μm and active region of 0.3 μm the power to be extracted from a mirror would therefore not be higher than 50 mWatt. Moreover, it was estimated [100] that for the same group of devices the threshold for the mirror catastrophic failure is about ten times higher in the case of pulse operation (t_{imp} = 100 nsec) than for CW operation. This demonstrates that power levels normally utilized in CW operation (from 5 to 10 mWatt) are not very far from those giving rise to the irreversible mirror failure. The detailed mechanism of such a failure is not yet clear, but it is probably connected to a very localized surface heating, which causes the material to melt. Consequently, it is not unlikely that, owing to local fluctuations in the electromagnetic field distribution on the optical cavity mirror, power densities capable of giving rise to surface failure can be obtained locally even in average safe operating conditions or that with time the quality and, therefore, the failure resistance of the same surface will deteriorate because of external contamination. Thus the probability of this degradation occurring may clearly be reduced by passivating these surfaces through the deposition of appropriate dielectric layers. Furthermore, the study of a structure capable of distributing the optical power over the largest possible mirror surface [101], together with a more in-depth analysis of the mechanism itself (so as to permit the prediction of the maximum power level utilizable by a certain structure in safe conditions) may also lead to a reduction in degradation.

1.6.2 *Present limits and development forecasts*

From what has been stated in the preceding section, it cannot be concluded that the reliability problems of semiconductor laser devices can be considered as solved as regards data communication requirements.

The only encouraging indication to emerge from the intensive work in this field is that the operational life of these devices appears to be getting steadily longer. Furthermore, numerous laboratories have by now succeeded in implementing laser devices, generally with different technologies and structures, that have largely overcome the limit of one year continuous operation at room temperature with limited degradation (an average of 5% in the emitted optical power or in the threshold current according to the criterion adopted for the life test) during the test time. This allows, though in a purely qualitative manner as compared to what would be possible using accelerated aging tests, operational lives between three and six years to be extrapolated. In the case of LED devices, to which accelerated aging methods can be applied with greater safety, it is generally accepted that the relevant operational lives are at present between 10^5 and 10^6 hours and this limit appears clearly satisfactory also for the more demanding applications. Consequently, as regards laser devices, two problems must be tackled and solved to give greater consistency to the encouraging prospects already described. Firstly, criteria for carrying out extensively accelerated aging tests based on the use of the Arrhenius diagram must be established; secondly, parameters must be identified for selecting a priori through a direct measurement the most degradation-resistant devices. From this point of view the criterion introduced by A. R. Goodwin *et al.* [102, 103], through the ratio

$$R = \frac{J_{th}(70\ ^\circ\mathrm{C})}{J_{th}(15\ ^\circ\mathrm{C})}$$

between the threshold current at 70 °C and the threshold current at 15 °C, and the experimental proof that the devices for which $R \leqslant 1.65$ have statistically a greater degradation resistance illustrate clearly the possibility of an a priori selection of devices through simple and direct methods. All the preceding considerations concern devices constructed using GaAs/GaAlAs ternary material, which are the only ones for which a sufficient quantity of experimental data on reliability problems is available. In the case of ternary and quaternary materials for emission at wavelengths higher than 1 μm, it is only possible to have at present preliminary indications of a purely qualitative nature.

However, even in these conditions, it emerges surprisingly that the first devices, both LED and laser, to be used at $\lambda > 1$ μm, already described in Sections 1.3 and 1.4, have a degradation resistance greater than that of the devices constructed with the ternary alloy of the GaAlAs type [39]. This, in spite of the fact that, it could be reasonably expected that the greater difficulty in the preparation of materials made of GaInAs and GaInAsP alloys would involve considerable complications from the reliability standpoint. Further confirmation of this fact would certainly contribute to the construction, earlier than planned, of components suitable for the optical fibres transmission region between 1 and 1.3 μm and therefore would have a considerable effect on system choices for the near future.

A quite large variety of degradation mechanisms has been described in this section. A satisfactory understanding of the physical cause of such mechanisms still remains to be established. However, the careful experimental observation of the degradation effects during the heterostructure laser operation has allowed the definition of a few empirical rules on the heteroepitaxial wafers growth and on device fabrication and mounting processes by which the most detrimental degradation mechanisms (the catastrophic mirror failure and the fast development of the dark line defect) have been eliminated in practice. A more subtle slow degradation mechanism, which seems to limit the useful life of the best selected laser devices available today to 10^5 hours, still remain largely undefined. Even if this figure could be considered adequate for optical communication purposes, large scale use of such long-lasting laser devices will remain uncertain as long as a reasonable production yield will not be achieved outside the laboratory environment. Nevertheless, a point worthy of note in conclusion: after ten years of intensive experimental work on degradation a fundamental limit to semiconductor laser operating life has not been identified.

REFERENCES

[1] E. W. Williams and R. Hall: *Luminescence and the light emitting diode*, Science of the Solid State, Vol. 13, Pergamon Press (1978), p. 72.

[2] L. R. Dawson: *Liquid phase epitaxy*, Progress in Solid State Chemistry, Vol. 7, Pergamon Press (1972), p. 117.

[3] C. M. Wolfe, G. E. Stillman and E. B. Owen: *Residual impurities in high purity epitaxial GaAs*, J. Electrochem. Soc., No. 117 (1970), p. 129.

[4] A. Y. Cho: *Impurity profiles of* GaAs *epitaxial layers doped with* Sn, Si *and* Ge *grown with molecular beam epitaxy*, J. Appl. Phys., Vol. 46 (1975), p. 1733.

[5] M. B. Panish and M. Ilegems: *Phase equilibria in ternary III-V systems*, Progress in Solid State Chemistry, Vol. 7, Pergamon Press (1972), p. 39.

[6] G. A. Antypas: *Liquid phase epitaxy of* $In_xGa_{1-x}As$, J. Electrochem. Soc., No. 117 (1970), p. 1393.

[7] C. J. Nuese: *III-V Alloys for opto-electronic applications*, J. of Electron. Mat., Vol. 6 (1977), p. 253.

[8] R. E. Nahory, M. A. Pollack, D. W. Taylor, R. L. Fork and R. W. Dixon: *Room-temperature operation and threshold temperature dependence of LPE-grown* $In_xGa_{1-x}As$ *homojunction lasers*, J. Appl. Phys., Vol. 46, (1975), p. 5280.

[9] C. J. Nuese, G. H. Oesen, M. Ettenberg, J. J. Gannon and T. J. Zamerowski: *CW room-temperature* $In_xGa_{1-x}As/In_yGa_{1-y}P$ 1.06 μm *lasers* Appl. Phys. Lett., Vol. 29 (1976), p. 807.

[10] J. J. Hsieh, J. A. Rossi and J. P. Donnelly: *Room-temperature CW operation of* GaInAsP/InP *double-heterostructure diode lasers emitting at* 1.1 μm, Appl. Phys. Lett., Vol. 28 (1976), p. 709.

[11] M. A. Pollack, R. E. Nahory, J. C. De Winter and A. A. Ballman: *Room-temperature operation and threshold temperature dependence of LPE-grown* $In_xGa_{1-x}As$ *homojunction lasers*, Appl. Phys. Lett., Vol. 33 (1978), p. 314.

[12] J. Pankove: *Optical Process in Semiconductors*, Prentice Hall, N. Jersey (1971), p. 107.

[13] S. M. Sze: *Physics of Semiconductor Devices*, J. Wiley, N. York (1969), p. 140.

[14] J. Pankove: *Optical Process in Semiconductors*, Prentice Hall, N. Jersey (1973), p. 160.

[15] S. M. Sze: *Physics of Semiconductor Devices*, J. Wiley, N. York (1969), p. 96

[16] S. M. Sze: *Physics of Semiconductor Devices*, J. Wiley, N. York (1969), p. 688.

[17] M. G. A. Bernard and G. Duraffourg: *Laser Conditions in Semiconductors*, Physica Status Solidi, Vol. 1 (1961), p. 699.

[18] F. Stern: *Effects of band tails on stimulated emission of light in semiconductors*, Phys. Rev., No. 148 (1966), p. 186.

[19] J. R. Biard, W. N. Carr and B. S. Reed: Trans. AIME, Vol. 230 (1964), p. 286.

[20] C. A. Burrus and R. W. Dawson: *Small-area high-current density* GaAs *electroluminescent diodes and a method of operation for improved degradation characteristics*, Appl. Phys. Lett., Vol. 17 (1970), p. 97.

[21] R. C. Goodfellow: Proc. Electro-Optics Int. Conference, Brighton (1974), p. 168

[22] F. D. King, A. J. Springthorpe and O. I. Szentesi: *Technical digest*, Int. Elect. Devices Meeting IEEE, N. York (1975), p. 480.

[23] R. C. Goodfellow and A. W. Mabbit: *Wide-bandwidth high-radiance gallium-arsenide light-emitting diodes for fibre-optic communications*, Electron. Lett., Vol. 12 (1976), p. 50.

[24] J. C. Dyment, A. J. Springthorpe, F. D. King and J. Straus: *Proton bombarded double heterostructure LED's*, J. of Electron. Mat., Vol. 6 (1977), p. 173.

[25] C. P. Basola, V. Michi, G. Randone and R. Rocak: *LED ad alta radianza ottenuti con diffusione planare: tecnologia e prestazioni*, Alta Frequenza, Vol. 47 (1978), p. 208.

[26] T. P. Lee, C. A. Burrus and B. I. Miller: *A stripe-geometry double-heterostructure amplified-spontaneous-emission (superluminescent) diode*, IEEE J. Quantum. Electron., Vol. QE-9 (1973), p. 820.

[27] Y. Horikoski and T. Seki: *High radiance light emitting diode as optical fiber transmission line sources*, Rev. Electr. Commun. Lab., Vol. 24 (1976), p. 187.

[28] J. P. Wittke, M. Ettenberg and H. Kressel: *High radiance LED for single-fiber optical links*, RCA Rev., Vol. 37 (1976), p. 159.

[29] W. P. Dumke: *The angular beam divergence in double-heterojunction lasers with very thin active regions*, IEEE J. Quantum Electron., Vol. QE-11 (1975), p. 400.

[30] A. G. Dentai, T. P. Lee and C. A. Burrus: *Small-area, high-radiance C. W.* InGaAsP *L.E.D.'s emitting at* 1.2 *to* 1.3 μm, Electron. Lett., Vol. 13 (1977), p. 484.

[31] A. C. Carter, R. C. Goodfellow and I. Griffith: GaInAsP/InP *high speed, high radiance LED's for the wavelength region* 1.05-1.3 μm, Fourth European Conference on Optical Communication, Genoa (September 12-15, 1978), Supplement to Conference Proceedings, p. 85.

[32] M. Abe, I. Umebu, O. Hasegawa, S. Yamakoshi, T. Yamaoka, T. Kotani, H. Okada and H. Takanashi: *High-efficiency long-lived* GaAlAs *LED's for fiber-optical*, IEEE Trans. Electron. Devices (July, 1977), p. 990.

[33] C. P. Basola and G. Chiaretti: *Design criteria for optimum coupling between a high radiance LED and a bulb-ended fiber*, Fourth European Conference on Optical Communication, Genoa (September 12-15, 1978), p. 313.

[34] H. F. Lockwood, H. Kressel, H. S. Sommers and F. Z. Hawrylo: *Low-threshold* LOC GaAs *Injection lasers*, Appl. Phys. Lett., Vol. 18 (1971), p. 43

[35] M. Ettenberg: *Very low-threshold double-heterojunction* $Al_xGa_{1-x}As$ *injection lasers*, Appl. Phys. Lett., Vol. 27 (1975), p. 652.

[36] J. C. Dyment: *Hermite-gaussian mode patterns in* GaAs *junction lasers*, Appl. Phys. Lett., Vol. 10 (1967), p. 84.

[37] G. H. B. Thompson and P. A. Kirkby: (GaAl)As *lasers with a heterostructure for optical confinement and additional heterojunction for extreme carrier confinement*, IEEE J. Quantum Electron., Vol. QE-9 (1973), p. 311.

[38] M. Nakamura, K. Aiki, A. Katzir, A. Yariv and H. W. Yen: GaAs-GaAlAs *double-heterostructure injection lasers with distributed feedback*, IEEE J. Quantum Electron., Vol. QE-11 (1975), p. 436.

[39] C. C. Shen, J. J. Hsieh and T. A. Lind: *1500-h continuous CW operation of double-heterostructure* GaInAsP/InP *lasers*, Appl. Phys. Lett., Vol. 30 (1977), p. 353.

[40] B. W. Hakki: *Carrier and gain spatial profiles in* GaAs *stripe geometry lasers*, J. Appl. Phys, Vol. 44 (1973), p. 5021.

[41] D. Marcuse: *Theory of Dielectric Optical Waveguides*, Academic Press, N. York (1974).

[42] H. Kogelnik and T. Li: *Laser beams and resonators*, Appl. Opt., Vol. 5 (1966), p. 1550.

[43] T. H. Zachos and J. C. Dyment: *Resonant modes of* GaAs *junction lasers*-III: *propagation characteristics of laser beams with rectangular symmetry*, IEEE J. Quantum Electron., QE-6 (1970), p. 317.

[44] T. L. Paoli: *Waveguiding a stripe-geometry junction laser*, IEEE J. Quantum Electron., Vol. QE-13 (1977), p. 662.

[45] P. A. Kirkby, H. R. Goodwin, G. H. B. Thompson and P. R. Selway: *Observations of self-focusing in stripe geometry semiconductor lasers and the development of a comprehensive model, of their operation*, IEEE J. Quantum Electron., Vol. QE-13 (1977), p. 705.

[46] J. E. Ripper, F. D. Numes and N. B. Patel: *Filaments in semiconductor lasers*, Appl. Phys. Lett., Vol. 27 (1975), p. 328.

[47] W. T. Tsang: *The effects of lateràl current spreading, carrier outdiffusion and optical mode losses on the threshold current density of* GaAs-$Al_xGa_{1-x}As$ *stripe-geometry DH lasers*, J. Appl. Phys., Vol. 49 (1978), p. 1031.

[48] L. A. D'Asaro: *Advances in* GaAs *junction lasers with stripe-geometry*, J. of Luminescence, Vol. 7 (1973), p. 310.

[49] J. C. Dyment: *Proton-bombardment formation of stripe-geometry heterostructure lasers for* 300 K *CW operation*, Proc. IEEE, Vol. 60 (1972), p. 726.

[50] J. M. Blum, J. C. Mc Grooddy, P. G. Mc Mullin, K. K. Shih, A. W. Smith and J. F. Ziegler: *Oxygen-implanted double-heterojunction* GaAs/GaAlAs *Injection lasers*, IEEE J. Quantum Electron., Vol. QE-11 (1975), p. 413.

[51] M. Takusagawa: *An internally planar laser with* 3 μm *stripe width oscillating in tranverse single mode*, Proc. IEEE, Vol. 61 (1973), p. 1758.

[52] T. Tsukada, R. Ito, M. Nakashima and O. Nakada: *Mesa-stripe-geometry double-heterostructure injection lasers*, IEEE J. Quantum Electron., Vol. QE-9 (1973), p. 356.

[53] K. Ito, H. K. Asahi, M. Inoue and I. Teramoto: *Embedded-stripe* GaAs-GaAlAs *double-heterostructure lasers with polycrystalline* GaAsP *layers*-I: *Laser with cleaved mirrors*, IEEE J. Quantum Electron., Vol. QE-13 (1977), p. 623.

[54] T. P. Lee, C. A. Burrus and R. A. Logan: $Al_xGa_{1-x}As$ *double-heterostructure rib-waveguide injection laser*, IEEE J. Quantum Electron; Vol. QE-11 (1975), p. 432.

[55] T. Tsukada: GaAs-$Al_xGa_{1-x}As$ *buried-heterostructure injection lasers*, J. Appl. Phys., Vol. 45 (1974), p. 4899.

[56] I. Hayashi: U. S. Patent, No. 3691476 (1972).

[57] C. J. Hwang: *Properties of spontaneous and stimulated emission in* GaAs *junction lasers. Part II*, Phys. Rev., Vol. 13, No. 2 (1970), p. 4126.

[58] M. C. Casey, M. B. Panish, W. O. Schlosser and T. L. Paoli: GaAs-$Al_xGa_{1-x}As$ *heterostructure laser with separate optical and carrier confinement*, J. Appl. Phys., Vol. 45 (1974), p. 322.

[59] G. H. B. Thompson and P. A. Kirkby: *Low threshold-current density in 5-layer-heterostructure* (GaAl) As/GaAs *localised-gain-region injection lasers*, Electron. Lett., Vol. 9 (1973), p. 295.

[60] G. H. B. Thompson and P. A. Kirkby: *Liquid phase epitaxial growth of six-layer* GaAs/(GaAl)As *structures for injection lasers with* 0.04 μm *thick centre layer*, J. of Cryst. Growth, Vol. 27 (1974), p. 70.

[61] J. A. Rossi, S. R. Chinn and H. Heckscher: *High-power narrow-linewidth operation of* GaAs *diode lasers*, Appl. Phys. Lett., Vol. 23 (1973), p. 25.

[62] H. Kogelnik and C. Shank: *Coupled-wave theory of distributed feedback lasers*, J. Appl. Phys, Vol. 43 (1972), p. 2327.

[63] S. Somekh, H. C. Casey and M. Ilegems: *Preparation of high-aspect ratio periodic corrugations by plasma and ion etching*, Appl. Opt., Vol. 15 (1976), p. 1905.

[64] C. J. Hwang: *Doping dependence of hole lifetime in n-type* GaAs, J. Appl. Phys., Vol. 42 (1971), p. 4408.

[65] H. C. Casey, S. Somekh and M. Ilegems: GaAs-GaAlAs *distributed-feedback diode lasers with separate optical and carrier confinement*, Appl. Phys. Lett., Vol. 27 (1975), p. 145.

[66] M. Nakamura, K. Aiki, J. Umeda and A. Yariv: *CW operation of distribution-feedback* GaAs-GaAlAs *diode lasers at temperature up to* 300 K, Appl. Phys. Lett., Vol. 27 (1975), p. 403.

[67] K. Aiki, M. Nakamura and J. Umeda: *Frequency multiplexing light source with monolithically integrated distributed-feedback diode lasers*, Appl. Phys. Lett., Vol. 29 (1976), p. 506.

[68] H. Kressel and J. K. Butler: *Semiconductor Lasers and Heterojunction LED's*, Academic Press, N. J. (1977), Chapter 11.

[69] R. E. Nahory, M. A. Pollack, E. D. Beebe, J. C. De Winter and R. W. Dixon: *Continuous operation of* 1.0 μm *wavelength* $GaAs_{1-x}Sb_x/Al_yGa_{1-y}As_{1-x}Sb_x$ *double-heterostructure injection lasers at room temperature*, Appl. Phys. Lett., Vol. 28 (1976), p. 19.

[70] J. J. Hsieh and C. C. Shen: *Room-temperature CW operation of buried-stripe double-heterostructure* GaInAsP/InP *diode lasers*, Appl. Phys. Lett., Vol. 30 (1977), p. 429.

[71] G. M. Olsen, C. J. Nuese and M. Ettenberg: *Low-threshold* 1.25 μm *vapor-grown* InGaAsP CW *lasers*, Appl. Phys. Lett., Vol. 34 (1979), p. 262.

[72] K. Nakagawa, T. Ito, K. Aida, K. Takemoto and K. Suto: 32 Mb/s *optical fiber transmission experiment with* 53 km *long repeater spacing*, Fourth European Conference on Optical Communication, Genoa (September 12-15, 1978), Supplement to Conference Proceedings, p. 102.

[73] W. W. Krühler, J. P. Jeser and H. G. Danielmeyer: *Properties and laser oscillation of the (Nd, Y) pentaphosphate system*, Appl. Phys., Vol. 2 (1973), p. 329.

[74] T. C. Damen, H. P. Weber and B. C. Tofield: NdLa *pentaphosphate laser performance*, Appl. Phys. Lett., Vol. 23 (1973), p. 519.

[75] S. R. Chinn and H. Y-P. Hong: *Fluorescence and lasing properties of* $NdNa_5(WO_4)_4$, $K_3Nd(PO_4)_2$ *and* $Na_3Nd(PO_4)_2$, Opt. Commun., Vol. 18 (1976), p. 87.

[76] T. Oreo, M. Morioka, K. Ito, A. Tachibana and K. Kurata: Hitachi Rev., Vol. 25 (1976), p. 129.

[77] K. Kubodera and K. Otsuka: *Diode-pumped miniature solid-state laser: design considerations*, Appl. Opt., Vol. 16 (1977), p. 2747.

[78] S. R. Chinn, J. W. Pierce and M. Heckscher: *Low-threshold transversely excited* $NdP\text{-}O_{14}$ *laser*, Appl. Opt., Vol. 15 (1976), p. 1444.

[79] W. W. Krühler and R. D. Plättner: *Miniature neodymium lasers as optical transmitters: requirements relating to material selection, laser properties LED pumping*, Siemens Forsch-.u. Entwickl.-Ber., Bd 7 (1978), p. 291.

[80] T. Kimura, S. Uehara, M. Sarowatari and J. Yamada: 800 Mbit/s *optical fiber transmission experiments at* 1.05 μm, 1977 International Conference on Integrated Optics and Optical Fiber Communication, Tokyo (July 18-20, 1977), C 72.

[81] T. P. Lee: *Effect of junction capacitance on the rise Time of LED's and on the turn-on delay of injection lasers*, Bell System Tech. J., Vol. 45 (1975), p. 53.

[82] J. E. Carroll and J. G. Farrington: *Short-pulse modulation of gallium-arsenide lasers with trapatt diode*, Electron. Lett., Vol. 9 (1973), p. 166.

[83] M. Chown, A. R. Goodwin, D. E. Lovelace, G. H. B. Thompson, and P. R. Selway: *Direct modulation of double-*

heterostructure lasers at rates up to 1 Gbit/s, Electron. Lett., Vol. 9 (1973), p. 34.

[84] M. J. Adams: *Rate equations and transient phenomena in semiconductor lasers*, Opto-Electron Vol. 5 (1973), p. 201.

[85] T. Suematsu, S. Akila and T. Hong: *Measurement of spontaneous-emission factor of* AlGaAs *double-heterostructure semiconductor lasers*, IEEE J. Quantum Electron., Vol. QE-13 (1977), p. 596.

[86] W. Harth and D. Siemsen: *Modulation characteristics of injection laser including spontaneous emission*-1. *Theory*, A.E.Ü., Vol. 30 (1976), p. 343.

[87] R. Lang and K. Kobayashi: *Suppression of relaxation oscillations in semiconduc/or lasers*, Conference on Laser Engineering and Applications, Washington D.C. (1975).

[88] P. Russer and S. Schulz: *Direkte Modulation eines Doppelheterostrukturlasers mit einer Bitrate von* 2,3 Gbit/s, A.E.Ü., Vol. 27 (1973), p. 193.

[89] P. R. Selway and A. R. Goodwin: *Effect of D.C. bias level on the spectrum of* GaAs *lasers operated with short pulses*, Electron. Lett., Vol. 12 (1976), p. 25.

[90] T. Ozek and T. Ito: *Pulse modulation of* DH-(GaAl)As *lasers*, IEEE J. Quantum Electron., Vol. QE-9 (1973), p. 388.

[91] A. Brosio, P. L. Carni, A. Moncalvo and V. Seano: *Level control circuit for injection laser transmitters*, Fourth European Conference on Optical Communication, Genoa (September 12-15, 1978), p. 438.

[92] N. G. Basov: *Dynamics of injection lasers*, IEEE J. Quantum Electron., Vol. QE-4 (1968), p. 855.

[93] T. P. Lee and D. Gloge: *Signal structure of continuously self-pulsing* GaAs *lasers*, IEEE J. Quantum Electron, Vol. QE-7 (1971), p. 43.

[94] B. I. Miller, E. Pinkas, I. Hayashi and R. J. Capik: *Reproducible liquid phase-epitaxial growth of double heterostructure* GaAs-$Al_xGa_{1-x}As$ *lasers diodes*, J. Appl. Phys., Vol. 43 (1972), p. 2817.

[95] R. L. Hartman and R. W. Dixon: *Reliability of DH* GaAs *lasers at elevated temperatures*, Appl. Phys. Lett., Vol. 26 (1975), pp. 239.

[96] S. Yamakoshi, O. Hasegawa, H. Hamaguchi, M. Abe and T. Yamaoka: *Degradation of high-radiance* $Ga_{1-x}Al_xAs$ *LED's*, Appl. Phys. Lett., Vol. 31 (1977), p. 627.

[97] R. L. Hartman and A. R. Hartman: *Strain-induced degradation of* GaAs *injection lasers*, Appl. Phys. Lett., Vol. 23 (1973), p. 147.

[98] P. M. Petroff and R. L. Hartman: *Defect structure introduced during operation of heterojunction* GaAs *lasers*, Appl. Phys. Lett., Vol. 23 (1973), p. 469.

[99] G. A. Rogozny and M. B. Panish: *Stress compensation in* $Ga_{1-x}Al_xAs_{1-y}P_y$ *LPE layers on* GaAs *substrates*, Appl. Phys. Lett., Vol. 23 (1973), p. 533.

[100] H. Kressel and I. Ladany: *Reliability aspects and facet damage in high-power emission from* (AlGa)As *CW laser diodes at room temperature*, RCA Rev., Vol. 36 (1975), p. 230.

[101] G. P. Henshall: *Gallium aluminium arsenide heterostructure lasers: factors affecting catastrophic degradation*, Solid-State Electron., Vol. 20 (1977), p. 595.

[102] A. R. Goodwin, J. R. Peters, M. Pion, G. H. B. Thompson and J. E. A. Whiteaway: *Threshold temperature characteristics of double heterostructure* $Ga_{1-x}Al_xAs$ *lasers*, J. Appl. Phys., Vol. 46 (1975), p. 3126.

[103] A. R. Goodwin, M. Pion and S. R. Baulcomb: *Enhanced degradation rates in temperature-sensitive* $Ga_{1-x}Al_xAs$ *lasers*, IEEE J. Quantum Electron, Vol. QE-13 (1977), p. 696.

[104] M. B. Panish: *Heterostructure injection lasers,* Proc. IEEE, Vol. 64, No. 10 (1976), p. 1512.

CHAPTER 2

PHOTODETECTORS

by Pietro Luigi Carni

2.1 General remarks

In the present state of the art, the detection and/or the amplification of optical signals takes place after their conversion into electrical signals.

The photodetectors used for this purpose operate, in principle, according to the following mechanisms:

— generation of electrical carriers by incident light
— transport and possible multiplication of these carriers
— current interaction with an external circuit.

The photodetectors employed in optical communication systems must meet the following requirements:

— high optical-electrical conversion efficiency
— high speed
— low noise
— small dimensions, reliability, reduced feeding power.

The components suited for these requirements are photodiodes, working under photovoltaic conditions, with or without a gain mechanism. They consist essentially of an inversely biased *p-n* junction, the current of which is modulated by electron-hole pairs generated by light absorption. PIN diodes, PN diodes and Schottky barrier diodes (semiconductor-metal junctions) operate according to this principle [1].

PIN diodes, being the most suitable for the applications of interest, will be considered further on.

As will be shown later, the device efficiency is connected to the absorption coefficient α [cm^{-1}] of the material. The behaviour of α for silicon, germanium and gallium arsenide is shown in Fig. 1 as a function of the energy gap.

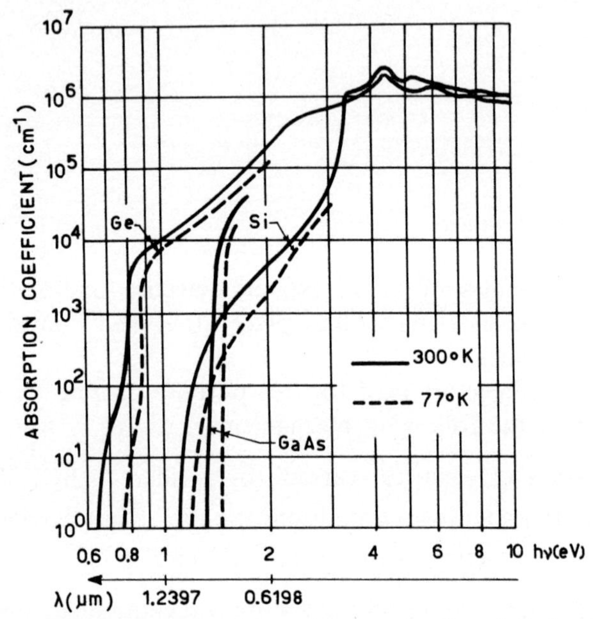

FIG. 1 – Measured absorption coefficients for pure Ge, Si, and GaAs. (After [1]).

The upper cutoff wavelength λ_c is determined by the energy gap of the material E_g ($E_g = 1.12$ eV for Si and $E_g = 0.803$ eV for Ge at 300 °K). The absorption coefficient α for $\lambda > \lambda_c$ becomes very small, and to achieve a significant absorption of the radiation the material must be of a certain thickness. Si can therefore be used in the first region ($\lambda \leqslant 1$ μm), while Ge would be also appropriate for values of $\lambda > 1$ μm. However, the use of Ge detectors constitutes a problem because of their noise factors.

2.2 Unity-gain photodetectors

2.2.1 *Efficiency and spectral response*

The PIN photodiode is the most widely used detector type, since its depletion layer depth can be planned so as to maximize

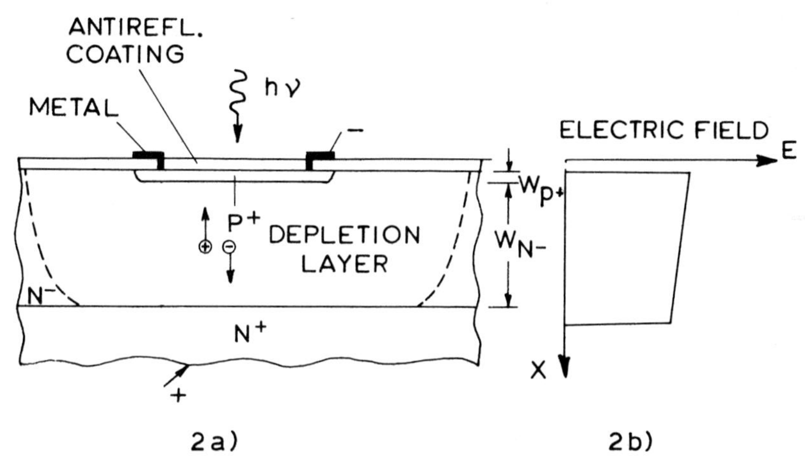

FIG. 2 – *a*) Structure of a PIN junction. *b*) Electric field vs. depth.

efficiency and speed. Figure 2*a* shows the structure of a PIN junction. Figure 2*b* shows the electrical field E versus x. The electromagnetic radiation incident on the surface is partly reflected and partly enters the semiconductor. The energy $h\nu$ associated with the photon is transferred to the electrons in the valence band if $h\nu \geqslant E_g$; these electrons pass into the conduction band, while holes are formed in the valence band. Carriers are kept apart from the electrical field E and drift towards the electrodes.

If a power $P = n_f h\nu/\Delta t$ impinges on the photodiode, a current $I = \varrho P$ is obtained at the output. The responsivity ϱ [AW^{-1}] can then be written as:

$$\varrho = \frac{q\lambda}{hc}\eta \tag{1}$$

where: q electronic charge ($1.602 \cdot 10^{-19}$ C)
λ wavelength of incident light
h Planck constant ($6.625 \cdot 10^{-34}$ J·s)
c velocity of light in vacuum
η efficiency
n_f number of incident photons.

The efficiency η is a non-dimensional number equal to the ratio of generated electron-hole pairs to the number of incident photons.

The responsivity depends on two factors: *a*) reflection at the semiconductor/outer medium interface; *b*) geometric and physical characteristics of the device.

a) If the semiconductor has a refractive index n_3 and the outer medium has a refractive index n_1, the reflectivity at the interface for a plane wave is [2]

$$R_1 = \left(\frac{n_3 - n_1}{n_3 + n_1}\right)^2 \qquad n_3 > n_1 . \tag{2}$$

To minimize reflection, a coating layer with a suitable refractive index n_2 ($n_1 < n_2 < n_3$) can be interposed.

It can be demonstrated that if the thickness h of this layer is given by

$$h = \frac{\lambda_0}{4n_2} \cdot \frac{1}{m} \tag{3}$$

with integer and odd m, the total reflectivity between the medium 1 and the medium 3 can be written

$$R_1 = \left[\frac{n_1 n_3 - n_2^2}{n_1 n_3 + n_2^2}\right]^2 . \tag{4}$$

Then, if $n_2 = \sqrt{n_1 n_3}$, R_1 is zero at λ_0 wavelength, and may be kept small for a narrow wavelength range around λ_0. Figure 3 shows the behaviour of R_1 as a function of λ for a SiO layer with $h \simeq 0.115$ μm.

In the case of Si detectors ($n_3 = 3.45$) with air interface ($n_1 = 1$) the antireflective coating must have $n_2 \simeq 1.86$. Therefore, the materials suitable for this application are SiO ($n = 1.9$), SiO_2 ($n = 1.46$), Si_3N_4 ($n = 2.06$), Al_2O_3 ($n = 1.76$).

Ignoring the absorption in the antireflective layer, the efficiency due to reflection is:

$$\eta_R = 1 - R_1 . \tag{5}$$

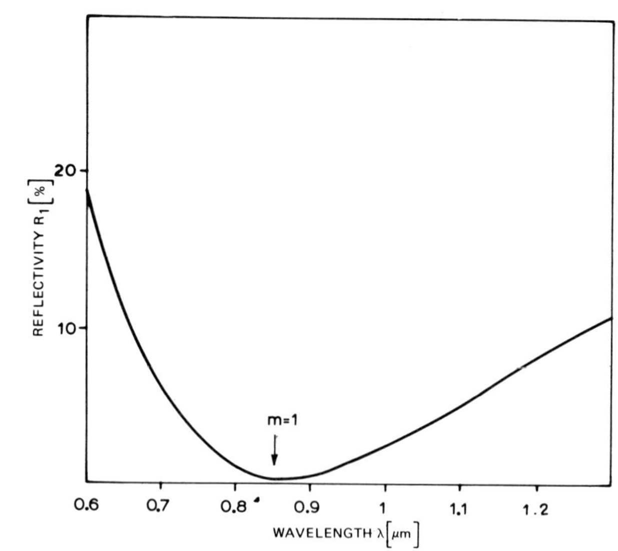

FIG. 3 – Behaviour of the reflectivity as a function of λ, for a SiO layer.

b) Inside the semiconductor the radiation is absorbed according to an exponential law

$$P(x) = P_0 \exp\left[-\alpha(\lambda)\, x\right]. \tag{6}$$

If $h\nu < E_g$, α is very small and the material does not absorb; on the other hand, if $h\nu \geqslant E_g$ there is a high probability of photon absorption and, therefore, of generation of electron-hole pairs.

In the heavily doped region P^+ the lifetime of minority carriers is short (i.e. there is a high probability of recombination before diffusion occurs in the high electrical field layer), therefore the radiation absorption in the region P^+ affects the device efficiency only slightly. Under the assumption that the overall radiation absorbed in the P^+ layers does not contribute to the current, the efficiency of the layer P^+ is equal to

$$\eta_{P^+} = \exp\left[-\alpha(\lambda)\, W_{P^+}\right] \tag{7}$$

where W_{P^+} is the P^+ layer thickness.

On the other hand, the thickness W_{N^-} of the N^- layer must be large enough to absorb a high level of radiation; its efficiency becomes

$$\eta_{N^-} = 1 - \exp\left[-\alpha(\lambda)\, W_{N^-}\right]. \tag{8}$$

The radiation which is not absorbed in the N^- layer partly penetrates the layer N^+ and is partly reflected in the N^-N^+ interface.

The N^- layer is generally designed so as to absorb most of the radiation in the wavelength range of interest. A reach-through structure with a fully depleted layer seems to be the best suited for this purpose. The explicit expression of the device efficiency can be written as:

$$\eta = \eta_{R_1} \cdot \eta_{P^+} \cdot \eta_{N^-}. \tag{9}$$

If the back reflectivity R_2 at the N^-N^+ interface is taken into account the device efficiency is [5]:

$$\eta = \frac{(1-R_1)[1-\exp[\alpha W_{N^-}]\{1+R_2 \exp[-\alpha(W_N + 2D)]\}\exp(-\alpha W_{P_+})}{1-R_1 R_2 \exp[-2\alpha(W_{N^-} + W_{P^+} + D)]} \tag{10}$$

where: R_2 reflectivity of the back surface (i.e. the interface between N^+ and N^- layers)

D thickness of the back region.

2.2.2 *Frequency response*

a) Transit time

The carriers generated in the N^- layer reach the N^+ contact in a time T depending on the material, on the device structure and on the electric field E. If the N^- layer is intrinsic (i.e. not intentionally doped) [1]

$$E = \frac{V}{W_{N^-}}, \qquad V = V' + V_{bi} \tag{11}$$

where V' is the applied reverse voltage and V_{bi} is the built-in voltage.

Then the transit time is:

$$T = \frac{W_{N^-}}{v} = \frac{W_{N^-}}{\mu_e E} = \frac{W_{N^-}^2}{\mu_e V} \tag{12}$$

μ_e [cm V^{-1} s^{-1}] being the electron mobility, v the drift velocity.
Note that for high electric fields, mobility tends to saturate and reaches the value $\simeq 1400$ cm^2 V^{-1} s^{-1} ($v \simeq 10^7$ cm s^{-1}, see Fig. 4).

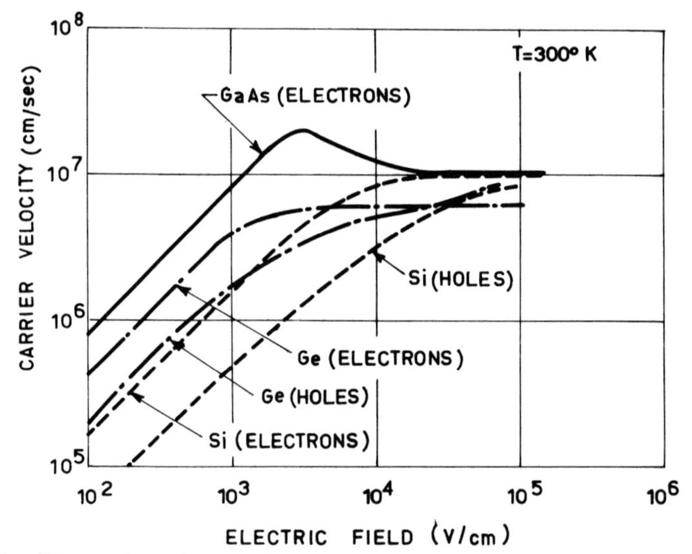

FIG. 4 – Measured carrier velocity vs. electric field for high-purity Ge, Si, and GaAs. (After [1]).

If the N^- layer is not intrinsic a reach-through voltage V_{RT}, given by,

$$V_{RT} = \frac{q}{2\varepsilon} N_d W_{N^-}^2, \tag{13}$$

is required to fully deplete the N^- layer. $(qN_d W_{N^-})/\varepsilon$ is the effective electric field and N_d is its doping level. Hence [3]:

$$T = \frac{W_{N^-}^2}{2\mu_e V_{RT}} \ln\left(1 + \frac{2V_{RT}}{V - V_{RT}}\right) \tag{14}$$

fits for $(V - V_{RT}) > 0$.

Note that, other conditions being equal, the time T calculated by (14) is larger than the time calculated by (12). In order to improve the device speed it is therefore convenient to use a material with high resistivity.

The device speed cannot be increased beyond a certain limit by only increasing the electric field, since the mobility tends to be saturated. Furthermore, a $P^+N^-N^+$ structure is to be preferred to a complementary $N^+P^-P^+$ structure since the hole mobility $\mu_h \simeq \frac{1}{3}\mu_e$ and the P^+ layer is very thin; in the latter case, the holes generated near the P^+ electrode should cover a longer path.

If the incident power is sinusoidally modulated, i.e. $P(t) = P_0 \exp[j\omega t]$, the conduction current at point x is:

$$I_{\text{cond}}(x) = I_0 \exp\left[j\omega\left(t - \frac{x}{v}\right)\right] \tag{15}$$

where $I_0 = \varrho P_0$.

By integration over the N^- region we obtain

$$I_{\text{cond}} = \frac{1}{W_{N^-}} \int_0^{W_{N^-}} I_{\text{cond}}(x)\, dx = I_0 \left(\frac{1 - \exp[-j\omega T]}{j\omega T}\right) \exp[j\omega T] \tag{16}$$

where $T = W_{N^-}/v$.

From (16) it may be seen that for $\omega T = 2.4$ the current amplitude is reduced by $\sqrt{2}$. The 3 dB cutoff frequency due to transit time with respect to the low frequency level is then given by

$$f_{3\text{dB}} \geqslant \frac{2.4}{2\pi T}. \tag{17}$$

Note that in most detectors for optical communications $T \sim 10^{-9}$ s, so that bandwidths in the order of 1 GHz are available.

b) Capacitance

In addition to the intrinsic limitations due to the transit time, another limiting factor of device speed is related to the device capacitance. This can be evaluated with reference to a fully depleted device, assuming that this can be schematically modelled as a plane capacitor. Since a PIN detector operates at a voltage $V > V_{RT}$

its capacitance is given by [1]

$$C = \varepsilon \frac{A}{W_{N^-}} \tag{18}$$

where A is the diffusion layer area and ε is the silicon dielectric constant ($\varepsilon_{Si} = 1.04$ pF/cm). Therefore, to reduce C it is convenient to limit the diffused layer area.

The cylindrical capacitance C_c due to the curvature of the diffused layer must also be taken into account. It may be shown [1] that

$$C_c = 1.5\pi\varepsilon r \tag{19}$$

where r is the curvature radius.

Finally, if the device has a field plate to shield the outer region from incident light, a further capacitance is introduced at low electric fields.

The total photodetector capacitance associated to the preamplifier input resistance gives rise to a pole in the frequency response, thus limiting the receiver bandwidth.

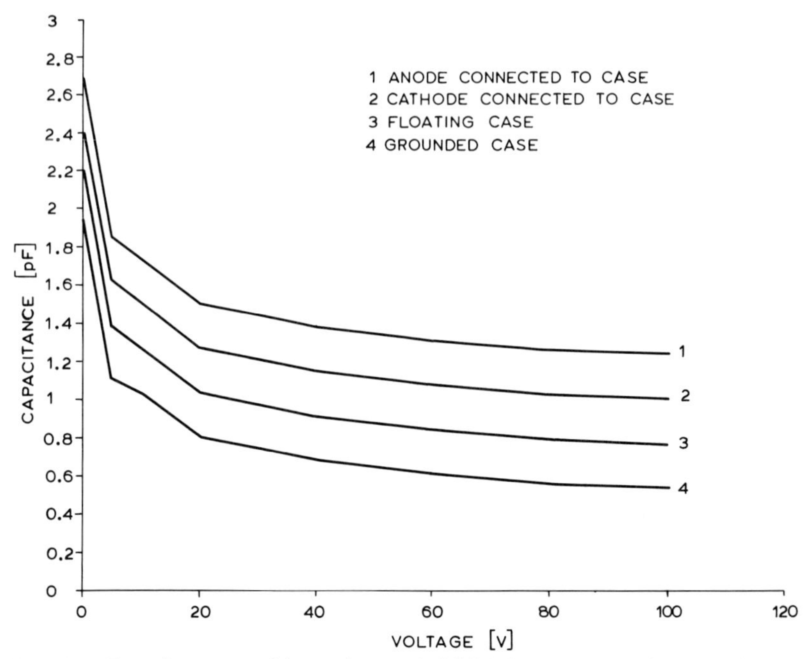

FIG. 5 – Capacitance vs. bias voltage of SGS OC PD 01 PIN detectors in various electrical configurations.

Typical values of the overall capacitance of PIN detectors range from 1 to 2 pF; Fig. 5 shows C versus V for a typical TO-18 packaged PIN detector (SGS OCPD 01) in various electrical configurations.

2.2.3 *Quantum noise*

A limiting characteristic of photodetector performance is the quantum noise due to the statistical nature of photon-electron conversion. It has been demonstrated [4] that this noise follows a Poissonian distribution with a variance

$$\langle i_n^2 \rangle = 2qIB \tag{20}$$

where I is the current average value and B is the noise equivalent bandwidth.

The current I is obtained as the sum of the signal average current and the dark current I_d. The latter comes from spontaneous generation phenomena occurring in the depleted layer and on the surface of the photodetector. The values of the two dark current terms may be expressed as

$$I_b = \frac{1}{2} q \frac{n_i}{\tau} A \qquad (I_b = I_{\text{bulk}}) \tag{21}$$

$$I_s = q n_i S_0 A_s \qquad (I_s = I_{\text{surface}}) \tag{22}$$

where: n_i carrier concentration in the intrinsic layer
τ minority carrier lifetime
A_s peripheric surface of the depleted layer
S_0 surface recombination speed.

The total dark current $I_d = \sqrt{I_s^2 + I_b^2}$ is shown in Fig. 6 as a function of V_p for ten photodiodes manufactured from the same material wafer. Since the dark current and the signal current are uncorrelated, the noise spectral density in the presence of an average signal current I_S is:

$$\frac{\langle i_n^2 \rangle}{B} = 2q\sqrt{I_S^2 + I_d^2}\,. \tag{23}$$

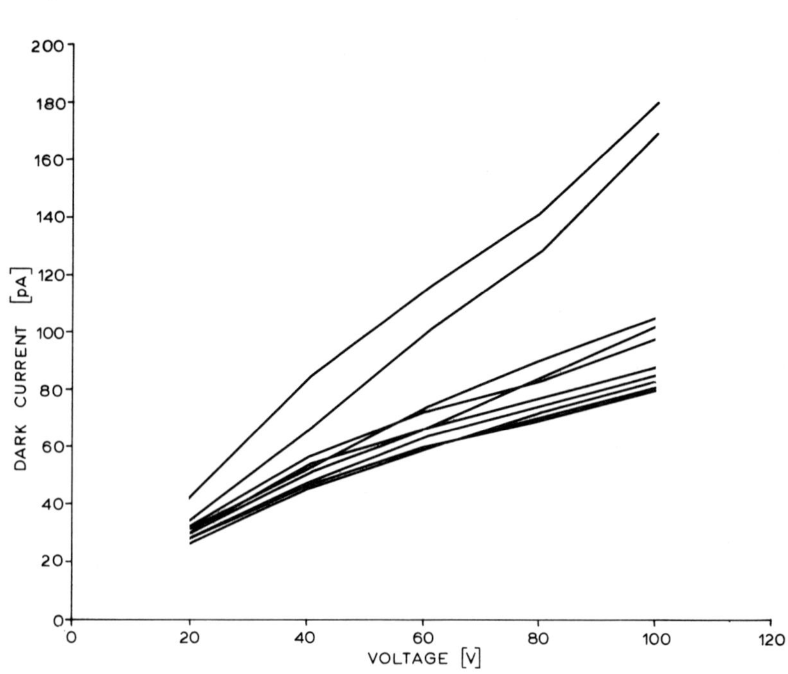

FIG. 6 – Total dark current vs. bias voltage for ten samples of SGS OC PD 01 PIN detectors.

2.2.4 *Equivalent circuit*

Figure 7 shows the current-voltage characteristics of a typical PIN photodetector. It should be noted that for values of the bias voltage higher than ~ 10 V, these characteristics are quite similar to those of a current generator with an internal resistance of the order of $10^{10}\ \Omega$.

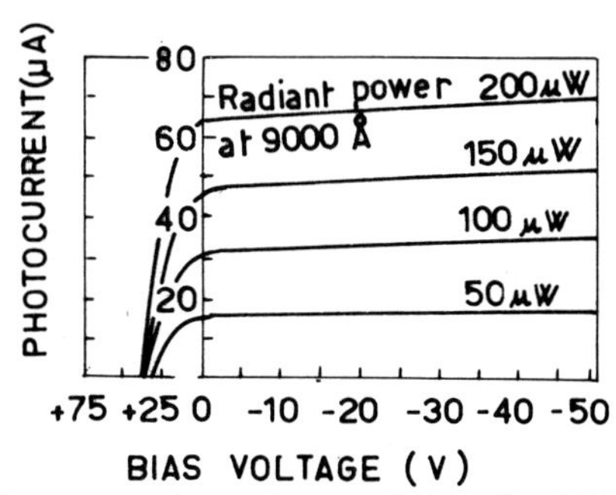

FIG. 7 – Typical current-voltage characteristics of a PIN photodetector.

Therefore, on the basis of these considerations the equivalent circuit of a PIN photodetector can be represented schematically as in Fig. 8, where

i_S	signal current ($\langle i_S \rangle = I_S$)
i_n	quantum noise current
R_P	reverse junction resistance
R_S	series resistance
C_P	total capacitance
Z_L	load impedance.

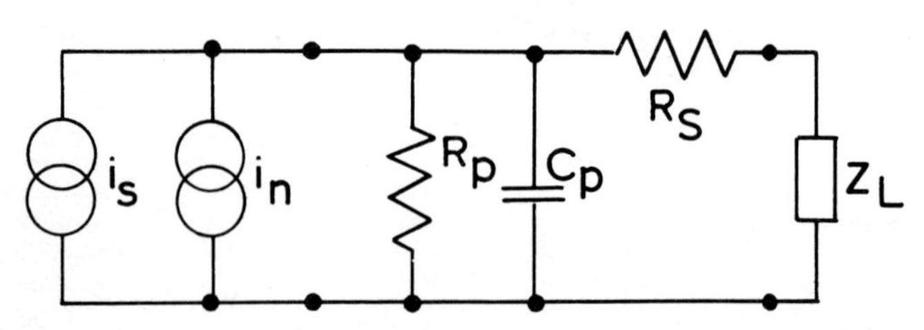

FIG. 8 – Schematical equivalent circuit of a PIN photodetector.

Note that in order to take into account the real performances of the device, effective values considering the physical effects which cannot simply be represented in terms of circuit parameters, must be assigned to the above parameters. In particular the capacitance value, if needed, has to include also the transit-time effect.

2.3 Avalanche photodetectors

2.3.1 *The avalanche mechanism*

In semiconductor photodetectors the signal current can be amplified by means of the avalanche multiplication process. As a matter of fact, if a high electric field is applied to a *p-n* junction, the hole-electron pairs generated in the depletion layer acquire enough energy to give rise to ionization of the lattice atoms. New carriers are thus generated which may produce new carrier pairs.

The avalanche process characteristics may be expressed in terms of the coefficient $\alpha(E)$ and $\beta(E)$ which represent the probability per unit length that electrons and holes respectively are produced as a result of collision ionization. Generally, the electric field is a function of position, so that by assuming a monodimensional path $\alpha = \alpha(x)$, $\beta = \beta(x)$.

Figure 9 shows a possible electric field distribution for an avalanche photodiode.

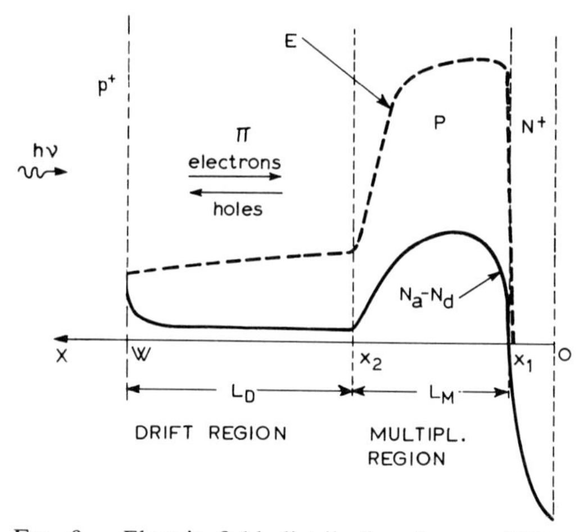

FIG. 9 – Electric field distribution for an APD.

If only one type of carrier (e.g. electrons) can undergo ionization, ($\alpha \neq 0$, $\beta = 0$) the continuity equation has the form

$$-\frac{dJ_n}{dx} = \alpha J_n \tag{24}$$

and integrating on the high electric field region we obtain

$$J_n(x) = J_n(x_2) \exp\left[\int_x^{x_2} \alpha \, dx\right]. \tag{25}$$

By defining the electron multiplication M_e as the ratio $J_n(x_2)/J_n(x_1)$ we have

$$M_e = e^{\delta}, \qquad \delta = \int_{x_1}^{x_2} \alpha \, dx . \tag{26}$$

In most materials $\beta \neq 0$. In this case the continuity equation must be solved in the form

$$\frac{dJ}{dx} = -\alpha J_n - \beta J_p - g(x) \tag{27}$$

where the optical generation term $g(x)$ in the high electric field region is taken into account. The solution to the Eq. (27) is given by:

$$\bar{M} = \frac{\int_0^W g(x) M(x)\, dx}{\int_0^W g(x)\, dx} \tag{28}$$

where $\bar{M}$ is the average multiplication for the carriers generated in the diode and $M(x)$ is the multiplication of a pair generated at x. This can be expressed in the form:

$$M(x) = \frac{\exp\left[-\int_x^W (\alpha - \beta)\, dx'\right]}{1 - \int_0^W \alpha \exp\left[-\int_x^W (\alpha - \beta)\, dx'\right] dx} \tag{29}$$

By defining the ratio k_0 and the weighted ratios k_1 and k_2 of the ionization coeffiecients:

$$k_0 = \frac{\int_{x_1}^{x_2} \beta\, dx}{\int_{x_1}^{x_2} \alpha\, dx}, \qquad k_1 = \frac{\int_{x_1}^{x_2} \beta M(x)\, dx}{\int_{x_1}^{x_2} \alpha M(x)\, dx}, \qquad k_2 = \frac{\int_{x_1}^{x_2} \beta M^2(x)\, dx}{\int_{x_1}^{x_2} \alpha M^2(x)\, dx} \tag{30}$$

$M(x)$ may then be expressed as

$$M(x) = \frac{G(x)(1 - k_1)}{G(x_1) - k_1} \tag{31}$$

where

$$G(x) = \exp\left[-\int_x^{x_1} (\alpha - \beta)\, dx'\right]. \tag{32}$$

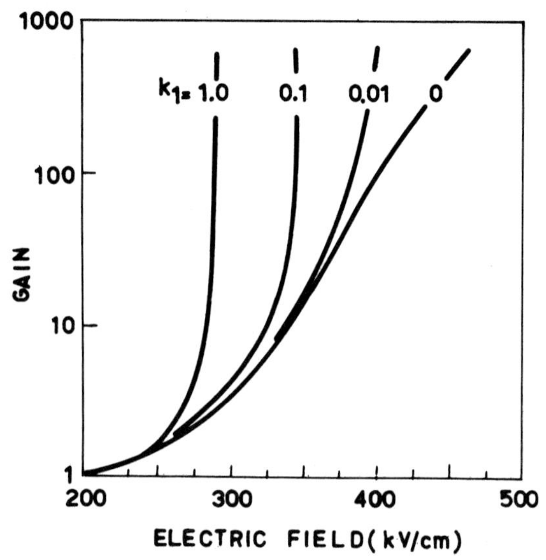

FIG. 10 – Electron multiplication as a function of electric field, with ionization ratio K_1 as a parameter. (After [5]).

Figure 10 [5] shows that for each value of $k_1 \neq 0$ the slope of $M(E)$ is determined to a large extent by k_1.

It appears that a low value of k_1 is required to obtain values of dM/dE that permit an automatic gain control by changing the bias point.

2.3.2 *Noise associated with the avalanche process*

Due to its statistical nature, the avalanche process gives rise to a noise contribution which adds to quantum noise, determined by photon-electron conversion. As a matter of fact, the value of $M(x)$ referred to in Eqs. (29) or (31) is an average value, whereas each hole-electron pair undergoes a multiplication M which is randomly distributed.

We recall that for a non-avalanche detector

$$\Phi = \langle i_n^2 \rangle = 2qI\,df\,. \tag{33}$$

The noise current spectral density, after the multiplication process can be written:

$$\Phi = \langle i_n^2 \rangle = 2qI\,df\,M^2F(M) \tag{34}$$

where F (excess noise factor) is defined as

$$F = \frac{\langle M^2 \rangle}{\langle M \rangle^2}. \tag{35}$$

To calculate F, we assume that $\alpha = \alpha(E(x))$, $\beta = \beta(E(x))$. The noise current spectral density due to either hole or electron current can be written

$$\langle (dI_{p,n} - \langle dI_{p,n} \rangle)^2 \rangle = 2q dI_{p,n} df. \tag{36}$$

Therefore, in the linear element dx inside the multiplication region:

$$d\Phi(x) = 2qM^2(x)\, dI_{p,n}(x) \tag{37}$$

and taking into account the hole current

$$\Phi = 2q \left[I_p(0)\, M^2(0) + I_n(W) M^2(W) + \int_0^W \frac{dI_p}{dx} M^2(x)\, dx \right] \tag{38}$$

$dI_{p,n}/dx$ can be obtained from the continuity equation:

$$I_p(x) = I_p(0) \exp\left[-\int_0^x (\alpha - \beta)\, dx' \right] + \\ + \int_0^x (\alpha I + g) \exp\left[-\int_x^W (\alpha - \beta)\, dx' \right] dx \tag{39}$$

where

$$I = I_p(0) M(0) + I_n(W) M(W) + \int_0^W g(x) M(x)\, dx. \tag{40}$$

Since $I = I_p(x) + I_n(x)$ under steady-state condition, differentiating (39) and substituting the result into (38) yields

$$\Phi = 2q \left\{ 2 \left[I_p(0) M^2(0) + I_n(W) M^2(W) + \int_0^W g(x) M^2(x)\, dx \right] + \\ + I \left[2 \int_0^W \alpha M^2(x)\, dx - M^2(W) \right] \right\}. \tag{41}$$

The general expression for noise spectral density is quite complex. A case of considerable interest can be obtained by considering a photodetector where *a*) $\alpha \gg \beta$, *b*) the injected current is an electron current ($I_p = 0$) and *c*) no carrier generation occurs within the high electric field region ($g(x) = 0$). We obtain

$$\Phi = 2qI_n M^3 \left[1 - (1-k)\left(\frac{M-1}{M}\right)^2\right] \tag{42}$$

$$F = M\left[1 - (1-k)\left(\frac{M-1}{M}\right)^2\right] \tag{43}$$

where $k = \alpha/\beta$.

Although the hypothesis $\alpha/\beta = k$ is not valid for all semiconductors, it can be demonstrated [8] that assuming for k an effective value defined by:

$$k_{\text{eff}} = \frac{k_2 - k_1^2}{1 - k_2} \qquad \text{for } M \geqslant 10 \tag{44}$$

the theoretical curves obtained from (44) are consistent with experimental values. Figure 11 [11] shows the behaviour of F as

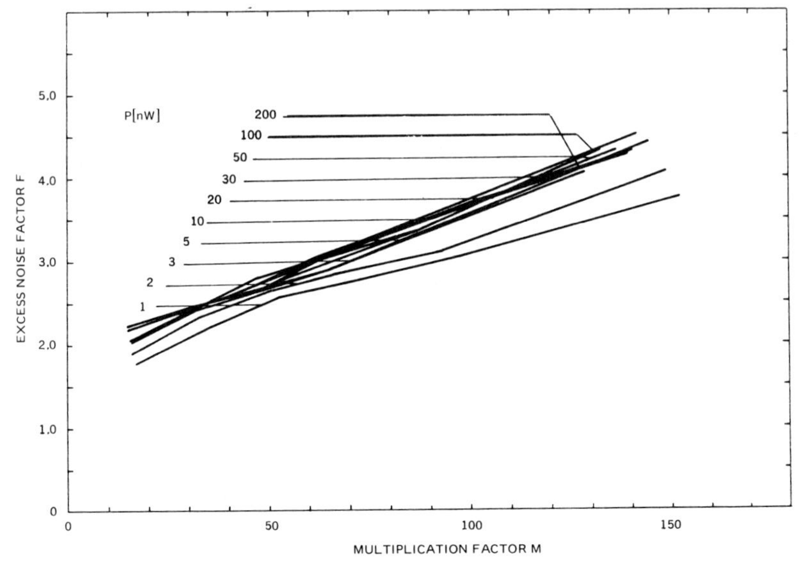

FIG. 11 – Excess noise factor F vs. multiplication factor M with incident optical power as a parameter.

a function of M for RCA C 30817 photodiodes for different values of incident power; a value $k_{eff} = 0.017$ inserted in (43) fits the experimental curves very well.

Note that in the range of values $M = (10 \div 150)$ the linear approximation proves to be satisfactory. Therefore the widely-used expression:

$$F = M^{2+x} \qquad (x = 0.3 \div 0.5) \tag{45}$$

should be excluded for $\alpha = 0$.

On the other hand the latter is suitable for devices with $k = 1$ (see Fig. 12).

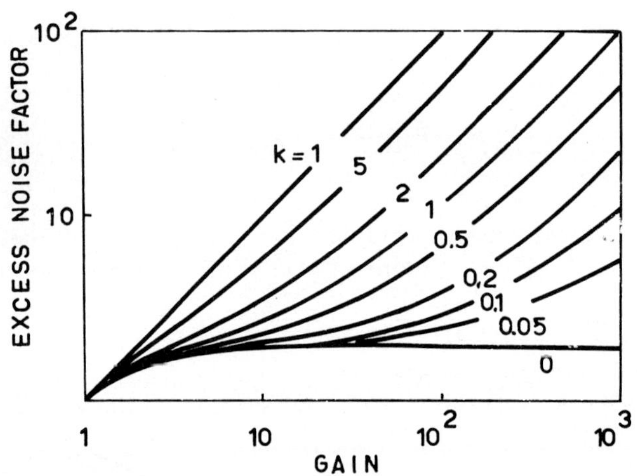

FIG. 12 – Excess noise factor F vs. M with K_{eff} as a parameter. (After [5]).

Finally it may be noted that in (41) the term $\int_0^W g(x)\, M^2(x)\, dx$ is included. As

$$g(x) \propto \exp\,[-\,\alpha(\lambda)\, x] \tag{46}$$

where $\alpha(\lambda)$ is the absorption coefficient, a contribution to quantum noise depending on the radiation wavelength is introduced if significant carrier generation occurs in the high field region [6].

For a complete characterization of noise in avalanche photodetectors the statistics of the multiplication process must be described in terms of its moment generating function (M_g). The calculations leading to the above mentioned function will not be carried out in detail here. Only the basic hypotheses will be indicated. Starting from these assumptions, the second order moment (i.e. the noise spectral density) has been obtained.

The basic hypotheses in the model are the following:

1) a probability of collision ionization per unit length (denoted by α for electrons and β for holes) is associated to each carrier type;

2) the subsequent ionization collisions of a carrier are statistically independent, i.e. the interval (both in time or distance) between two subsequent collisions is sufficiently large that the statistics of a single collision is not affected by previous collisions. Under these hypotheses the number of secondary carriers generated by a given carrier has a Poissonian distribution.

The analytical developments [7, 8] lead to the following implicit form for the moment generating function:

$$M_g(s, 0) = \exp \cdot \left\{ s + \frac{1}{1-k} \ln [M_g(s, 0) - \exp [(k-1)\,\delta][M_g(s, 0) - 1]] \right\} \quad (47)$$

where s is the complex variable of the Laplace transform. These results are consistent with those obtained in [9, 10] where a different computational method is applied, leading to the following form of the probability density of the random gain:

$$p_g(n, 0) = \frac{\Gamma(n/(1-k)+1)\, e^{-\delta}(e^{-k\delta} - e^{-\delta})^{n-1}}{n!\,\Gamma(n/(1-k)+2-n)} \quad (48)$$

$p_g(n, 0)$ is the probability that from one pair injected into the high electric field region at $x = 0$, n pairs are obtained as final result. For $k = 0$, i.e. for a photodetector with a unilateral gain, Eq. (48) becomes:

$$p_g(n, 0) = \frac{\Gamma(n+1)\, e^{-\delta}(1 - e^{-\delta})^{n-1}}{n!\,\Gamma(2)} \quad (49)$$

$$p_g(n, 0) = e^{-\delta}(1 - e^{-\delta})^{n-1} \quad (50)$$

and the moment generating function is

$$M_{\varrho}(s, 0) = \frac{1}{1 - \langle G \rangle [1 - \exp(-s)]} \tag{51}$$

where $\langle G \rangle = e^{\delta}$. The same result may be obtained from (47).

2.3.3 *Linearity and frequency response*

Both linearity and frequency response of avalanche photodetectors are affected by the avalanche mechanism. With regard to linearity, two effects have to be taken into account: *a*) saturation due to high signal level, *b*) thermal effects.

a) The electric field strength in the avalanche region depends on the current I flowing through the diode. Drifting of mobile carriers reduces the electric field by an amount

$$\Delta E = \frac{IW}{2\varepsilon v A} \tag{52}$$

where: v is the average drift velocity of carriers
A is the area of the device cross-section.

This corresponds to an effective bias voltage $v_{\text{eff}} = w(E - \Delta E)$ and therefore this effect can be represented by a space-charge resistance

$$R_{sc} = \frac{W^2}{2\varepsilon v A} \cdot \tag{53}$$

It must be outlined that as the space-charge resistance is not a physical one, it does not affect the time constant of the device, but only its gain.

b) The power dissipated by the photodetector at high current levels gives rise to considerable temperature variation, which modify the ionization coefficients. A general analysis of the thermal behaviour of the device is not possible because of its critical dependence on the device structure and package.

Figure 13 shows an equivalent circuit of an avalanche photodetector [5]. The presence of capacitances C_{th1} and C_{th2} should

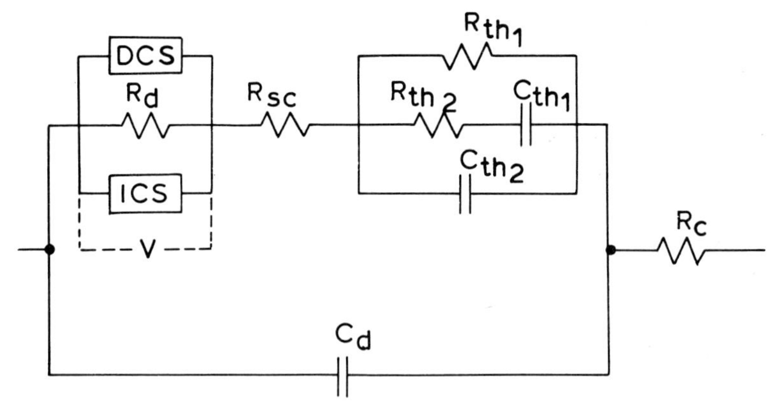

FIG. 13 – Equivalent circuit of avalanche photodiode. (After [5]).

C_d = Diode Capacitance
R_c = Diode Contact Resistance
R_{sc} = Effective Space Charge
$R_d = (dI_{ds}/dV)^{-1}$

R_{th1}, R_{th2}, C_{th1}, C_{th2}, are the effective thermal resistances and capacitances, respectively

be noted. This means that in principle the device speed performance is also affected by thermal effects, due both to high signal levels and to room temperature variations. As a matter of fact, Z_{th} has a very small value in comparison to R_{sc}. Thus, it can be ignored for the frequency range of interest.

For the frequency response the multiplication time effect must also be considered. Following the calculation developed for PIN detectors and substituting the transit time with the effective multiplication time we obtain:

$$M(\omega) = \frac{M(0) \cdot \exp\,[-j\omega M(0)(\tau_m/2)]}{1 + j\omega M(0)\,\tau_m} \tag{54}$$

where $\tau_m \sim 10^{-12}$ s is the effective multiplication time. For any practical purpose the avalanche process can be considered as instantaneous.

2.4 Materials and structures for photodetectors

2.4.1 *Photodetectors for* $\lambda < 1$ μm

The previous considerations are independent of the material employed for the construction of the device. It is to be noted, moreover, that all detectors used for the range $< 1\ \mu m$ are in silicon. Silicon technology is well-known, quite reliable and permits high values of quantum efficiency, speed and low noise when suitable structures are adopted.

The PIN structure already shown in Fig. 2 is the most widely used for non-avalanche detectors. It has the advantage over a *p-n* structure in that a thicker depletion layer can be obtained. This enables the optimum trade-off in terms of efficiency and speed to be achieved. Some structures used for avalanche detectors are shown in the following.

The structure of Fig. 14*a* is a *p-n* junction made of very high resistivity material, so that a wide depletion layer is also formed in the presence of weak bias voltages. In this structure the electric field in the *P* layer, where carriers are generated, is weak. Consequently, a high response time and a high gain may result for short wavelengths, whereas for longer wavelengths the APD shows a faster response but a lower effective gain.

The structure in Fig. 14*b* has a high electron multiplication factor only for short wavelengths which are absorbed in the *P* layer of the junction. Carriers generated by longer wavelengths in the

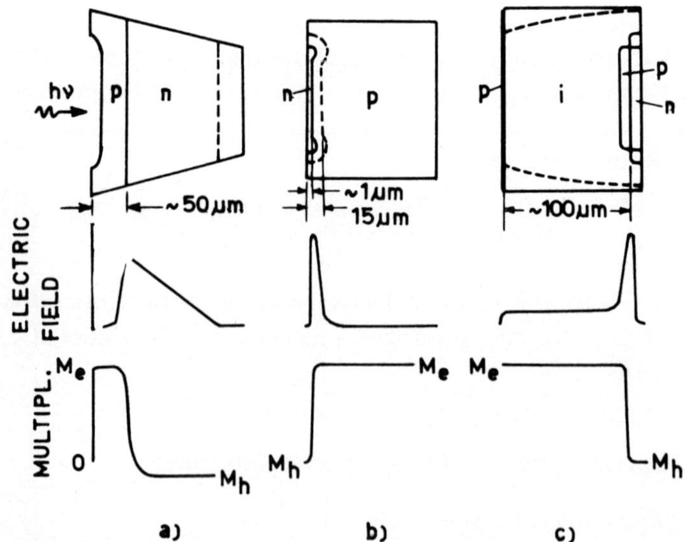

FIG. 14 – Structures of three commercially available silicon avalanche photodiodes: *a*) beveled edge diode; *b*) guard ring structure, and *c*) reach-through structure. (After [5]).

undepleted layer diffuse and are collected slowly. Therefore they do not contribute to high frequency response and introduce further noise.

A high performance is obtained with the reach-through structure of Fig. 14*c* where the depleted layer consists of a wide drift region, where carriers are generated, and a narrower high field region, where they undergo multiplication. Although the electric field in the intrinsic layer is not very high, carrier velocities allow rise and fall times (less than 2 ns) that are suitable for high bit rate [up to 140 Mbit/s]. High quantum efficiencies are also permitted due to the thick ($\sim 70\ \mu m$) intrinsic layer. In addition, effective values as low as 0.02 are obtained for the ionization coefficient ratio k.

2.4.2 *Photodetectors for* $\lambda > 1\ \mu m$

In order to realize the full potential of optical fibres in communication systems for very high bit rates and very long hauls, new materials are under investigation. In fact, silicon can still be employed up to 1.1 μm wavelength if thick depletion layers are employed to compensate the lower absorption coefficient. Unfortunately thick depletion layers limit detector speed. A structure of the type shown in Fig. 15 would probably allow faster transit times without any impairment of quantum efficiency.

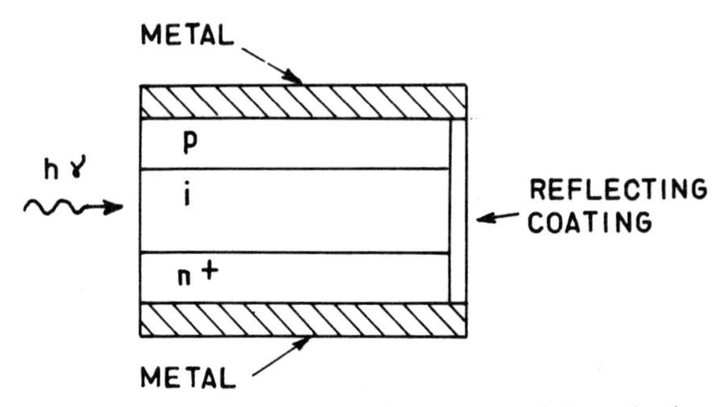

FIG. 15 – Photodetector with illumination parallel to the junction.

For wavelengths in the region 1.1-1.6 μm or longer, detectors must be constructed using materials with narrower bandgaps.

Due to a lack of theoretical studies, no guidance exists to choose a material suitable to be used for production of high efficiency, high speed and low noise detectors. Each material with a suitable bandgap must be experimented. Studies have been carried out on InGaAs, GaAsSb, HgCdTe and more recently on GaInAsP/InP. See, for instance, Section 1.3. An interesting heterojunction III-V alloy detector for $\lambda \simeq 1.06$ μm has been developed by R. C. Eden [17]. The structure of this photodetector is shown in Fig. 16. The GaAsSb/GaAs heterojunction allows light to be

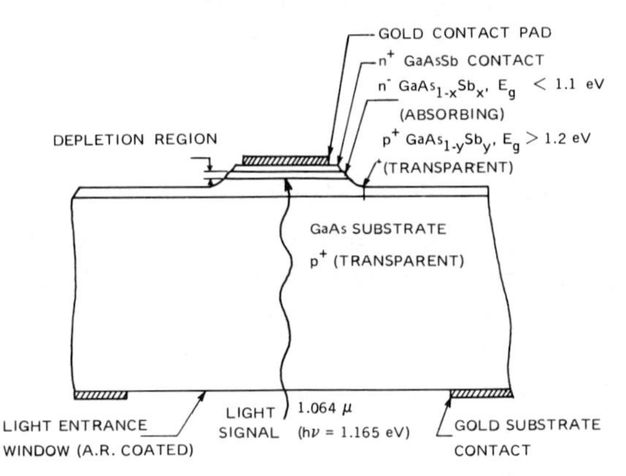

FIG. 16 – Inverted heterojunction 1.06 μm mesa photodiode. Light enters from the bottom in the figure, passes through the transparent p^+GaAs substrate and the transparent $p^+GaAs_{1-y}Sb_y$ buffer layer, and is absorbed in the n^-GaAs_{1-x} active layer (part of which is depleted). (After [17]).

absorbed within the depletion region since the p^+ GaAs substrate and the p^+ $GaAs_{1-y}Sb_y$ buffer layers are transparent for 1.06 μm: a very high quantum efficiency is thus achieved. The GaAs substrate absorption gives rise to a lower cutoff frequency for $\lambda \leqslant 0.9$ μm while for wavelengths longer than 1.2 μm the active layer becomes transparent. In this way a selective efficiency as high as 0.95 can be obtained together with a very low capacitance (0.1 pF), a wide bandwidth ($\sim$15 GHz) and low dark currents ($\sim$3 nA). The detector was not designed for avalanching, but the heterojunction mesa structure is protected against surface breakdown and avalanche gain operation is allowed. In fact gain factors over 100 were experimentally obtained.

Schottky barrier $In_xGa_{1-x}As$ alloy avalanche photodiodes have also been investigated [18]. Detectors with $x = 0.17$ show a quantum efficiency at $\lambda = 1.06\ \mu m$, which is nearly twice that of a silicon detector. Moreover, the responsivity peak can be slightly « tuned » by changing the In mole fraction (Fig. 17).

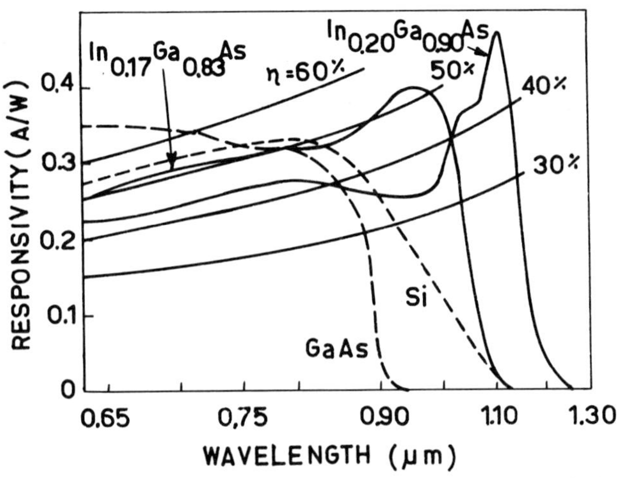

FIG. 17 – Spectral responsivity of two $In_xGa_{1-x}As$ avalanche photodiodes at unity gain. (After [18]).

A good gain uniformity (within ~20 %) across the photosensitive area was also obtained and no surface defects in the Schottky barrier seemed to occur.

Using a mode-locked laser as a light source and a 50 Ω sampling oscilloscope, rise and fall times of about 175 ps were observed. Preliminary tests gave a electron/hole ionization ratio of about 0.02, leading to a very low excess noise factor.

One of the most suitable materials for longer wavelengths appears to be the quaternary GaInAsP alloy. LED's and lasers with good lifetime performance and photodetectors have been developed. The GaInAsP bandgap can be varied continuously from 0.9 μm to about 1.6 μm by changing the mole composition. Avalanche photodetectors have been recently developed by epitaxial growth of GaInAsP layers on InP substrates [19]. Figure 18 shows a diode structure where a graded *p-n* junction is formed by Zn diffusion onto a GaInAsP layer.

In these devices quantum efficiency of 0.45 and gain up to 12

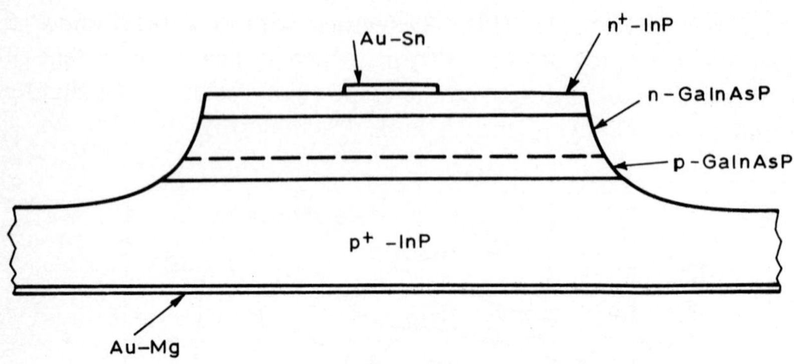

FIG. 18 – GaInAsP/InP APD structure. (After [19]).

for relatively low bias voltages (~ 70 V) were measured. Small capacitances and transit times may be achieved allowing rise time of about 150 ps, but a slower component ($\simeq$ 1 ns) is exhibited by the fall time due to the hole generated at the gap surface. The dark current is about 70 nA for $V_p \simeq 30$ V. Performance improvements can be achieved with the introduction of a passivated layer and a guard ring.

REFERENCES

[1] S. M. Sze: *Fisica dei Dispositivi a Semiconduttore*, Tamburini editore, Milano (1973).

[2] M. Born and E. Wolf: *Principles of Optics*, Pergamon Press, Oxford (1965).

[3] M. Conti, G. Corda and M. De Padova: *Tempi di commutazione, di fotorivelatori al silicio tipo PIN: teoria ed esperimenti*, SGS/ATES Techn. Rep. OPT/6/77.

[4] B. M. Oliver: *Thermal and quantum noise*, Proc. IEEE (May 1965), pp. 436-454.

[5] P. P. Webb, R. J. McIntyre and J. Conradi: *Properties of avalanche photodiodes*, RCA Review, Vol. 35 (June 1974), pp. 234-278.

[6] H. Kanbe, T. Kimura and Y. Mizushima: *Wavelength dependence of multiplication noise in silicon avalanche photodiodes*, IEEE Trans. Electron. Devices, Vol. ED-24, No. 6 (June 1977), pp. 713-716.

[7] S. D. Personick: *New results on avalanche multiplication statistics with applications to optical detectors*, Bell. Syst. Tech. J. Vol. 50, No. 1 (January 1971), pp. 167-189.

[8] S. D. Personick: *Statistics of a general class of avalanche detectors with applications to optical communication*, Bell. Syst. Tech. J., Vol. 50, No. 10 (December 1971), pp. 3075-3095.

[9] R. J. McIntyre: *Multiplication noise in uniform avalanche diodes*, IEEE Trans. Electron. Devices, Vol. ED-13, No. 1 (January 1966), pp. 164-168.

[10] R. J. McIntyre: *The distribution of gains in uniformly multiplying avalanche photodiodes: theory*, IEEE Trans. Electron. Devices, Vol. ED-19, No. 6 (June 1972), pp. 703-713.

[11] P. L. Carni, A. Moncalvo and M. Perino: *Misure di rumore su fotorivelatori a valanga*, CSELT Internal Report 78.03.159.

[12] C. C. Timmermann: *The statistic and dynamic multiplication factor in avalanche photodiode optical receivers*, IEEE Trans. Electron. Devices, Vol. ED-24, No. 12 (December 1977), pp. 1317-1322.

[13] H. Kanbe, T. Kimura, Y. Mizushima and K. Kajiyama: *Silicon avalanche photodiodes with low multiplication noise and high-speed response*, IEEE Trans. Electron. Devices, Vol. ED-23, No. 12 (December 1976), pp. 1337-1343.

[14] J. J. Goedbloed and J. Joosten: *Responsivity of avalanche photodiodes in the presence of multiple reflections*, Electron. Lett., Vol. 12, No. 14 (July 1976), pp. 363-364.

[15] K. Nishida: *Avalanche noise dependence on avalanche photodiode structure*, Electrop. Lett., Vol. 13, No. 14 (July 1977), pp. 419-421.

[16] S. Hata, K. Kajiyama and Y. Mizushima: *Performance of PIN photodiode compared with avalanche photodiode in the longer wavelength region of* 1 *to* 2 μm, Electron. Lett., Vol. 13, No. 22 (October 1977), pp. 668-669.

[17] R. C. Eden: *Heterojunction III-V alloy photodetectors for high sensitivity* 1.06 μm *optical receivers*, Proc. IEEE, Vol. 63, No. 1 (January 1975), pp. 32-37.

[18] G. E. Stillman, C. M. Wolfe, A. G. Foyt and W. T. Lindley: *Schottky barrier in* $In_xGa_{1-x}As$ *alloy avalanche photodiodes for* 1.06 μm, Appl. Phys. Lett., Vol. 24, No. 1 (January 1974), pp. 8-10.

[19] C. E. Hurwitz and J. J. Hsieh: GaInAsP/InP *avalanche photodiodes*, Appl. Phys. Lett., Vol. 32, No. 8 (April 1978), pp. 487-489.

PART III

CABLES AND CONNECTIONS

CHAPTER 1

OPTICAL FIBRE CABLES

by Giuseppe Galliano and Federico Tosco

1.1 Introduction

The practical implementation of optical fibres in the telecommunication field requires the manufacture of optical cable having a structure which facilitates laying and jointing operations and maintains stable transmission characteristics in operational conditions.

The design criteria and material used for optical cable construction are described in the following paragraphs.

Although the individual optical fibre is protected with a thin layer of organic material such as kynar or epoxy acrilate during drawing (primary coating), it does not have adequate mechanical characteristics for direct cabling. Therefore, it is necessary to protect the optical fibre with a secondary coating consisting of an extrusion of plastic material. The criteria for the implementation of the secondary coating are examined in detail in Section 1.2.

One or more strength members may be inserted in an optical cable to withstand tensile strength, thus preventing damage to fibres during cabling and, more especially, during laying of the cable. The principal materials used for the strength members are examined in Section 1.3.

The optical cable is then completed, in a similar manner to the conventional cable, with fillers, cushions etc. and an external sheath (see Section 1.4).

In Section 1.5 the most widely used fabrication type for experimental optical cables is described.

Some experimental sections of optical cables have been subjected to mechanical and environmental tests to verify their utilization limits. The various test types and measuring equipment are illustrated in Sections 1.6 and 1.7.

Section 1.8 describes the laying of experimental optical cable sections. Finally, Section 1.9 presents a review of worldwide developments in the design and manufacture of optical cables.

1.2 Secondary coating

Secondary coating of the fibre, also called jacketing or buffering, is obtained by applying a plastic covering over the primary coated fibre. Its main purpose is to improve the tensile strength of the fibre and to provide radial reinforcement. In conformity with these requirements, it is necessary to utilize plastic coatings of sufficient thickness, limited mainly by the need to retain good flexibility and to minimize the space occupied in the cable [1].

Generally, to provide this secondary coating a thermoplastic extrusion is used, obtained by machinery similar to that used in a conventional wire insulating line. During these operations it is important to apply a very low tension to the fibre and to maintain it constant under normal running conditions.

If the extrusion technique and cooling procedures of the extruded products are not carefully controlled, degradation of the fibre transmission characteristics caused by thermal contraction may occur.

Although many types of secondary coatings are at present used by optical cable manufacturers, they can be basically divided into two broad categories:

loose type: the fibre lies in a plastic tube, having an inner diameter considerably larger than the fibre diameter. Normally these jacketing tubes have an inner diameter in the range 0.4-1 mm and an outer diameter in the range 0.6-2 mm;

tight type: a jacket of 0.4-1.5 mm outer diameter is tightly extruded on the fibre.

The loose type jacketing permits, if suitable techniques are used, the fibre transmission characteristics (in particular attenuation) to be maintained practically unchanged. To achieve this objective, it is important to use jacketing tubes having smooth internal surfaces and a perfectly circular cross-section, to avoid microbendings of the fibre and the consequent attenuation increase.

With this extrusion technique the fibre is subjected to low stresses during cable stranding. In fact these stresses are only partially transmitted to the fibre by the friction forces and, as a consequence, the axial stresses are greatly reduced (1).

If the fibre is fixed to both ends of the jacketing, a large attenuation increase can in some cases occur at low temperatures (—20 - 0 °C). This phenomenon is caused by the different thermal contraction of the fibre and plastic jacket which produces microbendings in the fibre.

Measurements of attenuation (for $\lambda = 870$ nm) were carried out by several cable manufacturers at different temperatures (in the range -50 - $+60$ °C) with both ends of the jacket fixed to the fibre by an epoxy-type adhesive, using nylon-12 jacketing tubes of various thicknesses and inner diameters. Figure 1 shows one

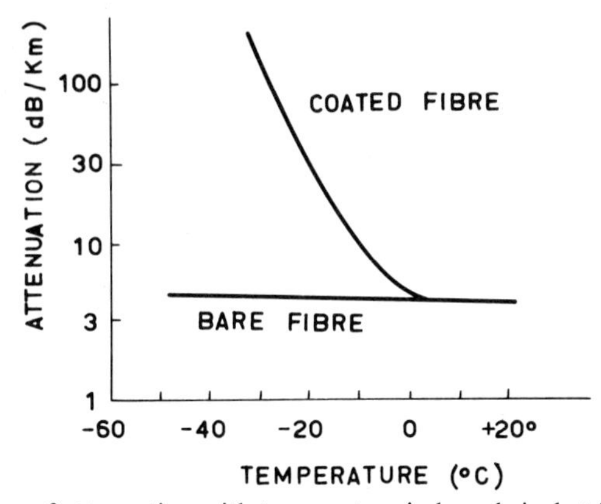

FIG. 1 – Variation of attenuation with temperature in loosely jacket fibre. (After [2]).

of the worst attenuation vs. temperature behaviours measured: at high temperatures the attenuation variations are very low, whereas an abnormal attenuation variation (up to more than 100 dB/km) occurs at low temperatures [2].

The attenuation increase is dependent on fibre-to-jacket clearance and jacket thickness. The larger the clearance, the more

(1) To reduce these friction forces some manufactures use silicone oil.

the attenuation increases. Presumably this is attributable to the fact that the fibre is more deformable within the jacket when the clearance is large than when it is small, with the result that local bending takes place with a smaller radius of curvature.

Attenuation variation with temperature also increases when jacket thickness is increased. The reason for this lies with the increase in the jacket contracting force: when the jacket is thin, its contracting force exceeds the fibre bending force only at very low temperatures, whereas with a thick jacket a slight temperature difference will cause a contraction having a force sufficient to bend the fibre.

With the tight type jacketing, a more compact fibre-jacket system is obtained; however, in this case, the coating always produces a certain fibre attenuation increase, since it is very difficult to eliminate the microbending phenomenon completely.

To limit the microbending effect it is very important to carefully control the jacket cooling phase during extrusion. When the cooling conditions are not appropriate, the loss increases at low temperatures, also in tightly jacketed fibres. The reason for this is that with over-rapid cooling the jacket surface layer solidifies, while the interior of the jacket remains in a molten state.

With the subsequent solidification of the jacket interior, voids are created around the fibre, due to the density difference between the molten and the solid state of the fibre. This can cause microbendings, particularly at low temperatures, when contraction of the jacket takes place.

The centricity of the fibre in the jacket is also very important for reducing microbendings in tightly jacketed fibres.

Thermal contraction due to cooling of an eccentric extruded jacket leads to bending of the fibre, and it can be observed that the greater the eccentricity, the more the bend radius decreases. Applying the elasticity theory, the curvature C of the fibre can be calculated (considering the jacket as an elastic tube and ignoring the plastic effects) by the following expression [3]:

$$C = \frac{1}{R} = \frac{4e\varepsilon[E_2(1-H) + E_1 H]}{4e^2[(E_1 - E_2)(K-1)^2 + E_2 R_2^2 K^2/R_1^2] + (E_1 - E_2)^2 R_1^2 + E_2 R_2^4/R_1^2} \tag{1}$$

where: $H = E_2(R_2^2 - R_1^2)/[E_1 R_1^2 + E_2(R_2^2 - R_1^2)]$

$K = R_1(E_1 - E_2)/(E_2 R_2^2 + R_1^2(E_1 - E_2))]$

e = eccentricity of the fibre

ε = relative shrinkage of the jacket

E_1, E_2 = Young's modulus of the fibre and jacket respectively

R_1, R_2 = radius of the fibre and jacket respectively.

Curvature C as a function of jacket radius R is shown in Fig. 2, considering different eccentricities and using nylon-12 for the jacket.

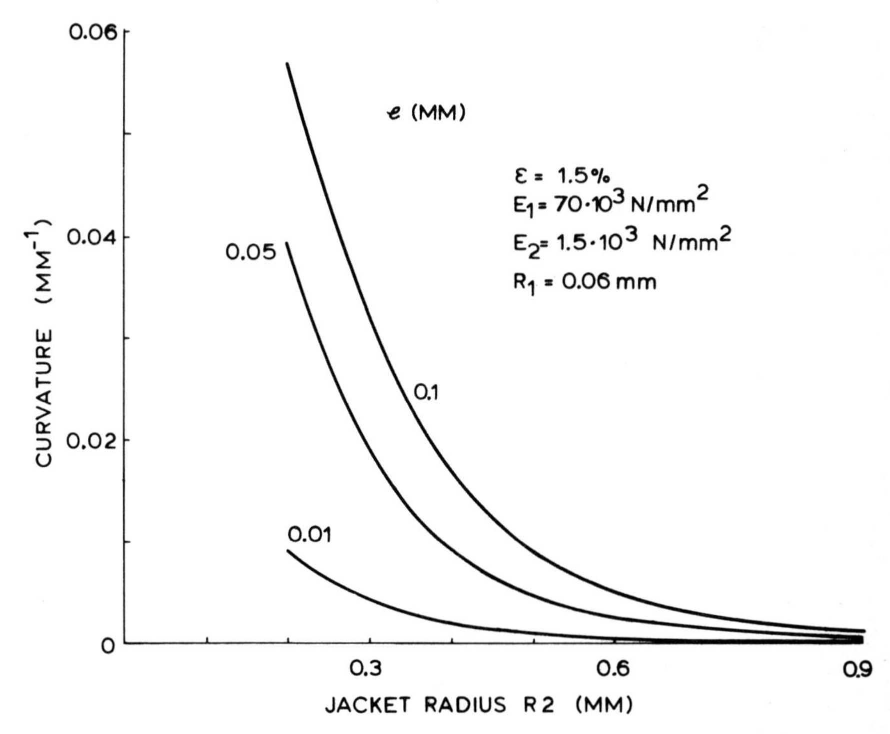

FIG. 2 – Curvature of the fibre as a function of jacket radius of eccentric jacket. (After [3]).

It can be observed in Fig. 2 that if the fibre eccentricity in the plastic jacket is large (e.g. 0.1 mm), variations in jacket thickness lead to a large variation in fibre curvature, causing an increase in fibre attenuation. This phenomenon is less evident for low eccentricity (e.g. 0.01 mm).

If the above-indicated parameters are carefully controlled, it is possible to obtain, in the case of tight type jacketing, attenuation variations vs. temperature lower than in the case of loose type jacketing: attenuation variation less than 0.2 dB/km in the temperature range −45 °C - +70 °C have been achieved.

In the case of tight type jacketing, a soft material is sometimes interposed between the fibre and the jacket in order to better protect the fibre from radial compression.

A number of high Young's modulus plastics have been used successfully for the construction of both loose and tight jacketings.

This list contains:

- High density polyethylene (HDPE);
- Polypropylene (PP);
- Nylon-12;
- Polytheylene-terephthalate (Arnite-PEPT).

All of these combine a fairly high Young's modulus for small elongation ($< 1\%$) with an adequate flexibility to permit a small bending radius. They also have high resistance to the abrasion and low friction coefficients, which are favourable to cable construction and performance.

Under completely dry conditions nylon has a higher Young's modulus (3000 N/mm^2) than the other materials, but when exposed to normally moist atmosphere the Young's modulus is reduced to about 1500 N/mm^2. The other polymers considered have a Young's modulus of the same order, ranging from about 1500 N/mm^2 for polyethylene to about 2000 N/mm^2 for arnite.

The tensile properties of extruded coating are greatly dependent upon the extrusion conditions, which influence the degree of molecular orientation. Orientation in the fibre axis direction can be used to improve the mechanical strength and Young's modulus [4, 5].

1.3 Strength member

The strength member may be defined as a cable component which is introduced to increase the tension which may be applied to the cable without risk of optical fibre break, particularly during laying, when the cable is pulled into ducts. Moreover, it can also play an important role during cable fabrication, ensuring the dimensional stability of the cable and avoiding the formation of microbendings in the fibres.

Optical fibres have a relatively low elongation per unit length to break under tension (0.01-0.02), and care must be taken to ensure that the working strain is comfortably below this value.

The tension that an optical cable can withstand for a given strain S (below 0.01-0.02) ([2]), is given as a first approximation, by the expression:

$$T = S \sum_i (E_i A_i) \tag{2}$$

where E_i, A_i are the Young's modulus and the cross-section of a cable component (fibre, strength member, sheath or filler).

The summation is extended over all the cable components.

The expression (2) is valid for a strain below the elasticity limit and shows that it is necessary for the strength member to have a high Young's modulus and a sufficient cross-section (compatible with the cable size and flexibility) to bear the main part of the tension. Therefore, the main requirements for the strength member are:

1) high Young's modulus;
2) strain at yield greater than the maximum desired cable strain;
3) low weight per unit length;
4) flexibility, to ensure good cable bending properties.

Additional features may be relevant, depending upon the cable design: friction coefficient against adjacent cable components, and transverse hardness and stability of properties over a specified range of temperatures.

High Young's modulus materials are generally stiff, but their flexibility can be improved by employing a stranded or bunched assembly of units of smaller cross-section, generally with an outer coating of extruded plastic.

([2]) The optical cable is considered as a system of parallel elements of uniform cross section and equal length.

Five main types of materials are employed for the construction of strength members:

1) steel wires;
2) plastic monofilaments;
3) multiple textile fibres;
4) glass fibres;
5) carbon fibres.

– *Steel wires*: these are widely used in conventional cables both for armouring and/or longitudinal reinforcement, due to their low cost and high ultimate tensile stress; for the same reasons they are also employed by many manufacturers of optical cables. Various grades of steels are available, with a high Young's modulus ($20 \cdot 10^4$ N/mm^2).

Although steel is one of the best materials for optical cable strength members and thus is probably at present the most widely used, it has some disadvantages, the main ones of which are: the high specific weight (this substantially increases the cable weight), the stiffness (if used in a single wire) and the need to provide an adeguate protective cushion to avoid fibre damage (in particular if stranded wires are used). Moreover, in some applications it may be desirable to have a completely dielectric optical cable.

– *Plastic monofilaments*: a special extrudable grade of polyethylene terephthalate (arnite) is often used, which is first extruded as a continuous rod with an amorphous isotropic structure. In this form its Young's modulus is relatively low but, with special procedures producing molecular orientation, it can be transformed into a form with a higher Young's modulus along the axial direction. This material is dimensionally stable at temperatures up to 200 °C, a valuable feature in the extrusion of the thermoplastic sheath. The surfaces can be smooth and perfectly cylindrical. The Young's modulus at low strain is for this material of the order of $1.6 \cdot 10^2$ N/mm^2.

This type of strength member is of particular interest where low weight or absence of metals are primary requirements for the cable, but its performance cannot compete with that of steel in applications requiring high tension strength, e.g. when long cable lengths (500 to 1000 m) must be pulled into telephone ducts. Also polyethylene or nylon could be employed as monofilaments, but they are not temperature stable, particularly during cable manufacturing.

– *Textile fibres*: in commercial forms these are normally available in assemblies of many small-diameter fibres (about 10 μm), arranged in parallel or in twisted configurations. These small fibres are highly molecularly oriented during manufacture, resulting in a high longitudinal strength and Young's modulus. Typical examples of textile materials are polyamides (nylon) or polyethylene therephatalate (terylene, dacron). These materials have an elastic modulus of $1.4 \cdot 10^4$ N/mm^2 for individual fibres. These materials are seldom used in optical cables, because they are more bulky than plastic monofilaments of equal strength. A material of this class, employed by many optical cable manufacturers, is an aromatic polyester, denominated kevlar and developed by Du Pont de Nemours.

The individual fibres have an exceptionally high Young's modulus ($13 \cdot 10^4$ N/mm^2) which, coupled with a specific weight of 1.45, makes this material competitive with steel wires.

Commercial forms of kevlar, suitable for use as optical cable strength members, consist of large numbers of filaments, assembled by stranding and/or resin bonding, which retain a high proportion of the single filament Young's modulus.

– *Glass fibres*: in some special applications, where high strength is not required, it is possible to use optical fibres themselves as strength members. If higher strength is required, non-active fibres are generally used, in a similar manner to textile fibres. The Young's modulus of glass fibres is high, typically $9 \cdot 10^4$ N/mm^2.

– *Carbon fibres*: carbon fibres, used particularly in the aircraft industry to obtain rigid compound, have a high Young's modulus ($20 \cdot 10^4$ N/mm^2) in single filaments.

The use of these fibres is at present limited to short optical cable lengths, because long lengths have yet to be manufactured. Furthermore, the price of this material is very high. The main mechanical characteristics of the above materials are summarized in Table I.

TABLE I

Properties of strength member materials (After [7]).

Material	Specific gravity	Young's modulus (N/mm²)	Yield stress (N/mm²)	Elongation at yield (%)	Tensile strength (N/mm²)	Elongation at break (%)
Steel wire	7.86	$20 \cdot 10^4$	$4\text{-}15 \cdot 10^2$	0.2-1	$5\text{-}30 \cdot 10^2$	25-2
Arnite monofil.	1.38	$0.6\text{-}1.4 \cdot 10^4$	$0.8\text{-}2 \cdot 10^2$	1.5	$5\text{-}30 \cdot 10^2$	25-10
Nylon yarn	1.14	$0.6\text{-}1.3 \cdot 10^4$	$> 8 \cdot 10^2$	> 6	$10\text{-}15 \cdot 10^2$	15-20
Terylene yarn	1.30	$0.6\text{-}1.3 \cdot 10^4$	$> 8 \cdot 10^2$	> 6	$10\text{-}15 \cdot 10^2$	15-20
Kevlar 49 fibre	1.44	$13 \cdot 10^4$	$30 \cdot 10^2$	2	—	2
Kevlar 29 fibre	1.44	$6 \cdot 10^4$	$7 \cdot 10^2$	1.2	$30 \cdot 10^2$	4
Glass fibre	2.48	$9 \cdot 10^4$	$30 \cdot 10^2$	—	$30 \cdot 10^2$	3
Carbon fibre	1.5	$10\text{-}20 \cdot 10^4$	$150\text{-}200 \cdot 10^2$	—	$15\text{-}20 \cdot 10^2$	1.5-1

The strength member may be placed at the centre or at the periphery of the cable cross-section. The main advantage of arranging it at the centre of the cable (where a steel wire is often employed) is the greater flexibility. A disadvantage consists in the compression given by the strength member on the fibres stranded around it when the cable changes direction.

The peripheral arrangement, which is normally used in textile fibres, provides protection against crushing and allows a reduction in cable dimensions.

The function of the strength member is less important when the cable is laid in a trench, rather than pulled into a duct. In this case it is very important to protect the cable against compression and shocks. For example it may be useful to employ metallic sheaths or armourings [1, 6, 7].

1.4 Other components used in optical cables

Other components used in optical cables are here briefly examined. They are generally similar or identical to those used in conventional telecommunication cables.

– *Insulated conductors*: insulated copper or aluminium conductors may be incorporated in the core assembly for the power feeding of repeaters.

The presence of conductors in the optical cable may be dangerous in the presence of interference due to voltages induced by electric power lines or due to lightning, which may damage the cable structure (e.g. sheath perforations). Moreover, the presence of conductors increases the cable weight. Sometimes, therefore, power is supplied locally or by remote feeding via a conventional cable laid beside the optical cable.

The types of insulation generally used are: polyvinylchloride (PVC), polyethylene, polypropylene, etc.

– *Fillers and cushion layers*: the cable core assembly is completed by the introduction of fillers, usually plastic, between the jacketed optical fibres and insulated conductors.

The materials for the fillers are chosen on the basis of the required hardness or softness: typical plastic materials used are PVC, polyethylene, polypropylene.

Cushion layers are used to protect the cable core against radial compression. Generally they are based on plastic material tapes wound helically or longitudinally around the cable core. The most widely used materials are low density cellulose paper and plastic synthetic paper (e.g. nylon or spun-bonded polyester).

– *Tapes around the cable core*: these have two purposes: firstly to hold the assemblies together and secondly to provide a heat barrier during the extrusion of the external sheath. Usually mylar tapes a few microns thick are employed, but cellulose paper or other material may be used.

– *External sheath*: the cable is completed by extruding a plastic sheath over the core. During extrusion the cable core is subjected to a pulling tension, while during cooling the thermal retraction of the plastic material produces a relatively strong compressive stress on all the cable elements.

Plastic materials used for external sheaths are: PVC, polyethylene, and polyurethane.

PVC possesses good mechanical characteristics and flexibility, is not inflammable, but has a high humidity permeability.

Polyethylene (3) has a humidity permeability about 100 times

(3) For the cable sheaths high density polyethylene is used. In this form a high quantity (80-95 %) of the plastic mass is in a crystalline state.

lower than PVC, good mechanical and chemical characteristics, but it is inflammable and less flexible than PVC. The coefficient of friction between polyethylene and PVC is lower than between two PVC surfaces; thus polyethylene is often preferred for laying cables in long PVC ducts.

Polyurethane can be used to achieve a very high flexibility, at the cost of a higher coefficient of friction and of reduced mechanical properties.

– *Water barrier*: although at the present time there is very little information available on the long term effect of water and humidity on the transmission and mechanical properties of optical cables, it is certain that these factors present less of a threat for optical cables than for conventional telephone cables and that probably their practical effect is minimal. However, most manufacturers and users prefer to provide optical cables with protection against water and humidity.

Some users adopt optical cable pressurization, in general with dry air, as a protection against humidity; however, this protection often does not offer sufficient guarantee because of the high pneumatic resistance of optical cables (due to their compactness). Thus special solutions have been studied for pressurization: for example, Bell in an experimental installation in Chicago have pressurized a PVC pipe containing an optical cable.

Many optical cable manufacturers employ sheaths with a water barrier included.

The most widely used type of sheath is that known under the trade names of polylam or LAP (aluminium-laminated polyethylene) which is also used in conventional buried cables.

It consists of an aluminium ribbon, a few tenths of a millimetre thick, wrapped longitudinally around the cable, with superposed edges, and situated directly under the sheath.

This barrier can be obtained in the most efficient way by employing an aluminium sheet covered on both sides by a polyethylene layer, a few tenths of a millimetre thick. During the sheath extrusion, the outer polyethylene sheet melts and is soldered to the sheath; the polyethylene of the superimposed edges is also soldered and seals the superimposed edges space.

Finally, some manufacturers use a metallic sheath (aluminium or steel) which has mainly a mechanical reinforcement function, but is also an excellent water barrier.

– *Radial reinforcement*: in some cases it may be necessary to ensure that optical cables are very well-protected against radial strains, in particular when the cables are not in ducts but directly buried. This can be obtained using a metallic sheath (generally extruded or welded aluminium or corrugated steel) or applying armouring wires or tapes around the cable core steel. An external plastic sheath is then generally applied as a protection against corrosion [1, 6, 7].

1.5 Optical fibre cabling

Optical fibre cabling is a very delicate process, because the only fibre protection in this stage are the plastic jackets, which have a limited mechanical resistance.

Throughout the world many optical cables are being manufactured using many different technologies. These cables are examined in Section 9.

This section examines only the most widely used fabrication type used in experimental optical cables. This fabrication type is directly derived from those used in conventional telecommunication cables and is based on stranding the optical fibre around a central element, usually the strength member. For this operation machinery similar to that employed in conventional cable is used.

The fibres, during cabling, are subjected to three types of stress: tension, flexure, and torsion [8].

The tensile stress (σ_t) to which the single tightly jacketed fibre is subjected, is given by [9]:

$$\sigma_t = \frac{4T}{\pi D_1^2} \cdot \frac{1}{[1 + (E_2/E_1)(D_2^2/D_1^2 - 1)]} \tag{3}$$

where: T = pulling tension

D_1 = outer fibre diameter

D_2 = outer coating diameter

E_1 = Young's modulus of the fibre

E_2 = Young's modulus of the coating material.

Since the fibre is bent around the strength member, flexural stress is induced. The maximum flexural stress (σ_f) suffered by the fibre is given by [10]:

$$\sigma_f = \frac{E_1 D_1}{2R} \tag{4}$$

where R is the bending radius, which is a function of the helic pitch of the cable.

Stranding the fibre around the strength member can also induce torsion. This stress can be considered a shearing stress (τ) to which the fibre is subjected, and can be calculated by the expression [10]:

$$\tau = D_1 G_1 \frac{\theta}{1} \tag{5}$$

where: D_1 = outer fibre diameter
G_1 = shear modulus of the fibre
$\theta/1$ = twisted angle per unit length.

The composition [4] of the three stress types yields the maximum stress [10]:

$$\sigma_{\max} = \frac{\sigma}{2} + \sqrt{\left(\frac{\sigma}{2}\right)^2 + \tau^2} \tag{6}$$

where: $\sigma = \sigma_t + \sigma_f$.

Therefore, the effect of the torsion is to increase the tensile stress to which the fibre is subjected. This effect can be reduced to a negligible amount if a cabling anti-torsional machine is used.

1.6 Mechanical tests on optical fibre cables

Many optical cable manufacturers and users have defined mechanical tests to be performed on short or long cable gauges [7, 12, 13].

The purpose of these tests is to estimate the stress to which the optical cables can be subjected without a deterioration in their

[4] The problem of superimposed shearing stress τ and tensile stress σ can be solved with the aid of Mohr's circle [11].

transmission characteristics. Consequently it is necessary to control the fibre transmission characteristics during and/or after the mechanical tests. At present these tests have not been standardized; therefore, in this section some of the more significant tests are considered.

– *Tensile test*: the purpose of this test is to determine the maximum tensile strength which can be applied to the cable without producing fibre breaks. The attenuation of the cable is often measured during the test execution. For this test it is very important to select gauge length, speed of load application, and method of gauge fixing.

Figure 3 shows schematically an apparatus for tensile strength measurement on long gauge lengths.

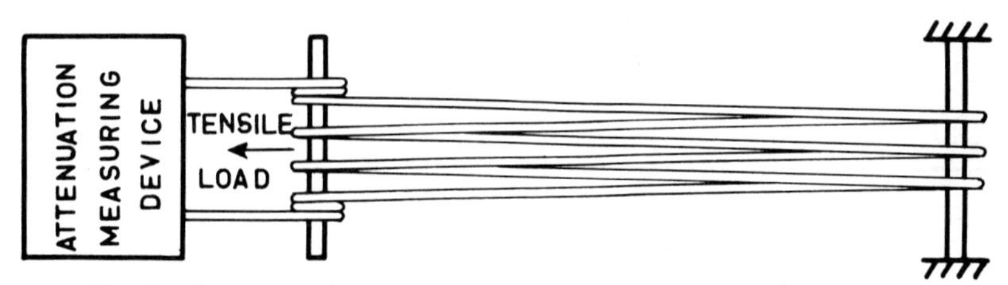

FIG. 3 – Apparatus for the tensile strength measurement. (After [12]).

– *Flexure test*: a cable section some metres long is passed through a series of rollers. Figure 4 shows a typical apparatus: the cable gauge is fixed, at one end, eccentrically to one roller and is moved backwards and forwards; a load P is applied at the other end. The roller movement can be motor-controlled.

The result of this measurement is the determination of the number of flexures reached without a fibre break with a given tensile strength and a given roller diameter. The attenuation is measured during the test execution. The test simulates a situation that is encountered during cable laying in ducts.

– *Bending test*: it is useful to determine the minimum bending radius of the cable. The cable to be tested is connected to an optical attenuation measuring device. A section of this cable is wound many times around mandrels of different diameter, under

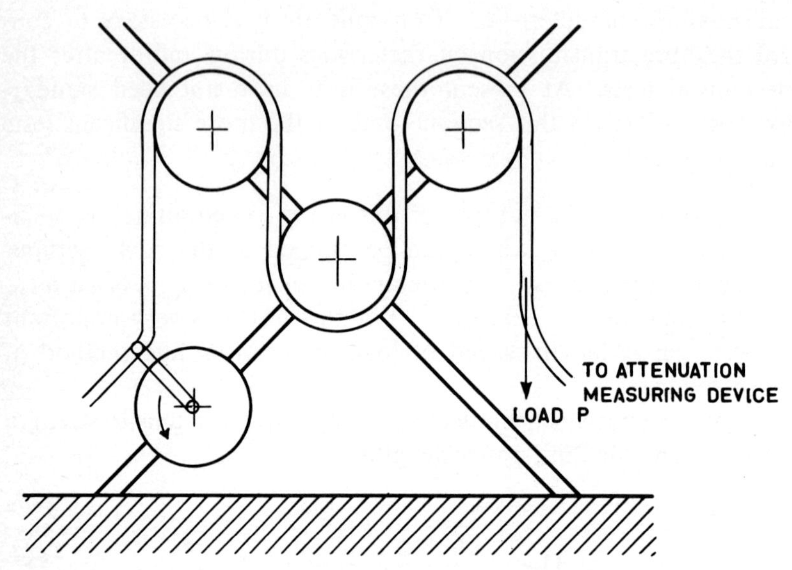

FIG. 4 – Apparatus for the flexure test.

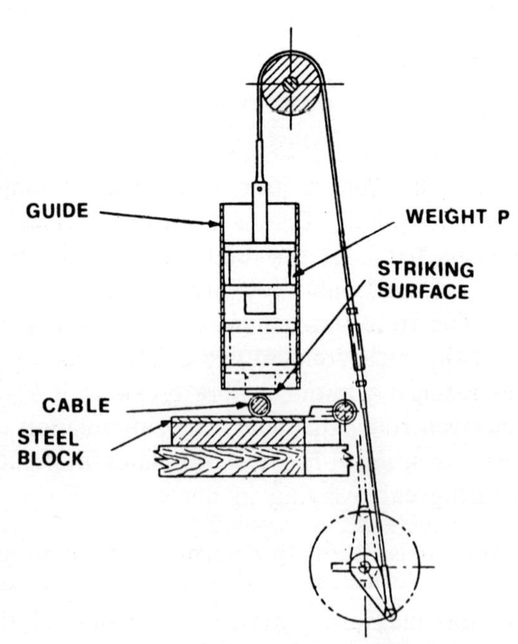

FIG. 5 – Impact test apparatus. (After [12]).

a low tensile strength. Decreasing the mandrel diameter step by step, there is a point where attenuation increases, following which the fibre starts to break. The minimum bending radius is where the attenuation starts to change.

– *Impact resistance*: a hammer of weight P and radius R falls on the cable from a certain height. The cable is supported by a stiff plane (e.g. a steel block). The scheme of this apparatus is shown in Fig. 5. The test is considered to have been passed if the specified number of impacts on the same cross-section of the cable is reached without fibre breakage.

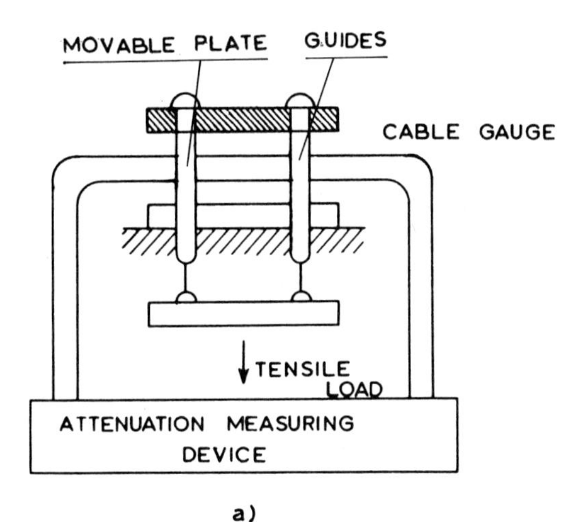

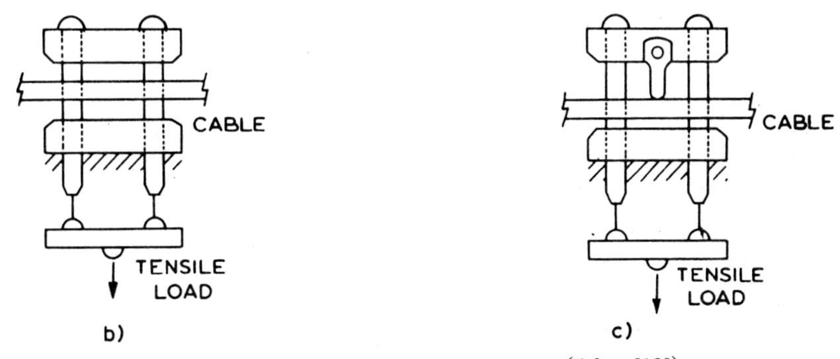

FIG. 6 – Compression test apparatus. (After [12]).

– *Compression test*: this test consists in the compression of the cable gauge between two hard surfaces. The gauge is placed in a testing apparatus according to Fig. 6*a* and loaded with different weights.

By varying the press type it is possible to distinguish two test types: crush test (Fig. 6*b*) and shear test (Fig. 6*c*).

For both tests it is necessary to measure the attenuation variation of the fibres, varying the compression strength.

– *Vibration test*: this is performed by subjecting the cable gauge to many vibration cycles. The vibration amplitude and frequency are controlled. No fibre breakage must occur.

– *Torsion test*: this test is carried out by subjecting the cable gauge to many torsion cycles. In this case it is necessary to specifiy the torsion angle and the cycle numbers. No fibre breakage must occur.

It is important to perform all these tests in various temperatures and relative humidity conditions, viz. simulating the conditions of optical cable use.

1.7 Behaviour of optical fibre cables as a function of environmental conditions

Studies are being carried out by many optical cable manufactures and users to test the behaviour of optical fibres in the presence of the natural phenomena that may be encountered during the life of an optical cable.

A very important test consists in the analysis of optical cable behaviour against temperature variations.

As stated in Section 1.2, temperature variation may cause microbendings in the optical fibres, and may limit the mechanical characteristics of the cable [2].

This test may be executed either by submitting cable gauges to temperature variations (e.g. in a temperature cabinet) and measuring the variations of attenuation, or by laying sections of experimental cable and recording over an extended period (e.g. some years) attenuation in relation to temperature variations [14-17].

Another important test is the measurement of the deterioration of optical fibre placed in a moist environment or in water.

For this reason, some manufacturers have immerged optical cable sections in water for a long period and controlled the cable transmission and mechanical characteristics before and after the test.

Other manufacturers have laid experimental sections of optical cable and monitored the transmission characteristic of the optical fibres over a long period [15, 16].

In both cases no appreciable variations in the mechanical and transmission characteristics were observed.

Tests were carried out in some laboratories to verify optical fibre resistance to damage from long-term exposure to environmental nuclear radiations. These test were performed by irradiating some types of optical fibre with gamma ray sources and neutron sources of a much higher intensity than that produced by natural radiation (5) [18-20].

A comparision of this data leads us to conclude that long-term exposure to gamma rays and neutrons in the natural environments of telecommunication cables does not pose a serious problem for optical fibres.

In addition, optical cable gauges may be subjected in the laboratory to corrosion and aging test, and the results compared to those obtained on conventional metallic cables.

1.8 Optical cable laying

At present a number of experimental optical cables have been laid in many countries, using procedures similar to those used for conventional cables; however, in these first trials more care has been taken [15, 21-24].

It is necessary to distinguish two fundamental laying methods: aerial laying and buried laying.

Generally aerial cable laying is achieved by clasping (with

(5) Gamma rays and neutrons are important constituents of the natural environmental radiation. Recent measurements of the intensity of the environmental radiation indicate values in the range from 0.1 to 0.2 rad/year (rad is a unit of absorbed dose) in normal regions.

metallic hooks) the optical cable to a steel rope (having a diameter of some millimetres), supported by poles. However, some optical cables have been manufactured having a steel rope incorporated in their structure and therefore not requiring the auxiliary metallic rope.

For buried laying it is necessary to distinguish two fundamental methods: trench laying (used mainly on trunk routes) and duct laying (used mainly in local networks).

In the first case, the optical cable is laid in a trench at a certain depth (from about 0.6 to 1 m); suitable protection such as concrete housing is provided.

Although this type of laying does not involve subjecting the cable to particular tension strengths, it may however be necessary to protect it against radial compression.

In the second method the cable is pulled into a duct. The pulling tension depends on the weight of the pulled cable, i.e. on its length and specific weight (which can be largely determined by the strength member); it also depends on the friction coefficient between the duct and the cable, and on the changes in direction or level of the ducts. Each cable can withstand a well-defined max. pulling tension and it is convenient to pull long cable sections (to minimize the number of splices); thus it is suitable to minimize the factors which determined the pulling tension. To reduce the friction, lubricants are often used; moreover, suitable equipment (e.g. pulleys) is used in the manholes where changes in duct direction or level are encountered.

1.9 Review of optical cables

This section is devoted to a review of optical cables produced by many manufacturers. This review is not intended to be exhaustive, but an effort has been made to include the main types of cables described in technical literature or presented at optical communication conferences. Although most these types of cables are in the experimental stage, many of them are currently being employed in field trials.

In the following description the various cables are grouped according to their country of origin: France, Germany (Federal

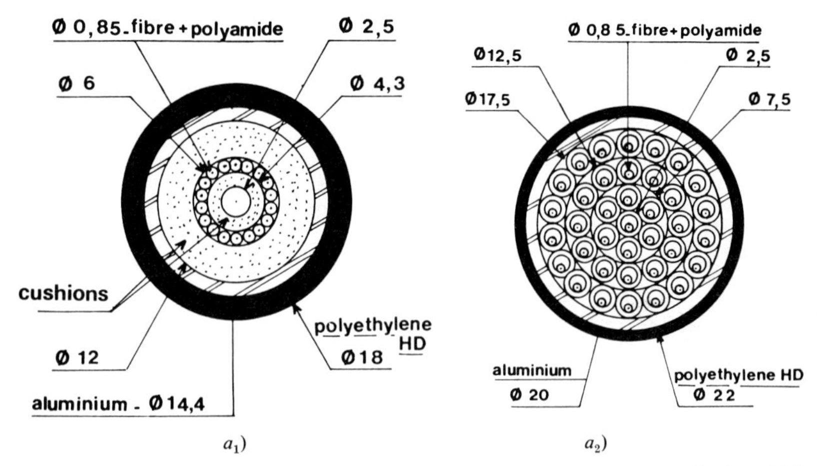

a_1) a_2)

FIG. 7 *a*) – Review of optical cables: Les Câbles de Lyon cables. (After [25]).

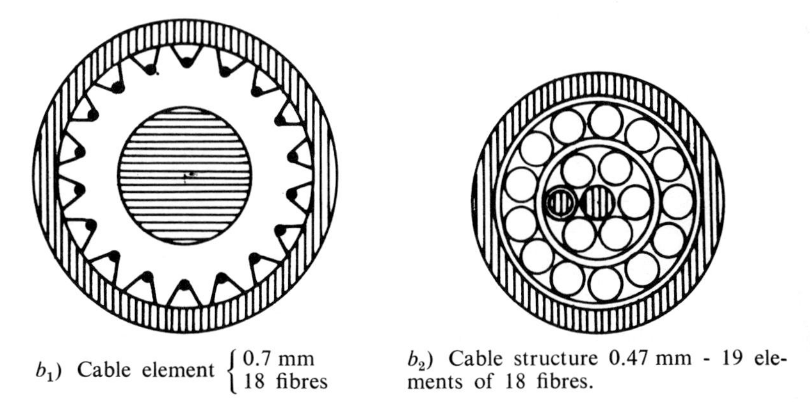

b_1) Cable element $\left\{ \begin{array}{l} 0.7 \text{ mm} \\ 18 \text{ fibres} \end{array} \right.$

b_2) Cable structure 0.47 mm - 19 elements of 18 fibres.

FIG. 7 *b*) – Review of optical cables: CNET cables. (After [26]).

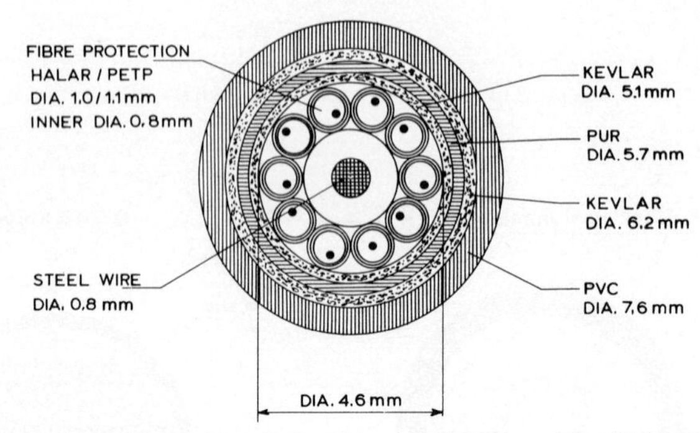

FIG. 7 *c*) – Review of optical cables: SIECOR cable. (After [13]).

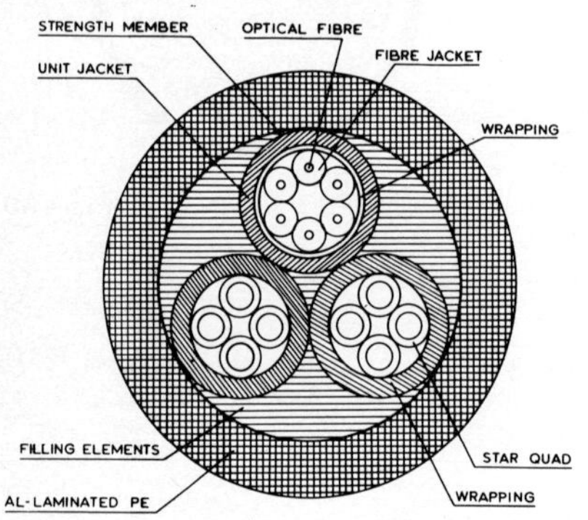

FIG. 7 *d*) – Review of optical cables: Felten & Guilleaume cable. (After [33]).

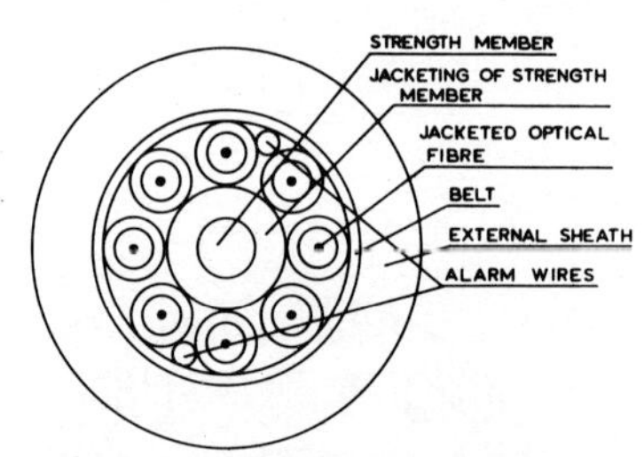

FIG. 7 *e*) – Review of optical cables: Pirelli cable. (After [23]).

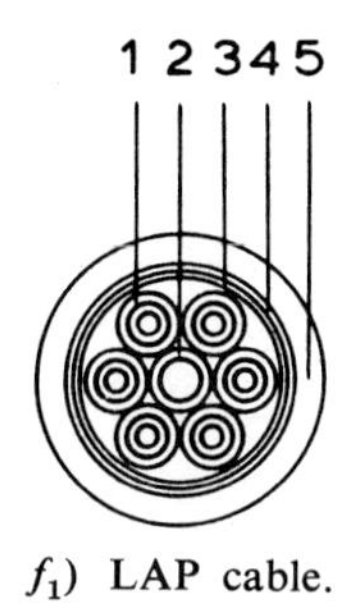

f_1) LAP cable.

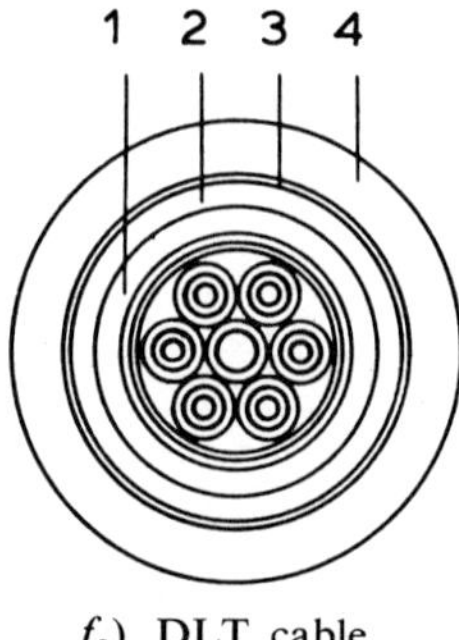

f_2) DLT cable.

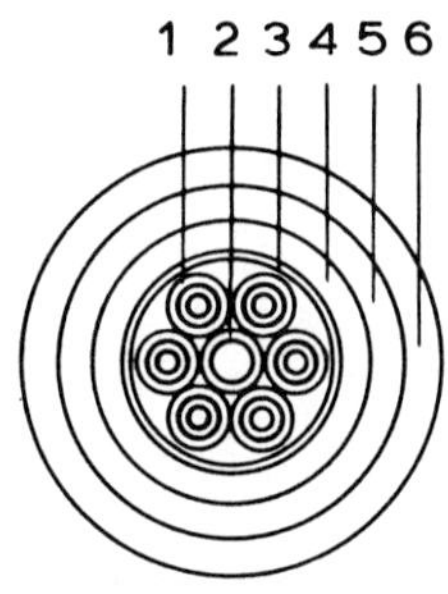

f_3) NMS cable.

Fig. 7 f) – Review of optical cables: Dainichi Nippon cables. (After [2]).

1) Jacketed fibre.
2) Dummy fibre.
3) Core wrap.
4) Aluminium tape.
5) Polyethylene.
6) Corrugated steel.
7) Bituminous compound.
8) Glass reinforced nylon compound.

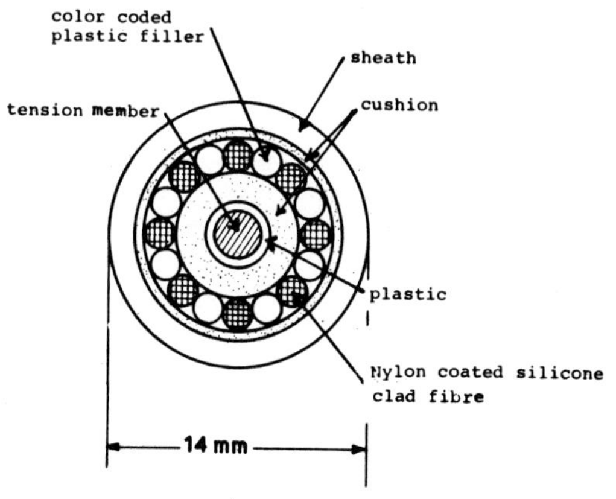

FIG. 7 g) – Review of optical cables: Fujikura cable. (After [14]).

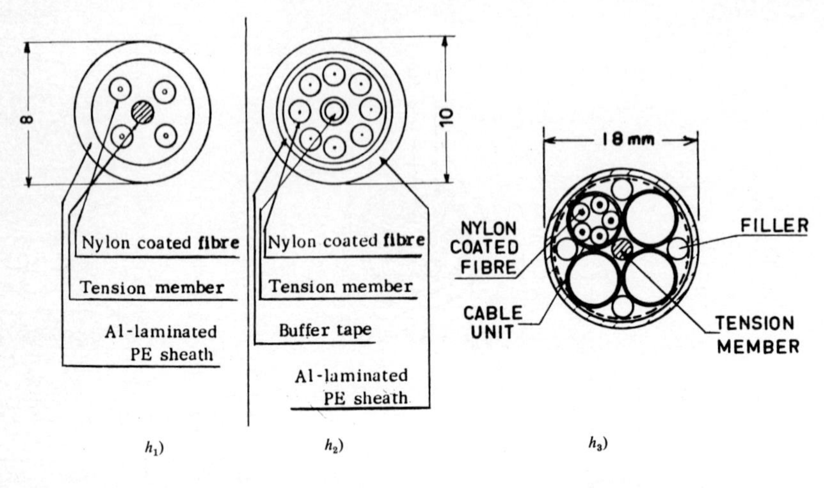
8
Nylon coated fibre
Tension member
Al-laminated PE sheath
10
Nylon coated fibre
Tension member
Buffer tape
Al-laminated PE sheath
18 mm
NYLON COATED FIBRE
CABLE UNIT
FILLER
TENSION MEMBER

h_1) h_2) h_3)

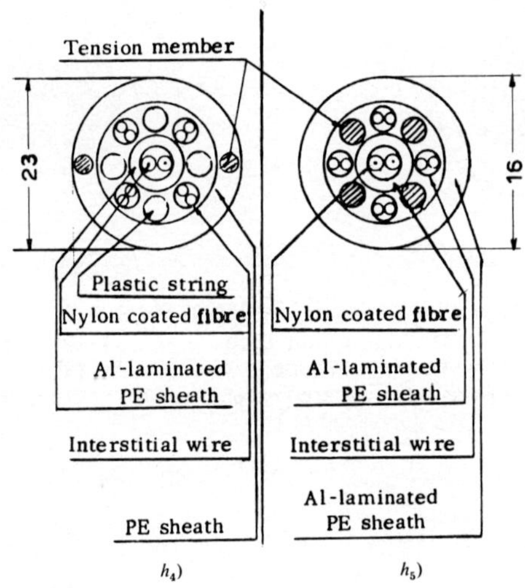
Tension member
23
Plastic string
Nylon coated fibre
Al-laminated PE sheath
Interstitial wire
PE sheath
16
Nylon coated fibre
Al-laminated PE sheath
Interstitial wire
Al-laminated PE sheath

h_4) h_5)

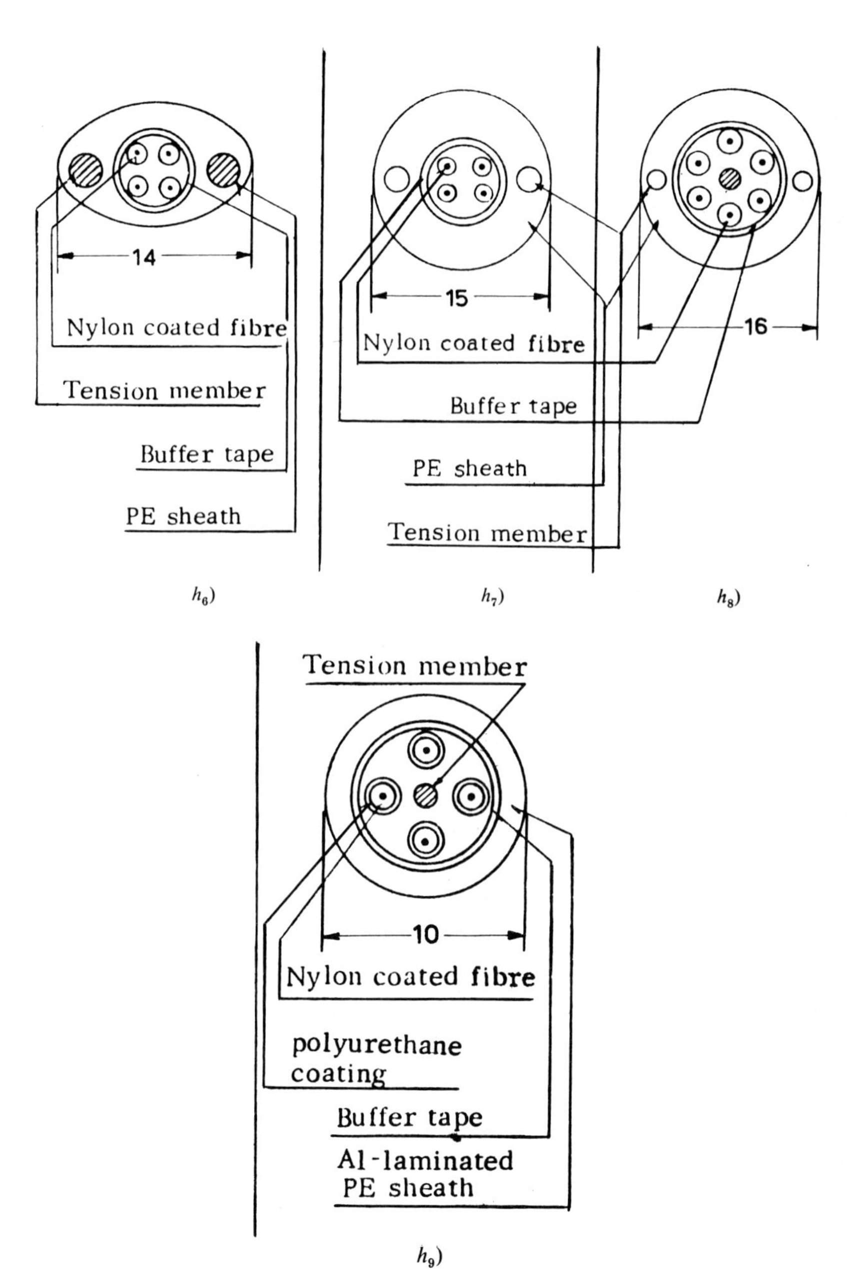

FIG. 7 *h*) – Review of optical cables: Furukawa cables. (After [34-36]).

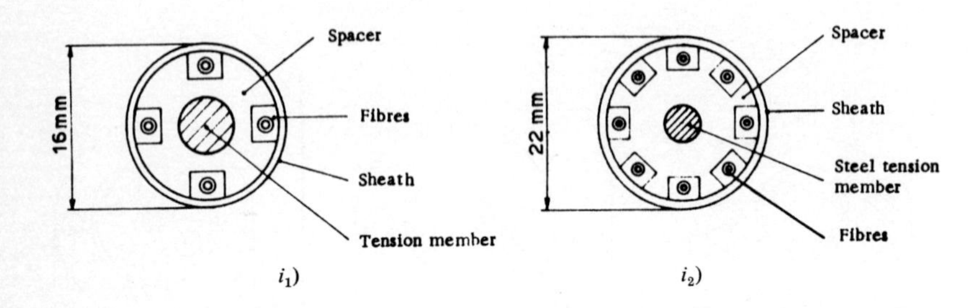

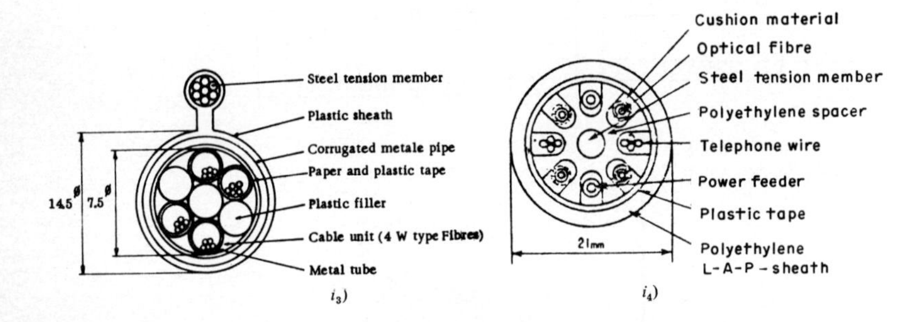

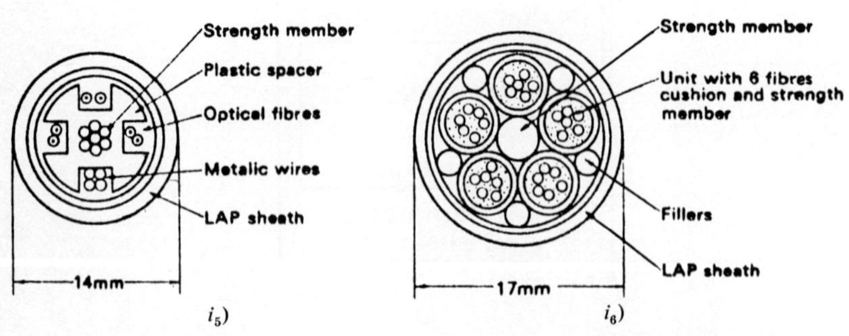

FIG. 7 i) – Review of optical cables: Hitachi cables. (After [37-39]).

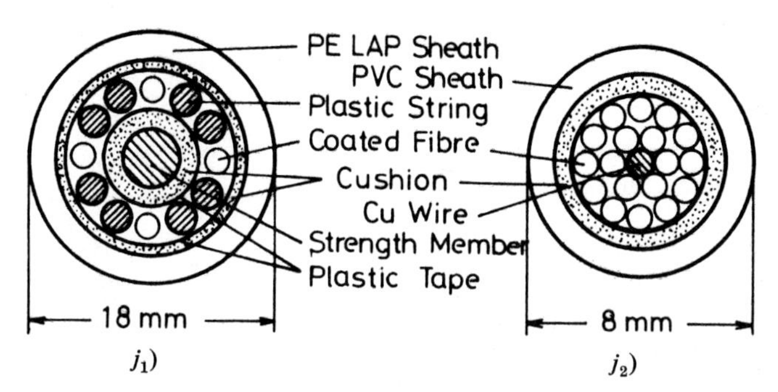

FIG. 7 *j*) – Review of optical cables: Sumitomo cables. (After [40]).

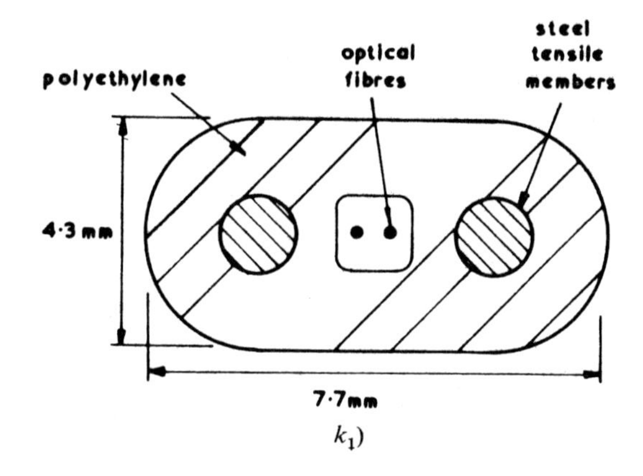

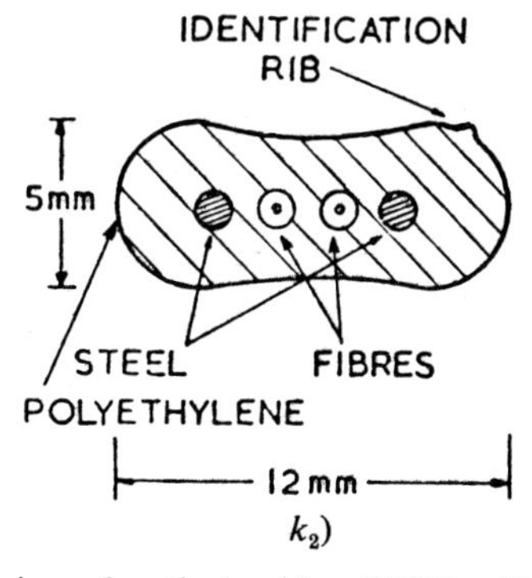

FIG. 7 *k*) – Review of optical cables: BICC cables. (After [42], [43]).

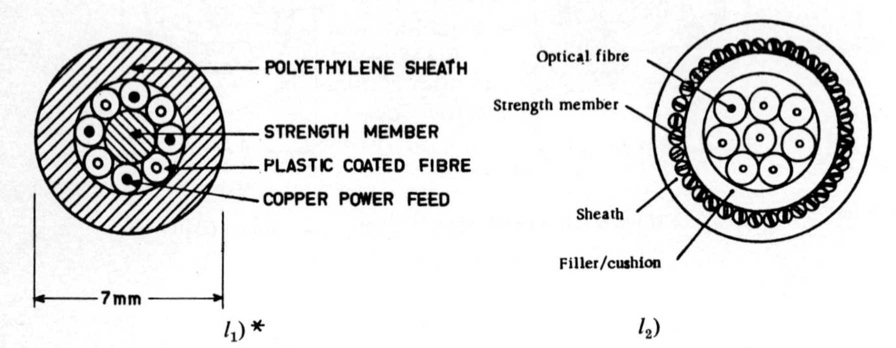

FIG. 7 *l*) – Review of optical cables: STL cables. (After [7], [45]).

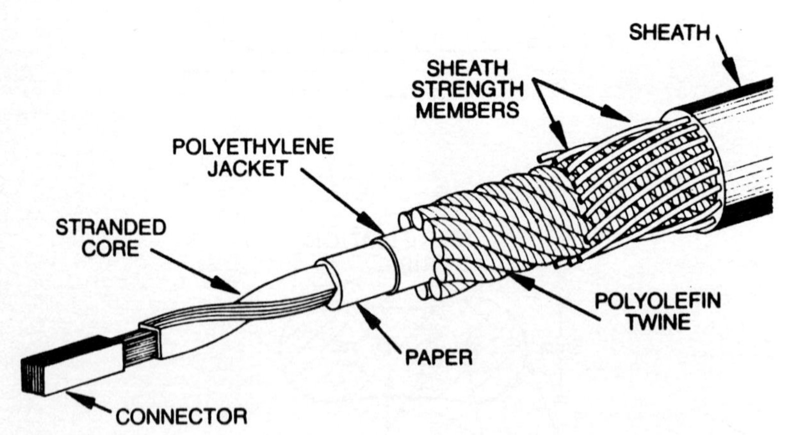

FIG. 7 *m*) – Review of optical cables: Bell cable. (After [51]).

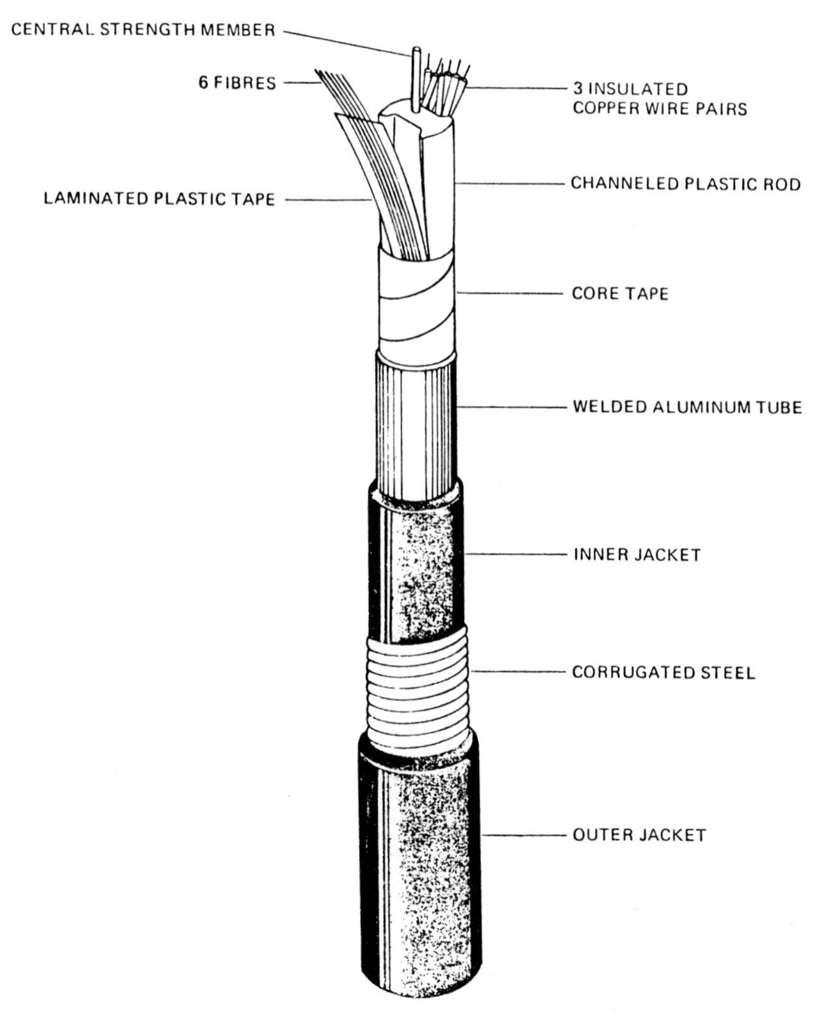

FIG. 7 *n*) – Review of optical cables: GTE cable. (After [52]).

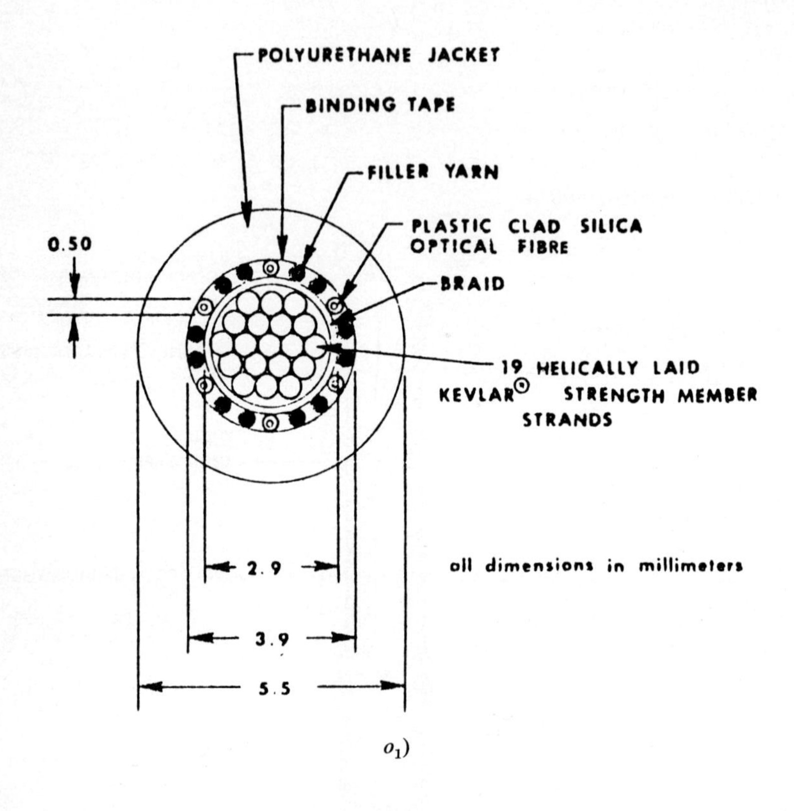

o_1)

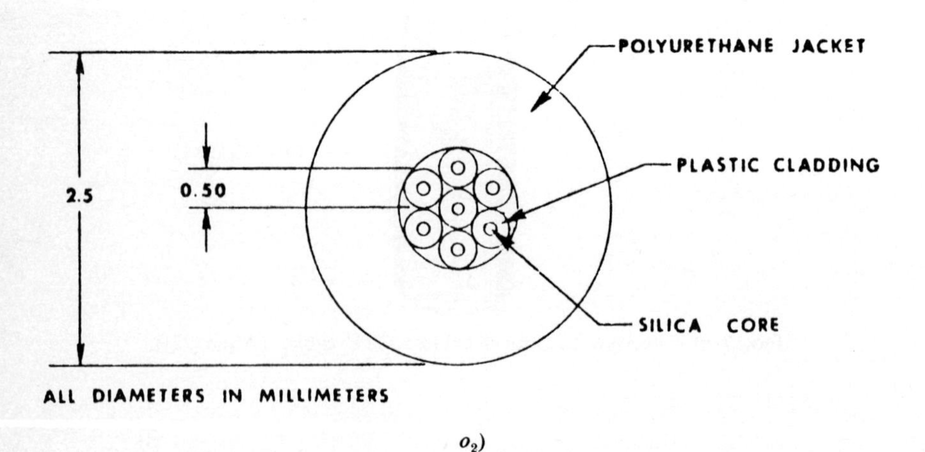

o_2)

FIG. 7 *o*) – Review of optical cables: ITT cables (After [54]).

Republic), Italy, Japan, United Kingdom and United States of America.

All the cables examined are shown in Fig. 7, where they are represented by a letter (the same for all cables produced by a manufacturer) and a number (to identify different cables of the same manufacturer).

France

In France optical cables have been developed by Les Câbles de Lyon and by CNET (Centre National d'Etudes de Telecommunications).

Les Câbles de Lyon have developed two types of optical cable structures.

In the first type 6, 12 or 18 fibres, tightly jacketed with polyamide 6.10 (outer diameter 0.85 mm) are assembled in a layer around a central strength member (aromatic polyamide); this layer is inserted between two cushions, based on helical tapes of high compressibility material (e.g. foamed polyurethane). The outer protection can be given by a polyethylene sheath extruded on an aluminium tape or by a plastic covered aluminium tube. a_1 is an example of a 18 fibre cable of this type.

In the second type of structure, the tightly jacketed fibres are provided with a second protection (a plastic loose tube) and are stranded in layers to obtain an element of 18-19 fibres or of 37 fibres (a_2); the cable can consist of one or more such elements. In this case, the sheath, for example a plastic covered aluminium tube, is the strength member [35].

Field trials are being carried out on such types of cables.

CNET has developed an unconventional type of cable, having a modular structure. The cable is obtained by the arrangement of elements of the type shown in b_1: each element is a cylindrical core presenting 18 helicoidal V-grooves, each containing a jacketed fibre. Many elements can be assembled to obtain a cable; b_2 is a possible structure for a cable containing 19 elements, i.e. 342 fibres.

This cable, currently under development, is being studied to permit an easy and quick jointing of the 18 fibres of each element, using special connection devices that are being studied by CNET [26-28].

Germany (*Federal Republic*)

Optical cables have been manufactured in Germany by SIECOR (a joint venture by Siemens AG and Corning Glass Works) and by Felten & Guilleaume.

In the SIECOR cables the fibres are protected by a loose plastic jacket and are stranded around a central strength member (plastic jacketed steel wire). A 4-layer sheath is then provided: a first layer of kevlar fibres, a second of polyurethane, another of kevlar fibres and an external sheath of PVC. Cables with 4, 6, 8 or 10 fibres are available; Fig. 7*c*) is an example of a 10-fibre cable.

SIECOR has also developed cables with only 1 or 2 fibres; in this case the central strength member is not provided: the fibres are in the centre of the cable, protected by the same 4-layer sheath described above.

Some SIECOR cables are undergoing field trials; in particular a 2.1 km cable (10 fibres) was installed in the summer of 1976 [13, 29-32].

The Felten & Guilleaume cables consist of a basic plastic sheathed unit, containing 6 loosely jacketed fibres (outer diameter about 1.3 mm) stranded around a kevlar-49 strength member. An optical cable can include several basic units, some of which contain conventional cable components, such as symmetrical or coaxial pairs instead of optical fibres. Figure 7*d*) is an example of a cable developed for a field trial in Berlin: three units are provided, one containing 6 fibres and two containing a copper quad and sheathed with Al-laminated polyethylene [33].

Italy

In Italy optical cables have been manufactured by Industrie Pirelli. They have a structure derived from conventional telecommunication cables: the fibres, jacketed with a plastic loose tube (outer diameter about 2 mm) are stranded around a central strength member (plastic jacketed steel wire). A plastic sheath is then applied; to provide a moisture barrier an inner Al-laminated sheath can be employed. At present, cables have been developed with a capacity of 8, 12 or 18 fibres. Figure 7*e*) is an example of an 8-fibre cable employed in a field trial (COS 2 experiment) in which about

5 km of optical cable have been installed in ducts (september 1977) between two telephone exchanges in Turin. This cable has only a polyethylene sheath; thus two bare copper wires have been inserted to obtain a water detection circuit. A similar cable, 1 km long, with a capacity of 12-fibres has been installed (March '76) in concrete boxes on a private site [15, 16, 23, 24].

Japan

In Japan optical cables have been developed by many manufacturers.

Dainichi has manufactured three types of cables (f_1, f_2, f_3), each containing six fibres cabled around a glass fibre strength member. f_1 has a LAP (Al-laminated polyethylene) sheath, f_2 a corrugated steel sheath, f_3 a plastic sheath. Various loose and tight fibre jacketings have been experimented, leading to the choice of a nylon-12 tight jacketing [2].

Fujikura has developed an optical cable (Fig. 7*g*) containing 6 fibres inserted between plastic fillers and cabled around a central strength member. The fibres are protected by a tight nylon jacket [14].

Furukawa has designed many optical cables with different characteristics and for different uses. In all these cables the fibres are tightly nylon jacketed.

h_1 is a 4-fibre cable with a FRP (fibre reinforced plastic rod) strength member. h_2 is a similar cable containing 8 fibres. The cable h_3 consists of 4 units, each containing 5 fibres, assembled with fillers; the strength member is central. h_4 and h_5 are 2-fibre cables: the fibres are in the central position and are surrounded by fillers and copper pairs; in h_4 the strength members are inserted in the external sheath, in h_5 they substitute the fillers. h_6 is a 4-fibre cable with strength members in the external sheath. h_7 is an all-dielectric cable used for data transmission between electric power stations. It contains 4 fibres; the non-metallic strength members are in the external sheath. h_8 is a similar cable containing 6 fibres. In the cable h_9 the fibre have a second polyurethane jacket on the nylon tight jacket.

h_1, h_2, h_3, h_5 and h_9 have a LAP sheath; h_4, h_6, h_7 and h_8 have a polyethylene sheath.

All these types of Furukawa cables have been employed in experimental installations, in ducts, in tunnels or aerial [9, 34-36].

Hitachi has developed 4 optical cable types (i_1, i_2, i_4, i_5) in which 4, 6 or 8 jacketed optical fibres are inserted in grooves obtained on a plastic central element; the strength member can be metallic (one or more steel wires) or non-metallic. In i_4 and i_5 metallic wires are also inserted.

Another type of cable (i_3) has been developed for aerial installation, in which several jacketed fibres are inserted in corrugated metallic tubes; the strength member (steel wires) is external. Finally i_6 is a unit type multicore cable. In each unit 6 or 10 fibres are inserted. The fibres are protected by strength members, cushion materials and a sheath. Several units are stranded with a strength member and fillers.

The sheaths of all these cables are in LAP. Some sections have been experimentally installed [37-39].

Sumitomo has developed two types of cable: one (j_1) for buried installation and another (j_2) for indoor installation. In both these cables the fibres are stranded around a central strength member (FRP for j_1, copper wire for j_2). j_1 has a LAP sheath, j_2 a PVC sheath. Some short lengths of these cables have been experimentally installed [40, 41].

United Kingdom

In the United Kingdom optical cables have been manufactured by BICC and STL (Standard Telecommunication Laboratories).

BICC has developed an original type of cable (PSP cable) based on a polyethylene element with 1 (k_1) or 2 (k_2) cavities in which one or more fibres can be installed. Two steel wires inserted in the polyethylene element have the function of strength members. This cable has been used in experimental installations; in particular 21 km have been laid in ducts on a 13 km ruote between Martlesham and Ipswich [42-44].

STL have developed a cable (l_1) containing 4 fibres with a tight polypropylene jacket (outer diameter 1 mm) and 4 copper wires with polypropylene insulation (outer diameter 1 mm) stranded around a central strength member (a cord of brass-plated steel wires, coated with a smooth layer of polypropylene). A cable of this type, with an outer sheath of black polyethylene has been installed in ducts on a 9 km route between Hitchin and Stevenage. Other types of plastic sheath have also been experimented. Another version of this cable contains 8 fibres and no copper wires.

STL have also developed another type of cable (l_2) consisting of 8 jacketed fibres inserted in a plastic tube and surrounded by a strength member (helically arranged steel wires); the external sheath is in polyethylene [7, 45-47].

United States of America

Many manufacturers in the USA have produced optical cables.

Bell Laboratories have developed a cable (Fig. 7*m*) having a non-conventional structure, based on a 1/8 inch wide plastic ribbon containing 12 fibres.

Several ribbons are superposed and twisted (to obtain improved bending properties) to form the cable core. A paper tape is wrapped around this core and a polyethylene jacket is then loosely extruded on it. The cable is completed by a polyolefin twine (for mechanical and thermal protection), strength members and an outer sheath.

This type of cable has been specially studied to facilitate fibre jointing, for which a special connector has been developed by Bell Laboratories. Moreover, the cable structure is such that very high capacity and very low dimensions can be obtained: with an outer diameter of 12 mm it can contain up to 12 ribbons (144 fibres). This type of cable has been used in field experiments in Atlanta (658 m of cable installed in ducts on the Bell Laboratories site) and in Chicago (about 2.6 km of cable installed in ducts). The Atlanta cable contains 12 ribbons, the Chicago cable only 2 ribbons [48-51].

Corning Glass Works, in addition to the activity in manufacturing of optical fibres, has developed a few types of optical cables in the past years. At present the activities of Corning in the field of optical cables are carried out by SIECOR, whose production is briefly illustrated above in the section dedicated to cables manufactured in Federal Republic of Germany.

An original cable has been studied by GTE and manufactured by General Cable Corporation. The cable of Fig. 7*n*) is based on an extruded plastic core having on its surface two opposite helical grooves and incorporating a central strength member. In one of these grooves 6 fibres, encapsulated in a flat plastic tape, are inserted, in the other three copper pairs. The cable has an inner sheath (a welded aluminium tube covered by black polyethylene) and an outer sheath (a corrugated steel tube covered by black polyethylene). This last sheath serves as the main strength member. This cable has been installed between Long Beach and Artesia (California) on a 9 km route [52, 53].

The cables of ITT are of two main types. The first one (o_1) contains six fibres stranded around a central support and encapsulated in an extruded polyurethane jacket. This cable is used where the mechanical performance required is not very severe. In the second type (o_2) optical fibres and filler yarns are stranded around a central strength member (kevlar cord). The external sheath is in polyurethane [54].

REFERENCES

[1] S. G. Foord, W. E. Simpson and A. Cook: *Some design principles for fibre optical cables*, 23rd International Wire and Cable Symposium, Atlantic City (December 3-5, 1974), pp. 276-280.

[2] M. Rokunoe, T. Shintani, M. Yajima and A. Utsumi: *Stability of transmission properties of optical fiber cables*, Second European Conference on Optical Fibre Communication, Paris (September 27-30, 1976), pp. 183-188.

[3] B. Hillerich, D. S. Parmar and P. Schlang: *Criteria for the jacketing of optical fibres*, Third European Conference on Optical Communication, Munich (September 14-16, 1977), pp. 86-89.

[4] D. Gloge: *Optical fiber packaging and its influence on fiber straightness and loss*, Bell Syst. Tech. J., Vol. 54, pp. 243-260.

[5] W. B. Gardner: *Microbending loss in optical fibres*, Bell. Syst. Tech. J., Vol. 54, pp. 457-465.

[6] R. E. J. Baskett and S. G. Foord, *Fiber optics cables*, Electrical Communication, Vol. 52, No. 1 (1977), pp. 49-53.

[7] S. G. Foord and M. A. Lees: *Principles of fibre-optical cable design*, Proc., IEEE Vol. 123, No. 6 (June 1976), pp. 597-602.

[8] E. Occhini: *Mechanical properties of optical fibres for cables*, Third European Conference on Optical Communication, Munich (September 14-16, 1977), pp. 77-82.

[9] H. Murata: *Broadband optical fiber cable and connecting*, Second European Conference on Optical Fibre Communication, Paris (September 27-30, 1976), pp. 167-174.

[10] H. Liertz and U. Oestreich: *Application of Weibull distribution to mechanical reliability of optical waveguide for cables*, Siemens Forsch Bd., Vol. 5, No. 3 (1976), pp. 129-135.

[11] W. A. Nash: *Strength of Materials*, 2nd ed., Schaum's Outline Series, McGraw Hill (1972), Chap. 17.

[12] Military Standard Fiber Optics Test Methods and Instrumentation Provisional Specifications MIL-STD-1678.

[13] U. Oestreich, G. Boscher, G. Shöber and H. Liertz: *Optical waveguide cables*, Third European Conference on Optical Communication, Munich (September 14-16, 1977), pp. 44-46.

[14] S. Tanaka, T. Naruse, H. Osakai, K. Inada and T. Akimoto: *Properties of cabled low-loss silicone clad optical fiber*, Second European Conference on Optical Fibre Communication, Paris (September 27-30, 1976), pp. 189-192.

[15] B. Catania, L. Michetti, F. Tosco, E. Occhini and L. Silvestri: *First Italian experiment with buried optical cable*, Second European Conference on Optical Fibre Communication, Paris (September 27-30, 1976), pp. 315-322.

[16] L. Michetti and F. Tosco: *One year results on a buried experimental optical cable*, Third European Conference on Optical Communication, Munich (September 14-16, 1977), pp. 56-58.

[17] P. W. Black, A. Cook, A. R. Gilbert, M. M. Ramsay and J. R. Stern: *The manufacture, testing and installation of rugged fibre optic cables*, Third European Conference on Optical Communication, Munich (September 14-16, 1977), pp. 50-52.

[18] J. Shah: *Effects of environmental nuclear radiation on optical, fiber*, Bell Syst. Tech. J., Vol. 54, No. 5 (September 1975), pp. 1207-1213.

[19] E. J. Friebele and G. H. Sigel Jr.: *In situ measurements of growth and decay of radiation damage in fibre optic waveguide*, Optical Fiber Transmission II, Williamsburg (February 22-24, 1977).

[20] E. J. Friebele, G. H. Sigel and M. E. Gingerich: *Enhanced low dose radiation sensitivity of fused silica and high silica core fiber optic waveguide*, Third European Conference on Optical Communication, Munich (September 14-16, 1977), pp. 72-74.

[21] A. L. Hale and M. R. Santana: *Interfering duct-run geometry from cable-tension data: a case history*, 25th International Wire and Cable Symposium, Cherry Hill, New Jersey (1976), pp. 152-157.

[22] A. R. Meier: *Real-world of Bell fiber optics system begin test*, Telephony (April 1977), pp. 35-39.

[23] G. Cocito, B. Costa, S. Longoni, L. Michetti, L. Silvestri, D. Tibone and F. Tosco: *COS 2 experiment in Turin: field test on optical cable in ducts*, IEEE Trans. Commun., Vol. COM-26, No. 7 (July 1978), pp. 1028-1036.

[24] L. Silvestri and C. Marchesi: *COS 2 - Cavo ottico sperimentale. Posa e giunzione in tubazione nella città di Torino*, SIRTI, Monography (1977).

[25] R. Jocteur: *Optical fiber cable for digital transmission systems*, Second European Conference on Optical Fibre Communication, Paris (September 27-30, 1976), pp. 193-199.

[26] G. Le Noane: *Optical fibre cable and splicing techniques*, Second European Conference on Optical Fibre Communication, Paris (September 27-30, 1976), pp. 247-252.

[27] M. Treheux and R. Bouillie: *Influence of baseband frequency responses of optical components on the transmission system design*, Optical Fiber Transmission II, Williamsburg (February 22-24, 1977). Paper Th A1.

[28] R. Bouillie: *Application of optical fibres to existing communication system*, Third European Conference on Optical Communication, Munich (September 14-16, 1977), pp. 231-236.

[29] *Optisches mehrfaserkabel, dämpfungsarm und robust*, NTZ, Vol. 29, No. 3 (1976), p. 232.

[30] Siecor Data Sheet: *Optical waveguide cables. Series BE.*

[31] P. Bark, G. Boscher, J. Gier, H. Goldmann and G. Zeidler: *Installation of an experimental optical cable link and experiences obtained with the transmission of TV and telephone signals*, Third European Conference on Optical Communication, Munich (September 14-16, 1977), pp. 243-245.

[32] H. Pascher: *Status of fiber transmission technique in Germany*, Optical Fiber Transmission II, Williamsburg (February 22-24, 1977).

[33] W. Weidhaas: *Optical fibre cables and accessories*, Third European Conference on Optical Communication, Munich (September 14-16, 1977), pp. 47-49.

[34] H. Murata, S. Inao and Y. Matsuda: *Step index type optical fiber cable*, First European Conference on Optical Fibre Communication, London (September 16-18, 1975), pp. 70-72.

[35] T. Sekizawa, F. Aoki, M. Nishida and H. Murata: *Optical systems, for electric power companies*, Optical Fiber Transmission II, Williamsburg (February 22-24, 1977). Paper Th B3.

[36] H. Murata, T. Nakahara and S. Tanaka: *Recent development of optical fiber cable for public communication*, International Conference on Integrated Optics and Optical Fiber Communication, Tokyo (July 18-20, 1977), pp. 281-284.

[37] T. Mizukami, T. Hatta, S. Fukuda, K. Mikoshiba and Y. Shimohori: *Spectral loss performance of optical fiber cables using plastic spacer and metal tube*, First European Conference on Optical Fibre Communication, London (September 16-18, 1975), pp. 191-193.

[38] F. Aoki, K. Ando, M. Nishida, K. Fukatsu, K. Mikoshiba, Y. Shimohori, S. Okada and K. Takeda: *Performance of optical, fiber cables using plastic spacer*, Second European Conference on Optical Fibre Communication, Paris (September 27-30, 1976), pp. 307-309.

[39] Hitachi, Monography - 1977 International Conference on Integrated Optics and Optical Fiber Communication, Osaka (July 22, 1977).

[40] T. Nakahara, M. Hoshikawa, S. Suzuki, S. Shiraishi, S. Kurosaki and G. Tanaka: *Design and performance of optical fiber cables*, First European Conference on Optical Fibre Communication, London (September 16-18, 1975), pp. 81-83.

[41] N. Uchida, M. Hoshikawa, M. Yoshida, Y. Hattori and S. Yoneji: *Transmission characteristics of step index fiber and optical fiber cable*, Optical Fiber Transmission II, Williamsburg (February 22-24, 1977). Paper Tu E6.

[42] R. J. Slaughter, A. H. Kent and T. R. Callan: *A duct installation of 2-fibre optical cable*, First European Conference on Optical Fibre Communication, London (September 16-18, 1975), pp. 84-86.

[43] D. Chadwick, K. W. Plessner and J. E. Taylor: *Production experience with PSP optical fibre cable for the Martlesham/Ipswich link*, Optical Fiber Transmission II, Williamsburg (February 22-24, 1977). Paper PD4.

[44] D. Brace and K. Cameron: BPO 8448 kbit/s *optical cable feasibility trial*, Third European Conference on Optical Communication, Munich (September 14-16, 1977), pp. 237-239.

[45] A. Cook, S. G. Foord, M. M. Ramsay and A. R. Gilbert: *The manufacture of hybrid multiple strand optical fibre cables*, International Conference on Integrated Optics and Optical Fiber Communication, Tokyo (July 18-20, 1977), pp. 297-299.

[46] K. C. Byron, G. J. Cannel, I. S. Few and R. Worthington: *Field measurement of fibre optic cables*, Third European Conference on Optical Communication, Munich (September 14-16, 1977), pp. 34-36.

[47] D. R. Hill, A. Jessop and P. J. Howard: *A* 140 Mbit/s *field demonstration system*, Third European Conference on Optical Communication, Munich (September 14-16, 1977), pp. 240-242.

[48] M. I. Schwartz, R. A. Kempf and W. B. Gardner: *Design and characterization of an exploratory fiber optic cable*, Second European Conference on Optical Fibre Communication, Paris (September 27-30, 1976), pp. 311-314.

[49] M. I. Schwartz, W. A. Reenstra and J. H. Mullins: *The Chicago lightwave communication project*, International Conference on Integrated Optics and Optical Fiber Communication, Tokyo (July 18-20, 1977), pp. 53-56.

[50] A. R. Meier: *Real-world aspects of Bell fiber optics system begin test*, Telephony (April 1977), pp. 35-39.

[51] I. Jacobs: *Lightwave communications passe its first test*, Bell Lab. Record (December 1976), pp. 291-296.

[52] *GTE first to carry public's calls*, Electronics (May 1977), pp. 34-36.

[53] *Technical details on GTE optical fiber communication systems*, Monography GTE (April 1977).

[54] ITT - Electro-Optical Products Division Data Sheet, (1976).

CHAPTER 2

CONNECTING AND SPLICING TECHNIQUES

by Feliciano Esposto and Emilio Vezzoni

2.1 Introduction

Two jointing techniques are adopted for optical cables or fibres for two different applications. A permanent splice is needed, for instance, whenever single fibres are joined to form a link. The other application concerns the case in which two optical fibres or groups of optical fibres (also sources and detectors) have to be connected or disconnected. This problem typically arises at the interface of an optical cable with a receiver or a transmitter.

The essential objective in making a connection is to join two optical fibres with minimum insertion losses.

Insertion losses may be divided into two groups: *a*) intrinsic losses; *b*) extrinsic losses.

The former are due to the characteristics of both optical fibres at the joining point, while the latter originate from inaccuracies in the execution of a joint. The most significant parameters contributing to intrinsic losses in a fibre-to-fibre connection include the difference between core diameters, the difference between the maximum value of the refractive index (that is, of the numerical, aperture) of the two fibres, the difference between the α profile and refractive index of the two fibres.

An additional source of intrinsic loss is due to signal distortion produced by the joint itself, the extent of which is a function of the core-cladding dimensions and the refractive index of the optical fibres. All these factors are fundamental in a link established by connecting a number of optical fibres, consisting of non-screened commercial fibres. Therefore in defining a joint an assessment must be made of the extent of the above losses with respect to the total loss.

Extrinsic losses are those arising from: *a*) displacement, separation and misalignment between the ends of the coupled optical fibres (reciprocal positioning errors of the two ends); *b*) quality of the two surfaces with respect to an ideal condition, with reference to smoothness, flatness, orthogonality to their longitudinal axes (face tilt), roughness, concavity and convexity.

The assessment of losses described above has been made by means of various theoretical (see [1-5]) as well as experimental models. In the case of theoretical models, a number of approaches have been adopted, which differ according to the type of optical fibres used (i.e. step, or graded index profile) and working condition, more or less approaching the steady-state at different levels of approximation. Furthermore, several experimental results have been reported, which are not, however, always comparable due to different experimental conditions, such as different lengths of optical fibre, different sources, use of a mode stripper, etc.

To keep the above losses to a minimum, researchers have concentrated on the construction of uniform fibres, the preparation of the ends to be joined and the execution of joints which allow the highest possible reduction in geometric errors. See Part I, Chapter 3 for the first two topics. Joint execution will be discussed in the following.

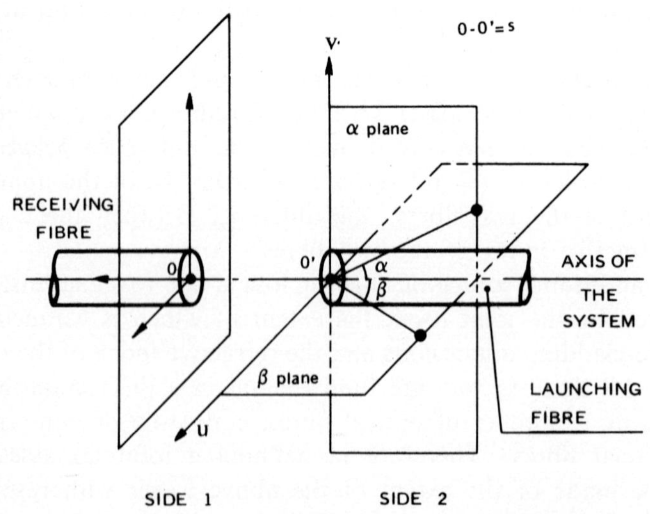

FIG. 1 – Degrees of freedom of the two coupling fibres.

To correctly assess the effects of geometric errors in the reciprocal positioning of two fibres, several researchers have carried out measurements of the coupled power as a function of the above errors. A method has been adopted which singles out the influence of extrinsic losses from intrinsic losses. This has been achieved by coupling two ends obtained from an optical fibre through one single cut. In so doing, any rotation of the two ends around the axis of the coupled optical fibres must be carefully prevented. The result of these precautions is the elimination of intrinsic loss effects. Losses assessed in this way consist in turn of reflection losses and losses due to fibre ends quality and positioning. The latter losses may be singled out from previous losses through a device performing rotational α and linear displacements u, v, s as schematized in Fig. 1.

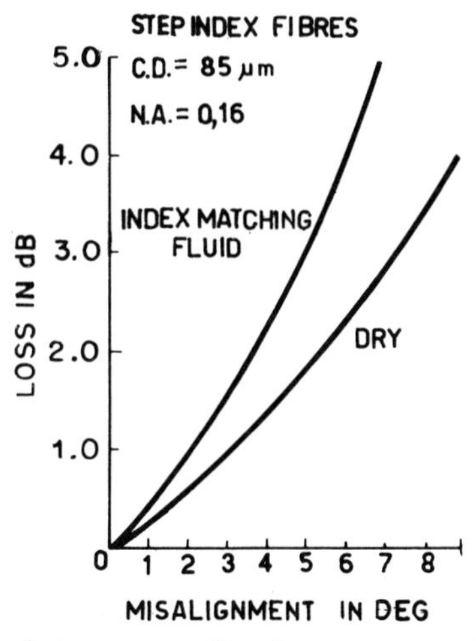

FIG. 2 – Comparison between coupling losses vs. angular misalignment, with and without index matching fluid. (After [6]).

We define the positioning loss as

$$P = -10 \log \frac{W(u, v, s, \alpha)}{W_M} \tag{1}$$

where $W(u, v, s, \alpha)$ is the actual value of the power measured at the output of the receiving fibre and W_M the value relative to the optimal positioning of the two fibres.

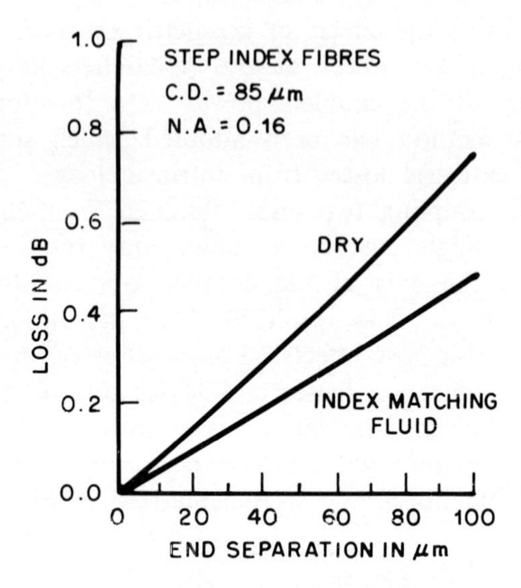

FIG. 3 – Comparison between coupling losses vs. separation, with and without index matching fluid. (See Part I, Chap. 2, Sec. 2.2)

These losses may be assessed either by introducing an index matching fluid between the two fibres or in dry condition (in this case reflection losses are also measured). In this context it should

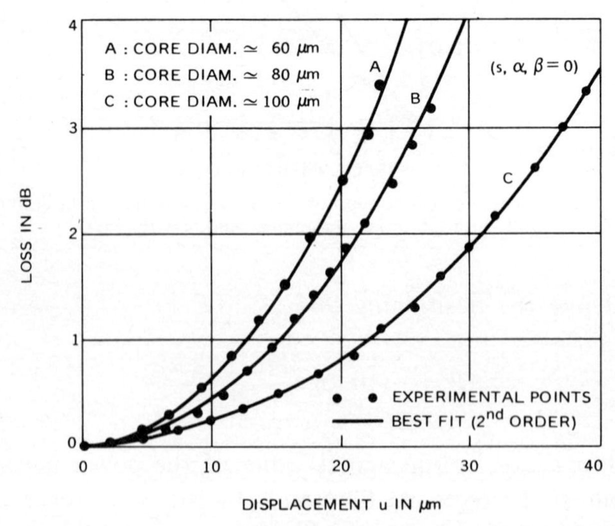

FIG. 4 – Coupling losses vs. displacement u for three fibre core diameters.

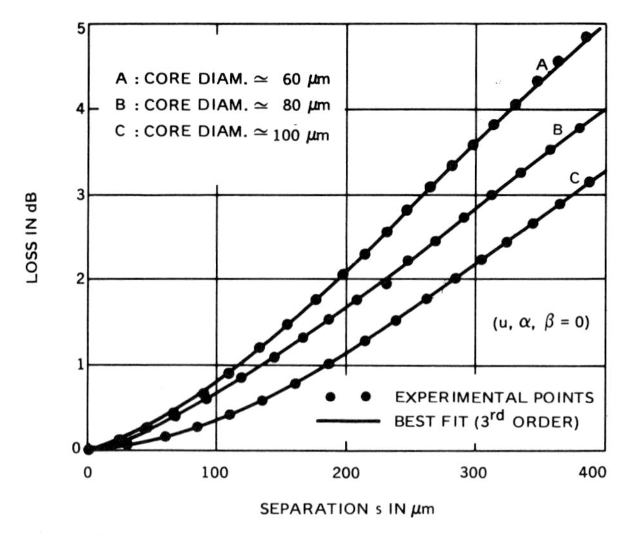

FIG. 5 – Coupling losses vs. separations s for three fibre core diameters.

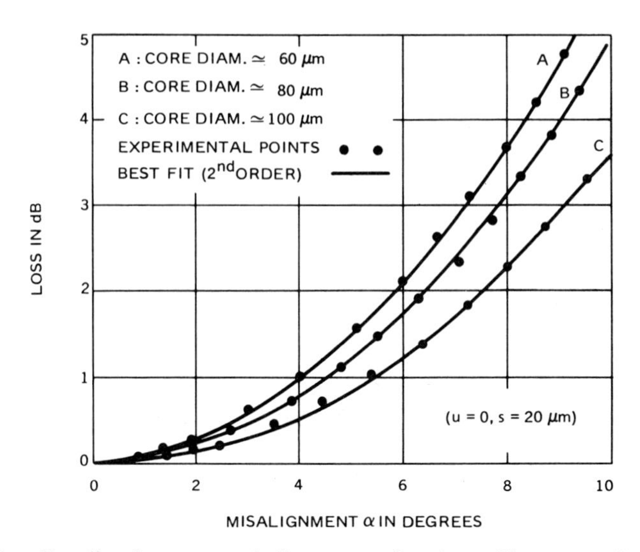

FIG. 6 – Coupling losses vs. misalignment α for three fibre core diameters.

be noted that the presence of index matching fluid affects both the absolute coupled power value (reflection loss being nil) and loss values [6] as a function of error in the two different situations.

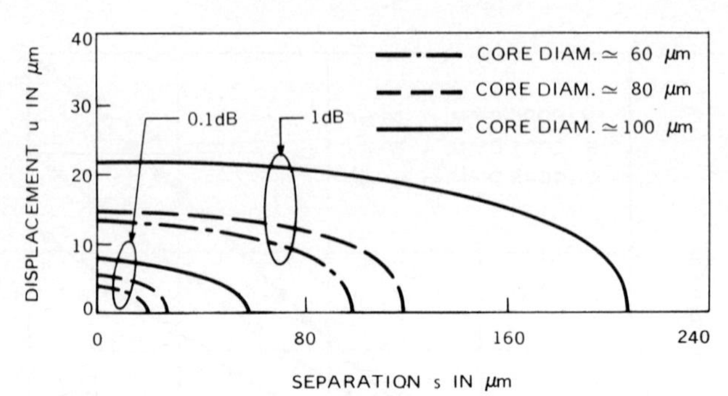

FIG. 7 – Comparison among curves at constant coupling loss for different core diameters.

This difference may be ascribed to the divergence of the emitted cone at the output end and the different acceptance of the receiving fibre in both cases. Figures 2 and 3 illustrate losses due to misalignment α and separation s for step-index fibres with and without index matching fluid between the two fibre ends. Several researchers have measured loss as a function of errors for step-index fibres [7] and graded-index fibres [8].

In Figs. 4, 5, 6 are plotted losses due to displacement u, separation s and misalignment α, for CGW graded-index optical fibres without index matching fluid. A comparison has been made of

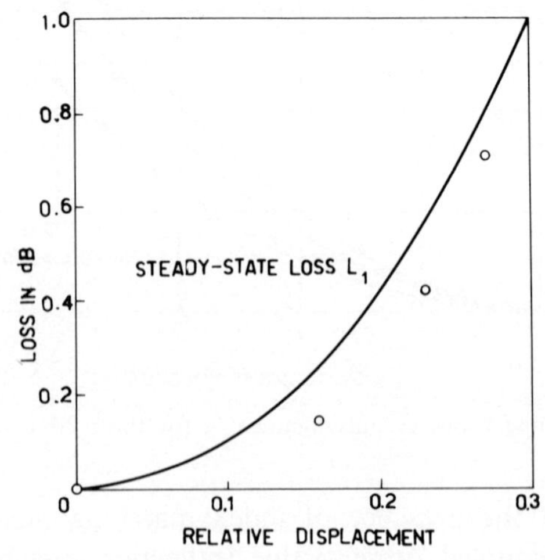

FIG. 8 – Steady-state loss vs. relative displacement. Points show loss measured for displacement between two ½ km fibres. (After [3]).

coupling losses for fibres with core diameters of 60, 80, 100 μm, respectively [8]. The above results have been obtained by means of an automatic bench shown in Fig. 120 of Part I, Chapter 3. The same bench was used to obtain constant coupling loss curves; an example is plotted in Fig. 7.

Similar measurements have been described in papers which will be referred to later on, whereas an example is depicted in Fig. 8 [3].

2.2 Splices

Several optical fibre lengths have to be jointed together to establish long-distance links. In most cases, these junctions must be permanent, since no repeated connection or disconnection is needed. This type of connection is called a splice.

Different types of splices have been developed to optimize performance, which is governed mainly by the alignment of the two optical fibres and fibre end preparation.

Splices must be divided into the following categories, according to their principle of operation [9]:

a) fibre fusing splices;

b) tube splices;

c) groove splices;

d) other techniques.

2.2.1 *Fibre fusing splices*

With this technique, the two ends to be jointed are heated to the fusing point while adequate axial pressure is applied between the two optical fibres. Thus, continuity of the transmission medium at the junction point is ensured.

Reciprocal positioning can be accomplished either by checking through an inspection microscope or by optimization of the transmitted power.

One of the first techniques applied to glass optical fibres uses a heated nichrome wire as the heating medium, which allows temperatures sufficient for material fusing to be reached.

Splices so obtained show losses as low as 0.48 dB (step index glass fibres; CD (1) = 10.8 and 20 μm; OD (1) = 75 and 150 μm respectively) [10].

Before heating, the two fibre ends must be disjointed by some μm, due to the high thermal expansion coefficient of glass. This is to prevent high pressure, which would cause fibre distortion and misalignment.

A similar jointing technique has been adopted in [11]. Suitable preparation of the optical fibres to be jointed together considerably reduces coupling losses. Ends prepared with scribing, bending, and pulling techniques show the requisite properties.

This technique has been further improved using a small electric arc, which brings the two ends to the required melting temperature. This method is used for jointing silica fibres and is described with a two-stage melting process in [12]. During the first stage fibres are softened and lightly pressed, whereas in the second stage they are melted together by a stronger arc.

Positioning is through a micromanipulator and a microscope. Figure 9 illustrates the schematics of the device.

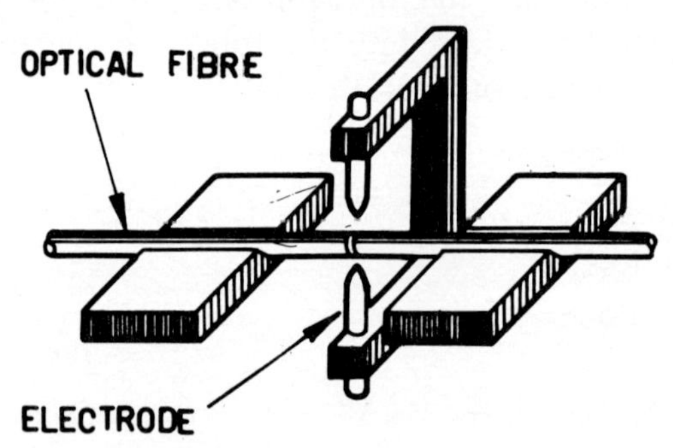

FIG. 9 – Welding fibres with an electric arc. (After [9]).

Duration of the two operations is selected so that coupling losses are reduced.

With this technique, mean losses of 0.26 dB have been measured on 19 samples (multimode step index optical fibres, N.A. = = 0.14, CD = 85 μm, OD = 125 μm, L_1 (2) = 1 m, L_2 (2) = 1 m, laser source wavelength = 632.8 nm).

Trials have also been conducted on single-mode optical fibres,

(1) CD and OD mean core diameter and outer diameter, respectively.
(2) L_1 and L_2 mean launching fibre length and receiving fibre length, respectively.

showing mean losses of 0.6 dB (CD = 7.3 μm, OD = 254 μm, $L_1 = L_2 = 1$ m).

In similar techniques, described in [13], the arc is displaced across the junction while the optical fibres are held in a vertical position to prevent gravity effects during melting. In this case, a 0.14 dB mean loss has been measured for step index fibres (NA = 0.16, CD = 75 μm, OD = 125 μm, $L_1 = 20$ m, $L_2 = 200$ m, laser source wavelength = 632.8 nm).

Two alternative heat sources have been experimented in [14], namely a micro-plasma torch (Fig. 10) and an oxhydrix micro-burner (oxygen, hydrogen and alcohol vapours). Positioning is by means of an optical system.

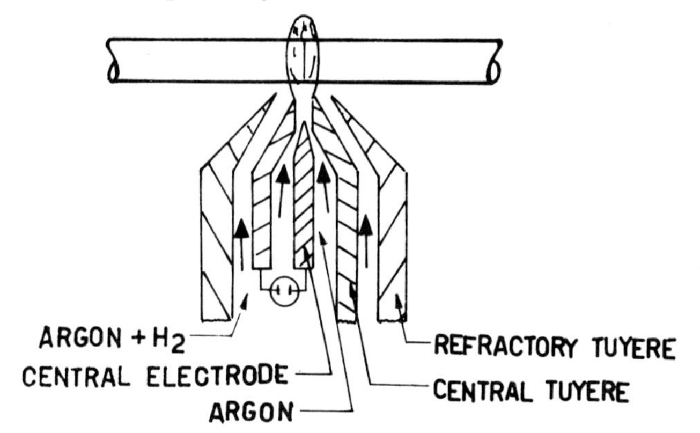

FIG. 10 – Welding fibres with a micro-plasma torch. (After [14]).

A 0.16 dB mean loss has been achieved on step index fibres (S (3) = 59, CD = 85 μm, OD = 125 μm, $L_1 = 1$ km, $L_2 = 2$ m, LED source) using the oxhydric micro-burner.

The micro-plasma torch has allowed the implementation of junctions with three different types of optical fibre: step index fibres have shown 0.21 dB mean losses ($S = 40$, CD = 85 μm, OD = 125 μm, $L_1 = 1$ km, $L_2 = 2$ m, LED source); step index fibres having a smaller core (62 μm, other characteristics being unchanged) showed mean losses of 0.2 dB; finally, with graded index fibre (25 samples, CD = 75 μm, other characteristics being unchanged), mean losses of 0.10 dB have been obtained.

(3) S means number of samples.

Obstacles to a successful fusing splice are represented by alignment difficulties and imperfect preparation of fibre ends.

A solution of the first problem, namely the « self-alignment » effect, is illustrated in [15]. This consists in the capacity of the two melted ends to align due to surface tension, which occurs in the stages illustrated in Fig. 11.

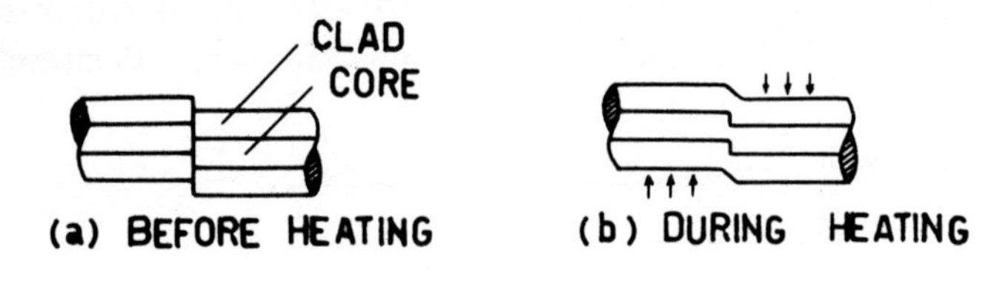

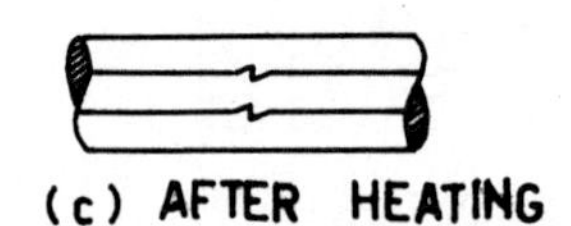

FIG. 11 – Self-alignment phenomenon. (After [15]).

This allows very good results to be achieved, even with displacement-sensitive fibres such as single mode fibres.

In fact, it may be noticed that, given initial displacement, the fusing splice results in a loss reduction higher than the Fresnel loss. This is due to the self-alignment effect, described above.

Mean losses of 0.2 dB have been obtained with single-mode fibres ($S = 40$, CD $= 7$ μm, OD $= 140$ μm, laser source wavelength $= 0.85$ μm).

The second problem involved in the implementation of fusing splices is associated with imperfect preparation of fibre ends, namely surfaces roughness, burrs, or lips, resulting in core distortion and small air bubbles. To overcome these obstacles, a 1 second prefusing has been suggested [16]. This allows the two fibre faces to be flattened by fusing any imperfections; soon after, fibres are pressed together and, finally, fusing is carried out with the usual technique.

In this way good quality splices may be obtained without sophisticated preparation of fibre ends; the time required for field operations is thus greatly reduced.

With step index optical fibres ($S = 51$, CD $= 60\ \mu$m, OD $=$ $= 150\ \mu$m, $L_1 = 1$ km) mean losses of 0.09 dB and a worst value of 0.3 dB have been measured.

An alternative heat source for fusing splices is the CO_2 laser [17], even with low power (a few watts), which is focused through a lens on the splicing area. With step index fibres 90% of the obtained splices have shown losses below 0.12 dB ($S = 40$, CD $=$ $= 56\ \mu$m, OD $= 154\ \mu$m, $L_1 = 30$ m, $L_2 = 3$ m, laser source wavelength $= 632.8$ nm).

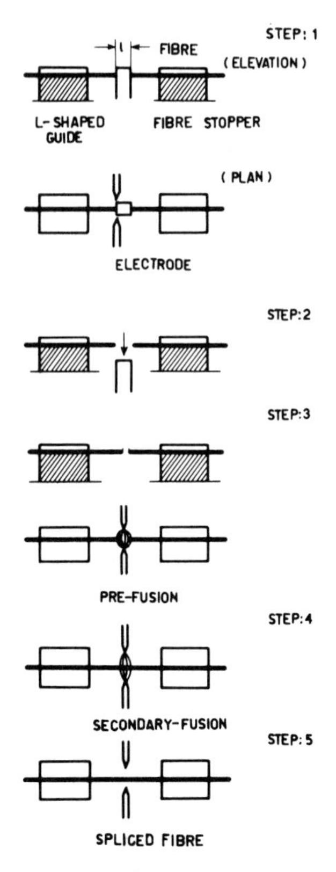

FIG. 12 – Operating stages of the splicing process. (After [18]).

The problem of fibre cutting and jointing has been thoroughly dealt with in [18], which also contains a description of an automatic set-up for field pre-fusion and fusion of optical fibres.

This technique is based on pre-fusion: its operating stages are illustrated in Fig. 12. Following manual pre-alignment of the two fibres, located at a given distance, successive stages are automatically executed. Their time sequence and duration is reproduced in Fig. 13.

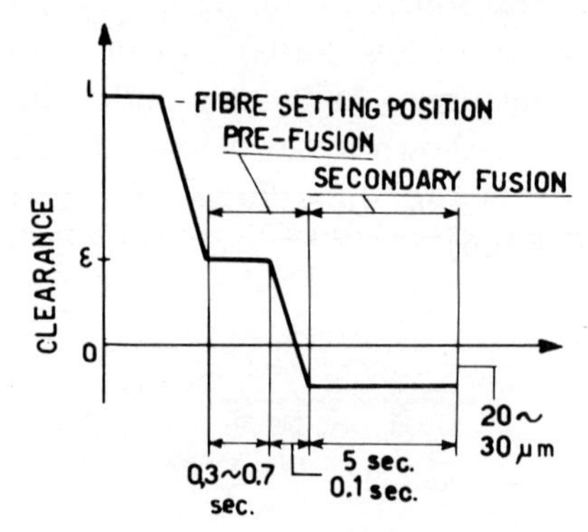

FIG. 13 – Time schedule of the splicing process. (After [18]).

The statistical distribution of losses (Fig. 14) on graded index fibres shows a mean value of 5/100 dB ($S = 100$, OD $= 150$ μm, $L_1 = 100$ m, LED source).

From the above it becomes apparent that fusing splices represent one of the most promising techniques currently available.

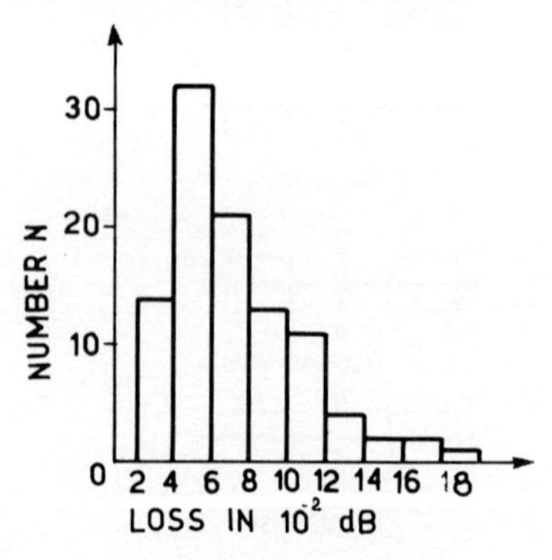

FIG. 14 – Splice losses by the automatic splicing technique. (After [18]).

This is particularly true when they are fabricated by means of fully automatic equipment. Small splices may thus be obtained relatively simply and with field losses in the region of or less than 0.1 dB.

The mechanical properties of splices thus obtained can be considered good, since the mechanical strength is approximately 60 % of the original fibre strength.

No significant loss variation has been detected during heat and mechanical tests conducted on this type of splice.

2.2.2 *Tube splices*

In this technique, alignment of the two fibre ends is through capillaries. These are generally made of glass and their inner diameter must be extremely accurate (snug-fitting). Figure 15 shows the schematics of their principle of operation. As may be seen, there is a transverse bore, located in the capillary middle, through which transparent adhesive is injected to ensure mechanical sealing and index matching.

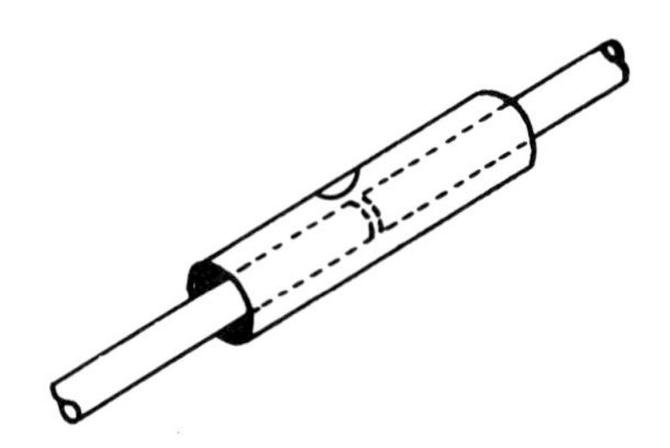

FIG. 15 – Snug tube splice. (After [9]).

In-field application of this procedure has already been discussed in [19, 20]. A description of the same is given in [21, 22].

The sequence of operations for splice fabrication is shown in Fig. 16.

The ends of the capillary tube are slightly flared to facilitate the insertion of the two optical fibres.

Mean losses of 0.25 dB have thus been obtained with step index fibres ($S = 20$, CD = 85 μm, OD = 125 μm) and of 0.21 dB ($S = 20$, CD = 85 μm, OD = 125 μm) with graded index fibres.

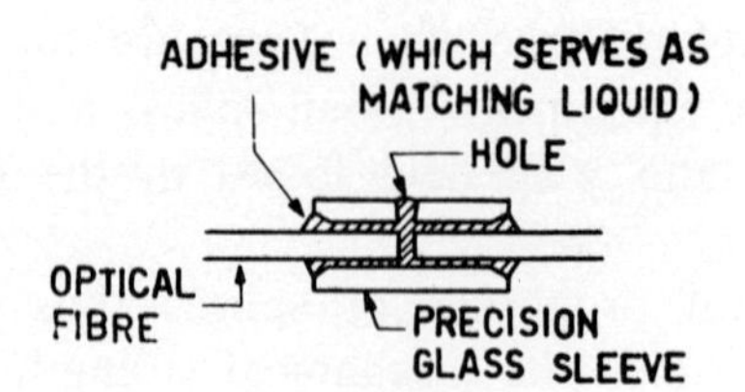

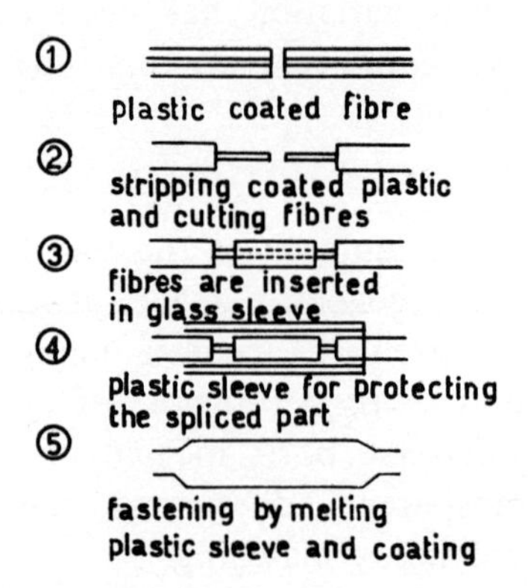

FIG. 16 – Tube splicing process. (After [21]).

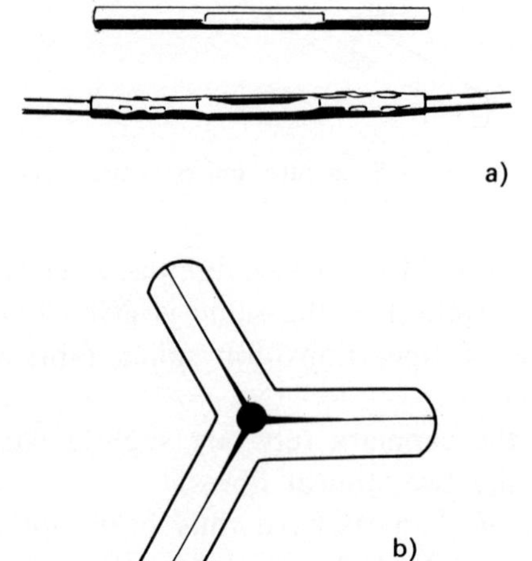

FIG. 17 *a*) – Splicing element and completed splice. (After [23]). *b*) – Cross-section of splicing element. (After [23]).

Another version of this technique, suitable for fibres with plastic coating, adopts a short length of steel tubing which has an alignment bore slightly wider than the fibre cladding diameter [23].

The two plastic coated fibres are inserted and then pressed until they contact.

Using appropriate pliers, the sleeve is crimped so as to block the fibre (see Fig. 17) by gripping its plastic coating. In this way, insertion losses between 0.85 and 1 dB have been obtained without index matching and between 0.25 and 0.35 dB with index matching ($S = 3$, NA $= 0.64$, CD $= 125\ \mu$m, OD $= 165\ \mu$m).

In order to minimize fibre play in the capillary tube a different solution has been proposed [24]. One of the two fibres is inserted into a pyrex sleeve, which has a lower melting point than quartz. The assembly is then heated so that the sleeve—due to tension—tends to collapse around the fibre (Fig. 18). Gripping of the fibre is obtained by injecting adhesive from the rear.

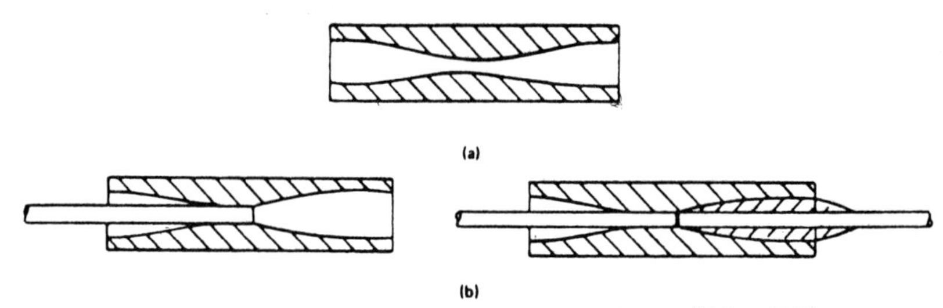

FIG. 18 - Collapsed sleeve splicing technique. (After [24]).

A second fibre is then inserted into the socket thus obtained, and the assembly is bounded with epoxy, which also serves as index matching.

If necessary, this process may compensate for possible diameter differences between the two fibres. In such a case it is sufficient to collapse the sleeve around the larger fibre so that it lips over the larger fibre end to form a socket snug fitting to the smaller diameter.

The whole set is then enclosed in a protective metal ferrule.

The splice shows losses of 0.2-0.3 dB with graded index fibres (CD $= 30\ \mu$m, OD $= 100\ \mu$m, $L_1 = 100$ m, $L_2 = 100$ m).

From the above example, it may be noted that the tube serves a double function, namely, it ensures alignment and guarantees splice protection.

Another attractive and simple technique of fabricating good quality splices [25] uses a loose fitting, square cross-section tube. Two techniques are here adopted: the sleeve-technique for rough alignment and protection, and the groove-technique for precision alignment (Fig. 19). Neither microscope nor micromanipulator are used with this method.

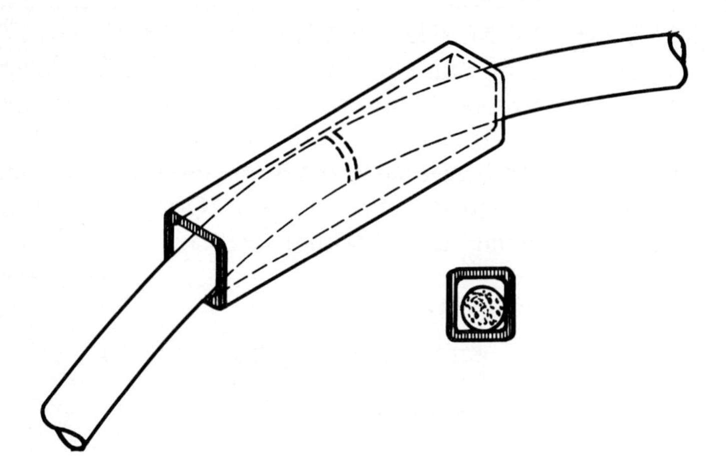

FIG. 19 – Square cross section, loose tube splice. (After [25]).

The two fibre ends are biased against one of the four inner edges of the tube by bending the fibres externally, and are held in position by means of an appropriate epoxy, which fills the tube.

To ensure a high level of accuracy, the curvature radius of the inner edges must be small. It is worth noting that this splice is self-aligning, i.e. externally bent fibres automatically position on the same edge.

Mean losses of 0.07 dB have been achieved with graded index optical fibres ($S = 8$, CD $= 68$ μm, $L_1 = 1.5$ m, $L_2 = 1.5$ m).

2.2.3 *Groove splices*

Another technique, which also permits the precision joining of two single-mode fibres, uses [26] a plain plexiglass substrate, on which an embossed precision groove is obtained (Fig. 20).

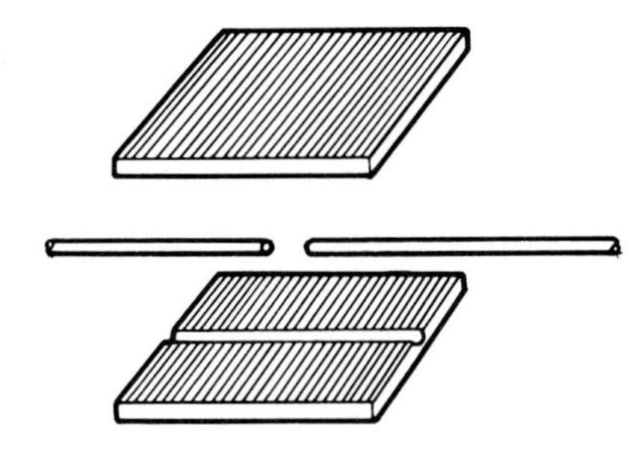

FIG. 20 – Grooved substrate splice.

The two fibre ends to be jointed are then inserted in the groove and secured by the cover.

With splices of this type mean losses of 0.5 dB have been achieved on single mode fibres ($S = 12$, CD = 3.7 μm, OD = = 254 μm, laser source wavelength = 632.8 nm), and 0.8 dB for a different source ($S = 24$, CD = 3.7 μm, OD = 254 μm, LED source wavelength = 0.9 μm).

In a similar technique [27], the alignment is accomplished by inserting the fibres into a groove and pressing them to the groove base (Fig. 21). 50% of the splices fabricated with step index fibres have shown an insertion loss of less than 0.2 dB.

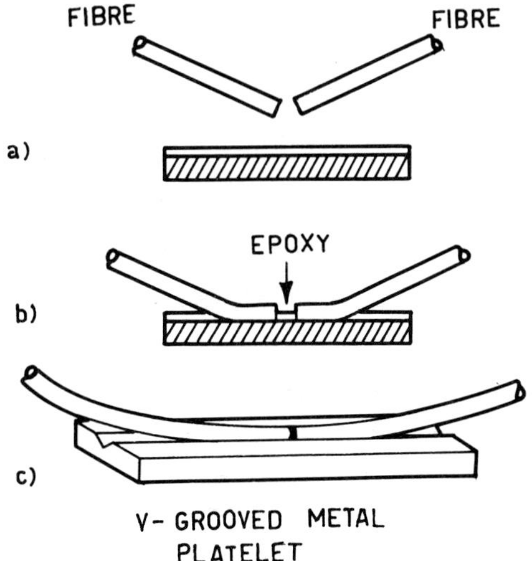

FIG. 21 – Splicing process using a *V*-groove substrate.

A splicing technique which detects and identifies losses due to the splice only and those due to differences in fibre geometrical characteristics (CD, OD, eccentricity, ellipticity) has been developed. It uses a *V*-groove and a retaining plate, both made of glass, and obtained through a drawing process [28].

A special technique has been employed to accurately measure losses which are only due to the jointing technique, i.e. not resulting from other factors.

In fact, two fibres obtained with a single cut have been joined together so as to avoid any rotation of the fibre ends, while the interposed index matching fluid does not act as an adhesive and is particularly transparent and bubble free. Therefore, resulting losses are a function of the mechanical accuracy and of the blocking technique.

Optical power losses due to the splice which have been measured with an integrating sphere placed just across the splice are not in excess of 1/100 dB (CD = 8 μm, OD = 130 μm, $L_1 = 1$ m, $L_2 = 1$ m).

FIG. 22 – Finished Springroove® joint.

A splicing method which adopts a precision groove, called the « Springroove » [29] (4), consists of a bracket incorporating two small cylinders, which serve as a guide system for the two fibres, and of an elastic element for fibre end alignment (Fig. 22).

(4) Patent Pending.

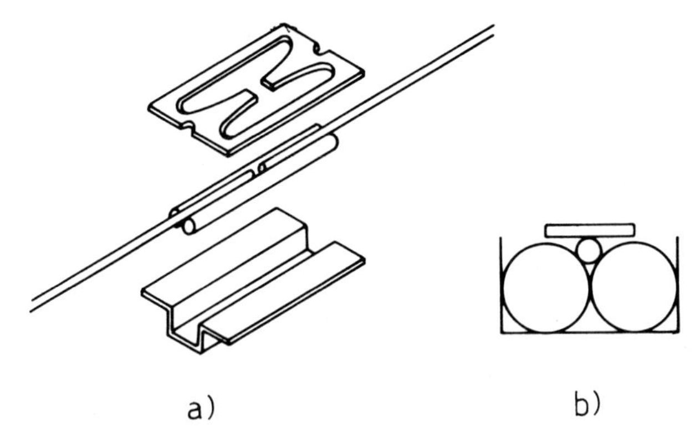

FIG. 23*a*) – Exploded view of the Springroove® joint. *b*) Schematic joint cross-section.

Cylinder size is selected so that the upper rim of the fibres is protruding above the cylinders and can be retained by a spring of simple construction, as shown in Fig. 23.

Figure 24 shows the statistical distribution of losses over 94 samples, indicating a mean loss of 0.05 dB with graded index fibres (S = 94, N.A. = 0.14, CD = 80 μm, OD = 125 μm, stabilized lamp source).

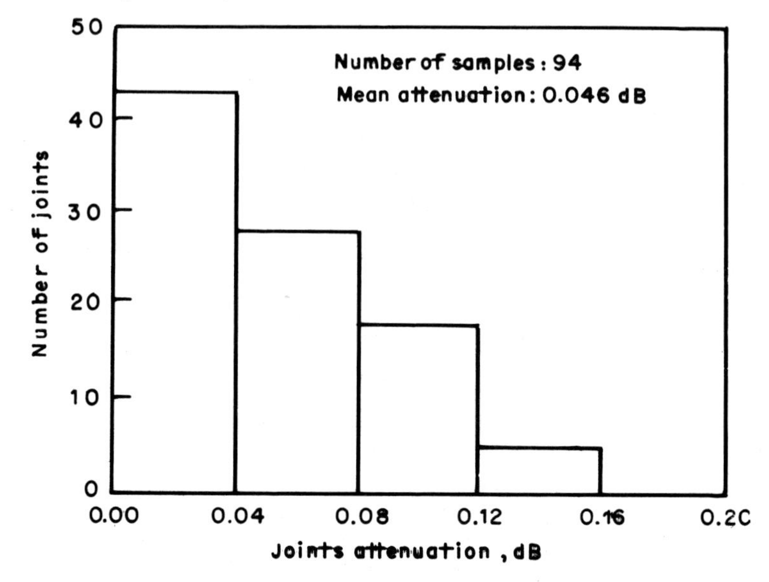

FIG. 24 – Histogram of the laboratory tests on Springroove® joint attenuation.

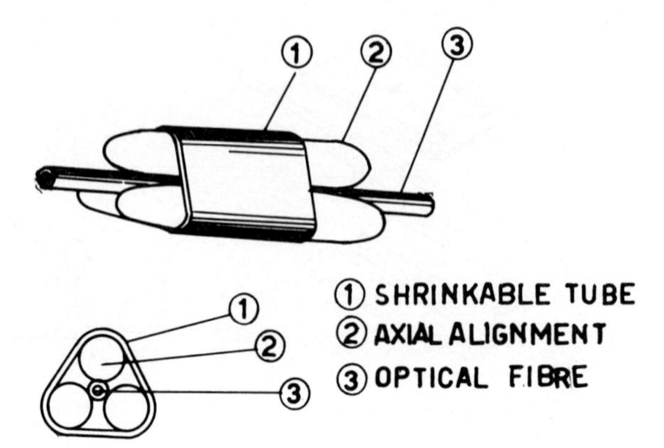

FIG. 25 – The three cylinder splice. (After [30]).

In a similar technique [30], the spring is replaced by a third cylinder, which is pressed against the fibre by a heat shrinkable tube (Fig. 25). With this method main losses of 0.2 dB have been achieved using multimode fibres (CD = 80 μm).

The groove technique is particularly suitable for obtaining substrates with multiple guides, required to splice several fibres contained in a cable.

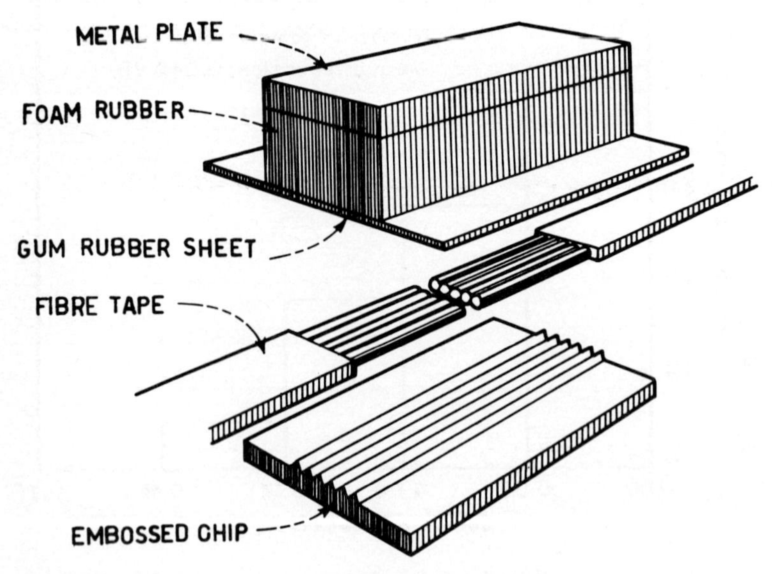

FIG. 26 – Sketch of splice arrangement before assembly. (After [31]).

A technique of this type has been described in [31]: six grooves are obtained on a metal plate, through embossing by means of a steel head; fibre-ribbon ends are removed from their plastic coating and then simultaneously prepared by a suitable tool using the « scribe, bend and pull » technique. As a next step, the fibres are introduced into the grooves and held in position through the pressure of a gum rubber sheet (Fig. 26). 90 % of all splices have shown a loss of less than 0.15 dB ($S = 90$, CD = = 80 μm, OD = 120 μm, $L_1 = L_2 = 2$ m).

A splice similar to that above [32, 33] has been obtained by embossing 12 parallel grooves on a plastic substrate with a specially designed tool. Once prepared, the two ends of the fibre ribbon are inserted into the prealignment slots on the splice and then pressed into the groove area. Their positioning is automatic and does not require the use of a microscope.

A cover plate bonded to the top side ensures the required mechanical stability. Glycerine is used as index matching.

Mean losses are 0.12 dB ($S = 120$, CD = 92 μm, OD = = 130 μm, $L_1 = 1$ m, $L_2 = 1$ m, laser source wavelength = = 632.8 nm). This splice is shown in Fig. 27.

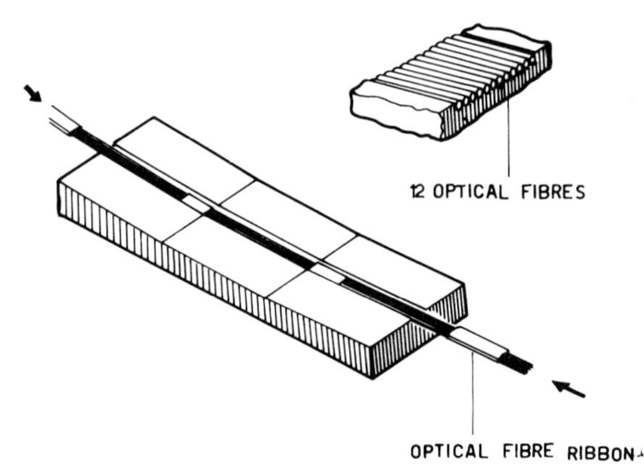

FIG. 27 – Illustration of embossed splicing tool and assembly technique. (After [32]).

A splice for multistripe cables, which consists of 12 sets of 12 grooves each (Fig. 28) lying in the same plane, is presented in [34]. It is an injection moulded plastic connector, obtained by means of a metal master, which allows the moulding of a

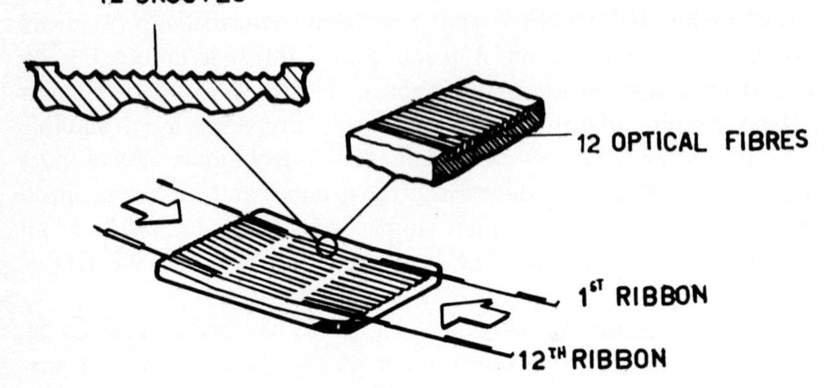

FIG. 28 – Precision-moulded splice for 12 ribbon each accomodating 12 optical fibres. (After [34]).

polycarbonate substrate ensuring highest accuracy and repeatibility.

The method employed here is similar to that previously described.

Once the fibre ends have been prepared, each of the 12 ribbons is inserted into its respective groove and then secured by a cover plate.

Suitable index matching is injected to complete the splice. Mean losses of 0.2 dB have been measured ($S = 425$, CD = = 55 μm, OD = 90 μm).

All 12 fibres are required to achieve an efficient alignment using this type of groove.

As an alternative, deeper grooves may be adopted whenever there is a risk of fibre breaking during splicing.

The advantages of the two above methods may be offered by an additional type of multiple splice [35]. The body and cover of the splice are made of polycarbonate by injection moulding and can be easily mass-produced. A silicon chip with a precision groove which is preferably produced using an etching technique is housed at the inside.

After preparation, the two fibre ends are inserted into the grooves and butted.

Fibres are secured by a cover plate, which is fitted with a central hole for index matching injection.

The two splice ends are then placed in a mould, into which

an encapsulant compound is poured, thereby forming the ribbon strain relieves at both ends.

Mean losses with graded index fibres are 0.22 dB ($S = 111$, $CD = 55$ μm, $OD = 110$ μm).

A detailed technique of fabricating silicon groove arrays for splices is dealt with in [36, 37]. The grooves, depicted in Fig. 29,

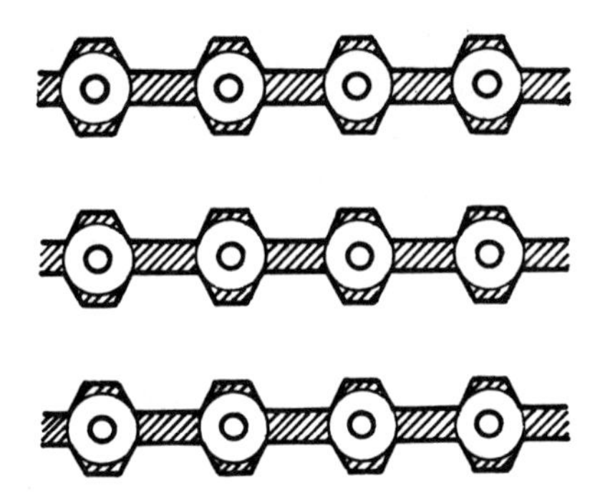

FIG. 29 – Cross section of a fibres sandwich, showing the grooves obtained by a photolithographic technique.

are obtained from a silicon plate with a photolithographic technique, which exploits the different etching rate along the characteristic planes of the crystal.

In-width tolerance for the grooves thus obtained is ± 0.7 μm. In this case the maximum fibre vertical positioning error is slightly less than 2 μm.

A splice for a multistripe cable, shown in Fig. 30, is based on a similar principle.

The fibres of one cable end are aligned so as to form a uniform matrix and then secured with adhesive. Ends are simultaneously prepared.

Alignment is by means of precision groove chips. Fibres are stacked in layers between the chips and the residual space is filled with epoxy.

The upper and lower faces of the sandwich thus obtained do not carry fibres, so that their grooves can act as guides for connecting the two parts.

A different splicing technique may be successfully applied if a special type of cable is used, as suggested in [38, 39].

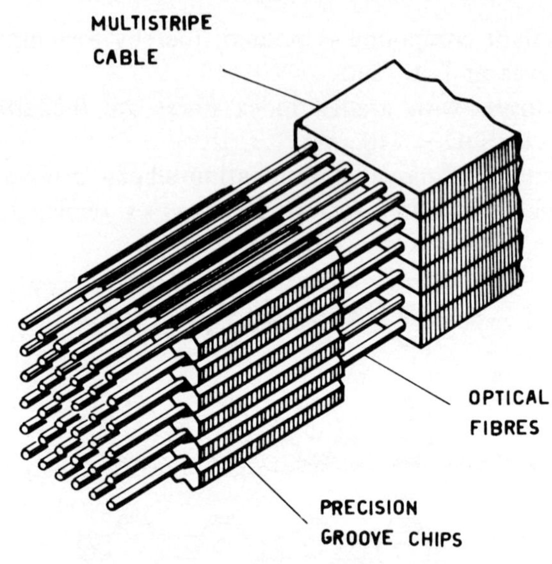

FIG. 30 – Multistripe cable splice. (After [37]).

Unlike the preceding example, this cable has a circular cross-section and the fibres are arranged in the grooves which have been obtained on the outer surface of a cylindrical core (Fig. 31).

Once the fibre has been removed from the cable sheath, a short length of the central core is cut in the form of a star and replaced by an equal length, fitted with precision grooves and perfectly adjusted.

Fibres are inserted into the precision grooves and secured by adhesive; then their ends are simultaneously cut together with the support, and covered with index matching.

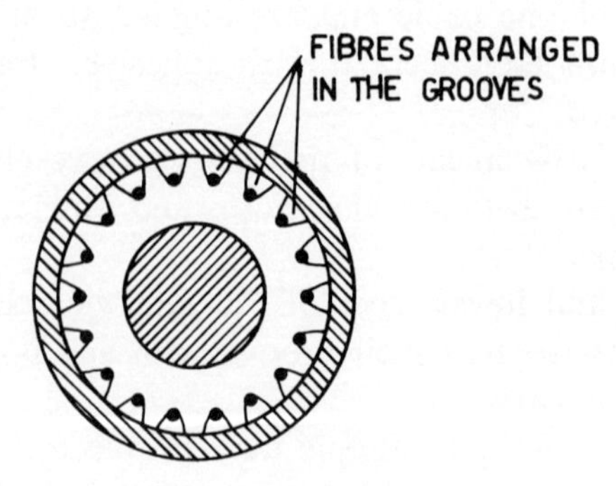

FIG. 31 – Cross section of the star-core cable. (After [38]).

Then the two ends of the cable to be jointed are brought together and fastened by semi-cylindrical shells and pins.

Improved accuracy may be achieved by obtaining the two grooved substrates from a single alumina piece. In this case, mean losses of 0.3 dB have been achieved with step index fibres ($S = 40$, CD = 65 μm, OD = 110 μm, $L_1 = 500$ m, $L_2 = 30$ m, laser diode source) and of 0.55 dB ($S = 40$, CD = 85 μm, OD = = 125 μm, $L_1 = 500$ m, $L_2 = 30$ m, laser diode source).

2.2.4 *Other techniques*

When particularly critical alignment conditions affect operation as in the case of single mode fibres or non-screened multimode fibres (where, for instance, core/cladding eccentricity errors are likely), one can resort to a technique of the « adjust-to-maximize power » type (ATMP), like that described in [40]. The ends of the two fibres to be jointed are inserted into special ferrules and then prepared and fastened at the rear.

Next, the two ferrules are mounted on a device which allows their relative movement in three directions, and are then micropositioned until the maximum transmitted power is attained. A suitable epoxy injected between the two ensures joint stability and serves as an index matching medium.

The whole set is covered with a heat-shrinkable tubing, which has a protective function (Fig. 32).

Measurements conducted on fibres with a relatively small core (10-20 μm) have demonstrated the efficiency of this method, showing losses restricted to within 0.5 dB.

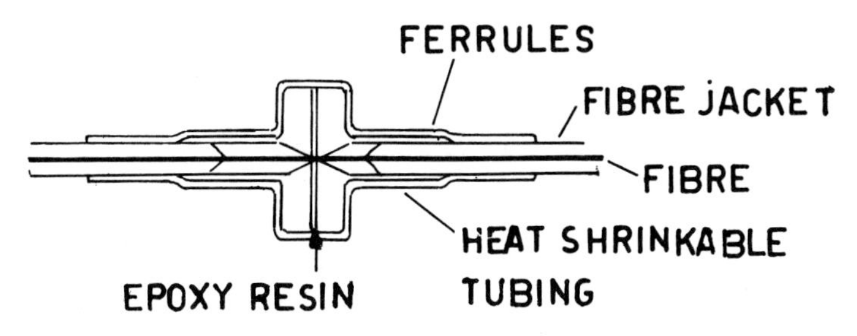

FIG. 32 – Section of a joint made with ATMP technique. (After [40]).

ATMP techniques allow good results to be achieved. However, they imply some drawbacks in actual operating conditions, owing to the fact that is difficult to check fibre output down the joint, especially at a distance of several hundred metres.

To overcome these difficulties, a suitable method has been proposed [41].

Although still based on optimum alignment detection, it uses an auxiliary fibre, the ends of which are readily accessible, in place of the direct fibre.

The sequence of the operations to be carried out is shown in Fig. 33.

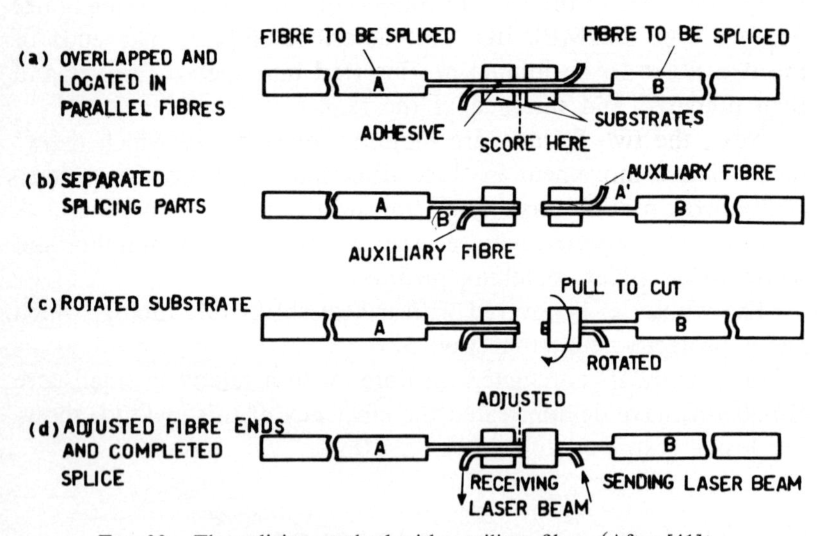

FIG. 33 – The splicing method with auxiliary fibres (After [41]).

The two fibres to be spliced are bounded in such a way that an overlapping zone is created (*a*).

A cut is then made at the centre of this area (*b*) and one of the two ends is rotated by 180° (*c*). The optimization can be accomplished by using only two short fibre lengths (*c*). In the case of a rotation of exactly 180°, the symmetry of the geometrical position makes it possible to perform optimization with the auxiliary fibre, achieving at the same time the optical positioning of the main fibre. The rotation by 180° requires high accuracy.

By applying this splicing method, mean losses of 0.04 dB can be achieved with step index fibres ($S = 20$, CD = 60 μm, OD = 150 μm, $L_1 = 10$ m, $L_2 = 10$ m, laser source wavelength = = 632.8 nm) and of 0.03 dB with graded index fibres ($S = 20$, CD = 50 μm, OD = 130 μm, $L_1 = 10$ m, $L_2 = 10$ m, laser source wavelength = 632.8 nm).

Ease of fabrication and automation possibilities are the main features of another splicing technique [42].

First, the two ends to be jointed are prepared. Next, after deposition of an adhesive drop on the tips, they are rolled into a transparent loop of polyester. Before the loop is fully tightened, the fibres are pressed together. Finally the loop is pulled by means of two drums (Fig. 34), whereby the alignment is accomplished.

A small heater facilitates adhesive grip.

The mean value of losses measured with step index fibres is 0.15 dB ($S = 6$, NA = 0.15, $L_1 = 1$ m, $L_2 = 1$ m, LED source wavelength = 820 nm).

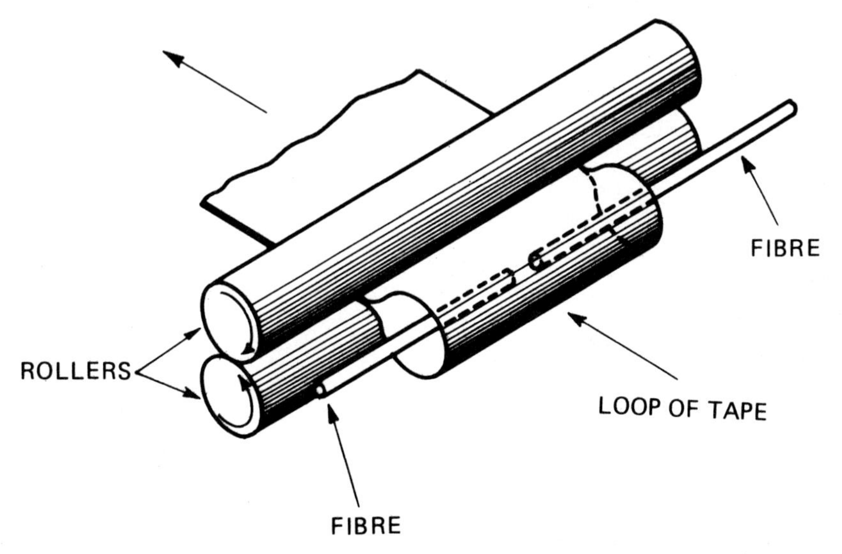

FIG. 34 – The tape joint. (After [42]).

Summary table of the characteristics of some single-fibre splices.

Splicing method	Reference (Example of application)	Typical loss in dB (average)	Fibre structure (step index: st graded index: gr)
Fibre fusion 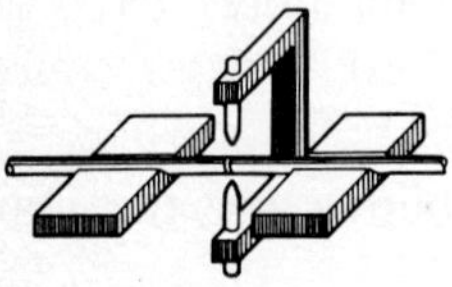	[18] Sumitomo, NTT (Japan)	0.05	gr - OD = 150 μm
Snug tube 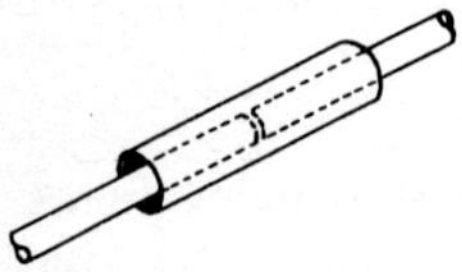	[21] Furukawa (Japan)	0.2	gr - CD = 85 μm OD = 125 μm
Collapsed sleeve 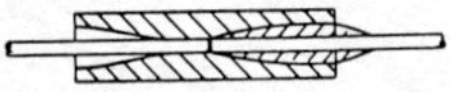	[24] STL (United Kingdom)	0.2/0.3	gr - CD = 30 μm OD = 100 μm

Summary table of the characteristics of some single-fibre splices.

Splicing method	Reference (Example of application)	Typical loss in dB (average)	Fibre structure (step index: st graded index: gr)
Loose square tube 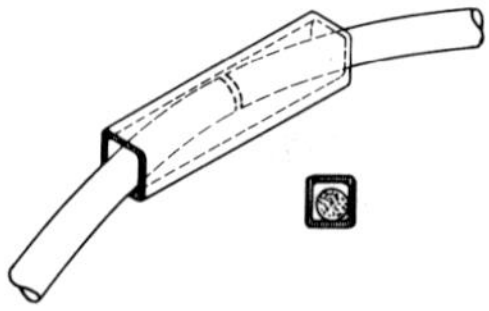	[25] Bell Lab. (U.S.A.)	0.07	gr - CD = 68 μm
V-groove 	[27] Siemens (West Germany)	0.2	st - CD = 80 μm
Springroove® 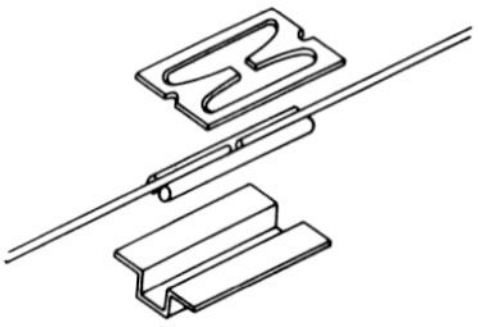	[29] CSELT (Italy)	0.05	gr - CD = 80 μm OD = 125 μm

Summary table of the characteristics of some multi-fibre splices.

Splicing method	Reference (Example of application)	Typical loss in dB (average)	Fibre structure (step index: st graded index: gr)
Precision moulded groove 	[34] Bell Lab. (U.S.A.)	0.2	gr - CD = 55 μm OD = 90 μm for 12×12 fibres
Etched groove 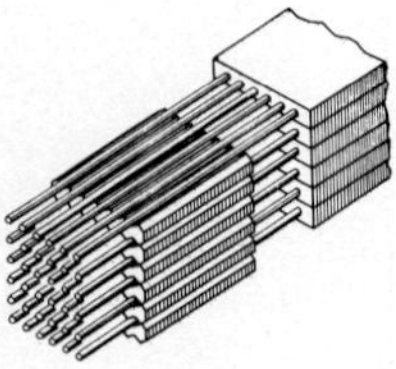	[37] Bell Lab. (U.S.A.)	0.2-0.3 (with index matching)	gr - CD = 55 μm OD = 110 μm for 12×12 fibres
Grooved cylindrical core 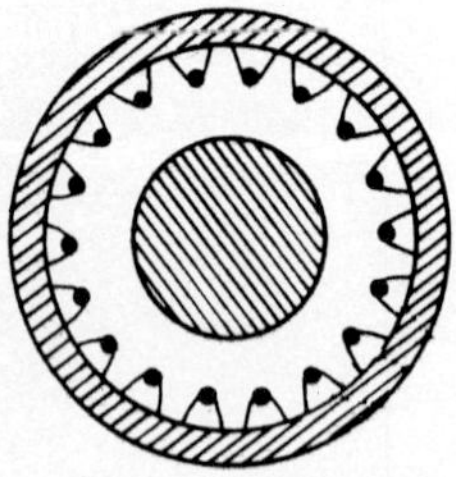	[38] CNET, LTT (France)	0.3	st - CD = 65 μm OD = 110 μm

2.3 Connectors

The aim of this section is to describe the most common types of removable connectors for optical fibres and cables. From a general standpoint, connectors may be divided into different categories according to their principle of operation, namely: *a*) watch jewel ferrule connectors, *b*) connectors with different types of groove, *c*) concentric sleeve connectors, *d*) eccentric sleeve connectors, *e*) moulded connectors, *f*) expanded beam connectors, *g*) other types of connectors.

The main characteristics for the performance evaluation of a connector are: low insertion losses, ease of construction and mounting, small variations in insertion losses even after a large number of connections and disconnections, interchangeability with different samples of the same connector, insensitivity to environmental factors (temperature and dust), low cost, and low crosstalk for multiple connectors. We shall now give a detailed description of the above connectors.

2.3.1 *Watch-jewel ferrule connectors*

This type of connector relies on a simple principle: it consists of a watch-jewel mounted on the front end of a metal sleeve, called a ferrule. Its accuracy depends on that of the watch jewel hole, through which the fibre is centered with respect to the ferrule. This serves as a guide for the optical fibre, which is bonded to the ferrule itself by means of cement. The two ferrules are then introduced into a metal sleeve for alignment. After attachment, the two terminations are prepared by means of a simple manual operation. Both the ferrule and the outer sleeve are readily available, with tolerances producing an average loss of 1 dB in the case of step index fibres with 75-100 μm core diameter. Figure 35 gives a schematic view of this connector [43, 44].

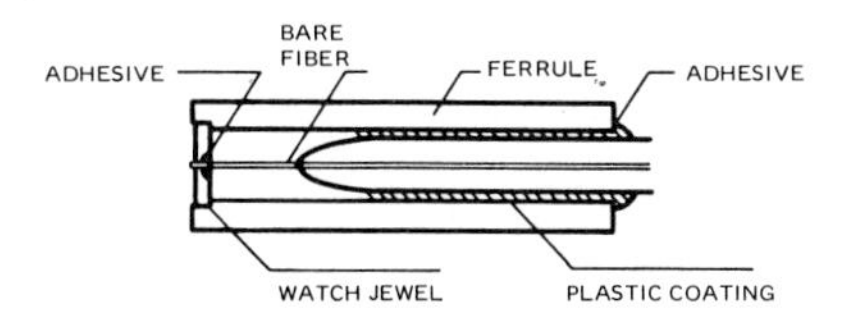

FIG. 35 – Jewelled ferrule termination. (After [44]).

2.3.2 *Groove connectors*

The principle of operation of the groove for optical fibre jointing is described in the section devoted to splices. This principle has also been successfully applied to the development of connectors. Various implementation techniques are available. In principle, however, they may be sorted into two broad categories according to whether the grooves are obtained through mechanical or chemical processes (etching).

In the mechanical process case, optical fibre grooves may be constructed with different geometries. A connector for two optical fibres developed on the above principle [45] is depicted in Fig. 36.

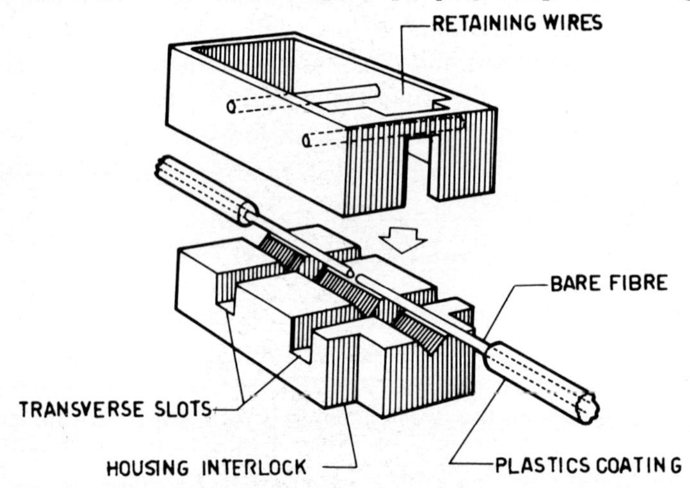

FIG. 36 – Principle of a groove connector. (After [45]).

The two prepared fibre ends are fixed to the groove bottom by two wires. In practice, the connector consists of two interlocking housings and a sleeve which holds the assembled housings together. The fibres are secured through a spring-loaded collar which grips the plastic coating, thus preventing the fibres from sliding. An 0-ring seal allows the use of an index matching fluid, preventing contaminations. The connectors assembled with step index fibres show losses of 1.1 dB (CD = 125 μm, OD = 165 μm), loss changes within ± 0.1 dB in the range – 40 °C to + 60 °C.

The following connector accomplishes the groove function by means of three small metal cylinders as schematized in Fig. 25.

This principle has been adopted to construct a multiple connector (Fig. 37*a*) capable of simultaneously aligning and connecting the whole set of optical fibres of a cable [6, 46]. The guide assembly consists of two arrays of small discrete cylindrical rods, each fixed to a bracket (Fig. 37*b*), or of two arrays of contiguous half-rods which are obtained directly from a monolithic structure, as dis-

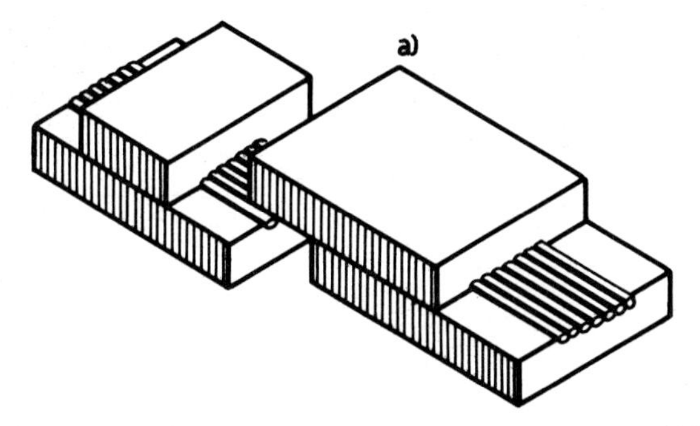

FIG. 37 *a*) – Illustration of the overlap connector concept. (After [6]).

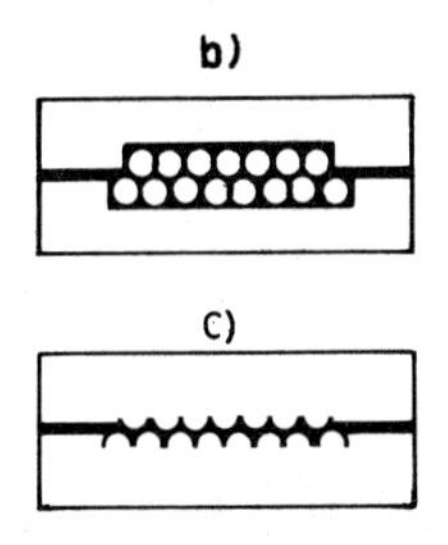

FIG. 37 *b*, *c*) – Two linear arrays of alignment rods arranged to provide a multiplicity of three-rod arrays obtained by two different techniques. (After [6]).

played in Fig. 37*c*. The rods are made of plastic or plastic-coated material, the elastic strain of which allows the fibres to be gripped and aligned. In its actual implementation, this connector is fitted with a check system for the separation of fibre ends, which uses an accumulation region where the fibre can bend easily, as shown in Fig. 38. In this way, when the two connector halves are clamped together, fibre terminations exert a pressure against each other, due to the above-mentioned spring action. The connector is also provided with a suitable flexible system for cable clamping. A

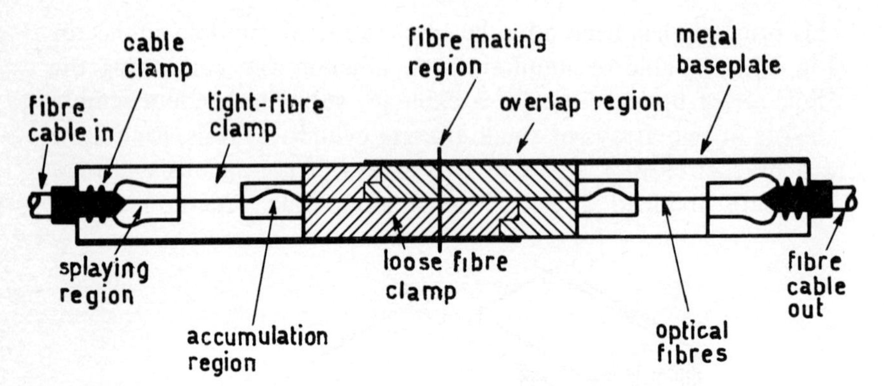

FIG. 38 – Longitudinal section of the overlap connector. (After [46]).

drawback of this connector (and generally of all those in which the fibre is strongly clamped) is due to micro-bending, which eliminates high order modes and produces a different modal distribution. The connector mean loss is 0.57 dB, the best value is 0.37 dB, the worst 0.77 dB; index matching and step index fibres are used (S = 44, NA = 0.18, CD = 85 μm, OD = 125 μm).

The next connector is also based on the groove principle and allows the connection of many optical fibres. On the basis of design data, insertion losses should not exceed 0.1 dB. If displacement is considered the only source of errors, this implies for a step index optical fibre a displacement error less than 1.8 % of the core diameter. The connector consists of only one cylinder, with grooves machined parallel to its axis along the surface. It is in these grooves that the optical fibres are housed [47].

To provide a perfect matching between the two connectors, the connectors are obtained from an individual grooved cylinder, which is then cut in two. The fibres to be coupled are bounded into these grooves, so that the fibre end planes and the cylinder plane face coincide. There is one groove deeper than the others which houses a steel needle. This aligns the two parts, pressing them against two additional supporting needles, as shown in Fig. 39. Mean losses of 0.1 dB have been measured with step index optical fibres (NA = 0.56, CD = 60 μm).

Insertion losses are unaffected by temperature values below 100 °C. By virtue of low inserion losses, this connector might also be used successfully for connecting single mode fibres.

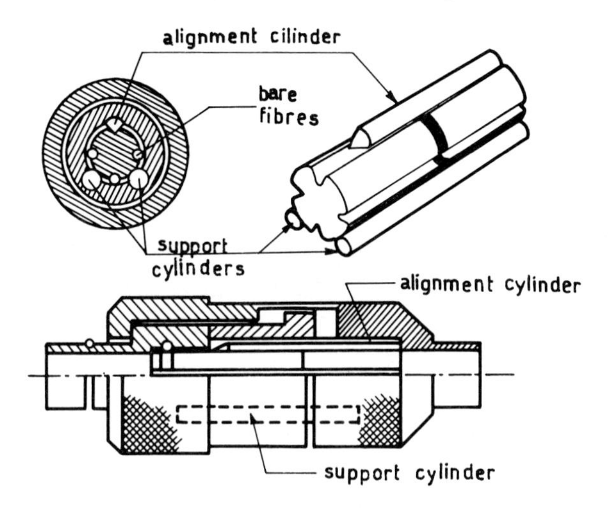

FIG. 39 – Multiple *V*-groove connector with cylindrical cross-section. (After [47]).

Another connector worth reviewing has the optical fibre located in a groove formed by two balls and held in place by a third ball (Fig. 40*a*). The three balls [48] gripping the fibre are housed in a bush. By lightly pressing together the two sets of three balls, the balance point is reached corresponding to a relative rotation of 60° of these sets. In this way the automatic alignment of the two fibres is brought about (Fig. 40*b*).

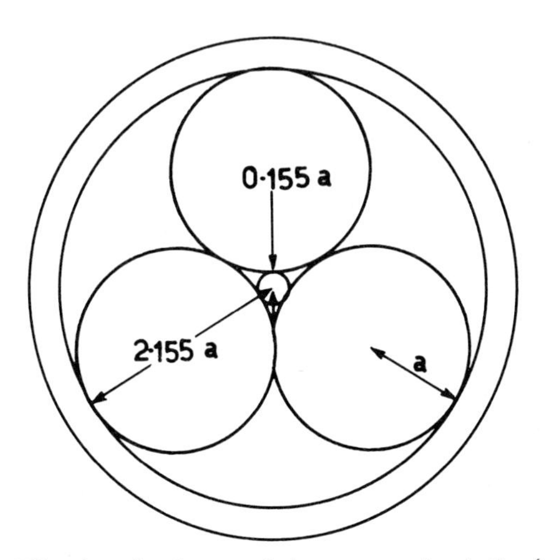

FIG. 40 *a*) – Fibre location by set of three contacting balls. (After [48]).

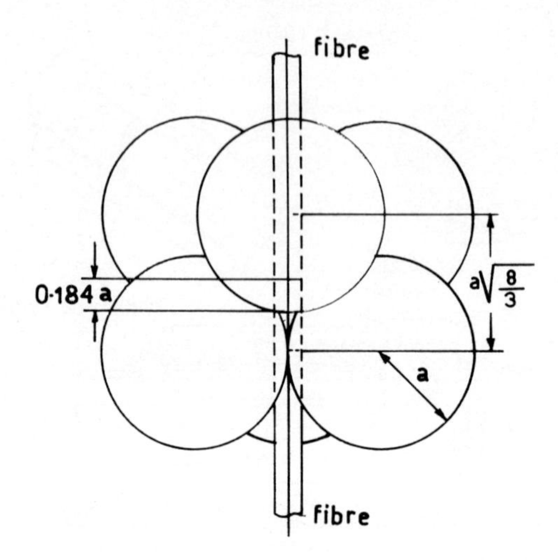

FIG. 40 *b)* – Two interlocking sets of three balls, showing position of fibre ends. (After [48]).

In order to ensure a correct location of the two fibre ends, the two fibres are recessed, with respect to the end plane of the three balls, slightly more than 0.184 a (where *a* is the ball radius). The geometrical characteristics linking together fibre diameter and ball radius are shown in the figure. Obviously, the longitudinal positioning of the fibres requires the use of a microscope. It should also be noted that the only high-precision components are tungsten-carbide balls, readily available with accuracies lower than one μm. The fibre is attached to the centering set through adhesive. The mean loss of this device is 0.49 dB, the best and worst values are 0.36 dB and 0.85 dB, using step index optical fibres (NA = 0.56, CD = 80 μm, OD = 125 μm, $L_1 = L_2 = 2$ m).

The connector we are going to consider next allows the connection, on the groove principle, of all the fibres of two multi-ribbon cables by jointing the two connector halves previously prepared on each cable end [49]. In particular, the two connector halves must ensure coupling between corresponding optical fibres. All fibres of one cable end are aligned to form a uniform matrix, the configuration of which is retained by means of cement. In this way, all ends of the optical fibres can be simultaneously prepared.

At first, the aligning operation was accomplished by high precision grooves machined on metal chips (Fig. 41). Afterwards, the chips forming the fibre ribbon guides were obtained [50] by photolithographic techniques on silicon layers cut along the 100 axis

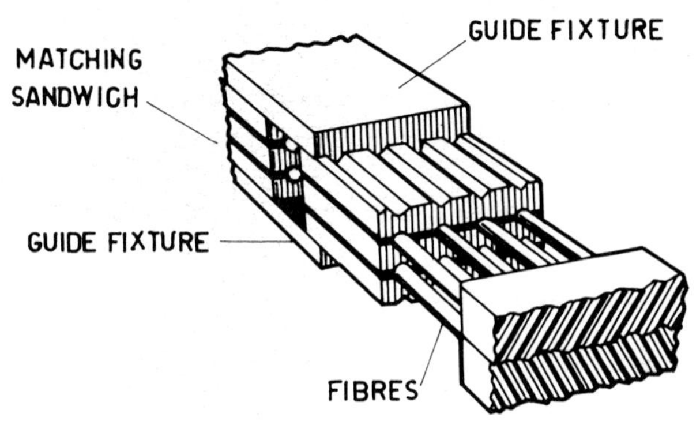

FIG. 41 – Splicing of two double-row sandwich arrays.

(Fig. 28). In both cases the fibres are stacked between the chips and the interspaces between the chips are filled with epoxy. The upper and lower parts of the laminate sandwich are not occupied by fibres so that the grooves can act as guides for the connection of the two parts.

The connector using metal chips shows a crosstalk of less than -65 dB between two continuous joints. Another relevant characteristic of this device is the loss, falling in the range 0.2-0.3 dB, using index matching and a 12×12 graded index fibres ribbon (CD $= 55$ μm, OD $= 110$ μm). The alternative method appears to offer higher fibre fabrication accuracy, keeping the periodicity error, width and vertical alignment, and thickness variation in individual chips within 1 μm, wheras guide angle variations are virtually non-existent.

The fabrication of the connector examined in the following is based on etching processes [51]. From thick layers of light sensitive plastic sheets rectangular grooves of given width and depth are obtained, in which the fibres will be placed (Fig. 42). If a proper exposure time is chosen, grooves considerably narrower ($\simeq 10$ μ) than the fibre diameter can be produced. In such a way, the fibres are elastically retained in the groove by means of a special locking cover. The two connector terminations are then lapped (or, more

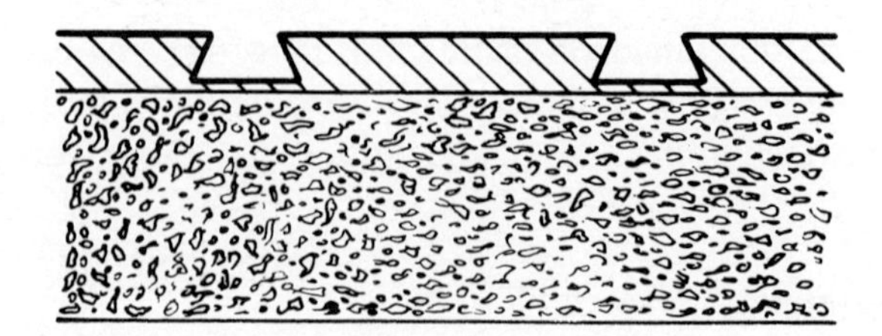

FIG. 42 – Plastic alignment grooves with undercut profile on top of glass substrate. (After [51]).

simply, cut with conventional techniques) and inserted into the two parts to be butt jointed that are aligned by pressing against an angle.

Statistical figures relative to insertion losses give a theoretical average displacement error of about 3 μm. By displacing the two connector halves by an integer multiple of the lateral fibre separation, it is possible to keep insertion losses within 0.2 dB, with step index optical fibres (NA = 0.14, CD = 92 μm, OD = = 132 μm, short fibres with mode stripper). Crosstalk has been estimated to less than — 20 dB.

Another connector adopting a photolithographic etching technique is depicted in Fig. 43 [52]. The fibre clamping system is

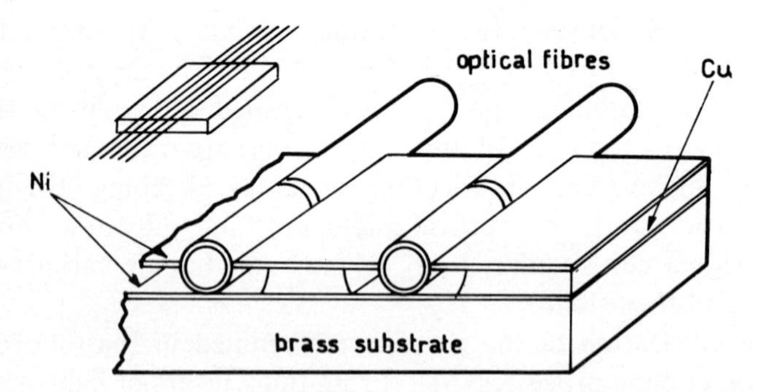

FIG. 43 – Cross-section and oblique view of aligning grooves with inserted optical-fibre ends. (After [52]).

based on low elastic deformation of the nickel strips, which grip the fibres slightly above their centre (≈ 5 μm) and hold them in place. The thickness of the Ni layers is ~ 5 μm, whereas gaps

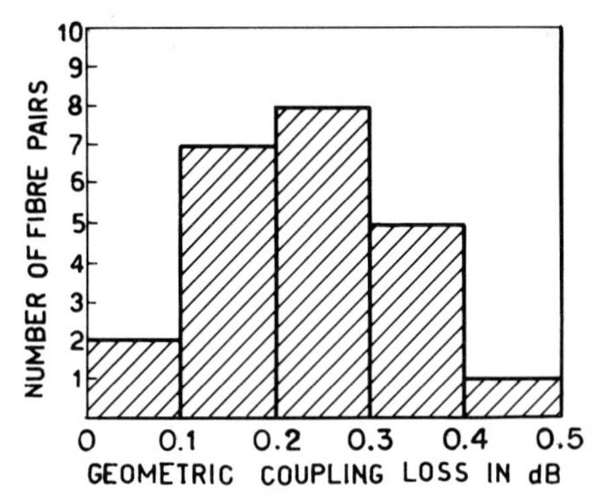

FIG. 44 – Distribution of geometric coupling loss related to twice 25 aligned fibres. (After [52]).

are slightly smaller than the optical fibre diameter. This system is particularly suitable for connecting stripe cables. Statistical data of insertion losses are reported in Fig. 44. Transverse positioning of the fibres can be reproduced within 5 μm. The mean loss is 0.26 dB, the worst value 0.42 dB with index matching and step index fibres ($S = 25$, CD $= 70$ μm, OD $= 82$ μm).

2.3.3 *Concentric sleeve connectors*

This category comprises all those connectors in which the fibre is fixed inside the concentric sleeve. Thus, losses of this

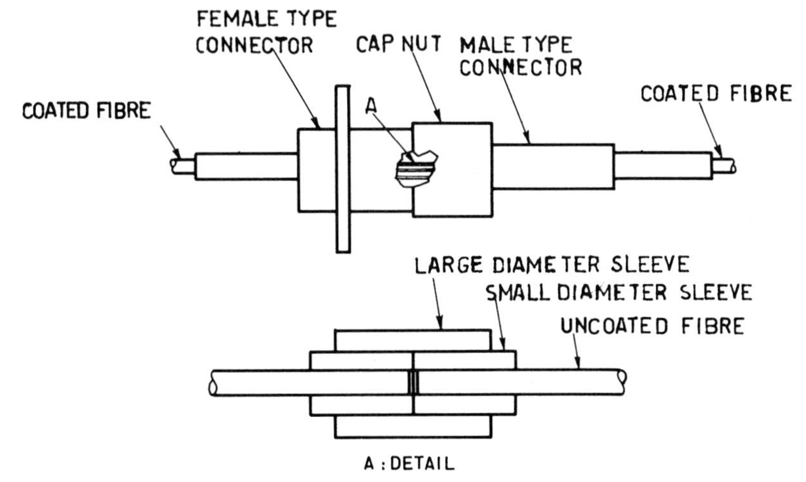

FIG. 45 – Concentric sleeve connector. (After [53]).

system depend on the accuracy with which the fibre is positioned at the centre of the outside sleeve. Although these connectors require alignment through a microscope, they offer the advantage of being unaffected by any small difference in optical fibre outside diameters. We first consider a sleeve connector [53] of particularly simple design (Fig. 45). After being cut and prepared, the fibre is inserted into a small diameter special metal pipe and bonded. The two small sleeves housing fibre ends are then introduced into an outer metal sleeve which acts as a guide. The mean loss is 0.32 dB, the best and worst values are 0.03 dB and 0.69 dB, with index matching fluid ($S = 20$).

In Figs. 46*a*) and *b*) loss variations are given as a function of temperature and of rotation angle of the two connector halves.

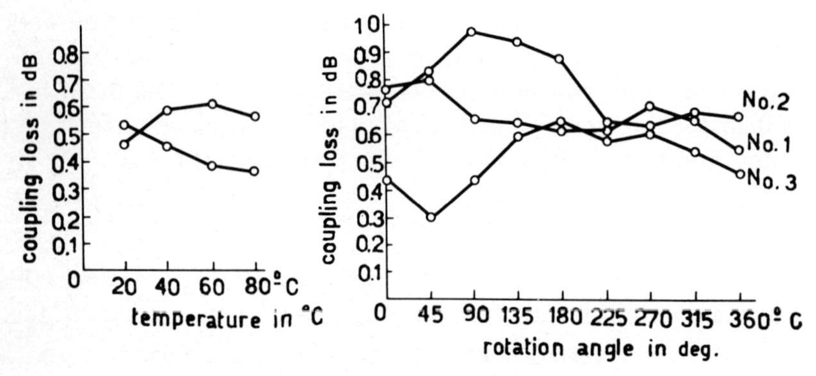

FIG. 46 – Loss variations as a function of: *a*) temperature, *b*) rotation angle. (After [53]).

A connector based on the same concept is depicted in Fig. 47 [54]. The fibre is first aligned in a plug connected to an *x-y-z* micro-manipulator, which optimizes its position with respect to a reference plug through optical checking. The connector is then cemented in the adjusted position in a sleeve, which serves as a guide for its connection with the other part. This connector is also suitable for connections between fibre and source by fixing the source to an appropriate socket. Reported losses are within 1 dB, with index matching. The sleeve for the alignment and fixing of the parts may be replaced by a groove, on which the two plugs will be aligned and coupled. An example is shown in

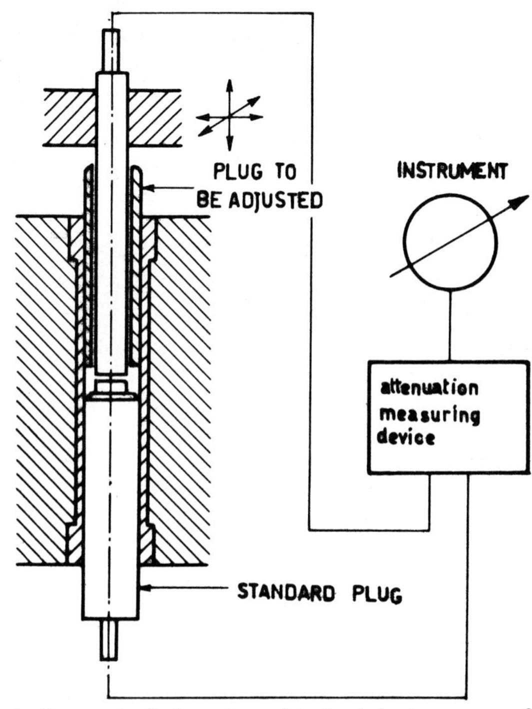

FIG. 47 – Central alignment of plug pin and outer tube by means of a standard plug. (After [54]).

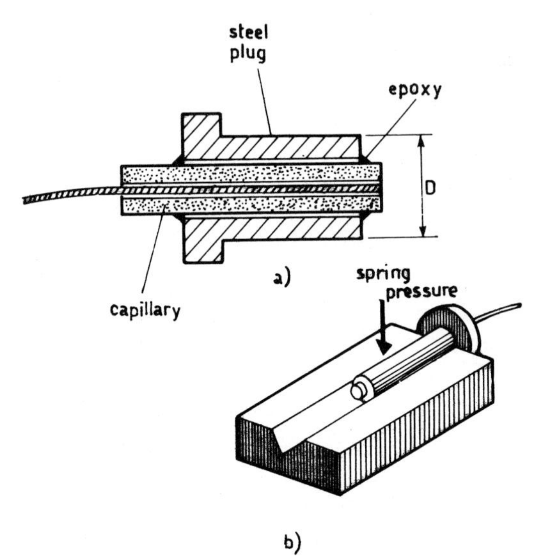

FIG. 48 – *a*) Fibre connector; *b*) coupler with only one connector shown. (After [55]).

Fig. 48 [55]; the mean loss is 0.3 dB, the worst value is 1.3 dB, with graded index fibre ($S = 19$, CD = 100 μm, $L_1 = 2$ m, $L_2 = 11$ m, LED source wavelength = 900 nm).

An example of a connector designed on experimental data [56] is illustrated in Fig. 49. The aim was to implement a connector

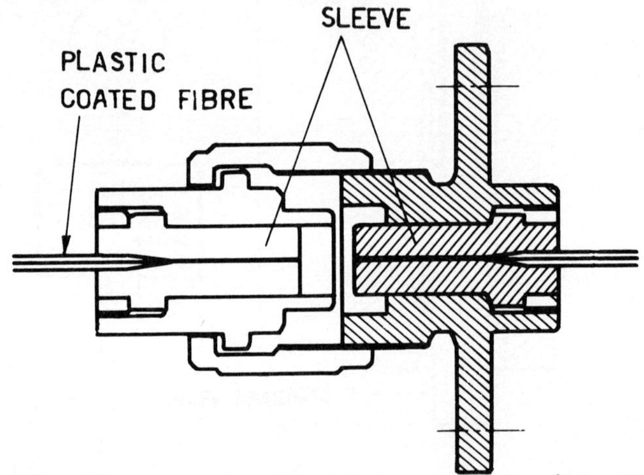

FIG. 49 – The cross-section of metal sleeve connector. (After [56]).

with losses below 1.5 dB (including reflection). This means that coupling errors must be kept within the following limits: the separation s must be less than 10 μm to keep the loss below 0.1 dB, the displacement u must be less than 8 μm to keep the loss below 1 dB, and the misalignment α must be less than 1° to keep the loss below 0.1 dB, with fibres N.A. = 0.14 and core diameter of 85 μm.

The connector is obtained by fixing the fibre in a central position by means of adhesive in a high-precision sleeve, which is then joined to the other part. Characteristics of this device show an average loss of 0.5 dB with index matching ($S = 50$, NA = = 0.14, CD = 85 μm, $L_1 = L_2 = 0.5$ m, NA = 0.14, LED source wavelength = 820 nm). Figure 50 shows variations in coupling losses for the two connector halves. Another concentric sleeve connector is presented in Fig. 51 [57]. The design target is a connector with total losses below 0.7 dB, of which 0.3 dB are due to reflection and 0.4 dB to connector inaccuracy. Of the latter loss value, 0.3 dB are ascribed to displacement u and 0.1 dB to misalignment α. With

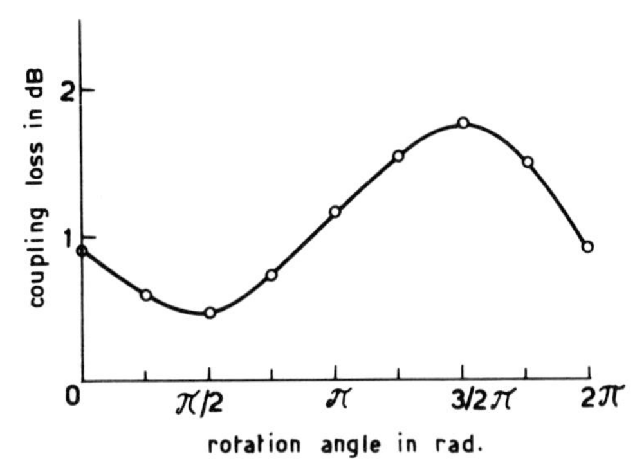

FIG. 50 – Coupling loss vs. rotation angle, without index-matching. (After [56]).

coupling nut

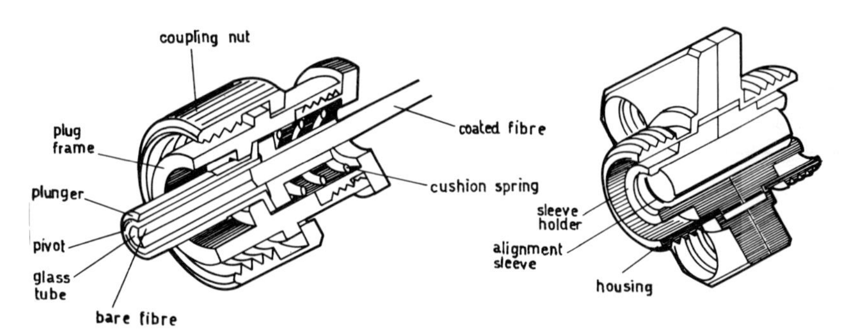

FIG. 51 – Optical fibre concentric connector. (After [57]).

step index fibres of CD = 60 μm, OD = 150 μm and NA = = 0.17, mechanical specifications imposed are: displacement u less than 5 μm, misalignment α less than 1°.

This connector consists of two plugs inserted in an adapter, which serves as a guide and fixing element (Fig. 51). The fibre is bonded in a glass tube and this in turn is fixed in a stainless steel pipe (called pivot). Then fibre ends are polished and the pivot is centered and fixed with adhesive in a plunger with the aid of a reference plunger, a microscope and a TV monitor for alignment check. Next adhesive is injected into the gap between plunger and pivot. Finally, during the coupling operation, an adapter joins and aligns the two plugs.

Further characteristic of this connector are: an average loss of 0.49 dB and a worst value of 1.20 dB, without index matching and with graded index fibres ($S = 140$, NA $= 0.17$, CD $=$ $= 60\ \mu$m, OD $= 150\ \mu$m). 150 samples were taken and core eccentricity with respect to plunger was evaluated, showing 2.1 μm as average value and 5.5 μm as worst value. Heat cycling, vibration and shock test were conducted, yielding good results. Statistically processed values for loss and eccentricity errors are supplied in Figs. 52*a*, *b* respectively.

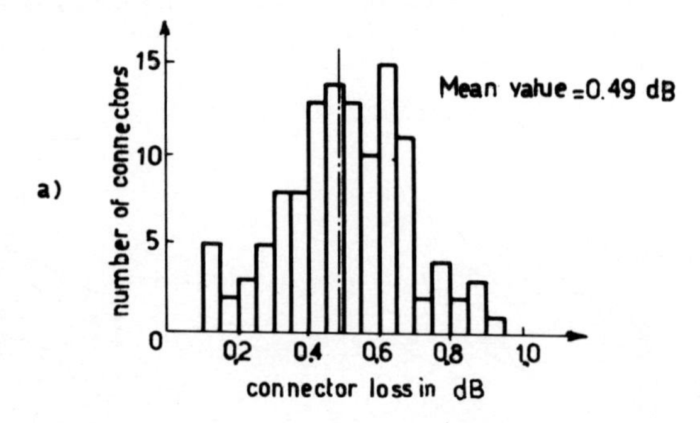

FIG. 52 *a*) – Coupling loss statistic. (After [57]).

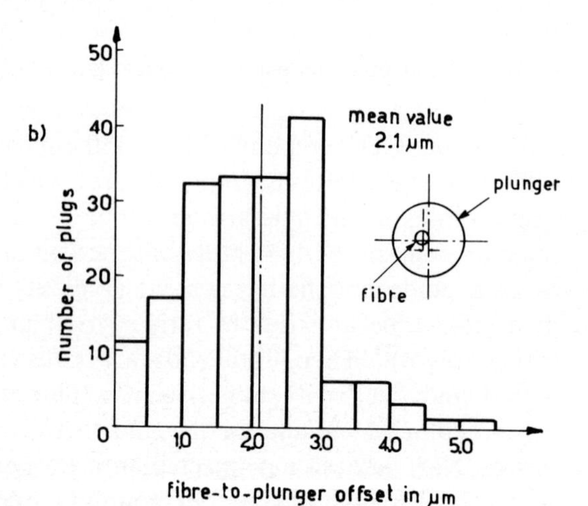

FIG. 52 *b*) – Fibre-to-plunger offset. (After [57]).

A connector based on a principle of operation similar to the previous example is described in [58]. It does, however, differ in two aspects. The adapter is in fact formed by a chucking sleeve, which also allows gripping and alignment of two plungers of slightly different diameter (Fig. 53*a*, *b*, *c*). Additionally, the con-

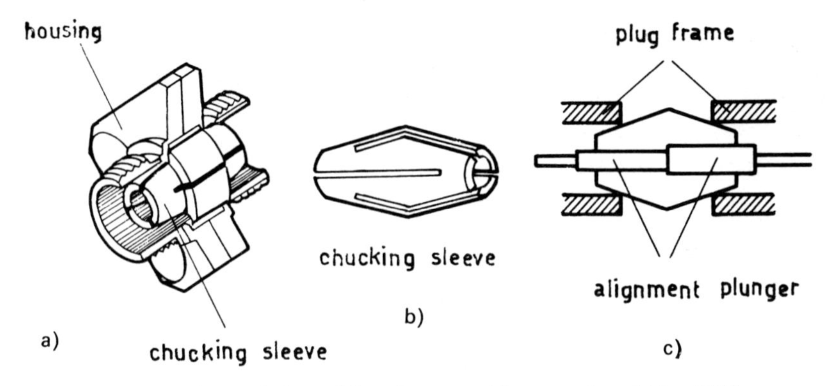

FIG. 53 – Dissected view of the adapter and its operation. (After [58]).

nector is no longer mounted in the plunger by means of core centering and then bonded. With this new technique, the glass tube housing the fibre is first lapped and then bonded in the plunger. The plunger is then mounted on the spindle of a grinding machine and observed through a microscope mounted as a tailstock. The chuck holding the plunger can be positioned so that the core centre is perfectly aligned with the spindle rotation axis. Next, the outher surface of the plunger is machined, ensuring concentricity of plunger surface and fibre core within the working tolerances of the tool. The design target was to implement a connector with losses less than 0.3 dB, excluding reflection. Consequently, this necessitated working with stringent tolerances, such as 6 μm displacement, 1° misalignment and 10 μm separation (for 65 μm core diameter). The mean loss is 0.4 dB, the best and worst values are 0.2 dB and 0.7 dB, with graded index fibres ($S = 40$, CD $= 65$ μm). Variations of individual characteristics are illustrated in Fig. 54(*a* to *f*).

A different adapter, which contains ball bearing arrays in the cylindrical inner surface of the sleeve (Fig. 55), allows both smooth detachment of the plug and alignment accuracy.

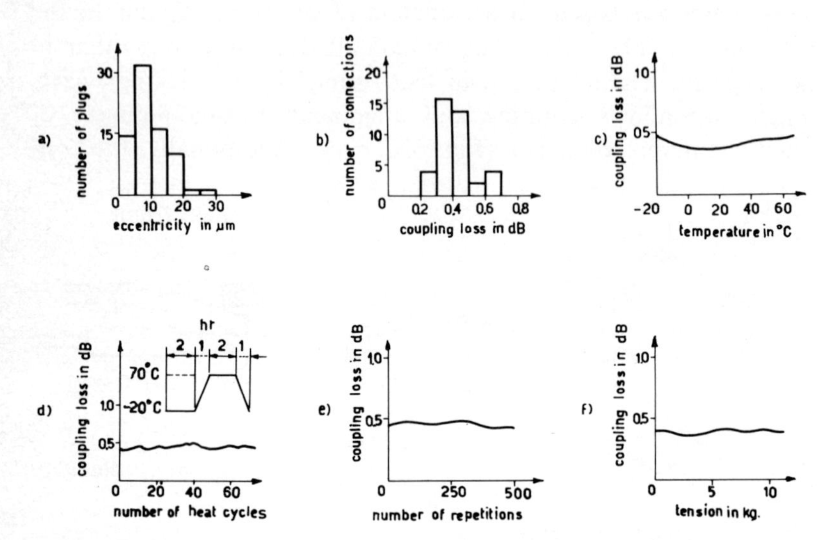

FIG. 54 *a-f*) – Relevant operation data about the concentric connector. (After [58]).

This structure can be used to obtain single mode connectors. Centreing of fibres in plugs is either through an eccentric tube alignment method [59] or a precision hole directly made at the plug centre by a discharge drilling machine [60]. This solution, although requiring fine machining of the hole on the plug tip, allows the assembling of the connector without any special equipment.

Measured losses show a 0.47 mean value, while the best and worst results are 0.3 dB and 0.7 dB; a single mode fibre ($S = 78$, CD $= 8.6\ \mu$m, OD $= 150\ \mu$m, laser diode source wavelength $=$ $= 830\ \mu$m) without index matching is used.

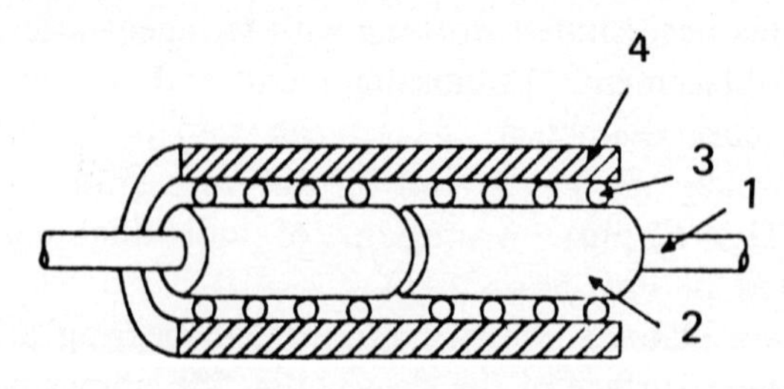

FIG. 55 – Single-mode optical fibre connector structure. Connector sectional view: 1) optical fibre; 2) fibre plug; 3) ball bearing; 4) sleeve. (After [59]).

A simpler method of centering a multimode fibre in the middle of the plug [61] makes use of an aligning jig as shown in Fig. 56; the fibre is heated to form a bead near its tip, and is inserted loosely into the plug.

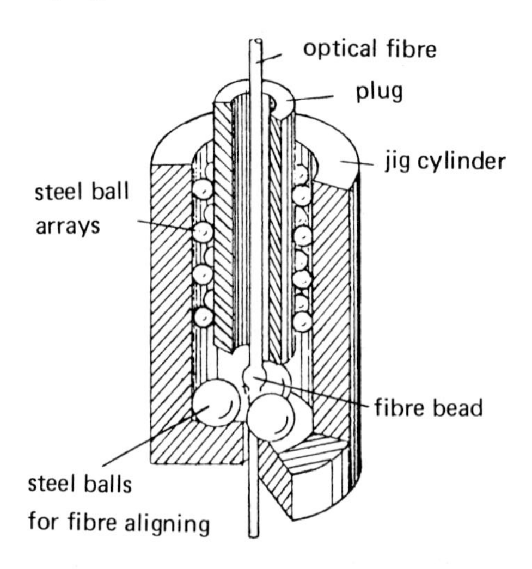

FIG. 56 – Cut away illustration showing the schematic structure of a fibre-aligning jig (After [61]).

The fibre tip is led into the jig and goes through the gap between three steel balls lying on the bottom, until it stops because of the bead, which is too large to pass through. Eventually the plug is filled with adhesive and, after hardening, the bead is broken off and the fibre end polished.

The mean loss is 0.5 dB, the best and worst values are 0.3 dB and 0.8 dB; a step index fibre is used ($S = 10$, CD $= 60$ μm, OD $= 150$ μm, $L_1 = 1$ km) and no index matching.

2.3.4 *Eccentric sleeve connectors*

The principle of operation of these connectors resembles that of the previous examples, in that it allows the direct, reciprocal centering of cores, regardless of any possible eccentricity between core and cladding. In a connector of this type [62-64] the two fibres are fastened eccentrically in two parallel and offset cylindrical pins. When the displacement of the two pins is lower than the sum of the two eccentricities, two optimum coupling positions are possible. Detection of these positions is carried out manually by monitoring the output power of the receiving fibre and by mechanically blocking the optimum position so obtained. The device consists of two pins mounted in two grooves having the same aperture, but different depth and vertex (Fig. 57*a*, *b*).

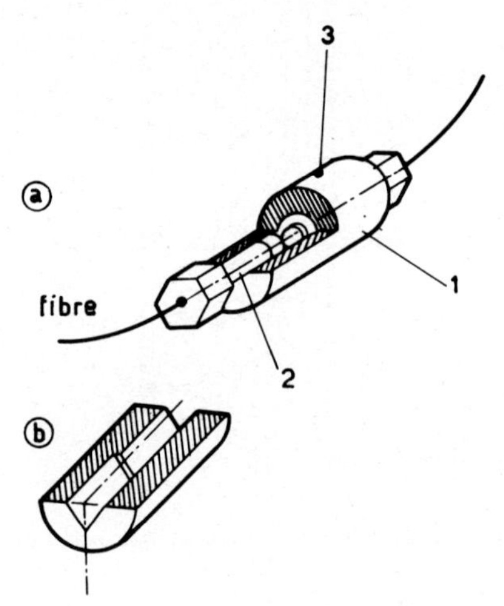

FIG. 57 – Eccentric connector: 1: housing; 2: pin; 3: plastic screw. (After [64])

The fibre is inserted eccentrically in one groove, fitted with a screw mounted at the front side of each pin and bonded with epoxy resin.

Coupling efficiencies between 0.5 and 0.2 dB were achieved using index matching liquid and single mode fibre with a 2 μm core.

Although still based on the eccentric sleeve principle, the next connector utilizes an original alignment system [65]. The fibres are mounted eccentrically on two displaced cylinders, which in turn are held in position in a groove by a spring coil. As shown in Fig. 58*a*, the groove consists of two transversally shifted parts. The fibre is housed in a capillary tube whose terminations facing the connector are cut at 45° (Fig. 58*b*). In this way, scattered light causing an unsatisfactory initial alignment can be collected perpendicularly to the fibre axis through an appropriate detector inserted into the connector itself. Thus, to achieve optimum

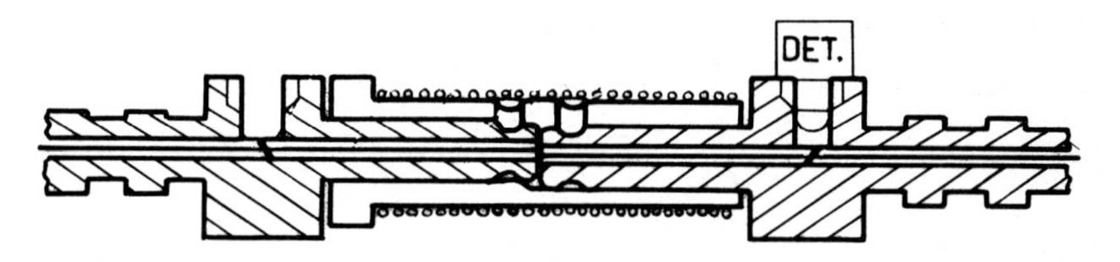

FIG. 58 *a*) – Mechanical outline of eccentric connector with scattered light monitor. (After [65]).

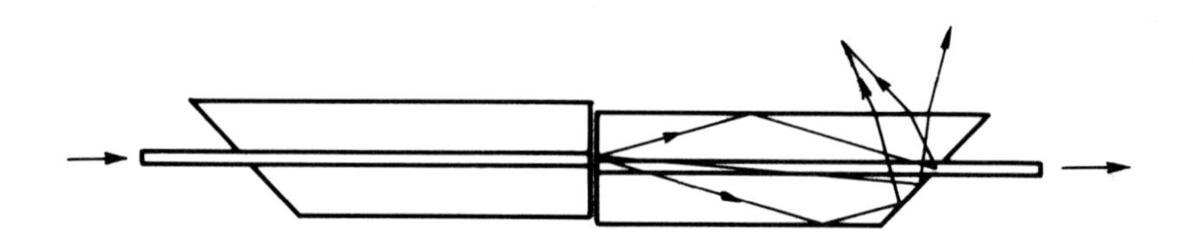

FIG. 58 *b*) – Principle of collecting scattered light. (After [65]).

coupling it is only necessary to rotate the two connector parts against each other until a minimum signal, picked up by the detector, is obtained. This adjustment appears to be much more sensitive (Fig. 59) than that based on the detection of the signal at the second fibre output and ensures reduced centering inaccuracy.

Indeed, with step index fibres (NA = 0.48, CD = 85 μm, OD = 105 μm, LED source wavelength = 800 nm), this technique gives a 0.087 dB mean loss, against a 0.124 dB value obtained with the usual technique; with graded index fibres (NA = 0.18, CD = 63 μm, OD = 130 μm) the new technique gives a 0.098 dB mean value, against a 0.115 dB value of the old one.

Finally, we will consider a connector which adopts the double eccentric principle [7] not for field adjustment but for mounting and initial alignment (through a microscope). Its diagram (Fig. 60) shows three eccentric tubes: an outer tube (1), an intermediate tube (2) and an inner tube (3). The optical fibre is mounted eccentrically in the inner tube. Denoting by l_1, l_2 and l_3 the eccen-

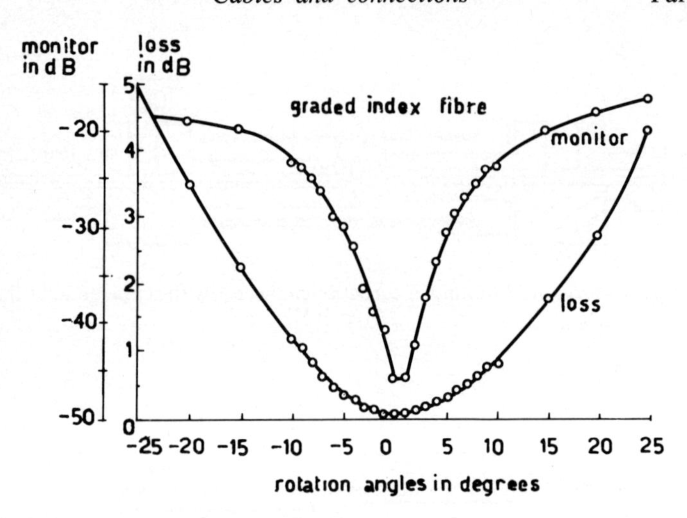

FIG. 59 – Adjustment sensitivity and monitor output level. (After [65]).

tricities of the above tubes with respect to the core centre, the area within which the core centre is always positioned as the tubes rotate is described by the following relationship:

$$|l_1 - l_2| \leqslant l_3 \leqslant l_1 + l_2 .$$

If inside and outside diameters of tube 3, in which the fibres are housed, have been suitably chosen, it is always possible (through rotations) to make the core centre coincide with the outer tube centre. This operation is monitored through a microscope with

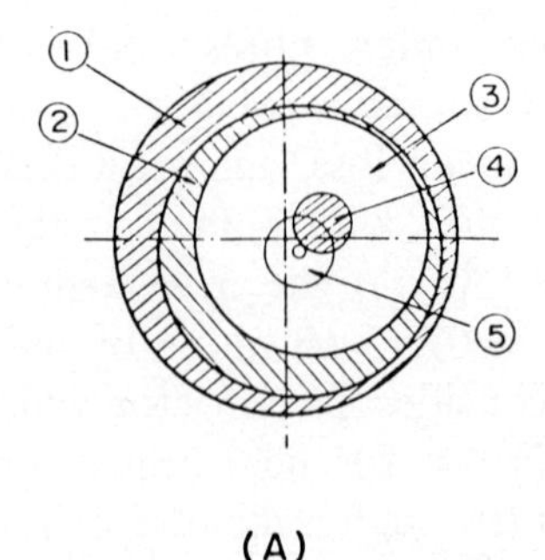

FIG. 60 – Double eccentric optical fibre connector structure. Connector plug sectional view: 1, outer eccentric tube; 2, intermediate eccentric tube; 3 inner tube; 4, optical fibre core; 5, movable core center area. (After [7]).

the aid of a reference tube. Once the two plugs have been centered and fixed mechanically, they are inserted into a jack to form the connector.

Figures 61*a, b* show the histograms of connecting losses with and without index matching. The connector has a 0.2 dB mean loss, with index matching, which becomes 0.48 dB without, using graded index fibres ($L_1 = 1$ m, $L_2 = 500$, $\Delta n = 0.7$ ‰, laser source wavelength = 632.8 nm). The residual eccentricity between core centre and plug has been measured to 2.3 μm, showing 1 μm as best and 6 μm as worst value.

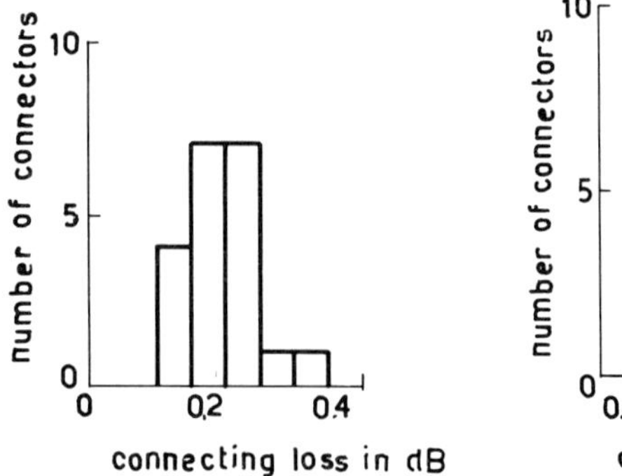

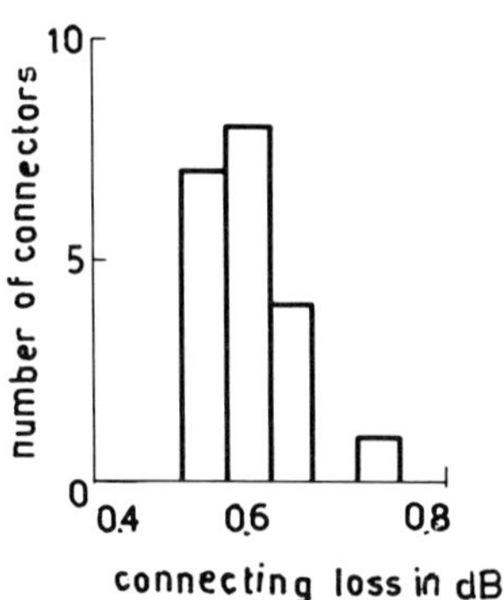

FIG. 61 – *a*) Histogram showing connecting loss with index matching (After [7]). *b*) – Histogram of the connecting loss without matching material. (After [7]).

2.3.5 *Moulded connectors*

These connectors are fabricated by directly moulding two tapers around the two fibres, which are then aligned by means of a central adapter [66-68]. The design goal is to implement a connector showing a displacement error less than 3 μm and a separation error less than 40 μm when fibres with 55 μm core diameter and 111 μm outer diameter are used. The material employed for transfer moulding the plug is epoxy with quartz powder additive, which does not damage the fibre during application. Furthermore, material, manufacture and assembly are inexpensive. The precision element for alignment is represented by the two conical surfaces, which are obtained through precision dies, machined with an accuracy better than 1 μm. Also, the central biconical adapter (Fig. 62) is moulded onto the precision metal die and

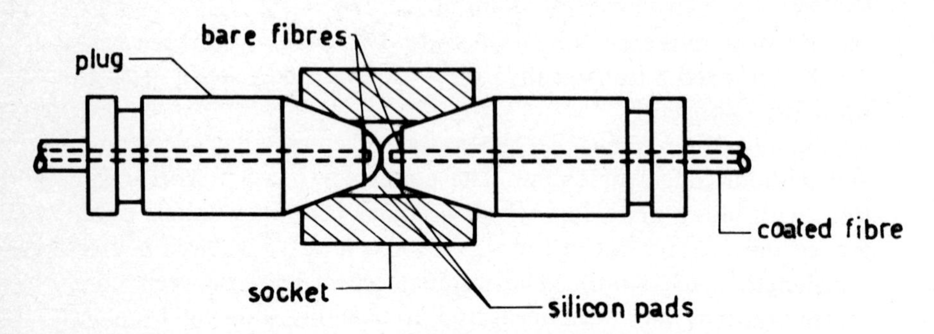

FIG. 62. – Molded-plug connector. (Reprinted from: Telephony, January 31, 1977, Issue copyright 1977 by Telephony Publishing Corp., 55 East Jackson BLVD., Chicago, IL 60604 USA).

allows a concentricity between the two connector tapers better than 1 μm. All operations involved are routinely performed. Measurements on 50 samples have shown an average loss of 0.38 dB, with index matching; a 360° rotation of one plug yields a loss variation of about ± 0.1 dB. Repetition tests have been conducted: for 100 connection-disconnection operation the initial loss has been measured in the region of ± 0.02 dB in 90% of all connections. Temperature changes between 0 °C and 60 °C have virtually no effect on connectors. Figure 63 illustrates a histogram of losses for about 50 samples. Graded index fibres have been used (CD = 55 μm, OD = 110 μm).

Another connector, suitable for optical fibre tapes and cables, is based on the structure depicted in Fig. 64. The plastic sheet

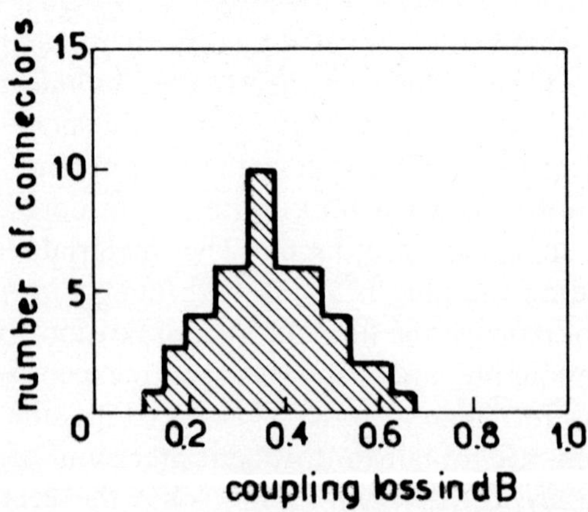

FIG. 63 – Coupling losses statistic. (After [66]).

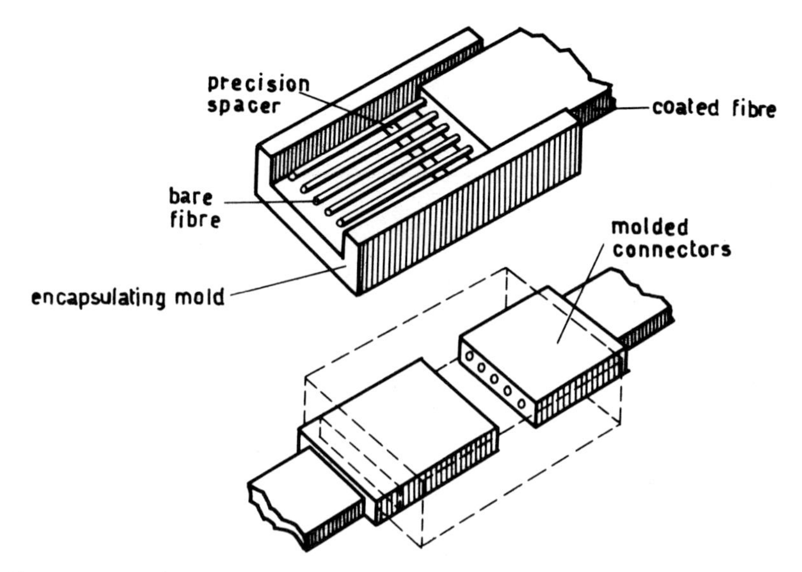

Fig. 64. – Molded plastic multiple connector. (Reprinted from: Telephony, January 31, 1977, Issue copyright 1977 by Telephony Publishing Corp., 55 East Jackson BLVD., Chicago, IL 60604 USA).

is removed from one tape end, which is then polished. Next, fibres are aligned on a precision spacer located in a mould. An appropriate plastic material [69] is moulded into the above mould and the resulting assembly is then scored and pulled over a curved surface with the procedure commonly adopted for single fibres. This process is repeated until the two ends are ready for assembly into a connection by means of the outer guide sleeve. A connector for multitape cable is obtained by stacking a number of these tapes. The implementation principle of this connector is depicted in Fig. 65. The connector shows losses in the range 0.2-0. 32dB, obtained with graded index fibres (CD = 55 μm, OD = 110 μm).

2.3.6 *Interposed optics connectors*

Another jointing technique collimates the fibre output beam, which is normally diverging, and then focuses it on the core of the receiving fibre by means of a second optical system. This procedure requires less stringent transverse and longitudinal tolerances than those examined so far. This is at the expense of a more stringent angular alignment, which, however, can be checked

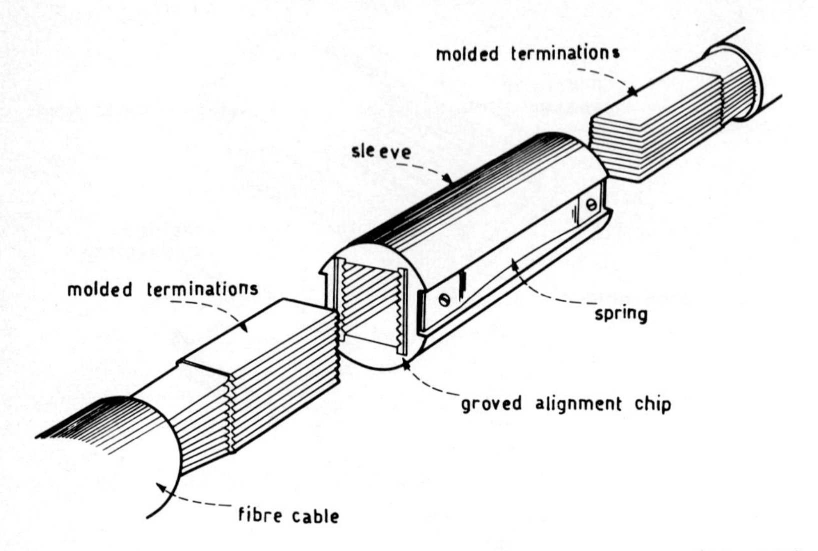

FIG. 65 – Grooved alignment sleeve for cable splicing and connectors. (After [69]).

more easily. Expanded beam connectors may have terminations built with different methods, i.e. tapers, lenses, holographic couplers. An example of a taper connector is shown in Fig. 66.

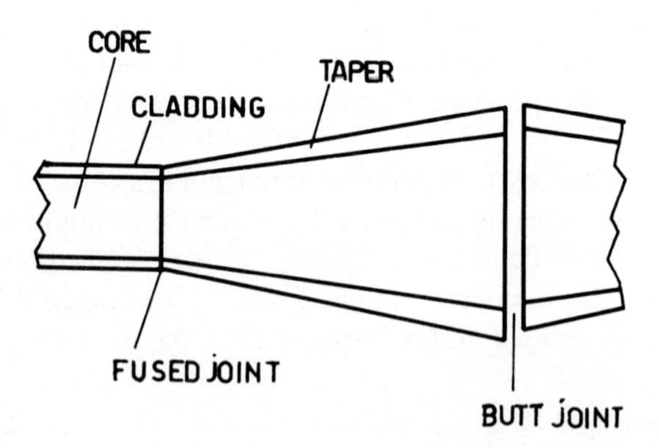

FIG. 66 – Expanded beam joint using tapers. (After [70]).

A cladded fibre taper is fused with the fibre to be coupled, forming a core of, say, 4 times the core size [70]. However, a low-cost, high-precision production of these taper sections is not yet possible.

A lens connector [70-72] has, however, been implemented. The fibre end requires accurate positioning at the lens focus. The lens is housed in an appropriate ferrule. Glass sphere lenses as well as low cost moulded plastic lenses can be used. The two lenses are bonded to their fibres with epoxy, which also serves as index matching. By virtue of the larger beam size, this connector proves also to be less sensitive to dust. Theoretical evaluations suggest that an additional 20 μm displacement may still ensure a 94% coupling efficiency in the case of a 50 μm optical fibre core, N.A. = 0.2, with 2 mm sphere lenses, and antireflection coating.

Another connector adopting two spherical lenses, making a 1-1 image of the emitting fibre upon the receiving one (Fig. 67),

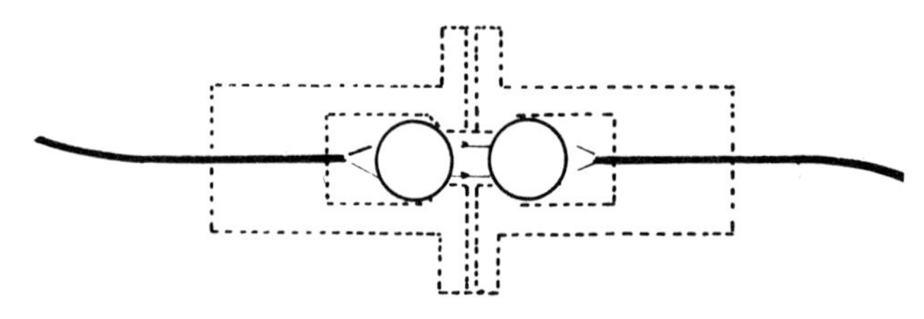

FIG. 67 – Demountable connector (schematic) with two lenses making a 1-1 image of the emitting fibre upon the receiving one. (After [73]).

is reported in [73]. The main feature of this connector is its insensitivity to lens surface contamination, because of the large width of the two coupled beams. Such a connector shows a critical angular alignment but, at the same time, the displacement and the separation of the two halves need less pressing tolerance.

Average losses of 1 dB are obtained which can be reduced to 0.7 dB using lenses with anti-reflection coating and graded index fibres ($S = 20$, CD = 50 μm, NA = 0.2). Sapphire balls have 1.5 mm radius and 1.77 refractive index.

A similar connector, using selfoc lenses instead of spherical lenses, is described in [74]. Like the one mentioned above, this technique strongly reduces problems of surface contamination.

Also in this case the critical factor is the angular misalignment, while the tolerance on the lateral displacement becomes greater.

A comparison between the losses of a butt joint and lens type connector is shown in Figs. 68, 69. The measured connection loss values of the lens type and of the butt joint type are 1.1 dB and

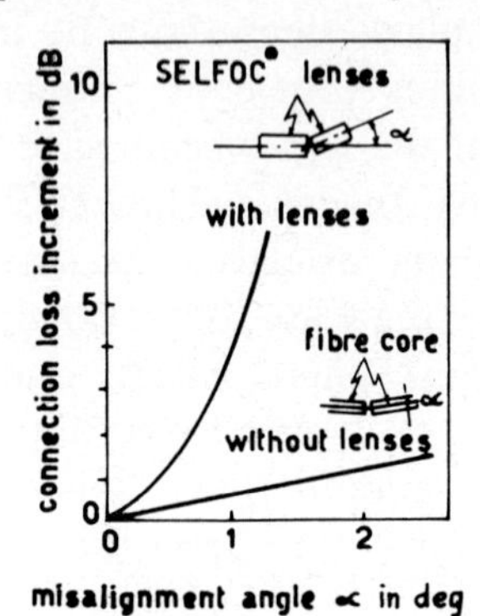

FIG. 68 – Measured losses vs. misalignment angle with and without selfoc lenses. (After [74]).

0.5 dB, respectively. In the lens type 1.8 mm diameter selfoc lens attached to a step fibre (CD = 60 μm, NA = 0.17) are used.

Connection loss variation (on 100 tests) are reduced to within

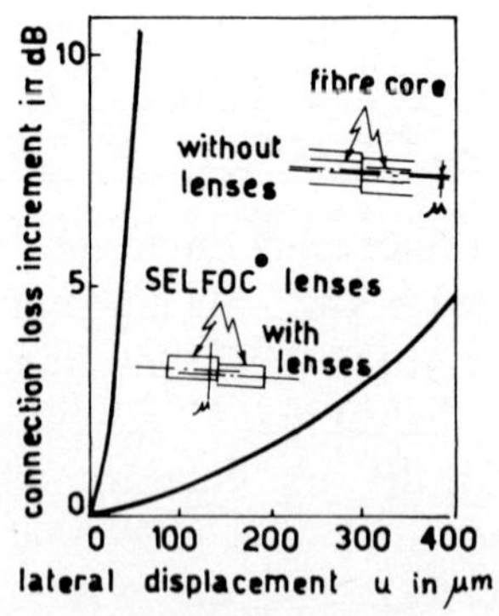

FIG. 69 – Measured lossed versus lateral displacement with and without selfoc lenses. (After [74]).

0.1 dB using the lenses, while conventional butt joint connectors showed variations of 0.25 dB.

A connector for two-fibre coupling can also be developed through holograms recorded on light-sensitive material [75]. It is based on the capacity of a hologram to convert wave front properties into those required. As shown in Fig. 70*a*, a holographic recording is made of the object spherical wave radiated by the transmitting fibre, using a plane wave as reference. A similar procedure is then followed for the receiving fibre. The two plates are developed with standard photographic procedures.

By illuminating the first hologram with the emitting fibre, a plane collimated outgoing beam is obtained, which the second hologram converts into a spheric wave, showing a divergence equal to that of the receiving fibre. The correct assembly of the

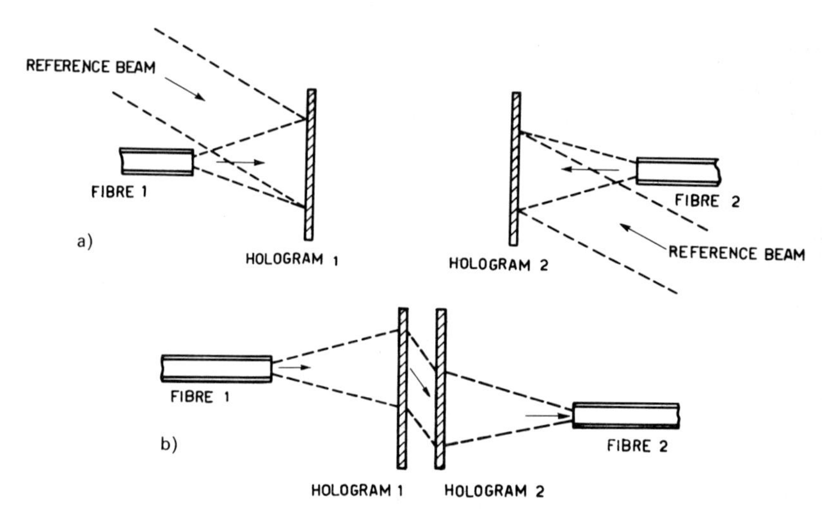

FIG. 70 – Principle of the holographic coupler: *a*) The two holograms are separately recorded. *b*) The coupling between two fibres is carried out by the compound holographic coupler.

fibres and holograms (Fig. 70*b*) does not call for a very accurate adjustment of lateral displacement with respect to the recording geometry.

2.3.7 *Other types of connectors*

A connector for individual fibres or multifibre cables is reported in [76]. The two optical fibres are mounted in a ring core held in place by two pressure plates (Fig. 71). A force applied to these plates causes the radial deformation of the elastic element clamping the fibres and the automatic alignment of the fibres themselves.

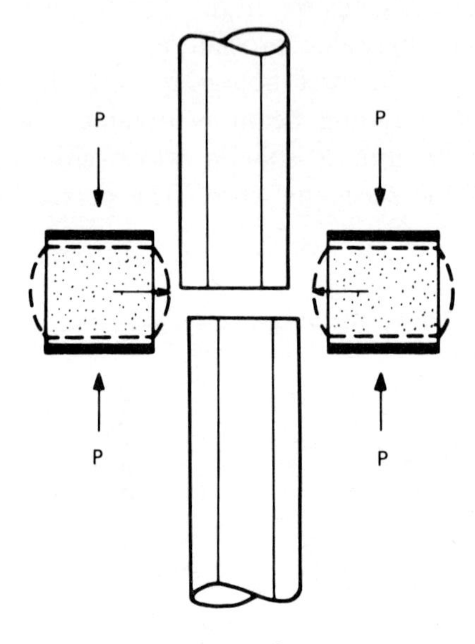

FIG. 71 – Elastic ring connector.

To simplify insertion and ensure efficient blocking, accurate dimensioning of the hole diameter is required. Insertion can be facilitated by suitably shaping the retainer rings. The coupling system may be extended to a number of fibres contained in a cable.

Finally, mention should be made of a magnetic connector for optical fibres [77]. The cylindrical surface of the two fibre ends is first covered with ferromagnetic material. Next the fibre ends are inserted into the connector, where their alignment has been prearranged. The final alignment should then be brought about through a magnetic field with force lines parallel to the fibre axis. In principle, its main advantage lies in its handy plugging and unplugging. Fibre end preparation, on the other hand, appears to be somewhat difficult.

Summary table of the characteristics of some single-fibre connectors.

Connecting method	Reference (Example of application)	Typical loss in dB (average)	Fibre structure (step index: st graded index: gr)
Watch jewel ferrule 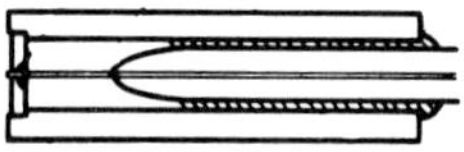	[43] I.T.T. (United Kingdom)	1	st - CD = 75 μm
Triple ball 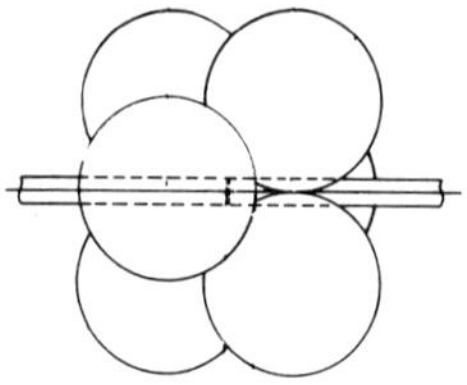	[48] Post Office (United Kingdom)	0.50	st - CD = 80 μm OD = 125 μm
Plug tubes 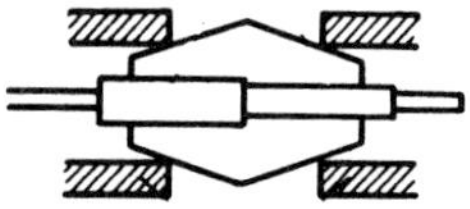	[58] NEC (Japan)	0.4	gr - CD = 65 μm

Summary table of the characteristics of some single-fibre connectors.

Connecting method	Reference (Example of application)	Typical loss in dB (average)	Fibre structure (step index: st (graded index: gr)
Eccentric tubes 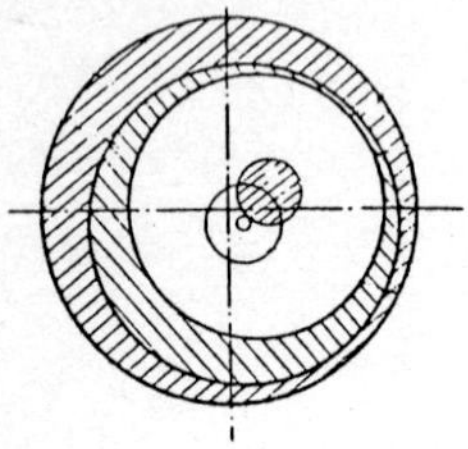	[7] NTT (Japan)	0.5	gr -
Eccentric tubes (adjustable with monitor) 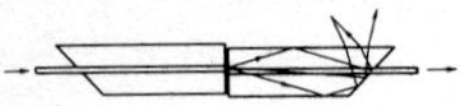	[65] Ericsson (Sweden)	0.1 (with index matching)	gr - CD = 63 μm OD = 130 μm
Moulded 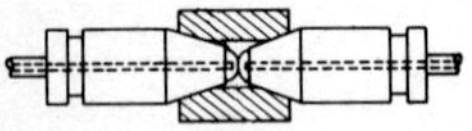	[68] Bell Lab., Western Electric (U.S.A.)	**0.38** (with index matching)	gr - CD = 55 μm OD = 110 μm

Summary table of the characteristics of some single-fibre connectors.

Connecting method	Reference (Example of application)	Typical loss in dB (average)	Fibre structure (step index: st graded index: gr)
Interposed lens 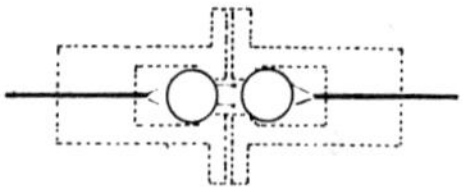	[73] Philips (Netherlands)	1 (without A.R. coating)	gr - CD = 50 μm

Summary table of the characteristics of some multi-fibre connectors.

Connecting method	Reference (Example of application)	Typical loss in dB (average)	Fibre structure (step index: st graded index: gr)
Rod array 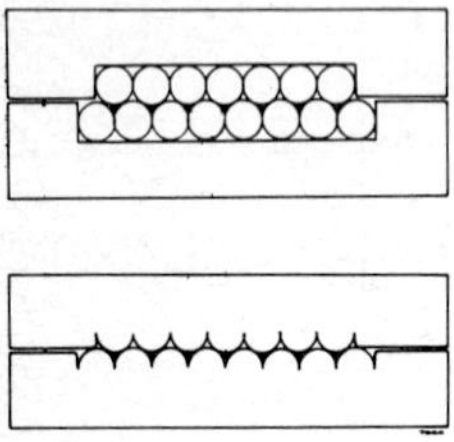	[6] Corning Glass Works (U.S.A.)	0.6 (with index matching)	st - CD = 85 μm OD = 125 μm
Multipole 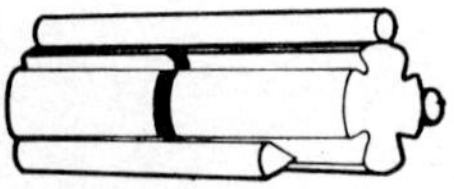	[47] (AEG Telefunken) West Germany	0.1 (with index matching)	st - CD = 60 μm for 3 fibres
Etched groove array (silica) 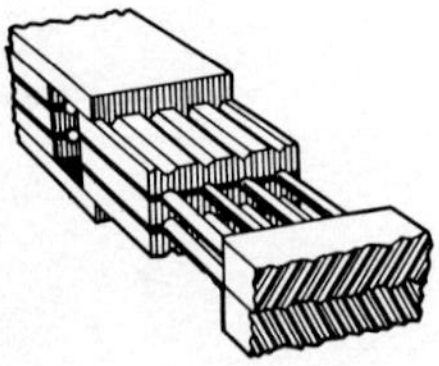	[37] Bell Lab. (U.S.A.)	0.2-0.3 (with index matching)	gr - CD = 55 μm OD = 110 μm for 12×12 fibers

Summary table of the characteristics of some multi-fibre connectors.

Connecting method	Reference (Example of application)	Typical loss in dB (average)	Fibre structure (step index: st graded index: gr)
Multiple etched groove (plastic) 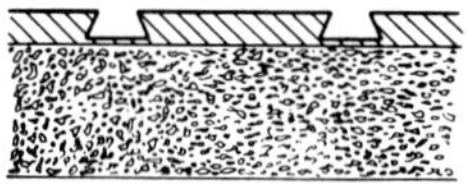	[51] Siemens (West Germany)	0.2 (with index matching)	st - CD = 92 μm OD = 132 μm

2.4 Source-fibre connections

2.4.1 *General*

The problem of jointing the source (LED or semiconductor laser) to the optical fibre is of particular importance for the implementation of an optical fibre communication system. The connection must basically ensure the highest possible power transfer from the source into the fibre. Thus, the launching efficiency is defined as

$$\eta_c = \frac{P_{\text{inj}}}{P_{\text{out}}} \cdot 100\,\% \tag{2}$$

where P_{inj} is the power injected into the fibre and P_{out} the output power of the source. This section will review the different jointing techniques suitable for efficiency optimization.

It should be noted that there is a tendency for researchers and manufacturers to supply the device already equipped with a short fibre (pig-tail) attached to the source itself. Consequently, the problem for the user will concern the connection between two fibres.

Source-fibre coupling can be through direct approaching (butt coupling) or through the interposition of optical systems to increase its efficiency. A theoretical discussion of the problems of direct coupling between a source and a fibre has been presented in Chapter 2 of Part I. Further information can be obtained from [78, 79], which describe the practical procedure for direct coupling a fibre to any source, in particular an LED or laser. Different models, specially designed for LED-fibre coupling, have been dealt with, for instance, in [80, 81], whilst for laser-fibre connections we recall [82].

These mostly theoretical studies supply approximated values of the direct coupling efficiency equal to a few percentage ($\leqslant 2\%$) for a Burrus type LED and 15-30% for a double heterostructure laser. These figures closely depend on the geometric characteristics of the source and fibres. The above values refer to standard types of multimode graded index fibre and currently used optoelectronic components. When the emitting surface of the source is smaller than the fibre core, the launching efficiency can be improved through the interposition of suitable optical systems [83]. This is possible in practice with stripe geometry lasers ($\sim 2 \times 15$ μm in dimension) and with the latest type of LEDs, with emitting area diameter of 30-40 μm, by using multimode optical fibres with core diameter of 60-80 μm and $\mathrm{NA} \simeq 0.15$. The practical implementation of source-fibre coupling, which unlike fibre-to-fibre jointing is usually permanent, calls for an accurate positioning of the fibre with respect to the source so as to keep coupling losses to a minimum by restricting mechanical tolerances to specified limits. These tolerance limits have been derived from experimental investigations on adopted components. Numerical values are presented in individual cases. In general, displacement appears critical. This is true in particular of a semiconductor laser for the displacement in the plane normal to the junction. Adoption of index matching may reduce reflection losses at the system interface.

2.4.2 *Semiconductor laser-optical fibre coupling*

The problems relative to the coupling between a semiconductor laser and an optical fibre are strictly dependent on the emission geometry of the source, which is exhibited by the near- and far-field curves. An example of a far-field curve is illustrated in

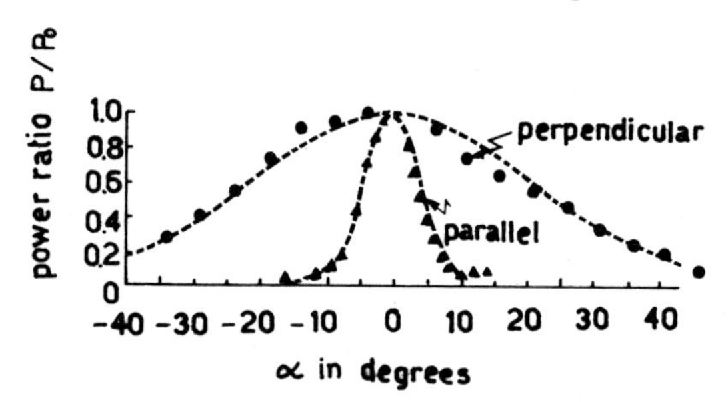

FIG. 72 – Far field intensity patterns from a junction laser measured in the planes parallel and perpendicular to the junction. The dashed lines are Gaussian fits to the intensity patterns. (After [84]).

Fig. 72, given in [84] for a double heterostructure laser having a 13 μm wide stripe: the emitted beam diverges much less in the plane parallel to the junction than in the normal one.

The need to couple an asymmetrically diverging beam to a circularly symmetric fibre makes optimization and design of interposition lenses rather difficult.

Direct coupling between a laser and an optical fibre is in any case much easier to accomplish, as only two objects need to be aligned. An example of this type of connection is presented in [85] taking into account implementation and manufacturing techniques. The laser is supplied with a short fibre (F_s), the end of which has already been prepared, and then bonded to a holder, to which the chip has been connected (Fig. 73). The assembly is hermetically

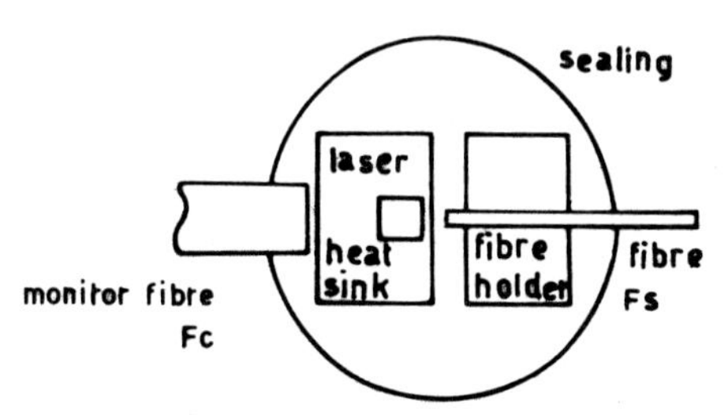

FIG. 73 – Configuration of a laser chip with main fibre (F_s) and monitor fibre (F_c). (After [85]).

sealed in a nitrogen atmosphere, to prevent contamination of the emitting surface. Figure 74 shows coupling losses as a function of displacement in the planes parallel and perpendicular to the junction. In this case the coupling efficiency ranges from 10%

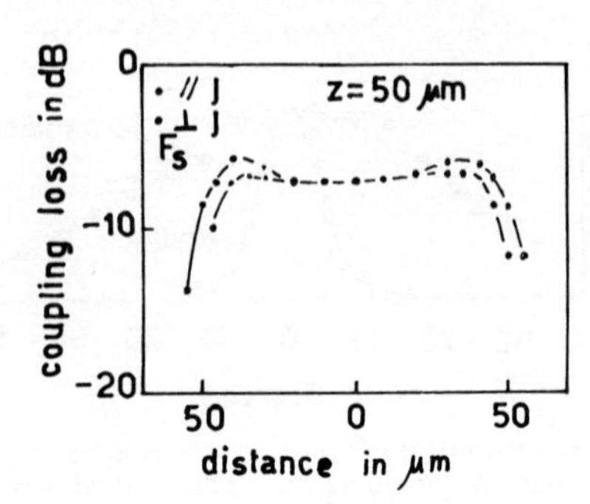

FIG. 74 – Coupling loss as a function of displacement between optical axes of a laser and a fibre. (After [85]).

to 20%, corresponding to losses between 10 and 7 dB. With the type of laser adopted here, this means 1-2 mW at the output of the fibre (step index, multimode, NA = 0.14, core diameter = = 85 μm). The DH laser emits in the region of $\lambda = 0.80\text{-}0.85$ μm with emitting stripe of 15-30 μm and oscillates on the first or second mode.

The package also contains another fibre (F_c) which detects a reference signal from the rear facet of the laser. As already mentioned, the coupling efficiency can be improved by interposing a converging optical system between the emitting surface and the fibre. This can be accomplished in different ways:

a) with a spherical lens, bonded to the fibre end or directly obtained by rounding the end through melting (bulb-ended fibre);

b) with a cylindrical lens, consisting of a glass rod of suitable size and refractive index, or obtained by direct etching on the fibre surface;

c) coupling using selfoc and cylindrical lenses;

d) tapering the fibre end so as to modify its acceptance (and the geometric configuration of its termination).

a) Coupling between laser and bulb-ended fibre.

The efficiency increase made possible with this techniques (i.e. fibre termination as shown in Fig. 75) is demonstrated by different papers. See in particular [86]. It can be shown that angle θ_1 within which all rays are picked up by the core is given by the

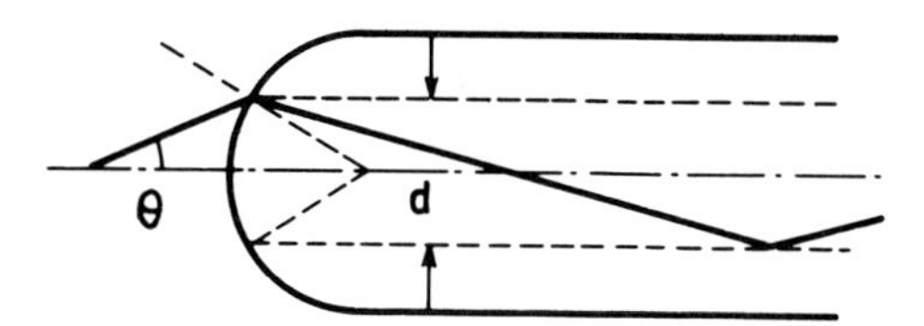

FIG. 75 – Spherical-ended fibre. (After [86]).

following relation [87]:

$$\theta_1 = \sin^{-1}\{n_1 \sin [\sin^{-1}(d/2r) + \cos^{-1}(n_2/n_1)]\} - \sin^{-1} d/2r \quad (3)$$

where: d = core radius;
r = radius of curvature;
n_1 = core refractive index;
n_2 = cladding refractive index.

The acceptance angle and coupling efficiency as functions of the percent difference of indices n_1 and n_2 are given in Fig. 76. The theoretical study has confirmed the possibility of increasing coupling efficiency to 60 % for spherical ended fibres.

Efficiency has been evaluated experimentally by also taking

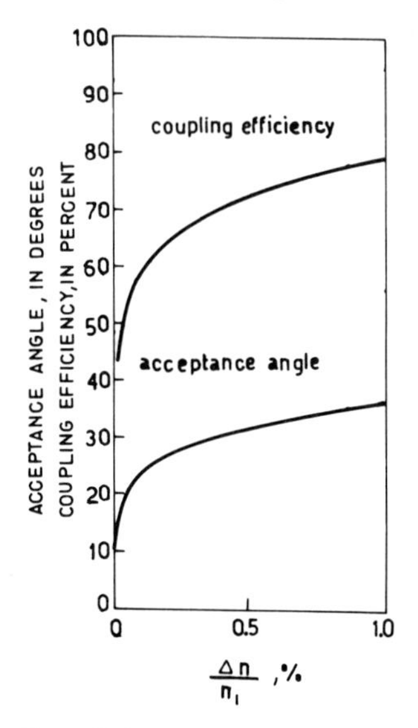

FIG. 76 – Maximum coupling efficiency and acceptance angle for spherical-ended fibre as a function of the core-cladding index difference $\Delta n/n_1$. (After [87]).

into account the effect of the different curvature radii of the lens. The efficiency of a glass fibre having CD = 80 μm and $\Delta n/n_1 = 0.3\%$ has been measured as a function of the curvature radius. The most favourable results have been obtained with radius values of 65-80 μm using a DH laser with Gaussian beam and half apertures of $6° \times 45°$ at point $1/e^2$.

Another technique for fabricating a lens at a fibre end is illustrated in [84]. The fabrication of a microlens is schematized in Fig. 77*a*, *b*. After thorough cleaning, the surface is coated with

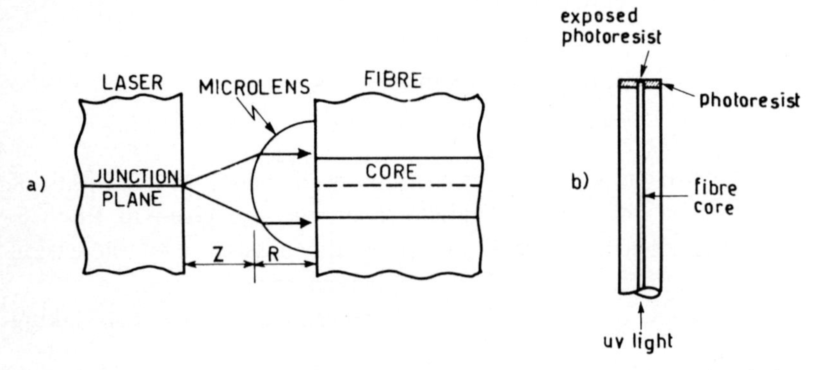

FIG. 77 – *a*) Experimental arrangement for automatically exposing a hemispherical lens over the core of a single mode fibre. (After [84]). *b*) Arrangement for coupling light power from a junction laser across a small air gap into the core of a single mode fibre. The fibre has a microlens over its core. (After [84]).

transparent photoresist ($n = 1.6$) and then illuminated through the core of the fibre. The illuminated area further undergoes an etching operation, whereas the remaining part is removed.

The Gaussian distribution of the beam guided through the core and a final baking make it possible to achieve spherical surfaces with the required accuracy. Using single mode fibres having a CD = 4 μm and a pulsed DH laser having an emitting stripe of 0.6×4 μm it has been possible to obtain—with the aid of the above-hemispherical lens—a 23% efficiency as against the 8% achieved through direct coupling. An improvement of about 3 dB in coupling efficiency may be obtained by forming an epoxy resin drop (transparent at 850 nm) at fibre end [88] simply by dipping the optical fibre into the resin.

A check of drop diameter can be made indirectly by varying the density of the epoxy used. Using this method with optical fibres having core and cladding diameter of 100 μm and 130 μm respectively, end curvature radii betxeen 70 and 120 μm can be obtained.

Figure 78 shows the power coupled into the fibre (with or without lens) as a function of lens-laser surface separation [88].

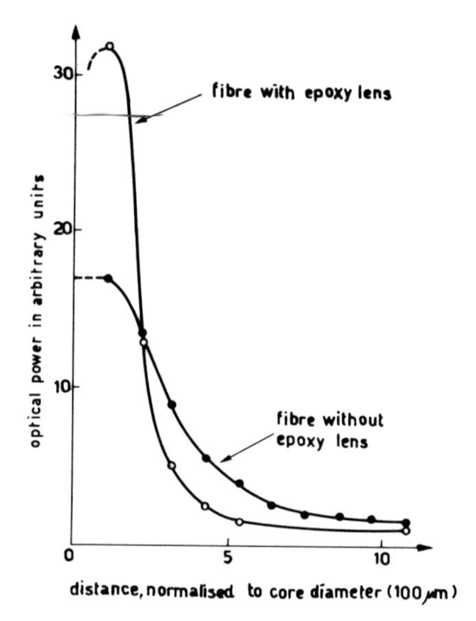

FIG. 78 – Power coupled into the fibre vs. fibre-laser surface separation (with and without lens). (After [88]).

The work reported in [89] describes in some detail the relationship between the chemical composition of the fibre and its behaviour when heated as regards curvature radius and cladding deformation. The fibre is first cut and the fibre tip is then melted by means of a microtorch (natural gas + oxygen). Fibre end distortion varies with type of fibre used, as shown in Fig. 79.

With the *a*) type fibre (the best as far as this problem is concerned) the cladding has a lower melting point and withdraws from the fibre end, leaving the core free to assume the shape of a high acceptance hemispherical lens; efficiency increases from 24 % to 63 %. This is also confirmed by the theoretical study [87], which predicts maximum coupling efficiency when lens curvature

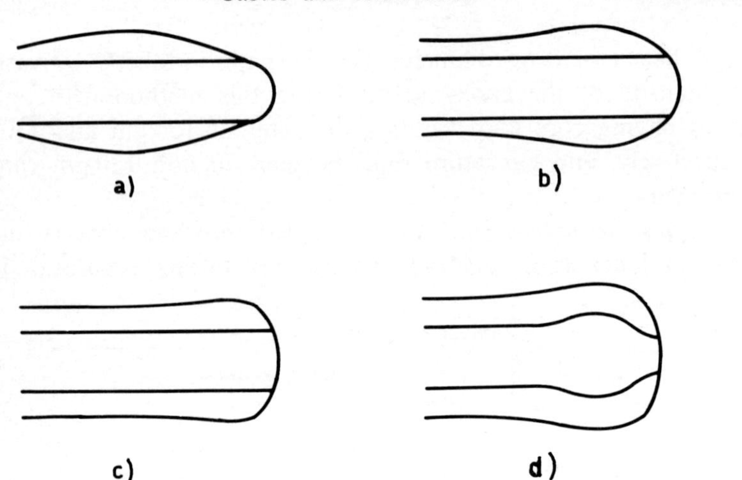

FIG. 79 – Lenses formed by melting of four different fibres.

radius equals core radius. This is in fact what happens with the above fibre. In the *b*) case, on the other hand, the core curvature radius is too small. In cases *c*) and *d*) the cladding shows a higher melting point than the core and tends to overlap the core itself. In the last three cases, efficiency improvement is less than in the first case.

A system using a CO_2 laser for the fabrication of a spherical lens on the fibre end is described in [90] and depicted in Fig. 80.

The problem relating to laser positioning on the heat sink is described in [91]. As a matter of fact, efficient coupling between a semiconductor laser and a flat-ended fibre is only possible if the laser is placed on the heat sink rim itself. However, as this is not usually possible, owing to cladding thickness, which is generally greater than the distance between active area and heat sink, a considerable loss is produced through displacement along the plane perpendicular to the junction. In the work cited above, this disadvantage has been avoided by removing all or part of the cladding by etching (Fig. 81). Coupling efficiency can be further improved by rounding the fibre ends. Moreover, lens materials with a low melting point are recommended to overcome the problem of the difference in melting point between core and cladding. In the case of quartz fibres, the material used is epoxy or glass (particularly suitable for high temperatures), whilst for glass fibres

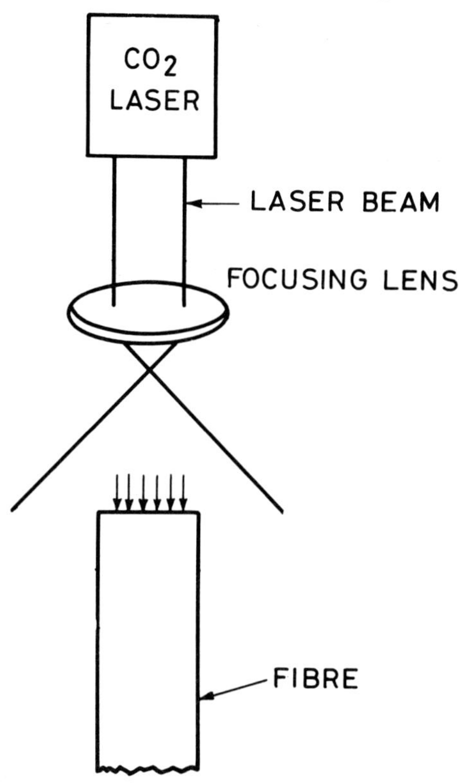

FIG. 80 – Fabrication of a spherical lens on the fibre end using a CO_2 laser. (After [90]).

epoxy is adopted. Whole or part cladding removal also facilitates the production of lenses of the required diameter through glass or liquid epoxy dipping. For instance, with a parabolic fibre (NA = 0.13, CD = 80 μm, OD = 125 μm) a 80% efficiency has been reported.

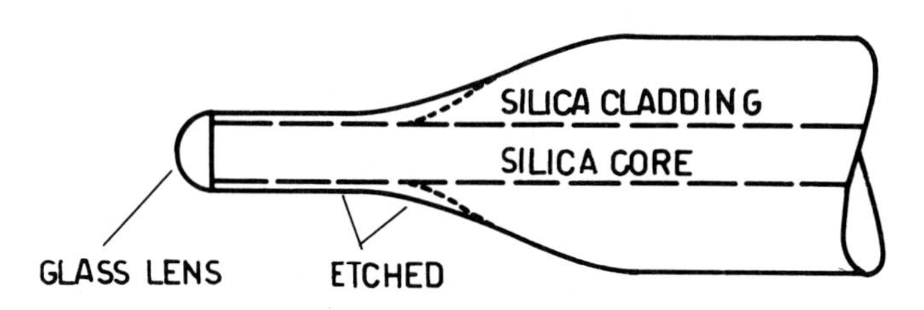

FIG. 81 – Etched silica fibre with glass or epoxy lens: —— step index, - - - graded index. (After [91]).

b) Coupling between laser and optical fibre through cylindrical lens.

A further increase in launching efficiency can be achieved by improving the matching between the beam emitted by the DH laser and the fibre acceptance. As is well-known, the former has a much higher divergence ($\sim 20°$) in the plane perpendicular to the junction than in the parallel plane ($\sim 4°$), while the latter has circular symmetry. The use of a cylindrical lens to make the beam converge only in the plane perpendicular to the junction may considerably enhance this situation. In fact, unlike the spherical lens which acts in the same way in all planes, the cylindrical lens can be adopted in one plane only, if this is so required, leaving unchanged the beam in the plane parallel to the junction. Here the divergence of the semiconductor laser is already below the fibre acceptance.

A first example of this technique is given in [84], where a hemicylindrical lens of transparent photoresist ($n \simeq 1.6$) is prepared directly at the fibre end, Fig. 82, using a photolithographic technique (similar to that already described for the fabrication of hemispherical lenses). With a single mode optical fibre (CD = 4 μm, $n\% = 0.3$) an efficiency of more than 30% can be achieved compared to 8% without a lens and 20% with a hemispherical lens.

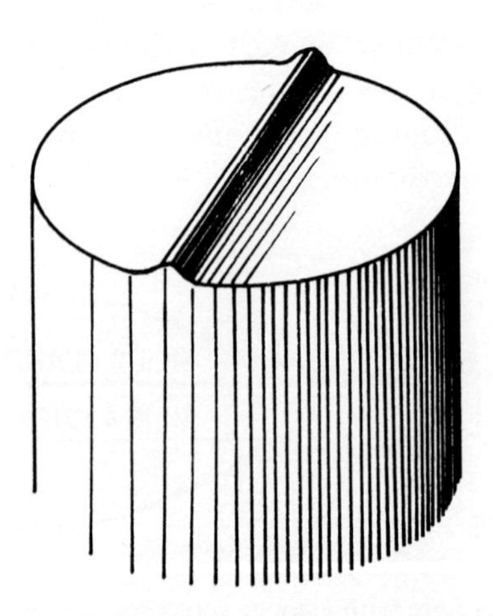

FIG. 82 – Cylindrical lens grown on a fibre end.

In [92] the hemispherical lens is obtained directly from the fibre glass through etching. The required roundness and smoothness of the lens surface is achieved by heating the almost trapezoidal preform obtained. With a glass fibre the efficiency goes up from 16 % (without the lens) to 40 % (with the lens). On the other hand, the fibre-lens system appears to be more sensitive to displacement than the fibre only system. In fact, to keep coupling efficiency better than 90 % of the maximum possible efficiency, it is necessary to keep the displacement within 1 μm for the fibre without a lens and within only 0.2 μm for the fibre with a lens.

The dependence of launching efficiency on displacement and separation is shown in Figs. 83*a*, *b*, *c*. It should be noted that the peak in Fig. 83*c* corresponds to the lens focal distance (~ 6 μm).

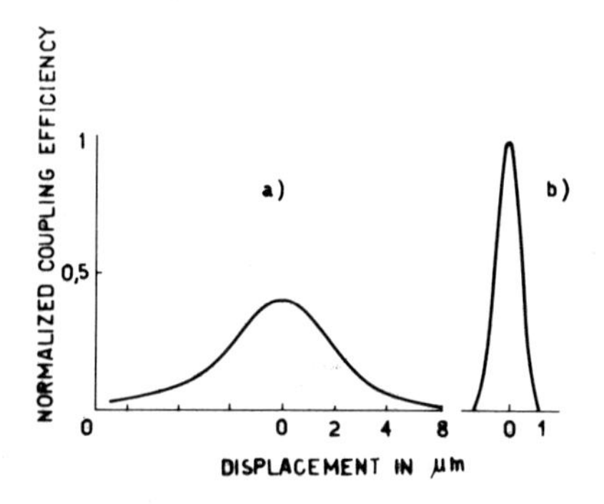

FIG. 83 *a*, *b*) – Normalized coupling efficiency vs. transverse displacement perpendicular to the *pn*-junction: *a*) plane fibre end face, *b*) fibre with a cylindrical lens on the end face. (After [92]).

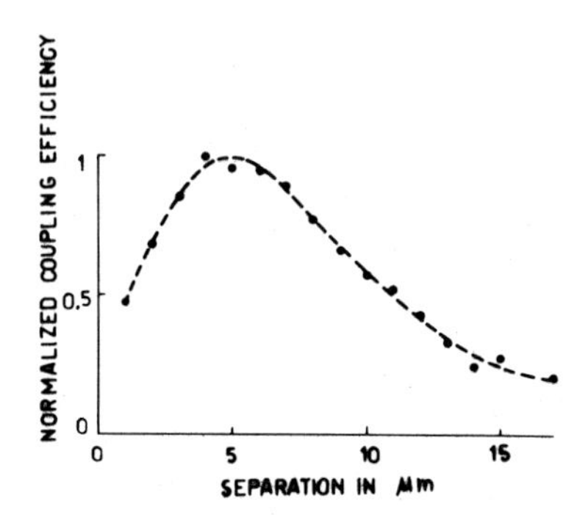

FIG. 83 *c*) – Normalized coupling efficiency vs. laser-cylindrical lensed fibre separation. (After [92]).

This lens type should show better resistance to optical and mechanical damage than similar plastic lenses.

The construction is simplified further with the structure of Figs. 84*a*, *b*, after [93]. This lens consists of a simple glass rod, which

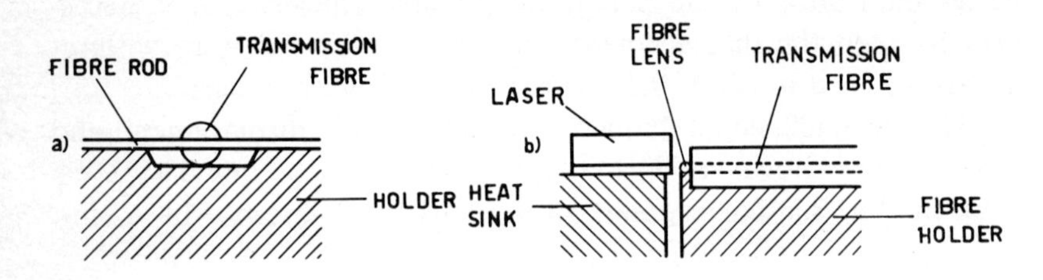

FIG. 84 – *a*) Location of the glass rod cylindrical lens (After [93]). *b*) Cross-section of coupling device with fibre lens. (After [93]).

has the same function as the other types of cylindrical lens described above. For a given transmitting fibre and laser the critical parameters of the lens are diameter, refractive index and laser-lens distance. These values can be selected so that a beam with a virtually circular section is obtained on the transmitting fibre flat face. The reflectivity of this face produces a Fabry-Perot effect, which depends on the position and reflectivity of the laser output mirror. This causes a damped oscillating behaviour of the output power as a function of the distance of the lens from the laser, with maximum values located at half wavelength intervals. Coupling more than 90% of the maximum power coupled by the laser into a single mode fibre (CD = 4.5 μm, $n\%$ = 0.53) requires an axial positioning tolerance of 0.05 μm. A 3 μm tolerance must be introduced, on the other hand, to couple more than 50%. Now, since these values are extremely critical, it is preferable to apply an anti-reflection coating to the fibre face which offers a 70% efficiency and avoids the Fabry-Perot effect. Particularly high also is the sensitivity to transverse displacement. If the above fibre is used, keeping the efficiency within 30% of the optimum value will mean that the displacement in the plane perpendicular to that of the junction must not exceed 0.2 μm. A similar coupling system between laser and multimode optical fibres is described in [94].

For a step-index optical fibre the core diameter of which is 50 μm, a theoretical direct coupling efficiency of about 20% is reported, against an experimental value of 15%. By interposing a cylindrical lens, variable efficiencies between 55 and 73% have been experimentally measured.

Using a graded optical fibre having a 25 μm core diameter 31-35% of the power of the laser used has been coupled without interposition of optics, against the theoretical evaluation of 35-40%.

The Fabry-Perot effect is shown in Fig. 85, in terms of coupled power as a function of the distance between laser and fibre face.

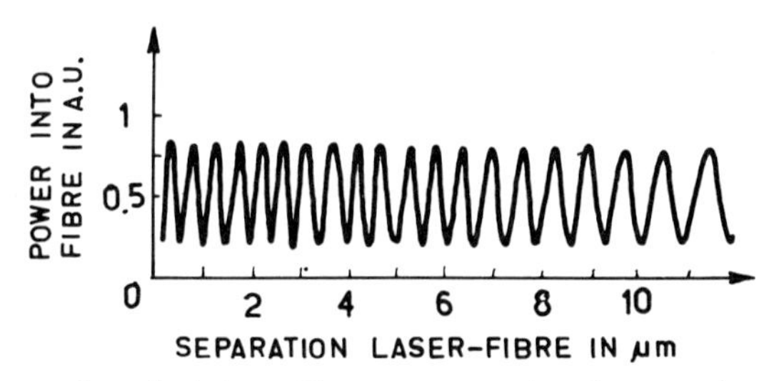

FIG. 85 – Power launched into fibre vs. separation between laser mirror and fibre end face; the Fabry-Perot effect is visible. (After [94]).

Figure 86 shows the coupled power curve in a step index fibre (CD = 50 μm) connected to a laser as a function of the laser-fibre separation, with or without lens. The dotted area highlights the Fabry-Perot effect over short distances. To keep the launching efficiency within 90% of the optimum value, in case of a coupling with a cylindrical transverse lens of 60 μm diameter, laser position tolerances with respect to the lens-fibre system are: ± 5 μm for separation and ± 3 μm displacement in the plane perpendicular to that of the junction and ± 10 μm on the parallel plane. Further, misalignment values are given which must be kept within 1° and 2.5° in the perpendicular and parallel planes with respect to the junction plane, so as to keep efficiency within 90% of the highest efficiency obtained.

In [95] is described the implementation of a coupler between a laser and an optical fibre using a cylindrical lens. The influence of geometric parameters (laser-lens distance, lens radius, different values of the offset between laser axis and fibre axis, as shown in Fig. 87), and measured coupling efficiencies are displayed in Fig. 88. The above diagrams also supply the optimum value for

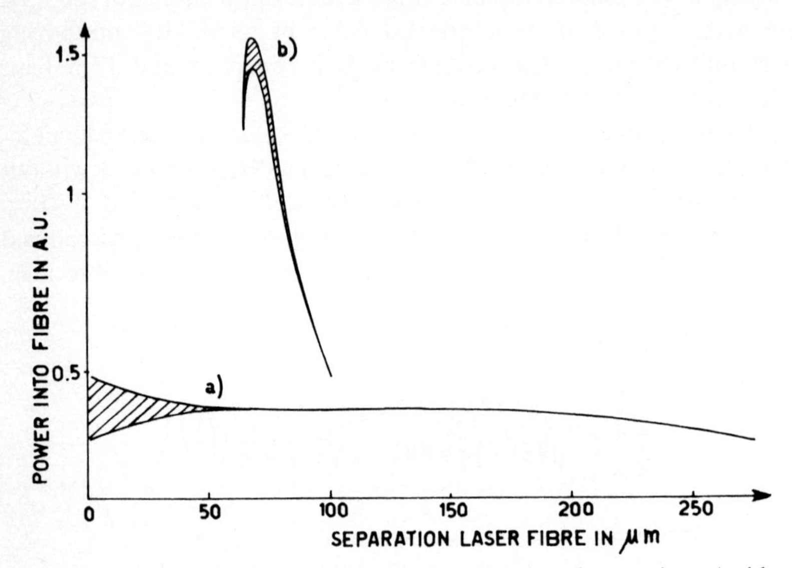

FIG. 86 – Light power launched into fibres as a function of separation: *a*) without lens; *b*) with lens. The Fabry-Perot effect is also visible (shadowed areas). (After [94]).

laser-lens separation, which may reach 80% for $R = R_c$ and offset $y =$ nil,whereas the experimental data shows a slightly lower value (70%).

Figures 89*a*, *b* illustrate an assembly schematic for a suggested

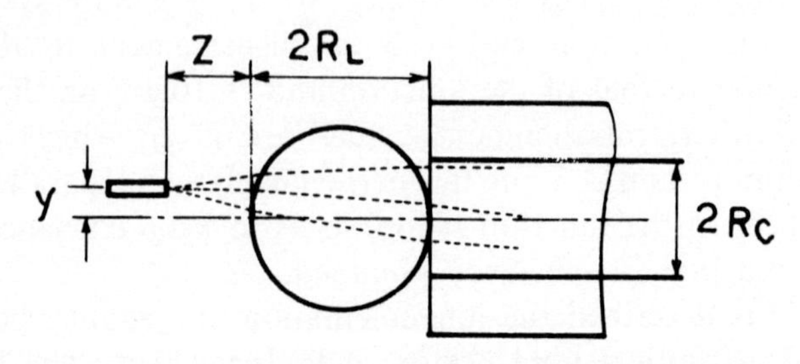

FIG. 87 – Schematic of a laser-to-fibre coupling using a cylindrical lens. Offset *y* and separation *z* are shown. (After [59]).

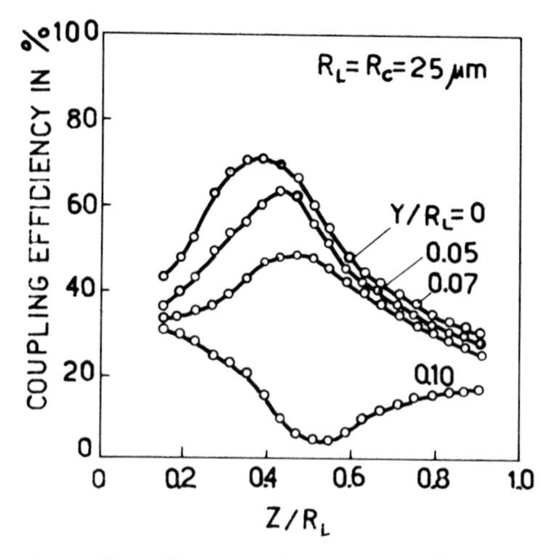

FIG. 88. – Measured coupling efficiencies for a laser-to-fibre coupling scheme with a cylindrical lens. They are shown with the misalignment parameters of the laser junction with respect to the lens axis defined in Fig. 87. (After [95]).

laser-fibre coupling with a cylindrical lens. The histogram of Fig. 90 shows measured coupling efficiencies with a 50 % average value.

A coupler between an array of 13 lasers and an equivalent number of fibres coupled simultaneously through a single cylindrical lens has been reported in [96]. The basic diagram is depicted

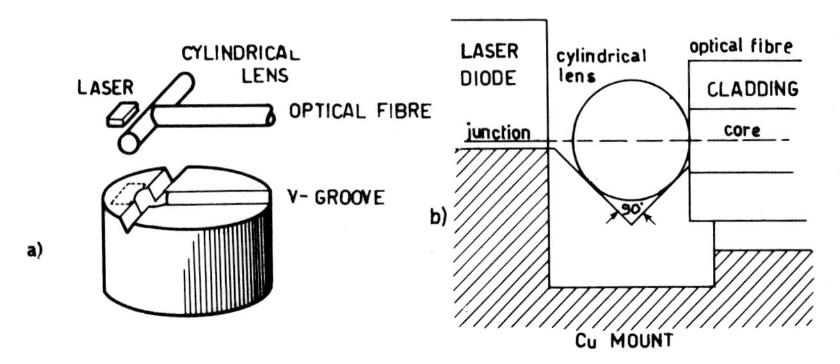

FIG. 89 – *a*) Schematic of a proposed mounting structure of the laser-to-fibre coupling scheme with a cylindrical lens (After [95]). *b*) Magnified diagram of the coupling section of the proposed laser-to-fibre coupler. (After [95]).

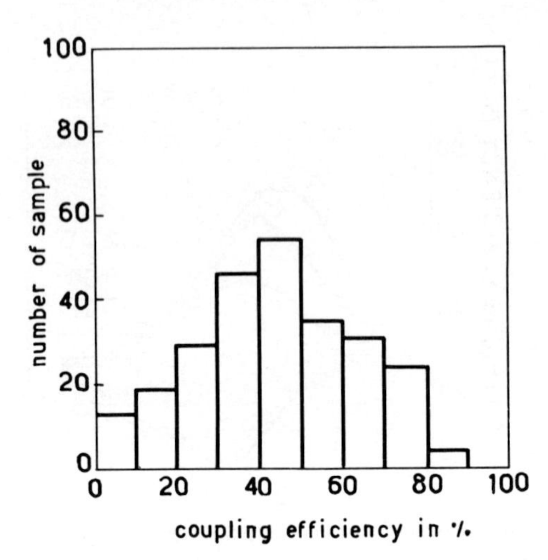

FIG. 90 – Histogram of measured coupling efficiencies for the couplers fabricated according to Fig. 89 (After [95]).

in Fig. 91. One cylindrical lens (70 μm in diameter, with 1.6 refractive index) can couple graded optical fibres (with NA = 0.15 and CD = 55 μm) to GaAs lasers having a 0.2-0.3 thick emis-

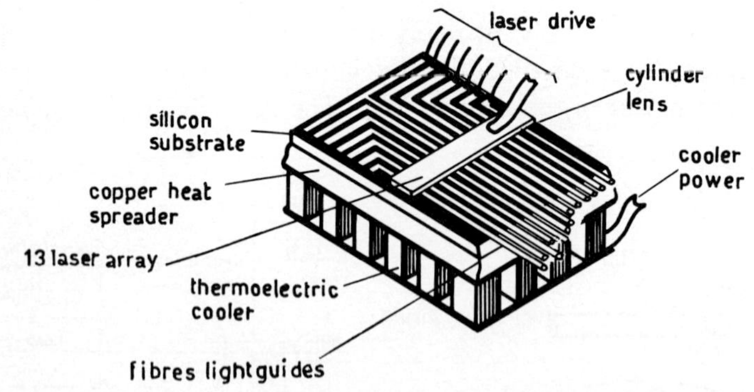

FIG. 91 – Schematic of the assembled laser array pakage, using a thirteen-laser array, as an example. (After [96]).

sion region and a 200 μm spacing. Fibres and lens must be first aligned through a precision groove and then bonded with epoxy as a package. Figure 92 shows the collimating effect of the cylindrical lens, i.e. the reduction in divergence of the emitted beam from 23° to 4°, half-width, half maximum, obtained using a cylindrical lens with 36 μm radius, placed 10 μm away from the laser. In the optimum coupling configuration, efficiencies of 70 % have

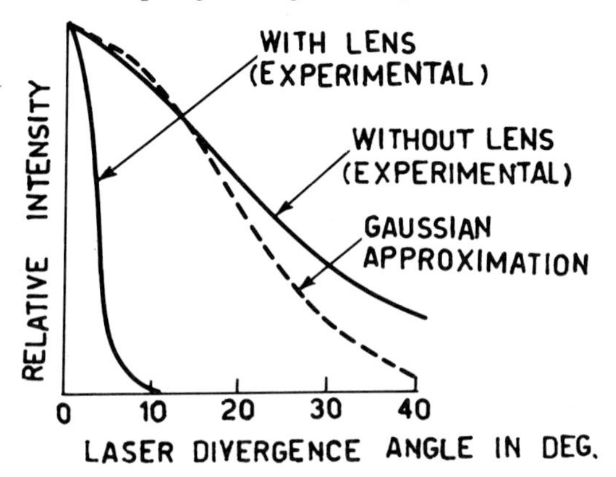

FIG. 92 – The experimentally measured far-field intensity profile of a laser with and without a 36 μm diam lens placed ≃ 10 μm away. (After [96]).

been reported. This value is particularly sensitive to the polarization current. As may be seen, configurations with a number of components of this type are subject to severe limitations in terms of band (5 MHz, — 3 dB) owing to the presence of relatively long and near biased metallized lines.

The use of spherical lenses shows some drawbacks [97]. The first is due to the fact that coupling efficiency increase appears to be small when the core radius is small compared to the lens curvature radius (which is actually equal to the fibre outside radius). Further, the spherical lens excites high-order modes, which are slow and loss producing. Lastly, the spherical lens is less efficient when used with graded fibres. The use is proposed of a hyperbolic profile lens, which should be free from these defects. Figure 93 illustrates the end profile of the fibre used. Efficiencies predicted for this type of lens are in the region of 70 %; experimental results have shown somewhat lower values.

c) Coupling between laser and optical fibre through combined lenses

In [98] a description is given of the use of selfoc lenses for adapting laser emission geometry to fibre acceptance. First, the use of only one small, easily alignable circular selfoc lens is suggested. However, it has been stressed that the use of a circular lens with an elliptical source does not bring about a sufficient increase in the launching efficiency. Thus, another configuration

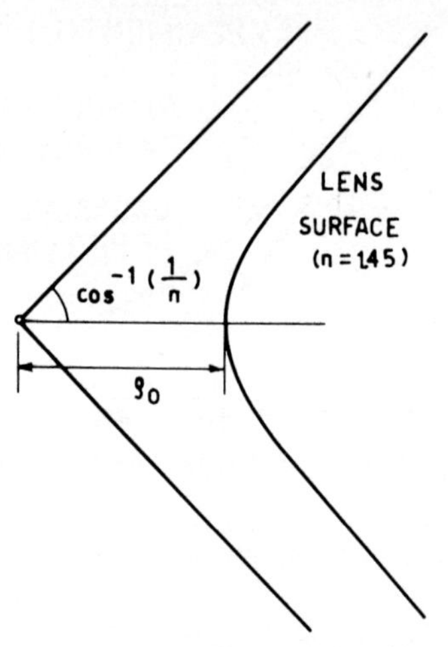

FIG. 93 – Hyperbolic lens profile. (After [97]).

is proposed using a slab-selfoc lens (the function of which is equivalent to that of a cylindrical lens), followed by a selfoc circular lens, which produces the spot size-matching. In this system lens positioning appears to be particularly crucial. A third proposal (Fig. 94) is based on two slab-selfoc lenses, one of which has a

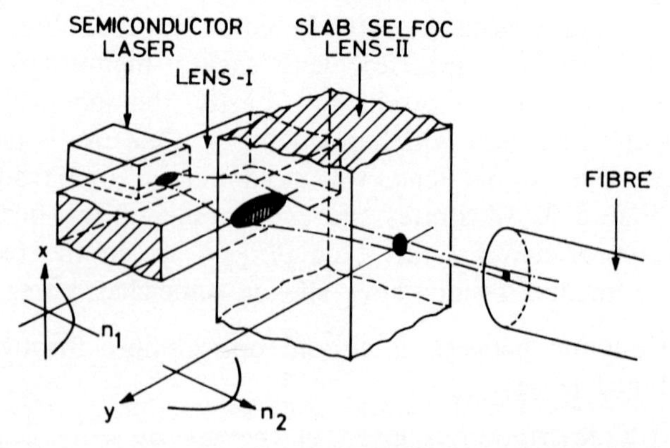

FIG. 94 – Schematic diagram of a coupling configuration using two perpendicularly crossed slab lenses. (After [98]).

much shorter focal length than the other. In this way the ellipticity of the beam emitted by the laser has been eliminated and the spot size matching required for the fibre used can be achieved. In Table I a comparison is made of the coupling losses for the first and the third systems, considering two different divergence values of the beam outgoing from the laser. These losses also include reflection losses: 1 dB in the case of two lenses and 0.5 dB in the case of one lens. The use of anti-reflection coatings on the facets may improve this value.

TABLE I (After [98])

Radiation angle	Slab Selfoc Lenses	Circular Selfoc Lens
16 deg.	1.8 dB	3.2 dB
40 deg.	2.8 dB	6.6 dB

Residual losses are due to a slight mode mismatching between laser and fibre and to the low numerical aperture of selfoc lenses. A similar configuration is presented in [99]. It is based on the adoption of a selfoc lens, preceded by a cylindrical lens, which is designed to reduce beam divergence by matching the beam to the low numerical aperture of a selfoc lens (Fig. 95). Coupling is performed with two different types of fibre, namely a multimode graded fibre (CD = 50 μm, NA = 0.17, 860 m in length)

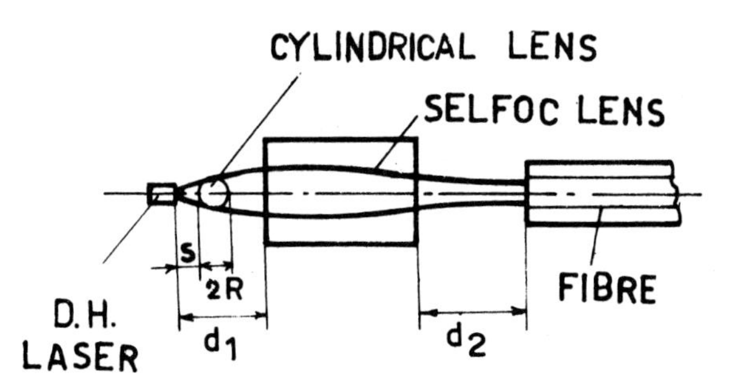

FIG. 95 – Coupling configuration with cylindrical lens and Selfoc lens. (After [99]).

and a single mode fibre (CD = 4.5 μm, Δn = 0.005%, 665 m in length). A 30 μm dia. glass rod is used as the cylindrical lens.

Selfoc lens characteristics are: NA = 0.5, diameter = 1 μm, length = 3.8 nm. The optimization of the configuration, i.e. of the choice of distances s, d_1, d_2, and R, results in coupling losses of 1 dB for the graded index fibre and 4 dB for the single mode fibre. Graded-index fibre offsets of 4 μm normal to the junction plane or 7 μm along the fibre axis lead to a loss increase of 0.5 dB.

d) Coupling between laser and tapered optical fibre

Another technique of matching laser emission geometry to fibre acceptance consists in flattening one end of the fibre in such a way that the core cross-section becomes elliptical [100]. The flattening process is performed gradually until the fibre end is transformed into a taper (see Fig. 96*a*). Further improvement is achievable by fabricating approximately cylindrical lenses on the elliptical core surface by means of photolithographic techniques,

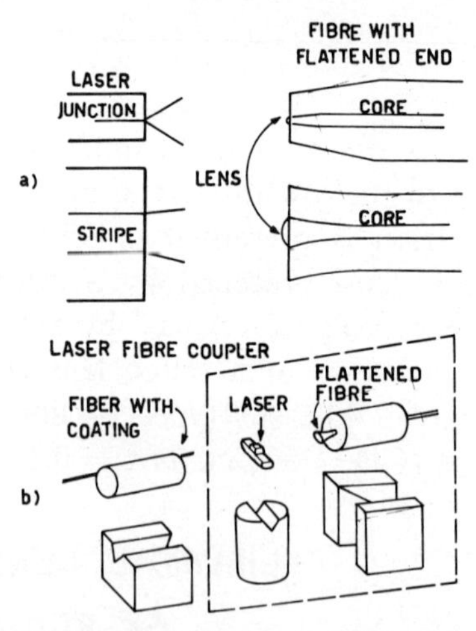

FIG. 96 – *a*) Gradually flattened fibre end with lens for mode adjustment (After [100]). *b*) Exploded view of a coupler with one laser and two fibres. (After [100]).

as described in [84]. The basic diagram of a coupler using a system of grooves and cylinders in shown in Fig. 96*b*. The implementation of a circular symmetry taper is presented in [101]. The taper may be prepared either by etching with HF or by heating and pulling over a microtorch. Figure 97 reproduces two tapers obtained in this way, from a fibre having CD = 80 μm, OD = = 120 μm and NA = 0.18. It should be noted that only the latter technique allows the cladding to be kept over the whole

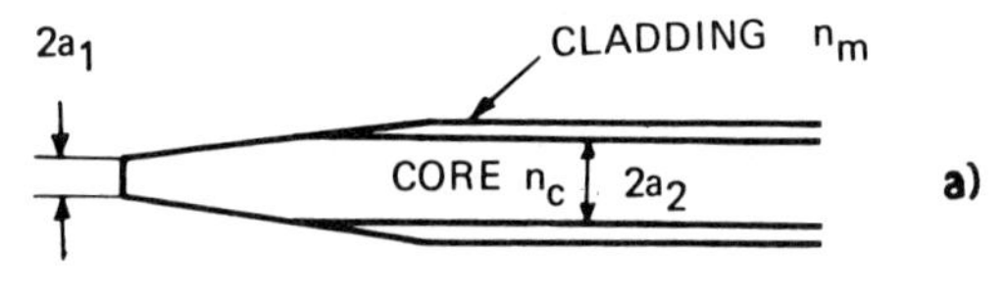

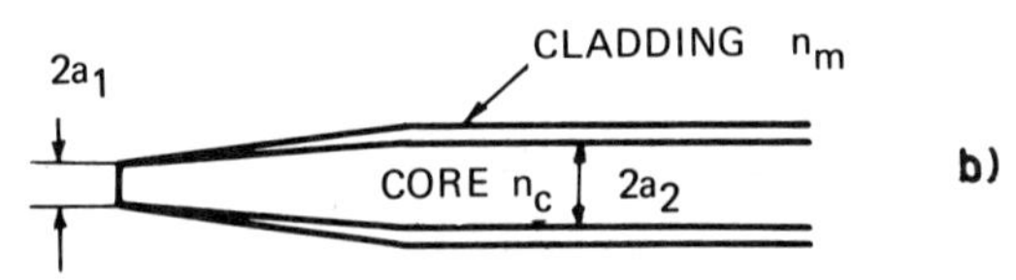

FIG. 97 – Structures of tapered launchers: *a*) graded index fibre; *b*) step index fibre. (After [101]).

taper length. Such a structure obviously increases the numerical aperture of the fibre, and in fact with the two tapers efficiencies of 87 % and 97 % respectively are obtained. The difference between these two values is due to scattering losses caused by imperfections on the outer surface of the etched taper. The coupling between a common laser beam and single mode optical waveguide may also be performed with integrated optics techniques as in [102]. Figures 98*a*, *b*, *c* show a basic diagram of the system. The device consists of a grating for coupling the laser beam to the guide which is obtained with photolithographic or interferometric techniques (the latter allows a periodicity of 0.57 μm to be achieved) and commercially available photoresist material. As can be seen in Fig. 98*b*, the coupled beam propagates in the waveguide (silicon exynitride, $n = 1.62$, 2 μm thick) and is then collimated by means of a lens to the guide core. Figure 98*c* shows the fibre fixed to the substrate (silicon) of the dielectric guide by means of a groove obtained through etching. The optical fibre used here is a single mode fibre having a 10 μm core and a 83 μm outside diameter. Experimental evaluations of the coupled power carried out with a He-Ne laser have shown an overall coupling efficiency for the system of 10 %. The major losses may be attributed to the difficulty

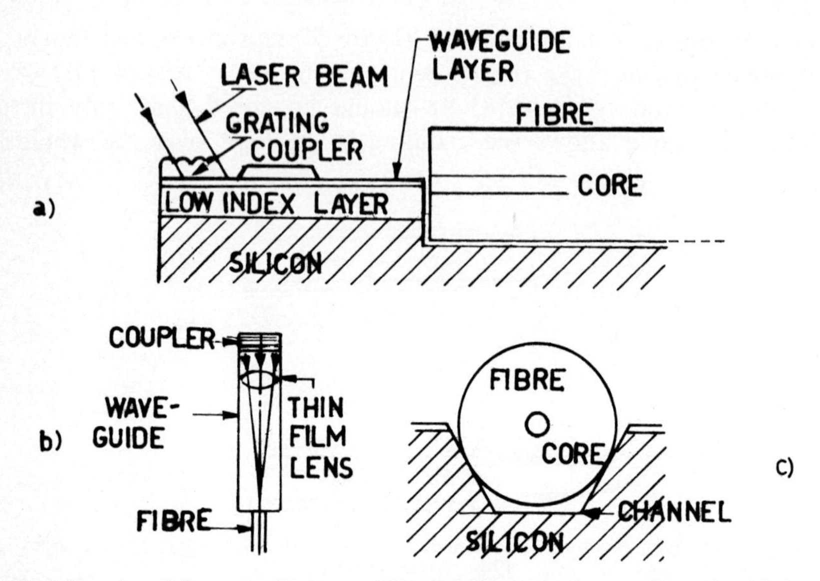

FIG. 98 *a-c*) – Schematic diagram of laser-to-fibre coupler, performed with integrated optics techniques *a*) side view; *b*) top view; *c*) section through fibre and channel. (After [102]).

in fabricating a grating of high coupling efficiency and a low attenuation waveguide. Considerable losses also originate at the optical fibre-dielectric guide interface, owing to the high divergence of the beam emerging from the guide.

A method of increasing the power emitted by a laser on the fundamental mode (which is more stable than the subsequent ones) has been proposed in [103]. This mode is excited by increasing locally the reflectivity of one of the two laser mirrors, namely at the point where the mode itself supplies highest output. The mirror of the laser output face is replaced by a suitable, partially reflective coating of metal on the core of the monomode fibre. In this way a new oscillating cavity is obtained (Fig. 97) which allows a good direct coupling to the optical fibre. To reduce reflection on the laser output face to a minimum, this face has to be coated with an antireflective layer.

The same principle has been applied in [100]. Figure 99 shows the set-up for the coupling of a laser to an optical fibre, the core of which is coated with partially reflective material.

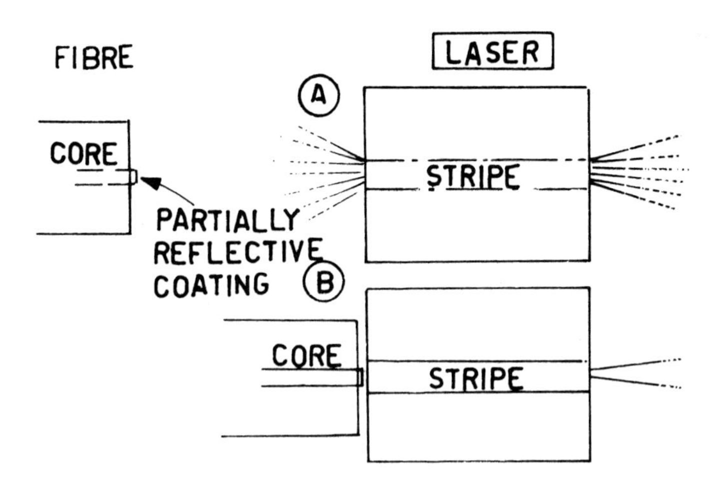

FIG. 99 – Zero-order transverse mode stimulated by increased local mirror reflectivity. (After [100]).

In [104] a traditional laser package is shown combined with a spherical lens to maximize the power injected into an optical fibre.

In Fig. 100 the cross-section of the package is shown; the spherical lens, the glass window and the source are visible.

A pull relief is pressed on the above mentioned mounting, from which the pig tail comes out. A special-purpose optical set was

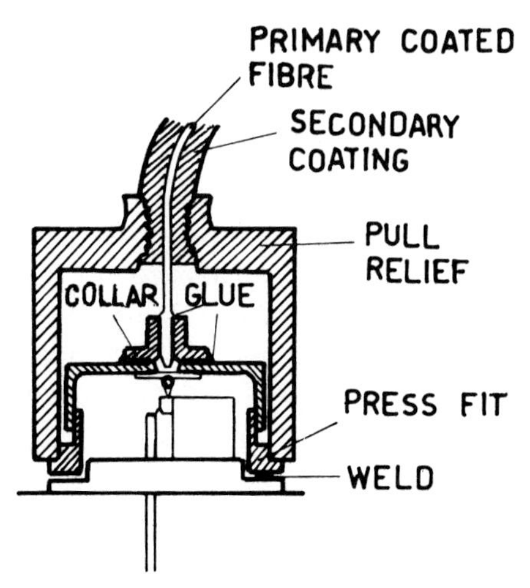

FIG. 100 – Hermetic packaged, lens-coupled laser diode: cross-section. (After [104]).

designed to carry out the optical alignment. Several packages were coupled using graded index fibres (CD = 50 μm, NA = 0.2) and micro beads with diameters in the range 70-120 μm and 1.9 refractive index.

A coupling efficiency of 56% was obtained; temperature variation between -10—80 °C did not influence the coupling efficiency.

2.4.3 *Light emitting diode-optical fibre coupling.*

Losses caused by positioning errors have also been measured for LED-optical fibre couplings [105]. Distribution is shown in Figs. 101 and 102. They refer to the coupling losses between

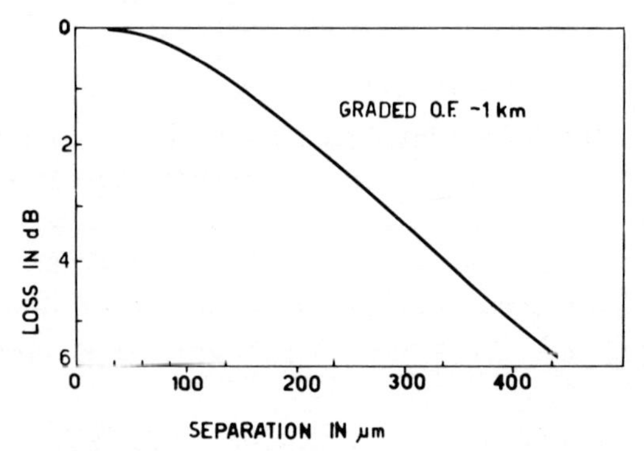

FIG. 101 – Coupling losses vs. separation s between 1 km graded index fibre and an LED.

Burrus-type LED with emitting area of 50 μm in diameter and a 1 km long graded-index CGW optical fibre with CD = 80 μm. As can be seen, here again the displacement error is much more critical than the separation and misalignment errors. By interpolating these values, one can plot the isoloss curves shown in Fig. 103, which can be helpful to the connector designer.

Figure 23, Chapter 1, Part II illustrates the schematics for direct coupling between a LED and a fibre, described in [106] and [107]. The launching efficiency of the most recent LED's having emitting areas smaller than the core of standard multimode fibres can, as in the case with lasers, be improved by interposing

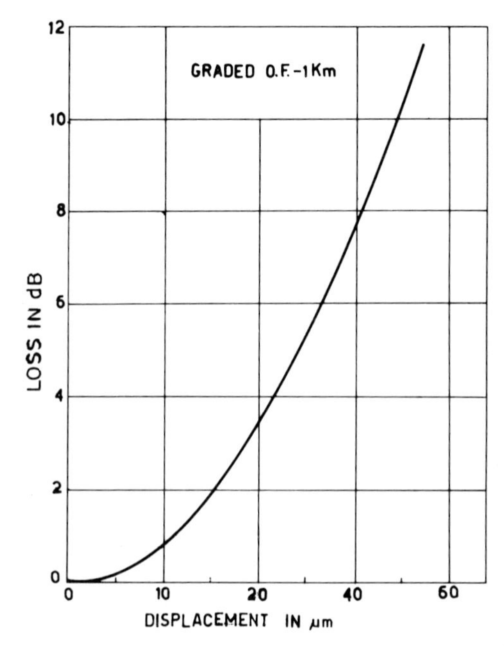

FIG. 102 – Coupling losses vs. displacement v between 1 km of graded index fibre and an LED.

a suitable optical system. This may be accomplished in different ways, namely by rounding the fibre end, introducing a small spherical lens of suitable radius between LED and fibre or using integrated techniques.

The theoretical aspects of the first case have already been

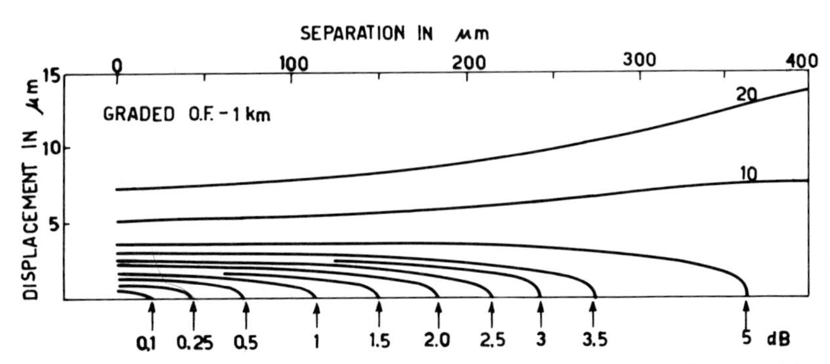

FIG. 103 – Output power losses vs. radial and axial coupling errors between 1 km graded index fibre and an LED.

discussed in the previous section. An experimental solution for the LED case has been thoroughly investigated in [108]. The same analysis also presents an LED assembled in an appropriately designed package: the output power emerges at the end of a pigtail, which is attached to the LED. Using an optical fibre (CD = 85 μm, OD = 125 μm, NA = 0.14, with a 75 μm radius spherical end), it may be shown that there is a value of the emitting area (Fig. 104) for which the maximum coupled power is achiev-

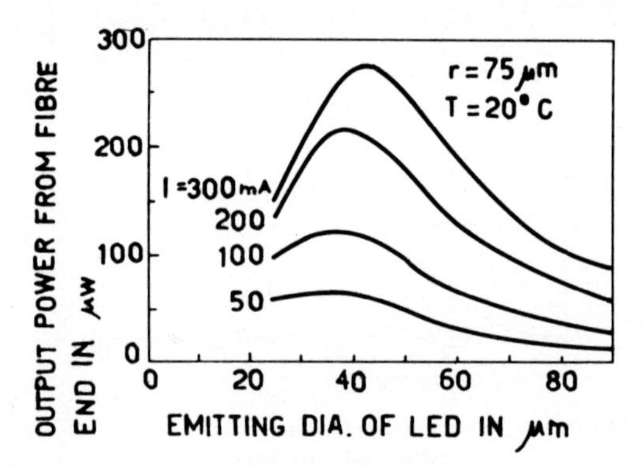

FIG. 104 – Calculated curves of the output power from the fibre end as a function of the emitting diameter, taking the forward current I as a parameter. (After [108]).

able. Different curves have been plotted as a function of drive current. Measurements have been conducted using an LED whose emitting area has the optimum dimensions required by the theory.

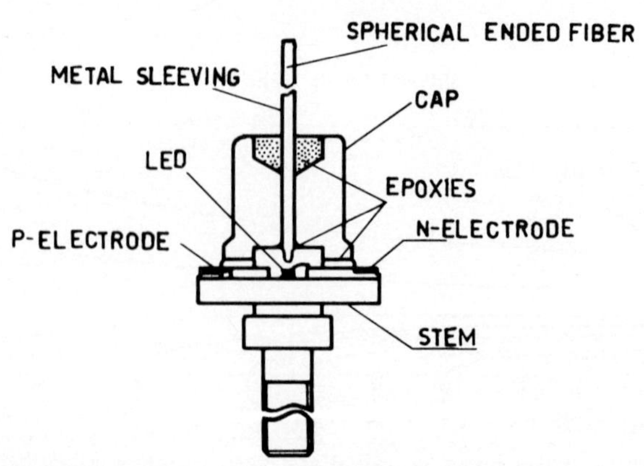

FIG. 105 – Cross-sectional view of an LED-fibre mounting package. (After [108]).

This may be obtained by adopting an electrode of 35 μm in diameter. The efficiencies obtained are 7-8 %, i.e. four times those achievable through direct coupling.

Figure 105 shows the cross-section of the above-mentioned package, and Fig. 106 the structure of the LED and the spherical-ended fibre.

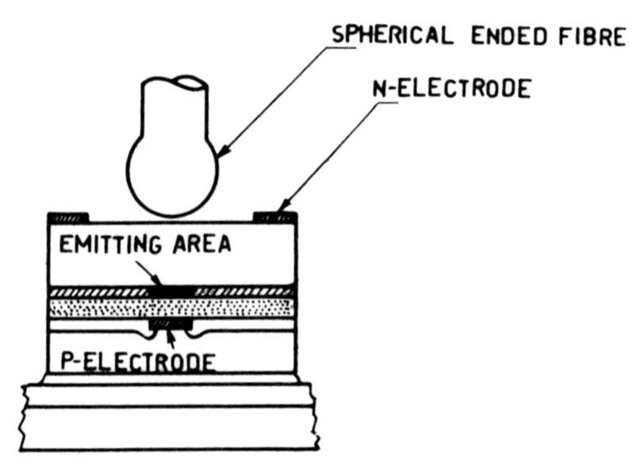

FIG. 106 – Structure of the GaAlAs LED with bulb-ended fibre. (After [108]).

Similar results are reported in [109], where bulb ended fibres allow a four-fold increase in direct coupling efficiency.

The second method of increasing the coupling efficiency is based on the use of a spherical lens [110]. This positions and aligns itself at the centre of the photoemitting surface (as shown

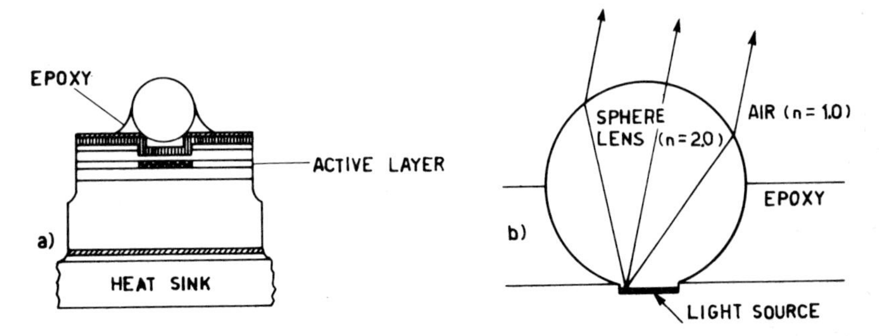

FIG. 107 – *a)* A schematic cross-sectional drawing of the GaAs-GaAlAs double heterostructure LED with a self-aligned sphere lens (After [110]). *b)* A simple geometrical configuration of the LED with a sphere lens. (After [110]).

in Figs. 107*a*, *b*) and is then bonded with epoxy. This lens (100 μm dia. and $n \simeq 2$ refractive index) allows a 3-4-fold increase in direct coupling efficiency, achieving values higher than 9% with LED's of 35 μm-diameter emitting area and fibres of CD =

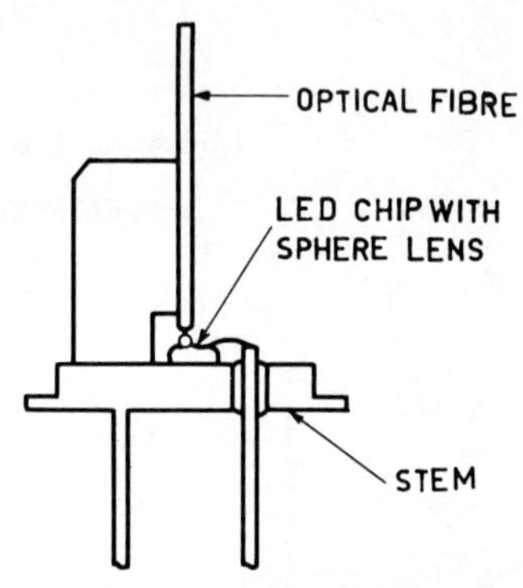

FIG. 108 – Schematic diagram of a mounted device with a fibre for the measurement of the coupling efficiency. (After [110]).

= 80 μm and NA = 0.14. The power coupled into fibres by means of a device such as that shown in Fig. 108 is about 200 μW with a LED drive current of 100 mA.

In [111] a device is presented which is fitted on the diode emitting surface, with a set of small lenses allowing a considerable

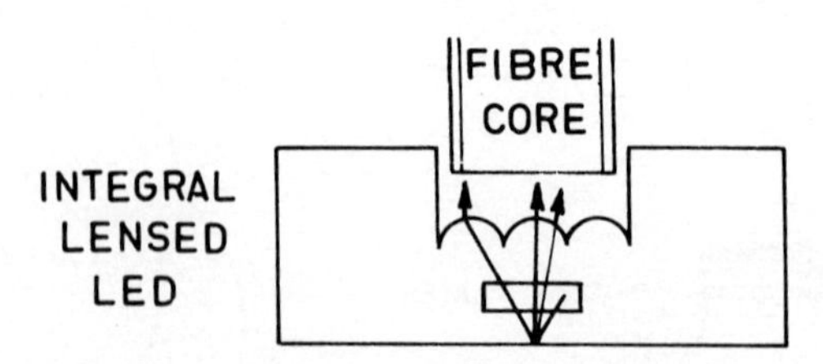

FIG. 109 – Integral lensed LED. (After [111]).

increase in the launching efficiency. The basic diagram of this set-up is illustrated in Fig. 109.

For the sake of completeness, mention should be made of the

coupling characteristics of the edge emitting LED with double heterostructure [112]. The source having a FWHM of 30° in the plane perpendicular to the junction allows the coupling of up to 800 μW power into a graded index, spherical ended fibre (C.D. = 90 μm, N.A. = 0.14) with an efficiency of about 8 %. Coupling through a spherical-ended fibre gives an efficiency 2.5 greater than that of direct coupling (Fig. 110). The same technique already presented

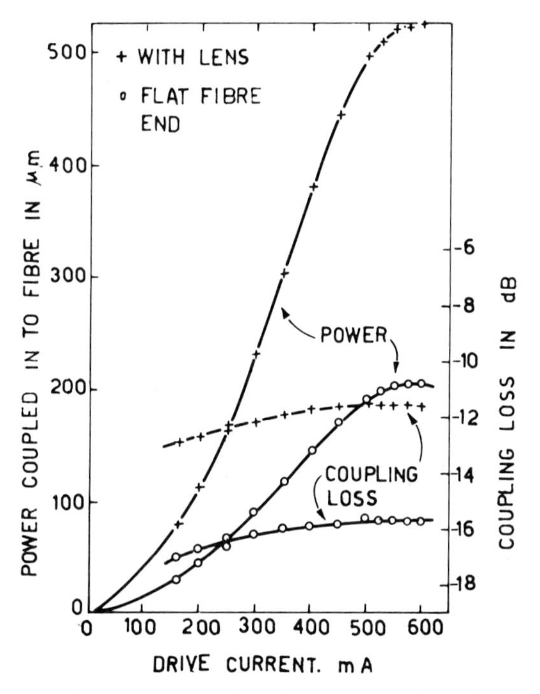

FIG. 110 – Effect of melting an optimal lens onto a silica fibre of NA = 0.14. The lens increases the power coupled into the fiber by a factor of about 2.5(4 *dB*) (After [112]).

for laser-to-fibre couplings, based on the use of a tapered fibre, is applied to coupling between a fibre and a multi-heterostructure LED [113]. Using a tapered end graded index CGW fibre of NA = 0.18, it is possible to achieve launching efficiencies of 53 % with taper ratio of about four.

Summary table of the characteristics of DH laser-optical fibre connections.

Technique	Reference (Example of application)	Efficiency range (%)	Fibre structure (step index: st graded index: gr)
Butt coupling 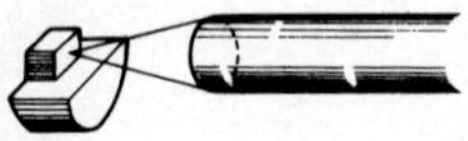	[85] Fujitsu Lab. (Japan)	10-20	st - CD = 85 μm NA = 0.14
Microbead lens 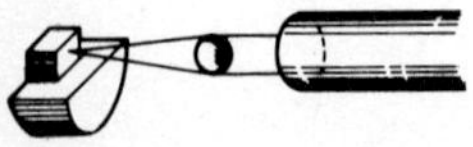	[104] Philips (Netherlands)	50-80 with A.R. coating	gr - CD = 50 μm NA = 0.2
Cylindrical lens 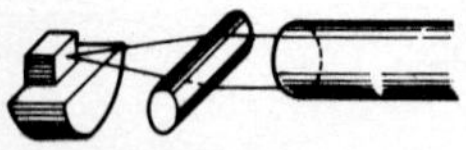	[95] Hitachi (Japan)	40-50	gr - CD = 50 μm NA =

Summary table of the characteristics of DH laser-optical fibre connections.

Technique	Reference (Example of application)	Efficiency range (%)	Fibre structure (step index: st graded index: gr)
Selfoc lens 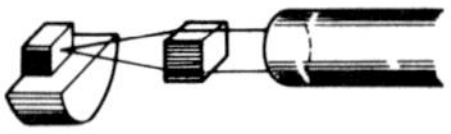	[98] NEC (Japan)	25-50	gr
Bulb-ended fibre 	[82] Bell Lab. (U.S.A.)	25-60	st - CD = 55 μm NA = 0.17
Tapered fibre end 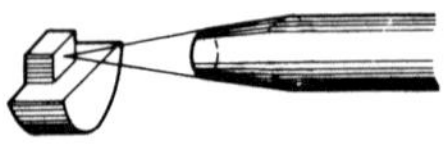	[101] C.R.C. (Canada)	80-90	st - OD = 80 μm NA = 0.18

Summary table of the characteristics of DH laser-optical fibre connections.

Technique	Reference (Example of application)	Efficiency range (%)	Fibre structure (step index: st graded index: gr)
Bulb-ended etched fibre 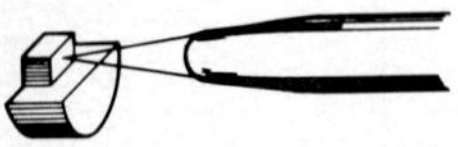	[91] R. Bosch (West Germany)	80	gr - OD = 80 μm NA = 0.13

Summary table of the characteristics of LED-optical fibre connections

Technique	Reference (Example of application)	Efficiency range (%)	Fibre structure (step index: st graded index: gr)
Butt coupling 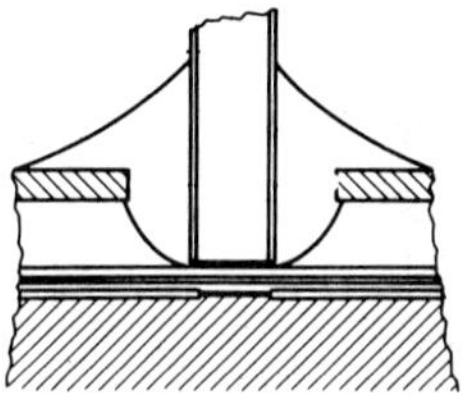	[107] Bell Lab. (U.S.A.)	2	st - CD = 85 μm NA = 0.14
Spherical lens 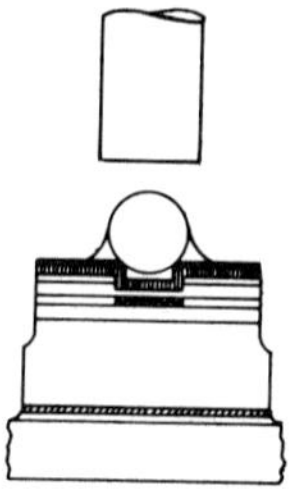	[110] Mitsubishi (Japan)	8-10	CD = 80 μm NA = 0.14
Bulb-ended fibre 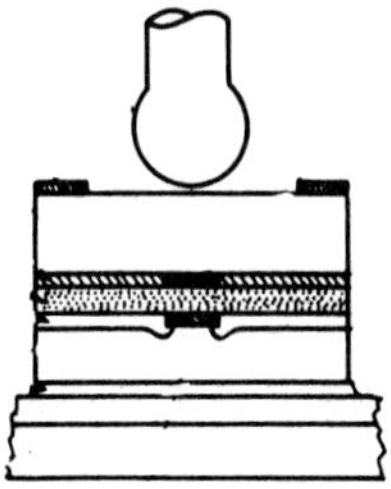	[108] Fujitsu (Japan)	6-8	st - CD = 80 μm NA = 0.14

REFERENCES

[1] P. Di Vita and R. Vannucci: *Geometrical theory of coupling errors in dielectric optical waveguides*, Opt. Commun., Vol. 14, No. 1 (1975), pp. 139-143.

[2] P. Di Vita and U. Rossi: *Theory of power coupling between multimode optical fibres*, Opt. Quantum Electron., Vol. 10 (1978), pp. 107-117.

[3] D. Gloge: *Offset and tilt loss in optical fiber splices*, Bell. Syst. Tech. J., Vol. 55, No. 7 (1976), pp. 905-916.

[4] C. M. Miller: *Transmission vs. transverse offset for parabolic profile fiber splices with unequal core diameters*, Bell Syst. Tech. J., Vol. 55, No. 7 (1976), pp. 917-927.

[5] M. Young: *Geometrical theory of multimode optical fiber-to-fiber connectors*, Opt. Commun., Vol. 7, No. 3 (1973), pp. 253-255.

[6] F. L. Thiel and R. M. Hawk: *Optical waveguide cable connection*, Appl. Opt., Vol. 15, No. 11 (1976), pp. 2782-2791.

[7] H. Tsuchiya *et al.*: *Double eccentric connectors for optical fibres*, Appl. Opt., Vol. 16, No. 5 (1977), pp. 1323-1330.

[8] G. Cocito *et al.*: *Coupling loss for three different fibre core diameters*, CSELT Rapporti Tecnici, Vol. VI, No. 2 (1978), pp. 113-119.

[9] C. M. Miller: *Optical fiber splicing*, Optical Fiber Transmission II, Williamsburg (February 22-24, 1977), pp. WA3, 1-2.

[10] D. L. Bisbee: *Optical fiber joining technique*, Bell Syst. Tech. J., Vol. 50, No. 10 (December 1971), pp. 3153-3158.

[11] R. B. Dyott, J. R. Stern and J. H. Stewart: *Fusion junctions for glass-fibre waveguides*, Electron. Lett., Vol. 8, No. 11 (June 18, 1972), pp. 290-292.

[12] Y. Kohanzadeh: *Hot splices of optical waveguide fibers*, Appl. Opt., Vol. 15, No. 3 (March 1976), pp. 793-795.

[13] D. L. Bisbee: *Splicing silica fibers with an electric arc*, Appl. Opt., Vol. 15, No. 3 (March 1976), pp. 796-798.

[14] R. Jocteur and A. Tardy: *Optical fiber splicing with plasma torch and oxydric microburner*, Second European Conference on Optical Fibre Communication, Paris (September 27-30, 1977).

[15] M. Tsuchiya and I. Hatakeyama: *Fusion splices for single-mode optical fibers*, Optical Fiber Transmission II, Williamsburg (February 22-24, 1977), pp. PD1, 1-4.

[16] M. Hirai and N. Uchida: *Melt splice of multimode optical fibre with an electric arc*, Electron. Lett., Vol. 13, No. 5 (March 1977), pp. 123-125.

[17] K. Egashira and M. Kobayashi: *Optical fiber splicing with a low-power* CO_2 *laser*, Appl. Opt., Vol. 16, No. 6 (June 1977), pp. 1636-1638.

[18] K. Sakamoto *et al.*: *The automatic splicing machine employing electric arc fusion*, Fourth European Conference on Optical Communication, Genoa (September 12-15, 1978), pp. 296-303.

[19] P. M. Buhite and D. A. Pinnow: *Optical splicing device and technique*, U.S.Patent 3,810,802 (May 1974).

[20] R. M. Derosier and J. Stone: *Low-loss splices in optical fibers*, Bell Syst. Tech. J., Vol. 52, No. 7 (September 1973), pp. 1229-1235.

[21] M. Murata *et al.*: *Connection of optical fiber cable*, Optical Fiber Transmission, Williamsburg (January 7-9, 1975), pp. WA5, 1-4.

[22] M. Murata *et al.*: *Splicing of optical fiber cable on site*, First European Conference on Optical Fibre Communication, London (September 16-18, 1975), pp. 93-95.

[23] J. F. Dalgleish *et al.*: *Splicing of optical fibres*, First European Conference on Optical Fibre Communication, London (September 16-18, 1975), pp. 87-89.

[24] D. G. Dalgoutte: *Collapsed sleeve splices for field jointing of optical fibre cable*, Third European Conference on Optical Communication, Munich (September 14-16, 1977), pp. 106-108.

[25] C. M. Miller: *Loose tube splice for optical fibers*, Bell Syst. Tech. J., Vol. 54, No. 7 (1975), pp. 1215-1225.

[26] C. G. Someda: *Simple low-loss joints between single-mode optical fibers*, Bell Syst. Tech. J., Vol. 52, No. 4 (April 1973), pp. 583-597.

[27] D. Kunze *et al.*: *Jointing technique for optical cables*, Second European Conference on Optical Fibre Communication, Paris (September 27-30, 1976), pp. 257-260.

[28] A. R. Tynes and R. M. Derosier: *Low-loss splices for single-mode fibres*, Electron. Lett., Vol. 13, No. 22 (October 27, 1977), pp. 673-674.

[29] G. Cocito *et al:* COS 2 *Experiment in Turin: Field Test on an optical cable in ducts,* IEEE Trans. on Comm., Vol. COM-26, No.7 (1978), pp. 1028-1036.

[30] T. Tatsuya and S. Ohara: *Recent advance in optical fiber transmission technology*, Japan Telecommunications Review (January 1976), pp. 3-10.

[31] E. L. Chinnock *et al.*: *Preparation of optical fiber ends for low-loss tape splices*, Bell Syst. Tech. J., Vol. 54, No. 3 (March 1975), pp. 471-477.

[32] A. H. Cherin and P. J. Rich: *Multigroove embossed-plastic splice connector for joining groups of optical fibers*, Appl. Opt., Vol. 14, No. 12 (December 1975), pp. 3026-3030.

[33] A. H. Cherin and P. J. Rich: *A splice connector for joining linear arrays of optical fibers*, Optical Fiber Transmission, Williamsburg (January 7-9, 1975), pp. WB3, 1-4.

[34] A. H. Cherin and P. J. Rich: *An injection molded plastic connector for splicing optical cables*, Bell Syst. Tech. J., Vol. 65, No. 8 (October 1976), pp. 1057-1067.

[35] A. H. Cherin and P. J. Rich: *An injection molded splice connector with silicon chip insert for joining optical fiber ribbons*, Optical Fiber Transmission, Williamsburg (February 22-24, 1977), pp. WA7, 1-4.

[36] C. M. Schroeder: *Accurate silicon space chips for an optical fiber cable connector*, Bell Syst. Tech. J., Vol. 57, No. 1 (January 1978), pp. 91-97.

[37] C. M. Miller: *Fiber optic array splicing with etched silicon chips*, Bell Syst. Tech. J., Vol. 57, No. 1 (January 1978), pp. 75-90.

[38] G. Le Noane: *Optical fibre cable and splicing technique*, Second European Conference on Optical Fibre Communication, Paris (September 27-30, 1976), pp. 247-252.

[39] G. Le Noane: *Further developments on compartmented fiber optic cable structure and associated splicing technique*, 1977 International Conference on Integrated Optics and Optical Fiber Communication, Tokyo (July 18-20, 1977), pp. 355-358.

[40] J. H. Stewart and P. Hensel: *Technique for jointing small-core optical fibres*, Electron. Lett., Vol. 12, No. 21 (October 1976), p. 570.

[41] H. Takimoto *et al.*: *A splicing method with auxiliary fibers*, 1977 International Conference on Integrated Optics and Optical Fiber Communication, Tokyo (July 18-20, 1977), pp. 343-346.

[42] P. Hensel: *A new jointing technique suitable for automation*, Third European Conference on Optical Communication, Munich (September 14-16, 1977), pp. 103-105.

[43] J. D. Archer and B. Tech: *Fibre optics interconnection components*, Optical Fibres, Integrated Optics and their Military Application, London (May 16-20, 1977), pp. 52, 1-8.

[44] J. F. Dalgleish: *A review of optical fiber connection technology*, Telephony (January 31, 1977), pp. 25-31.

[45] J. F. Dalgleish and H. H. Lucas: *Optical fibre connectors*, Electron. Lett., Vol. 11, No. 1 (1975), pp. 24-26.

[46] *Fiberoptic cable getting connector for use in field*, Electronics (August 21, 1975), pp. 29-30.

[47] J. Guttman *et al.*: *Multi-pole optical fibre-fibre connector*, First European Conference on Optical Fibre Communication, London (September 16-18, 1975), pp. 96-98.

[48] P. Hensel: *Triple ball connector for optical fibres*, Electron. Lett., Vol. 13, No. 24 (1977), pp. 734-735.

[49] C. M. Miller: *A fiber-optic-cable connector*, Bell Syst. Tech. J., Vol. 54, No. 9 (1975), pp. 1547-1555.

[50] C. M. Schroeder: *Accurate silicon spacer chips for an optical fiber cable connector*, Optical Fiber Transmission, Williamsburg (January 7-9, 1975), pp. WA6, 1-4.

[51] F. Auracher and K. H. Zeitler: *Multiple fiber connectors for multimode fibres*, Opt. Commun., Vol. 18, No. 4 (1976), pp. 556-558.

[52] K. Höllerl *et al.*: *Retention grooves for optical fibre connectors*, Electron. Lett., Vol. 13, No. 3 (1977), pp. 74-76.

[53] H. Murata: *Broad band optical fiber cable and connecting*, Second European Conference on Optical Fibre Communication, Paris (September 27-30, 1976), pp. 167-174.

[54] D. Kunze *et al.*: *Jointing techniques for optical cables*, Second European Conference on Optical Fibre Communication, Paris (September 27-30, 1976), pp. 257-260.

[55] M. L. Dakss and A. Bridger: *Plug-in fibre-to-fibre coupler*, Electron. Lett., Vol. 10, No. 14 (1974), pp. 280-281.

[56] K. Miyazaki *et al.*: *Theoretical and experimental considerations of optical fiber connector*, Optical Fiber Transmission, Williamsburg (January 7-9, 1975), pp. WA4, 1-4.

[57] N. Suzuki *et al.*: *A new demountable connector developed for a trial optical transmission system*, 1977 International Conference on Integrated Optics and Optical Fiber Communication, Tokyo (July 18-20, 1977), pp. 351-354.

[58] N. Kuroki *et al.*: *A development study on design and fabrication of an optical fiber connector*, Third European Conference on

Optical Communication, Munich (September 14-16, 1977), pp. 97-99.

[59] N. Shimizu and H. Tsuchiya: *Single-mode fibre connectors*, Electron. Lett., Vol. 14, No. 19 (September. 1978), pp. 611-613.

[60] N. Shimizu *et al.*: *Low-loss single-mode fibre connectors*, Electron. Lett., Vol. 15, No. 1 (January 1979), pp. 28-29.

[61] I. Hatakeyama: *Multimode fibre connector fabricated with a self-aligning jig*, Electron. Lett., Vol. 15, No. 1 (January 1979), pp. 34-35.

[62] M. Börner *et al.*: *Lösbare Steckverbindung für Ein-Mode-Glasfaserlichtwellenleiter*, A.E.Ü., Vol. 26, No. 6 (1972), pp. 288-289.

[63] O. Krumpholz: *Fiber coupler*, Appl. Opt., Vol. 11, No. 8 (1972), pp. 1679-1680.

[64] J. Guttman *et al.*: *A simple connector for glass fibre optical waveguide*, A.E.Ü., Vol. 29, No. 1 (1975), pp. 50-52.

[65] V. Vucins: *Adjustable single-fiber connector with monitor output*, Third European Conference on Optical Communication, Munich (September 14-16, 1977), pp. 100, 102.

[66] P. K. Runge *et al.*: *Precision transfer molded single fiber optic connectors and encapsulated connectorized devices*, Optical Fiber Transmission, Williamsburg (February 22-24, 1977), pp. WA4, 1-4.

[67] J. S. Cook *et al.*: *An exploratory fiber guide interconnection system*, Second European Conference on Optical Fibre Communication, Paris (September 27-30, 1976), pp. 253-256.

[68] P. K. Runge and S. S. Cheng: *Demountable single-fiber optic connectors and their measurement on location*, Bell Syst. Tech. J., Vol. 57, No. 6 (July-August 1978), pp. 1771-1790.

[69] P. W. Smith *et al.*: *A molded-plastic technique for connecting and splicing optical-fibre tapes and cables*, Bell Syst. Tech. J., Vol. 54, No. 6 (1975), pp. 971-984.

[70] M. A. Bedgood *et al.*: *Demountable connectors for optical fiber systems*, Elect. Commun., Vol. 51, No. 2 (1976), pp. 85-91.

[71] R. W. Berry and R. C. Hooper: *Practical design requirements for optical fibre transmission systems*, First European Conference on Optical Fibre Communication, London (September 16-18, 1975), pp. 153, 155.

[72] *British Post Office to field-test fiber-optic telephone transmission*, Electronics (September 18, 1975), pp. 14E-18E.

[73] A. Nicia: *Practical low-loss lens connector for optical fibres*, Electron. Lett., Vol. 14, No. 16 (1978), pp. 511-512.

[74] Y. Koyama *et al.*: *Optical devices for optical fiber communication*, NEC Research and Development, No. 49 (April 1978), pp. 51-57.

[75] H. Nishihara *et al.*: *Holocoupler: a novel coupler for optical circuits*, IEEE J. Quantum Electron (1975), pp. 794-796.

[76] F. R. Trambarulo: *Optical fiber connector*, U.S.Patent 3,734,594 (May 22, 1973).

[77] P. Di Vita: *Connettori magnetici per fibre ottiche*, Pat. pending.

[78] P. Di Vita and R. Vannucci: *Optical waveguide with graded refractive index: theory of power launching*, Appl. Opt., Vol. 15, No. 11 (1976), pp. 2765-2772.

[79] G. P. Kidd: *The effect of excitation conditions on fibre launching and coupling losses*, Australian Telecommunication Research, Vol. 11, No. 3 (1977), pp. 4-12.

[80] M. K. Barnoski: *Coupling component for optical fiber waveguides*, Fundamental of Optical Fibre Communications, Acc. Press, New York (1976), Chapter 3.

[81] M. C. Hudson: *Calculation of the maximum optical coupling efficiency into multimode optical waveguides*, Appl. Opt., Vol. 13, No. 5 (1974), pp. 1029-1033.

[82] L. G. Cohen: *Power coupling from* GaAs *injection lasers into optical fibers*, Bell Syst. Tech. J., Vol. 51, No. 3 (1972), pp. 573-594.

[83] P. Di Vita and R. Vannucci: *The radiance law in radiation transfer processes*, Appl. Phys., Vol. 7 (1975), pp. 249-255.

[84] L. G. Cohen and M. V. Schneider: *Microlenses for coupling junction lasers to optical fibers*, Appl. Opt., Vol. 13, No. 1 (1974), pp. 89-94.

[85] T. Yamaoka *et al.*: *Fiber-mounted semiconductor device for optical communication system*, Optical Fiber Transmission, Williamsburg (January 7-9, 1975), pp. WD2, 1-4.

[86] D. Kato: *Light coupling from a stripe-geometry* GaAs *diode laser into an optical fiber with spherical end*, J. Appl. Phys., Vol. 44, No. 6 (1973), pp. 2756-2758.

[87] C. A. Brackett: *On the efficiency of coupling light from stripe-geometry* GaAs *laser into multimode optical fibers*, J. Appl. Phys., Vol. 45, No. 6 (1974), pp. 2636-2637.

[88] J. Wittmann: *Contact-bonded epoxy-resin lenses to fibre end-faces*, Electron. Lett., Vol. 11, No. 20 (1975), pp. 477-478.

[89] W. W. Benson *et al.*: *Coupling efficiency between* GaAlAs *laser and low-loss optical fibers*, Appl. Opt., Vol. 14, No. 12 (1975), pp. 2815-2816.

[90] U. C. Paek and A. L. Weaver: *Formation of a spherical lens at optical fiber end with a* CO_2 *laser*, Appl. Opt., Vol. 14, No. 2 (1975), pp. 294-297.

[91] C. C. Timmerman: *Highly efficient light coupling from* GaAlAs *lasers into optical fibers*, Appl. Opt., Vol. 15, No. 10 (1976), pp. 2432-2433.

[92] E. Weidel: *Light coupling from a junction laser into a monomode fibre with a glass cylindrical lens on the fibre end*, Opt. Commun., Vol. 12, No. 1 (1974), pp. 93-97.

[93] E. Weidel: *New coupling method for* GaAs-*laser-fibre coupling*, Electron. Lett., Vol. 11, No. 8 (1975), pp. 436-437.

[94] E. Weidel: *Light coupling problems for* GaAs *laser-multimode fibre coupling*, Opt. Quantum Electron., Vol. 8, (1976), pp. 301-307.

[95] M. Maeda *et al.*: *Hybrid laser-to-fiber coupler with a cylindrical lens*, Appl. Opt., Vol. 16, No. 7 (1977), pp. 1966-1970.

[96] J. D. Crow *et al.*: *Gallium arsenide laser-array-on-silicon package*, Appl. Opt., Vol. 17, No. 3 (1978), pp. 479-485.

[97] K. Kurokawa *et al.*: *Laser fibre coupling with a hyperbolic lens*, IEEE Trans. on Microwave Theory and Technique (March 1976), pp. 309-311.

[98] S. Sugimoto *et al.*: *Light coupling from a DH laser into a selfoc fiber using slab selfoc lenses*, Optical Fiber Transmission, Williamsburg (January 7-9, 1975), pp. WD1, 1-4.

[99] Y. Odagiri *et al.*: *High-efficiency laser-to-fibre coupling circuit using a combination of a cylindrical lens and a selfoc lens*, Electron. Lett., Vol. 13, No. 14 (1977), pp. 395-396.

[100] G. D. Khoe: *Power coupling from junction lasers into single mode optical fibers*, First European Conference on Optical Fibre Communication, London (September 16-18, 1975), pp. 114-116.

[101] T. Ozeki and B. S. Kawasaki: *Efficient power coupling using taper-ended multimode optical fibers*, Electron. Lett., Vol. 12, No. 23 (1976), pp. 607-608.

[102] L. P. Boivin: *Thin-film laser-to-fiber coupler*, Appl. Opt., Vol. 13, No. 2 (1974), pp. 391-395.

[103] R. B. Dyott: *Direct coupling from a* GaAs *laser into single-mode fiber*, Electron. Lett., Vol. 11, No. 14 (1975), pp. 208-209.

[104] G. D. Khoe *et al.*: *Fiberless hermetic packaged lens-coupled laser diode for wide band optical fibre transmission*, Optical Fiber Communication, Washington, D. C. (March 6-8, 1979), pp. 94-97.

[105] G. Cocito *et al.*: *Theory and experiments on LED/FIBRE optic coupling*, CSELT Rapporti Tecnici, Vol. 6, No. 1 (March 1978).

[106] C. A. Burrus and B. I. Miller: *Small-area double-heterostructure alluminum-gallium sources for optical fiber transmission lines*, Opt. Commun., Vol. 4, No. 4 (1971), pp. 207-209.

[107] C. A. Burrus and R. W. Dawson: *Small-area, high-current density* GaAs *electroluminescent diodes and a method of operation for improved degradation characteristics*, Appl. Phys. Lett., Vol. 17, No. 3 (1970), pp. 97-99.

[108] M. Abe *et al.*: *High-efficiency long-lived* GaAlAs *LED's for fiber optical communications*, IEEE Trans. Electron. Devices, Vol. ED-24, No. 7 (1977), pp. 990-994.

[109] G. P. Basola and G. Chiaretti, *Design criteria for optimum coupling between a high radiance LED and a bulb-ended fiber*, Fourth European Conference on Optical Communication, Genoa (September 15-18, 1978), pp. 313-317.

[110] S. Horiuchi *et al.*: *A new LED structure with a self-aligned sphere lens for efficient coupling to optical fibres*, IEEE Trans. Electron. Devices, Vol. ED-24, No. 7 (July 1977), pp. 986-990.

[111] F. D. King and A. J. Springthorpe: *The integral lens coupled LED*, J. of Electronic Materials, Vol. 4, No. 2 (1975), pp. 243-253.

[112] J. P. Wittche *et al.*: *High radiance LED for single-fibre optical links*, RCA Review, Vol. 37 (1976), pp. 159-183.

[113] Y. Uematzu and T. Ozeki: *Efficient power coupling between a MH LED and a multimode fiber with tapered launcher*, 1977 International Conference on Integrated Optics and Optical Fiber Communication, Tokyo (July 18-20, 1977), pp. 371-374.

PART IV

SYSTEMS

CHAPTER 1

TOPICS IN OPTICAL FIBRE COMMUNICATION THEORY

by Angelo Luvison

INTRODUCTION

This chapter deals with some theoretical and statistical issues related to optical fibre communication systems. It depends to a large extent on the previous discussion about components and may be considered as an introduction to the following, more system-oriented chapters. Since we are concerned almost exclusively with an overview of communication theory principles, Part A of this chapter outlines some results pertinent to the Poisson random process and its generalizations. Just as the Gaussian process plays a key role in conventional transmission systems, the Poisson process is fundamental to optical communications. In particular, the photoelectron counting process is modelled and its statistical features derived. For this purpose mean value, variance and power spectral density are given.

In Part B, we apply this background to the canonical direct-detection optical communication system. The overall performance criterion is stated in terms of signal-to-noise ratio and is illustrated for analogue baseband intensity or subcarrier-intensity modulations.

Digital systems are covered in Part C, where, again, we limit ourselves to the direct-detection approach. Performance is measured according to the average bit error probability. Finally, typical linear and nonlinear equalization structures are reviewed and their relative merits are discussed. Moreover, problems related to signalling formats and synchronization strategies are briefly recalled.

Due to space limitations, there are many topics of current interest that will be either touched over or omitted entirely: we mention quantum detection, heterodyne receivers and parameter or waveform estimation.

1.1 The optical fibre communication system

The subject is best understood by considering a system model whose characteristics are attractive enough and not far from the present state of the art. It seems advisable to consider optical sources such as light-emitting diodes (LED's), injection laser diodes or miniature Nd-YAG lasers pumped by LED's, multimode fibres and ordinary or avalanche photodiodes as optical detectors. These assumptions entail that emphasis be given to the electrical viewpoint, whereas the optical signal processing aspects are only touched on. One of the consequences is that we deal with *direct detection* of photons in an optical fibre. Direct detection refers to the processing of the electrical signal at the output of a photodetector, as opposed to detection schemes based on optimum processing of the electromagnetic field considered as a quantum system.

The functional elements of a fibreoptic communication system can be represented as in Fig. 1. The diagram is composed of three

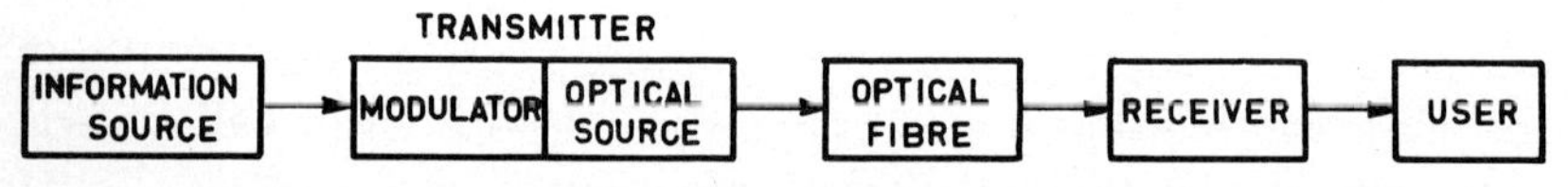

FIG. 1 – General model of optical fibre communication system.

standard blocks—a transmitter, the channel and a receiver—for information transfer between remotely located points. The wide range of possible information sources results in many different message forms. Regardless of their exact form, however, messages may be waveforms in time, digital symbols, etc. Before conveying the message over the fibre channel, it is transformed by modulating an optical carrier in order to cast it into a form suitable for propagation. The optical fibre is coupled to the transmitter and receiver by input and output couplers, respectively. The receiver's function is to extract the desired message from the received field. In direct-detection receivers (Fig. 2), this is performed by a photodetection that transfers the instantaneous optical power, as it

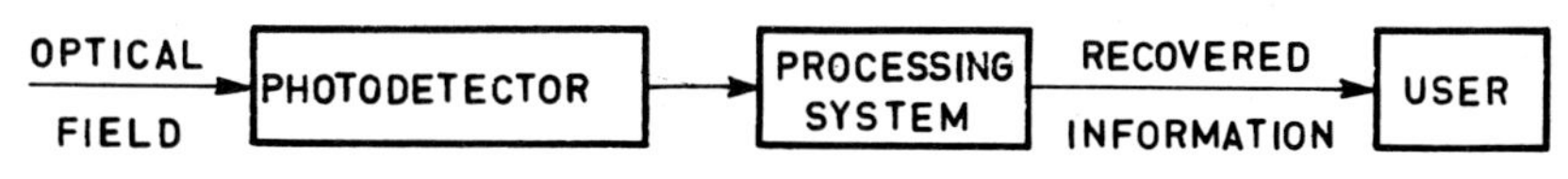

FIG. 2 – Optical receiver.

arrives at the receiver, into an electrical signal, which is then processed (e.g., amplified, filtered, equalized, etc.) in order to combat the various noise, loss and distortion sources.

A. Poisson and related processes

Many different physical phenomena have essentially the same probability characteristics as the *shot-noise process*, e.g., a current generated by the randomness of the flow of electrons from the photocathode to the anode in a *vacuum photodiode.* In this case, the electrons are found to arrive at the anode independently of one another, viz., the number of electrons, which arrive in any given time interval, is independent of the number arriving in any other nonoverlapping time interval. Furthermore, the average intensity of emitted electrons is assumed to be constant, say, λ electrons per unit time, and the probability that a single electron is emitted in a very short time interval of duration Δt is essentially equal to $\lambda \Delta t$, while the probability that more than a single electron will be emitted in such a short time interval is essentially zero.

This and other physical phenomena of the same type can be modelled as a *Poisson random process.*

1.2 The Poisson random process

The emission of electrons from a photoemissive material may be described by a *counting process* $\boldsymbol{N} := \{N_t,\ 0 \leqslant t < \infty\}$ (1,2),

(1) In the first part of this chapter we will use boldface capital letters (say $\boldsymbol{X}$) to denote random processes, capital letters with a time subscript (say X_t) to denote a random variable at time t, and small letters (say $x(t)$) to denote a member of the process. For notation simplicity, this use will be relaxed from Section 1.5 onwards: the meaning of each symbol will be obvious from the context.

(2) Symbol $:=$ means *equal by definition.*

which is a family of non-negative-integer-valued random variables N_t, nondecreasing in time t. N_t changes value by unity every time an electron is emitted; therefore, it *counts* the accumulated number of *events* in time interval $[0, t)$. It is usual to set $N_0 = 0$, assuming that initially there are no counted events. A typical counting function might be that shown in Fig. 3. If $0 \leq u < v$,

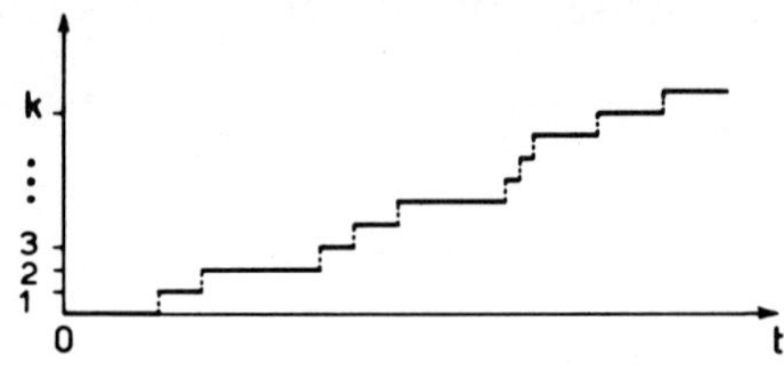

FIG. 3 – A typical counting function.

the *increment* of N_t over the interval $[u, v)$ is the random variable

$$N_{u,v} := N_v - N_u \, . \tag{1}$$

The average number of events in $[0, t)$ (the *expected value* of N_t) is called the *principal function* of the process $\boldsymbol{N}$ and is denoted by

$$E[N_t] := \Lambda(t) \, . \text{ (3)}$$

The electron emission discussed previously can be modelled by means of a counting random process, which satisfies the following basic properties:

a. By definition, N_t assumes only non-negative integer values and

$$N_0 := 0 \, .$$

b. The process $\boldsymbol{N}$ has stationary and independent increments.

c. The probability that a single event will occur in a very short interval Δt is essentially proportional to the length of the interval

$$P[N_{t+\Delta t} - N_t = 1] := \lambda \, \Delta t + 0(\Delta t) \, . \text{ (4)} \tag{2}$$

(3) $E[\ ]$ denotes statistical average (expectation) of the quantity between brackets. When necessary, a subscript of E emphasizes the random variable over which the expectation is taken.

(4) $0(\Delta t)$ is *any* function such that $\lim_{\Delta t \to 0} \frac{0(\Delta t)}{\Delta t} = 0$.

d. The probability that in the same interval more than one event will occur is practically negligible, viz.,

$$P[N_{t+\Delta t} - N_t > 1] := 0(\Delta t) . \tag{3}$$

It follows immediately from hypotheses (*c*) and (*a*) that

$$\begin{aligned} P[N_{t+\Delta t} - N_t = 0] &= 1 - \{P[N_{t+\Delta t} - N_t = 1] + P[N_{t+\Delta t} - N_t > 1]\} \\ &= 1 - \lambda \Delta t + 0(\Delta t) . \end{aligned} \tag{4}$$

More generally, a random process which satisfies hypotheses (*a*) through (*d*) is called a Poisson counting process, as it can be proved that the number of events occurring in any given interval has the Poisson probability distribution:

$$p_k(t) := P[N_t = k] = \frac{(\lambda t)^k}{k!} \exp[-\lambda t] . \tag{5}$$

In Fig. 4, the family of curves $p_k(t)$ as a function of k and λt is plotted.

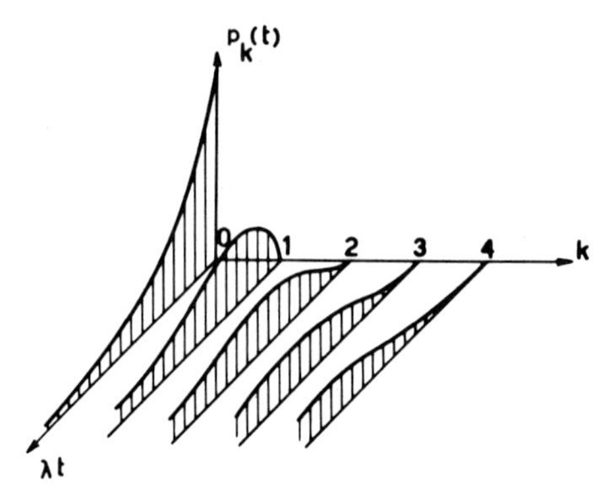

FIG. 4 – The Poisson distribution.

The expected value, or principal function, of the Poisson process is

$$\begin{aligned} \Lambda(t) = E[N_t] &= \sum_{k=0}^{\infty} k p_k(t) \\ &= \sum_{k=0}^{\infty} k \frac{(\lambda t)^k}{k!} \exp[-\lambda t] = \exp[-\lambda t]\, \lambda t \sum_{k=1}^{\infty} \frac{(\lambda t)^{k-1}}{(k-1)!} \\ &= \lambda t \exp[-\lambda t] \sum_{m=0}^{\infty} \frac{(\lambda t)^m}{m!} = \lambda t \exp[-\lambda t] \exp[\lambda t] \\ &= \lambda t . \end{aligned} \tag{6}$$

Hence $\lambda = E[N_t]/t$ is the average number of events occurring per unit time; λ, accordingly, is often called the *intensity* or the *rate* of the Poisson counting process. It also follows from (5) after some algebra that

$$\operatorname{var}(N_t) := E[(N_t - E[N_t])^2] = \lambda t; \tag{7}$$

in other words, the variance increases linearly with time.

The Poisson process has been generalized in many ways. It is worth taking note of the following examples because of their importance in assessing suitable models in optical communication systems.

1.3 Shot noise and filtered Poisson process

Let us next reconsider the shot-noise process generated during photodetection. As the radiation impinges upon the detector surface, electrons are released and travel to the collecting surface. The total current flow through such a detector is the superposition of the current pulses produced by the individual electrons emitted randomly in time and passing through the photodiode. In general, the duration of any individual electronic current pulse will be the transit time of the electron passage through the photodetector. For example, typical electronic current pulses $h(t-u)$ caused by the passage of an electron through an optical detector have the forms shown in Fig. 5, where an electron of charge

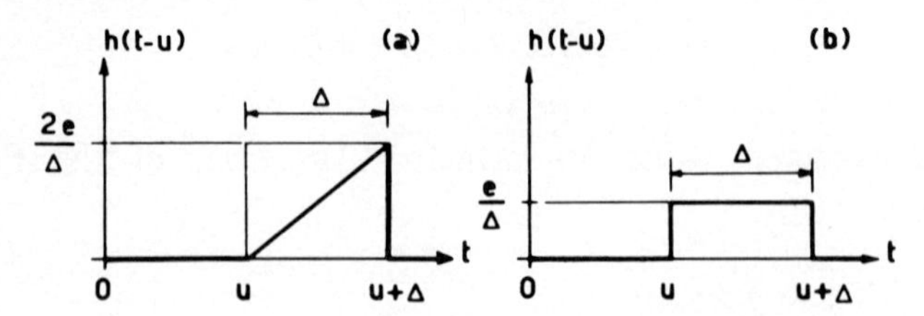

FIG. 5 – Photodetector current pulses.

$e \simeq 1.601 \times 10^{-19}$ C was emitted at time $t = u$ and where Δ is the transit time.

We can therefore represent the output process $X := \{X_t,$

$0 \leqslant t < \infty\}$ by means of

$$X_t := \sum_{j=1}^{N_t} h(t - U_j)\,, \tag{8}$$

where N_t is the random variable counting the number of electrons emitted during the interval $[0, t)$, U_j is the emission random time of the jth electron, and $h(t - U_j)$ is the contribution to X_t due to the jth electron emitted. More generally, the detailed form of h is difficult to predict; however, the physical model allows us to impose some global constraints. Specifically, h must be everywhere non-negative and its area must be equal to the electronic charge and the duration of h is about the same as the travel time, typically of the order of 10^{-7}-10^{-9} s. With the finite-duration assumption, the counting variable N_t in (8) can be replaced by the increment of N_t over the interval $[t - \Delta, t)$:

$$X_t = \sum_{j=1}^{N_{t-\Delta,t}} h(t - U_j)\,. \tag{9}$$

If h has the rectangular form shown in Fig. 5*b*, i.e.,

$$h(t) = \begin{cases} e/\Delta & \text{for } 0 \leqslant t \leqslant \Delta \\ 0 & \text{otherwise;} \end{cases} \tag{10}$$

then, X_t can be further simplified as follows

$$X_t = \frac{e}{\Delta} \underbrace{N_{t-\Delta,t}}_{\text{number of counts in } [t-\Delta,\, t)} \tag{11}$$

and the shot-noise process becomes proportional to the counting process controlled by electron emissions.

If N is a Poisson counting process, a process of the form (8) is usually referred to as a *filtered Poisson process*. In the following, we will outline the main statistical properties of this process. Remember that assumptions (*a*) through (*d*) of the preceding section are still valid. The rate parameter λ in particular is to be considered a constant.

1.3.1 *Occurrence times*

First, by looking at (8) we see that the output process $\boldsymbol{X}$ depends not only on the count statistics but also on the occurrence times $U_1, U_2, \ldots,$ which are a sequence of random *points* in time. The random variable U_j is the time at which the *object* (e.g., the electron) labelled with j causes the event of interest to occur (e.g., the electron emission). Therefore, U_j is not necessarily the same as the jth arrival time. In this sense, the random variables U_j are also said to represent *unordered arrival times.*

Suppose that it is known that exactly k events of a Poisson process occur during the interval $[0, t)$; in other words, suppose that $N_t = k$, where N_t is a Poisson counting random variable. It can be shown that *the occurrence times* $U_1, U_2, \ldots, U_k$ *of those* k *events are mutually independent random variables, each of which is uniformly distributed over the interval* $[0, t)$:

$$f_{U_j}(u|N_t = k) = \begin{cases} 1/t & \text{for } 0 \leqslant u < t \\ 0 & \text{otherwise} \end{cases} \tag{12}$$

for all $j = 1, 2, \ldots, k$.

1.3.2 *Counting statistics*

The results given in the previous section can be used to evaluate the statistical moments of X_t. We note from (8) that X_t is a random sum of random variables. With this fact in mind, the suggested procedure is to compute the conditional moment of X_t by first assuming a particular value (say k) of N_t, then conditionally averaging the resultant fixed sum with respect to the k occurrence times, and, finally, averaging the result over all possible values k of N_t.

– *Average value.* The statistical mean of the shot-noise process can be defined as

$$E[X_t] = \sum_{k=0}^{\infty} P[N_t = k] E[X_t|N_t = k], \tag{13}$$

where N_t has the Poisson probability distribution and $E[X_t|N_t = k]$ is obtained by averaging the sum

$$\sum_{j=1}^{k} h(t - U_j)$$

with respect to $U_1, U_2, \ldots, U_k$ that are mutually independent random variables, each uniformly distributed over the interval $[0, t)$.

With some algebra it is easy to show that

$$E[X_t|N_t = k] = E\left[\sum_{j=1}^{k} h(t - U_j)\right]$$

$$= \frac{k}{t}\int_0^t h(\tau)\, d\tau; \tag{14}$$

then

$$E[X_t] = \sum_{k=0}^{\infty} P[N_t = k]\, E[X_t|N_t = k]$$

$$= \lambda \int_0^t h(\tau)\, d\tau\,. \tag{15}$$

For values of t greater than the duration Δ of an electronic current pulse, this last expression applied to the shot noise yields

$$E[X_t] = e\lambda\,. \tag{16}$$

– *Characteristic function.* It is possible to determine the characteristic function of the filtered Poisson random variable X_t:

$$\Phi_{X_t}(v) := E[\exp(jvX_t)]$$

$$= E\left[\exp\left\{jv \sum_{j=1}^{N_t} h(t - U_j)\right\}\right]. \tag{17}$$

We start from

$$\Phi_{X_t}(v) = \sum_{k=0}^{\infty} E[\exp(jvX_t)|N_t = k]\, P[N_t = k] \tag{18}$$

and, paralleling our calculation of $E[X_t]$, we arrive at

$$\Phi_{X_t}(v) = \exp\left[\lambda \int_0^t (\exp[jvh(\tau)] - 1)\, d\tau\right]. \tag{19}$$

This expression is generally difficult to invert to determine the probability distribution for X_t, unless the electron response function h is particularly simple. It is useful, however, for evaluating the *moments* of X_t through the *cumulants* (or *semi-invariants*)

γ_n, $n = 1, 2, \ldots$, of X_t itself. By using the relationship

$$j^n \gamma_n = \{\partial^n \ln \Phi_{X_t}(v)/\partial v^n\}_{v=0}\,, \tag{20}$$

we deduce from (19) that

$$\gamma_n = \lambda \int_0^t h^n(\tau)\, d\tau\,. \tag{21}$$

It is known that moments $m_n := E[X^n]$, for $n = 1, 2, \ldots$, are related to cumulants by the following relationship

$$\begin{aligned} m_1 &= \gamma_1\,, \\ m_2 &= \gamma_2 + \gamma_1^2\,, \\ m_3 &= \gamma_3 + 3\gamma_1\gamma_2 + \gamma_1^3\,, \\ m_4 &= \gamma_4 + 3\gamma_2^2 + 4\gamma_1\gamma_3 + 6\gamma_1^2\gamma_2 + \gamma_1^2\,, \\ &\vdots \end{aligned} \tag{22}$$

In compact form we have the recurrent relation

$$m_{n+1} = \gamma_{n+1} + \sum_{j=0}^{n-1} \binom{n}{j} \gamma_{j+1} m_{n-j}\,, \qquad n \geqslant 1 \tag{23}$$

with $m_1 = \gamma_1$. From the expressions above, we see that the first cumulant γ_1 gives the expectation of X_t, and that the second cumulant $\gamma_2 = m_2 - m_1^2$ equals the variance of X_t. Therefore, according to (21) the variance of a filtered Poisson process is

$$\operatorname{var}(X_t) = \gamma_2 = \lambda \int_0^t h^2(\tau)\, d\tau\,. \tag{24}$$

For the shot-noise case, the response function is nonzero only over a finite interval. Therefore, we can write

$$\exp[jvh(\tau)] - 1 = 0 \tag{25}$$

for $\tau < 0$ and $\tau > \Delta$ where Δ is the duration of h. With this fact in mind, Eq. (19) may be written in the form

$$\Phi_{X_t}(v) = \exp\left[\lambda \int_{-\infty}^{\infty} (\exp[jvh(\tau)] - 1)\, d\tau\right]. \tag{26}$$

It should be noted in this case that the characteristic function (and hence the probability distribution of X_t) is *not* a function of t. Since the same result is obtained when the joint-characteristic function is determined for any n time instants, it follows that such a filtered Poisson process is *stationary in the strict sense.*

For rectangular electron function in particular, we have $h(t) = e/\Delta$, $0 \leqslant t \leqslant \Delta$, and the variance reduces to

$$\operatorname{var}(X_t) = e^2 \lambda/\Delta . \tag{27}$$

This equation shows that $\operatorname{var}(X_t) \to \infty$ as $\Delta \to 0$; thus *infinite bandwidth filtered Poisson processes do not correspond to bounded mean square processes.*

– *Limiting form as* $\lambda \to \infty$. We have noted that it may be quite difficult to invert the characteristic function of a filtered Poisson process to obtain the corresponding probability density:

$$f_{X_t}(x) = \int_{-\infty}^{\infty} \Phi_{X_t}(v) \exp[-jvx] \frac{dv}{2\pi} . \tag{28}$$

Nonetheless, it can be proved that Φ_{X_t} (26) tends to the characteristic function of a Gaussian random variable as $\lambda \to \infty$. Since both the mean and the variance of a Poisson process increase linearly with λ, it is convenient to consider the normalized random variable

$$Y_t := \frac{X_t - E[X_t]}{[\operatorname{var}(X_t)]^{\frac{1}{2}}} , \tag{29}$$

which has a zero mean and a unit variance. By defining for notational convenience the parameters

$$\alpha := \int_0^t h(\tau)\, d\tau \qquad \text{and} \qquad \beta^2 := \int_0^t h^2(\tau)\, d\tau , \tag{30}$$

from (15) and (24) we can write

$$E[X_t] = \alpha\lambda \qquad \text{and} \qquad \operatorname{var}(X_t) = \beta^2 \lambda . \tag{31}$$

Hence, applying this result to (29), we have

$$Y_t = \frac{X_t}{\beta\sqrt{\lambda}} - \frac{\alpha}{\beta}\sqrt{\lambda} \tag{32}$$

and

$$\Phi_{Y_t}(v) = E\left[\exp\left\{jv\left(\frac{X_t}{\beta\sqrt{\lambda}} - \right)\frac{\alpha}{\beta}\sqrt{\lambda}\right\}\right]$$

$$= \exp\left(-j\frac{\alpha}{\beta}\sqrt{\lambda}\, v\right)\Phi_{X_t}\left(\frac{v}{\beta\sqrt{\lambda}}\right), \tag{33}$$

which gives

$$\Phi_{Y_t}(v) = \exp\left\{-j\frac{\alpha}{\beta}\sqrt{\lambda}v + \lambda\int_0^t\left[\exp\left(j\frac{v}{\beta\sqrt{\lambda}}h(\tau)\right) - 1\right]d\tau\right\}. \tag{34}$$

Using a power-series expansion of the exponential inside the integral in (34) and putting $\lambda \to \infty$ yields

$$\lim_{\lambda\to\infty}\Phi_{X_t}(v) = \exp[-v^2/2], \tag{35}$$

provided that integrals $\int_0^t h^n(t)dt$, $n \geqslant 3$, all exist.

Thus, we see that *the characteristic function of the normalized filtered Poisson random variable Y_t tends to that of a Gaussian random variable with a zero mean and a unit variance as the average number of events occurring per unit time increases without limit.*

The same results can be obtained by considering the cumulants $\hat{\gamma}_n$ of the normalized random variable Y_t, which can be written on the basis of definition (29) as

$$\begin{aligned} \hat{\gamma}_1 &= 0, \\ \hat{\gamma}_n &= \frac{\gamma_n}{[\text{var}(X_t)]^{n/2}}, \qquad n \geqslant 2, \end{aligned} \tag{36}$$

where the cumulants γ_n of X_t are given by (21). Hence

$$\begin{aligned} \hat{\gamma}_1 &= 0, \\ \hat{\gamma}_2 &= 1, \\ &\vdots \\ \hat{\gamma}_n &= \frac{\lambda\int_0^t h^n(\tau)\,d\tau}{\left[\lambda\int_0^t h^2(\tau)\,d\tau\right]^{n/2}}. \end{aligned} \tag{37}$$

Only the first two cumulants (mean and variance) of Y_t are non-zero as $\lambda \to \infty$, thus Y_t approaches a Gaussian random variable.

A similar derivation for the limiting form of the joint-characteristic function of any n random variables $Y_{t_1}, Y_{t_2}, \ldots, Y_{t_n}$ also results in a Gaussian characteristic function. *It therefore follows that a filtered Poisson process tends to become a Gaussian random process as* $\lambda \to \infty$.

1.3.3 *Autocorrelation function and spectral power density*

By resorting again to the averaging technique, it may be shown that the autocorrelation function of the shot-noise process is given by

$$R_X(t_1, t_2) := E[X_{t_1} \ X_{t_2}] = \lambda \int_{-\infty}^{\infty} h(z) h(t_1 - t_2 + z) \, dz + (e\lambda)^2, \qquad (38)$$

or, because of stationarity, by

$$R_X(\tau) = \lambda \int_{-\infty}^{\infty} h(z) h(\tau + z) \, dz + (e\lambda)^2. \qquad (39)$$

In the frequency domain, the power spectral density is

$$S_X(\omega) := \int_{-\infty}^{\infty} R_X(\tau) \exp[-j\omega\tau] \, d\tau$$

$$= \lambda |H(\omega)|^2 + 2\pi (e\lambda)^2 \, \delta(\omega) \, . \qquad (40)$$

Note that the spectrum always appears as the sum of a continuous portion and a delta function. The former is the same as if the filter h were excited by a random process with constant spectral power density λ. Such a property is widely used in models for devices exhibiting shot noise. The contribution $\lambda |H(\omega)|^2$ becomes flat for an ideal photodetector, i.e., having impulse response $h(t) = e\delta(t)$.

1.3.4 *Random gain photodetectors and marked Poisson process*

It has been pointed out in Chapter 2 of Part II that an ideal photomultiplier with gain G effectively multiplies $h(t)$ by G in the shot-noise model. If the gain of the multiplier is itself a random variable, statistically independent from electron to electron, the component functions are multiplied by independent random variables, producing a random amplitude from one h to the next. Then a more general representation for filtered Poisson processes, e.g., modelling shot-noise processes in avalanche photodiodes (APD's), becomes

$$X_t = \sum_{i=1}^{N_t} G_j h(t - U_j), \tag{41}$$

where the gains G_j (or *marks* in the Poisson process nomenclature) are mutually independent, independent of N, identically distributed random variables.

We can obtain Φ_{X_t} by a straightforward application of the procedure outlined for deducing (19), that is

$$\Phi_{X_t}(v) := E[\exp[jvX_t]] = \sum_{k=0}^{\infty} E[\exp[jvX_t]|N_t = k] P[N_t = k], \tag{42}$$

where the expectation is over $U_1, U_2, \ldots, U_k$ *and* $G_1, G_2, \ldots, G_k$. Using (5) and (12), we can repeat the calculations in (18) to (19) with an additional average over G. We obtain

$$\Phi_{X_t}(v) = \exp\left[\lambda \int_0^t E[\exp[jvGh(\tau)] - 1]\, d\tau\right]. \tag{43}$$

The present version of cumulants corresponding to (21) is

$$\gamma_n = \lambda E[G^n] \int_0^t h^n(\tau)\, d\tau; \tag{44}$$

therefore, for a shot-noise process with random gains we have

$$E[X_t] = \gamma_1 = \lambda E[G] \int_0^t h(\tau)\, d\tau = e\lambda E[G] \tag{45}$$

and

$$\operatorname{var}(X_t) = \gamma_2 = \lambda E[G^2] \int_0^t h^2(\tau)\, d\tau. \tag{46}$$

Similarly, the power density (40) is modified according to

$$S_X(\omega) = \lambda E[G^2] |H(\omega)|^2 + 2\pi (e\lambda)^2 E^2[G]\, \delta(\omega). \tag{47}$$

Note that in our case

$$\int_0^\infty h(\tau)\,d\tau = H(0) = e; \tag{48}$$

hence, more generally,

$$S_X(\omega) = \lambda E[G^2]|H(\omega)|^2 + 2\pi\lambda^2 E^2[G]\,H^2(0)\,\delta(\omega), \tag{49}$$

which is sketched in Fig. 6.

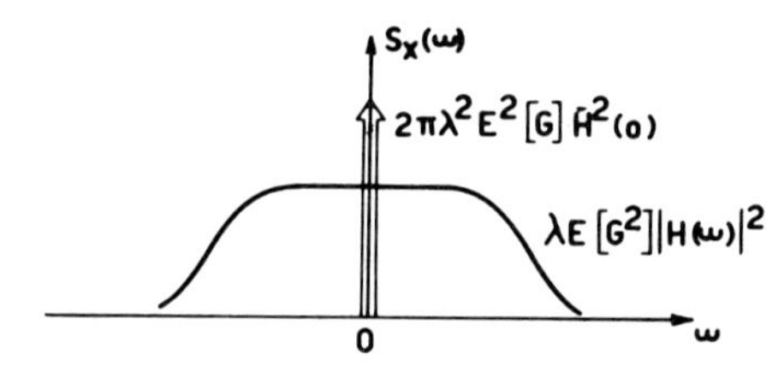

FIG. 6 – Power spectrum of filtered homogeneous Poisson process with gain.

1.4 Nonhomogeneous Poisson process

Our previous methods and results can be extended to the case of the nonhomogeneous (or inhomogeneous) Poisson process, i.e., the case in which the intensity λ is a function of time.

In this case, we change assumption (*b*) of Section 1.2 to:

b'. The counting random process $\boldsymbol{N} := \{N_t,\ 0 \leqslant t < \infty\}$ has independent (but nonstationary) increments,

and we replace assumption (*c*) by:

c'
$$P[N_{t+\Delta t} - N_t = 1] := \lambda(t)\,\Delta t + 0(\Delta t), \tag{50}$$

the assumptions (*a*) and (*d*) remaining unchanged. It then follows that $\boldsymbol{N}$ is a *nonhomogeneous Poisson process* whose probability distribution function may be shown to be

$$p_k(t) := P[N_t = k] = \frac{1}{k!}\left(\int_0^t \lambda(\tau)\,d\tau\right)^k \exp\left(-\int_0^t \lambda(\tau)\,d\tau\right), \tag{51}$$

for $k = 0, 1, 2, \ldots$. More generally,

$$p_k(t_1, t_2) := P[N_{t_1,t_2} = k] = \frac{1}{k!}[\Lambda(t_2) - \Lambda(t_1)]^k \exp(-[\Lambda(t_2) - \Lambda(t_1)])$$

for $k = 0, 1, 2, \ldots$, where $\Lambda(t) = \int_0^t \lambda(\tau)\, d\tau$. A simple computation shows that

$$\Lambda(t) = E[N_t] = \int_0^t \lambda(\tau)\, d\tau \tag{52}$$

and

$$\operatorname{var}(N_t) = \int_0^t \lambda(\tau)\, d\tau\, . \tag{53}$$

$\Lambda(t)$, the principal function of the process $\boldsymbol{N}$, is nonnegative, non-decreasing, and right continuous with $\Lambda(0) = 0$. Its derivative $\lambda(t)$ represents the *instantaneous* intensity of the nonstationary process.

In the following, we will extend the results given in the preceding sections to inhomogeneous processes without getting involved in proofs. Analogies and, when appropriate, discrepancies with the homogeneous case will be emphasized.

– *Occurrence times.* Again, the events times $U_1, U_2, \ldots, U_k$ conditioned to $N_t = k$ are statistically independent, identically distributed random variables. However, they are no longer uniformly distributed over the interval $[0, t)$, the common distribution being

$$f_{U_j}(u|N_t = k) = \begin{cases} \lambda_u \left(\int_0^t \lambda(\tau)\, d\tau \right)^{-1} & \text{for } 0 \leqslant u < t \\ 0 & \text{otherwise} \end{cases} \tag{54}$$

for all $j = 1, 2, \ldots, k$.

– *Filtered Poisson process.* The model for a filtered Poisson process, and, hence, for the shot-noise, becomes

$$X_t = \sum_{i=1}^{N_t} G_j h(t - U_j)\, , \tag{55}$$

where $\boldsymbol{N} := \{N_t,\ 0 \leqslant t < \infty\}$ is an *inhomogeneous* counting process and the marks, or gains, G_j are statistically independent and identically distributed random variables. The usual approach based upon conditional averages gives the following relationships.

Characteristic function

$$\Phi_{X_t}(v) = \exp\left\{\int_0^t \lambda(\tau) E[\exp(jvGh(t-\tau)) - 1]\, d\tau\right\}. \tag{56}$$

Cumulants

$$\gamma_n = E[G^n]\int_0^t \lambda(\tau)\, h^n(t-\tau)\, d\tau\,. \tag{57}$$

Mean

$$E[X_t] = \gamma_1 = E[G]\int_0^t \lambda(\tau)\, h(t-\tau)\, d\tau\,. \tag{58}$$

Variance

$$\mathrm{var}(X_t) = \gamma_2 = E[G^2]\int_0^t \lambda(\tau)\, h^2(t-\tau)\, d\tau\,. \tag{59}$$

Autocorrelation function

$$R_X(t, t-\tau) = E[G^2]\int_{-\infty}^{\infty} \lambda(z)\, h(t-z)\, h(t-\tau-z)\, dz + E[X_t]E[X_{t-\tau}]\,. \tag{60}$$

For mnemonic purposes, Fig. 7 shows how the cumulants can be interpreted according to a *linear system* model driven by the intensity function $\lambda(t)$.

For the shot-noise process with an infinite bandwidth $h(t) = e\,\delta(t)$ and (58) becomes

$$E[X_t] = eE[G]\,\lambda(t); \tag{61}$$

hence, the mean-value function is proportional to the instantaneous count rate. In this case, however, the variance grows without limits as implied by relation (59); this is in agreement with the result found for the homogeneous case.

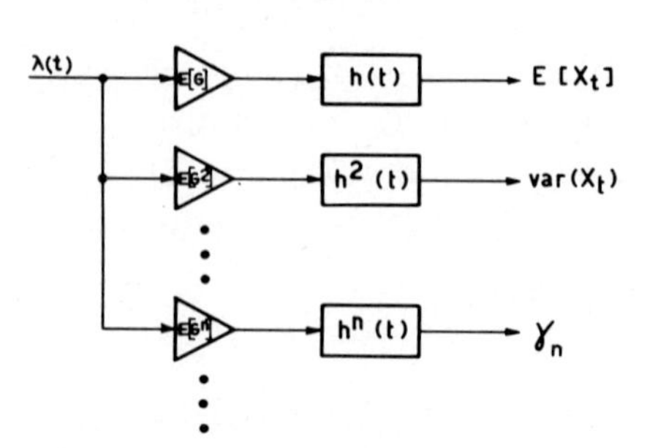

FIG. 7 – Generation of shot-noise cumulants by linear filtering.

– *Power spectral density.* The statistical averages (56)-(59) exhibit an explicit dependence on time t and (60) depends on both t and τ. Hence the general filtered Poisson process we are dealing with is *nonstationary.* A first troublesome consequence is that time averages are *not* equivalent to statistical averages. Most importantly, the general recommended approach in computing the power spectrum of a stationary random process by Fourier transforming its statistical autocorrelation function becomes unfeasible. To overcome this drawback, it is possible to consider the power spectral density as a *time average.* That is, we consider the power spectrum of process $\boldsymbol{X} := \{X_t, -\infty < t < \infty\}$ (5) as

$$S_X(\omega) := \lim_{T\to\infty} \frac{1}{2T} E\left[\left|\int_{-T}^{T} X_t \exp[-j\omega t]\, dt\right|^2\right], \tag{62}$$

where $\left|\int_{-T}^{T} X_t \exp[-j\omega t]\, dt\right|^2$ is a random variable at each ω representing the energy density at each ω. It must therefore be averaged over the statistics of the process. Thus, $S_X(\omega)$ *is the time average of the statistical average of the energy density at each* ω.

The approach for computing (62) is usual, since it is based on conditioning and then averaging. As an intermediate result we get

$$E\left[\left|\int_{-T}^{T} X_t \exp[-j\omega t]\, dt\right|^2\right] = |H(T,\omega)|^2 \{E[G^2]L(T) + E^2[G]L^*(T,\omega)L(T,\omega)\}, \tag{63}$$

(5) We assume that the events of interest are counted starting from time $t_0 \to -\infty$. This assumption means that we have the infinite part available to operate on in order to make our estimate of $S_X(\omega)$.

where

$$H(T, \omega) := \int_{-T}^{T} h(t) \exp[-j\omega t]\, dt\,, \tag{64}$$

$$L(T) := \int_{-T}^{T} \lambda(t)\, dt\,, \tag{65}$$

$$L(T, \omega) := \int_{-T}^{T} \lambda(t) \exp[-j\omega t]\, dt\,. \tag{66}$$

Taking the limit yields the desired result

$$S_x(\omega) = |H(\omega)|^2 \{E[G^2]\,\bar{\lambda} + E^2[G]\, G_\lambda(\lambda)\}\,, \tag{67}$$

where $\bar{\lambda} = \lim_{T\to\infty} (1/2T) \int_{-T}^{T} \lambda(t)\, \mathrm{d}t$ is the time-averaged count intensity and $G_\lambda(\omega) = \lim_{T\to\infty} |L(T, \omega)|^2/(2T)$ is the power spectral density of $\lambda(t)$.

Two remarks are appropriate:

1. $G_\lambda(\omega) = 0$ when $\lambda(t)$ is a finite-energy function.
2. When $\lambda(t)$ is a constant, $S_x(\omega)$ reduces to

$$S_x(\omega) = \lambda E[G^2] |H(\omega)|^2 + 2\pi\lambda^2 E^2[G]\, H^2(0)\, \delta(\omega)\,, \tag{68}$$

which coincides with (49). As a matter of fact, $\lambda = \bar{\lambda}$ and

$$G_\lambda(\omega) = 2\pi\lambda^2\, \delta(\omega); \tag{69}$$

thus (67) becomes (68) since

$$|H(\omega)|^2\, \delta(\omega) = H^2(0)\, \delta(\omega)\,. \tag{70}$$

In this respect, (68) is a generalization of (49) in the case of non-homogeneous Poisson counting processes.

1.5 Extensions to doubly stochastic filtered Poisson process

There are many physical situations for which the intensity function $\lambda(t)$ is a sample of a random process $\boldsymbol{\lambda}$ [6] and is no longer a deterministic time function. These are typical in optical com-

[6] See note [1] on page 649.

munication systems; such as when the intensity of a detected optical field corresponds to the modulating (analogue or digital) process. In this case the results given in the foregoing sections are still valid for a specific sample function of the random intensity process. The effect of this conditioning is that *an additional average over the intensity process* $\boldsymbol{\lambda}$ *is required* in order to derive the desired statistics.

Thus, for the doubly stochastic Poisson process the occurrence times are *not* independent. In other words, given a member $\lambda(t)$ of $\boldsymbol{\lambda}$, the k event times $U_1, U_2, \ldots, U_k$ are generated independently with a joint distribution that is a k-fold product of the marginal probability densities (54). However, the additional average over the randomness of $\boldsymbol{\lambda}$ does produce nonindependent event times.

– *Moments*. The moments of X_t can be obtained [see (22)] as

$$\begin{aligned} E[X_t] &= E_\lambda[\gamma_1]\,, \\ E[X_t^2] &= E_\lambda[\gamma_2] + E_\lambda[\gamma_1^2]\,, \\ E[X_t^3] &= E_\lambda[\gamma_3] + 3E_\lambda[\gamma_1\gamma_2] + E_\lambda[\gamma_1^3]\,. \\ &\vdots \end{aligned} \tag{71}$$

In particular, we see from (58) and (59) that

$$E[X_t] = E[G]\int_0^t E[\lambda_\tau]\,h(t-\tau)\,d\tau \tag{72}$$

and

$$\begin{aligned} \operatorname{var}(X_t) = {} & E[G^2]\int_0^t E[\lambda_\tau]\,h^2(t-\tau)\,d\tau \\ & + E^2[G]\int_0^t\int_0^t h(t-\tau)\,K_\lambda(\tau, u)\,h(t-u)\,d\tau\,du\,, \end{aligned} \tag{73}$$

where K_λ is the covariance function for $\boldsymbol{\lambda}$.

– *Time averaged power spectral density*. The power spectral density given by (67) depends on a *given* intensity function. Therefore, the spectrum of the doubly stochastic filtered Poisson process is obtained by performing the average of (67) with respect to **λ**, i.e.,

$$S_x(\omega) = |H(\omega)|^2\{E[G^2]E[\bar{\lambda}] + E^2[G]\,S_\lambda(\omega)\}\,, \tag{74}$$

where

$$E[\bar{\lambda}] = \lim_{T\to\infty}\frac{1}{2T}\int_{-T}^{T}E[\lambda_t]\,dt \tag{75}$$

and

$$S_\lambda(\omega) = \lim_{T\to\infty}\frac{1}{2T}E[|L(T,\omega)|^2]\,. \tag{76}$$

Note also that for an intensity process whose statistical mean is time independent, i.e., $E[\lambda_t] = E[\lambda]$, it results that

$$E[\bar{\lambda}] = E[\lambda_t] = E[\lambda]\,,$$

and if the power spectrum of **λ** is ergodic it can be computed as the Fourier transform of the autocorrelation function:

$$S_\lambda(\omega) = \int_{-\infty}^{\infty}E[\lambda_t\,\lambda_{t-\tau}]\exp[-j\omega\tau]\,d\tau\,. \tag{77}$$

Remember also that in photodetection processes $H(0) = e$ in general, and $H(\omega) = e$ for all ω under the infinite bandwidth assumption. Thus, the information bearing random process **λ** is essentially undistorted if $S_\lambda(\omega)$ is significant only in the region where $H(\omega) \cong e$ (Fig. 8). The spectral density (76) reveals the

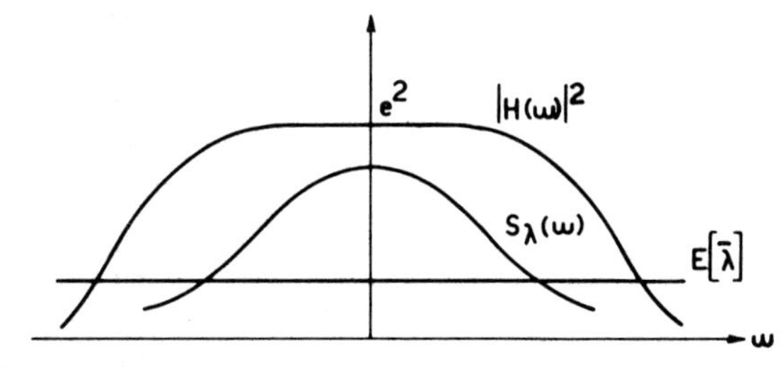

FIG. 8 – Spectral power components of shot-noise process.

inevitable noise level associated with $E[\bar{\lambda}]$ and is due to the discrete nature of the detector model. The corresponding noise power will be referred to as *quantum-noise power*.

B. Analogue communications

The statistical background of the preceding part is now used for giving an analysis of direct-detection optical communication receivers. System performance is measured in terms of signal-to-noise ratio by taking into account the shot noise (dark current included) and the thermal noise, whereas all other sources are considered negligible. The emphasis is on information signals continuous in time, that is, on *analogue* communications.

The typical model of a baseband direct-detection receiver is illustrated in Fig. 9. The desired information modulates the intensity of an optical source and is transmitted through the fibre.

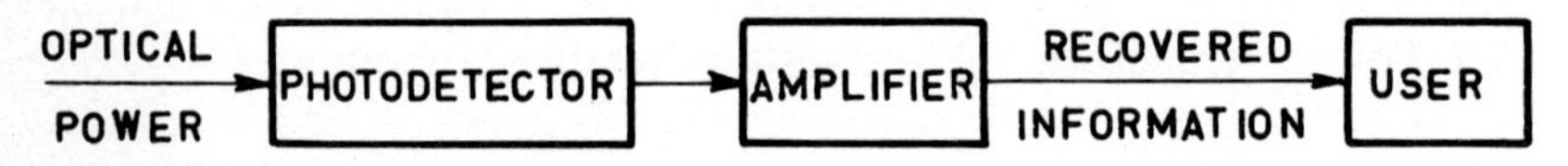

Fig. 9 – Baseband direct-detection optical receiver.

The received field impinges on the surface of the photodetector, thereby giving an output current whose count intensity depends on the received optical power.

We also consider subcarrier systems, where the optical carrier is intensity modulated by a high frequency subcarrier wave which is itself modulated by the information signal.

1.6 Optical direct detection

A receiver model which emphasizes the signal and noise contributions is shown in Fig. 10. The photodetector output process (e.g., a current) $\boldsymbol{I} := \{I_t,\ 0 \leqslant t < \infty\}$ can be written as a sum

$$I_t = I_t^g + I_t^s + I_t^d; \tag{78}$$

hence, $\boldsymbol{I}$ account for the conversion of the optical power $p(t)$ into a signal current. The optical power $p(t)$ is associated with a

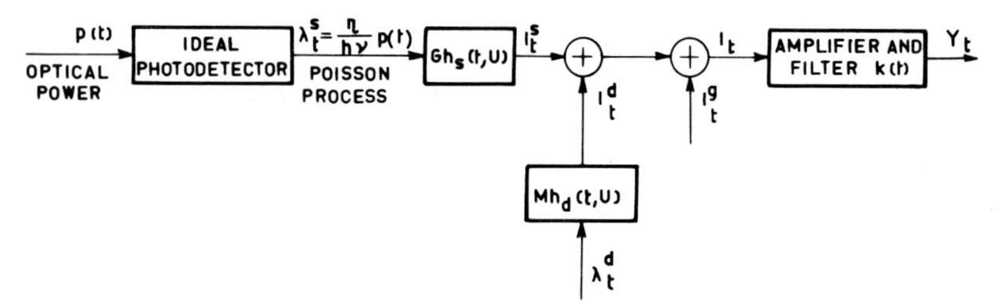

FIG. 10 – Statistical model for the receiver.

narrow-band optical signal field

$$z(t, \boldsymbol{r}) = \sqrt{2}\,\mathrm{Re}\,[s(t, \boldsymbol{r})\exp(j2\pi\nu t)]\ (^7) \tag{79}$$

incident on the active surface $\mathcal{A}$ of the detector according to

$$p(t) = \int_{\mathcal{A}} |s(t, \boldsymbol{r})|^2\, d\boldsymbol{r}\,, \tag{80}$$

$s(t, \boldsymbol{r})$ being the complex envelope of the received signal, $\boldsymbol{r}$ the points of a planar coordinate system on $\mathcal{A}$ and ν the optical carrier frequency. Of course, $p(t)$ is related to the modulation format. The optoelectrical conversion performed by the photodetector is affected by disturbances caused either by the detector itself or by the amplifier or other circuits following the detector. Hereafter the physical quantities are explained and modelled from a statistical viewpoint.

As usual, the noise process caused by the amplifier is assumed to be a thermally generated current. As a consequence, it can be represented as a stationary zero-mean Gaussian process $I^g := \{I^g_t, 0 \leqslant t < \infty\}$ with constant bilateral spectral height $\mathcal{N}_0/2 = 2k\mathcal{T}/R$, where $k \simeq 1.38 \times 10^{-23}$ J/°K is Boltzmann's constant, $\mathcal{T}$ is the receiver temperature in degrees Kelvin and R is the effective output resistance of the photodetector.

In (78), I^s_t is the photodetector useful output in response to $s(t, \boldsymbol{r})$, whereas I^d_t is the so-called *dark current* noise contribution caused by extraneous, internally generated thermoelectrons even

(7) When an incoherent source such as a light-emitting diode is used, the field representation has to account for the source bandwidth. The beat between spectral components within the spectral width of the source gives rise to a noise term that is, however, negligible in the frequency range in which we are interested. If a coherent source is used, the beat noise is not present at all.

in the absence of a driving signal. According to the model of Fig. 10, $\boldsymbol{I}^s := \{I_t^s, 0 \leqslant t < \infty\}$ is a shot-noise process

$$I_t^s = \sum_{j=1}^{N_t^s} G_j h_s(t - U_j) . \tag{81}$$

The intensity function of the counting process $\boldsymbol{N}^s$ results from the conversion of an incident photon stream of power $p(t)$ and optical frequency ν into a primary current $\boldsymbol{I}^s$; thus (see Chapter 2 of Part II)

$$\lambda_t^s = \frac{\eta}{h\nu} p(t) , \tag{82}$$

where η is the quantum efficiency of the detector, $h = 6.626 \times \times 10^{-34}$ J·s is Planck's constant, and $h\nu$ is the energy in a photon.

The representation (81) resembles expression (55); therefore, I_t^s is a filtered (inhomogeneous and marked) Poisson process (with respect to the notation used in Section 1.4, symbols N_t^s, G_j, h_s and U_j are obvious). More generally, $\boldsymbol{I}^s$ is doubly stochastic since its intensity λ_t^s is a sample of a random process in information transmission systems.

The dark current $\boldsymbol{I}^d := \{I_t^d, 0 \leqslant t < \infty\}$ is again a shot-noise process:

$$I_t^d = \sum_{j=1}^{N_t^d} M_j h_d(t - U_j) \tag{83}$$

and accounts for extraneous electrons generated during $[0, t)$. The intensity function λ_t^d is the instantaneous rate at which such spontaneous events occur.

The model described previously is quite general and appropriate simplifications can be carried out when prompted by physical considerations. For this purpose it is generally assumed that λ_t^d is a constant equal to λ^d and that $M_j h_d = G_j h_s$. In these cases, we can consider the overall counts $N_t = N_t^s + N_t^d$ during $[0, t)$ with an intensity function

$$\lambda_t = \lambda_t^s + \lambda^d = \frac{\eta}{h\nu} p(t) + \lambda^d. \tag{84}$$

With these premises the amplifier output $\boldsymbol{Y} := \{Y_t, 0 \leqslant t < \infty\}$

should clearly be written as a filtered version of the shot-noise process $\boldsymbol{I}$ given by (78); i.e.,

$$Y_t = \sum_{j=1}^{N_t} G_j h(t - U_j) + \int_0^t I_\tau^g k(t-\tau)\, d\tau\,, \tag{85}$$

where $k(t)$ is the amplifier impulse response and

$$h(t) = h_s(t) * k(t)\,. \tag{86}$$

(Symbol $*$ denotes convolution). Hereafter we assume infinite bandwidth photodetectors; thus $h_s(t) = e\delta(t)$, and (85) reduces to

$$Y_t = \sum_{j=1}^{N_t} eG_j k(t - U_j) + \int_0^t I_\tau^g k(t-\tau)\, d\tau\,. \tag{87}$$

Relationships (58) and (59) together with the assumptions made in this section allow the computation of the statistical mean (conditioned to a member of $\boldsymbol{\lambda}$)

$$E[Y_t] = eE[G]\int_0^\infty \lambda_\tau k(t-\tau)\, d\tau \tag{88}$$

and conditional variance

$$\operatorname{var}(Y_t) = e^2 E[G^2]\int_0^\infty \lambda_\tau k^2(t-\tau)\, d\tau + \frac{\mathcal{N}_0}{2}\int_0^\infty k^2(t-\tau)\, d\tau\,, \tag{89}$$

where the integration upper limit has been extended to ∞ according to the hypothesis that $k(t)$ is *causal*, i.e., $k(t) = 0$ for $t < 0$.

In the absence of dark current and thermal noise, (88) and (89) result, respectively, in

$$E[Y_t] = \frac{\eta e}{h\nu} E[G]\int_0^\infty p(\tau)\, k(t-\tau)\, d\tau \tag{90}$$

and

$$\operatorname{var}(Y_t) = \frac{\eta e^2}{h\nu} E[G^2]\int_0^\infty p(\tau)\, k^2(t-\tau)\, d\tau\,. \tag{91}$$

To summarize, the photodetector output appears as a shot-noise process whose count intensity function is proportional to the collected power. The photodetector is a square-law device in which the desired signal corresponds to the statistical mean of the shot-noise (dark current excluded), but the randomness inherent to the process yields a variance that must be considered.

One particular type of linear filter $k(t)$ is the averager in a time interval of duration T_a. Such a device is called a *short-time*

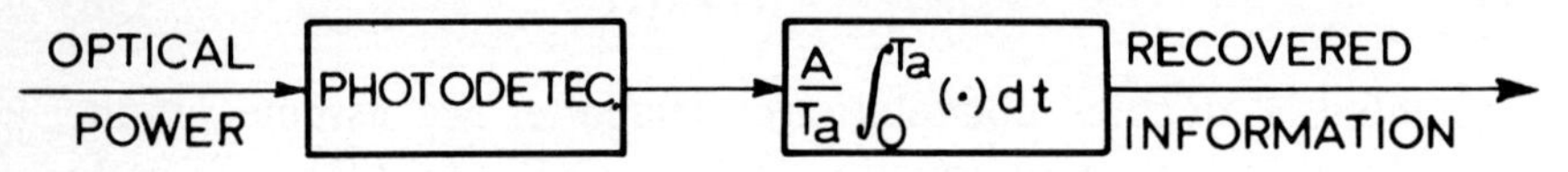

FIG. 11 – Baseband short-time averager with gain.

averager (Fig. 11) and its impulse response (including a constant gain A) is

$$k(t) := \begin{cases} A/T_a & \text{for } 0 \leqslant t \leqslant T_a \\ 0 & \text{otherwise;} \end{cases} \tag{92}$$

hence the output becomes

$$Y_t = \frac{A}{T_a} \int_{t-T_a}^{t} I_\tau^s \, d\tau + \text{dark current} + \text{thermal noise}\,, \tag{93}$$

where, according to (81) written without avalanche gain, we have

$$I_t^s = \sum_{j=1}^{N_t^s} e\delta(t - U_j)\,. \tag{94}$$

Putting (94) into (93) gives

$$Y_t = \frac{eA}{T_a} \underbrace{[N_t^s - N_{t-T_a}^s]}_{\text{number of counts in } [t-T_a,\, t)} + \text{dark current} + \text{thermal noise}\,. \tag{95}$$

Now

$$E[N_t^s - N_{t-T_a}^s] = \int_{t-T_a}^{t} \lambda_\tau \, d\tau = \frac{\eta}{h\nu} \int_{t-T_a}^{t} p(\tau)\, d\tau\,. \tag{96}$$

If the integration time T_a is small compared to the highest frequency component in $p(t)$, we have

$$E[N_t^s - N_{t-T_a}^s] \cong \frac{\eta}{h\nu} p(t) T_a , \tag{97}$$

and the expectation of Y_t becomes *proportional* to the information-bearing signal $p(t)$.

1.7 Signal-to-noise ratio

A fundamental tradeoff in the design of any communication system is that of complexity versus performance. A major determining factor of both system complexity and performance is the choice of modulation, demodulation and signal processing techniques used. In this section, the performance of optical fibre communication systems employing *analogue* modulation technique is investigated. The receiver performance is measured by the signal-to-noise ratio (SNR). For digital communication systems, which will be considered in Part C, we will resort to the average probability of error.

As already emphasized in the preceding sections, a photodetector gives rise to a signal-dependent noise, which, as a consequence, can be nonstationary. Hence a SNR measure requires exact definition for its meaning to become clear.

We may see, e.g. from (95) and (97), that the average value of the detected signal in the absence of dark and thermal noises is related to the information message $p(t)$. It can be, therefore, interpreted as the *useful* or *desired* signal, whereas the fluctuations due to all the noises give rise to « disturbance ». A first obvious approach is then to define the *instantaneous* SNR at time t as

$$\mathrm{SNR}_t := \frac{E^2[Y_t | I_t^d = I_t^g = 0]}{\mathrm{var}(Y_t)} , \tag{98}$$

where unfortunately $\mathrm{var}(Y_t)$ is affected by the signal. By using relationships (88) and (89), we may write

$$\mathrm{SNR}_t = \frac{e^2 E^2[G][\int_0^\infty \lambda_\tau^s k(t-\tau)\, d\tau]^2}{e^2 E[G^2] \int_0^\infty [\lambda_t^s + \lambda^d] k^2(t-\tau)\, d\tau + (\mathcal{N}_0/2) \int_0^\infty k^2(t-\tau)\, d\tau} . \tag{99}$$

1.7.1 *Unmodulated optical carrier*

This expression for SNR_t can be simplified when the signal field corresponds to an unmodulated optical carrier. In this case we have $\lambda_t^s = \eta P_s/h\nu$, where P_s is the total carrier power incident on the active area of the detector. P_s may be easily computed in terms of transmitted power and link losses, which must account for both total loss in the fibre and all input/output coupling losses. We then have a *time-independent* SNR, i.e.,

$$\mathrm{SNR} = \frac{\left(\dfrac{\eta P_s}{h\nu}\right)^2}{2B\dfrac{E[G^2]}{E^2[G]}\left(\dfrac{\eta P_s}{h\nu} + \lambda^d\right) + 2B\dfrac{\mathcal{N}_0}{2e^2E^2[G]}}, \tag{100}$$

where $2B$ is the noise (frequency) bandwidth of $k(t)$:

$$2B := \frac{\int_0^\infty k^2(t-\tau)\,d\tau}{\left(\int_0^\infty k(t-\tau)\,d\tau\right)^2} = \frac{\int_{-\infty}^{\infty}|K(\omega)|^2(d\omega/2\pi)}{K^2(0)}, \tag{101}$$

[$K(\omega)$ is the Fourier transform of $k(t)$]. We note that for detectors without gain, i.e., $G = 1$, the last term at the denominator in SNR usually dominates the first term. The converse holds for avalanche photodiodes; i.e. the shot-noise is predominant, and SNR becomes

$$\mathrm{SNR} \cong \frac{E^2[G]}{E[G^2]}\left[\frac{\left(\dfrac{\eta P_s}{h\nu}\right)^2}{2B\left(\dfrac{\eta P_s}{h\nu} + \lambda^d\right)}\right]. \tag{102}$$

It should be emphasized that (99), (100) and (102) can be computed using only second moment properties, and complete shot-noise statistics are not necessary. In particular, $E^2[G]/E[G^2]$ plays a key role in showing how the shot-noise-limited SNR is affected by the multiplication mechanism. Note that the parameter

$$F := \frac{E[G^2]}{E^2[G]} = 1 + \frac{\mathrm{var}(G)}{E^2[G]} > 1 \tag{103}$$

behaves as a photodetector *excess noise factor* since it indicates the factor by which the shot-noise-limited SNR is reduced with avalanche gains. F is unity for an ideal multiplication mechanism; however, the randomness of the multiplication causes a gain spread and $F > 1$. For avalanche photodiodes the following has been proposed

$$F = E[G]\{1 - (1 - k_I)(E[G] - 1)^2/(E^2[G])\}$$
$$= k_I E[G] + (2 - 1/E[G])(1 - k_I)\,, \qquad (104)$$

where k_I is the ionization coefficient ratio (see, once again, Chapter 2 of Part II) which is near zero for silicon photodiodes. When the average gain is large enough, the excess noise factor can be approximated as

$$F \cong k_I E[G] + 2(1 - k_I)\,, \qquad (105)$$

and F is an increasing function of $E[G]$. Figure 12 exhibits F versus $E[G]$ according to law (104), with k_I as a parameter.

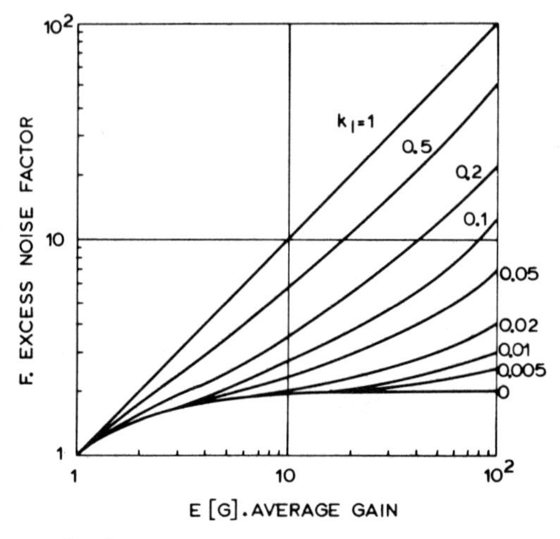

FIG. 12 – Excess noise factor as a function of average gain and ionization ratio.

A quantity often used in the discussion of optical photodetectors is the *noise-equivalent power* (NEP), i.e., the amount of optical power P_s per unit bandwidth needed to produce an output power equal to the detector output noise power, thus giving a unity SNR. We see that, for fixed thermal noise power, the NEP can be minimized by properly adjusting the average detector gain $E[G]$. Physically, the average signal increases with respect to the thermal noise if $E[G]$ increases. However, if $E[G]$ is increased too far, the randomly multiplied shot noise begins to dominate since F increases with $E[G]$.

1.7.2 *Modulated optical carrier*

The most common method of transmitting information by means of an optical carrier is to associate the information with the *intensity* of the carrier. This technique is called *baseband* or *direct intensity modulation* and has been proposed in optical fibre systems because it is both inexpensive and easy to implement. In this case, the received power waveform is [8]

$$p(t) = P_s[1 + m(t)] , \tag{106}$$

where P_s is, again, the unmodulated carrier power and $m(t)$ is the intensity modulating signal, which is proportional to the source message $a(t)$. If $S_m(\omega)$ denotes the spectral density of $m(t)$ in a frequency bandwidth $[-B_m, B_m]$, the total power of $m(t)$ is

$$P_m = \int_{-2\pi B_m}^{2\pi B_m} S_m(\omega) \frac{d\omega}{2\pi} . \tag{107}$$

We further assume $m(t)$ is normalized so that $|m(t)| \leqslant 1$ with a very high probability in order to prevent overmodulation.

In this case the detected signal is a doubly stochastic filtered Poisson process and the instantaneous SNR_t as given by (98) is not very meaningful. It is more convenient to follow the frequency approach outlined in Part A. The detected shot noise has the time-averaged power spectral density derived in (74). With the assumption of infinite bandwidth detectors the signal is given

[8] For the introductory features of this and the next paragraphs, we assume that the received signal is essentially undistorted, even though attenuated. In addition to other degradations, other disadvantages include nonlinearity of the LED input-output characteristic, imperfections of the electronics of the LED driving circuit, nonideal characteristics of fibre, photodetector and amplifier, etc.

by (87), and, starting from (74), its power spectral density can be written as

$$S_Y(\omega) = e^2|K(\omega)|^2\{E[G^2]E[\bar{\lambda}] + E^2[G]S_\lambda(\omega)\} + \frac{\mathcal{N}_0}{2}|K(\omega)|^2, \quad (108)$$

where $S_\lambda(\lambda)$ is the power spectral density of $\boldsymbol{\lambda} := \{\lambda_t,\ 0 \leqslant t < \infty\}$ and

$$E[\bar{\lambda}] = \lim_{T\to\infty} \frac{1}{2T}\int_{-T}^{T} E[\lambda_t]\,dt\,. \quad (109)$$

Now, (106) into (84) gives

$$\lambda_t = \frac{\eta P_s}{h\nu}[1 + m(t)] + \lambda^d, \quad (110)$$

and, for $m(t)$ having a zero average value,

$$E[\bar{\lambda}] = \frac{\eta P_s}{h\nu} + \lambda^d. \quad (111)$$

Moreover, according to (110), the power spectral density of $\boldsymbol{\lambda}$ is

$$S_\lambda(\omega) = \left(\frac{\eta P_s}{h\nu} + \lambda^d\right)^2 2\pi\delta(\omega) + \left(\frac{\eta P_s}{h\nu}\right)^2 S_m(\omega)\,, \quad (112)$$

where the second term is the spectral component due to the desired message, whose total output power becomes

$$P_{so} = \left(\frac{\eta e}{h\nu}E[G]P_s\right)^2 \int_{-\infty}^{\infty} |K(\omega)|^2 S_m(\omega)\frac{d\omega}{2\pi}\,. \quad (113)$$

If $|K(\omega)| = 1$ for $-2\pi B_m \leqslant \omega \leqslant 2\pi B_m$, and $|K(\omega)| = 0$ for $|\omega| > > 2\pi B_m$, then

$$P_{so} = \left(\frac{\eta e}{h\nu}E[G]P_s\right)^2 P_m\,. \quad (114)$$

Similarly, the overall output noise power associated with the quantum and thermal noises has the evaluation

$$P_{N_0} = 2B_m e^2 E[G^2]\left(\frac{\eta P_s}{h\nu} + \lambda^d\right) + 2B_m\frac{\mathcal{N}_0}{2}\,. \quad (115)$$

The output signal-to-noise ratio may be evaluated as

$$\mathrm{SNR} := \frac{P_{so}}{P_{No}} = \frac{P_m\left(\frac{\eta P_s}{h\nu}\right)^2}{2B_m\left(\frac{E[G^2]}{E^2[G]}\right)\left(\frac{\eta P_s}{h\nu} + \lambda^d\right) + 2B_m\frac{\mathcal{N}_0}{2e^2E^2[G]}} \tag{116}$$

that, for a sinusoidal modulating signal

$$m(t) := m_a \cos\omega_m t\,, \tag{117}$$

becomes

$$\mathrm{SNR} = \frac{\frac{1}{2}\left(m_a\frac{\eta P_s}{h\nu}\right)^2}{2B_m\frac{E[G^2]}{E^2[G]}\left(\frac{\eta P_s}{h\nu} + \lambda^d\right) + 2B_m\frac{\mathcal{N}_0}{2e^2E^2[G]}}\,. \tag{118}$$

1.7.3 *Subcarrier intensity modulation*

The SNR expressions derived previously are related to a fibre-optic system in which the desired information signal $a(t)$ is directly modulated onto the optical carrier. In this paragraph alternative modulation formats are analyzed. That is, the baseband message $a(t)$ modulates a sinusoidal signal (named subcarrier) using standard modulation methods. The resulting signal $m(t)$ modulates the intensity of the optical carrier. Thus the received power is again

$$p(t) = P_s[1 + m(t)]\,, \tag{119}$$

where $m(t)$ has to be defined by the subcarrier modulation format.

Figure 13 is a block diagram of the so-called *Subcarrier/Analogue IM* fibreoptics communication system. The system now differs from the previous *Baseband/Analogue IM* system, in that the photodetector is followed by a bandpass filter of (one-sided) bandwidth B_m equal to the bandwidth of $m(t)$, rather than a lowpass filter of bandwidth B_a equal to the bandwidth of $a(t)$. It is then

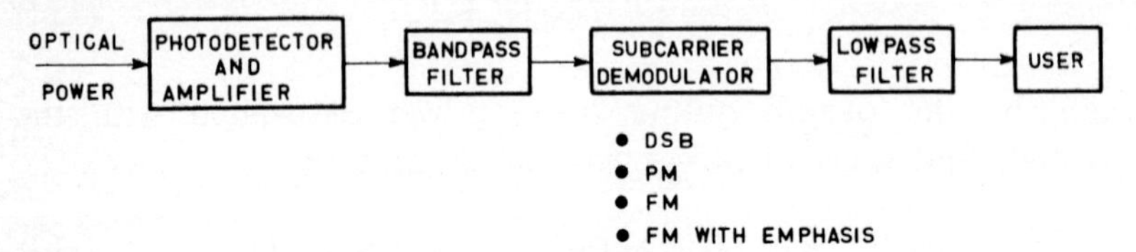

FIG. 13 – Receiver for Subcarrier/Analogue IM system.

followed by an electrical demodulator that recovers the desired signal. The primary advantage of such an alternative operation is the possible SNR improvement obtained during the subcarrier demodulation. In the following, we examine some possible candidate inputs for such modulation formats.

– *DSB/Analogue IM*. The double-sideband (DSB) amplitude-modulation is achieved by setting

$$m(t) = a(t) \cos \omega_0 t \,, \text{ [9]} \tag{120}$$

where ω_0 is the subcarrier angular frequency. Then we have

$$p(t) = P_s[1 + a(t) \cos \omega_0 t] \,. \tag{121}$$

Again the numerical value of $a(t)$ is normalized, so that $|a(t)| \leqslant 1$, to prevent overmodulation and, as a consequence, the value of its power P_a is $\leqslant 1$. The subcarrier signal occupies a bandwidth $B_m = 2B_a$ and $P_m = P_a/2$. The $(\text{SNR}_i)_\text{DSB}$ that occurs in the subcarrier bandwidth at the input to the demodulator is

$$(\text{SNR}_i)_\text{DSB} = \frac{\left(\dfrac{\eta P_s}{h\nu}\right)^2 P_a/2}{2B_m \mathcal{N}_\text{tot}} \,, \tag{122}$$

where for notation simplicity we have defined

$$\mathcal{N}_\text{tot} := \frac{E[G^2]}{E^2[G]}\left(\frac{\eta P_s}{h\nu} + \lambda^d\right) + \frac{\mathcal{N}_0}{2e^2 E^2[G]} \,. \tag{123}$$

An ideal DSB demodulator would then yield an output signal-to-noise ratio

$$(\text{SNR}_o)_\text{DSB} = 2(\text{SNR}_i)_\text{DSB} = \frac{\left(\dfrac{\eta P_s}{h\nu}\right)^2 P_a/2}{2B_a \mathcal{N}_\text{tot}} \,. \tag{124}$$

Thus a degradation of 3 dB is obtained with the DSB format even though P_s, the message, the thermal noise and the photo-

[9] To be precise, $m(t)$ should be equal to the *numerical value* of $a(t) \cos \omega_0 t$.

detector are the same [see (116)]. This fact partly explains why DSB is of little interest with respect to the following subcarrier systems.

– *PM/Analogue IM.* In phase modulation (PM) systems, we have

$$m(t) = C \cos[\omega_0 t + k_p a(t)], \tag{125}$$

where k_p is now the *deviation constant* in radians per unit of $a(t)$, and the unmodulated carrier amplitude C must be $\leqslant 1$ to prevent intensity overmodulation. The bandwidth of the PM signal is given by Carson's rule:

$$B_m \cong 2(D_p + 1) B_a, \tag{126}$$

where D_p is the frequency deviation ratio

$$D_p = \Delta\omega / 2\pi B_a, \tag{127}$$

$\Delta\omega$ being the peak frequency deviation in a PM signal

$$\Delta\omega = k_p \left| \frac{da(t)}{dt} \right|_{\max}. \tag{128}$$

The signal-to-noise ratio at the input to the PM demodulator is

$$(\mathrm{SNR}_i)_{\mathrm{PM}} = \frac{\left(\frac{\eta P_s}{h\nu}\right)^2 C^2/2}{2B_m \mathcal{N}_{\mathrm{tot}}}. \tag{129}$$

An ideal PM demodulator operating above threshold gives an output signal-to-noise ratio

$$(\mathrm{SNR}_o)_{\mathrm{PM}} = D_p^2 \frac{P_a \left(\frac{\eta P_s}{h\nu}\right)^2 C^2/2}{2B_a \mathcal{N}_{\mathrm{tot}}}, \tag{130}$$

which is maximized when $C = 1$.

– *FM/Analogue IM.* The subcarrier is frequency modulated (FM) with the data

$$m(t) = C \cos\left[\omega_0 t + k_f \int_0^t a(\tau)\, d\tau\right], \tag{131}$$

where k_f is the *angular-frequency deviation constant* in radians per second per unit of $a(t)$, and, again, we assume $C \leqslant 1$. The bandwidth given by Carson's rule is

$$B_m \cong 2(D_f + 1)B_a\,, \tag{132}$$

where D_f is the *frequency deviation ratio*

$$D_f = \Delta\omega/2\pi B_a \tag{133}$$

and $\Delta\omega$ is the peak frequency deviation in a FM signal

$$\Delta\omega = k_f|a(t)|_{\max}\,. \tag{134}$$

The signal-to-noise ratio at the input to the FM demodulator is

$$(\mathrm{SNR}_i)_{\mathrm{FM}} = \frac{\left(\dfrac{\eta P_s}{h\nu}\right)^2 C^2/2}{2B_m \mathcal{N}_{\mathrm{tot}}}\,, \tag{135}$$

and at the output

$$\begin{aligned}(\mathrm{SNR}_o)_{\mathrm{FM}} &= 6D_f^2(D_f + 1)\,\frac{P_a\left(\dfrac{\eta P_s}{h\nu}\right)^2 C^2/2}{2B_m \mathcal{N}_{\mathrm{tot}}}\\ &= 3D_f^2\,\frac{P_a\left(\dfrac{\eta P_s}{h\nu}\right)^2 C^2/2}{2B_a \mathcal{N}_{\mathrm{tot}}}\,.\end{aligned} \tag{136}$$

A comparison of FM/IM performance with Baseband/IM transmission shows that the former is far superior from the SNR viewpoint. Similarly, a comparison of (136) and (130) shows that FM/IM is three times as good, or 4.76 dB better than PM/IM under conditions of equal transmission bandwidth.

– *FM/Analogue IM with emphasis.* Preemphasis and deemphasis can be exploited with a certain advantage in the presence of noise. The deemphasis filter is usually a first-order *RC* low-pass filter placed directly at the demodulator output to reduce the total output noise power. Prior to modulation, the signal is passed through a preemphasis filter with a system function which is the reciprocal of the receiver deemphasis filter system function. Assume that the deemphasis filter has an amplitude response

$$|H_{de}(\omega)| = \frac{1}{1 + \left(\frac{\omega}{\omega_c}\right)^2}, \qquad (137)$$

where $\omega_c = 2\pi f_c = 1/RC$. The transmitter preemphasis filter will necessarily have a system function $H_{pe}(\omega) = 1/H_{de}(\omega)$, so that there is no overall message distortion. The subcarrier modulation becomes

$$m(t) = C \cos\left[\omega_0 t + k_f \int_0^t h_{pe}(t - \tau)\, a(\tau)\, d\tau\right]. \qquad (138)$$

In a typical situation $f_c \ll B_a$; in this case it can be shown that the emphasis gives an improvement in signal-to-noise ratio approximatively equal to

$$R_{\mathrm{FM}} \cong \frac{1}{3}\left(\frac{B_a}{f_c}\right)^2, \qquad (139)$$

and the output SNR becomes

$$(\mathrm{SNR}_o)_{\mathrm{FM}} = \left(\frac{D_f B_a}{f_c}\right)^2 \frac{P_a \left(\frac{\eta P_s}{h\nu}\right)^2 C^2/2}{2B_a \mathcal{N}_{\mathrm{tot}}} \qquad (140)$$

with a significant improvement in noise performance with respect to standard FM/IM.

Comparisons. Table I summarizes performance characteristics of the five analogue subcarrier modulation techniques under the assumption of equal received optical power P_s, equal degrees of intensity modulation and the same value of $\mathcal{N}_{\mathrm{tot}}$. In this table, the postdetection SNR gains are given for each modulation format, as well as the required transmission bandwidth. The amplitude C of PM or FM subcarrier was set to unity to maximize SNR_o. The gain is obtained by referring the output signal-to-noise ratio to direct IM signal-to-noise ratio. The tradeoff between postdetection SNR and transmission bandwidth is evident.

TABLE I

Noise performance characteristics

System	Modulation format $m(t)$	One-sided transmission Bandwidth B_m (Hz)	Postdetection SNR Gain (dB)
Baseband IM	$m(t) = a(t)$	$B_m = B_a$	0
DSB/IM	$m(t) = a(t) \cos \omega_0 t$	$B_m = 2B_a$	-3
PM/IM (above threshold)	$m(t) = \cos [\omega_0 t + k_p a(t)]$	$B_m \cong 2(D_p + 1) B_a$	$-3 + 20 \log D_p$
FM/IM (above threshold)	$m(t) = \cos \left[\omega_0 t + k_f \int_0^t a(\tau (d\tau\right]$	$B_m \cong 2(D_f + 1) B_a$	$1.76 + 20 \log D_f$
FM with emphasis/IM (above threshold)	$m(t) = \cos \left[\omega_0 t + k_f \int_0^t h_{pe}(t-\tau) a(\tau) d\tau\right]$	$B_m \cong 2(D_f + 1) B_a$	$-3 + 20 \log D_f + 20 \log (B_a/f_c)$

C. Digital communications

All the modulation methods discussed so far involve transmission of analogue information waveforms from the source to the transmitter with as little distortion as possible. In this part, we consider *digital modulation systems* in which the information message is generated as binary symbols. Whereas the signal-to-noise ratio was the performance criterion for analogue modulations, here, we will measure performance by means of the average symbol error probability P_e.

We give a general method which takes into account the statistical properties of the main impairments that affect the information signal, i.e., shot-noise (dark current included), thermal noise and intersymbol interference. Unlike other methods, the analysis of shot-noise statistics does not entail the second-moment characterization alone, but uses a number of moments sufficient to evaluate the error probability with a high degree of accuracy. For this purpose, it is shown how the cumulants of shot noise and intersymbol interference of any order can be computed. The baseband pulse response and the statistical distribution of the avalanche-photodetector gain are assumed to be available.

1.8 Digital system model

A digital optical fibre communication system can be represented as in Fig. 14. The information source generates a sequence of symbols $b(i)$ at rate $1/T$ taking values from the binary alphabet $\{0, 1\}$ at every time i. The baseband signal modulates the intensity of the optical source and the output power is transmitted through

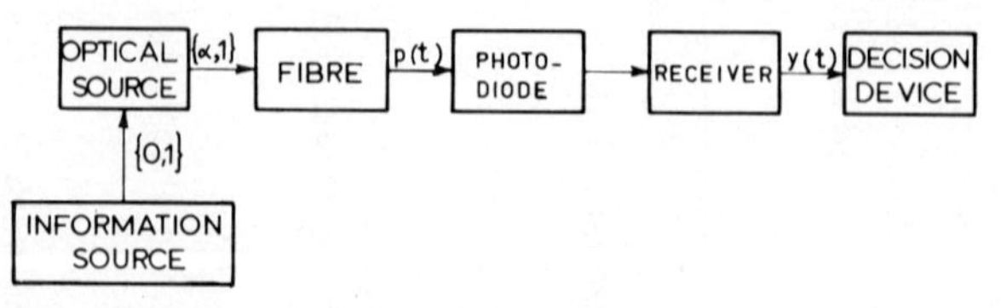

Fig. 14 – Block diagram of digital optical fibre transmission system.

an optical fibre. As a result of imperfect modulation, the optical line signal cannot be completely extinguished. Therefore, the sequence $b(i)$ is mapped into a stream of line symbols $a(i)$ belonging to the set $\{\alpha, 1\}$ with $\alpha \ll 1$. Hereafter, for the sake of generality the values of $a(i)$ will be denoted by $q_0(=\alpha)$ and $q_1(=1)$. Note that a more satisfactory exploitation of the available fibre bandwidth can be achieved resorting to multilevel operations.

The received pulses are smeared out in time because of imperfect channel characteristics, due to differential time delay, material dispersion, mode coupling, etc. From a geometrical viewpoint, an optical pulse launched into the fibre drives rays with different paths, hence, at the output of the fibre, the pulse is broadened owing to different travelling times. As a consequence, the information signal suffers a distortion that takes the form of an overlap between adjacent pulses, thereby causing what is called *intersymbol interference* (ISI).

A typical receiver for digital signalling schemes consists of a photodetector, an amplifier, an equalizer, which compensates for distortion effects, and, finally, a decision device. Under fairly mild conditions, the fibre can be treated as linear in power, viz.,

$$p(t) = \sum_{i=-\infty}^{\infty} a(i) h_p(t - iT) . \tag{141}$$

In (141), $p(t)$ is the optical power driving the photodetector and $h_p(t)$ is the fibre response to the *basic pulse shape.*

Looking at the receiver model of Fig. 10, we remember that the intensity function of the detected random process was related to the incident optical power by

$$\lambda_t = \frac{\eta}{h\nu} p(t) + \lambda^d, \tag{142}$$

where we recall that η is the quantum efficiency of the APD and $h\nu$ is the energy in a photon, viz., h is Planck's constant and ν is the unmodulated optical carrier frequency. Note that in (142) λ^d accounts for the constant rate at which spontaneous but extraneous electrons are generated during $[0, t)$. As a consequence, the detected current $\boldsymbol{I}^s$ is affected by the dark current disturbance, which is usually negligible provided that $p(t)$ is not too low.

Substituting (141) into (142) gives the following equation, which emphasizes the relationship between the counting process and the information signal, i.e.,

$$\lambda_t = \frac{\eta}{h\nu} \sum_{i=-\infty}^{\infty} a(i)\, h_p(t - iT) + \lambda^d. \tag{143}$$

As a matter of fact, the detected current is a filtered doubly stochastic Poisson process, since its intensity depends on the information sequence $S = \{a(i)\}$. If S is given, we will refer to λ_t as an intensity *conditioned* by S. Similarly, the counting random variable N_t may entail a conditioning with respect to the information squence.

The analysis performed in Section 1.6 is still valid. Therefore, for infinite bandwidth detectors, the signal at the input of the decision device is given again by (87), viz.,

$$y(t) = \sum_{j=1}^{N_t} eG_j k(t - \tau_j) + \int_0^t i_\tau^g k(t - \tau)\, d\tau\,, \tag{144}$$

where $k(t)$ is the impulse response of amplifier plus equalizer. On the other hand, the signal can be also written as

$$y(t) = s(t) + \left\{ \sum_{j=1}^{N_t} eG_j k(t - \tau_j) - s(t) \right\} + \int_0^t i_\tau^g k(t - \tau)\, d\tau\,, \tag{145}$$

where $s(t) := E[y(t)|S,\ i_t^d = i_t^g = 0]$ represents the data signal. Since [see, e.g., (88)]

$$E[y(t)|S] = eE[G] \int_0 \lambda_\tau k(t - \tau)\, d\tau\,, \tag{146}$$

we may write

$$E[y(t)|S] = \frac{\eta e}{h\nu} E[G] \sum_{i=-\infty}^{\infty} a(i) \int_0^t h_p(\tau - iT)\, k(t - \tau)\, d\tau$$

$$+ \lambda^d eE[G] \int_0^t k(t - \tau)\, d\tau\,. \tag{147}$$

The use of (145) together with (147) gives

$$y(t) = \sum_{i=-\infty}^{\infty} a(i) r(t - iT) + n_s(t) + n_g(t) \,. \tag{148}$$

In (148), $r(t)$ is the overall baseband pulse response

$$r(t) = \frac{\eta e}{h\nu} E[G] h_p(t) * k(t) \tag{149}$$

and $n_s(t)$ is the shot-noise contribution given by

$$n_s(t) = \sum_{j=1}^{N_t} eG_j k(t - \tau_j) - s(t) \,. \tag{150}$$

Finally, the thermal noise

$$n_g(t) = \int_0 i_\tau^g k(t - \tau)\, d\tau \tag{151}$$

can be assumed, as usual, to be a sample from a stationary zero-mean Gaussian process.

The decision device samples $y(t)$ given by (148) every T seconds to determine the amplitude of the transmitted symbols. At the decision time t_0, we have

$$y(t_0) = a(0) r_0 + x(t_0) + n_s(t_0) + n_g(t_0) \,, \tag{152}$$

where

$$r_0 = \frac{\eta e}{h\nu} E[G] \int_0^{t_0} h_p(\tau) k(t_0 - \tau)\, d\tau \,, \tag{153}$$

$$x(t_0) = \sum_{i=-\infty}^{\infty}{}' a(i) r(t_0 - iT)$$

$$= \sum_{i=-\infty}^{\infty}{}' \frac{\eta e}{h\nu} E[G] a(i) \int_0^{t_0} h_p(\tau - iT) k(t_0 - \tau)\, d\tau \,, \tag{154}$$

$$n_s(t_0) = \sum_{j=1}^{N_t} eG_j k(t_0 - \tau_j) - s(t_0) \,, \tag{155}$$

$$n_g(t_0) = \int_0^{t_0} i_\tau^g k(t_0 - \tau)\, d\tau \,, \tag{156}$$

and $\sum'$ does not include the term $i = 0$. In (152), $a(0)r_0$ denotes the desired signal, and the other terms represent the intersymbol interference, the shot-noise and the thermal noise, respectively. Therefore, the received signal can be considered as a superposition of a desired component related to the transmitted symbol to be detected and three types of impairments, i.e., intersymbol interference, shot noise and thermal noise.

Equation (152) represents the starting point for computing the average error probability P_e, which is the best technical criterion for analyzing the performance of a given system. A procedure for computing P_e will be fully explained and some results reported in the following section. Note that such a procedure does not entail the commonly used hypothesis about the Gaussianity of the filtered Poisson process $y(t)$ defined by means of (144); as a matter of fact, the process is nonGaussian and has an amplitude probability density function with larger tails than the Gaussian density, especially for low Poisson intensities.

1.9 Error probability evaluation

For the sake of simplicity the method is outlined for binary systems, the extension to multilevel operation being straightforward. For binary signalling, the symbol $a(0)$ takes the values $q_0(= \alpha)$ with probability p_0 and $q_1(= 1)$ with probability $p_1 = 1 - p_0$. Henceforth the sequence of transmitted symbols will be denoted by S_c where the subscript c means that the symbol $a(0)$ to be detected is given. Simple computations show, therefore, that the error probability $P_e(S_c)$ conditioned by a given sequence S_c is

$$P_e(S_c) = p_1 D[(d + q_0 - q_1)r_0 - z_1] + p_0[1 - D(dr_0 - z_0)], \quad (157)$$

z_1 and z_0 being the values of the random variable (RV)

$$z(t_0) = x(t_0) + n_s(t_0), \quad (158)$$

in hypotheses $a(0) = q_1$ and $a(0) = q_0$, respectively. Moreover, $D(\cdot)$ denotes the probability distribution function of the Gaussian RV $n_g(t_0)$ and dr_0 is the distance of the detector slicing level from $q_0 r_0$. Of course, it is straightforward to apply such an approach to multilevel digital transmissions.

The evaluation of the average error probability P_e may be performed by taking the expectation of $P_e(S_c)$ with respect to z_1 and z_0 in their ranges of definition R_1 and R_0:

$$P_e = p_1 \int_{R_1} D[(d + q_0 - q_1) r_0 - z_1] \, dF_{z_1}(z_1)$$

$$+ p_0 \int_{R_0} [1 - D(dr_0 - z_0)] \, dF_{z_0}(z_0) , \tag{159}$$

where F_{z_1} and F_{z_0} are the probability distribution functions of z_1 and z_0, respectively. Since $n_g(t_0)$ is a zero-mean Gaussian RV with variance σ_g^2, $D(\cdot)$ becomes

$$D(u) = 1 - \tfrac{1}{2} \operatorname{erfc} [u/(\sqrt{2}\sigma_g)] , \tag{160}$$

where

$$\operatorname{erfc}(u) = (2/\sqrt{\pi}) \int_u^\infty \exp(-t^2) \, dt . \tag{161}$$

Substitution of (160) into (159) yields the average error probability for Gaussian thermal noise

$$P_e = \frac{1}{2} \left\{ p_1 \int_{R_1} \operatorname{erfc} \left[\frac{(q_1 - (d + q_0)) r_0 + z_1}{\sqrt{2}\sigma_g} \right] dF_{z_1}(z_1) \right.$$

$$\left. + p_0 \int_{R_0} \operatorname{erfc} \left[\frac{dr_0 - z_0}{\sqrt{2}\sigma_g} \right] dF_{z_0}(z_0) \right\} . \tag{162}$$

Note that the right-hand side of (162) relies on the probability distribution function of z, which is the sum of two *non-independent* RV's, i.e., $x(t_0)$ and $n_s(t_0)$.

1.9.1 *Exhaustive method*

An exact computation of the average error probability P_e can be performed by resorting to a method which is exhaustive with respect to all possible patterns of intersymbol interference.

P_e, as expressed by (162), is comprised of two contributions, each implying an average with respect to the probability distribution function of $x(t_0) + n_s(t_0)$ under hypothesis q_1 or q_0. Since both $x(t_0)$ and $n_s(t_0)$ depend on the information sequence S_c, it is possible to evaluate these integrals for each S_c and then to average over all the conditional error probabilities $P_e(S_c)$, i.e.,

$$P_e = p_1 E_{S_c}[P(S_c, q_1)] + p_0 E_{S_c}[P(S_c, q_0)]$$
$$= \frac{1}{2^{M+1}} \left\{ p_1 \sum_{i=1}^{2^M} \int_{R_1} \operatorname{erfc}\left[\frac{(q_1 - (d + q_0))\, r_0 + x_i + n_{s_{1i}}}{\sqrt{2}\sigma_g}\right] dF_{N_{1i}}(n_{1i}) \right.$$
$$\left. + p_0 \sum_{i=1}^{2^M} \int_{R_0} \operatorname{erfc}\left[\frac{dr_0 - x_i - n_{s_{0i}}}{\sqrt{2}\sigma_g}\right] dF_{N_{0i}}(n_{0i}) \right\}. \qquad (163)$$

In writing (163), it has been assumed that both $x(t_0)$ and $n_s(t_0)$ are dependent on a finite length $M+1$ of the sequence S_c; moreover, the subscript i means that S_c is given and is the ith sequence.

The computation of P_e as expressed by (163) may be performed according to the following steps:

1) evaluation of the current value x_i of $x(t_0)$ for each of the 2^M possible sequences;

2) average with respect to the quantum noise RV $n_{s_{1i}}$ or $n_{s_{0i}}$ for the considered sequence;

3) average over all possible sequences.

The key item for P_e lies in the second step, i.e., in averaging the probability of error conditioned by a fixed S_c and a given value of RV $n_{s_{1i}}$ or $n_{s_{0i}}$ with respect to this last RV, that is to solve the integrals in (163). This averaging can be performed by resorting to numerical integration rules, since both integrals in (163) are in the form

$$Q = \int_R f(u)\, dF(u), \qquad (164)$$

where $f(u)$ is a function of the shot-noise disturbance $n_s(t_0)$ depending on a given sequence S_c. Q can be approximated by a Gauss quadrature rule with N terms as

$$Q \simeq \sum_{k=1}^{N} w_k f(u_k). \qquad (165)$$

A detailed discussion of how to construct a suitable integration formula $\{w_k, u_k\}_{k=1}^N$ is beyond the scope of this presentation. However, it is worth emphasizing that high accuracy integration methods, which can be generated on a computer, are currently available. Beside this advantage, the set $\{w_k, u_k\}_{k=1}^N$ can be generated without an explicit knowledge of the probability distribution function $F(u)$. A statistical description through moments (or cumulants) is sufficient; therefore, in the following discussion, attention will be devoted to the moments computation.

1.9.2 *Shot-noise and avalanche-gain moments*

The characteristic function for a conditional Poisson process with intensity λ_t given by (143) and filtered through $k(t)$ was given in (56), that is,

$$\exp\left\{\int_0^{t_0} \lambda_\tau E[\exp(jveGk(t-\tau)) - 1]\, d\tau\right\}. \tag{166}$$

The cumulants γ_l' are then obtained by exploiting relationship (57) together with expression (143) for the intensity function. Therefore, at the decision time t_0, we have

$$\gamma_l' = E[G^l]\frac{\eta e^l}{h\nu}\sum_{i=-\infty}^{\infty} a(i)\int_0^{t_0} h_p(\tau - iT)\,k^l(t_0-\tau)\, d\tau + \lambda^d e^l E[G^l]\int_0^{t_0} k^l(t_0-\tau)\, d\tau\,. \tag{167}$$

Note that the cumulants γ_l' are computed for a *given* sequence S_c'. The cumulants $\gamma_l^{(n_s)}$ for $n_{s_i}(t_0)$ are easily evaluated [see (147) and (155)], since

$$\gamma_1^{(n_s)} = \lambda^d eE[G]\int_0^{t_0} k(t_0-\tau)\, d\tau\,, \tag{168}$$

$$\gamma_l^{(n_s)} = \gamma_l'\,, \qquad l \geqslant 2\,. \tag{169}$$

Finally, for a RV u the *central moments* μ_m can be obtained by recurrence starting from its cumulants γ_m with an expression resembling (23):

$$\mu_{l+1} = \gamma_{l+1} + \sum_{j=1}^{l-1} \binom{l}{j} \gamma_{j+1} \mu_{l-j}, \qquad l \geqslant 2 \tag{170}$$

with $\mu_1 = 0$ and $\mu_2 = \gamma_2$. The use of central moments is for numerical convenience, since the computation of $\{w_k, u_k\}_{k=1}^N$ is less sensitive to roundoff errors when the integration region is symmetrical with respect to the origin. The abscissae u_k computed in this way are translated with the quantity $E[u]$, whereas the weights w_k are unchanged.

Note that each γ_l' depends on the moment $E[G^l]$ of the avalanche gain. The APD moments can be computed starting from the physical description of the avalanche-gain mechanism given in Chapter 2 of Part II, where it has been stated that the moment generating function $M(s)$ of G satisfies the following differential equation

$$M^{(1)}(s) = \frac{\varrho M(s)}{1 - \beta[M(s) \exp(-s)]^{\varepsilon}} \tag{171}$$

with

$$\begin{aligned} \varrho &= (k_I - 1)/k_I, \\ \beta &= 1 - (\varrho/E[G]), \\ \varepsilon &= k_I - 1, \\ M(0) &= 1, \end{aligned} \tag{172}$$

where $E[G]$ is the average gain and k_I is the ionization ratio. Remember that k_I varies between 0 and 1. In particular, $k_I \simeq 0$ corresponds to silicon photodiodes, whereas larger values of k_I are typical for germanium photodiodes. For $k_I = 0$ or $k_I = 1$, (171) is still meaningful, provided that a limit operation is introduced. By differentiating (171) and exploiting the equivalent expression:

$$1 - \beta[M(s) \exp(-s)]^{\varepsilon} = \varrho[M(s)/M^{(1)}(s)],$$

it can be shown that $M(s)$ is the solution to the differential equation

$$M^{(2)} = a_3 \frac{(M^{(1)})^3}{M^2} + a_2 \frac{(M^{(1)})^2}{M} + a_1 M^{(1)}, \tag{173}$$

where

$$a_1 = \varepsilon\,, \qquad a_2 = 1 - \varepsilon - \varepsilon/\varrho\,, \qquad a_3 = \varepsilon/\varrho\,. \tag{174}$$

A straightforward application of (173) yields

$$\left[\frac{(M^{(1)})^l}{M^{l-1}}\right]^{(1)} = la_3\,\frac{(M^{(1)})^{l+2}}{M^{l+1}} + \\ + (la_2 - l + 1)\,\frac{(M^{(1)})^{l+1}}{M^l} + la_1\,\frac{(M^{(1)})^l}{M^{l-1}}\,. \tag{175}$$

Differentiating (173) iteratively and applying (175) gives the lth derivative of $M(s)$ as

$$M^{(l)} = \sum_{k=1}^{2l-1} A_{k,l}\,\frac{(M^{(1)})^k}{M^{k-1}}\,, \tag{176}$$

where the $A_{k,l}$ are coefficients which can be easily computed with the aid of (174). Finally, the moments of G can be deduced from

$$E[G^l] = \left\{\frac{d^l M(s)}{ds^l}\right\}_{s=0} = \sum_{k=1}^{2l-1} A_{k,l} E^k[G] \tag{177}$$

using $M(0) = 1$ and $M^{(1)}(0) = E[G]$.

It is worth mentioning that with $s = 0$ (173) gives for $E[G^2]$ an expression in complete agreement with the APD second order statistics summarized in relationships (103) and (104).

1.9.3 *The proposed method*

The exhaustive procedure illustrated previously becomes unfeasible because of computer time when dealing with a large number M of interfering samples. This drawback can be overcome if in (162) a GQR approach is used in which weights w_k and abscissae u_k are obtained starting from the moments of RV z (z_1 and z_0) defined by (158). The steps are the following: 1) elaborate on cumulants and moments of $z(t_0)$ *conditioned* to S_c and 2) average the conditional moments.

As regards the steps above, the conditional shot-noise cumulants (167)-(169) are still useful. Let

$$r_l(t) = \frac{\eta e^l}{h\nu}\,E[G^l]\int_0^t h_p(\tau)\,k^l(t-\tau)\,d\tau \tag{178}$$

and

$$c_l(t) = \lambda^d e^l E[G^l] \int_0^t k^l(t-\tau)\, d\tau\,. \tag{179}$$

Then the conditional cumulants of $z(t_0)$ from (158) and (167)-(169) become

$$\gamma_1^{(z)} = \sum_i{}' a(i) r_1(t_0 - iT) + c_1(t_0) \tag{180}$$

and

$$\gamma_l^{(z)} = \gamma_l^{(n_s)} = \sum_i{}' a(i) r_l(t_0 - iT) + a(0) r_l(t_0) + c_l(t_0)\,, \quad l \geqslant 2\,. \tag{180}$$

The conditional moments $m_l^{(z)}(S_c)$ are obtained from (180) and (181) by recurrence, and, finally, are averaged to give

$$m_l^{(z)} = \int_R m_l^{(z)}(S_c)\, dF_{x_l(t_0)}\,, \tag{182}$$

where

$$x_l(t_0) = \sum{}' a(i) r_l(t_0 - iT) \tag{183}$$

resembles the intersymbol interference RV. This kind of definition permits the moments of each $x_l(t_0)$ to be computed with fast methods for either statistically independent line symbols or correlated data sequences. In any case, each $m_l^{(z)}$ in (182) may be evaluated by means of a Gauss quadrature rule.

For example, we outline the cumulants computation of $x_l(t_0)$ for *independent* symbols. In this hypothesis, its mth cumulant is the M-term sum of the mth cumulants of each independent component $u_l(i) = a(i) r_l(t_0 - iT)$. The cumulant generating function $\Psi(s)$ of $u_l(i) - E[u_l(i)]$ for $p_1 = p_0 = \frac{1}{2}$ is

$$\Psi(s) = \ln \tfrac{1}{2} \{\exp(s\alpha_i) + \exp(-s\alpha_i)\}\,, \tag{184}$$

where

$$\alpha_i = \frac{r_l(t_0 - iT)}{2}(q_1 - q_0)\,. \tag{185}$$

It is easy to show that $\Psi(s)$ satisfies the following differential equation

$$\Psi^{(2)}(s) = -[\Psi^{(1)}(s)]^2 + \alpha_i^2 , \tag{186}$$

where $\Psi^{(k)}(s)$ means kth derivative with respect to s, and $\Psi^{(1)}(0)=0$.

Equation (186) allows all the cumulants of $u_l(i)-E[u_l(i)]$ to be computed iteratively since an extension of (186) by induction yields

$$\Psi^{(m+1)} = P_2(\Psi^{(1)})\left[\frac{d}{d\Psi^{(1)}} P_m(\Psi^{(1)})\right], \tag{187}$$

where $P_k(\cdot)$ is a polynomial of degree k in the quantity between brackets and

$$P_2(\Psi^{(1)}) = \Psi^{(2)}(s) = -[\Psi^{(1)}(s)]^2 + \alpha_i^2 , \tag{188}$$

$$P_m(\Psi^{(1)}) = \Psi^{(m)}(s) . \tag{189}$$

Since the cumulants $\Gamma_{m+1}(i)$ of $u_l(i)-E[u_l(i)]$ are given by

$$\Gamma_{m+1}(i) = \left\{\frac{d^{m+1}}{ds^{m+1}} \Psi(s)\right\}_{s=0} , \tag{190}$$

we obtain by recalling that $\Psi^{(1)}(0) = 0$

$$\Gamma_{m+1}(i) = \left\{P_2(\Psi^{(1)})\left[\frac{d}{d\Psi^{(1)}} P_m(\Psi^{(1)})\right]\right\}_{\Psi^{(1)}=0} \tag{191}$$

or

$$\Gamma_{m+1}(i) = \alpha_i^2 P_m^{(1)}(0) . \tag{192}$$

Finally, the mth cumulant $\gamma_m^{(x_l)}$ of $x_l(t_0)$ can be evaluated:

$$\gamma_m^{(x_l)} = \sum_{i=1}^{M} \Gamma_m(i) . \tag{193}$$

1.9.4 *Gaussian approximation of shot noise*

In computing P_e, a Gaussian model for the filtered Poisson process is often suggested. Thus it is assumed that shot noise corresponds to increasing the thermal-noise variance σ_g^2 by a quantity $\sigma_{n_s}^2$ equal to the second cumulant $\gamma_2^{(n_s)}$ of $n_s(t_0)$. The probability of error becomes

$$P_e = \frac{1}{2}\left\{p_1 \iint_{R_1} \operatorname{erfc}\left[\frac{(q_1 - (d + q_0))\, r_0 + x}{\sqrt{2}(\sigma_g^2 + \sigma_{n_{s1}}^2)^{\frac{1}{2}}}\right] dF_{X,N_{s1}}(x, n_{s1}) + p_0 \iint_{R_0} \operatorname{erfc}\left[\frac{dr_0 - x}{\sqrt{2}(\sigma_g^2 + \sigma_{n_{s0}}^2)^{\frac{1}{2}}}\right] dF_{X,N_{s0}}(x, n_{s0})\right\}. \tag{194}$$

Equation (194) shows the need of a general algorithm for evaluating two-dimensional integrals of the form

$$Q = \iint_R f(u, v)\, dF(u, v)\,, \tag{195}$$

where $f(u, v)$ is a known function and $F(u, v)$ is always non-negative, monotone increasing and with bounded variations in the integration region R. It is assumed that the joint moments

$$E[u^h v^k] = m_{h,k} = \iint_R u^h v^k\, dF(u, v) \tag{196}$$

exist and are known. In fact, u coincides with $x(t_0)$ as given by (154) and v is $\gamma_2^{(n_s)} = \sigma_{n_s}^2$ [see (167) and (169), and remember that $a(0)$ is fixed]. Therefore, u and v resembles the in-phase and quadrature components of a two-channel pulse amplitude modulation (PAM) system, and their joint moments can be easily computed by recurrence.

There exist appropriate methods for approximating (195) that use a linear combination of the integrand:

$$Q \simeq \sum_{i=1}^{N_1} \sum_{j=1}^{N_2} A_{ij} f(u_i, v_j)\,, \tag{197}$$

where the nodes u_i and v_j are points of a rectangular lattice but are not necessarily distributed uniformly. As in the one-dimensional case, the quadrature formula (197) can be obtained by resorting to a suitable interpolation of $f(u, v)$.

Expression (194) may be simplified in the absence of intersymbol interference. In such a case, it is no longer necessary to know the joint moments of random variables $x(t_0)$ and $\sigma_{n_s}^2$. In general, it is still necessary to compute the integrals of (194) with respect to N_{s1} and N_{s0}, for the no-intersymbol-interference assumption does not always entail the statistical independence of cumulants $\gamma_l^{(n_s)}$ and information symbol sequence. Further, if P_e is evaluated for a given pattern of information symbols, $\sigma^2_{n_{s1}}$ and $\sigma^2_{n_{s2}}$ are numbers, and (194) may be simplified in

$$P_e = \frac{1}{2}\left\{p_1 \operatorname{erfc}\left[\frac{(q_1 - (d + q_0))\, r_0}{\sqrt{2}(\sigma_g^2 + \sigma^2_{n_{s1}})^{\frac{1}{2}}}\right] + \right. \\ \left. + p_0 \operatorname{erfc}\left[\frac{d r_0}{\sqrt{2}(\sigma_g^2 + \sigma^2_{n_{s0}})^{\frac{1}{2}}}\right]\right\}. \tag{198}$$

This expression has been adopted by many authors even though it gives poor results. On the contrary, it is possible to cope with the computation effort needed in the exact method outlined in the previous paragraph by resorting to *fast standard routines*. Moreover, the advantage of dealing with real disturbances altogether outweighs the small increase in computation complexity.

1.9.5 *An example*

A typical approximation $h_\delta(t)$ to the measured fibre responses to a Dirac delta function is given in Fig. 15. In the figure, the

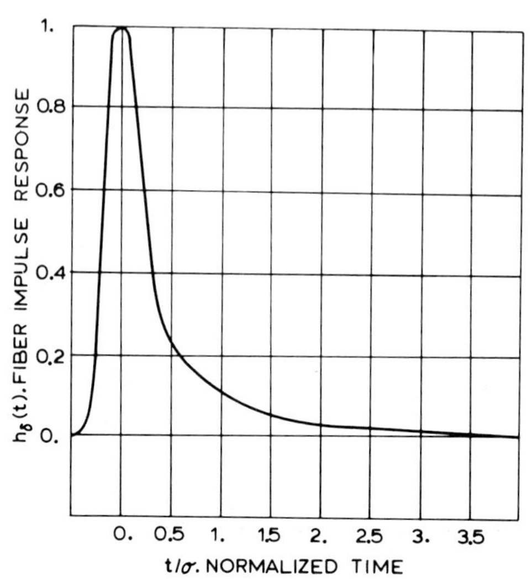

FIG. 15 – Fibre response $h_\delta(t)$ to Dirac impulse.

time variable is normalized to the fibre root-mean-square dispersion defined as

$$\sigma = \left[A^{-1}\int_0^\infty h_\delta(t)(t - t_m)^2\, dt\right]^{\frac{1}{2}}, \tag{199}$$

where

$$A = \int_0^\infty h_\delta(t)\, dt \tag{200}$$

and

$$t_m = A^{-1}\int_0^\infty t h_\delta(t)\, dt\,. \tag{201}$$

We have evaluated the performance of a 34-Mbit/s PCM system with a dispersion $\sigma \simeq 15$ ns, which corresponds to $\sigma/T = 0.5$. A rectangular pulse transmission has been considered with sta-

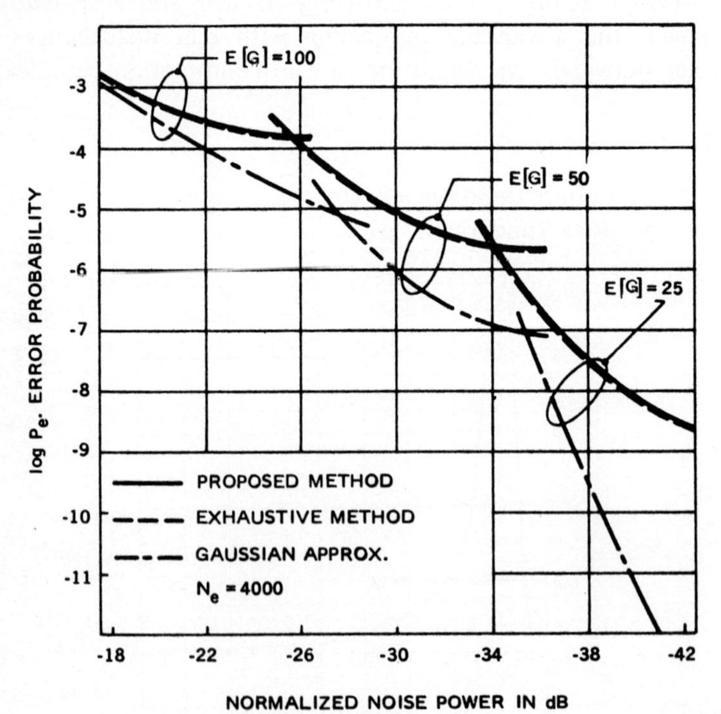

FIG. 16 – Comparison of error probability computational methods for different values of $E[G]$. Thermal noise power is normalized to reference signal power corresponding to $N_e E[G] = 4 \times 10^5$. N_e is equal to 4000 and threshold level is central. (Courtesy of IEEE Trans. Inform. Theory, Vol. IT-25, pp. 170-178).

tistically independent, equally likely symbols $q_0 = 0.1$ and $q_1 = 1$ having a duty cycle equal to 0.5. For computation convenience, the parameter

$$N_e = \frac{\eta}{h\nu} h_p(t_0) T, \tag{202}$$

has been introduced, where $h_p(t_0)$ represents the received optical power for an isolated symbol $q_1 = 1$.

The overall impulse response of the receiver has been assumed to be

$$h(t) = (1/C) \exp(-t/RC), \tag{203}$$

with $RC = 0.2T$, C and R being respectively the total load capacitance and resistance of the photodiode. Since silicon APD's seem to be the most suitable for telecommunication systems, the following results refer to a silicon photodiode with ionization

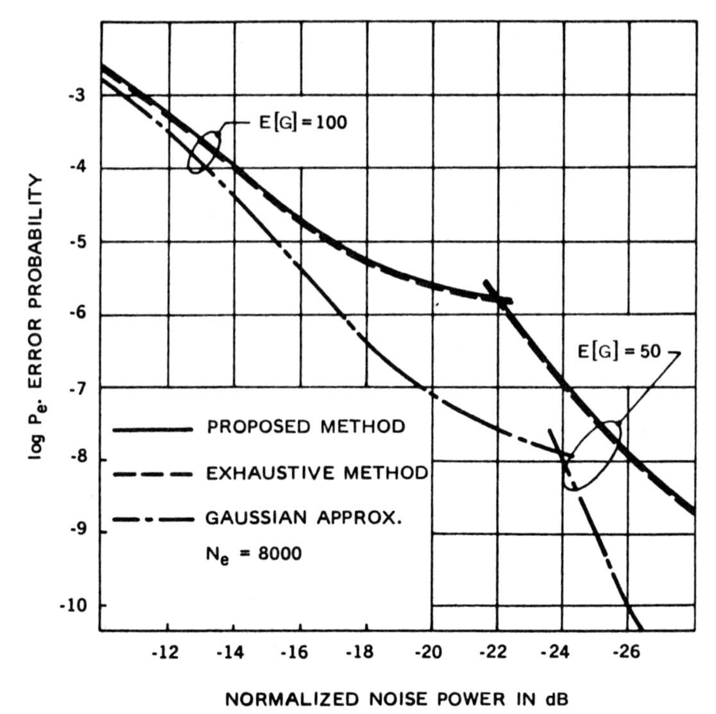

FIG. 17 – Comparison of error probability computational methods for $E[G]=100$ and 50. (Same as Fig. 16 except that $N_e = 8000$). (Courtesy of IEEE Trans. Inform. Theory, Vol. IT-25, pp. 170-178).

ratio k_I of 0.16 and average gain $E[G]$ considered as a parameter. Currently, lower values of the ionization ratio, e.g., between 0.02 and 0.1, seem to be more realistic.

In Figs. 16-19, the abscissa is the thermal-noise power normalized to a reference signal power corresponding to $E[G]N_e = 4\times10^5$. Figures 16 and 17 compare the illustrated methods for evaluating P_e with parameter $N_e = 4000$ and $N_e = 8000$, respectively. The proposed technique, i.e., P_e given by (162) is represented by the solid line. The exhaustive approach gives a curve, which effectively coincides with the preceding one. Another curve shows a Gaussian approximation to shot noise; this assumption gives less satisfactory results than the proposed approach.

The curves mentioned above are obtained by assuming that the threshold level dr_0 is equidistant from q_0r_0 and q_1r_0. The presence of intersymbol interference and shot noise means that this choice is by no means optimal. This is emphasized in Fig. 18.

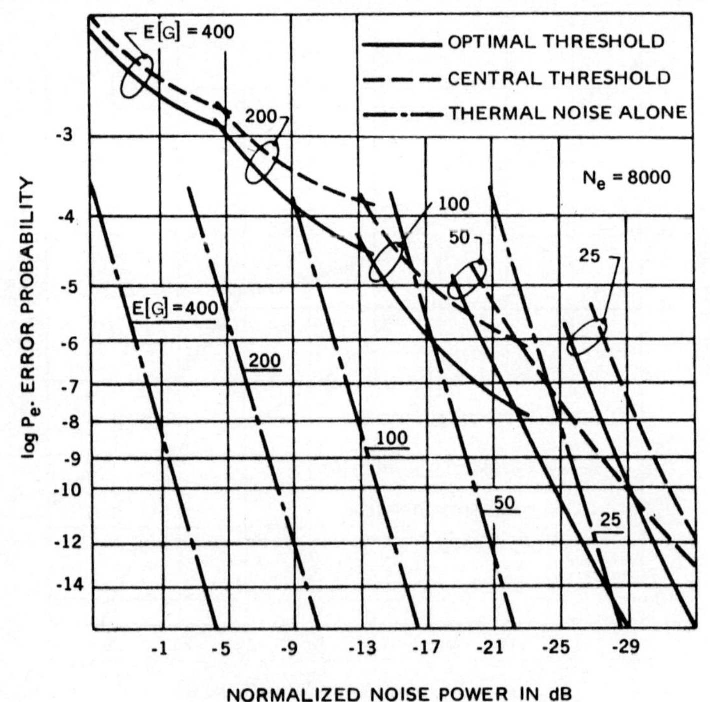

FIG. 18 – Comparison of error probabilities for central and optimal threshold levels with different values of $E[G]$. Parameter N_e is 8000. (Courtesy of IEEE Trans. Inform. Theory, Vol. IT-25, pp. 170-178).

The solid line is based on a choice of d, which is optimal in the sense that the average probabilities of error on hypotheses q_1 and q_0 become equal. The dashed lines, on the other hand, are related to a central threshold. In this figure, we also give the P_e curves obtained by neglecting shot noise and intersymbol interference effects; only thermal noise and ideal gain $E[G]$ are taken into account. Of course, in multilevel systems, the question of a favourable spacing of signal levels and decision thresholds is even more important.

Finally, Fig. 19 gives P_e for different values of N_e and $E[G]$ with optimal threshold levels.

The proposed method, while being approximatively ten times faster than the exhaustive technique, requires about the same computational time as the Gaussian approximation. It is an exact method, except for numerical errors, and, as such, is an improvement on previous approaches.

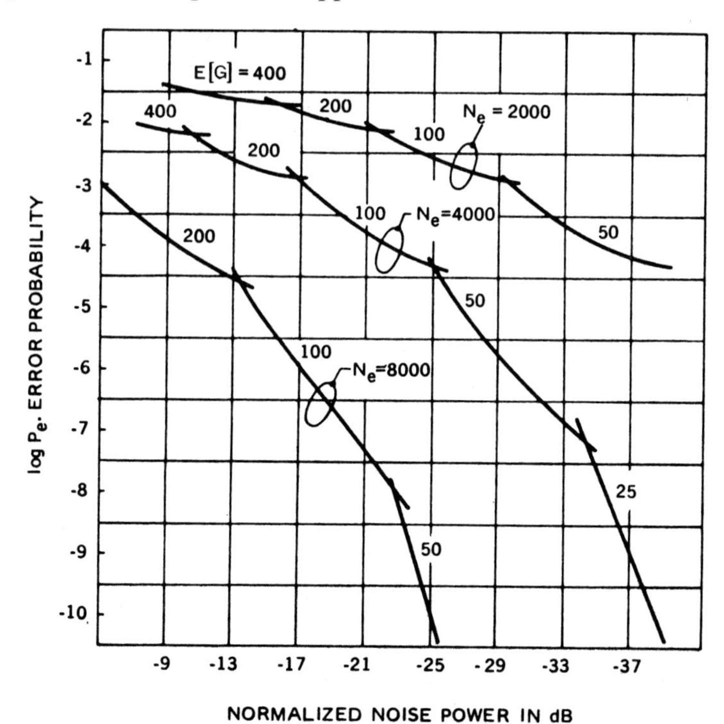

FIG. 19 – Error probability for different values of N_e and $E[G]$. Threshold level is optimized. (Courtesy of IEEE Trans. Inform. Theory, Vol. IT-25, pp. 170-178).

1.10 Communication systems

We now return to the system model of Fig. 14 with emphasis on the basic receiver and trasmitter blocks, as depicted in Fig. 20.

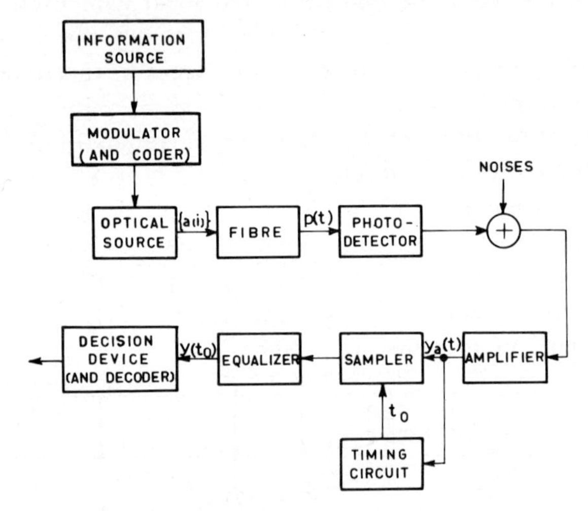

FIG. 20 – Model of digital fibreoptic transmission system.

The fundamental design problems of an optical communication system can be summarized according to this figure. The system designer is interested in processing the detected signal in order to estimate the transmitted symbols. Therefore, it is advisable to operate both in the transmitting section and in the receiving part of the system. With reference to the transmitter, attention should be paid to the *line encoding*, i.e., choice of the alphabet of $a(i)$ for easy time synchronization, error monitoring and bandwidth economy. Moreover, another item is the *pulse waveform* design that affects $h_p(t)$ and, therefore, the received optical power. However, the peak-power limitation of physical LED's makes questionable the opportunity of pulse shaping.

Direct processing of the electrical signal implies a hierarchy of problems. First, the sequence of symbols $a(i)$ must be detected in the presence of intersymbol interference, shot noise and thermal noise. This can be accomplished by means of a linear or non-linear equalizer following a low-noise amplifier. Moreover, the receiver must be provided with a proper synchronization circuit, which gives the appropriate instant t_0 to a sampler. In summary, the receiver performs three basic functions; i.e., pulse equalization, timing and detection, which will be surveyed in the next paragraphs.

1.10.1 *Pulse equalization*

For high signalling rates, the main impairment degrading system performance, i.e., the average probability of error, is due to the presence of intersymbol interference that is caused by an overlapping between adjacent symbols. As we have seen, the band-limitation of a fibreoptic system gives essentially a spread in the optical pulses travelling along the fibre; hence at the output of the fibre a pulse is broadened and distorted.

Most of the theoretical work of any area in digital communications has been devoted to the problem of detecting pulses in the presence of distortion. This objective in optical fibre systems can be attained by means of two different approaches. The first is accomplished by inserting an optical device in the step-index fibre in order to modify the angular distribution of power. In this way a tradeoff between optical paths of various rays should be achieved. It must be stressed that these equalizers can be fully exploited in the absence of scattering, when a biunivocal correspondence between ray delay-time and angle exists. Likewise, the relative losses introduced by the optical equalizer are approximatively between 1 dB and 8 dB, according to the type of the equalizer. The rather sophisticated technology required in the implementation to guarantee, for example, low tolerances, no-aberrations optics, etc., suggests that these devices will probably be convenient from an economic viewpoint in connection with advanced integrated-optic systems (Part V).

The second method of equalization is to resort to baseband systems processing the detected electrical signal and designed to match the Nyquist criterion or its extensions. As a matter of fact, there are several situations which require a frequency-domain analysis. As an example, the overall transmission channel is the resultant of transmitter, fibre and receiver characteristics; therefore, in the frequency domain it is easy to deal with their partition or summation. A number of techniques exist which give the exact characteristics of an overall system in order to compensate for intersymbol interference. However, with this approach it is difficult to design exact trasmitting and receiving filters, and therefore to obtain the optimal system response. Moreover, this approach takes into account intersymbol interference alone and suggests a possible solution for zero intersymbol interference. Sometimes, however, it is more appropriate to design equalizers that keep the intersymbol interference between tolerable bounds with respect to a desirable and achievable goal of error probability. Finally, the equalizer parameters obtained with this approach cannot be made adjustable by on-line techniques when the channel characteristics are not known in advance and possibly are time varying. These shortcomings can be reduced with the aid of other techniques.

In a conventional transversal equalizer (Fig. 21) the form is

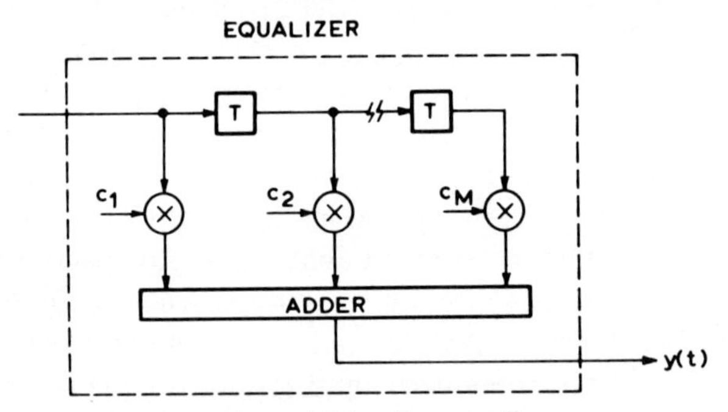

FIG. 21 – Tapped-delay-line equalizer.

chosen in advance, viz., the classical transversal filter with tap coefficients $\{c_i\}$. The intersymbol interference is reduced by filtering the amplified signal $y_a(t)$ and by computing the $\{c_i\}$ in order to satisfy a suitable criterion.

In the literature, two equalization criteria are used, viz., zero forcing and the minimization of the mean-square error (MMSE). The zero-forcing criterion is known to minimize the equalized peak distortion if the peak distortion of the unequalized pulse is < 1. In this case, the set $\{c_i\}$ is computed by solving a linear system. The second criterion minimizes the MMSE between the received sequence estimate and the transmitted sequence. From a mathematical programming viewpoint, this approach searches for the linear least squares solution and leads to the so-called *normal system* of equations. Incidentally, the matrix of normal equations is frequently ill-conditioned and influenced greatly by roundoff errors. Therefore, one must resort to appropriate numerical techniques for solving linear least squares problems in a highly accurate manner.

Generally speaking, the main drawback of the tapped delay line (TDL) results from the fact that the length of the equalizer tends to ∞, even though the channel has a finite memory. Some recent works consider the problem of equalization as a *sequence estimation* problem without imposing a given structure on the receiver. In this case, the received process is written according to its *state equations* and the optimum estimator is derived as a Kalman filter. The resulting equalizer is given in Fig. 22, where

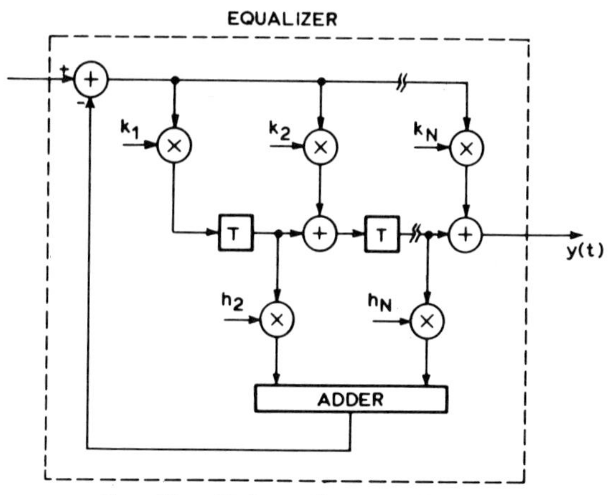

FIG. 22 – Kalman-filter equalizer.

the parameters $\{h_i\}$ are the values of the amplifier response to $h_p(t)$ sampled at the bit rate, and the gains $\{k_i\}$ satisfy an algebraic matrix Riccati equation. In Fig. 23, the functional blocks of two possible loops for estimating the unknown sets $\{h_i\}$ and $\{k_i\}$ are illustrated; usually, the adaptive algorithms are based upon a gradient minimization of the error.

If the transmission rate is pushed closer to the Nyquist rate, the linear approach tends to increase the noise power significantly

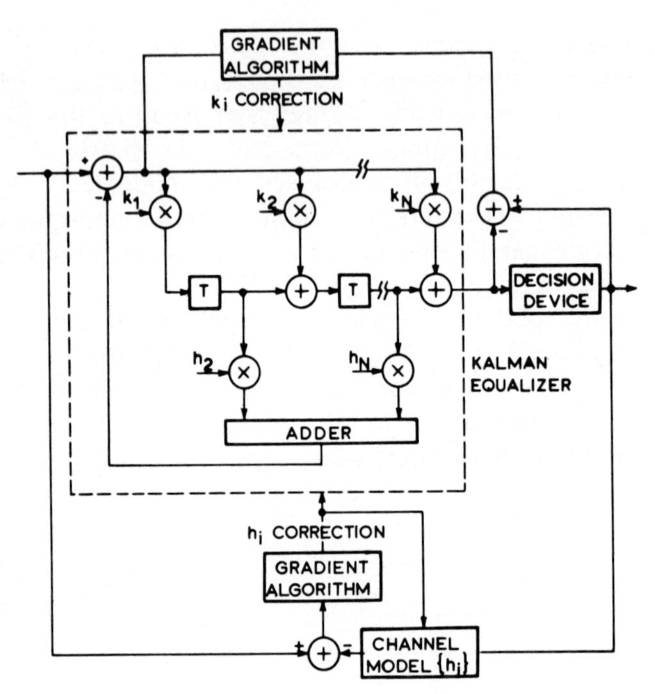

FIG. 23 – Adaptive Kalman-filter equalizer.

in attempting to equalize a very distorted pulse response. As an example, for the fibre response of Fig. 15 we have computed the signal-to-noise ratio SNR_i, defined at the equalizer input, necessary to obtain a given error probability with different MMSE equalization orders M (Fig. 24). SNR_i is defined as $10 \log (r_0^2 B/2\sigma_g^2)$ where B is the noise-enhancement factor introduced by an equalizer and r_0 is the signal evaluated at the decision time t_0 [see (153)]. Note that this definition implies that the noise at the input of the equalizer is a white noise process. This hypothesis is useful since it enables us to take into account the noise-bandwidth effects of the equalizer. Alternatively, one could operate with the signal-to-noise ratio defined at the output, thereby removing the assumption of white noise; however, it would be quite difficult to compare results obtained for various equalizers in different situations of σ/T, M, etc. If the unequalized signal is corrupted by an additive white Gaussian noise with two-sided spectral height $\mathcal{N}_0/2$, the noise power in the Nyquist frequency bandwidth $[-1/2T, 1/2T]$

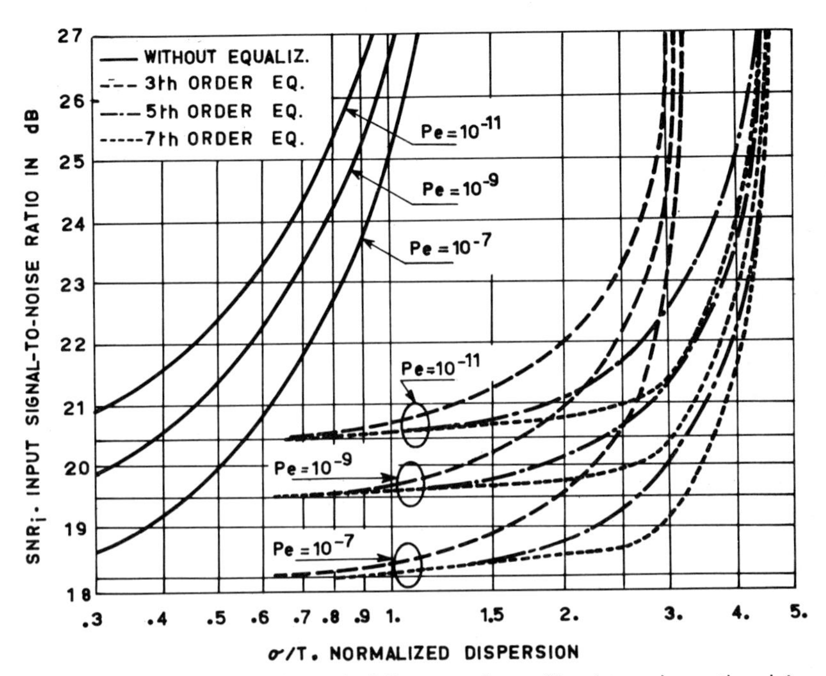

FIG. 24 – MMSE equalizer of different orders. Signal-to-noise ratio giving $P_e = 10^{-7}$, 10^{-9} and 10^{-11} versus σ/T. (Courtesy of Opt. Quantum Electron., Vol. 8, pp. 743-753).

is $\sigma_n^2 = \mathcal{N}_0/2T$. Therefore,

$$\sigma_g^2 = B\sigma_n^2 = B\mathcal{N}_0/2T\,, \tag{204}$$

with

$$B = \frac{T}{2\pi}\int_{-\pi/T}^{\pi/T} G(\exp[j\omega T])\,G(\exp[-j\omega T])\,d\omega\,, \tag{205}$$

where $G(\exp[j\omega T])$ is the frequency response of the equalizer.

In Fig. 24, the abscissa is σ/T and three kinds of curves can be recognized, each for a different error-probability objective, i.e., $P_e = 10^{-7}$, 10^{-9} and 10^{-11}. Moreover, every P_e can be reached with a given equalizer complexity, that is 0, 3, 5 and 7 equalizer taps. Of course, for a given tolerable SNR_i and for this equalizer complexity, the dispersion and the transmission rate tend to be upper-bounded by a ratio $\sigma/T \simeq 5$.

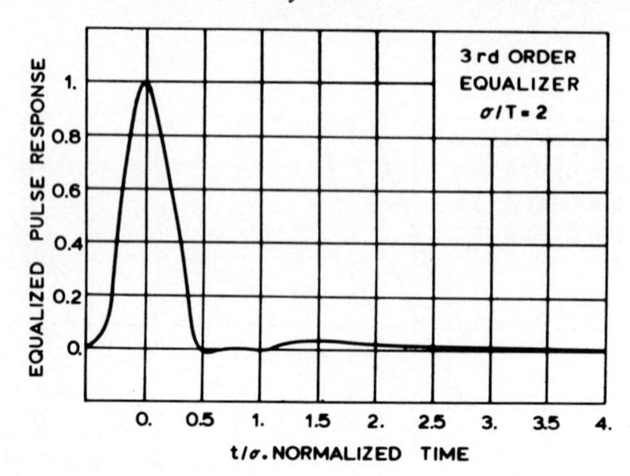

FIG. 25 – 3rd order MMSE equalizer for $\sigma/T = 2$. Equalized time response $r(t)$ to rectangular source pulse. (Courtesy of Opt. Quantum Electron., Vol. 8, pp. 743-753).

Therefore, when $\sigma/T < 1$, the objectives mentioned above can be reached without equalization, provided that SNR_i is high enough. On the other hand, in the zone defined by $\sigma/T > 1$ the resulting intersymbol interference dominates the error probability and any SNR_i increase cannot compensate for P_e degradations, unless a suitable equalizer is introduced.

As an example of the distortion effects, assume that $\sigma/T = 2$. This value corresponds to a *step-index* fibre with a length of 2 km

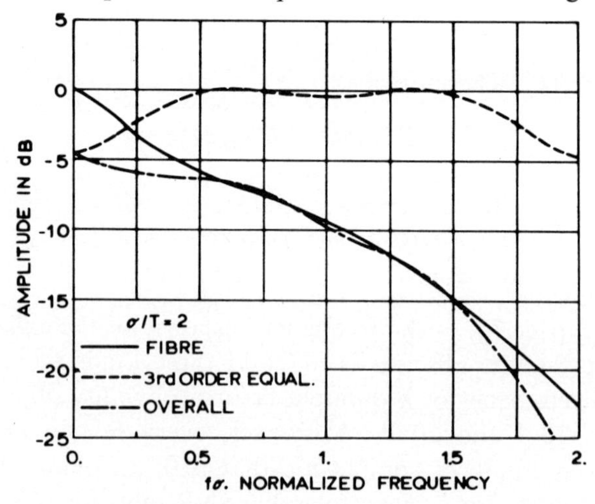

FIG. 26 – 3rd order MMSE equalizer for $\sigma/T = 2$. Amplitude of fibre, equalizer and overall system function versus normalized frequency. (Courtesy of Opt. Quantum Electron., Vol. 8, pp. 743-753).

and a half-value width of $\simeq$ 13.5 ns/km in a 34-Mbit/s PCM system. Otherwise, $\sigma/T = 2$ may correspond to a *graded-index* profile, a length of 4 km, and a half-value width of $\simeq$ 1.6 ns/km in a 140-Mbit/s PCM system. In both cases a third-order equalizer can be used to obtain $P_e = 10^{-9}$ with a signal-to-noise-ratio penalty lower than 1.5 dB.

For the same example, Fig. 25 gives the time response to the *rectangular pulse* after equalization, whereas Figs. 26 and 27 give the frequency behaviour of fibre, equalizer and overall system.

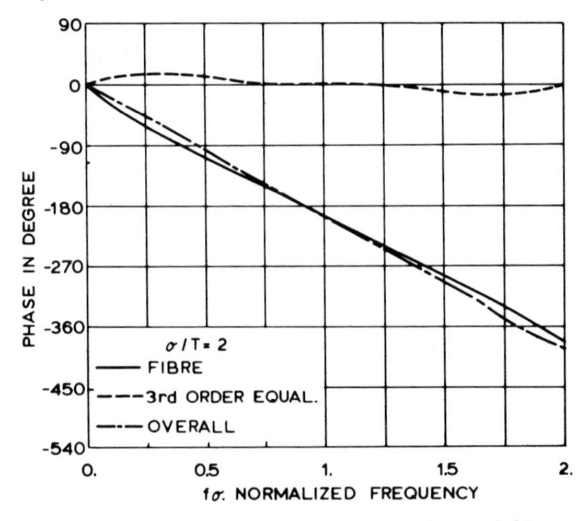

FIG. 27 – 3rd order MMSE equalizer for $\sigma/T = 2$. Phase of fibre, equalizer and overall system function versus normalized frequency. (Courtesy of Opt. Quantum Electron., Vol. 8, pp. 743-753).

Note that time and frequency responses are quite different from standard Nyquist responses. As a matter of fact, a desired performance can be reached with very simple equalizers.

In the last few years there have been two promising approaches towards improving on linear equalizers, i.e., the decision-feedback equalizer (DFE) and the Viterbi-algorithm (VA) detector.

In the DF equalizer the postcursors or tail of the distorted pulse are removed by cancellation, using the detected symbols. The scheme is given in Fig. 28, where the feedback filter compensates for the pulse tail. Since the nonlinear decision device is in the forward path, the DF equalizer does not suffer the noise enhancement drawback. On the other hand, two kinds of impair-

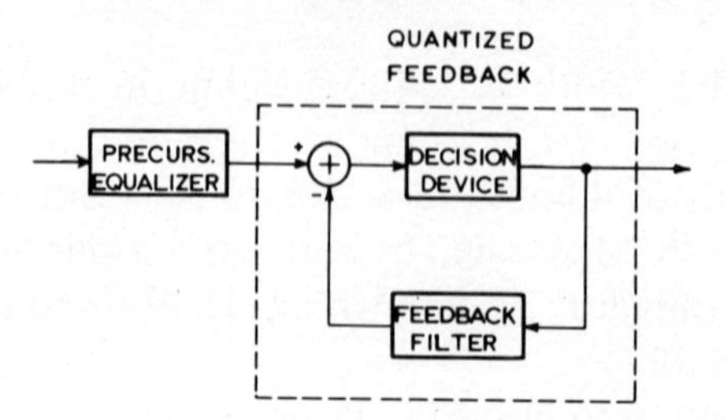

FIG. 28 – Block diagram of a decision feedback equalizer and linear compensation of precursors.

ments should be considered, i.e., the presence of precursors of the unequalized pulse, and the error propagation effects due to the feedback structure.

Usually, the precursors are reduced by resorting to a conventional transversal filter before the tail cancellation (see Fig. 28); in this way, however, a certain noise enhancement is caused by the TDL bandwidth. For this purpose, the new scheme of Fig. 29 has been studied and based upon a nonlinear iterative correction of *one* precursor, which is of particular significance in view of high-speed transmission.

The Viterbi receiver is based on a maximum-likelihood sequence estimator because it determines the transmitted sequence that maximizes the *a posteriori* probability density of the received signal.

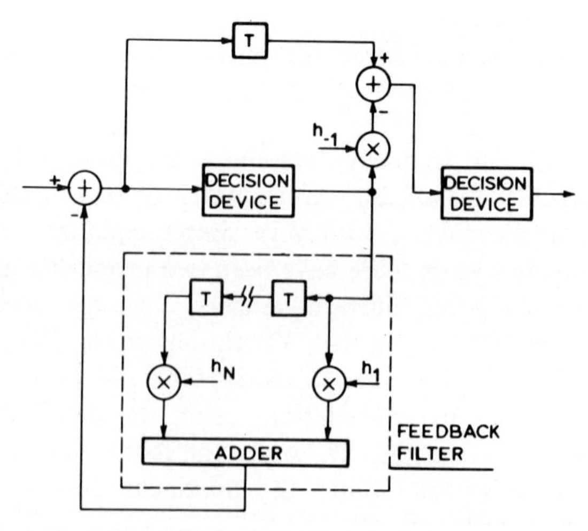

FIG. 29 – Block diagram of nonlinear equalizer with decision feedback and one precursor (h_{-1}) correction.

The Viterbi algorithm is a dynamic programming solution which can be implemented in iterative form, even though it is generally quite complicated. It is worthwhile mentioning that the Viterbi algorithm and the Kalman filter are both *sequence* estimators; the relevant difference lies in the linearity of the Kalman structure that, therefore, represents a suboptimal solution.

The VA detector is required to perform 2^{M+1} additions and squaring operations per received symbol, MT being the channel duration or memory. The problem with the VA is obviously the complexity; therefore, its use in practical optical fibre systems has not yet been investigated. Anyway, current research is centred on finding less complex but nearly equivalent systems.

It is worthwhile emphasizing once again that the practical use of nonlinear Viterbi detectors is questionable. However, the error probability curves presented in Fig. 24 show that, in many cases, satisfactory performance can be achieved with a relatively modest linear transversal equalizer, whose introduction, even though requiring an additional cost, increases the feasibility of higher capacity systems.

To conclude this brief review on pulse equalization, some performance comparison results are given in terms of the average probability of error P_e. Figure 30 shows the SNR_i, giving a $P_e = 10^{-9}$, for the decision-feedback equalizer compared with both

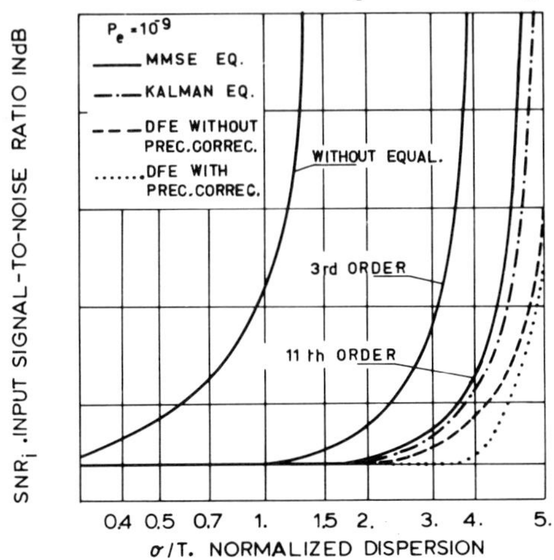

FIG. 30 – Signal-to-noise ratio giving $P_e = 10^{-9}$ for transversal MMSE, Kalman and *DF* (with and without precursor cancellation) equalizers. (Courtesy of Ann. Télécommun., Vol. 32, pp. 357-366).

Kalman and MMSE equalizers as the rate is increased over the bandlimited channel shown previously.

The theoretical and numerical aspects of error probability evaluation in linear equalization systems are discussed in the previous sections. The performance computation must be carried out differently when a DFE is used. This point has been developed only recently and a detailed explanation is beyond the scope of this brief review.

For the sake of simplicity the results of Figs. 24 and 30 are obtained under the hypothesis that the light detector is an ordinary phototiode, and, therefore, it is assumed that the shot noise is negligible.

1.10.2 *Synchronization*

In deriving our expressions for P_e we have assumed that the symbol timing is perfect. If a synchronization error occurs during a symbol period, system performance is degraded. In order to emphasize the timing effects, Fig. 31 shows the sensitivity to timing error of a 5-tap MMSE equalizer. The results refer to the same parameters and quantities shown in Fig. 24, which has been discussed in the previous paragraph. Moreover, the SNR_i for three situations of different timing error are presented. Note

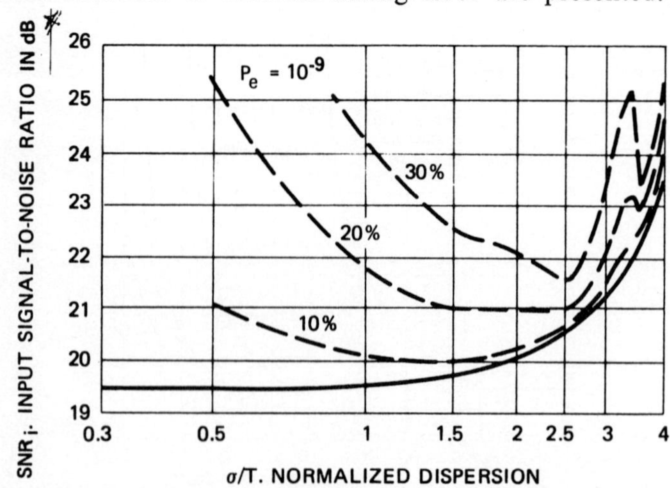

FIG. 31 – 5th order MMSE equalizer. Signal-to-noise ratio giving $P_e = 10^{-9}$ for different timing errors, i.e. 10%, 20% and 30% of symbol period. (Courtesy of Ann. Télécommun., Vol. 32, pp. 357-366).

that the oscillating behaviour beyond $\sigma/T \simeq 3$ is due to the appearance of a precursor in the unequalized pulse response.

We shall adopt the same model and the same notations used in Section 1.8, except for an unknown *timing parameter* φ to be tracked by the receiver. By focussing on this item, expression (143) becomes

$$\lambda_t = \frac{\eta}{h\nu} \sum_{i=-\infty}^{\infty} a(i) h_p(t - iT - \varphi) + \lambda^d, \tag{206}$$

while expression (144) for the signal at the amplifier output is still valid:

$$y_a(t) = \sum_{j=1}^{N_t} eG_j k(t - \tau_j) + \int_0 i_\tau^g k(t - \tau)\, d\tau\,, \tag{207}$$

provided that $k(t)$ is the impulse response of the amplifier *alone.*

The *unconditional* average of $y_a(t)$ in the absence of dark current is easily obtained as

$$E[y_a(t) | i_t^d = 0] = E[a(i)] \sum_{i=-\infty}^{\infty} r(t - iT - \varphi)\,. \tag{208}$$

Equation (208) emphasizes the synchronization problem, where the nonrandom signal $E[y_a(t) | i_t^d = 0]$ is usually referred to as *timing wave* since it bears the information on the unknown parameter φ. Note that the data pattern produces deviations from the timing wave, thereby causing what is called *pattern-noise.* In view of (208), we may look at $E[y_a(t) | i_t^d = 0]$ as the synchronization signal to be tracked by phase-lock loops or tuned resonant circuits (Fig. 32*a* and *b*). A possible option is to insert some form of nonlinearity before the timing circuit to enhance the spectral line at the pulse rate. This solution seems to be convenient for large pulse spreadings.

The timing synchronization at the receiver could be also carried out by an automatic tracking structure which operates by forming a zero-seeking error signal. This structure, generally called data-aided loop (Fig. 32*c*), yields superior performance by taking advantage of whatever is known at the receiver about the transmitted signal.

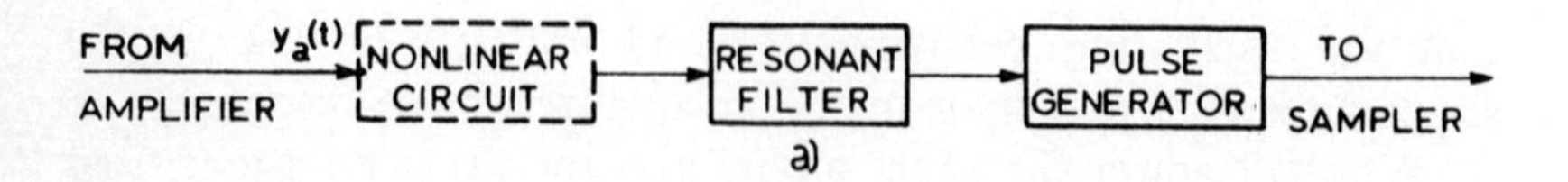

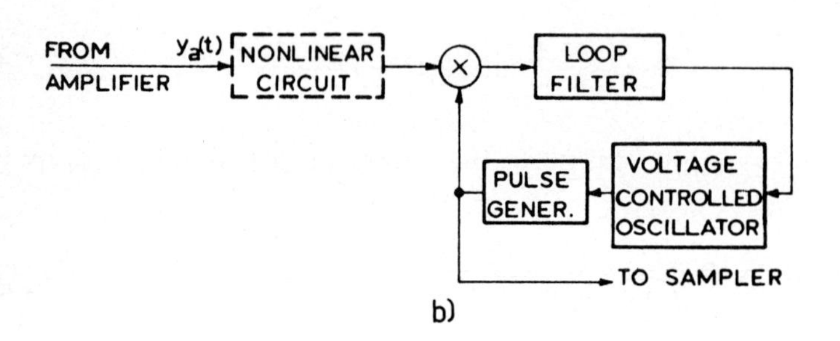

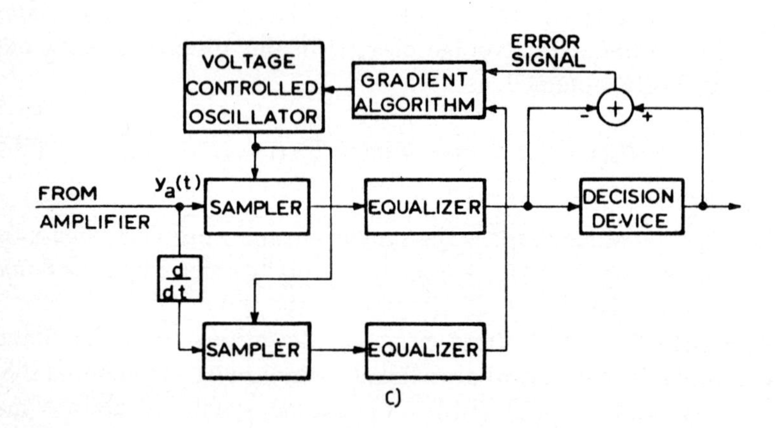

FIG. 32 – Data synchronization subsystem. (*a*) Conventional bandpass model; (*b*) Phase-lock loop; (*c*) Decision-directed tracking loop.

As a final comment, it is worth noting the these subjects are currently under investigation and that many aspects need further development. In particular, the influence of shot noise on synchronization circuits in the presence of a nonlinear element is still to be evaluated.

CONCLUSIONS

1.11 Comments and bibliographic notes

For comprehensive discussion of optical communication theory see Pratt [1], Hoversten [2], Gagliardi and Karp [3], and Saleh [4]. Snyder [5] gives an excellent background and interesting examples for the Poisson process and other hierarchically derived stochastic processes.

Section 1.5. See Gagliardi and Karp ([3], pp. 119-123) for details about the power spectral density in the case of (conditional) Poisson shot-noise processes.

Section 1.6. The statistical model of Fig. 10 for the photodetected signal has been proposed by Hoversten [2]. See also Snyder ([5], pp. 174-178).

Personick [6] suggests the use of *quantum amplifiers* in optical communication systems, either as optical analogue repeaters or as front ends of regenerative repeaters. The quantum amplifier gives rise to a spontaneous emission noise, which can be assumed to be a zero-mean Gaussian random field added to the information-bearing signal. Let us assume for simplicity a direct-detection receiver observing a single mode of the optical field. The received field can be modelled as a scalar field and the intensity of the detected signal becomes

$$\lambda_t^s = \frac{\eta}{h\nu} A_r |s(t) + b(t)|^2, \tag{209}$$

where A_r is the area of the receiving surface, $s(t)$ and $b(t)$ are the *complex envelopes* of the signal and the quantum-amplifier noise, respectively. The signal-to-noise ratio, e.g., given for direct intensity modulation, may account for the quantum-amplifier noise starting from (209) and paralleling the computations given in Section 1.7.

Section 1.7. Chapter 5 of Gagliardi and Karp [3] is very accurate in defining the various signal-to-noise ratios, and in applying them to optical communication systems.

Another method for analogue modulation suitable for fibre-optic systems is *pulse modulation*, which involves changing a parameter of a pulse train. Hubbard [7] considers pulse position modulation (PPM), which offers the advantage of improvement in noise immunity by exploiting the wide available bandwidth of optical systems. A variant of PPM is pulse frequency modulation (PFM) (see Timmermann [8]), whose major attractive feature with respect to PPM is the simplicity of modulation and demodulation equipments.

Section 1.8. For the digital system model and related problems see Personick [9], Catania and Sacchi [10], Dogliotti and Luvison [11], Bosotti and Pirani [12], and Luvison and Sacchi [13]. Personick [14] illustrates important situations, in which the fibre can be considered « linear in power », meaning that expression (141) holds. Muoi and Hullett [15], and Bosotti and Tamburelli [16] extend the analysis given herein to *multilevel* pulse amplitude modulation.

Situations may arise which require that digital data be transmitted by digital modulation of an electrical subcarrier modulating, in turn, the intensity of an optical source. At the receiver, the electrical subcarrier would be detected directly as shown in Fig. 13 and the digital data would be recovered by electrical demodulation of the subcarrier. As an example, a frequency-division-multiplex (FDM) system, which transmits both analogue and digital signals, would require the application of such a digital subcarrier. In addition, schemes using multiple optical carriers, the so-called wavelength division multiplexing (WDM), have been experimented (Ishio and Miki [17]) and analyzed (Bosotti and Laforgia [18]).

Section 1.9. The method for error probability evaluation has been proposed by Dogliotti, Luvison and Pirani [19], who give more numerical details. Preliminary results are given in Dogliotti *et alii* [20]. For a different approach see Cariolaro [21].

Fundamental issues in the analysis and design of optical fibre communication systems are the channel modelling and simulation. We have said that the linearity assumption is valid under physically reasonable hypotheses regarding, e.g., signalling format and modulation; this enables us to deal with a baseband model of the transmitting channel. As a consequence, the overall fibre behaviour is completely characterized by the Dirac impulse response or by the corresponding transfer function. A good mathematical model has been proposed by Gloge [22], who considers the theoretical baseband pulse response of multimode fibres. Currently, experimental results are also available, since a large number of measurements on fibreoptic experimental systems are carried out in many countries (see, e.g., Catania *et alii* [23]).

In modelling the fibre, it is convenient to deal with a lumped-parameter structure whose response approximates the behaviour illustrated previously. First of all, it is possible to design optimum transmitting and receiving filters, according to standard techniques provided that the channel system function is available in rational form. Furthermore, an analogue lumped-parameter system can be easily discretized and the resulting digital filter gives an accurate and flexible model for the numerical simulation. Finally, a time-domain exponential fitting is a prerequisite for carrying out analytic performance evaluation for nonlinear receivers (Cianci and Dogliotti [24]). Discussions and results about approximation are given by Dogliotti and Luvison [11], and Dogliotti, Luvison and Pirani [25].

Section 1.10. The pulse-equalization problem for fibre communication systems is considered by Dogliotti, Guardincerri and Luvison [26]. The *nonlinear* equalization of optical fibres implemented by quantized feedback, feedforward and shaping filter is approached by Tamburelli [27]. Tamburelli [28] proposed the decision-feedback equalizer with the precursor correction of Fig. 29.

The synchronization of pulse and digital optical systems is discussed by Forrester and Snyder [29], Gagliardi and Karp ([3], Chapters 9-11), Andreucci and Mengali [30], and Mengali and Pezzani [31]. A general theory concerning the timing jitter accumulation in a chain of regenerative repeaters for optical fibre transmission is presented by Mengali and Pirani [32].

For high signalling rates in fibre communications, suitable codes can be devised with synchronizing properties, that is, to give a reasonably constant timing content, besides guaranteeing bandwidth economy, error monitoring and tolerance to intersymbol interference. Transmission codes for fibreoptic digital systems are discussed by Rousseau [33], Takasaki *et alii* [34], Balicco and Iudicello [35], and Bosotti and Pirani [36]. The method for the performance computation given in Section 1.9 can be applied to evaluate the error probability for encoded optical communication systems and to compare different codes with the criterion of the error probability. If the data symbols are encoded, $x_i(t_0)$ given by (183) is no longer the sum of statistically independent random variables. However, a fast technique for computing its moments has been proposed by Cariolaro and Pupolin [37].

Foschini, Gitlin and Salz [38] report on the optimum (maximum likelihood) reception of digital signals transmitted over optical fibres. The optimum direct-detection processing is examined in the presence of quantum noise, dispersion and thermal noise.

REFERENCES

[1] W. K. Pratt: *Laser Communication Systems*, New York, Wiley (1969).

[2] E. V. Hoversten: *Optical Communication Theory*, Laser Handbook, F. T. Arecchi and E. O. Schulz-Du Bois, Eds. Amsterdam, North-Holland (1972), pp. 1805-1862.

[3] R. M. Gagliardi and S. Karp: *Optical Communications*, New York, Wiley (1976).

[4] B. Saleh: *Photoelectron Statistics with Applications to Spectroscopy and Optical Communication*, Berlin, Springer-Verlag (1978).

[5] D. L. Snyder: *Random Point Processes*, New York, Wiley (1975).

[6] S. D. Personick: *Applications for quantum amplifiers in simple digital optical communication systems*, Bell. Syst. Tech. J., Vol. 52 (January 1973), pp. 117-133.

[7] W. M. Hubbard: *Utilization of optical frequency carriers for low- and moderate-bandwidth channels*, Bell Syst. Tech. J., Vol. 52 (May-June 1973), pp. 731-765.

[8] C. C. Timmermann: *Signal-to-noise ratio of a video signal transmitted by a fiber-optic system using pulse-frequency modulation*, IEEE Trans. Broadcast., Vol. BC-23 (March 1977), pp. 12-16.

[9] S. D. Personick: *Receiver design for digital fiber optic communication systems, I*, Bell Syst. Tech. J., Vol. 52 (July-August 1973), pp. 843-874.

[10] B. Catania and L. Sacchi: *Design of optical fiber digital communication links*, Conf. Rec. XXIII Convegno Internazionale delle Comunicazioni, Genoa, Italy (October 9-11, 1975), pp. 564-573.

[11] R. Dogliotti and A. Luvison: *Signal processing in digital fiber communications*, Conf. Rec. Second European Conference on Optical Fibre Communication, Paris, France (September 27-30, 1976), pp. 331-341, and Ann. Télécommun., Vol. 32 (September-October 1977), pp. 357-366.

[12] L. Bosotti and G. Pirani: *Analysis and modeling of integrate and dump circuits in fiber optic communication receivers*, Conf. Rec. 1978 IEEE Nat. Telecomm. Conf., Birmingham, Ala (December 4-6, 1978), pp. 13.5.1-13.5.5.

[13] A. Luvison and L. Sacchi: *Topics in digital fiber communication systems. Theory and implementation*, Conf. Rec. 1979 Int. Symp. Circuits and Systems, Tokyo, Japan (July 17-19, 1979), pp. 515-518.

[14] S. D. Personick: *Baseband linearity and equalization in fiber optic digital communication systems*, Bell Syst. Tech. J., Vol. 52 (September 1973), pp. 1175-1194.

[15] T. V. Muoi and J. L. Hullett: *Receiver design for multilevel digital optical fiber systems*, IEEE Trans. Commun., Vol. COM-22 (September 1975), pp. 987-994.

[16] L. Bosotti and G. Tamburelli: *Performance evaluation in multilevel digital transmission on optical fibres*, Conf. Rec. Fourth European Conference on Optical Communication, Genoa, Italy (September 12-15, 1978), pp. 539-545.

[17] H. Ishio and T. Miki: *A preliminary experiment on wavelength division multiplexing transmission using LED*, Tech. Digest 1977 Int. Conf. on Integrated Optics and Optical Fiber Communication, Tokyo, Japan (July 18-20, 1977), pp. 493-496.

[18] L. Bosotti and D. Laforgia: *Performance evaluation of optical fibre digital systems employing wavelength division multiplexing (WDM)*, CSELT Rapporti Tecnici, Vol. 7 (March 1979), pp. 11-23.

[19] R. Dogliotti, A. Luvison and G. Pirani: *Error probability in optical fiber transmission systems*, IEEE Trans. Inform. Theory, Vol. IT-25 (March 1979), pp. 170-178.

[20] R. Dogliotti, A. Guardincerri, A. Luvison and G. Pirani: *Error probability in optical fiber transmission systems*, Conf. Rec. 1976 IEEE Nat. Telecomm. Conf., Dallas, Tx (November 29-December 1, 1976), pp. 37.6-1 - 37.6-5.

[21] G. L. Cariolaro: *Error probability in digital fiber optic communication systems*, IEEE Trans. Inform. Theory, Vol. IT-24 (March 1978), pp. 213-221.

[22] D. Gloge: *Impulse response of clad optical multimode fibers*, Bell Syst. Tech. J., Vol. 52 (July-August 1973), pp. 801-816.

[23] B. Catania, F. Tosco, L. Michetti, E. Occhini and L. Silvestri: COS 1: *first Italian experiment with buried optical cable*, Conf. Rec. Second European Conference on Optical Fibre Communication, Paris, France (September 27-30, 1976), pp. 315-322.

[24] C. Cianci and R. Dogliotti: *Calculation of exact error rates for decision-feedback equalizers*, unpublished work (1975).

[25] R. Dogliotti, A. Luvison and G. Pirani: *Pulse shaping and timing errors in optical fibre systems*, Conf. Rec. Fourth European Conference on Optical Communication, Genoa, Italy (September 12-15, 1978), pp. 510-520.

[26] R. Dogliotti, A. Guardincerri and A. Luvison: *Baseband equalization in fibre optic digital transmission*, Opt. Quantum Electron., Vol. 8 (July 1976), pp. 743-753.

[27] G. Tamburelli: *Nonlinear equalizer with shaping filter for optical fibres*, CSELT Rapporti Tecnici, Vol. 5 (September 1977), pp. 229-235.

[28] G. Tamburelli: *Decision feedback and feedforward receiver (for rates faster than Nyquist's)*, Alta Frequenza, Vol. 45 (October 1976), pp. 231E-238E.

[29] R. H. Forrester, Jr., and D. L. Snyder: *Phase-tracking performance of direct-detection optical receivers*, IEEE Trans. Commun., Vol. COM-21 (September 1973), pp. 1037-1039.

[30] F. Andreucci and U. Mengali: *Timing extraction in optical transmission*, Opt. Quantum Electron., Vol. 10 (November 1978), pp. 445-458.

[31] U. Mengali and E. Pezzani: *Tracking properties of phase-locked loops in optical communication systems*, IEEE Trans. Commun., Vol. COM-26 (December 1978), pp. 1811-1818.

[32] U. Mengali and G. Pirani: *Timing jitter in a chain of regenerative repeaters for optical fiber transmissions*, to be published.

[33] M. Rousseau: *Transmission code and receiver selection for optical fibres PCM communications*, Conf. Rec. First European Conference on Optical Fibre Communication, London, England (September 16-18, 1975), pp. 174-176.

[34] Y. Takasaki, M. Tanaka, N. Maeda, K. Yamashita and K. Nagano: *Optical pulse formats for fiber optic digital communications*, IEEE Trans. Commun., Vol. COM-24 (April 1976), pp. 404-413.

[35] E. Balicco and G. Iudicello: *Binary coding in fiber optic digital transmission with respect to timing extraction and error monitoring*, Conf. Rec. XXIII Congresso Internazionale per l'Elettronica, Rome, Italy (March 22-24, 1976), pp. 47-58.

[36] L. Bosotti and G. Pirani: *A PAM-PPM signalling format in optical fibre digital communications*, Opt. Quantum Electron., Vol. 11 (January 1979), pp. 71-86.

[37] G. L. Cariolaro and S. G. Pupolin: *Moments of correlated digital signals for error probability evaluation*, IEEE Trans. Inform. Theory, Vol. IT-21 (September 1975), pp. 558-568.

[38] G. J. Foschini, R. D. Gitlin and J. Salz: *Optimum direct detection for digital fiber-optic communication systems*, Bell. Syst. Tech. J., Vol. 54 (October 1975), pp. 1389-1430.

CHAPTER 2

SYSTEM CONSIDERATIONS

by Agostino Moncalvo and Luigi Sacchi

2.1 Summary

The aim of this chapter is to provide a systematic approach to the different problems related to the system design of optical fibre communication links.

Since fibre systems are more naturally suited to digital transmission applications and most of the development work on fibre systems is in this direction, the following considerations refer mainly to digital transmission. Comments on analogue transmission systems are given only when general relations, such as signal-to-noise ratio at the receiver, are considered.

In Section 2.2 the properties of optical fibres and optoelectronic components are reviewed in terms relevant to the system design.

Section 2.3 deals with the design criteria for optical fibre digital transmission systems: the required optical power to achieve a desired error rate and the repeater spacing for different line symbol rates are deduced on the basis of a simplified theory which gives quite precise results, thus allowing an easy understanding of the influence of the different factors.

Section 2.4 gives an overview of the line coding schemes most suited to digital transmission over optical fibres.

Finally, Section 2.5 is devoted to optical fibre applications on short distance systems: different system configurations are examined and some basic considerations on the design of such links are reported.

2.2 System model

2.2.1 *Optical fibre system block diagram*

A block diagram of an optical fibre digital communication system is given in Fig. 1, where only one regeneration section is indicated. Like any other transmission system it comprises three main blocks: a transmitter, a transmission medium or channel and a receiver.

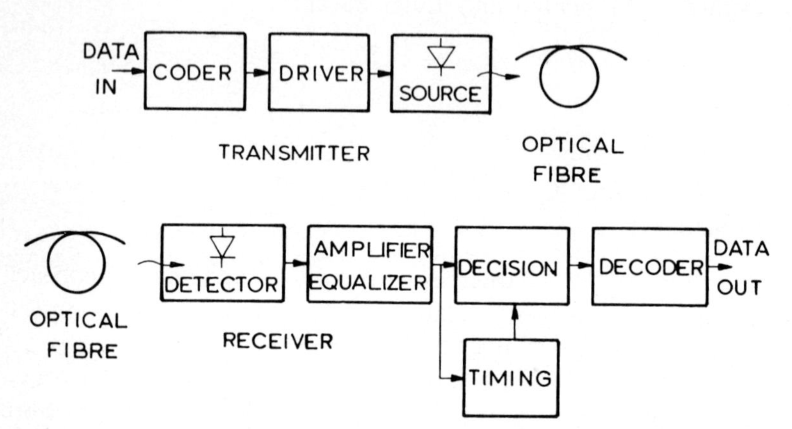

FIG. 1 – Block diagram of an optical fibre digital transmission system.

The transmitter contains a line coder which converts the input sequence of statistically independent binary symbols, with a rate f_0, into a sequence of symbols, with a rate f_r, suitable for the transmission medium and monitoring purposes. This signal modulates, through a driver circuit, the intensity of the output power of the optical source (typically a high current, low voltage device) which can be a light emitting diode (LED) or an injection laser diode.

Part of the emitted power is injected into the fibre and propagates along it according to the principles of total internal reflection. During propagation the optical signal pulses suffer attenuation and distortion.

At the receiving side the optical pulses are converted, through the photodetector (PIN diode or avalanche diode), into electrical current pulses, which are amplified and equalized in order to maximize the signal-to-noise ratio at the decision point. The equalized signal is then regenerated and decoded in order to deliver the original sequence.

For a better understanding of the different system design aspects, it is necessary to briefly recall the characteristics and properties of optical fibres and optoelectronic components already discussed in the preceding Sections.

2.2.2 *Optical fibre characteristics*

Multimode optical fibres are currently being considered for telecommunication applications. Transmission through such fibres introduces an attenuation on the signal proportional to the length and, because of imperfect channel and source characteristics, a distortion known as « pulse dispersion » that leads to a broadening of the transmitted pulse along the fibre.

In the following, the hypothesis of linearity in the power response of the fibre will be assumed. Such an assumption has been proved valid for performance evaluations for the systems of interest [1, 2].

The attenuation is caused by material absorption, namely conversion of optical power into heat, and by scattering, namely loss of light from the fibre caused by material impurities.

Considering the attenuation dependence on the wavelength it is easy to recognize two minima in the bands between 0.8 and 0.9 μm and above 1.05 μm, usually known as first and second « windows ». Attenuation values ranging between 3 and 5 dB/km and of about 1 dB/km are currently reported for the two windows respectively.

Pulse dispersion is a distortion corresponding to a band limitation of the fibre, and is caused by two main factors: modal and material dispersion.

Modal dispersion is due to the difference of group velocity of the different modes propagating along the fibre at a single wavelength. This effect, which is eliminated in monomode fibres, can be reduced in multimode fibres by grading the refractive index profile of the core.

Pulse broadening τ_m due to modal dispersion increases with the numerical aperture Δ of the fibre, as this parameter is related to the number of modes that can be excited. The pulse broadening versus Δ for different fibre index profiles is shown in Fig. 2 [3]. In practical low-loss fibres normal values of Δ range between 0.15 and 0.2, and values of τ_m as small as a fraction of 1 ns/km are currently reported.

Fibre or route irregularities and splices may cause mode mixing, that is an exchange of energy among different modes, with a tendency to concentrate the energy in the lower order modes.

For fibres without mode mixing, modal dispersion is propor-

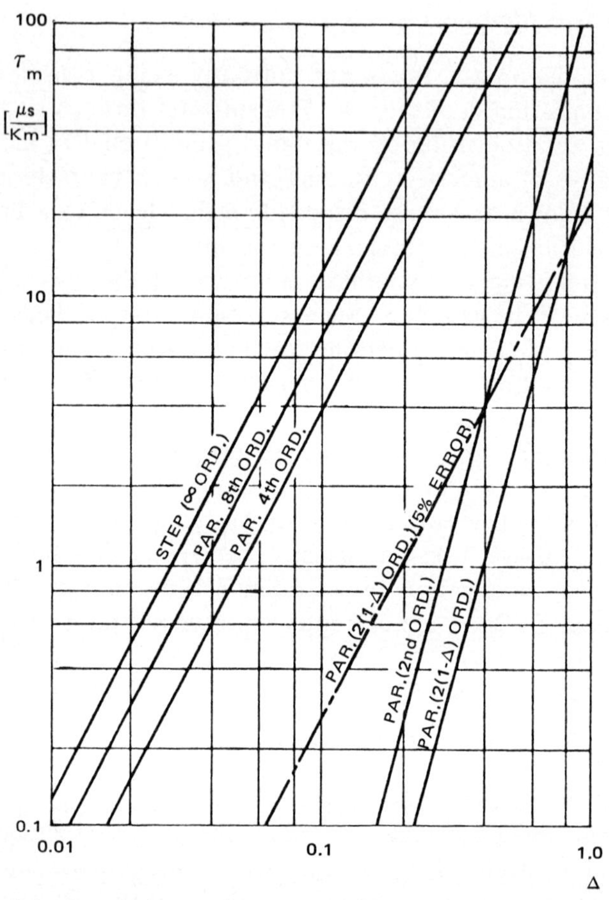

FIG. 2 – Pulse broadening τ_m due to modal dispersion versus numerical aperture Δ for different fibre index profile.

tional to the fibre length L. For fibres with mode coupling the pulse broadening τ_m is proportional to $\sqrt{L}$; in this case the fibre impulse response due to modal dispersion $h_m(t)$ can be considered, with a good approximation, Gaussian in shape. According to experimental results obtained on spliced graded-index fibres, τ_m will be assumed in the following to be proportional to $\sqrt{L}$. At any rate, such an assumption does not impair the general considerations that will be made.

Material dispersion is associated with the bandwidth of the optical source (typically much larger than the modulation bandwidth) and is caused by the variation of refractive index with the optical wavelength, which produces a difference in group velocity of a single mode at different wavelengths. The corresponding pulse broadening τ_λ is proportional to the source bandwidth $\Delta\lambda$ and to the fibre length L, and decreases with the source wavelength as indicated in Fig. 3 [4].

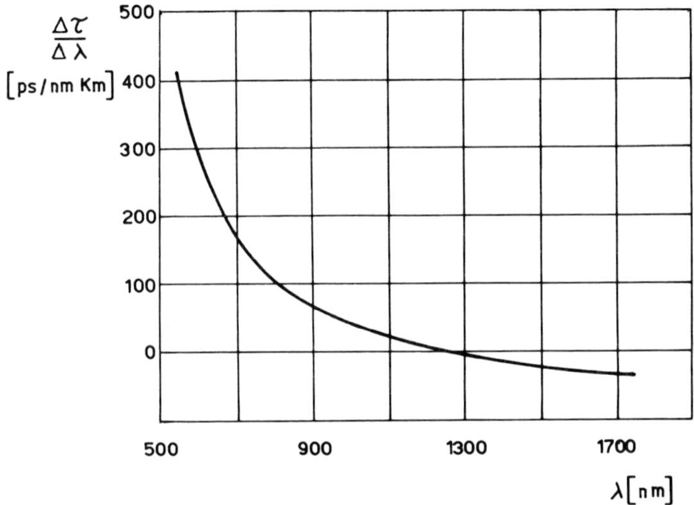

FIG. 3 – Pulse broadening $\Delta\tau_\lambda/\Delta\lambda$ due to material dispersion versus wavelength for a typical silica fibre.

Current values of material dispersion are around 100 ps/km$\cdot$ $\cdot$nm($\Delta\lambda$) and below 10 ps/km$\cdot$nm($\Delta\lambda$) in the first and second window respectively. The material dispersion effect is important, mainly in the first window, when LED sources ($\Delta\lambda = 20\text{-}50$ nm) are used.

In the second window, where material dispersion becomes negligible, a second order dispersion effect, known as waveguide dispersion, could become relevant. This distortion is due to the dependence of the guiding effect and then of the propagation constant on the wavelength, and is proportional to the length L.

If the differential delay δ can be considered linear with the wavelength over the optical source band, the fibre impulse response $h_\lambda(t)$ due to material dispersion has the same shape as the source spectrum, with a time scale $t = \delta\cdot\lambda$. In practical cases this shape can be considered, with a good approximation, Gaussian [5, 6].

Under the assumed hypothesis of linearity of the fibre in the power response, it can be shown [7, 8] that the fibre impulse response $h(t)$ is given by:

$$h(t) = h_m(t) * h_\lambda(t) \tag{1}$$

where $*$ denotes convolution and $h_m(t)$ and $h_\lambda(t)$ are the impulse responses due separately to modal and material dispersion. From the dispersion point of view, the fibre behaves like two quadripoles in tandem. Assuming $h_m(t)$ and $h_\lambda(t)$ as Gaussian in shape the total impulse broadening τ is given by:

$$\tau^2 = \tau_m^2 + \tau_\lambda^2 \tag{2}$$

Accordingly, for the following evaluations the fibre transfer characteristic $H(f)$ can be assumed, with a good approximation, valid in the presence of mode mixing, Gaussian and equal to:

$$H(f) = \exp[-2\alpha L]\cdot\exp[-\ln 2\cdot(f/f_t)^2] \tag{3}$$

where α is the attenuation per unit length (in Np/km), L is the length of the fibre and f_t is the 3 dB optical power cut-off frequency given by:

$$f_t = \frac{1}{\sqrt{\dfrac{1}{f_m^2}L + \dfrac{1}{f_\lambda^2}L^2}} \tag{4}$$

in which f_m and f_λ are the cut-off frequencies per unit length due to modal and material dispersion respectively, related to the corresponding pulse broadening by:

$$f_{m,\lambda} = \frac{\ln 4}{\pi\tau_{m,\lambda}} \simeq \frac{0.44}{\tau_{m,\lambda}} \tag{5}$$

In the first window, with laser sources ($\Delta\lambda \simeq 2$ nm) both dispersion effects can contribute to bandwidth limitation on long distance links; f_m and f_λ values of 800-1000 MHz·km (~ 0.5 ns/km) and of 2-2.5 GHz·km (~ 0.2 ns/km) respectively can be found in practical low loss graded-index fibres. With LED sources

($\Delta\lambda = 20\text{-}50$ nm), material dispersion is the predominant effect and values of f_λ ranging between 100 and 200 MHz·km (2-5 ns/km) have to be considered.

2.2.3 *Optical sources*

The two main optical sources suitable for telecommunication applications are the high radiance LED's and the semiconductor injection lasers LD. Both are easy to use since they can be directly modulated, the optical output power varying almost linearly with the input driving current.

LED's are the optical sources at present available with a reliability suitable for transmission system applications (mean lifetime higher than 100,000 hrs). They are isotropic semi-coherent light sources which, according to the semiconductor utilized, can operate either in the first or in the second window, with a spectral width ranging from 20 to 50 nm and from 40 to 100 nm in the two cases respectively. Radiance values can vary from 20 to 100 $W/sr \cdot cm^2$ and from 10 to 50 $W/sr \cdot cm^2$ for LED's operating in the first and in the second window respectively.

The power coupling efficiency with the fibre is an important factor in system design. It can easily be shown that the maximum power $\hat{P}_i$ injectable into a fibre with a core diameter d_f and numerical aperture Δ by a LED, with an emitting area diameter d_s and with a radiance in air R_a, is approximately given by [9]:

$$\hat{P}_i \simeq \left(\frac{x}{x+2}\right) R_a \left(\frac{\pi}{2} d\Delta\right)^2 \qquad (6)$$

where $d = \min(d_f, d_s)$, and x is the exponent of the index profile ($x = 2$ for parabolic profile, $x \to \infty$ for step-index profile).

The relationship (6) is given in Fig. 4 for the case of a step-index fibre ($x \to \infty$). The product $d\Delta$, otherwise known as the « acceptance factor », has been assumed as a parameter. For example, for Corning fibres, with $d_f = 62$ μm and $\Delta \simeq 0.2$, the acceptance factor is about 12 μm, when $d_s > d_f$.

From (6), it results that the injectable power is proportional to the radiance and to the square of the acceptance factor. As the number of guided modes increases with the same law as a function of the latter, it follows that the injectable power is pro-

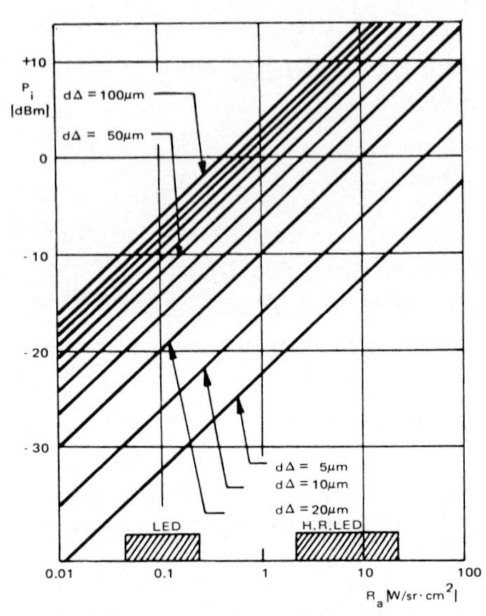

FIG. 4 – Power injected $\hat{P}_i$ into a step-index fibre versus LED radiance for different acceptance factors $d\Delta$.

portional to the number of modes and inversely proportional to the fibre bandwidth. Therefore, when a large bandwidth is needed, to increase the input power it is better to use high radiance LED's than high acceptance fibres. From Fig. 4 it can be seen that with high radiance LED's it is possible to inject into Corning-type fibres optical powers of 50 μW (— 13 dBm) or higher.

Due to non-linearity phenomena in the electrical to optical conversion, LED sources are better suited for digital transmission, for which it is advisable to use a two-level signal format. A small prebias can be useful to improve the linearity and also the rise time in on-off signalling.

LED's bandwidth is physically limited by the recombination time of carriers and can vary from 10 to 100 MHz; the cut-off characteristic is similar to that of a single pole *RC* network.

To overcome the LED's intrinsic bandwidth limitation and that due to the non-coherence of the light emission (material dispersion), laser diodes have to be used for high-frequencies signal transmission.

Laser sources have many advantages over LED's, as they have much higher emitted beam radiance and coherence. The total emitted optical power ranges from 1 to 10 mW for devices radiating in the first window ($\lambda \simeq 0.85\ \mu m$); lasers suitable for the second window are also reported in the literature with a maximum output power of about 4 mW. As the spectral width of the emitted light is about 1-2 nm, the use of these sources allows the material dispersion effects to be reduced. The radiation does not have an isotropic pattern and consequently the optical power injectable into the fibre can be, for small-diameter low numerical aperture fibres, up to two orders of magnitude higher than with LED's. With Corning-type fibres the power coupled into the fibre can vary from 0.5 to 5 mW (-3 to $+7$ dBm).

For a correct laser operation the driving current must be greater than a threshold ranging between 50 and 200 mA. Normally, laser diodes are biased slightly above the threshold to achieve a better linearity and a larger bandwidth. In these conditions, bandwidths up to 500-1000 MHz can be reached.

Due to non-linearity effects, laser sources are also more suitable for two-level digital transmission. Multilevel signals could be utilized only when considerable fibre bandwith limitations arise.

Laser behaviour is very sensitive to temperature and has to be stabilized by using suitable feedback, as will be shown in the chapter dedicated to circuit considerations.

At present, the main drawback of laser sources is their limited lifetime, about 10,000 hrs as a maximum, while for telecommunication applications the minimum requirement is 100,000 hrs. However, taking into account the very rapid technological progress in this area, it is expected that this problem will be overcome in the next few years.

2.2.4 *Optical detectors*

The most interesting optical detectors for optical fibre transmission are solid state Si PIN photodiodes and Si or Ge avalanche photodiodes APD.

The basic parameter of the photodetector is the responsivity, i.e., the output current per unit of incident power. For Si diodes this can vary from 0.4 to 0.6 A/W in the first window, and is below 0.2 A/W in the second window. Ge photodiodes can have a responsivity of about 0.6 A/W in the second window, but they are not suitable because of their high noise penalty.

PIN photodiodes require only moderate bias voltage (a few tens of Volts) in contrast with avalanche photodiodes. Their bandwidth is basically limited by the transit time effect; this can be reduced with the thickness of the junction, but in this way the responsivity is also reduced. Therefore, a compromise is necessary. Optimized diodes with a responsivity of about 0.6 A/W and a rise time of 0.1 ns have been reported.

Avalanche photodiodes are, of course, more sensitive than PIN diodes, but require high bias voltage (100-300 V) and possibly a stabilization of the operating point, which greatly influences the gain G. Normal values of G are about 40-100. A limitation of the maximum avalanche gain attainable at high frequencies is a result of a finite gain·bandwidth product, due to the transit time of the carriers through the multiplicative region. By careful design, Si and Ge avalanche diodes with gain·bandwidth products between 20 and 100 GHz can be obtained.

A good small signal model [10] of a photodetector is the equivalent circuit shown in Fig. 5, where C_d is the junction capacitance

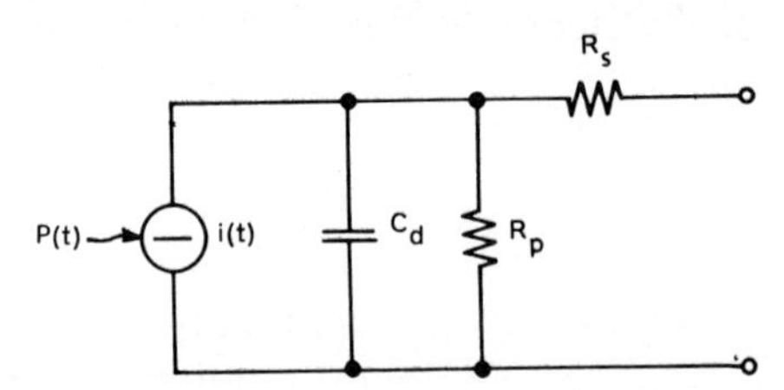

FIG. 5 – Equivalent circuit of an optical detector.

of the diode (1-3 pF), R_p is the equivalent parallel resistance ($\sim 100\ \text{M}\Omega$) and R_s is the contact resistance ($\sim 50\ \Omega$). The current generator $i(t)$ represents the generation of carriers by optical incident power.

In the more general case of avalanche photodiodes, the current generation process can be divided into two steps.

In the first step, free carriers (electrons) are generated by the incident light by means of the photoelectric effect with a statistical law corresponding to a Poisson process. The average rate of « primary » electrons is proportional to the incident optical power giving rise to the signal current. The types of noises introduced at this stage are quantum or shot noise due to the Poisson process, and dark current noise due to the spurious inverse current which flows even in the absence of illumination.

In the second step, each primary electron can produce many secondary electrons by the mechanism of collision ionization in the high field region of the junction. In this way, the signal current is multiplied by the average avalanche gain G. To take into account the excess noise introduced by the random nature of the avalanche process, this can be represented as an equivalent amplifier, with a gain G and a noise figure F_G.

The most important noise source at the output of an avalanche photodetector is the quantum current, with a mean square value given by:

$$\overline{i_q^2} = 2qi(t)\,bG^2 F_G \tag{7}$$

where q is the electron charge equal to $1.6 \cdot 10^{-19}$ C, $i(t)$ is the signal current due to primary electrons and b is the signal bandwidth.

A noise contribution of minor importance is given by the dark current noise that can be divided into two parts: the noise due to the surface current I_s, which is not involved in the avalanche process, and the noise due to the bulk current I_b, which flows through the avalanche gain region. The mean square values of these noises are given by:

$$\overline{i_s^2} = 2qI_s b \tag{8}$$

$$\overline{i_b^2} = 2qI_b G^2 F_G b \tag{9}$$

In avalanche diodes I_s and I_b are about 10^{-7} and 10^{-10} A respectively; in PIN diodes ($G = 1$) values of $I_s + I_b$ of 10^{-8} A are usual.

It has been shown theoretically [11] and confirmed by experimental results [12] that the noise figure of the avalanche process F_G is a function of the average gain G and of the hole to electron ionization ratio k, given by:

$$F_G = G\left[1 - (1-k)\frac{(G-1)^2}{G^2}\right] \tag{10}$$

Typical values of k range between 0.03 and 0.1 in Si diodes and are close to unity in Ge diodes. According to (10), F_G versus G

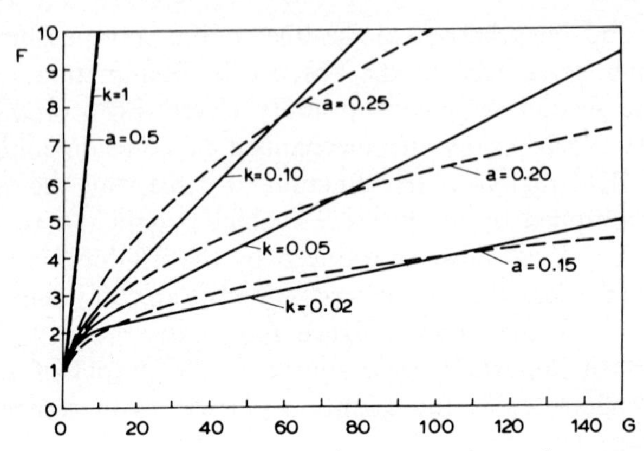

FIG. 6 – Noise figure F_G of the avalanche process versus the average gain G. Exact relation on solid curves; approximate relation on dashed curves.

for different k is plotted in Fig. 6 (solid lines). F_G is often approximated in the form:

$$F_G \simeq G^{2a} \tag{11}$$

with a varying between 0.15 and 0.25 for Si diodes and equal to about 0.5 for Ge diodes. This relation will be used in the following, as it is more suitable for analytical processing. It is also plotted in Fig. 6 (dashed lines), from which it can be seen that with a correct choice of a the approximation is satisfactory for usual values of G ($G = 20$-100).

2.3 Optical fibre system analysis

2.3.1 *General*

Having recalled the main characteristics of optical fibres and optoelectronic components, design aspects of optical fibre transmission systems are here considered. Reference will be made to digital transmission to which fibre systems are more suited.

As the optical transmitted power is proportional to the drive current and the output current at the receiver is proportional to the received power, an intensity modulated optical fibre system can be regarded as equivalent to a baseband transmission system.

For the evaluation of the repeater spacing it is necessary to relate the error probability in the signal detection with the mean received optical power (equal to the mean transmitted power attenuated by αL). In this evaluation it must be taken into account that the quantum noise given by (7) is a function of the signal amplitude and results from a filtered non-stationary Poisson process with non-Gaussian amplitude probability density.

A theory for the exact computation of the error probability in the presence of quantum noise has been developed in detail in the preceding Chapter. A calculation of the upper bound of the error probability generally applicable is reported in [13].

However, the results of both theories are given in the form of integral equations, which must be computed numerically.

To overcome this difficulty, approximate relationships are here derived which are based on the assumption that quantum noise is a stationary Gaussian process, independent of the signal level and proportional to the mean received optical power [14, 15]. This approximation makes possible an analytical approach, allowing a straightforward understanding of the influence of the different parameters in the system design. Moreover, it has been proved to be valid for practical purposes in the repeater spacing evaluation for most cases of interest [16], and is conservative with respect to the theories mentioned above.

For the reasons given above relating to the non-linearity of the optical sources, we shall be referring in the following section to two-level digital transmission, unless otherwise stated.

2.3.2 *Optical receiver*

Figure 7 shows the equivalent circuit of a typical optical receiver.

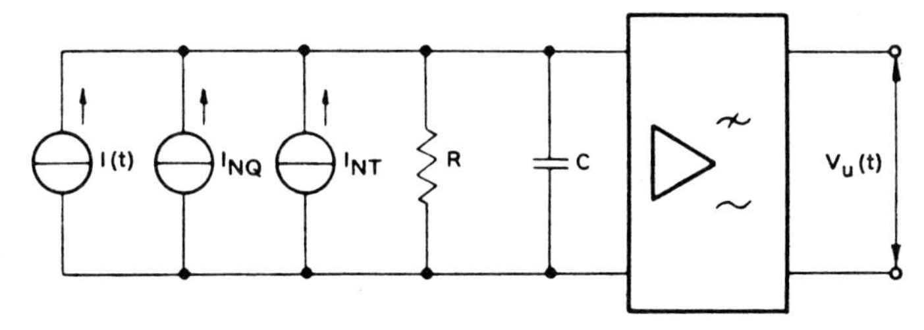

FIG. 7 – Equivalent circuit of an optical receiver.

It consists of:

— a signal current generator, with:

$$i(t) = \varrho G P(t) \tag{12}$$

where $P(t)$ is the optical input power, G is the avalanche gain ($G = 1$ for PIN diodes) and ϱ is the responsivity;

— an RC network, where R is the photodetector load resistance and C is the sum of the diode capacitance and the amplifier input capacitance, with a cut-off frequency f_c equal to a portion δ of the amplifier noise equivalent bandwidth B_N:

$$f_c = \frac{1}{2\pi RC} = \delta B_N \tag{13}$$

— an amplifier-equalizer which compensates for loss and distortion of the fibre and of the RC network in the signal band.

In Fig. 7 a voltage amplifier is indicated. Other amplifier schemes showing the same performance are possible and will be discussed later in the chapter on circuits;

— a quantum noise current generator, with a flat power spectral density equal, for the assumption made and taking into account (7) and (11), to:

$$W_{iq} = 2q\varrho G^{2(1+a)} P_r \tag{14}$$

where P_r is the mean received optical power, and q and a, as already defined, are the electronic charge and the excess noise factor of the avalanche process;

— a thermal noise current generator, with a power spectral density given by:

$$W_{it} = \frac{4KTF}{R} \tag{15}$$

where K is the Boltzmann constant equal to $1.38 \cdot 10^{-23} J \cdot {}^{\circ}K^{-1}$, T is the absolute temperature that can be assumed equal to 293 °K, and F is the amplifier noise figure referred to the noise produced by the input resistance R. A more detailed amplifier noise sources analysis is given in Chapter 3 on circuit considerations.

Taking into account (13), relation (15) can be written in the following form:

$$W_{it} = 8\pi KTB_N(\delta FC) \tag{16}$$

The product δFC is a very useful parameter which will be utilized in the following, as it indicates the accuracy of the receiver design, in the sense that the lower is δFC the better is the receiver. Values of δFC of about 1 pF and of about 10 pF can be obtained for medium and high capacity digital systems respectively.

The effect of the photodiode dark current can be, in a first approximation, ignored.

2.3.3 *Required optical power*

The amplifier-equalizer is generally designed in such a way as to produce output pulses without intersymbol interference; in the following it is assumed that the output pulse spectrum is a raised cosine type with a given roll-off β.

In these conditions, having assumed the noise to be additive Gaussian, the error probability is related to the peak signal-to-noise ratio σ at the decision point by the well known erfc function.

Let us assume that the transmitted optical pulses are rectangular with a peak value $\hat{P}_t$ and a duty cycle d_c, and that no optical power is emitted in the absence of pulses; in this case the peak signal current $\hat{I}$ at the decision point is given by:

$$\hat{I} = \varrho G d_c \hat{P}_t \exp[-2\alpha L] \tag{17}$$

Taking into account that the mean transmitted power P_t is related to $\hat{P}_t$ by:

$$P_t = \frac{1}{h} d_c \hat{P}_t \tag{18}$$

where h is the inverse of the probability of « 1 », and that:

$$P_r = P_t \exp[-2\alpha L] \tag{19}$$

$\hat{I}$ can be given, as a function of the mean received power, by:

$$\hat{I} = h\varrho G P_r \tag{20}$$

where $h = 2$ for equiprobable symbols. This condition is satisfied with the line codes usually adopted in optical fibre systems.

Relationship (20) is also valid for multilevel digital transmission and for baseband analogue transmission (e.g. TV signal) with h depending on the statistics of the signal and being generally higher than 2. Even for these cases h can be assumed conservatively equal to 2.

From (20), with $h = 2$, and from (14) and (16) the peak signal-to-noise ratio at the decision point is given by:

$$\sigma = \frac{2\varrho G P_r}{\sqrt{2q\varrho G^{2(1+a)} P_r B_N + 8\pi KTB_N^2(\delta FC)}} \tag{21}$$

As σ is generally fixed by the quality desired on the detected signal (e.g. error probability in digital transmission) it is better to invert (21) in order to have the mean optical power required for assuring a given σ. By doing so, we obtain:

$$P_r = \frac{qB_N G^{2a} \sigma^2}{4\varrho} \left[1 + \sqrt{1 + \frac{32\pi KT(\delta FC)}{\sigma^2 a^2 G^{2(1+2a)}}}\right] \tag{22}$$

From (22) the required power for a given σ results proportional to the noise equivalent bandwidth of the amplifier.

The required power per unit of bandwidth versus σ, for the case of a PIN diode ($G = 1$) with $\varrho = 0.5$ A/W, is plotted in Fig. 8 for different values of the product δFC.

From Fig. 8 it can be seen that high values of σ require high values of optical power which are related to high values of quantum noise. In this region the power tends to increase by 1 dB/dB with the desired σ (as is clear from the asymptote corresponding to $\delta FC = 0$ pF, i.e. in the absence of thermal noise). This behaviour, known as « quantum noise limited », is typical in the analogue transmission of TV signals, where a peak signal-to-noise ratio higher than 50 dB is required. In the case of low values of σ, typical of a digital transmission where for two level signals a σ value of about 23 dB corresponds to an error probability

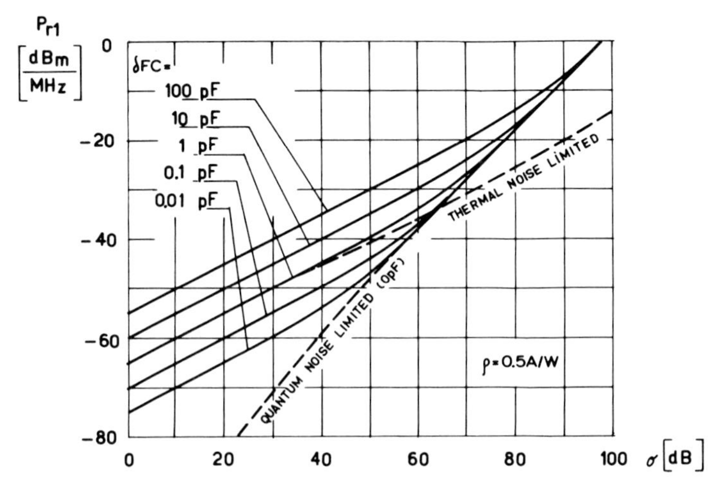

FIG. 8 – Optical mean power per unit of bandwidth P_{r1} versus signal-to-noise ratio σ at the decision point for a PIN diode.

lower than 10^{-10}, thermal noise is predominant and the required power increases by 0.5 dB/dB both with σ and with the parameter δFC.

In the case of avalanche photodiodes, from (22) it can be seen that for every value of σ there is an optimum value of G which minimizes the required optical power. Making equal to zero the derivative of (22) versus G, it can be deduced that:

$$G_{\text{opt}} = \left[\frac{8\pi KT(\delta FC)}{a(1+a)\,q^2\sigma^2}\right]^{1/2(1+2a)} \tag{23}$$

From (23) it results that the optimum gain is independent of bandwidth; moreover, for high values of σ, G_{opt} can be less than 1: this condition is obviously unrealizable, so (23) is valid only when $G_{\text{opt}} > 1$. This fact can be explained considering that while the signal current is proportional to G (12), the quantum noise current increases more rapidly with $G^{(1+a)}$ (14); so when quantum noise is predominant (high values of σ) it is not convenient to use avalanche photodiodes.

It is interesting to evaluate the maximum optical received power saving $\Delta P_{\max}$ that can be obtained with an avalanche photodiode with respect to a PIN photodiode. From (22), making the ratio between P_r computed for $G = 1$ and P_r computed for

$G = G_{opt}$, we obtain:

$$\Delta P_{max} = \frac{a^{a/(1+2a)}}{(1+a)^{(1+a)/(1+2a)}} \cdot \frac{\frac{1}{2} + \sqrt{\frac{1}{4} + \frac{8\pi KT(\delta FC)}{q^2 \sigma^2}}}{\left[\frac{8\pi KT(\delta FC)}{q^2 \sigma^2}\right]^{a/(1+2a)}} \tag{24}$$

Just as G_{opt} is independent of bandwidth, so is ΔP_{max}. In Fig. 9 ΔP_{max} and G_{opt} are plotted versus σ for different values of

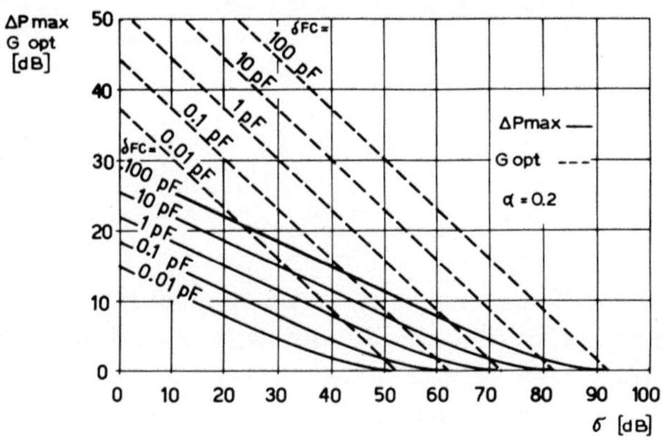

FIG. 9 – Maximum power saving ΔP_{max} and optimum gain G_{opt} attainable by an avalanche photodiode versus signal-to-noise ratio σ at the decision point.

δFC and for an excess noise factor $a = 0.2$ (typical value for Si diodes). For evaluating the influence of this latter factor in Fig. 10 ΔP_{max} and G_{opt} are plotted for different values of a in the typical case of $\delta FC = 1$ pF. From the preceding relations and figures it can be seen that the lower the amplifier quality (δFC high) or the better the photodiode (a small), the higher is the power saving attainable with avalanche diodes, even if higher G_{opt} values are required. As a consequence, the choice of a good low noise avalanche photodiode can then partly compensate for a low accuracy in the amplifier design. As already seen, the advantage of using avalanche photodetectors decreases with the required σ, and for high σ values is practically negligible.

For digital two-level transmission, where a σ of about 23 dB is required, the power saving attainable with a typical avalanche photodiode ($a = 0.2$) and a well-designed amplifier ($\delta FC = 1$ pF),

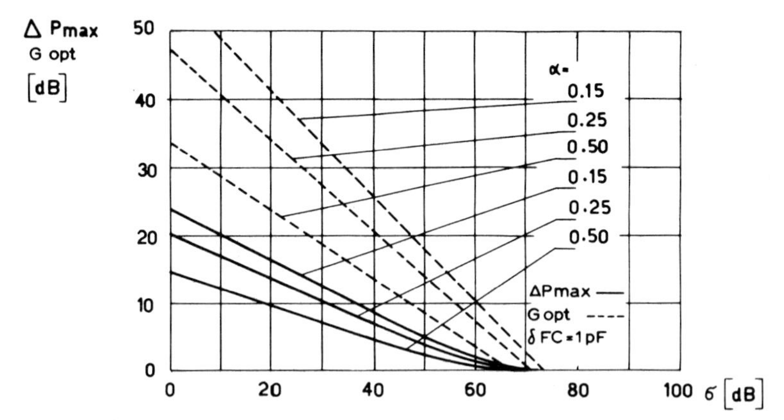

FIG. 10 – Comparison among three types of avalanche diodes as regards the maximum power saving ΔP_{max} and the optimum gain G_{opt}.

is about 13 dB; the necessary avalanche gain is about 34 dB, which is a value below the maximum attainable in commercial diodes (40-50 dB).

By substituting (23) into (22), the minimum received power P_{rm} required for a desired σ can be obtained, given by:

$$P_{rm} = \frac{B_N q^{1/(1+2a)}}{2\varrho} \left[\frac{8\pi KT(\delta FC)}{a}\right]^{a/(1+2a)} [(1+a)\sigma^2]^{(1+a)/(1+2a)} \quad (25)$$

The above relation is valid only for $G_{opt} \geqslant 1$, that is for:

$$\sigma^2 \leqslant \frac{8\pi KT(\delta FC)}{a(1+a)q^2} \quad (26)$$

In Fig. 11 the minimum required power per unit of bandwidth versus σ, computed from (25) (for $G_{opt} > 1$) and from (22) ($G = 1$) is given for different values of δFC in the case of $\varrho =$ $= 0.5$ A/W and $a = 0.2$. It can be seen that the avalanche diode behaviour is practically « quantum noise limited » even for low σ values, where the required optical power increases by 0.86 dB/dB with σ. Moreover, it can be noted that in this region accuracy in receiver design has far less effect in performance than in the case with PIN diodes; in fact, the required power increases with the parameter δFC only by 0.14 dB/dB.

Before concluding, it seems worthwhile to have an expression for the required power as a function of the equivalent input

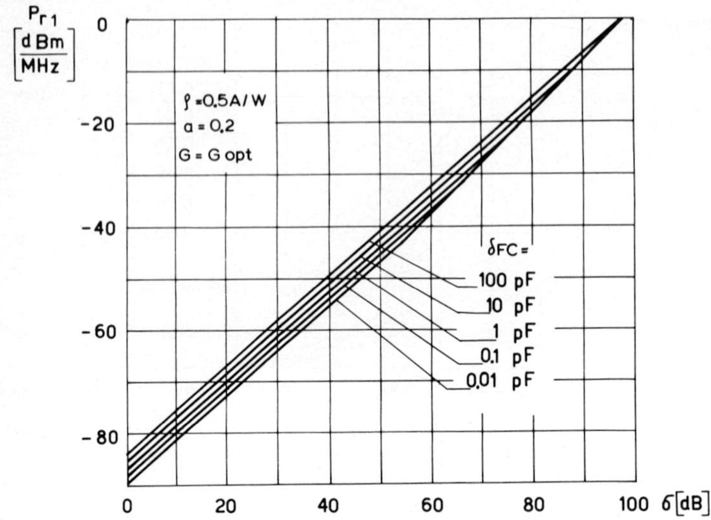

FIG. 11 – Optical mean power for unit of bandwidth P_{r1} versus signal-to-noise ratio at the decision point for an avalanche diode with $G = G_{opt}$.

thermal noise current power density, given by (15), which is a parameter of the receiving amplifier, and easier to measure than the δFC parameter. In this case (22) and (25) become: PIN diode

$$P_r = \frac{qB_N G^{2a} \sigma^2}{4\varrho} \left[1 + \sqrt{1 + \frac{4W_{it}}{B_N \sigma^2 q^2 G^{2(1+2a)}}}\right] \tag{27}$$

avalanche diode

$$P_{rm} = \frac{q^{1/(1+2a)}}{2\varrho} \left[\frac{W_{it}}{a}\right]^{a/(1+2a)} [(1+a)\sigma^2 B_N]^{(1+a)/(1+2a)} \tag{28}$$

a - *Effect of imperfect source modulation*

As already mentioned, the optical emitted signal can not be completely extinguished in the optical sources (LED's or lasers); a pedestal is also introduced in most cases (especially with lasers) in order to achieve a better linearity and a larger bandwidth.

Let us assume that in the absence of signal an optical power is emitted equal to a portion γ of the peak value $\hat{P}_t$. In this case the mean transmitted power P_t is given by:

$$P_t = \frac{d_c}{2}\left[1 + \frac{2\gamma}{d_c(1-\gamma)}\right] \hat{P}_t \tag{29}$$

The mean received signal power P_{rs}, proportional to the signal peak amplitude at the decision point, is related to the total mean power P_r, proportional to the mean square value of the quantum noise, by:

$$P_{rs} = \frac{(1-\gamma)\, d_c}{(1-\gamma)\, d_c + 2\gamma}\, P_r \tag{30}$$

From (21) taking into account (30), it follows that the relations and the figures of the preceding paragraph are still valid in the presence of a light pedestal provided that, instead of the required σ, a value σ' is considered, given by:

$$\sigma' = \left[1 + \frac{2\gamma}{d_c(1-\gamma)}\right] \sigma\,. \tag{31}$$

A usual value of γ is around 0.1; in the common case of $d_c = 0.5$ σ' is 3.2 dB higher than σ.

b - *Effect of dark current noises*

Let us now consider the effect of the dark current noises i_s and i_b that have been ignored in the preceding considerations.

Taking into account (8) and (9), relation (21) becomes:

$$\sigma = \frac{2\varrho G P_r}{\sqrt{2qG^{2(1+a)}(\varrho P_r + I_b)\, B_N + 8\pi KT(\delta FC)\, B_N^2 + 2qI_s B_N}} \tag{32}$$

Making equal to zero the derivative of (32) versus G the optimum gain that maximizes σ can be deduced:

$$G_{\text{opt}} = \left[\frac{4\pi KT(\delta FC)\, B_N + qI_s}{aq(\varrho P_r + I_b)}\right]^{1/(2(1+a))} \tag{33}$$

By substituting (33) into (32) we obtain the maximum signal-to-noise ratio σ_M for a given P_r:

$$\sigma_M = \sqrt{\frac{a^{a/(1+a)}}{1+a}}\,\cdot$$

$$\cdot \frac{2\varrho P_r}{[2q(\varrho P_r + I_b)\, B_N]^{1/(2(1+a))}\, [8\pi KT(\delta FC)\, B_N^2 + 2qI_s B_N]^{a/(2(1+a))}} \tag{34}$$

It must be noted that Eq. (34) also relates the minimum required power P_{rm} to the desired σ.

Typical values for avalanche photodiodes are $I_b = 10^{-10}$ A and $I_s = 10^{-7}$ A; for PIN diodes usually $I_b + I_s = 10^{-8}$ A.

Inserting these values in (32) and (34), the required power per unit of bandwidth versus σ is given in Fig. 12, both for a

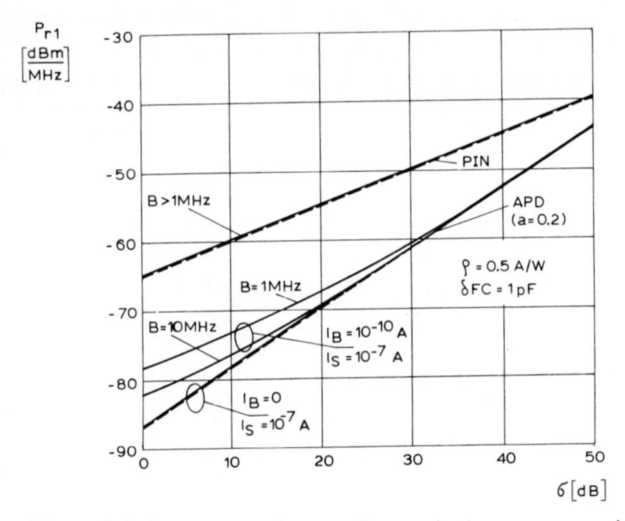

FIG. 12 – Effect of dark current noises on the required power per unit of bandwidth P_{r1} versus signal-to-noise ratio at the decision point, for PIN diode and avalanche diode ($G = G_{opt}$). Dashed lines refer to $I_D = I_L = 0$.

PIN diode ($G = 1$) and for an avalanche diode ($G = G_{opt}$) with $a = 0.2$, in the case of $\varrho = 0.5$ A/W and $\delta FC = 1$ pF. As a reference, the curves relative to the case of $I_b = I_s = 0$ are also plotted.

It can be seen that in PIN diodes the dark current effect is practically negligible: the required optical power increases by about 0.1 dB with a 1 MHz bandwidth and decreases for larger bandwidths. In avalanche photodiodes the effect of I_s is negligible; the effect of I_b is noticeable only at low σ and small B_N and becomes negligible for medium and high σ and large B_N. For $\sigma = 23$ dB the increase in the required optical power is about 1 dB with 1 MHz bandwidth and becomes negligible for larger bandwidths.

Accordingly the dark current effect can be ignored in the systems of interest.

2.3.4 *Optical fibre system repeater spacing*

Having deduced the mean required optical power P_r for a given σ (and then for a given error probability) both for PIN diodes ($G = 1$) (22) and for avalanche diodes ($G = G_{opt}$) (25), the repeater spacing can be evaluated from (19). However, the difficulty in this evaluation is that P_r depends on the receiver equivalent noise bandwidth B_N which, in the presence of fibre cut-off, is a function of the length L.

Let us first consider that no bandwidth limitation is caused by the fibre. In this case, known as « attenuation limited », the equivalent noise bandwidth depends only on the transmitted pulse duty cycle d_c and on the roll-off β of the signal spectrum at the decision point, and can be assumed equal to:

$$B_N = B_0 = k_b f_r \tag{35}$$

where f_r is the symbol rate and k_b is a constant equal to about 0.5 for usual values of d_c and β.

As the required power at the receiver is, in this condition, independent of the length L (1), this latter can be easily calculated from (19) and is given by:

$$L = \frac{1}{\alpha}\,[10 \log P_t - 10 \log P_{r0}] \tag{36}$$

where P_{r0} is the power at the receiver computed by (22) and (25) in the condition specified by (35), and α is the attenuation in dB/km.

In Fig. 13 the required power P_{r0} for a two level digital transmission with an error probability $P_e \leqslant 10^{-10}$ ($\sigma = 23$ dB) is plotted versus f_r for both PIN ($G = 1$) and avalanche ($G = G_{opt}$, $a = 0.2$) diodes, in the case of $\varrho = 0.5$ A/W and $\delta FC = 1$ pF and for

(1) In the presence of source cut-off, B_N is always independent of L and can be computed as a function of f_r.

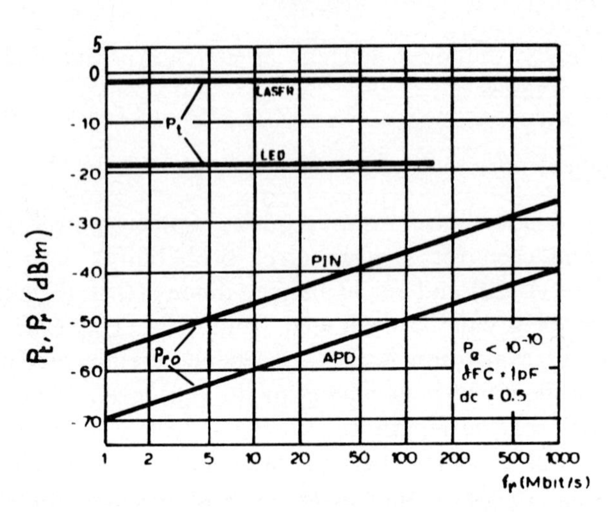

FIG. 13 – Transmitted and received optical power levels (P_t, P_{r0}) versus symbol rate f_r in absence of fibre cut-off for a two-level digital transmission with error probability less than 10^{-10}.

$B_0 = f_r/2$. In the same figure the levels of the mean power injected into the fibre by laser and LED sources are given, assuming a peak power value equal to $+3$ dB and -12 dBm in the two cases respectively and a duty cycle $d_c = 0.5$. LED sources cannot be used at high symbol rates for their intrinsic bandwidth limitation.

The difference between the transmitted and received power levels for the symbol rate of interest gives the total attenuation available on the fibre, which divided by α gives, according to (36), the length of the repeater section.

From (22) and (25) and from Fig. 13, the total available loss in the absence of fibre dispersion for a given set of optoelectronic components and for a fixed accuracy of the receiving amplifier has the following form (in dB):

$$P_t - P_{r0} = A_L - 10 \log f_r \tag{37}$$

where A_L represents the available loss for a unitary symbol rate. For example, in the case of Fig. 13, A_L is equal to 67 dB/Mbit/s for the laser-APD combination and equal to 52 dB/Mbit/s for the LED-APD combination.

It is worth noting that in the design of a link a certain margin must be kept for taking into account signal degradations (residual intersymbol interference, jitter, low frequency cut-off, etc.), device imperfections (incertainty in the decision level and instant) and fibre attenuation spread. A margin of 3-5 dB optical power should be adopted.

Let us now consider the more general case where a bandwidth limitation due to the fibre arises. Equation (19) can be written as follows:

$$\alpha L = 10 \log \frac{P_t}{P_r} = 10 \log \frac{P_t}{P_{r0}} - 10 \log \frac{P_r}{P_{r0}} \tag{38}$$

where the first term is the total attenuation in dB available in the absence of fibre cut-off (36), which can be read for each symbol rate directly from Fig. 13, while the second term represents the so called « power penalty » due to the fibre cut-off.

It can be shown by taking into account (22) and (25) that the power penalty η_P in dB is the sum of two terms and is given by:

$$\eta_P = 10 \log \frac{P_r}{P_{r0}} = 10 \log \frac{B_N}{B_0} + 10 \log M_q \tag{39}$$

where M_q is a correcting factor, which takes into account the dependence of the quantum noise power on the adjacent pulses when intersymbol interference is present on the optical signal at the input of the receiver.

The equivalent noise bandwidth referred to the Nyquist frequency is equal to:

$$B_N = \int_0^\infty \left[\frac{\pi d_c x}{\sin \pi d_c x}\right]^2 \exp\left[2 \cdot \ln 2 \cdot \left(\frac{f_r}{f_t}\right)^2 x^2\right] [C(x, \beta)]^2 \, dx \qquad x = f/f_r \tag{40}$$

where f_t is the cut-off frequency of the fibre, given by (4), and $C(x, \beta)$ is the well known raised cosine function with a roll-off β.

The correcting factor M_q can be evaluated empirically by comparing the present approximate theory with the more precise ones, such as that reported in the preceding chapter or that given in [13]. A good approximation has been found for M_q that leads to conservative results [2] with respect to the exact theory, and is given by:

$$M_q = 2 - \exp\left[-0.1\left(\frac{f_r}{f_t}\right)^2\right] \tag{41}$$

From (40) and (41) it can be seen that the power penalty given by (39) is a function of the ratio f_r/f_t between the pulse repetition frequency and the fibre cut-off frequency.

Power penalty versus f_r/f_t is shown in Fig. 14 for different values of the roll-off β of the raised cosine signal spectrum at the decision point, assuming a duty cycle $d_c = 0.5$ of the transmitted pulses and no bandwidth limitation in the optical source.

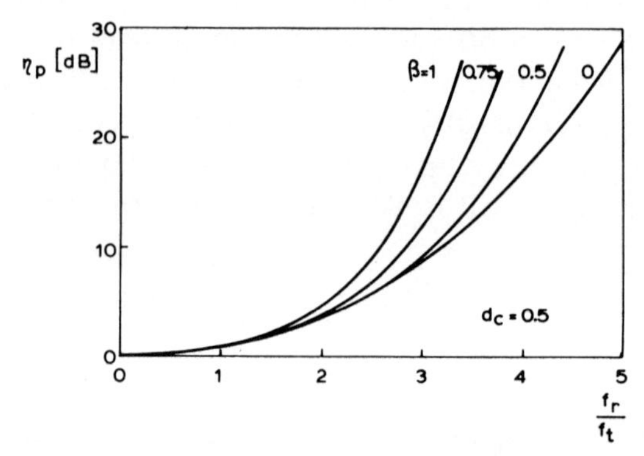

FIG. 14 – Power penalty η_P versus f_r/f_t, for different values of the roll-off of the raised cosine signal spectrum at the decision point, assuming a duty cycle $d_c = 0.5$ and no bandwidth limitation in the optical source.

It can be seen that for low values of f_r/f_t the power penalty is practically independent of β, while for high values of f_r/f_t the power penalty decreases as β decreases. On the other hand, it must be noted that the choice of β is also influenced by circuit complexity and timing recovery facility, which are less complex for high values of β. A good compromise for the value of β in the presence of fibre cut-off can range between 0.6 and 0.9.

Coming back to the repeater spacing evaluation, since the power penalty depends on the length L of the fibre through f_t, as results from (4), relation (38) has the form of an integral

[2] The discrepancy with respect to the theory of [13] is less than 1 dB.

equation. The difficulty of solving this equation can be overcome by means of a graphical method.

Let us rewrite relation (38) in the following form, where all quantites are expressed in dB:

$$P_t - \alpha L = P_{r0} + \eta_P(f_r/f_t) \tag{42}$$

Both left and right side terms of Eq. (42) give the received power P_r, and both depend on the fibre length L as f_t is a function of L according to (4). For a given symbol rate it is then possible to plot the two terms of (42) versus L on the same graph: the abscissa corresponding to the intersection of the two curves gives the achievable repeater spacing. This is shown in Fig. 15, which

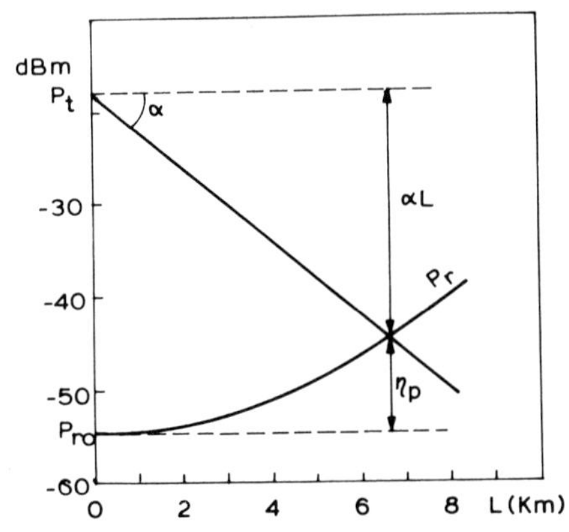

FIG. 15 – Required power P_r ($P_e = 10^{-10}$) and available power $P_t - \alpha L$ at the receiver versus link length for a 34 Mb/s digital transmission system using a LED ($\Delta\lambda = 50$ nm at 820 nm) an avalanche photodetector, and a fibre with $\alpha = 4$ dB/km.

refers to a 34 Mbit/s digital transmission system using a LED source with $\Delta\lambda = 50$ nm at 0.85 μm, an avalanche photodetector and a fibre with $\alpha = 4$ dB/km and with material and modal dispersions of 100 ps/km·nm($\Delta\lambda$) and of 0.5 ns/km respectively. The power penalty has been computed for a duty cycle $d_c = 0.5$ of the transmitted pulses and for a roll-off $\beta = 1$ of the equalized signal spectrum. The transmitted and received power P_t and P_{r0} have been taken, according to Fig. 13, equal to − 18 dBm and − 54.5 dBm respectively. The resulting repeater spacing for such a system is 6.6 km.

The graphical method described above allows an accurate evaluation of the repeater spacing and is very useful in the design of a well defined system at a particular symbol rate. For a straightforward evaluation of the repeater spacings achievable at different symbol rates with different optical fibres and components, explicit formulas, even if approximate, are better suited.

Looking at Fig. 14, it results that the power penalty behaviour versus f_r/f_t can be well approximated, at least up to a certain value of f_r/f_t, by a parabolic function. We can assume:

$$\eta_P \simeq c\left(\frac{f_r}{f_t}\right)^2 \tag{43}$$

where c is a constant depending on the roll-off β of the equalized signal spectrum. For example $c = 1.5$ for $\beta = 1$ and $c = 1$ for $\beta = 0.5$. The approximation error with respect to the corresponding exact curves is less than 1 dB for f_r/f_t values lower than 3-4.

From (43), taking into account (4), relation (38) becomes, with all quantities expressed in dB:

$$\alpha L = A_L - 10 \log f_r - c\left(\frac{f_r}{f_\lambda}\right)^2 L^2 - c\left(\frac{f_r}{f_m}\right)^2 L \tag{44}$$

where f_m and f_λ are the cut-off frequencies per unit length due to modal and material dispersion respectively and related to the pulse broadening by (5), and A_L is the available loss in the absence of fibre dispersion.

From the quadratic equation (44), we obtain:

$$L = \frac{1}{2}\left\{\sqrt{\left[\frac{\alpha}{c}\left(\frac{f_\lambda}{f_r}\right)^2 + \left(\frac{f_\lambda}{f_m}\right)^2\right]^2 + \frac{4(A_L - 10 \log f_r)}{c}\left(\frac{f_\lambda}{f_r}\right)^2} - \left[\frac{\alpha}{c}\left(\frac{f_\lambda}{f_r}\right)^2 + \left(\frac{f_\lambda}{f_m}\right)^2\right]\right\} \tag{45}$$

which gives the repeater section length as a function of f_r, α, f_m and f_λ for a certain choice of optoelectronic components and receiving amplifier design (on which A_L and c depend).

In many cases the bandwidth limitation of a fibre link is determined mainly by only one type of dispersion. For example, with LED sources operating in the first window, material dispersion is the predominant limiting factor, while in the second window, especially with laser sources, the limiting factor is due to modal dispersion. It is quite useful to have a simplified expression of (45) for the evaluation of the repeater spacing.

Putting in (45) $f_\lambda = \infty$ or $f_m = \infty$, according to whether modal or material dispersion effects are predominant, we obtain for the two cases respectively:

modal dispersion predominant

$$L = \frac{A_L - 10 \log f_r}{c\left(\dfrac{f_r}{f_m}\right) + \alpha} \tag{46}$$

material dispersion predominant

$$L = \frac{1}{2}\left[\sqrt{\frac{\alpha^2}{c^2}\left(\frac{f_\lambda}{f_r}\right)^4 + \frac{4(A_L - 10 \log f_r)}{c}\left(\frac{f_\lambda}{f_r}\right)^2} - \frac{\alpha}{c}\left(\frac{f_\lambda}{f_r}\right)^2\right]. \tag{47}$$

For example, Fig. 16 shows the repeater spacing computed by means of (45) for different symbol rates in the case of two-level digital transmission with an error probability $P_e \leqslant 10^{-10}$.

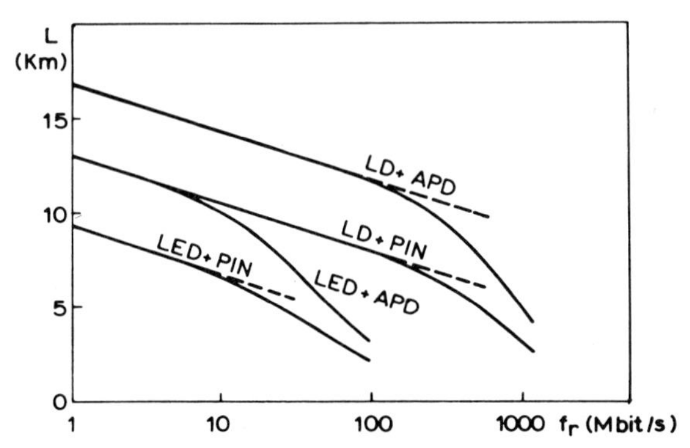

FIG. 16 – Repeater spacing L versus bit rate f_r for a two-level digital transmission and for different system configurations corresponding to an error probability less than 10^{-10}.

Both laser diodes ($\Delta\lambda \simeq 2$ nm) and LED's ($\Delta\lambda \simeq 50$ nm) operating in the first window ($\lambda \simeq 0.85$ nm) with no frequency cut-off are considered, giving pulses at the input of the fibre with a duty cycle $d_c = 0.5$ and a peak power of $+3$ dBm and -12 dBm respectively, as for Fig. 13; for lasers a pedestal with $\gamma = 0.1$ has been assumed. A multimode optical fibre has been considered with $\alpha = 4$ dB/km and with modal and material dispersion of 0.5 ns/km and 100 ps/km·nm($\Delta\lambda$). The corresponding cut-off frequencies are $f_m = 800$ MHz·km for modal dispersion and $f_\lambda = 2$ GHz·km or $f_\lambda = 100$ MHz·km for material dispersion with lasers or LED's respectively. As for Fig. 13, a detector responsivity $\varrho = 0.5$ has been taken, and for the avalanche photodiode an excess noise factor $a = 0.2$ and the optimum avalanche gain are assumed. The curves have been drawn for a receiver with $\delta FC = 1$ pF and a raised cosine signal spectrum with a roll-off $\beta = 1$ (corresponding to $c = 1.5$) at the decision point.

It can be seen that LED sources suffer considerably from the material dispersion effects which restrict the application of such devices to digital systems with a symbol rate below 34-40 Mbit/s. Laser diodes allow a much larger bandwidth and are suitable for use on digital transmission up to 800-1000 Mbit/s. With these sources, both material and modal dispersions affect system performance at high symbol rates.

TABLE I

Repeater spacing in km for European hierarchical bit rates.

bit rate [Mbit/s]		2.048	8.448	34.368	139.264
Optical fibre first window	LED + PIN	9	7	5	—
	LED + APD	12	10.5	7	—
	LD + PIN	12	10.5	9	7
	LD + APD	15.5	14	12.5	11
Copper cables	symm. pairs	2	—	—	—
	0.7-2.9 coax.	—	4	2	—
	1.2-4.4 coax.	—	—	—	2

The optical fibre repeater spacing attainable with different system configurations at the symbol rates of the different levels of the European digital hierarchy are given in Table I, together with the repeater spacing of the corresponding systems on conventional copper cables.

As can be seen, the repeater spacings attainable with optical fibres at the different symbol rates are always larger than in copper cable systems; this is one of the main factors contributing to the economy of fibre systems with respect to conventional systems.

Before concluding this section, it is worth noting that in the presence of a limited amount of intersymbol interference the power penalty to be paid for the fibre cut-off is practically independent of whether or not a compensation for the fibre dispersion is provided at the receiver. This fact is clearly shown in Fig. 17, where two

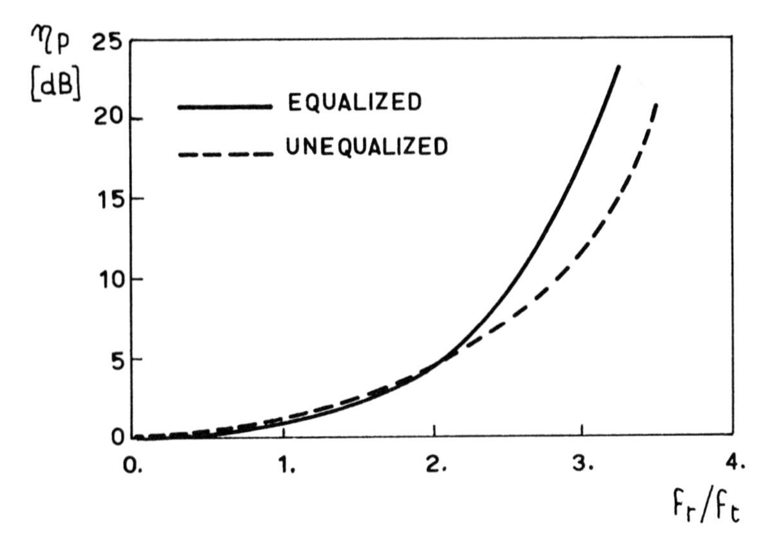

FIG. 17 – Comparison of power penalty η_P for equalized and unequalized optical links versus f_r/f_t.

conditions of power penalty have been reported according to whether the receiver is set up for delivering output raised cosine pulses with a unitary roll-off in any fibre cut-off condition or only in the absence of fibre dispersion. Figure 17 refers to a Gaussian fibre response and to transmitted pulses with a duty cycle $d_c = 0.5$. As can be seen, the receiver sensitivity loss is approximately the same in both cases for symbol rates up to 2-3 times the fibre cut-off frequency. This means that the power penalty due to the noise enhancement of the equalizing networks equals that due to the increase in the intersymbol interference when no equalization is provided. As a consequence, when a limited amount of fibre dispersion is foreseen, as in most practical situations, only a compromise fixed equalization can be envisaged for any system plant condition, provided that a flat automatic gain control can compensate for different section lengths.

From the above considerations it follows that, as pulse dispersion can vary from fibre to fibre and also sometimes be time dependent, it is advisable to limit the dispersion effects, in order to avoid the need for adaptive equalization, which would impair the advantages of optical fibre systems.

To this end, upper bounds to the repeater spacing due to fibre dispersion can be obtained by putting a limit on the maximum value of the ratio f_r/f_t between the symbol rate and the fibre cut-off frequency or of the ratio τ/T between the pulse broadening an the symbol time interval. On the basis of the preceding considerations and of simulations carried out on the computer, the maximum allowable value of f_r/f_t can be reasonably assumed equal to $\sqrt{10}$, which corresponds to a maximum pulse broadening $\tau = 1.4T$. The consequent power penalty ranges between 10 and 15 dB for the usual values of the roll-off (0.6-1) of the received signal spectrum. Moreover, the proposed figure also corresponds to the limit under which approximation (43), and consequently relation (45), are valid.

Imposing the constraint $f_r/f_t = \sqrt{10}$, from (4) we obtain:

$$\left(\frac{f_r}{f_\lambda}\right)^2 L^2 + \left(\frac{f_r}{f_m}\right)^2 L = 10 \tag{48}$$

Solving equation (48) the limit to the repeater spacing due to the fibre dispersion is given by:

$$L_{\text{Max}} = \frac{1}{2}\left[\sqrt{\left(\frac{f_\lambda}{f_m}\right)^4 + 40\left(\frac{f_\lambda}{f_r}\right)^2} - \left(\frac{f_\lambda}{f_m}\right)^2\right] \tag{49}$$

In a similar way as for (45), simpler expressions of (49) can be derived for the cases when only one type of dispersion is dominant, and are given by:

modal dispersion predominant

$$L_{\text{Max}} = 10\left(\frac{f_m}{f_r}\right)^2 \tag{50}$$

material dispersion predominant

$$L_{\text{Max}} = \sqrt{10}\,\frac{f_\lambda}{f_r} \tag{51}$$

For example, Fig. 18 shows the limiting curves for the repeater spacing at different symbol rates and for different system con-

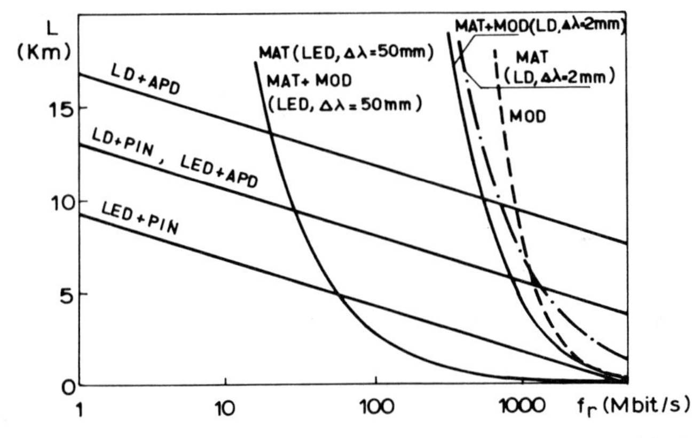

FIG. 18 – Limiting curves for the repeater spacing L for a two-level digital transmission and for different system configurations due to attenuation and pulse dispersion effects.

figurations due to pulse dispersion effects together with those due to attenuation, both computed with the same assumptions made for Fig. 16. The different limiting factors are much more evident than in Fig. 16. It can be seen that material dispersion strongly affects systems using LED sources, while with laser sources both dispersion effects contribute to limit the repeater spacing at high symbol rates.

2.4 Line coding for optical fibre systems

2.4.1 *General*

A last important item to be considered in the system design is the line coding, by which a redundancy on the information digital stream is introduced to ensure efficient timing recovery and error monitoring facilities at the receiver, and to suitably shape the signal line spectrum.

In the case of optical fibre communications, some peculiar factors must be taken into account.

Considering the large bandwidth available and, as already stated, the possible non-linear behaviour of the optical sources, it is advisable that the insertion of redundancy results in an increase of the line symbol rate, rather than in an increase of the number of transmitted levels. For this reason, two-level codes seem to be the most suitable; moreover, with such codes decision circuits are simple and gain instability effects are avoided.

As regards spectrum shaping, it must be noted that in optical fibre transmission the need to avoid d.c. components is less stringent than in wire communications, as the inter-device a.c. coupling in the receiver can be assured by means of capacitors. Unlike trasformers, these introduce a low frequency cut-off independent of the amplifier bandwidth, which can be easily compensated by suitably scrambling the digital signal.

In the following some classes of two level codes that seem attractive for transmission on optical fibres are presented.

2.4.2 *Parity check codes*

The most simple way of providing easy error monitoring facilities is to increase the symbol rate so as to provide a line signal frame structure, in which every m information bits n parity check bits are inserted.

Different frame formats can be envisaged. The n control bits can be grouped after the m information bits or distributed along the frame in order to be less affected by line errors. Moreover, the n additional bits can either carry the control information relative to the whole block of m information bits, or can be divided into p words each controlling the parity of the p interleaving subsequence in which the information block is supposed to be splitted. This allows adjacent errors to be detected.

At the receiver, the parity check bits are recognized and compared with those locally computed on the basis of the received information bits. An indication of the transmission quality is given by the parity violation rate V, which is related to the line error rate ε by the approximate relation:

$$V \simeq (m + n)\,\varepsilon \tag{52}$$

which is valid for ε less than 10^{-3}.

In order to identify and to remove the control bits at the receiver, a frame alignment circuit must be provided which operates in a similar way as that used in existing PCM terminals. Reframing time depends on the redundancy introduced into the line signal and, for redundancy value ranging from 5 to 1 per cent, it can vary from thousands to tens of thousands of bit time intervals.

The use of alignment circuits could hardly be justified, for economic and power consumption reasons, within the line repeaters, where simpler error monitoring facilities must be provided.

An attractive scheme for error detection in line repeaters has been recently proposed [17], based on the fact that the number of « ones » in a frame (including control bits) is even if no errors occur. A bi-stable circuit (flip-flop), which operates on each « one », turns to be in a particular state at specified instants when the parity bits are received. Over a period of time embracing many specified instants, a d.c. component is introduced in the flip-flop output. This d.c. component only changes when an error, or odd number of errors, occurs between two adjacent specified instants, because the flip-flop output changes to its alternative state at the specified instants after the error, this new state becoming the normal state. Error rate can then be monitored by the changes between the two levels of the d.c. component.

The main advantage of the parity control codes is that the increase in the line symbol rate required for a good error monitoring performance is very low (few percents). On the other hand, no power spectrum shaping is provided and a d.c. component is present. Moreover, a scrambler is needed on the information signal in order to assure a sufficient number of level transitions for an efficient timing recovery at the repeaters; this also reduces the possibility of d.c. wander.

A code of this kind with $m = 17$ and $n = 1$ is adopted in an STL experimental 140 Mbit/s optical fibre system [18]. A similar code with $m = 384$, $n = 4$ and $p = 2$ is also used, with d.c. restoring, in the ATT T4 coaxial line system at 274 Mbit/s [19].

2.4.3 *Block codes*

These codes, generally known as m B - n B codes, convert blocks of m binary digits into blocks of n binary digits where $n > m$.

The n bit words corresponding to the 2^m information words must, to be transmitted, be chosen in such a way as to facilitate clock extraction in the repeaters and to reduce low frequency components in the coded signal continuous spectrum.

The first requirement is satisfied if the all « 0 » and all « 1 » words are excluded in the coded signal.

As regards the second requirement, it is convenient to introduce the concept of the accumulated or running disparity D, that is, the difference between the number of « ones » and the number of « zeroes » which can be easily measured with an up-down counter. The d.c. component of the continuous spectrum can be cancelled, provided that the accumulated disparity is bounded. Moreover, the low frequency spectrum content is proportional to the disparity variation (i.e. the difference between the maximum and the minimum values of the accumulated disparity).

For the above reasons, the coded words must be chosen among those with the lowest disparity and the coding law must assure a limit in the disparity variation. Since for n odd there are no coded words with zero disparity, it is usually preferred to have n even; in this case the number of zero disparity words is equal to $\binom{n}{n/2}$.

If the number of n bit words with zero disparity is higher or equal to the number 2^m of words to be coded, a biunivocal correspondence can be set up between the information and the coded blocks. This is the case of the 6 B - 8 B code for which

$$\binom{n}{n/2} = 70 > 2^6 = 64 \tag{53}$$

When the above condition is not satisfied, in order to prevent the disparity from increasing without limit, a bimodal operation becomes necessary, by which some input words can generate two complementary (with opposite disparity) output words. The jump from one operating mode to the other is determined by the value of the accumulated disparity at the end of the preceding emitted word.

Examples of such codes are the 3 B - 4 B and the 5 B - 6 B codes. The coding table and state transition diagram of the 3 B - 4 B is shown, as an example, in Fig. 19.

a)

Binary word	State S_1 coded word	Word disparity	State S_2 coded word	Word disparity
0 0 0	0 1 0 1	0	0 1 0 1	0
1 0 0	0 1 1 0	0	0 1 1 0	0
0 1 0	1 0 0 1	0	1 0 0 1	0
1 1 0	1 0 1 0	0	1 0 1 0	0
0 0 1	0 0 0 1	−2	0 1 1 1	2
1 0 1	0 0 1 0	−2	1 0 1 1	2
0 1 1	0 1 0 0	−2	1 1 0 1	2
1 1 1	1 0 0 0	−2	1 1 1 0	2

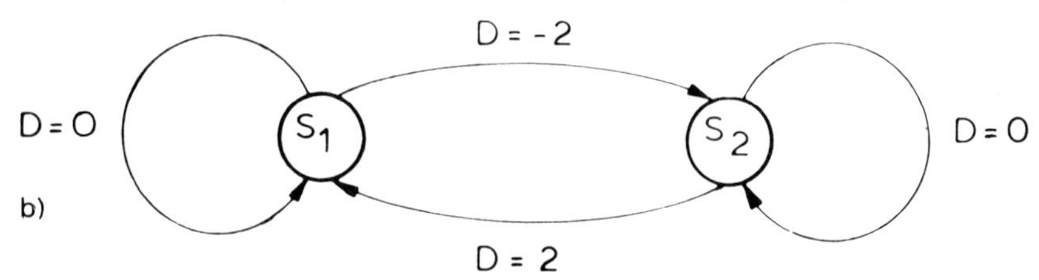

FIG. 19 – Coding table (*a*) and state transition diagram (*b*) for 3*B*-4*B* code.

It should be noted that in all block codes the two n bit zero disparity words with the first $n/2$ digits equal are excluded in order to reduce the bounds in the disparity variation.

The power continuous spectra [20] of the three above-mentioned codes are shown in Fig. 20; all codes present a zero at d.c. and at the repetition frequency.

The error detection in the block codes can be easily implemented by driving an up-down counter with the received coded sequence and controlling that the accumulated disparity does not exceed the upper or the lower limits relative to the adopted code. The lower the disparity variation of a code the more reliable is the error detection.

In order to identify and correctly decode the incoming signal at the receiver, a block alignment circuit must be provided which operates according to the coding rules. For example, in the 3 B - 4 B code a word with a certain disparity can not be followed by a word with the same disparity; moreover, some words are

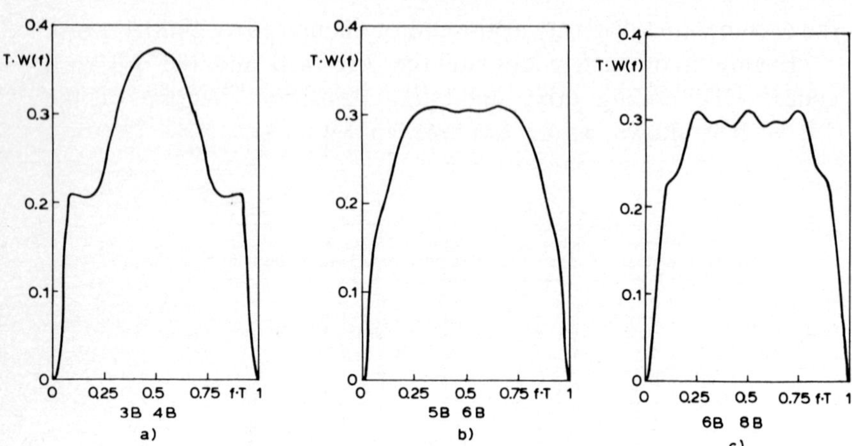

FIG. 20 – Continuous spectra of 3*B*-4*B* (*a*), 5*B*-6*B* (*b*), 6*B*-8*B* (*c*) codes.

prohibited and can not be received in the alignment state, unless errors occur.

Table II shows, for the three considered codes:

— the ratio n/m which gives the increase in the symbol rate and then the power penalty with respect to the uncoded binary signal;

— the longest number of N_{max} consecutive equal symbols which is related to the clock recovery efficiency;

— the bounds of the accumulated disparity D which are related to the low frequency spectral content and to the error monitoring facilities.

TABLE II

Properties of possible block codes

Code	n/m	N_{max}	D
6 B - 8 B	1.33	6	± 3
3 B - 4 B	1.33	4	± 3
5 B - 6 B	1.20	6	± 4

It should be noted that the lower n, the simpler are the coding and decoding circuits. From this point of view 3 B - 4 B is the most suitable code. However, as regards the bandwidth requirements the 5 B - 6 B code is the most advantageous. This code has been utilized in a B.P.O. experimental system at 8.448 Mbit/s [21].

2.4.4 1 B - 2 B *codes*

A particular class of block codes are the 1 B - 2 B codes which, due to the high redundancy, show a very good performance and require very simple circuits. The main drawback is that the symbol rate and then the required bandwidth is double (corresponding to a power penalty of 3 dB in the absence of fibre cut-off) with respect to that of the uncoded signal; so these codes are not well suited for high capacity digital transmission when bandwidth limitations due to the fibre can arise.

The most interesting codes of this family are CMI and biphase codes.

CMI (Coded Mark Inversion) code is internationally adopted for transmission at 140 Mbit/s digital interfaces. By this code digit « 0 » is transmitted as 01 and digit « 1 » alternatively as 11 or 00. The word 10 is forbidden in the coded signal and allows an easy alignment procedure at the receiver. Timing information is assured by the very frequent positive to negative transitions, which can happen only at the beginning of a word. Finally, the disparity variation is bounded to 3 allowing an easy and efficient error detection.

Biphase code utilizes only 2 bit balanced words: i.e. digits « 0 » and « 1 » are transmitted respectively as 01 and 10. It shows some minor advantages with respect to the CMI: the prohibited coded words are two (00 and 11) and disparity variation is equal to 2. On the other hand, it is a bit more sensitive to the high frequency cut-off (around Nyquist frequency), as can be seen from Fig. 21 where the power density spectral of the two codes are shown.

Another type of 1 B - 2 B code is the so called AMI-2 (Alternate Mark Inversion, 2 levels) code, that can be derived from the biphase code.

A precoding is applied to the information binary sequence $\{a_n\}$ in order to obtain an intermediate binary sequence $\{b_n\}$

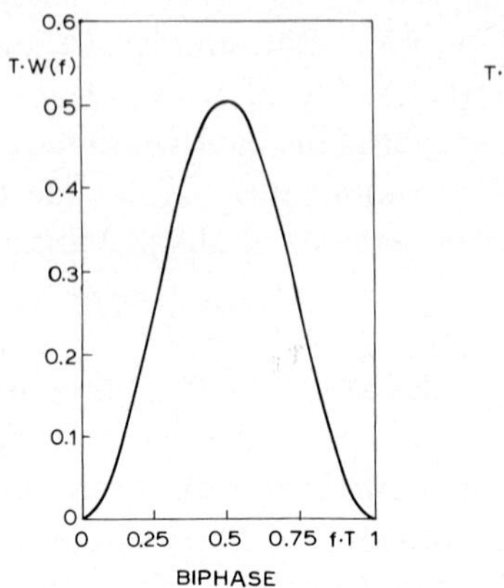

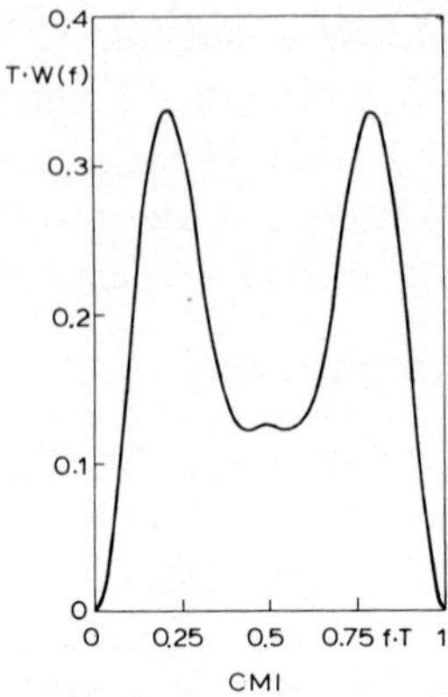

FIG. 21 – Continuous spectra of biphase and CMI code.

given by:

$$b_n = a_n \oplus b_{n-1} \tag{54}$$

where $\oplus$ denotes modulo 2 sum.

The precoded sequence $\{b_n\}$ is then coded according to the biphase coding law and sent to the line.

At the receiver, the signal is passed through a 2 tap delay line transversal filter which makes the sum of two line symbols $T/2$ apart and has a transfer function given by:

$$H_R(f) = 1 + \exp[-j\pi fT] \tag{55}$$

At the output of the filter there is a three level signal which corresponds to the information sequence coded according to the well known bipolar law.

In fact, as the power spectrum of the biphase code, shown in Fig. 21, is given by:

$$S_{bi}(f) = \frac{1}{4T}\sin^2\frac{\pi}{2}fT \tag{56}$$

the resulting spectrum, after filtering according to the transfer function given by (55) is equal to:

$$S_R(f) = \frac{1}{T}\sin^2 \pi fT \tag{57}$$

which is the spectrum of a bipolar coded signal.

The precoding is necessary in order to have an univocal correspondence between the information bits and the received pulses.

The relation between the information sequence and the precoded, the biphase, and the received signals is illustrated in Fig. 22.

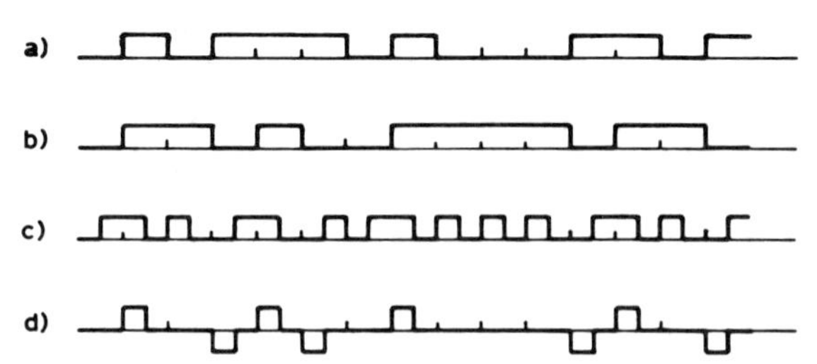

FIG. 22 – Sequence of signals for AMI-2 code: information sequence (*a*), precoded sequence (*b*), biphase coded sequence (*c*), received sequence (*d*).

Just as with the bipolar code, this code provides for easy error monitoring; moreover, timing information can be derived from the signal before filtering, where a high number of transitions is assured.

The drawbacks of this code are the considerable power penalty of about 5.2 dB in the absence of fibre cut-off and the higher complexity of the 3-level decision circuit.

2.5 Short distance systems

2.5.1 *General*

Besides the established advantages of low attenuation and large bandwidth, the dielectric nature and intrinsic physical characteristic of optical fibres (e.g. non-conductivity, absence of short circuits and ground loops, EMI and EMP immunity, small size, light weight, etc.) make them particularly well-suited to short distance applications.

Military applications [22, 23], with more severe system requirements, are the most suited to the utilization of optical fibres.

In airborne communications weight and size, electromagnetic

interferences and ground loops play a fundamental role. Optical fibres can be utilized in general for data transmission between different on-board equipment. This can be accomplished either by means of point to point links, or, as it is expected in the near future, by means of multipoint connections or data busses. The maximum length of such links ranges between 50 m and 100 m; the maximum required capacity is about 10 Mbit/s and the number of data terminals to be interconnected is of the order of 10 [24, 25].

In shipboard communications weight and size are less critical and the most serious problems are those connected with ground loops, crosstalk and interference. Also in this case optical fibres can profitably replace copper cables in the interequipment data transmissions, where the maximum link length is of the order of several hundred metres and the number of terminals to be connected is about 20. Optical fibres can be also used for voice or TV signal on-board transmission [26].

In civil engineering applications, optical fibres are particularly advantageous for transmission on links in the presence of adverse environmental conditions, such as exposure to high temperature or chemical agents, or near high voltages or high electromagnetic fields. Typical examples of possibile optical fibre applications are in power stations for the transmission of measurement data from high voltage points, where an electrical insulation between the transmitter and the receiver is needed [27], or in the connection between computers and peripherals in the presence of high level disturbances. In these applications the link length ranges between several tens and several hundreds of metres.

In the following, the problems related to optical fibre applications on short-distance links are examined and some basic considerations on the design criteria of such links are reported, with particular reference to the choice of the most suitable components.

2.5.2 *Components*

Taking into account length and required capacity, multimode step-index fibres, currently available at moderate prices, seem to be the most suitable for use in short-distance systems. Graded-index fibres are not considered, as they are expensive and their performance is much higher than required. According to the materials used and to their purity, different step-index fibres are available with low ($<$ 20 dB/km), medium (20-100 dB/km), or high ($>$ 100 dB/km) attenuation values; numerical aperture can range between 0.1-0.2 and 0.5-0.6: the corresponding dispersion values vary from 10-15 ns/km to 200-300 ns/km respectively; the core diameter is usually of the order of 50-80 μm. Of course, the higher the attenuation, the lower is the price of such fibres; thus, considering the distances to be covered, preference should be given to medium loss fibres.

The optical transmission channel can consist just as well of a single fibre or of a bundle of many fibres (from 7 to more than 100) closely packed with hexagonal symmetry.

Although the latter solution is more expensive, it is currently very much in use over short distances since it provides higher coupling efficiency with the optical source and reduces the risk of link interruptions due to fibre damage.

Even if considerable improvements are expected in single fibre reliability, the bundle of fibres is at present, as will be shown, the most suitable carrier for short-distance applications. The bundle cost is expected to fall below 1 $/m in the next few years.

The criteria governing the choice of optoelectronic components should be cost and reliability, considering the length and high number of links encountered in many applications (e.g. aircraft and ships).

From this point of view, the optical sources best suited for short distance applications are commercial GaAs LED's emitting in the first window with a radiance of about 0.1-1 $W/sr \cdot cm^2$, which present high reliability (more than 100,000 hrs of mean life) and moderate cost (about $ 10).

Si PIN diodes are generally preferred as optical detectors since they are cheaper and require simpler circuitry than avalanche diodes. Avalanche diodes, in fact, require high bias voltage and possibly a stabilization of the operating point and must be used only in special cases.

As far as the coupling elements are concerned, in the case of bundle connectors with about 1 dB loss, source couplers including suitable optical devices with about 10 dB loss and photodetector couplers with about 1 dB loss are available at present. Single fibre connectors with a loss of the order of 1 dB are envisaged in the near future, while source and photodetector coupling will continue to play a predominant role so as to avoid losses due to possible misalignments.

2.5.3 *Data busses*

It is envisaged that optical fibres will be used in the near future not only for point-to-point links, but also for the so-called data busses where a single transmission line carries many different multiplexed signals and serves a number of spatially distributed terminals [28]. Two fundamental arrangement are possible, the trunk configuration and the star configuration, as indicated in Fig. 23 [29]. Bundles of fibres are normally used in the experiments described in the literature to increase the power coupled into the optical cable and to facilitate signal extraction at the different terminals.

In the trunk configuration, the different terminals are connected to a common optical highway by means of *T* couplers. The insertion and the extraction of optical power can be easily obtained, by suitably furcating the optical bundle and by inserting a mixer element in order to equally distribute throughout the bundle the power injected into a small number of fibres, as indicated in Fig. 23 *a*). *T* coupler losses are quite high: about 8 dB in the signal extraction, about 2 dB in each mixer and less than 1 dB of insertion loss. While the advantage of such a solution is its flexibility, its limitation lies with the high losses of each tap. The maximum attenuation between terminals is proportional to the total number of terminals, which usually can not exceed 10.

In the star configuration [30], the signal coming from every terminal is sent to a single optical component and then distributed to all other terminals. A possible scheme of a star coupler is given in Fig. 23 *b*). The bundles from every terminal are packed together at one endface of a mixer rod with a reflecting opposite endface. The rays from a particular fibre are reflected from the mixer rod walls and mirrored endface and return to the input endface. In practice, the rod length must be chosen so that rays from any single fibre are uniformly distributed over the entire input endface. In this configuration all terminals are equivalent in the sense that the attenuation between any two terminals is equal. The insertion loss of a star coupler is about 8 dB and the power-splitting loss

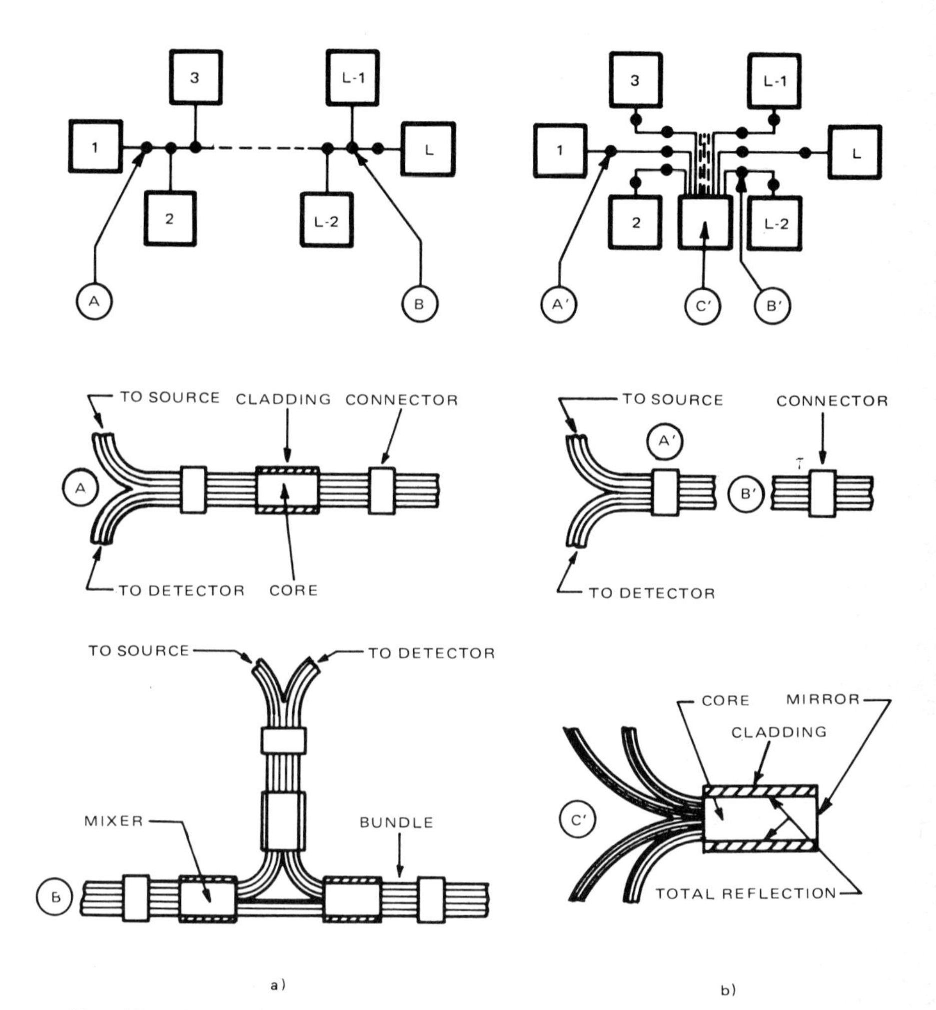

FIG. 23. – Trunk (*a*) and star (*b*) configurations for data busses and relevant components. (After [35]).

is proportional to the logarithm of the number of terminals. Therefore, a substantially greater number of terminals (20-50) can be supported by the star system than by the *T* system; moreover, its structure is such that the star system is less subject to outage from damage than a *T* system.

Recent literature has described experimental *T* and star couplers [31, 32], but further tests are needed before these components will be available. Coupling devices for single fibre systems have also been announced [33, 34], but they are only in an early experimental stage and require further research.

2.5.4 *System considerations*

Let us now consider some design aspects of short-distance optical fibre links which may be useful in deciding whether to use the fibre or the bundle in such applications [35]. Reference will be made mainly to point-to-point links; star type multipoint connections will also be considered.

The overall attenuation of a digital point-to-point link depends, as seen in Section 2.3, upon the power that can be injected into the optical channel and the power required at the receiver for assuring a given error probability.

Let us consider the case of binary digital transmission at a symbol rate f_r over a bundle of n fibres with core diameter d and numerical aperture Δ, where a LED source with radiance R_a and a PIN diode with responsivity ϱ are used. Taking into account (4), (16) and (15) the maximum allowable loss αL in dB for a given signal-to-noise ratio (and a given error probability) at the decision point is given, if no cut-off is introduced by the channel, by:

$$\alpha L = 10 \log \frac{\varrho d_c R_a n(\pi d \Delta)^2}{\sigma f_r \sqrt{32\pi KT(\delta FC)}} \tag{58}$$

where d_c is the duty cycle of the transmitted pulse, K is the Boltzmann constant, T the absolute temperature and δFC is the parameter that indicates the accuracy of the receiver (see Section 2.3). In (58) quantum noise has been ignored, since the PIN diode operates, as can be seen from Fig. 8, in the thermal noise limited region.

The overall loss given by (58) versus the acceptance factor $\sqrt{n} \cdot d \cdot \Delta$ of the bundle for different values of the ratio R_a/f_r between the LED radiance and the symbol rate is reported in Fig. 24. A PIN diode, with a typical value $\varrho = 0.5$ A/W, has been considered and a value of $\delta FC = 10$ pF, corresponding to a not sophisticated amplifier design, has been assumed. Curves have been drawn for a duty cycle $d_c = 0.5$ and for an error probability

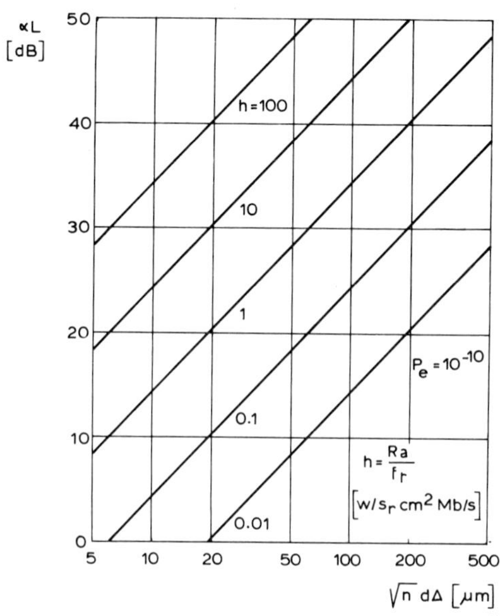

FIG. 24 – Bundle overall losses αL versus acceptance factor $\sqrt{n}d\Delta$ for different values of the ratio R_a/f_r between the LED radiance and the symbol rate.

$P_e = 10^{-10}$ with a margin of 5 dB (optical power) to allow for receiver imperfections and connector losses.

From Fig. 24, on the basis of the optical carrier and of the optical source adopted, it is easy to determine for different symbol rates the maximum link length. For example, with a Corning bundle of 19 fibres ($d = 85$ μm, $\Delta = 0.14$), having an acceptance factor of 52 μm, and a LED with $R_a = 0.5$ W/sr·cm², a very conservative value, the maximum link attenuation for a 10 Mbit/s digital signal is 18 dB, which corresponds to a maximum length of 700 m, with $\alpha \simeq 25$ dB/km.

Selection of the optical carrier is aided by Fig. 25, which shows fibre attenuation per unit length required for different typical link lengths and for different bundle structures versus fibre numerical aperture. Bundles with hexagonal symmetry have been considered. The distances considered are respectively 30 m (typical for aircraft and computer links), 100 m (typical for shipboard and remote control links) and 300 m (maximum shipboard and remote terminal links). Figure 25 has been drawn under the hypothesis of Fig. 24, assuming a typical value for the fibre core diameter

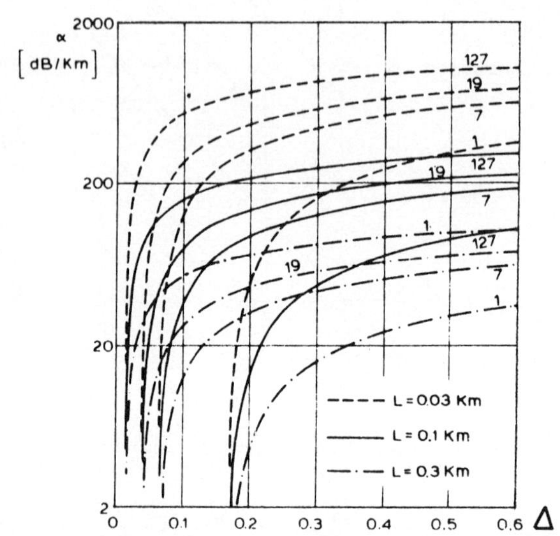

FIG. 25 – Fibre attenuation per unit length α, versus fibre numerical aperture Δ for hexagonal bundles in point to point links; fibre number as parameter.

$d = 50\ \mu m$ and a LED radiance $R_a = 0.5\ W/sr \cdot cm^2$; a symbol rate of 10 Mbit/s, that corresponds to the upper limit required for short distance digital transmission, has been considered.

From Fig. 25, considering the most critical length, that is 300 m, it can be seen that in the case of a single fibre it is necessary to utilize low loss fibres ($\sim$10 dB/km) if low numerical aperture values are required (0.2-0.3), while it is possible to accept medium attenuations of about 30 dB/km with high numerical aperture values (0.5-0.6). In the latter case, advantageous from a cost point of view, some bandwidth limitations could possibly arise due to the pulse dispersion, that for high value of Δ can be of the order of 200 ns/km. The bandwidth limitation problem can be resolved by using bundles with a limited number of fibres (7 or 19) which allow, as can be seen from the figure, the use of medium loss (30-50 dB/km) and low numerical aperture (0.2-0.3) fibres, available at moderate cost. The utilization of bundles with a high number (> 61) of fibres is neither necessary nor convenient as the advantage in attenuation is not proportional to the number of fibres and thus to the channel cost; moreover, the pulse dispersion increases, although slightly, with the number of fibres.

Generally, even if shorter length links (100 m and 30 m) are considered, the figure shows an equivalence, from the attenuation point of view, between bundles with a limited number (7 or 19) of fibres having low numerical aperture (0.2-0.3) and single fibres with high numerical aperture. Taking into account, as already mentioned, the possible bandwidth limitation related to high numerical apertures, the most suitable optical carriers for short distance links seem to be bundles with 9 or 17 fibres.

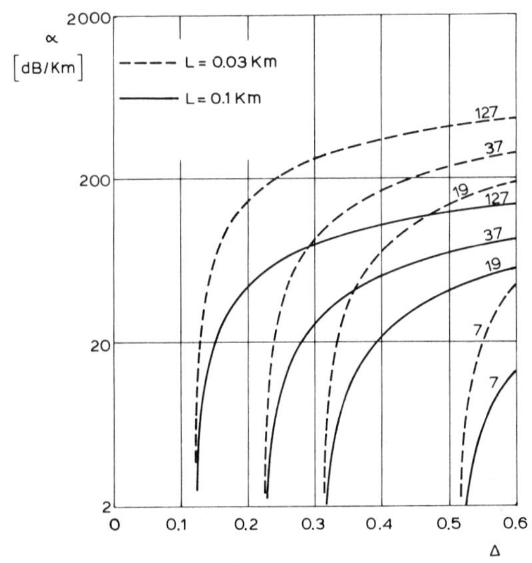

FIG. 26 – Fibre attenuation per unit length α versus numerical aperture Δ in a multipoint system with star configuration and $n = 10$ terminals; bundle fibre number as parameter.

Figure 26 is similar to Fig. 25 but refers to a multipoint system with a star configuration. Curves have been drawn under the hypothesis made for Fig. 25, assuming equal to 10 the number of terminals to be interconnected, and equal to 8 dB the star coupler insertion loss. The link lengths considered between two terminals are only 100 m and 30 m, which should cover the normal requirements for such applications, and only bundles have been considered, as a single fibre would require a too low attenuation.

From Fig. 26, taking 100 m as the most critical length, it can be seen that there is an equivalence, from the attenuation point of view, between bundles with 7 fibres having a high numerical aperture (0.5-0.6) and bundles with 19 (or 37) fibres having a low numerical aperture (0.2-0.3). Considerations similar to those made for point-to-point links lead to the use of bundles with 19 (or as a maximum 37) fibres having an attenuation of about 50 dB/km for data bus applications.

The remarks made concerning the choice between a single fibre and bundles of fibres refer to the present situation of optical component development and cost, which could change in the future. It is worth noting that a trend is emerging towards the use of single fibres also in short distance applications. Such a trend is justified by the continuous improvement in fibre mechanical and transmission characteristics, which has led in the last few years to a reduction of the number of fibres within bundles (from several hundreds to 19 or 7).

The opinion of single fibre advocators is that, due to the high production volume required for telecommunication applications, in the near future high quality fibres (with attenuation lower than 10 dB/km and dispersion lower than 10 ns/km) will be available at low cost (about 0.1 $/m). Thus, neither performance nor cost considerations would warrant the use of bundles.

In this connection, the cost of an optical link is not only determined by the carrier cost, but, especially in short distance applications, is also influenced by the cost of coupling elements and optoelectronic components, which are more expensive in the case of a single fibre. Moreover, considering multipoint connections (data busses), all coupling elements (T and star couplers, mixers, etc.) are available for bundles, while they are not yet available for single fibres. At any rate, it is extremely difficult to make estimates of the future, as the situation is continuously evolving.

REFERENCES

[1] S. D. Personick: *Baseband linearity and equalization in fiber optic digital communication systems*, Bell Syst. Tech. J., Vol. 52, No. 7 (September 1973), pp. 1175-1191.

[2] C. Vassallo: *Linear power responses of an optical fiber*, IEEE Trans. Microwave Theory Tech., Vol. MTT-25, No. 7 (July 1977) pp. 572-576.

[3] S. E. Miller, E. A. Marcatili and T. Li: *Research toward optical fiber transmission systems. - Part I: The transmission medium*, Proc. IEEE, Vol. 61, No. 12 (December 1973), pp. 1703-1726.

[4] D. N. Payne and A. M. Hartog: *Pulse delay measurement of the zero wavelength of material dispersion in optical fibres*, NTZ, Vol. 31, No. 2 (1978), pp. 130-132.

[5] C. C. Timmermann: *Bandwidth limitation of a fibre optic system with linear intensity modulation*, Opt. Quantum Electron., Vol. 9, No. 10 (1977), pp. 532-535.

[6] C. C. Timmermann: *Wavelength dependence of pulse broadening and distortion in glass fibres with dominant material dispersion*, NTZ, Vol. 31, No. 11 (1978), pp. 822-824.

[7] S. D. Personick: *Optical fiber, a new transmission medium*, IEEE Communication Society NewsLett., Vol. 13, No. 1 (1975), pp. 21-24.

[8] S. D. Personick: *Optimal trade-off of mode-mixing optical filtering and index difference in digital fiber optic communications systems*, Bell Syst. Tech. J., Vol. 53, No. 5 (May-June 1974), pp. 785-800.

[9] R. Di Vita and R. Vannucci: *Multimode optical waveguides with graded refractive index: theory of power launching*, Appl. Opt., Vol. 15, No. 11 (November 1976), pp. 2765-2772.

[10] T. Ozeki and A. Watanabe: *Equivalent circuit parameters of a silicon avalanche photodiode*, Electron. Lett., Vol. 12, No. 6 (March 1976), pp. 144-145.

[11] R. Mc Intyre: *The distribution of gains in uniformly multiplying avalanche photodiodes: theory*, IEEE Trans. Electron. Devices, Vol. ED-19, No. 6 (June 1972), pp. 703-713.

[12] J. Conradi: *The distribution of gains in uniformly multiplying avalanche photodiodes: experimental*, IEEE Trans. Electron. Devices, Vol. ED-19, No. 6 (June 1972), pp. 713-718.

[13] S. D. Personick: *Receiver design for digital fiber optic communicaion systems*, Bell Syst. Tech. J., Vol. 52, No. 7 (July-August 1973), pp. 843-886.

[14] B. Catania and L. Sacchi: *Progetto di collegamenti numerici in fibra ottica*, XXIII Congresso Internazionale delle Comunicazioni, Genoa (October 9-11, 1975), pp. 363-379.

[15] S. Longoni, E. Paladin and L. Sacchi: *Digital communication system over optical cables for medium and long distances*, Eurocon 1977. Conference Proceedings on Communications, Venice (May 3-7, 1977), pp. 523-531.

[16] A. Fausone, G. Rosso and G. Veglio: *Repeater for 34 Mb/s transmission system over optical fibre*, Second European Conference on Optical Fibre Communication, Paris (September 27-30, 1976), pp. 351-356.

[17] CCITT Period 1977-1980 - Comm. XVIII, Doc. n. 59: *In service error rate detection at dependent repeaters using mark parity control.*

[18] R. E. Epworth: *ITT 140 Mbit/s optical fibre system*, NTC '77, Conference Record, Los Angeles (December 5-7, 1977), pp. 13:3-1 to 14–3-6.

[19] M. I. Maunsell, R. B. Robrock and C. A. von Roesgen: *The M13 and M34 digital multiplexer*, ICC '75 - International Conference on Communications, San Francisco (June 16-18, 1975), pp. 48-5 - 48-9.

[20] D. Laforgia, I. Pilloni and G. Pirani: *Codici di linea per la trasmissione numerica su fibra ottica*, Rapporto Interno CSELT, No. 77.06.106.

[21] D. J. Brace and I. A. Ravenscroft: *Optical fibre transmission systems: the 8.448 Mb/s feasibility trial*, The Post-Office Electrical Engineer's Journal, Vol. 70, No. 3 (October 1977), pp. 146-153.

[22] L. U. Dworkin and J. R. Christian: *Army fiber optic program: an update*, IEEE Trans. Commun., Vol. COM-26, No. 7A (July 1978), pp. 999-1006.

[23] L. R. Weisberg: *DoD activities in optical fibres*, Optical Fiber Communication, Washington D.C. (March 6-8, 1979), paper TuA2.

[24] T. A. Hawkes and J. C. Reymond: *Systèmes de communication optique à bord d'avions*, Second European Conference on Optical Fibre Communication, Paris (September 27-30, 1976), pp. 309-409.

[25] R. A. Greenwell: *Fiber optics cost models for the A-7 aircraft*, Fiber and Integrated Optics, Vol. 1, No. 2 (1977), pp. 197-225.

[26] E. M. Hara and H. C. Frayn: *An experimental optical-fiber link for the command and control system 280*, Optical Fibres, Integrated Optics and their Military Applications, London (May 16-20, 1977), pp. 12.1-12.11.

[27] L. Thione and E. Viganò: *Electrooptic in electrical measurement: a review of experience in system design and applications*, XXIII Congresso Internazionale per l'Elettronica, Rome (March 22-24, 1976), pp. 124-131.

[28] M. K. Barnoski: *Data distribution using fiber optics,* Appl. Opt., Vol. 14, No. 11 (Nov. 1975), pp. 2571-2577.

[29] F. L. Thiel: *Coupling consideration in optical data buses*, Optical Fiber Transmission, Williamsburg (January 7-9, 1975), pp. WE 1-1 - WE 1-4.

[30] M. C. Hudson and F. L. Thiel: *The star coupler: a unique interconnection component for multimode optical waveguide communications systems*, Appl. Opt., Vol. 13, No. 11 (November 1974), pp. 2540-2545.

[31] L. D'Auria and A. Jacques: *Bidirectional central couplers for links with optical fiber bundles*, Optical Fibres, Integrated Optics and their Military Applications, London (May 16-20, 1977), pp. 47.1-47.11.

[32] L. Jeunhomme and J. P. Pocholle: *T-coupler for multimode optical fibers*, Optical Fibers, Integrated Optics and their Military Applications, London (May 16-20, 1977), pp. 48.1-48.13.

[33] C. A. Villaruel, T. G. Giallorenzi and A. F. Milton: *Couplers for multiterminal data distribution systems*. Optical Fiber Communication, Washington D.C. (March 6-8, 1979) pp. 76-80.

[34] E. G. Rawson and R. M. Metcalfe: *Fibernet: multimode optical fibers for local computer networks*, IEEE Trans. Commun., Vol. COM-26, No. 7 (July 1978), pp. 983-990.

[35] B. Catania, A. Di Lullo, S. Longoni and L. Sacchi: *Applicazioni delle fibre ottiche ai collegamenti a breve distanza*, XXIII Congresso Internazionale per l'Elettronica, Rome (March 22-24, 1976), pp. 75-92.

CHAPTER 3

CIRCUIT CONSIDERATIONS

by Alfredo Fausone

3.1 General

The purpose of this chapter is to illustrate the general design criteria of the typical circuits that constitute an optical repeater.

For the transmitting side, driving amplifiers for both LED and laser optical sources are discussed for both analogue and digital transmission.

For the receiver side, considerations are made on the low-noise amplifiers best suited to guarantee a good current transfer from the photodetector, to maximize the signal-to-noise ratio at the decision point and to minimize the intersymbol interference. The timing recovery circuits are also briefly discussed.

Since it is the accurate design of the preamplifier which essentially determines the optical repeater's overall performance, the criteria relating to the design of this device are considered in greater detail.

3.2 Driving circuits

Two types of semiconductor diode sources are widely used in optical communications systems: light emitting diodes (LED's) and injection lasers.

LED's have generally been considered to be more suitable than lasers for linear transmission systems, as they exhibit a linear power-current law, whereas the laser response can exhibit non linearities.

Ga As lasers, on the other hand, are preferred for digital transmission systems, in that they can be modulated directly on an ON-OFF basis with more brightness. This allows more optical power to be launched into a given fibre, and thus a greater distance between repeaters.

They are also characterized by a faster response with rise and fall times of less than one nanosecond, and by a lower value for the spectral spread of 2 nm or less between half-optical-power points; thus the maximum bandwidth available from fibres can be increased, which is an important factor in the highest capacity long haul systems.

The driving circuits for these semiconductor diode sources aim to convert the transmitting voltage signal into modulation current with peak values from a few tens to hundreds of milliamperes depending on the source used.

There are many circuits that can be used as optical source drivers, every one of which can show particular characteristics in a given design.

Accordingly, some kinds of circuits will be examined later on as typical examples for different solutions.

3.2.1 *Driving circuits for analogue systems*

In analogue transmission systems, particularly CATV and FDM, light-emitting-diodes (LED) are preferred as optical sources, as they have a better linearity than other sources.

For multichannel signal transmission over repeatered systems, however, the inherent non-linear distortion characteristics of the optical sources must be compensated.

In the literature, various techniques for distortion reduction are shown, namely:

— complementary distortion, in which the nonlinear distortion produced can be reduced if compensating additional non-linear devices are employed [1].

  This compensation technique is limited by the fact that only over a reduced range of modulation amplitudes a significant improvement in the linearity of the system can be expected.

  A diode network, whose distortion is complementary to the originally distorted signal, is commonly used as a non-linear device.

  Problems can arise with this method by the mismatching

of the non-linear device characteristics introduced and the relative optical source;

— negative feedback, in which a photodiode, joined with the optical source by a beam-splitter monitoring the optical signals, provides the necessary feedback signal.

This is a more general compensation technique because it consists of a feedback of the whole transmitting signal such that the amount of residual distortion is determined by the feedback loop gain.

Of course, as in every feedback circuit, difficulties due to the large bandwidth requirements of the feedback may arise at high frequencies;

— feedforward, in which the error signal, obtained by the comparison of the signal before and after the distorting device, is combined beyond the circuit output.

The advantage of using feedforward is that it does not require signal loops and therefore results in an inherently stable system.

Moreover, with this technique for the reduction of distortion, the resulting compensation does not depend upon device frequency limitations, as in the case with feedback;

— phase-shift modulation, in which a selective harmonic compensation can be obtained by using, as the optical signal, the combination of light emitted by a couple of optical sources having similar characteristics, modulated by the same phase shifted signal [2].

— The advantages of this technique are its simplicity and the good performance shown. However, a simultaneous reduction of second and third order distortion cannot be achieved, due to the different phase shift requirements imposed for the cancellation of various harmonics.

Moreover, for a single channel transmission system over short distances without repeaters, the total harmonic distortion obtained by a LED is acceptable. Therefore, it is sufficient to use a simple driving circuit without compensation for the inherent non linearity of the optical component.

A simple driving circuit which can be used for this kind of

system consists of a differential amplifier that directly modulates the optical source.

Its diagram is shown in Fig. 1.

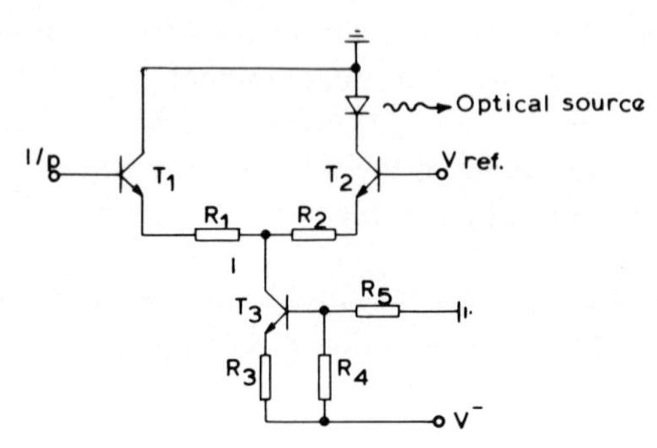

FIG. 1 – Schematic diagram of a differential driving amplifier.

The current generator *I*, realized by the transistor *T*3 feeding the differential stage *T*1 and *T*2, limits the top value of LED polarization.

Moreover, *V* ref voltage enables the most suitable polarization point of the same LED to be chosen on the basis of transmission efficiency and total distortion limitation.

The two equal resistors *R*1 and *R*2, in series with the emitters of *T*1 and *T*2, provide a current-series feedback by reducing to smaller values the transimpendance of the stage.

These feedback resistors can be used to equalize the transfer function of the driving circuit and optical source.

Other solutions for optical source driving can be chosen, employing 50 Ω or 75 Ω high band conventional amplifiers and increasing the dynamic impedance of the source to 50 or 75 Ω, respectively, for a good matching.

To reduce the output power supplied by the driver for a given optical power, a transformer coupled driver stage, which offers improved efficiency in comparison with *RC* couple circuits, can be used.

3.2.2 *Driving circuits for digital systems*

In digital transmission systems, the driving circuit transforms the logic voltage levels, available at its input, into current levels suitable for the optical source driving, and can assume different forms according to the data rate to be transmitted.

Since, above threshold, the dynamic impedance of both Laser and LED optical sources is typically 1-2 Ω, a single stage current amplifier using microwave transistors allows a direct modulation of the optical source with a typical peak current of up to 300 mA with a pulse-width of about 5-10 ns.

Therefore, it is necessary to use frequency compensations to obtain an excellent optical pulse with low values for the rise and fall time according to the data transmitted rate.

A simplified diagram of a driving circuit using bipolar microwave transistors is shown in Fig. 2, where the time constant R_1C

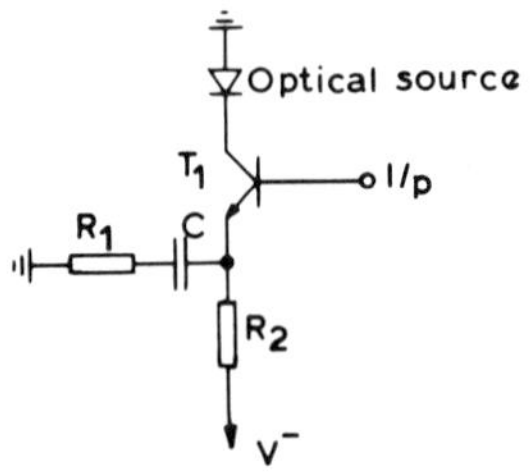

FIG. 2 – Schematic diagram of a digital driving circuit with compensation.

is chosen to optimize the optical pulse rise and fall time without ringing [3].

In fact, the delayed rise of the collector current is compensated by the emitter *RC* network, the value of which depends on the characteristics of the photodiode.

Another technique which can be useful in driving the optical source is the use of a transisfor emitter-follower circuit with a suitable matching network that can achieve fast, direct modulation of LED's or lasers at relatively low driving power.

The emitter-follower driving circuit with compensating matching network studied by [4], is shown in Fig. 3.

With the optimum values for the matching network, light output pulses without ringing, with a rise and fall time of about 2.5-3 ns at a 100 Mbit/s rate, can be obtained.

For systems at higher data rates, voltage amplifiers can be used for driving circuits to drive a 50 Ω load. In these cases the

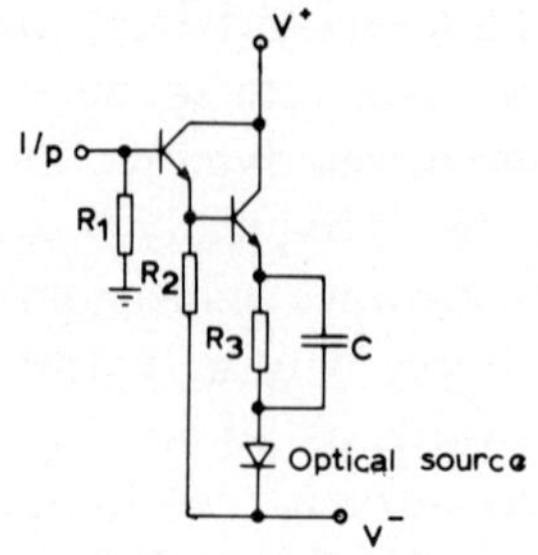

FIG. 3 – Schematic diagram of an emitter-follower driving circuit.

dynamic impedance of optical sources is raised to 50 Ω in the bandwidth considered to ensure a good matching.

In this case the performance, such as rise and fall time and relative ringing, is determined by the bandwidth and shape of the amplifier response.

At ultrahigh bit rates, step-recovery diode circuits have been proposed for direct modulation of laser diodes [5].

An attractive diode stage, in which the non ideal properties of the diodes are nearly eliminated, is the former circuit that contains four step-recovery and one Schottky-barrier diodes operating in a push-pull mode driven by pump voltages shifted by 180° with regard to each other.

The basic circuit is shown in Fig. 4.

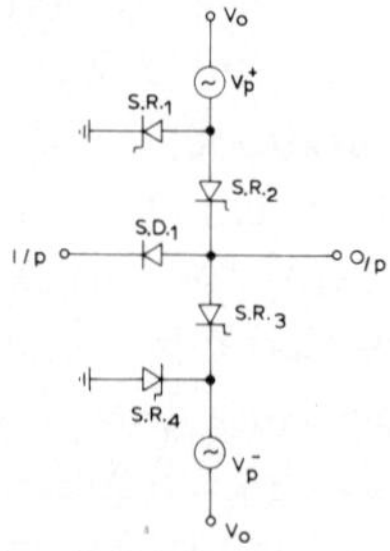

FIG. 4 – Schematic diagram of a step-recovery diode driving circuit. [After [5]].

The operating cycle consists of two separate time intervals. In the first, the input signal modifies the balance of the charge in the diffusion capacitances of the step-recovery diodes, while in the second, the power transfer from the clock source to the optical source is due to this unbalance.

3.2.3 *Automatic level control circuits*

The automatic level control circuit is used to ensure a constant output power and modulation depth in respect to variations between changes in temperature and long term degradation of optical sources.

Some of the techniques that will be discussed below may be applicable to high radiance LED's even if they are developed for Ga As lasers that show more critical characteristics from this point of view.

The lasers are usually modulated by d.c. biasing to the threshold level and by adding a modulation current.

Unfortunately, the threshold varies with temperature and life and can differ significantly from device to device. Therefore it is very important to maintain the correct bias and modulation current if optimum performance is to be obtained.

Any automatic level control systems compare the optical output of the source with the required output signal by monitoring a fraction of the light launched into the fibre.

The simplest feedback arrangement maintains the mean optical power constant by varying the threshold current level: this solution only requires a d.c. feedback loop [6].

For a complete control of the device behaviour, two items of information are needed. A signal related to the average power monitors the threshold current drift, while the efficiency of the source above threshold is monitored from changes in the peak to peak power.

As the average power is easily obtained with an integrator, the peak power can be directly measured using a high speed detector and an amplifier almost equal to the regenerator front end.

The peak to peak value of the optical signal can also be extracted from the spectral properties of a return to zero waveform.

A small fraction of the output optical signal is detected by means of a PIN photodetector and filtered with a low Q resonant circuit tuned at the pulse rate.

To reduce the sensitivity of this method to the statistics of the coded symbols, a scrambler must be inserted in the transmitting unit.

A voltage proportional to the monitored signal envelope is obtained with a coherent detector and, after a low-pass filtering, is used to control the modulation current.

A control technique that offers accurate threshold tracking and very little device dependence is the switch-on delay method. The system consists in measuring pulse width in order to maintain it constant.

The feedback signal is amplified and phase compared with the data pulses transmitted: the signal so obtained is low-pass filtered and fed back to the driving circuit.

3.3 Preamplifier circuits

Overall performance in optical repeaters is determined by the suitability of the opto-electronic components used (sources and photodetectors), and by a careful design of the preamplifier.

This latter is generally designed in such a way as to maximize the signal-to-noise ratio at the decision point with minimum intersymbol interference.

However, we can observe that if a Si PIN diode is used as a photodetector, the signal-to-noise ratio is mainly determined by the preamplifier, whereas, if an APD is employed, the same ratio is determined by the APD itself.

Therefore it is possible, by minimizing the preamplifier noise, to reduce the APD optimum gain, thus lowering its bias voltage and the corresponding gain sensitivity.

In optical receivers, two types of preamplifier circuit can be considered:

— the transimpedance preamplifier, which is the configuration, in the medium frequency range (below 100 MHz), most attractive from the design standpoint because an amplitude equalizer network is not necessary [7-9];

— the voltage preamplifier, both high input impedance and front end amplifier, which needs, on the contrary, a differentiating network for equalizing the integrating first stage [10-13].

Further advantages of the transimpedance over the voltage preamplifier without overall feedback are better response, better input and output impedance stability and reduced non-linear distortion with respect to the latter.

Some problems arise during the adjustment of the circuit owing to the broad bandwidth and feedback resistance value required for a high current gain; on the other hand, the voltage amplifier is a simpler and cheaper circuit.

3.3.1 *Noise analysis*

The sources of noise which must be considered to evaluate the limits of the performance of an optical receiver are: the shot noise associated with the signal current in the photo-diode with some excess noise due to the avalanche gain mechanism, if an avalanche detector is used, the thermal noise of the detector load resistor and the thermal noise introduced by the preamplifier.

The effect of the shot noise associated with the dark current of the photo-diode can be ignored in comparison with the contribution of the shot noise due to the received optical power.

To simplify the noise analysis of an amplifier, a simple noise model containing only the two noise parameters E_n and I_n, which are not difficult to measure, can be considered.

With respect to midband frequency, parameters E_n and I_n increase at low frequencies because of the $1/f$ noise source and increase at high frequencies because of reduced gain: this is true for any transistor, whether bipolar or field effect.

In particular, at midfrequencies the noise parameters of a bipolar transistor are given by:

$$E_n^2 = 4KTr_{bb'} + \frac{2K^2T^2}{q} \cdot \frac{1}{I_c},$$

$$I_n^2 = 2qI_b$$

which show that both E_n and I_n are highly dependent on the operating point.

Plots of E_n and I_n versus the collector current (I_c) for certain values of transistor parameters are shown in Figs. 5 and 6.

The noise voltage E_n rises steadily as collector current decreases whilst, at high collector currents, it is determined by the thermal noise of the base resistance.

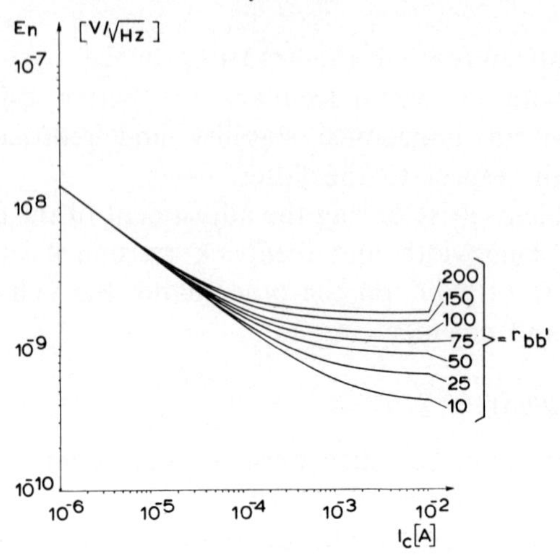

FIG. 5 – Plots of E_n versus collector current for certain values of $r_{bb'}$ (bipolar transistor).

On the other hand, the noise current I_n decreases with decreasing collector current and with increasing β. However, the effect of β is a minor penalty compared to the collector current contribution.

The main sources of noise in both FET types (JFET and MOSFET) are: thermal noise in the channel, shot noise in the

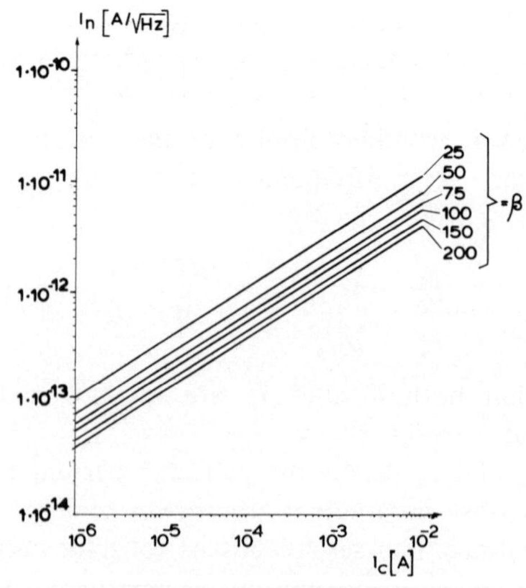

FIG. 6 – Plots of I_n versus collector current for certain values of current gain (β) (bipolar transistor).

gate to channel leakage, flicker noise, and thermal fluctuations in the drain circuit that are coupled into the gate.

The noise voltage generator at midfrequencies and high frequencies is due to the thermal excitation of carriers in the device channel and can be expressed by:

$$E_n = \sqrt{4KTR_n}$$

where: R_n = noise resistance that depends on the resistance of the channel by the rough expression:

$$\cong \frac{2}{3}\frac{g_{\max}}{g_m^2} \cong \frac{2}{3}\frac{1}{g_m},$$

with g_m, the transconductance of FET, equal to $g_{\max}$.

Therefore, to minimize the effect of this noise generator, it is necessary to operate the FET where the g_m value is large, that is, at high values of static drain current.

At lower frequencies the value of this noise increases due to the excess $1/f$ noise.

This noise source is dependent on the technology used. Its contribution is smaller in JFET, in that only the lower frequencies are involved, while in MOSFET it is significant at higher frequencies such as 1-10 MHz.

The equivalent noise current generator I_n represents either the shot noise of the current flowing through the gate or the capacitance-coupled thermal noise at the input, and its expression is:

$$I_n = \sqrt{2 \cdot Ig \cdot q + \frac{4 \cdot K \cdot T \cdot \omega^2 \cdot C_{gs}^2}{g_m}}$$

where: I_g = gate leakage current

C_{gs} = gate to source capacitance.

While the shot noise of the gate leakage is usually independent of frequency and is the higher factor at lower frequencies, the component of noise induced through the gate-source capacitance is dependent upon frequency and arises in particular as frequency increases.

By selecting a FET with a small enough leakage current, it

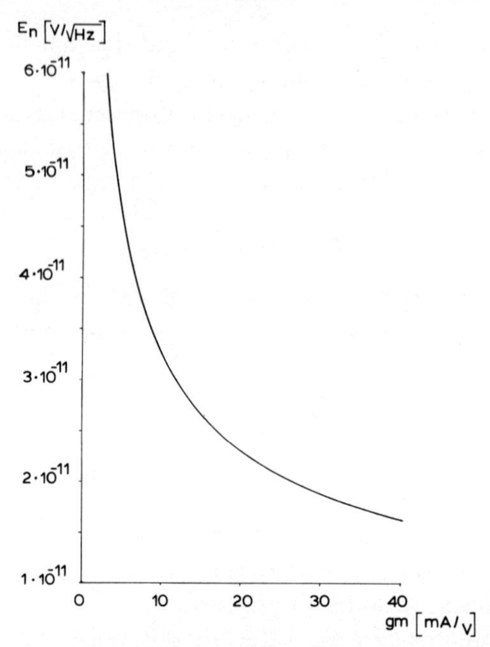

FIG. 7 – Plot of E_n versus the transimpedance g_m (FET devices).

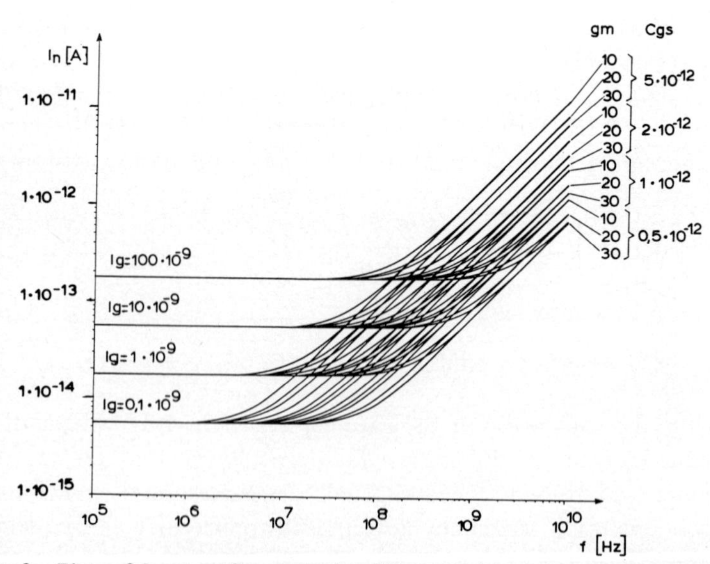

FIG. 8 – Plots of I_n versus frequency for different value of the transimpedance g_m, the gate-source capacitance C_{gs} and the current gain I_g (FET devices).

is usually possible to make this noise source negligible compared with channel noise and capacitance coupled thermal noise sources.

Plots of E_n versus transconductance and I_n versus frequency for certain values of FET parameters are shown in Figs. 7, 8 and 9.

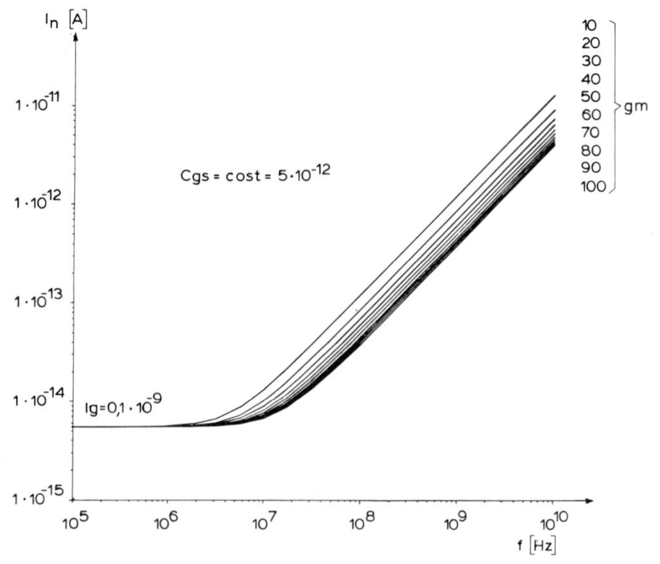

FIG. 9 – Plots of I_n versus frequency for different value of the transimpedance g_m with typical values for the gate-source capacitance C_{gs} and the current gate I_g (FET devices).

3.3.2 *Transimpedance and voltage amplifier comparison*

The receiver model for a transimpedance and a high input impedance voltage preamplifier including the noise sources is depicted in Figs. 10 and 11, respectively.

Figure 12 shows a schematic differentiating network for equal-

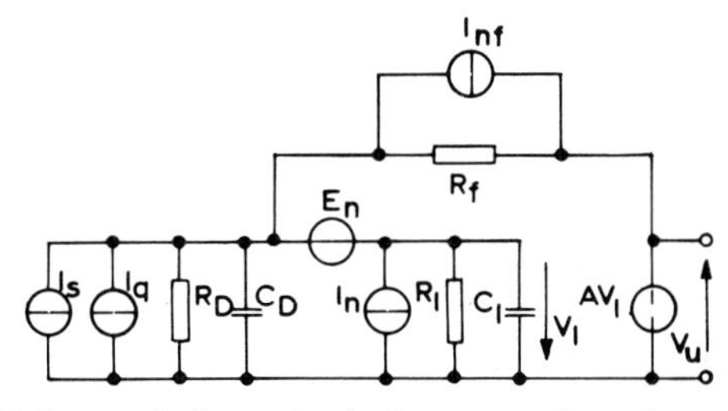

FIG. 10 – Noise equivalent circuit for a transimpedance preamplifier.

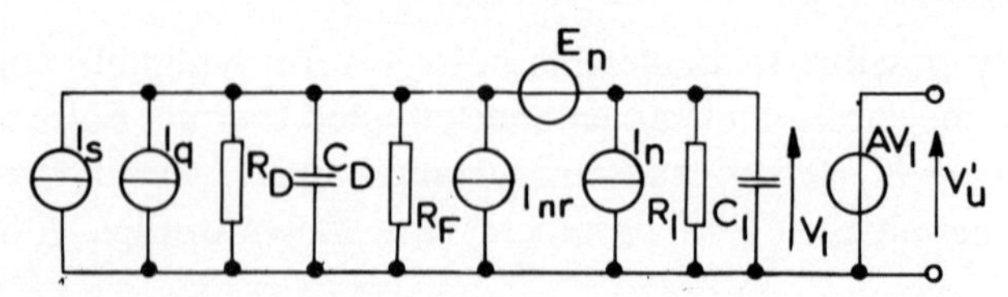

FIG. 11 – Noise equivalent circuit for a high input impedance voltage preamplifier.

izing the integrating input stage of the voltage preamplifier. It can be shown that the signal-to-noise ratio obtainable by both circuits is the same if the preamplifier equivalent noise generators are the same and the front end resistor R and the transimpedance R_F of the two amplifiers respectively are equal.

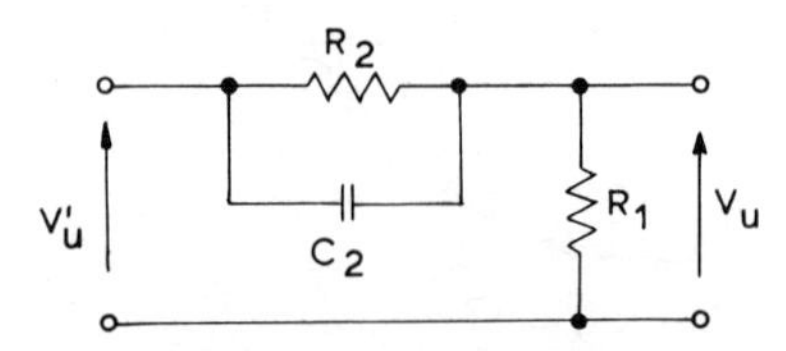

FIG. 12 – *RC* equalizer.

For the front end amplifier, the observations that will be made are valid apart from the absence of the input resistor.

Using the symbols shown in Fig. 10 where:

I_s = equivalent optical signal current generator

I_a = photodetector equivalent noise current generator

R_D = photodetector bias resistor

C_D = photodetector junction capacitance

E_n = amplifier equivalent noise voltage generator

I_n = amplifier equivalent noise current generator

C_I = amplifier input capacitance

R_I = first stage input resistance

R_F = amplifier transimpedance

I_{nF} = transimpedance equivalent thermal noise current generator

A = amplifier open loop voltage gain

for the transimpedance preamplifier, the optical signal transfer-function is represented by the following relation:

$$\frac{V_u}{I_s} = -\frac{R_F}{\dfrac{A+1}{A} + \dfrac{R_F}{A \cdot R_I} + s \cdot C \cdot \dfrac{R_F}{A}} \tag{1}$$

where: $C = C_D + C_I$.

The related transfer-function for the equivalent noise current is represented by the same relation as above:

$$\frac{V_u}{I_{q,n,nF}} = -\frac{R_F}{\dfrac{A+1}{A} + \dfrac{R_F}{A \cdot R_I} + s \cdot C \cdot \dfrac{R_F}{A}} \,. \tag{2}$$

On the contrary, the transfer-function for the equivalent noise voltage is given by the following relation:

$$\frac{V_u}{E_n} = -\frac{1 + s \cdot C_D \cdot R_F}{\dfrac{A+1}{A} + \dfrac{R_F}{A \cdot R_I} + s \cdot C \cdot \dfrac{R_F}{A}} \,. \tag{3}$$

For the high impedance input voltage amplifier, using the same symbols as in the previous equivalent circuit with the exception of:

R_F = preamplifier input resistance

I_{nr} = equivalent thermal noise current generator of input resistance

and including the differentiating network for equalizing the integrating input stage, the overall signal and equivalent noise current transfer functions are represented by the following relation:

$$\frac{V_u}{I_{s,q,n,nr}} = A \frac{\dfrac{R_F \cdot R_I}{R_F + R_I}}{1 + s \cdot C \cdot \dfrac{R_F \cdot R_I}{R_F + R_I}} \cdot \frac{R_1}{R_1 + R_2} \cdot \frac{1 + s \cdot C_2 \cdot R_2}{1 + s \cdot C_2 \dfrac{R_1 \cdot R_2}{R_1 + R_2}} \,. \tag{4}$$

Introducing the following relations:

$$C_2 = C = C_D + C_I$$

$$R_2 = \frac{R_F \cdot R_1}{R_1 + R_F} \tag{5}$$

$$\frac{R_1 \cdot R_2}{R_1 + R_2} = \frac{R_F}{A}$$

the following can be obtained:

$$\frac{V_u}{I_{s,q,n,nr}} = \frac{R_F}{\frac{A+1}{A} + \frac{R_F}{A \cdot R_I} + s \cdot C \cdot \frac{R_F}{A}}. \tag{6}$$

If $R_F \ll AR_I$ and $A+1 \simeq A$ the previous relation can be written:

$$\frac{V_u}{I_{s,q,n,nr}} = \frac{R_F}{1 + s \cdot C \cdot \frac{R_F}{A}}$$

The relating transfer-function, for the equivalent noise voltage generator, is represented by the following relation:

$$\frac{V_u}{E_n} = \frac{\frac{R_I}{1 + pC_I R_I}}{\frac{R_I}{1 + s \cdot R_I \cdot C_I} \cdot \frac{R_F}{1 + s \cdot C_D \cdot R_F}} \cdot A \cdot \frac{R_F}{A} \cdot \frac{R_F + R_I}{R_F R_I} \cdot$$

$$\cdot \frac{1 + s \cdot C \cdot \frac{R_F \cdot R_I}{R_F + R_I}}{1 + s \cdot C \cdot \frac{R_F}{A}} = \frac{1 + s \cdot C_D \cdot R_F}{1 + s \cdot C \cdot \frac{R_F}{A}} \tag{7}$$

From the comparison of the relations (1), (2), (6) and (3), (7), it can be said that, as the transfer-functions are alike, the signal-to-noise ratio obtainable by both circuits is equal.

Furthermore, considering the relation (3), we can say that the power density diagram for the signal and the noise due to E_n appear in the form shown in Fig. 13 where:

$$f_1 = \frac{1}{2\pi \cdot R_F \cdot C_D},$$

$$f_2 = \frac{1}{2\pi \cdot \frac{R_F}{A} \cdot C}.$$

Therefore, from the foregoing diagram, it is plain that, for a preamplifier in which the output noise due to E_n is predominant, the signal-to-noise ratio decreases more than proportionally with the bandwidth.

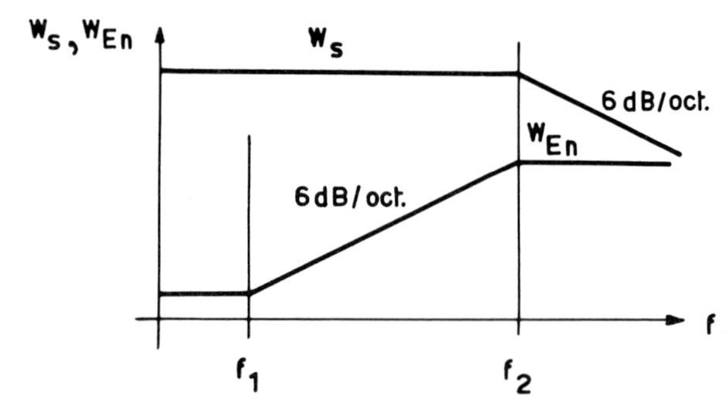

FIG. 13 – Spectral density diagram for the signal and the noise due to the E_n.

3.3.3 *Preamplifier design criteria*

Before going into detail of the preamplifier circuit design criteria to minimize the thermal noise contributions, some general observations are required.

There are three useful connections in which bipolar or field effect transistors can be operated: common emitter or source common base or gate, and emitter or source follower.

It can be shown that each configuration offers approximately the same power gain-bandwidth product and the same equivalent input noise.

Therefore, the circuit designer is allowed to operate at low-noise condition, and simultaneously to select the better configuration for meeting the requirements of gain, frequency response and impedance levels imposed by the amplifying system.

Moreover, it is very interesting for the designer to measure the contribution of two equivalent noise generators, and thus minimize the preamplifier thermal noise.

To determine I_n, we must measure the total output noise with a large source resistance R and divide this by the amplifier voltage gain A, as measured with this source resistance.

This gives the total input noise voltage, which is mostly the $I_n \cdot R$ term; dividing it by R, the I_n component is obtained.

On the contrary, a measurement of the total output noise under the $R = 0$ (short-circuit) condition therefore equals $A \cdot E_n$ where A is again the amplifier voltage gain.

The division of total output noise by A gives the value of E_n.

a) *Transimpedance preamplifier*

To minimize the transimpedance preamplifier noise, the feedback resistance value must be chosen on the ground of the following considerations:

— whereas the preamplifier output signal increases proportionally to the feedback resistance value, the thermal noise increases proportionally to the square root of its value, on the ground of the following espression:

$$E_{n R_F} = \sqrt{4 \cdot K \cdot T \cdot B \cdot R_F}$$

where: R_F = feedback resistor.

Therefore, to obtain a better signal-to-noise ratio it is necessary to select the highest possible value of this resistance;

— a lower value for the feedback resistance becomes advisable to obtain a low input impedance in such a way as to eliminate the amplitude equalization for the integrating input stage.

Moreover, the stray capacitance of the feedback resistance modifies the closed loop response, and must therefore be of a small and steady value.

Various versions of transimpedance preamplifiers employing either low noise bipolar and junction or MOS field-effect transistors can be implemented.

In any case, particular attention must be given to the first preamplifier stage, which has a leading role in minimizing the overall noise level.

In fact, to reduce the noise contribution due to the second and the following stages, the first stage gain must be increased by grounding the emitter or source; this causes the input capacitance to rise, both for the Miller effect and the base-emitter or gate-source capacitance, so rendering the amplifier adjustment more complex from the stability standpoint.

For the foregoing statements, in a bipolar transistor preamplifier, the output noise is chiefly due to the noise current generator I_n of the first stage. The thermal noise I_{nF} of the feedback resistor can be ignored with respect to I_n if the resistor value is very high.

As the I_n value is proportional to the base current, high current gain transistors at a low collector current must be used.

However, the bias current cannot be reduced below certain limits without an excessive lowering of the transistor cut-off frequency and its current gain.

The penalty due to the thermal noise I_{nF} of the feedback resistor rises as the preamplifier bandwidth increases due to the design difficulties associated with the circuit stability.

The contribution due to the transistor voltage noise generator E_n is negligible compared with the two foregoing cases.

The spectral density of the preamplifier output noise is « coloured » by E_n in that, as previously mentioned, the transfer function V_u/E_n shows a zero introduced by the combined effect of the photo-detector capacitance and the preamplifier transimpedance.

Today, commercial bipolar transistors with good performance characteristics are available, factoring the significant progress made in the field of bipolar transistors designed for use in low noise and small signal amplifiers.

Typical values of the most important parameters which influence the design of preamplifiers for commercial bipolar transistors can be summarized as follows:

— Cut-off frequency f_T at $I_c = 10$ mA	8-12 GHz
— minimum noise figure at $I_c = 2$-3 mA and $f = 1$ GHz	2-3.5 dB
— Forward current gain at $I_c = 0.2$-2 mA	50-100

Laboratory tests have shown that an equivalent input noise current generator value as low as 1.5 $\mathrm{pA}/\sqrt{\mathrm{Hz}}$ can be obtained for a transimpedance preamplifier with a bandwidth below 50 MHz and 80 kΩ for the feedback resistance. For a 100 MHz bandwidth and a 5 kΩ transimpedance, an equivalent input noise current generator of 5 $\mathrm{pA}/\sqrt{\mathrm{Hz}}$ can be achieved.

The experimental results obtained on a transimpedance preamplifier using FET as a first stage have not proved satisfactory because the corresponding output noise has always been higher than that obtained with the bipolar transistor model.

If the first stage is a FET having its source grounded, the input capacitance becomes high, thus reducing the feedback factor; hence the output noise will be increased by the same factor with which the feedback is reduced.

To reduce the input capacitance a source follower stage could be used. The corresponding output preamplifier noise is not lowered because of the noise effects of the second stage and the thermal noise of the feedback resistance inserted between the source and the ground of the first stage.

b) *Voltage preamplifier*

This type of amplifier is usefully employed in those systems where high bandwidth (above 100 MHz) is required. For these frequencies, the propagation delay existing in the closed loop of the feedback amplifier reduces the phase margin and becomes a significant design factor.

In particular, there are two criteria that can be followed in the design of a voltage amplifier, namely the development of the high input impedance and the front end amplifier.

For the first kind, a FET, showing a high input impedance, is usefully employed as the first stage.

This preamplifier, as already mentioned, is a simpler and cheaper circuit but it needs a differentiating network to equalize the integrating first stage due to the total input capacitance.

Also in this case, particular attention should be given to the design of the first-stage, both to minimize the output noise level and to facilitate the design of the subsequent equalization.

To avoid the thermal noise due to the input loading resistor, the physical input resistance is kept large. The detector capacitance must be kept small to provent the E_n contribution from increasing.

A circuit configuration which gives a good noise performance is a cascade of a first common source stage followed by a common drain stage to match the impedance. This increases the gain of the first stage and makes negligible the noise equivalent of subsequent stages.

The common emitter front end amplifier, on the other hand, employs bipolar transistors, since FET's have high input impedance which reduces the bandwidth obtainable.

This circuit is chosen because the emitter follower circuit exhibits a large input resistance, with consequent reduction in bandwidth due to the photodiode capacitance, and the common base circuit has an input resistance which gives insufficient power gain because of the high impedance of the PIN or APD.

To reduce the thermal noise contribution of this amplifier, it is necessary to choose, as the first stage, a transistor characterized by a high current gain at a low emitter current so as not to excessively reduce the cut-off frequency.

In this circuit a collector inductance is introduced as partial equalization of the integrating input stage.

Laboratory tests have shown that a value for the equivalent input noise current generator of 3-5 $pA/\sqrt{Hz}$ can be obtained for these amplifiers with a bandwidth below 500 MHz.

3.4 Timing acquisition in synchronous data-transmission systems

Whether in optical fibre or coaxial cable digital transmission systems, a regenerator is required to reconstitute the received data, contaminated with noise and with amplitude and delay distortions.

Since a statistically random data stream contains the information to syncronize the properly constructed retiming circuit, the timing waveform, at the bit rate frequency, is extracted from the same data without auxiliary pilot or extra signalling levels.

The power spectrum of a band limited NRZ scrambled binary signal does not show components at the bit rate. A non linear circuit in the timing information extractor must therefore be used to produce a spectral line, with coherent phase, at the data rate.

Good results can be obtained considering the different solutions shown in the literature if the non linear circuit is constituted by a differential network followed by a full-wave rectifier with clipping.

The overall effect is to produce a unipolar pulse coincident with every transition of the input data.

This treatment of the received data stream can be developed with analogue circuits that implement the single functions described above using discrete or integrated components.

Another easy circuit implementation of the functions in digital form consists in deriving from the data pulse zero-crossings a secondary pulse stream obtained with an exclusive-OR function between data pulses and pulses delayed by an approximately half bit period.

The Fig. 14 shows the basic zero-crossing half-width pulse generator.

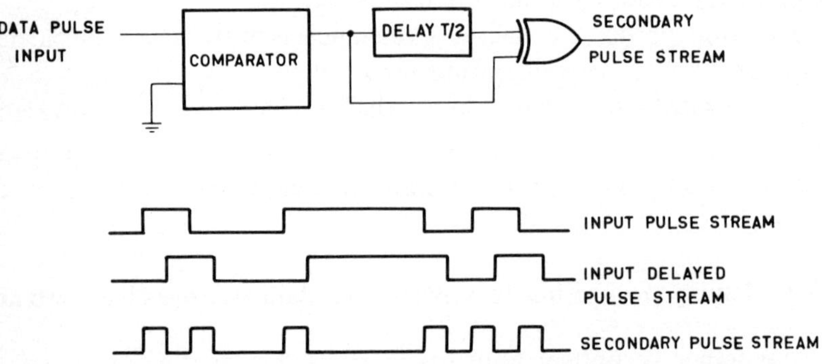

FIG. 14 – Diagram of a basic zero-crossing half width pulse generator.

In this solution, the main effect of the thermal noise and intersymbol interference is to perturb the zero-crossing odd data train causing a phase jitter in the recovered clock: this means that only the average position of pulses is at the data transition time.

As the mean square jitter is inversely proportional to the Q factor of the tuned circuit used for the timing extraction, the timing jitter may be reduced to a very low value by using a sufficiently narrow filter.

This requirement is also dictated by the consideration that a pseudo-random sequence can ensure a long term average transition density of 50 per cent, but there can still be long intervals without timing information.

In the literature, we can find two different criteria that are followed in the choice of the Q factor of the tuning circuit.

In fact, on the basis of general system considerations, the bandwidth of this tuned circuit can change from 10^{-2} to 10^{-4} of the bit rate.

While a large value for the bandwidth ($Q \simeq 100$) can be obtained with a traditional *LC* filter, for a very narrow band ($Q \gg 1000$), crystal filter or phase locked loops (PLL's), using as VCO's *LC* thermally stable or crystal oscillators, must be used to perform the filtering process.

While crystal filters are intrinsically selective filters with highly stable central frequencies, and are thus characterized by an acceptable static phase error, in a phase locked clock extractor the phase errors can be reduced by controlling the drift of the quiescent VCO frequency, the offset voltages of the phase detector and the d.c. loop amplifier.

In particular, the VCO choice is merely due to its ease implementation, and good thermal stability which minimizes the obtainable quasi static phase error.

Another important design factor is the difficulty of implementation of the phase detector, particularly because of the high frequency range.

In this case, a good balance, necessary for ensuring minimum offset voltage influence, is obtained by a double balanced multiplier equipped with Schottky-barrier diodes.

In this solution, the influence of offset voltages can be made smaller by increasing the level of the signal voltages and by the good matching available for the diodes diffused on a single chip. Of course, a d.c. loop amplifier with very low offset voltages should be used, so as not to increase the quasi static phase error.

REFERENCES

[1] K. Asatani and T. Kimura: *Non linear phase distortion and its compensation in L.E.D. direct modulation*, Electron. Lett., Vol. 13, No. 6 (1977), pp. 162-163.

[2] J. Straus, A. J. Springthorpe and O. I. Szentesi: *Phase-shift modulation technique for the linearisation of analogue optical transmitters*, Electron. Lett., Vol. 13, No. 5 (1977), pp. 149-151.

[3] M. Chown, A. R. Goodwin, D. F. Lovelace, G. H. B. Thompson and P. R. Selway: *Direct modulation of double-heterostructure lasers at rates up to* 1 Gb/s, Electron. Lett., Vol. 9, No. 2 (1973), pp. 34-36.

[4] G. White and C. A. Burrus: *Efficient* 100 Mb/s *driver for electroluminescent diodes,* Int. J. Electron., Vol. 35, No. 6 (1973), pp. 751-754.

[5] U. Wellens: *High-bit-rate pulse regeneration and injection laser modulation using a diode circuit*, Electron. Lett., Vol. 13, No. 18 (1977), pp. 529-530.

[6] S. R. Salter, D. R. Smith, B. R. White and R. P. Webb: *Laser automatic level control circuits for optical communications systems*, Third European Conference on Optical Communication, Munich (September 14-16, 1977).

[7] Y. Ueno, Y. Ohgushi and A. Abe: *A* 40 Mb/s *and a* 400 Mb/s *repeater for fiber optic communication,* First European Conference on Optical Fibre Communication, London (September 16-18, 1975), pp. 147-149.

[8] A. Fausone, G. Rosso and G. Veglio: *Repeater for* 34 Mb/s *transmission system over optical fibre*, Second European Conference on Optical Fibre Communication, Paris (September 27-30, 1976), pp. 351-356.

[9] J. L. Hullet and T. V. Muoi: *A feedback receive amplifier for optical transmission systems*, IEEE Trans. Commun. (October 1976), pp. 1180-1185.

[10] J. E. Goell: *An optical repeater with high-impedance input amplifier*, Bell Syst. Tech. J., Vol. 53, No. 4 (April 1974), pp. 629-643.

[11] J. E. Goell: *Input amplifiers for optical PCM receivers*, Bell Syst. Tech. J., Vol. 53, No. 9 (November 1974), pp. 1771-1793.

[12] P. K. Runge: *An experimental* 50 Mb/s *fiber optic PCM repeater*, IEEE Trans. Commun., Vol. Com-24, No. 4 (April 1976), pp. 413-418.

[13] T. Ogawa, T. Yamashita, Y. Mochida and K. Yamaguchi: *Low noise* 100 Mb/s *optical receiver*, Second European Conference on Optical Fibre Communication, Paris (September 27-30, 1976), pp. 357-363.

PART V

INTEGRATED OPTICS

INTRODUCTION TO INTEGRATED OPTICS

by Vittorio Ghergia and Antonio Scudellari

The term « Integrated Optics » (IO) appeared at the end of the Sixties, when research into the propagation of light in dielectric films had its first significant development. This term was invented by Miller [1], who merged « optical integrated data processors » [2] and « optical integrated circuits » [3] which signified the utilization of optical surface waves in data-processing networks. The main purpose of substituting electronic integrated circuits with optical integrated circuits was to obtain better performance: optical integrated circuits present large bandwidths, lower power dissipations and lower sensitivity to electromagnetic noises.

At the present time, the replacement of electronics by integrated optics in data processing networks seems to have become a less fundamental objective. In fact, with the increasing possibility of using long-distance optical communication links, integrated optics has also opened up new horizons for the generation, guiding, modulation and detection of light in miniaturized and reliable components having low power dissipation and high thermal and mechanical stability. The application of integrated optics to telecommunications entails the transition from lumped circuits to distributed circuits. A similar phenomenon was observed not long ago with the advent of microwave-integrated circuits techniques. Although microwave systems employ more complex components than those currently employed in fibre optical systems, many techniques can be transferred to the latter, such as those relating to periodic structures, directional couplers, ring resonant filters, bends and tapers.

Passive components for optical integrated circuits have therefore been developed as the equivalent of microwave integrated circuits. Active components such as sources, modulators and detectors have also been obtained. Semiconductor laser sources formed by groups III and V elements have proved to be of great interest for integrated optics, due to their dimensions and compactness. Several methods have been followed for modulators, by determining electro-optical, acousto-optical and magneto-optical effects in thin film structures. In detectors the experience acquired in silicon technology for electronic circuits has been applied.

As a first approach, the above passive and active components made with materials of different types have allowed a hybrid type of integration (on the analogy of electronic integrated circuits). In order to obtain complete monolithic integration, which is particularly suitable for future applications, it will be necessary to implement all the required processing functions on a single material. This has led to strong performance limitations due to the impossibility of choosing the most suitable material for each component. Consequently, major technological difficulties still have to be resolved.

At the present time, definite results have not yet been obtained. The basic material which appears to meet the requirements of optical monolithic integrated circuits is GaAs, which is likely to play the same role in integrated optics as silicon in electronic integrated devices. In fact, GaAs by itself or in ternary or quaternary compounds together with III and V elements is basic to all the functions required by each single component of an optical integrated circuit, such as light generation, light guiding, modulation and detection. Unfortunately, considerable technological difficulties have already been encountered in the preliminary development stage of integrated optics. The first difficulty related to the small dimensions of the devices which are bounded by the optical radiation. This problem was overcome by the use of submicron photolithographic technologies such as electron beam exposition and plasma etching. Another difficulty concerned the new materials, either in thin film or semiconductor form, and the development of special epitaxial growth technologies (for instance, molecular beam epitaxy). Another difficulty was encountered regarding electrical and optical measurement methods and refined techniques of chemical-physical characterization.

At present, production processes for integrated optics have

not yet been established. Several components have already been produced with good characteristics in the laboratory. Very good results have been obtained even in very sophisticated versions such as the parallel integration of various components performing the same function as multiplex and multi-switching networks.

However, the relatively slow development of such components and integrated optics technology in general may be observed. This is due mainly to the fact that the major advantage of integrated optics, that is the large bandwidth due to its monomode operation, is not fundamental to present and future projects. In fact, optical communication systems employing traditional electronics together with bulk optical devices and multimode fibres may currently be used at bit rates of several hundreds of Mbit/s. These values are largely sufficient for present and future requirements.

The great potential of integrated optics from a technological point of view must also be taken into account. It is sufficient to remember that some of the most interesting among discrete components, such as distributed feedback lasers, are currently being produced by integrated optics technologies.

In the following we will highlight the main aspects of this new technique. In particular, we will consider light propagation in dielectric guides, materials for manufacturing integrated optical devices, technologies and characterization methods, and active and passive devices which have been produced, as well as future development prospects.

REFERENCES

[1] S. E. Miller: *Integrated optics: an introduction*, Bell Syst. Tech. J., Vol. 48 (1969), pp. 2059-2069.

[2] R. Shubert and J. H. Harris: *Optical surface waves on thin films and their application to integrated data processors*, IEEE Trans. Microwave Theory Tech., Vol. MTT-16 (1968), pp. 1048-1054.

[3] D. B. Anderson: *Application of semiconductor technology to coherent optical transducer and spatial filters*, in *Optical and Electro optical Information Processing*, J. Tippet, M.I.T. Press, Cambridge (March 1965), p. 221.

CHAPTER 1

LIGHT PROPAGATION IN INTEGRATED OPTICS WAVEGUIDES

by Antonio Scudellari

All IO components are essentially made of optical waveguides, i.e. of structures which confine the electromagnetic field in one dimension (slab guides) or in two dimensions (strip guides). In fact, the miniaturization process in the case of IO is such that the configuration of the electromagnetic field inside the devices must be taken into account. To introduce IO waveguides it may be useful to compare them to microwave waveguides, waveguides for acoustic surface waves and optical fibres.

In regard to the comparison with microwave waveguides, two different cases must be considered corresponding to metallic boundary waveguides, such as rectangular waveguides, and to dielectric waveguides, such as those used for millimeter-wave integrated circuits. In the first case, the comparison is only qualitative, while in the second it may be quantitative.

In the metallic boundary case it must be considered that metallizations for optical frequencies are quite different from those for microwave frequencies. For microwave frequencies a metal may be considered as a perfect conductor. For optical frequencies it must be considered as an overdense plasma with a complex dielectric constant, whose real and imaginary parts are strongly influenced by frequency variations. Very often the real part of the dielectric constant is negative and the imaginary part is large, resulting in high losses. Therefore, while for microwaves the boundary conditions in metallic waveguides take a particularly simple form, in the case of IO waveguides the conditions are more complex. They must correspond to the evanescence condition in a medium which, as in the case of metallizations, may show losses and a negative dielectric constant. Obviously, however, the solution obtained for microwave configurations may be a useful reference for understanding the more complex optical problem. This is particularly true when electromagnetic fields are concentrated inside the guide and decay rapidly outside.

In regard to comparison with millimeter-wave dielectric guides, it should be noted that from the theoretical point of view any IO guide can also be used for microwaves. In this case, however, the typical phenomenon of poor guidability arises. Therefore, the use of dielectric guides is possible only in the millimeter-wave frequency range and only if the difference between the refractive indices of the waveguide and of the surrounding medium is very large. This quantitative comparison is also important from an experimental point of view. For instance, the only measurements of bent losses in IO guides have been performed by models obtained at millimetric wavelengths [1].

In the area of waveguides for acoustic surface waves, two types of guides are quite similar to IO guides, namely, the overlay type (strip or slot) and the diffused or implanted type. It should be noted that these analogies do not only apply to the theoretical model, but also to the practical implementation. It is useful to observe that the techniques used to obtain higher diffusion depths in acoustic waveguides may also be utilized for increasing the number of modes in optical diffused waveguides. The theoretical correspondence between acoustic and optical waveguides may be easily understood by taking the acoustic strip as a reference. The strip is a « slow » acoustic material lying on a « fast » substrate. It corresponds to a strip of optical material with a high refractive index on a substrate with a low refractive index. The fields which are equivalent to the magnetic and the electric fields in the optical waveguide are the « velocity » and the « stress tensor » fields. The concentration of these fields in the strip varies as a function of its thickness and decays exponentially in the surrounding medium [2].

The comparison between IO guides and optical fibres has two fundamental aspects: the difference in the propagation characteristics required and the difference in the geometries employed, so that different methods for propagation analysis are needed. As regards propagation characteristics, in the fibre optic field more attention is paid to attenuation and distortion of a modulated signal propagating over a long distance. In the case of IO, on the other hand, emphasis is placed on the following points: 1) the necessity of avoiding conversions to degenerate or higher-order modes which lowers the efficiency of devices working in a very narrow bandwidth around a well-defined guided mode wavelength; 2) the reduction of roughness which, perturbing the field intensity in the proximity of etched walls, represents the main loss mechanism; 3) the selection of a well-confined or a fringed mode configuration when active (or non-linear) light processing or coupling of light outside the guide are required. From the geometrical and propagation point of view it should be noted that the guiding structure is limited in both cases by a boundary of the open type, consisting of a dielectric discontinuity surface beyond which an electromagnetic field, albeit evanescent, still exists. This field must at the same time be matched to the internal field and decay very rapidly with distance. In the fibre optic case (circular geometry), polar coordinates are used since decay and dielectric discontinuity occur along one coordinate (radial variable r) only. On the contrary, in the case of IO guides having rectangular geometry, perpendicular Cartesian coordinates are particularly suitable for expressing the matching condition, while a radial coordinate r is necessary to impose the evanescent condition. As a consequence, an exact analytical expression for the electromagnetic field does not exist. Slab guides (see Fig. 1 of Section 1.1) are the only exception to the above considerations. In fact, the continuity condition must be imposed on the vertical direction of field evanescence. For limited geometry IO guides, only an approximate solution may be found to the problem of boundary conditions. For this purpose, different analysis methods may be followed, providing results the precision of which depends on the simplifying conditions introduced (these methods and pertinent references will be referred to in Section 1.2).

1.1 Slab waveguides

Slabs are waveguide types characterized by the absence of bounds in the horizontal direction. They represent a preliminary step, both from a theoretical and an experimental point of view, in the direction

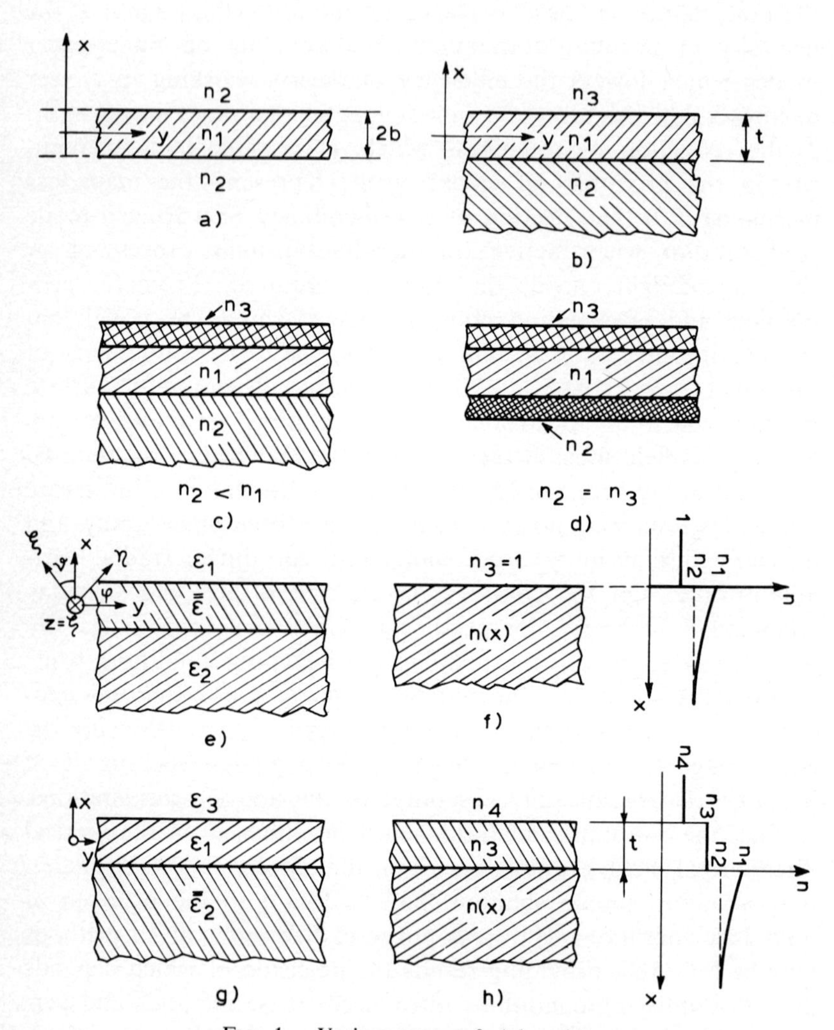

FIG. 1 – Various types of slab guides.

of more sophisticated guide types. The theoretical interest in slabs is due to the effective index method [3] which allows channel waveguides to be treated as slab waveguides. The experimental interest is due to the fact that propagation measurements are simpler on slab waveguides than on channel waveguides. In spite of these facts, some applications exist for slabs when horizontal confinement is unnecessary or undesired. For instance, in several types of semiconductor lasers lateral confinement is not required, while for acousto-optic deflection modulators confinement cannot be tolerated. Other examples are planar devices for optical data processors (lenses and prisms), where confinement is not desired, and systems for light coupling to narrow channel guides from the air. In this case coupling is obtained as a result of the following steps: 1) light coupling from the air to a large channel guide by means of prisms or gratings; 2) transition from a large channel guide to a narrow guide by means of a tapered section.

Figure 1*a* shows the simplest version of a symmetric slab. It is made of an infinite homogeneous dielectric plane slab surrounded by a medium with a lower refractive index. (The analogous microwave guide is the so-called parallel planes guide). As previously stated for the symmetric slab, an exact analysis is possible (see, for instance, [4]). Symmetry, which is achieved when the substrate and the superstrate have the same refraction index, finds practical applications in double heterostructure GaAs lasers.

Figure 1*b* represents one asymmetric slab, where the underlying and the overlying medium are different. The substrate must have a smaller refractive index ($n_2 < n_1$) in order to confine light inside the slab by total reflection. The cover is usually air and therefore $n_1 > n_2 > n_3$. An exact electromagnetic analysis exists [4] and several simplified methods may be applied (the zig-zag model [5] and the equivalent network for transverse section [2]). Among the various types of slab, the asymmetric one is the most studied and also the most experimented and implemented. In practical applications, the number of modes is usually fixed and the slab thickness and index difference must be properly chosen for light confinement. Thin slabs with a large refraction index difference show higher field intensity on the surface; therefore, they are suitable for surface excitation and interaction. On the other hand, thick guides with a smaller refraction index difference show lower losses, and are better suited for end coupling by lenses. As their modes may be better controlled by external fields, the implementation of modulators is made easier.

When metallizations are present, either Fig. 1*c* (slab on a lower refraction index substrate ad metallization on the top) or Fig. 1*d* (slab between two metallizations) may apply. Figure 1*c* and Fig. 1*d* are also known as asymmetric and symmetric, respectively. In practice, the thickness of the layers is much larger than the penetration depth of electromagnetic fields, so that metallizations may be considered to be infinite. In this way, the propagation characteristics are obtained by the same method used for slabs with optically transparent media at the boundaries. The only complication arises from the introduction of complex refractive indices with a negative real part [6]. As regards the utilization of these types of guides, metallizations produce larger losses. However, they are sometimes necessary, both for the application of electrical fields (electro-optic modulators), and for the implementation of polarizers and mode filters, since TE and TM modes are differently attenuated by metallizations.

Figure 1*e* shows an asymmetric slab similar to that of Fig. 1*b* but now the main propagation medium is an anisotropic material. ξ and η are the principal dielectric axes and θ is their rotation with reference to the x and y axes. Anisotropic slabs may be analyzed in a rigorous way. The procedure is complex, since it is not possible to separate transverse electric and magnetic solutions in the general case of $\theta \neq 0$. Hybrid modes are obtained, in the sense that a sort of coupling exists between the TE and TM configurations [7]. Only in some particular cases the anisotropic slabs show linear polarization. For instance, in the case of axial reference ($\theta = 0$) the TE and TM modes can be separated and are similar to those of isotropic slabs. The only exception with respect to isotropic slabs is that TE modes, with horizontal polarization, are affected by the principal dielectric constant along the y-axis, while TM modes, with vertical polarization, are prevalently influenced by the principal dielectric constant along the x-axis.

The most widely used types of anisotropic material (i.e. lithium niobate and tantalate) are uniaxial. Therefore they are characterized by an ordinary refractive index n_0 and an extraordinary index n_e. If a vertical crystal axis is chosen, the TE modes are prevalently influenced by n_0 and the TM modes by n_e. A further simplification is to consider the TE modes as depending only on n_0 and TM modes on n_e. This simplification is useful for overcoming further complications introduced by the presence of graded junctions in anisotropic materials.

Another type of slab guide is the graded index slab shown in Fig. 1*f*. In the graded index guide (Fig. 1*b*) the refractive index is not constant, but is a function $n(x)$ of x, reaching a maximum value n_1 in the upper interface ($x = 0$) and a minimum value n_2 in the lower interface. This type of waveguide (which is actually made of two adjacent half-spaces) is also given the male of slab guide since rays are bent by the index gradient in the same way as they are reflected by a step boundary. The technological processes used in the production of graded junctions are ion-exchange, ion-implantation and diffusion. Ion-exchange is employed mainly for glass guides. Ion-implantation finds wider applications in the production of guides on glass for passive devices (in place of sputtering) and of guides on semiconductors (such as GaAs) for active devices. Diffusion is the most widely used method for guides on semiconductor materials and single-crystal substrata. Devices produced by ion-implantation can be more easily repeated and controlled than those obtained by diffusion. In fact, the typical defect of diffusion is the difficulty of obtaining a correspondence between the technological process variables (type and method of doping, diffusion temperature, diffusion time) and the geometrical parameters which are of interest for electromagnetic analysis (shape and characteristic depth of function $n(x)$, values n_1 and n_2). In propagation analysis several $n(x)$ shapes have been assumed: half-Gaussian, asymmetric Gaussian, complementary error function and exponential. Furthermore, in order to simplify the propagation analysis, approximate shapes such as piecewise linear or truncated Taylor expansions have been introduced. It should be noted that the asymmetric Gaussian function is suitable for describing « buried guides », that is guides where the value of $n(x)$ corresponding to the maximum concentration light falls within the medium. In a buried configuration, roughness is less critical. Most studies concerning graded index slab guides deal with high diffusion depths (thick slabs) Thick guides correspond to graded fibres with a large radius. Theoretical analyses in this case are based on the WKB approximation [8] or the zig-zag model [9]. Thin graded slabs (monomode or quasi-monomode), on the other hand, require a more rigorous analysis. This may be obtained only by direct solution of the Helmotz equations. Exact solutions have been obtained for an exponential profile of $n(x)$ [10], while for a Gaussian profile only approximate solutions have been derived [11]. Several modifications have been made to the graded index slab, owing to the necessity of improving the correspondence between the technological processes and geometrical parameters that are essential for electromagnetic analysis.

Figure 1*g* shows an example of a graded slab with an overlayer having a refraction index $n_3 < n_1$. This layer is used to remove a small fraction of optical energy from the guide so that the phase velocity of light in the medium can be adjusted to the desired values. The layer with refraction index n_3 is obtained by means of liquids with a controlled refraction index or by deposition of a sequence of very thin films. In other cases, a type of slab similar to that in Fig. 1*g* is used, where the hypothesis $n_3 < n_1$ is changed to $n_3 > n_1$. In this case, the medium with index n_3 may become the main propagation medium if its thickness t is larger than the cut-off value [12]. Otherwise, the main propagation medium is still that with graded index. It appears that it is still less easy to implement $t < t_{\text{cut-off}}$ solutions than those with $t > t_{\text{cut-off}}$, since now the critical parameters are those relating to technological processes such as the overlay thickness t and the diffusion profiles. Furthermore, tuning of propagation characteristics is more difficult. As regards the problem of tuning, it should be mentioned that strip diffused channel guides (to be examined in Section 1.2) are often employed because propagation is controlled by width w rather than by thickness t, which is more difficult to control from a technological point of view.

Figure 1*h* shows an asymmetric slab guide with an isotropic propagation medium and an anisotropic substrate. This may be employed to produce waveguide modulators which are similar to those obtained by means of the anisotropic slab in Fig. 1*e*.

1.2 Strip waveguides

Strip waveguides are more widely employed than slabs in IO circuits, since they guide light, instead of simply confining it in one dimension, as do slabs. Figure 2 shows the most significant types of strip guide from the point of view of practical applications.

The simplest example of a strip waveguide is the rectangular dielectric waveguide in Fig. 2*a*. It corresponds to a symmetric slab with both vertical and horizontal confinement. The microwave equivalent structure is the rectangular metallic waveguide. When $b \simeq a$, the propagation characteristics of strip waveguides are quite similar to those of a step-index fibre with a radius equal to a. Owing to technological implementation difficulties, the interest in this type of guide is mainly theoretical. Its study is introductory to that of more complex geometries. The lack of an exact electromagnetic analysis has already been mentioned. For the various methods used for an approximate solution, see references [13-22]. A more commonly used waveguide is the raised strip guide shown in Fig. 2*b*. From a technological point of view it must be considered as the result of etching an asymmetric slab, Fig. 1*b*. From a theoretical point of view it may be considered as a rectangular dielectric waveguide coupled to the underlying substrate. Owing to its simplicity the raised strip waveguide has a wide utilization: all other strip guides are modifications of this fundamental type. Modifications are mainly dictated by the need to avoid its basic defect: attenuation due to roughness on the etched sides. Figure 2*c* shows a ridge guide (rib guide) which, from a technological point of view, is obtained in a similar way to the raised strip guide, limiting the removal of surrounding material. From a theoretical point of view, on the other hand, it may be considered as a composite structure obtained by coupling an asymmetric slab and a rectangular dielectric waveguide. Numerical analysis method, such as variational or field expansion, appeared only recently for the study of propagation characteristics [13, 21]. In the past, approximate methods were available as an alternative: the method of Marcatili [22], which is valid only for large transversal dimensions, and the effective index method,

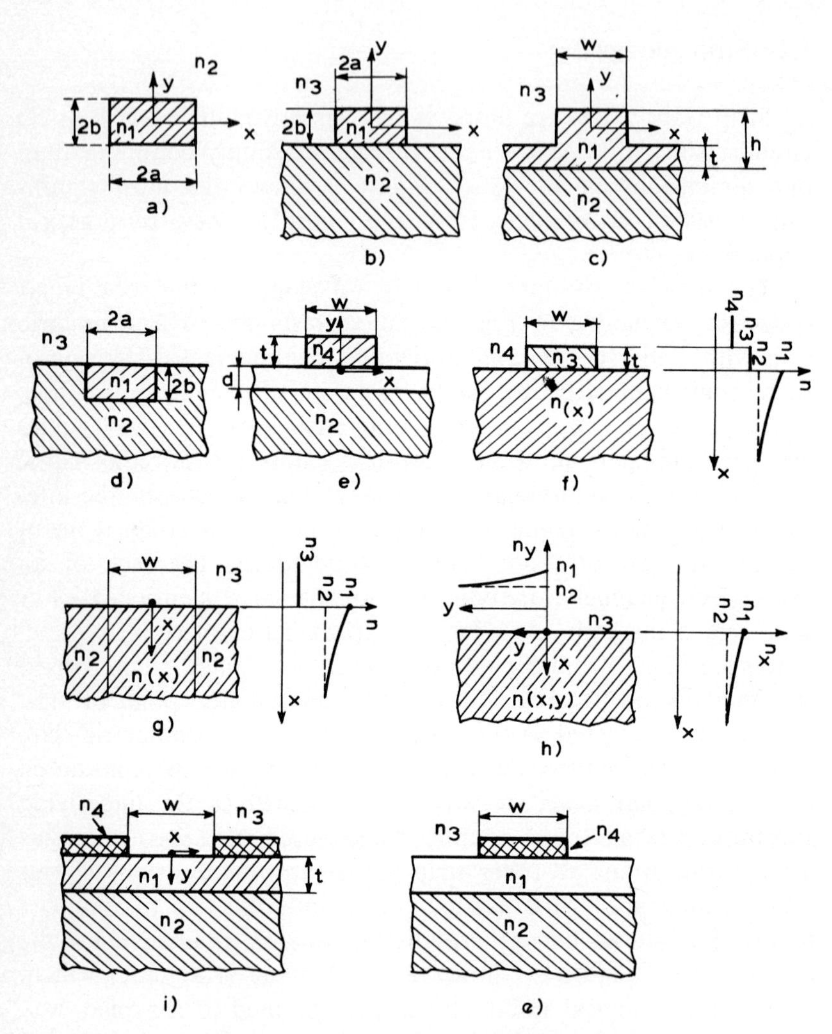

FIG. 2 – Various types of IO strip guides.

which is also useful to justify qualitatively the operation of this type of guide, were used. In fact, starting from the raised strip guide, in order to reduce roughness the index difference $(n_2 - n_1)$ could be reduced (see Section 1.5). However, if the index difference is lowered to avoit cut-off, it is necessary to produce thicker guides. Since the production of thicker guides increases technological difficulties, the configuration of the ridge guide may be used as an alternative. The parameters of the ridge guide, in fact, are chosen in order to keep the thinner portions of the slab guide with index n_1 below cut-off so that light propagates only in the thicker region (within W). In this way, only a part of the transmitted wave strikes the etched vertical walls which show the roughness defect. Another notable characteristic of this type of guide arises from the fact that the decay of electromagnetic fields in lateral slabs of thickness t is very slow. Coupling to adjacent guides is thus facilitated. A drawback lies in the fact that the minimum bent radius is larger (see Section 1.4). Furthermore, it should be noted that a ridge guide is closely related to the single material monomode fibre and the image ridge guide for millimiter-wave integrated circuits [23]. In fact, studies on single material fibres [22] often refer to the symmetric ridge guide in Fig. 3*a*, and the image ridge guide in Fig. 3*b* is obtained from the ridge guide of Fig. 2*c* by substituting the total-reflection lower boundary with an ideal conductor. The modes of the image ridge guide are identical those of a symmetric ridge guide having field components unsymmetries with respect to the vertical direction.

Another type of guide allowing a reduction of roughness problems is the embedded channel guide in Fig. 2*d*. From a theoretical

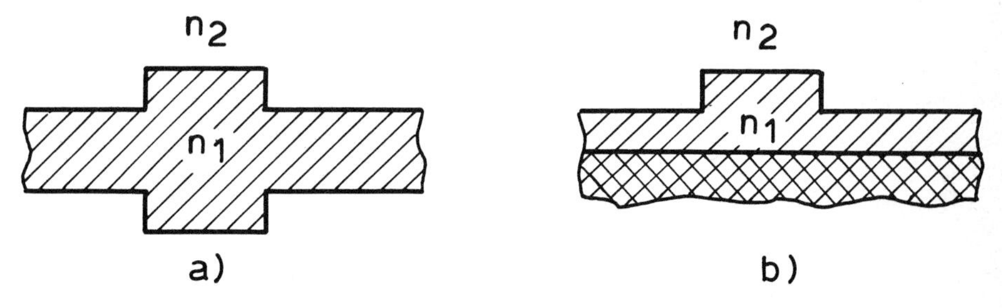

FIG. 3 – Symmetric ridge guide (*a*) and image ridge guide (*b*).

point of view this guide is similar to the raised strip in Fig. 2*c*. From a technological point of view it may be obtained by diffusion or ion-implantation through a mask, provided that the resultant transition are sufficiently step-like. Another process for obtaining embedded guides is embossing. In this case a channel is obtained when a plastic material is deformed by a medium with a larger refractive index (Fig. 2*e*). Roughness problems may also be reduced with strip loaded waveguides. Two versions of strip loaded waveguides are available: one where the greatest index is n_1 [9], and one where the medium with the greatest refractive index is that with index n_4 [24]. In both cases light propagates within the medium with index n_1. In the first case the thickness of this layer is lower than the cut-off value (a function of difference $n_1 - n_3$), except in the region underlying the dielectric bar with index n_4. In the second case the bar itself, owing to its geometrical parameters W and t, undergoes cut-off conditions. The microwave equivalent is the microstrip line. Both types of guide are less sensitive to roughness, allow a greater separation between side-coupled guides and wider strips for maintaining single mode operation. The drawback is, as usual, a greater tolerable bend radius. The version with a strip having refractive index n_4 greater than n_1, as opposed to that having a strip of refractive index n_4 lower than n_1, has the advantage of a greater light concentration within the medium of index n_1. However, the amount of confinement is very sensitive to strip thickness: this detail makes the construction of parallel guide couplers critical.

Still in the field of channel waveguides with weak horizontal confinement, guides with graded transitions must, be considered. One type, the strip-loaded diffused waveguide, is illustrated in Fig. 2*f*. It is produced by superimposing a strip of material with index $n_3 < n_2$ on a diffused substrate. A second type, a channel diffused waveguide with unidimensional diffusion of width W, is shown in Fig. 2*g*. Another type is represented in Fig. 2*h*, which shows a channel diffused waveguide with bidimensional diffusion showing a maximum of refraction index at $x = 0$, $y = 0$. Diffused channels are in general obtained by diffusion (or ion-implantation or ion-exchange, etc.) after masking. The diagram in Fig. 2*h* is more general but also more complex. It must be applied, however, when diffusion is obtained from a strip-like source material that has been deposited on the substrate. For all three types of diffused transition guides it must be noted that a graded boundary has a much lower guidance effect than a step-index boundary. For the same reason, a lower lateral confinement is obtained with a bidimensional diffused channel than with a unidimensional diffused channel. A still lower confinement is obtained in the case of the strip-loaded configuration, that is a slab (without confinement along y) with a superimposed strip. As regards losses, channel diffused waveguides in Fig. 2*g* and Fig. 2*h* have lower losses than the strip-loaded guide in Fig. 2*f*, because of their embedded configuration. For all three types of diffused transition guides, a widely employed method for electromagnetic study is that of the effective index [25].

In conclusion, two strip guides with metallization must be mentioned: the slot guide in Fig. 2*i* and the metal strip in Fig. 2*e* Guides with metallizations are seldom used at optical frequencies, since they are lossy. Nevertheless, they find applications in the field of modulators and active devices. In the production of electro-optic devices the slot guide is more important: a couple of electrodes, separated by a gap, have the configuration in Fig. 2*i*. Losses in slot guides are relatively small, since light is concentrated within the region under the gap without reaching metallizations. The metal strip, on the other hand, is a limited-geometry version of a metal slab (see Fig. 1*c*). The main characteristic is that of mode filtering. The electromagnetic analysis may be carried out by the effective index method [26, 27] or by the method of the equivalent network for the transversal sections [2]. Finally, it should be noted that the slot guide is not similar to the slot line used at microwaves, but resembles the slot guide for acoustic waves.

1.3 Bent guides

With strip guides the utilization of bent guides cannot be avoided, for instance, for dividing two previously coupled guides, and for producing ring resonators in branching filters (see Section 2.5). In the first application the calculation of radiation losses is important since they determine the minimum allowable bend radius.

For the second application the calculation of losses is important for determining the quality factor of the resonator. For design purposes the knowledge of the guided wavelength is also important.

Almost all available theoretical analyses are similar to those utilized for fibres. Among these methods the most important is of the perturbative type (Marcatili [28], Marcuse [29], Lewin [30]) where the external field is represented by functions which express radiation (Henkel functions of complex order). The field is matched to an internal field which is assumed equal to that of a straight guide (therefore bending is considered as a small perturbation). Other methods have appeared recently: that of Snyder [31], which is applicable only to small bends like those of fibres, and that of Heiblum-Harris [32], which is based on conformal mapping which may also be applied to graded structures.

Analytical expressions for bending losses are quite involved. Consequently, only the following typical form may be given:

$$A = C_1 \exp [- C_2 R(\beta^2 - \beta_c^2)^{\frac{1}{2}}]$$

where, R is the bend radius, β the propagation constant of the mode, β_c the cut-off value and C_1, C_2 constants. By means of this expression it may be shown that only near cut-off modes are strongly attenuated, and that losses are low if the bend radius is greater than a critical value and become very high if this value is exceeded. As an example, at $\lambda = 0.63$ μm, with an index difference of 0.01, a bend radius of 1 mm may be used if the guide is far from cut-off [28].

1.4 Coupled guides

The typical configuration of a directional coupler (see also Section 2.4) is represented in Fig. 4, where two parallel and symmetric optical waveguides are coupled by their evanescent fields. When coupling is absent, the two guides are single-mode and their respective modes have the same phase constant (condition of synchronism) so that they are in the condition of maximum coupling. Such a coupling occurs when the two guides are close together so that their external fields overlap. This distributed coupling is very weak. Therefore a forward directional coupler results [17]. The power exchange is characterized by spatial beats with complete transfer after a length given by $\lambda/2k$, when k is

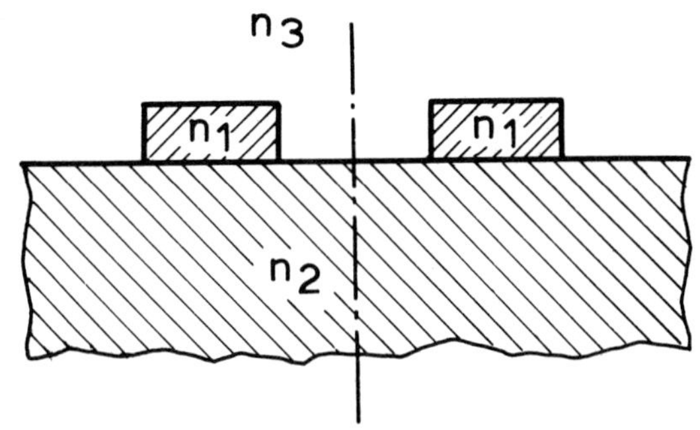

FIG. 4 – Cross-section of a parallel strip directional coupler.

the coupling coefficient. This coupling coefficient determines the period of spatial beats and the bandwidth above which coupling is maintained. With lateral decaying fields, coupling is weak and therefore bandwidth is very narrow. As a consequence, the phenomenon of spatial beats is very delicate, since it is quite sensitive to small deviations from synchronism. For this reason practical implementation must be very accurate: the two guides must be perfectly equal and devoid of roughness, materials must be uniform and the gap between guides must be constant. In order to relax such requirements it is necessary to design a stronger coupling by means of guides with very slow transversal decay (the drawback is represented by greater bending losses: see preceding sections).

From an analytical point of view, coupling constant k may be computed as an integral, known as the transverse overlap integral, containing the product of the transverse profiles of the two interacting modes. As an alternative, k may be expressed in terms of the difference $(\beta_E - \beta_0)$, between the propagation constants of even and odd modes (the even and odd modes are the two configurations obtained when a magnetic or an electric wall is set in the plane of symmetry). The design of a directional coupler requires the siution of an electromagnetic problem in a limited section geometry.

1.5 Losses in guides

Almost all types of losses in IO guides, such as material absorption, material scattering, etc., are similar to those of fibres. The only exception concerns roughness which is negligible in fibres while it is the main source of losses in IO guides. Owing to roughness (and the need to use special materials), losses in IO guides are much higher than those of fibres. On the other hand, losses do not cause serious problems in an IO circuit, as the overall length of guides in this case is small. Usually losses up to 10 dB/cm may be tolerated, even if satisfactory values are in the range of 1 dB/cm.

For all kinds of losses that are common to fibres analytical methods and experimental verification procedures exist. For roughness, on the contrary, only an approximate theory is available, developed by Marcuse [33], but not verified experimentally.

A noteworthy result obtained by Marcuse regards a symmetric slab (see Fig. 1*a*), where roughness is present on a single face and is transverse to the propagation direction: with a roughness of less than 5% high losses can be avoided. Furthermore, Marcuse shows that losses increase when the index difference between the core and the external medium increases.

1.6 Periodic structures

Periodic structures are very important in IO, since they are used in the production of several devices: input-output couplers (with laser beams, with fibre or other planar circuits), band-stop filters, modulators, non-linear devices, Distributed Feed-Back (DFB) and Distributed Bragg Reflectance (DBR) lasers, etc.

The theory of periodic structures has already been developed in other areas of applied physics (crystalline and atomic gratings, multilayer optical films, slow structures for microwaves, etc.). IO periodic structures give rise to the following new and more complex problems: 1) structures are limited by radiative boundaries, so that they are neither unlimited, as in the case of atomic gratings, nor limited by closed boundaries, as in the case of slow structures for microwave tubes; 2) there are various types of materials which may be active (lasers and optical amplifiers), non-linear (parametric devices) or non-reciprocal (isolators). For IO periodic structures the only simplifying hypothesis that may be added is that all structures have a planar geometry. Therefore, if the direction of propagation z coincides with the direction of non-uniformity, and if x is the direction of confinement of the slab, the geometry must be independent of the y-coordinate (Fig. 5).

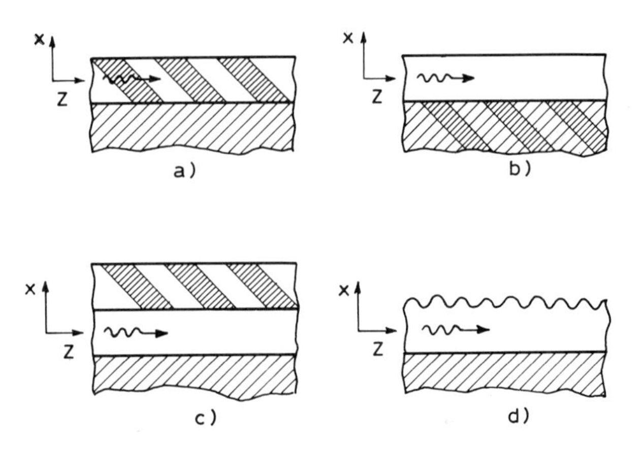

FIG. 5 – Various schemes of IO periodic structures.

The various schemes in Fig. 5 show the theoretical models that must be studied to solve all the possible cases that may arise in practice. These correspond to three distinct situations: 1) perturbation of the dielectric constant of the slab where propagation occurs (slab of periodic medium with open boundaries, Fig. 5*d*); 2) perturbation of the dielectric constant of the substrate, Fig. 5*b* or of the superstrate, Fig. 5*c* (slab of homogeneous material with a periodic medium at one boundary); 3) perturbation in the upper boundary of the slab where propagation occurs (slab of homogeneous material with a periodic boundary, Fig. 5*d*).

The more rigorous theory for studying periodic structures is that known as « Floquet's approach ». For small perturbations, as is the case in IO, a sufficiently exact theory is that of « coupled waves ». For a well-defined frequency (determined by the period of the grating), a propagating mode transfers a part of its power to another mode propagating in the same direction or in the opposite direction or to a radiation mode out of the plane. The following mode conversions may be obtained: conversion from an air mode to a guided mode (input coupler, Fig. 6*a*);

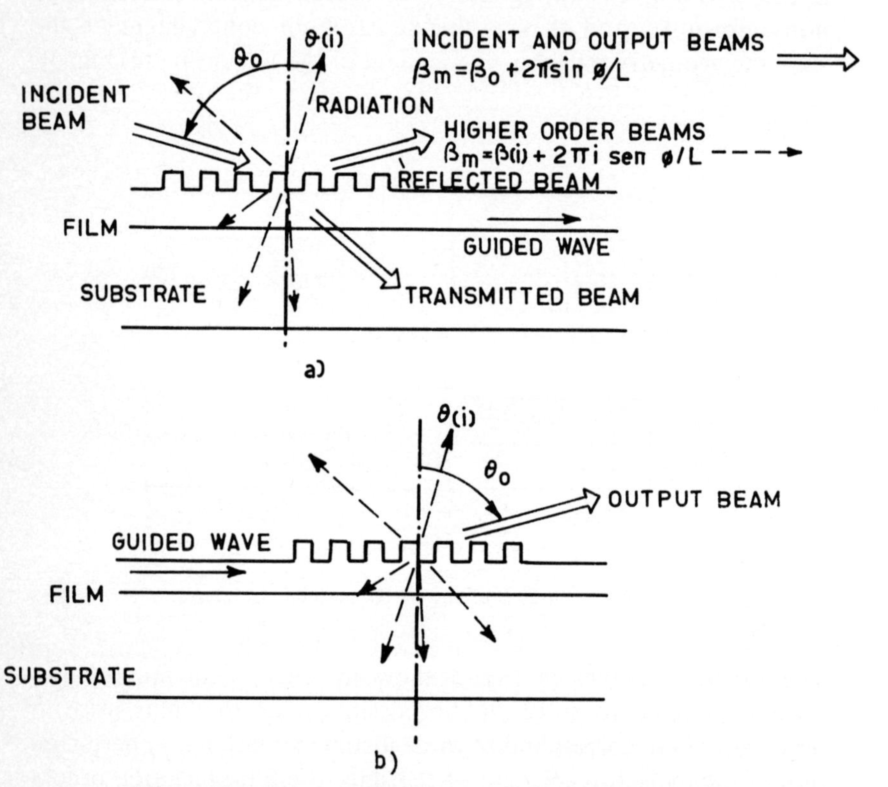

FIG. 6 – Diffracted beams in a grating coupler. *a*) Grating input coupler (excitation from the air), *b*) grating output coupler. (After [34]).

conversion from a guided mode to air mode (output coupler, Fig. 6*b*); forward conversion between guided modes, Fig. 7*a*; backward conversion between guided modes, Fig. 7*b*; reflection, Fig. 7*c*; and conversion between modes with different directions (Bragg deflection, Fig. 7*d*) [34].

The vectors $\boldsymbol{\beta}_n$, $\boldsymbol{\beta}_m$ appearing in Fig. 7 represent the propagation wavenumber along the direction of propagation of the

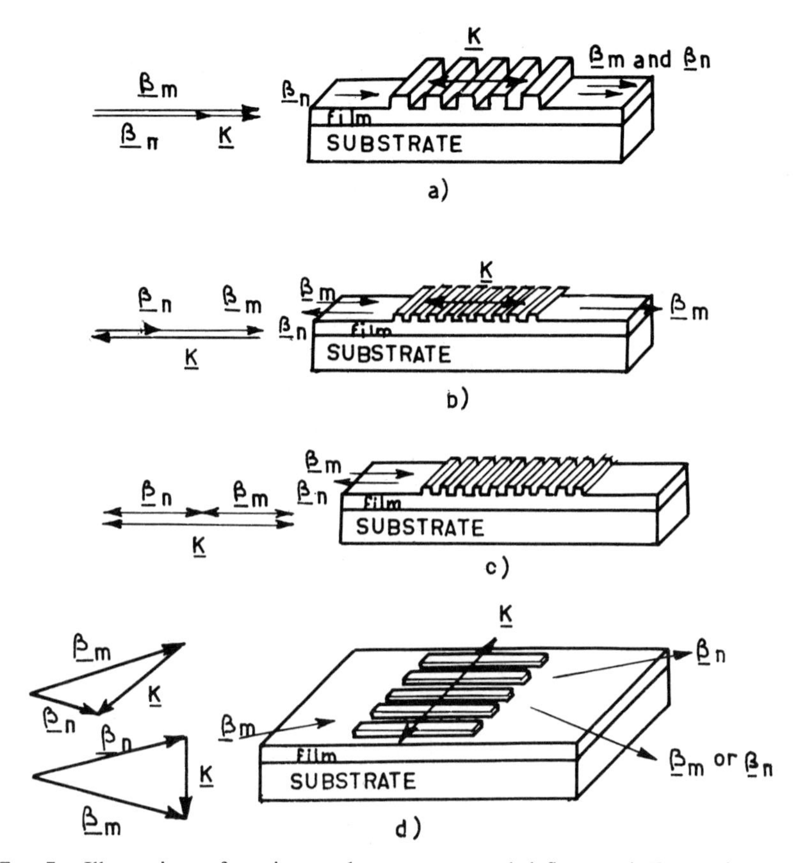

FIG. 7 – Illustrations of grating mode converters and deflector. *a*) Forward mode converter; *b*) backward mode converter; *c*) reflector; *d*) Bragg's deflector. (After [34]).

m-th and the *n*-th mode, while **k** is a vector representing the spatial frequency of the grating (it is perpendicular to the grating and has an amplitude $2\pi/L$, where L is the grating period). When the phase-matching condition $\pm\mathbf{k} = \boldsymbol{\beta}_m - \boldsymbol{\beta}_n$ is satisfied, the grating will convert the energy from the *m*-th mode to the *n*-th mode and vice versa. For the particular case of distributed reflection, Fig. 7*c*, the condition of phase-matching is $K = 2\beta$ (Bragg condition). Furthermore, while the grating period L determines the frequency of strongest coupling, the depth of the grooves (or more generally, the magnitude of the periodic variation of refraction index) determines the entity of coupling of the two interacting modes. In practical implementations perturbations are very small so that coupling is weak. As a consequence, the bandwidth above which coupling is maintained is very narrow. Therefore, I.O. gratings must be very accurate, as a small deviation from periodicity may cause a deviation from the phase-matching condition. For this reason IO gratings technology requires special processes, such as holography or electron-beam exposure.

REFERENCES

[1] E. G. Neumann and H. D. Rudolph: *Radiation from bends in dielectric rod transmission lines*, IEEE Trans. Microwave Theory Tech., Vol. MTT-23 (1975), pp. 142-149.

[2] A. A. Oliner: *Acoustic surface waveguides and comparison with optical waveguides*, IEEE Trans. Microwave Theory Tech., Vol. MTT-24 (1976), pp. 914-920.

[3] V. Ramaswamy: *Strip-film waveguide*, Bell Syst. Tech. J., Vol. 4 (1974), pp. 697-704.

[4] R. Collin: *Fields and Waves in Communication Electronics*, McGraw-Hill, New York (1960), p. 470.

[5] P. K. Tien: *Light waves in thin film and integrated optics*, Appl. Opt., Vol. 10 (1971), pp. 2395-2413.

[6] I. P. Kaminov *et al.*: *Metal-clad optical waveguides: analytical and experimental study*, Appl. Opt., Vol. 13 (1974), pp. 396-405.

[7] D. Marcuse: *Coupled-mode theory for anisotropic optical waveguides*, Bell Syst. Tech. J., Vol. 54 (1975), p. 985.

[8] D. Marcuse: *TE modes of graded-index slab waveguides*, IEEE J. Quantum Electron., Vol. QE-9 (1973), pp. 1000-1006.

[9] H. Furuta *et al.*: *Novel optical waveguide for integrated optics*, Appl. Opt., Vol. 13 (1974), pp. 322-326.

[10] E. Conwell: *Modes in optical waveguides formed by diffusion*, Appl. Phys. Lett., Vol. 23 (1973), pp. 328-329.

[11] R. D. Standley and V. Ramaswamy: Nb-*diffused* $LiTaO_3$ *optical waveguides: planar and embedded strip guides*, Appl. Phys. Lett., Vol. 25 (1974), pp. 711-713.

[12] J. F. Weller and T. G. Giallorenzi: *Indiffused waveguides: effects of thin film overlays*, Appl. Opt., Vol. 14 (1975), pp. 2329-2330.

[13] M. Matsuhara *et al.*: *Analysis of the guided modes in slab-coupled waveguides using a variational method*, IEEE J. Quantum Electron., Vol. QE-12 (1976), pp. 378-382.

[14] E. Voges: *Variational calculation of dielectric bulge guides and coupler for integrated optics*, A.E.Ü., Vol. 30, No. 4 (1976), pp. 176-179.

[15] J. B. Andersen *et al.*: *Propagation in rectangular waveguides with arbitrary internal and external media*, IEEE Trans. Microwave Theory Tech., Vol. MTT-23 (1975), pp. 555-560.

[16] W. Schlosser and H. G. Hunger: *Advances in Microwaves*, New York: Academic Press (1966), p. 319.

[17] E. A. Marcatili: *Dielectric rectangular waveguide and directional coupler for integrated optics*, Bell Syst. Tech. J., Vol. 48 (1969), pp. 2071-2102.

[18] J. E. Goell: *A circular-harmonic computer analysis of rectangular dielectric waveguides*, Bell Syst. Tech. J., Vol. 48 (1969), pp. 2133-2160.

[19] C. Yeh *et al.*: *Arbitrarily shaped inhomogeneous optical fiber or integrated optical waveguides*, J. Appl. Phys., Vol. 46 (1975), p. 2125.

[20] T. S. Bird: *A numerical analysis of optical guides with poligonal boundaries*, Proc. IEEE (1976), p. 235.

[21] E. Voges: *Field expansion analysis of dielectric rib-guides*, A.E.Ü., No. 4 (1977), pp. 170-173.

[22] E. A. Marcatili: *Slab-coupled waveguides*, Bell Syst. Tech. J., Vol. 53 (1974), pp. 645-674.

[23] R. M. Knox: *Dielectric waveguide microwave integrated circuits. An overview*, IEEE Trans. Microwave Theory Tech., Vol. MTT-24 (1976), pp. 806-814.

[24] N. Uchida: *Optical waveguide loaded with high refractive-index strip film*, Appl. Opt., Vol. 15 (1976), pp. 179-182.

[25] G. B. Hocker and W. K. Burns: *Mode dispersion in diffused channel guides by the effective index method*, Appl. Opt., Vol. 16 (1977), pp. 113-118.

[26] J. Hamasaki and K. Nosu: *A partially metal-clad-dielectric-slab waveguide for integrated optics*, IEEE J. Quantum Electron., Vol. QE-10 (1974), pp. 822-825.

[27] Y. Yamamoto *et al.*: *Propagation characteristics of partially metal-clad optical guide: metal-clad optical strip line*, Appl. Opt., Vol. 14 (1975), pp. 322-326.

[28] E. A. Marcatili: *Bends in optical dielectric guides*, Bell Syst. Tech. J., Vol. 48 (1969), pp. 2103-2132.

[29] D. Marcuse: *Bending losses of the asymmetric slab waveguide*, Bell Syst. Tech. J., Vol. 50 (1971), pp. 2551-2563.

[30] L. Lewin: *Radiation from curved dielectric slabs and filters*, IEEE Trans. Microwave Theory Tech., Vol. MTT-22 (1974), pp. 718-727.

[31] A. W. Snyder *et al.*: *Radiation from bent optical waveguides*, Electron. Lett., Vol. 11 (1975), pp. 332-333.

[32] M. Heiblum and J. H. Harris: *Analysis of curved optical waveguides by conformal transformation*, IEEE J. Quantum Electron., Vol. QE-11 (1975), pp. 75-83.

[33] D. Marcuse: *Mode conversion caused by surface imperfections of a dielectric slab waveguide*, Bell Syst. Tech. J., Vol. 48 (1969), pp. 3187-3215.

[34] W. S. C. Chang: *Periodic structures and their application in integrated optics*, IEEE Trans. Microwave Theory Tech., Vol. MTT-21 (1973), pp. 775-785.

CHAPTER 2

INTEGRATED OPTICS DEVICES

by Antonio Scudellari

Two types of integration can be considered: monolithic integration, where the optical signal is processed sequentially by devices of different kinds, and monofunctional integration, where the same function is repeated with several devices of the same kind. Both have been demonstrated with several implementations.

Examples of monolithic integration are combinations of a laser and an absorption modulator [1], of a laser and a frequency modulator [2], of a laser and a detector [3] and of a laser and a laser-amplifier [4]. Other interesting possibilities under development are the combination of a filter and a detector (see Section 2.2), and the combination of optical components with electrical components. As far as this last case is concerned, the combination of a microwave Gunn oscillator and a laser has been already demonstrated [5]: with this arrangement the laser has been modulated up to 1 GHz frequency.

Nowadays, although IO has scarcely reached the level that is necessary to provide monolithic integrated circuits, in the field of parallel integration laboratory products have reached an advan-

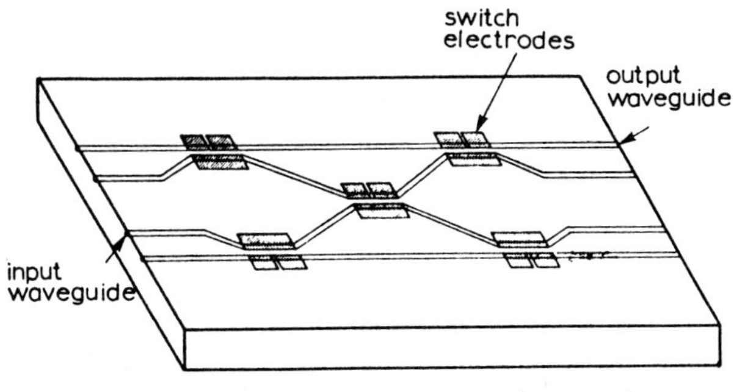

FIG. 1 – 4×4 switching network. (After [8]).

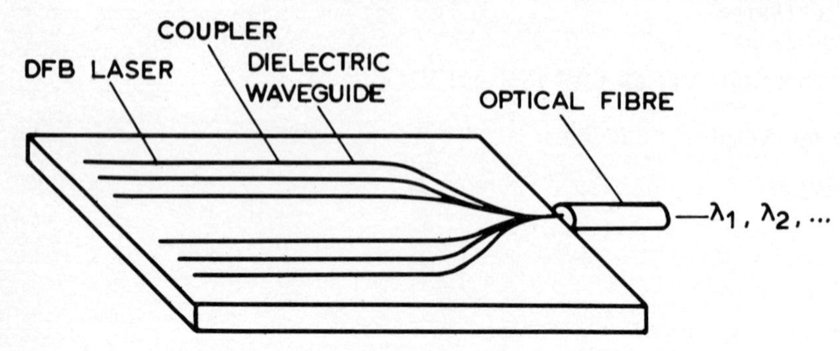

FIG. 2 – Scheme of optical wavelength multiplexer experimented at Hitachi laboratories. (After [6]).

ced stage of development. The most notable results have been the integration of several laser sources [6] to form arrays of sources for WDM (Wavelength Division Multiplexing) systems, silicon detectors [7] for optical data processing and electro-optical switches [8], Fig. 1.

The first case is probably the most surprising nowadays.

Several laser sources are arranged in the configuration of a complete WDM transmitter. In Fig. 2, from [6], six DFB lasers are integrated on a single output waveguide which may be coupled to an optical fibre. The structure of the laser device and of the transition to the corresponding waveguide is given in Fig. 3. The

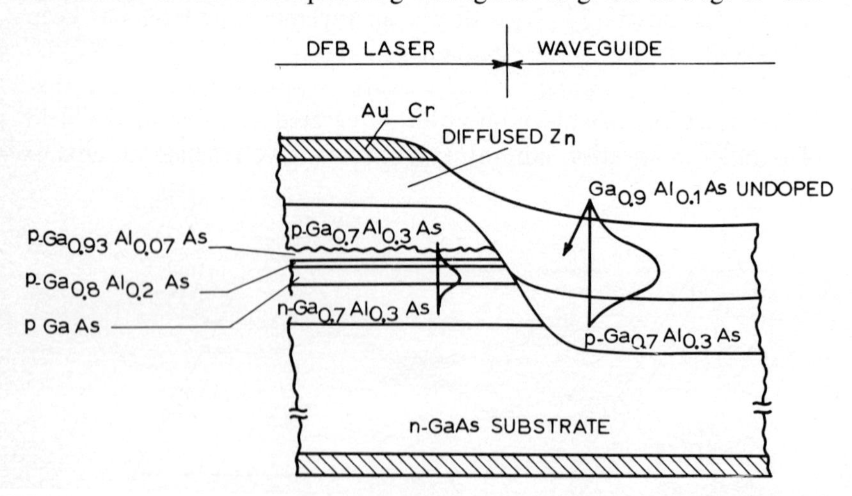

FIG. 3 – Cross-section of one of DFB lasers of the multiplexer in the region of coupling to dielectric waveguide. (After [6]).

laser devices in Fig. 2 operate with pulsed currents having a repetition rate of 1 kHz and a duration of 100 ns. The threshold current density varies from 3 to 6 kA/cm^2 at room temperature. The wavelength of light emission from the six lasers is varied adopting a different feedback grating period for every laser. Results are given in the diagram of Fig. 4, showing wavelength spacing of about 20 Å with a variation of the grating period of about 9 Å.

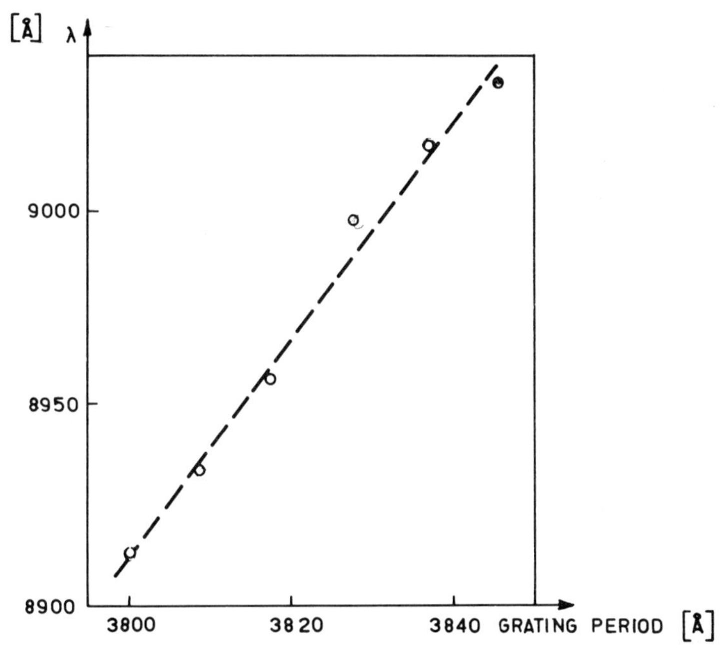

FIG. 4 – Wavelength of light emitted by six single lasers, correlated to grating period. (After [6]).

A WDM communication system should be completed by a receiver for demultiplexing and detecting the six optical carriers propagated along a fibre. A good approach seems to be that of chirped gratings proposed by Yariv [9]. At any rate, the problem of separating six frequencies, spaced a few tens of Ångstroms, expecially after propagation along a graded index fibre, is a serious technological problem.

Ignoring compound integrated circuits, a review of consolidated discrete implementations may be given.

2.1 Sources

Sources will be considered in Chapter 3. Considering mainly their suitability for integration, semiconductor lasers with double heterostructure are particularly interesting for both hybrid and monolithic integration; in particular, special Distributed Feed-Back (DFB) lasers may be employed for monolithic integration. In fact, owing to the substitution of cleaved semisilvered surfaces with an optical feedback grating, they can be coupled to planar optical waveguides and other integrated devices.

A further development of the grating laser is the Distributed Bragg Reflectance (DBR) laser (see also Section 2.5). The use of terminal grating reflectors overcomes some of the drawbacks due to the proximity of corrugations in the active region (degradation of recombination efficiency and shortening of operating life). Coupling of a complex integrated circuit to planar waveguides is also made easier with DFB lasers.

Recently, other types of lasers suitable for integrated optics have been experimentally demonstrated. They do not require the use of grating structures, so that coupling to output passive waveguide is even simpler. One of them makes use of an Integrated Interferometric Reflector (IIR) to obtain optical feed-back [10], Fig. 5. The operation of a IIR laser relies on an interferometric phenomenon which is analogous to that of microwave bridges and is also used to obtain amplitude modulators and switches (see Section 2.3). An alternative approach to the fabrication of integrated lasers utilizes reflectors made by etching [3]. Through

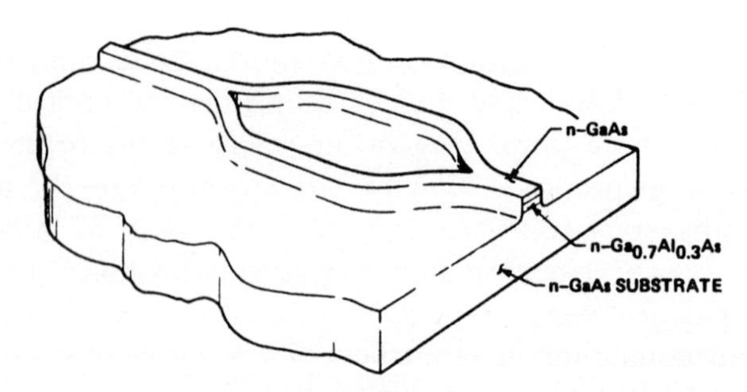

FIG. 5 – Schematic of the integrated interferometric reflector. (After [10]).

the same technological process used for etched reflector laser, a detector may also be obtained, so that the combination laser-detector is easily achieved (Fig. 6).

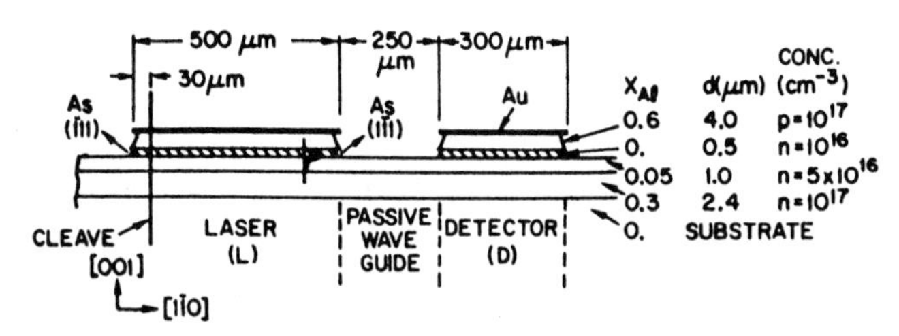

FIG. 6 – Schematic diagram of etched laser-waveguide-detector. (After [3]).

Looking towards the future, it is likely that the laser itself will find more applications than simply that of a source (or a detector). One possibility, that of using a laser as an injection locked amplifier, has been already experimentally demonstrated [4]. The same experiment showed that, when no electrical bias was applied to the laser amplifier, a detector was obtained (Fig. 7).

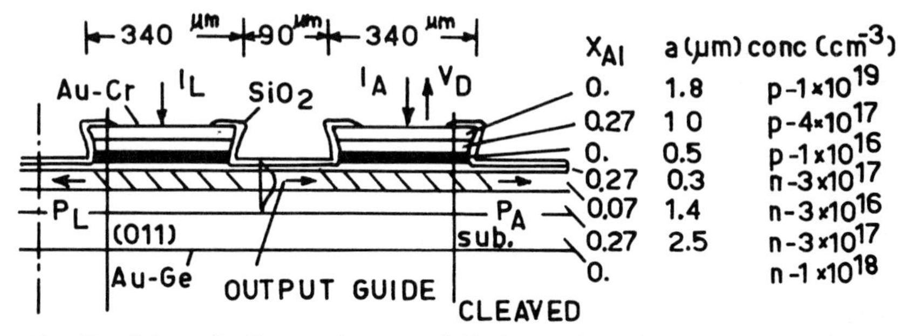

FIG. 7 – Schematic diagram for monolithic integration of laser and amplifier or detector. (After [4]).

2.2 Detectors

For detectors, integration is in a more transitory phase.

For wavelengths near 0.9 μm, which are common in optical communications, silicon detectors are usually employed; silicon is a material with high sensitivity for these wavelengths. These detectors cannot be integrated with sources in a monolithic form, since the latter are made of materials of the III-V groups (see Chapter 3).

On the other hand, long-distance optical communications favour the use of longer wavelengths (1.1-1.3 μm). In this case, owing to loss of sensitivity, silicon must be abandoned in favour of elements of the III-V groups (for instance $GaAs/GaAs_{1-x}Sb_x$ or InGaAsP/InP ... heterojunctions) which are suitable for monolithic integration.

As far as the problem of detection selectivity is concerned, a tuned detector may be obtained by combining a detector with a selective coupling elements, such as a grating coupler, so that radiation transfer to the photodiode active volume is obtained at a well-defined wavelength.

The structure of an integrated detector could be that shown in Fig. 8, from [11], even if the materials employed here are of

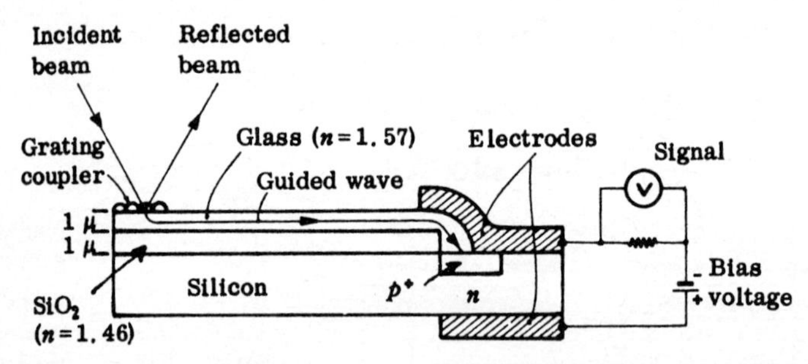

FIG. 8 – Opto-electronically integrated photodetector schematic. (After [11]).

different type (hybrid integration) and the input grating coupler operates on air modes rather than on propagation modes. Inspite of some implementations of the kind shown in Fig. 8, little experimental work has been done in this direction. More attention has been given to the problem of detector arrays for optical data processing. These devices also seem suitable for optical fibre multiplexing [7].

When detector arrays are considered, silicon is the preferrep substrate, since various integrated electronic circuits can be fabricated to perform complementary electronic processes. In Fig. 9, from [7], a waveguide array, made of channel waveguides, is used for a parallel injection of signals into a Charge-Coupled Device (CCD) linear imaging array.

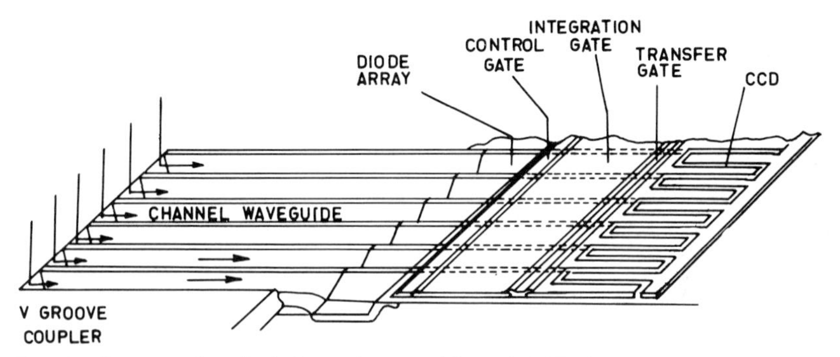

FIG. 9 – Integrated optical channel waveguide end CCD detector array along with *V*-groove end-coupler. (After [7]).

2.3 Modulators and switches

Laser sources may be modulated by an electrical signal superimposed on the drive current. Much effort is being concentrated on the direct modulation of the optical signal at the laser output. In this way higher modulation rates may be obtained with low power consumption.

Time division multiplexers and switching networks may be also obtained using switches that are proper modulators with two identical outputs.

Integrated optical modulators are generally made of crystalline thin films whose propagation characteristics are properly modified by an applied external field. The most widely used physical effects are the Electro-Optic (EO), the Acusto-Optic (AO) and Magneto-Optic (MO) effects.

In 1970 Yariv [12] first showed the possibility of obtaining an intensity EO modulator in the form of a Schottky Barrier on a high intensity GaAs waveguide, exploiting the propagation characteristics of the depletion layer of a reverse biased diode.

Other EO produced modulators were obtained using thin films

on $LiNbO_3$ and $LiTaO_3$ substrata. Owing to the high light concentration in the optical guide, the power demand of these modulators is much smaller than that of their bulk counterparts. A phase modulator (unsuitable for direct implementation in an intensity modulation system) was reported by Kaminov in 1973 [13], Fig. 10.

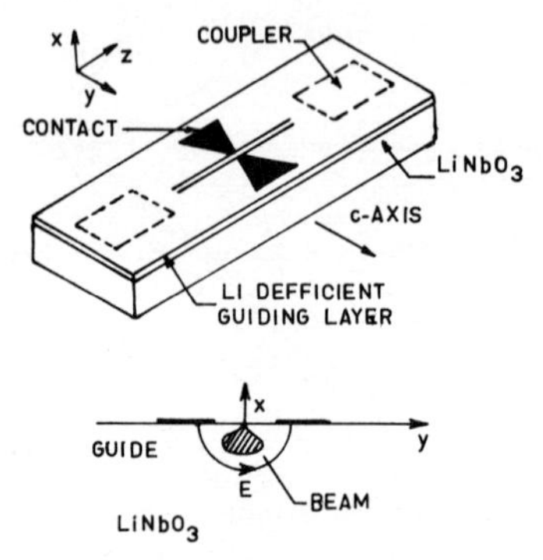

FIG. 10 – Thin film phase modulator on $LiNbO_3$: *a*) modulator configuration; *b*) field configuration. (After [13]).

Other EO modulators on lithium niobate are of the deflection type, with interdigital electrodes producing a Bragg-type phase grating. Modulation depths of about 70 % at 900 MHz are available from these modulators.

Another EO modulator, produced on GaAs by Texas Instrument Laboratories [14], consists of two channel optical guides of the metallic-gap type. The two guides are sidecoupled and the coupling conditions are varied by an external field applied to the metallizations. This modulator has a bandwidth of 100 MHz, a power-bandwidth ratio of 180 μW/MHz and a modulation depth of about 95 %, so that it is an almost perfect switch.

This type of EO switch, consisting basically of two-side-coupled channel waveguides, represents the most attractive type of IO produced to date (see also Section 2.4). It is sufficient to associate the two different values of the refractive index of the material to two synchronism conditions to obtain a complete transfer or a complete power cut-off after a length L. A first method of obtaining the two refractive index values is based on the use of parallel and external electrodes giving rise to opposite polarizations, and to variations of the refractive index in the waveguides. A more recent technique, known as COBRA [15] (see Fig. 11), requires only two electrodes lying on the waveguides,

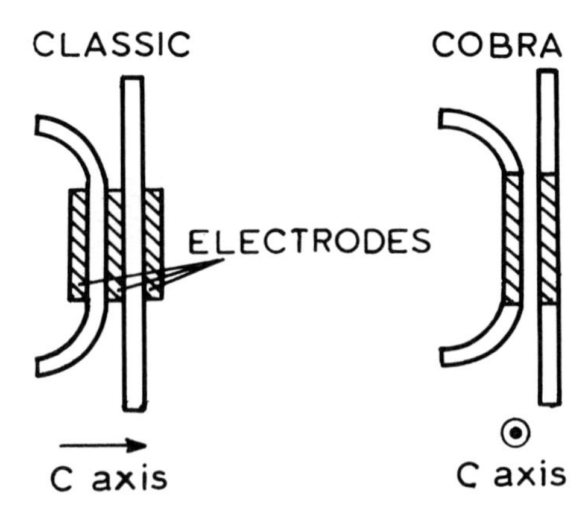

FIG. 11 – Two possible electrode configurations for electrically switched directional couplers: *a*) classical configuration (three electrodes); *b*) COBRA configuration (two electrodes). (After [15]).

so that the index variation occurs in the direction of decaying fields, which are opposite for the two guides. Another recent modification of EO switches is that of adopting more electrodes in cascade so that fabrication errors may be compensated [16] (see Fig. 12). Moreover, this switch configuration can also be arranged to be polarization insensitive. In fact, with a new design [17], the presence of two different electrode types has permitted a tuning for a polarization insensitive operation. The problem of obtaining polarization insensitive devices is very important for optical communication systems employing fibre optical transmission line, since two states of polarization normally propagate in

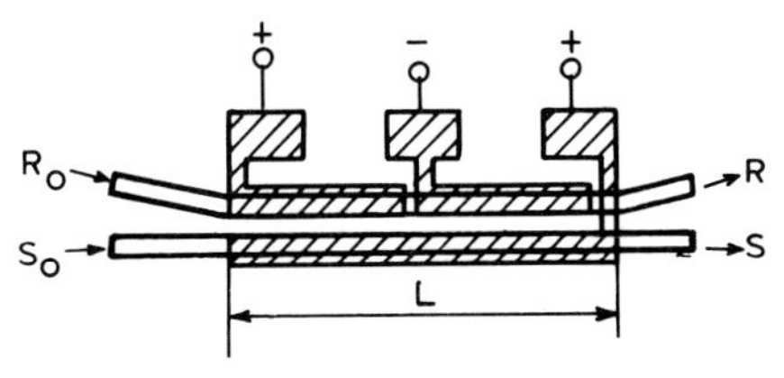

FIG. 12 – Schematic diagram of stepped $\Delta\beta$ reversal switch. (After [16]).

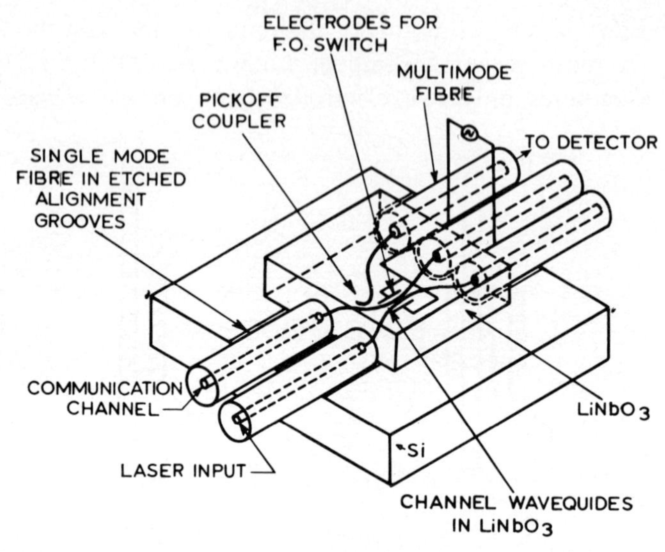

FIG. 13 – Four port data switching terminal module. (After [18]).

single mode fibres, while single-section switches are optimized for a single polarization (see Fig. 13 from [18]).

Another type of switch the balanced bridge modulator switch should present less problems on polarization sensitivity, (see Fig. 14 from [19]). This switch is based on the principle of bridge modu-

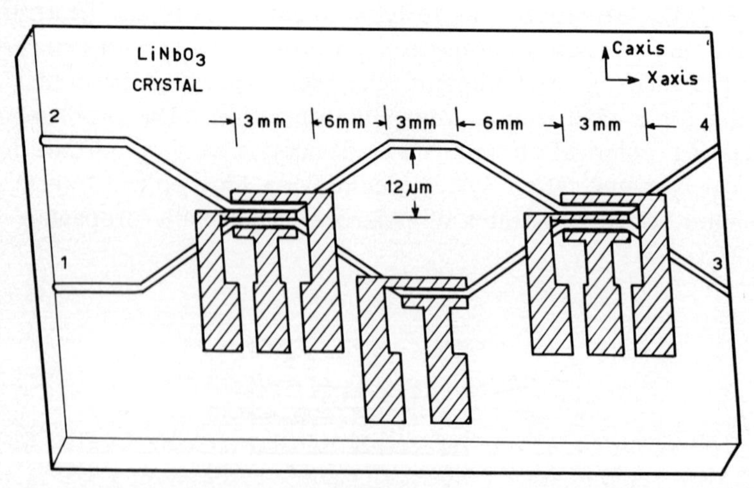

FIG. 14 – Balanced bridge modulator swith. (After [19]).

lators, which are analogous to the conventional microwave bridges. They consist of two 3 dB directional couplers interconnected via two paths, one of which is an electro-optically controllable phase-shifter.

Another electro-optical device, with wider applications than a simple switch, has recently been introduced [20]: it consists of a waveguide version of a non-linear Fabry-Perrot resonator, whose nonlinearity is produced by using the output of a photodetector, which samples the transmitted light, to drive the electro-optic element of the resonator (Fig. 15). Such a device may be used, not only as an optical switch, but also as a pulse shaper and limiter, as a differential amplifier and an « optical triode ».

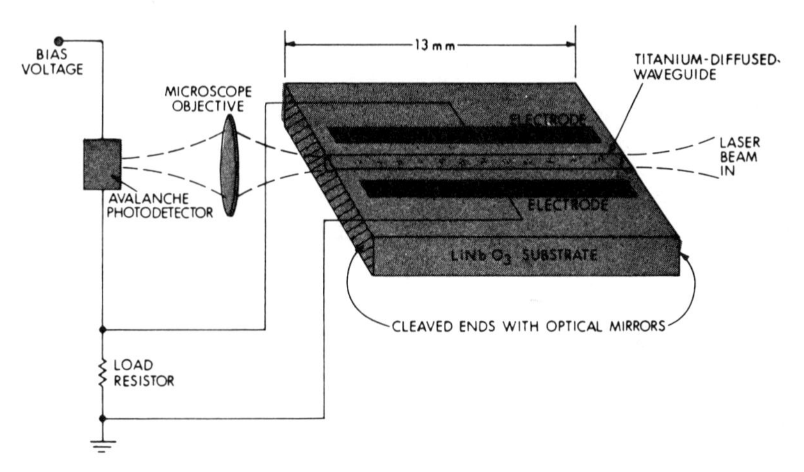

FIG. 15 – Schematic drawing of the integrated version of a non-linear Fabry-Perot resonator. (After [20]).

AO modulators are based on beam deflection by a grating created by the periodic index variations caused by an acoustic wave. AO deflections have been observed in thin films made of lithium niobate, As_2S_3, Ta_2O_5 and glass. Generally, the acousto-optic effect does not permit high data rates: the upper limit is strongly influenced by the difficulty of obtaining acousto-electric transducers for generating high frequency acoustic waves.

MO modulators utilize the Faraday effect within iron-garnet thin films which are grown by liquid epitaxy on garnet substrata.

The main problem in this case is that of transforming a polarization modulator into an amplitude modulator.

Few references are available on absorptive-type modulators, even if their operating mechanism often employs GaAs semiconductor material which is suitable for monolithic integration. These modulators consist of reverse biased GaAs *p-n* junctions, or of GaAs and GaAlAs heterostructure junctions. Under these conditions the electric field in the junction is sufficient to produce, by the Franz-Keldysh effect, a shift of the absorption characteristics of the material. Therefore, a variation of the light power crossing the junction results. Modulation bandwidth is larger than 1 Gbit/s and power requirements are less than 0.1 mW/MHz.

An example of integration of these modulators is that obtained by Reinhard and Logan [1].

2.4 Directional couplers

A family of interesting devices for IO, such as power dividers, frequency selectors and polarization selectors, is that of directional couplers. Moreover, when it is possible to control coupling by external fields, modulators and switches of the type already mentioned in Section 2.3 may be obtained.

A directional coupler, as described in Section 1.4, consists in the exchange of power between two guides which are coupled by their evanescent fields. Therefore, the two basic schemes are that of top coupling and that of side coupling.

In the case of top coupling (Fig. 16) two planar waveguides are separated by a layer which acts as a substrate for the upper guide and as a cover for the lower. This type of coupler is seldom

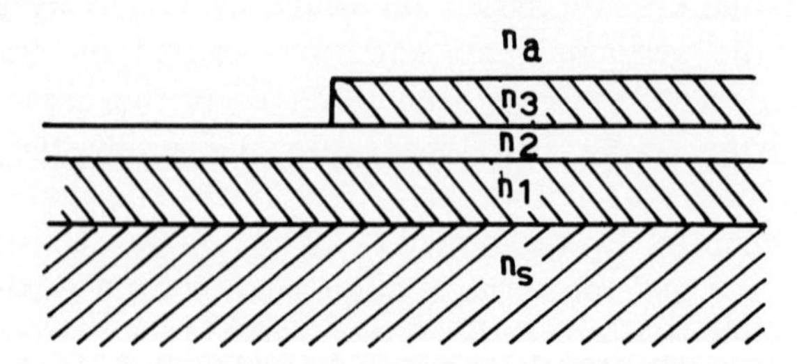

FIG. 16 – Top coupling between two slab guides.

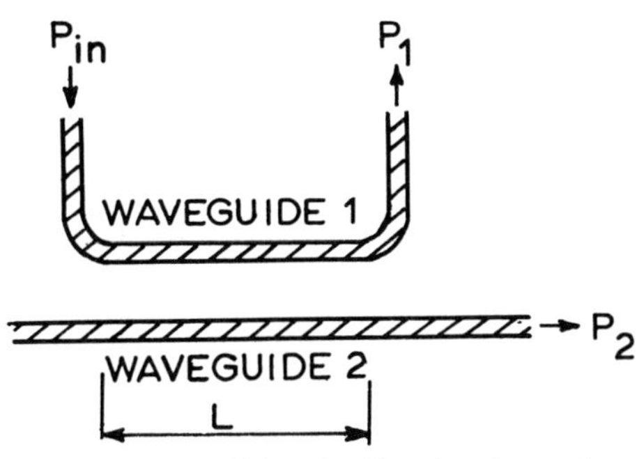

FIG. 17 – Parallel strip directional coupler.

used owing to the difficulty of controlling synchronism during coupling. Wilson [21] has tried to overcome the strict geometric tolerances required by synchronous coupling by varying the thickness of the coupling layers in a prescribed manner.

However, the most widely used type of coupling is that shown in Fig. 17, where two channel waveguides are parallel for a length L and then diverge. Experiments have been carried out with a great variety of materials.

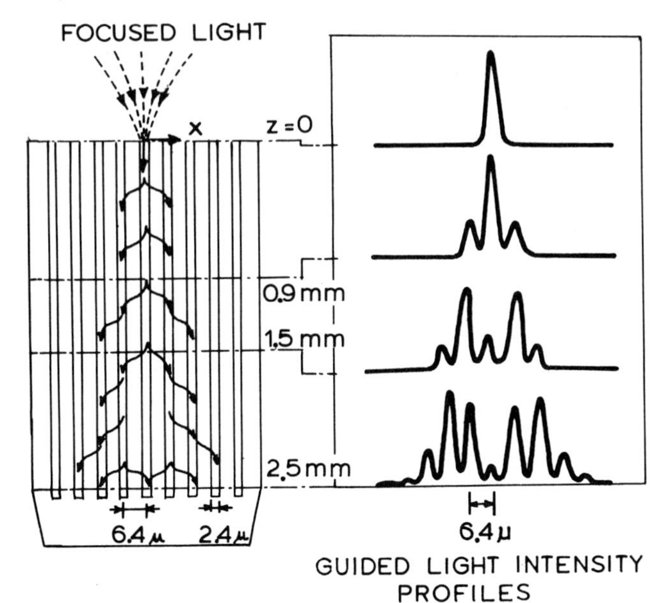

FIG. 18 – Experimental results which show directional coupling between guides on GaAs. (After [22]).

The possibility of the subsequent division of light energy by coupling several guides, particularly in the case of GaAs, (Fig. 18) [22] has been demonstrated. For lateral coupling a non-uniform version may also be tried, even though the increase in bandwidth is offset by technological difficulties in controlling the spacing variation (Fig. 19).

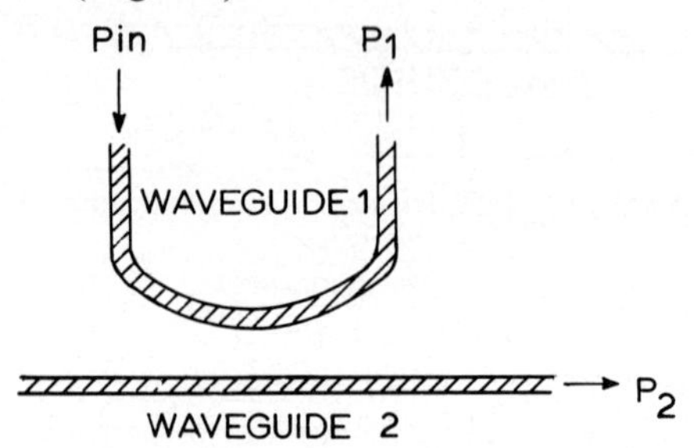

FIG. 19 – Parallel strip non uniform directional coupler.

2.5 Filters

Most of the filters produced are of the band-stop type. Corrugation gratings use Bragg reflection for a well-defined wavelength (see also Section 2.7).

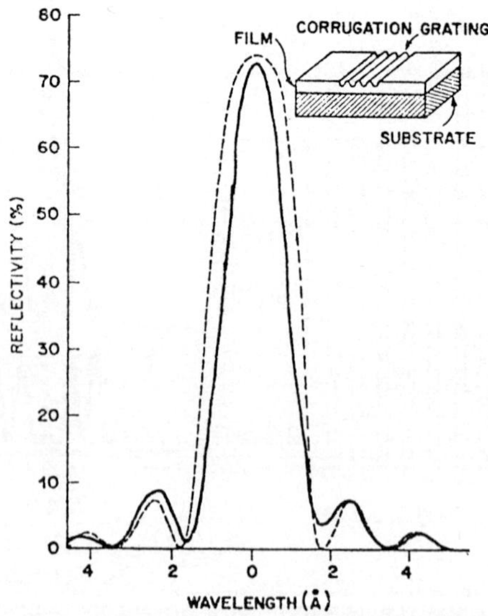

FIG. 20 – Frequency response of a grating band-stop filter. (After [23]).

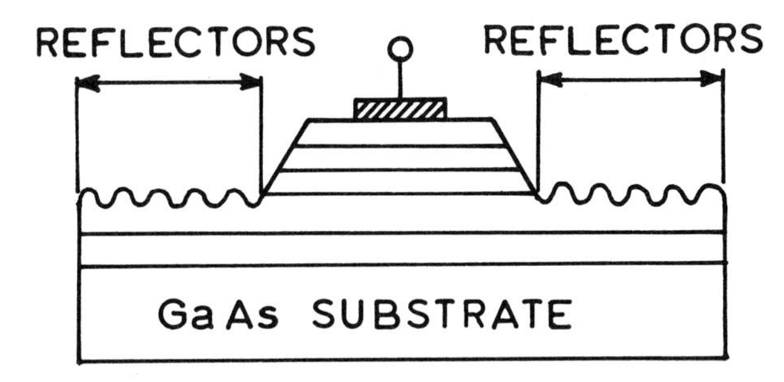

FIG. 21 – Schematic drawing of a DBR laser.

First examples of filters with a corrugated waveguide have been obtained on glass (by ion-beam etching through a photoresist mask after exposure to the interference pattern of two laser beams) at $\lambda = 0.57\ \mu m$ [23] (see Fig. 20). The bandwidth of the filter has passed from some Å in the first experiments to 0.1 Å in a more recent filter by Schmidt [24].

For improving the performance of grating filters, quasi-periodic structures have also been proposed. The amount of coupling between modes is gradually varied by a modification of corrugation depth. It has been shown that a quadratic taper with a coupling coefficient proportional to the square of distance can reduce lateral lobes of frequency response [25].

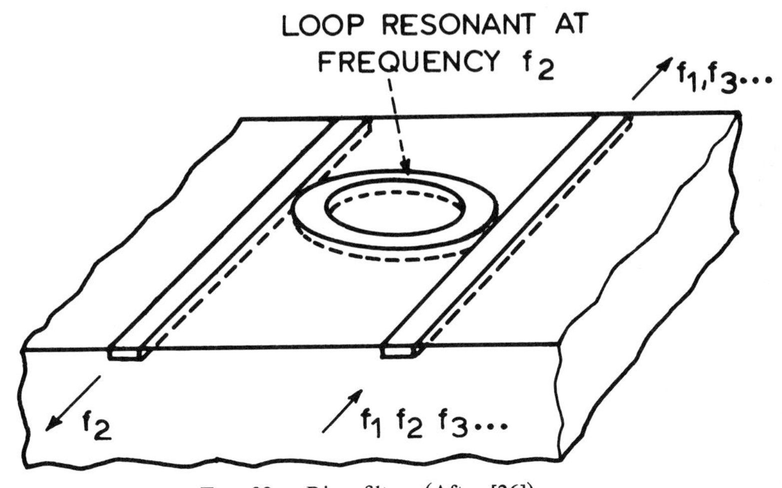

FIG. 22 – Ring filter. (After [26]).

These grating filters may be easily integrated in complex circuits. For instance, by closing a section of waveguide between two reflecting gratings a resonant circuit is formed so that with an active layer a DBR laser may result, Fig. 21 (see also Section 2.1).

Other filtering networks different than grating structures could also be employed: in general, a resonant circuit coupled to input and output lines can produce a frequency discrimination. Such a discrimination may be used for demultiplexing a WDM signal. As an example, the ring filter shown in Fig. 22 could be used for this purpose (see also Section 1.4).

REFERENCES

[1] F. K. Reinhardt and R. A. Logan: *Monolithically integrated* AlGaAs *double-heterostructure optical components*, Appl. Phys. Lett., Vol. 25 (1974), pp. 622-624.

[2] F. K. Reinhardt and R. A. Logan: *Integrated electro-optic intracavity frequency modulation of double-heterostructure injection laser*, Appl. Phys. Lett., Vol. 27 (1975), pp. 532-534.

[3] J. L. Merz and R. A. Logan: *Integrated* GaAs-$Al_xGa_{1-x}As$ *injection lasers and detectors with etched reflectors*, Appl. Phys. Lett., Vol. 30 (1977), pp. 530-533.

[4] K. Kishino *et al.*: *Twin-guide laser monolithically integrated with amplifier or detector*, Integrated and Guided Wave Optics, Salt Lake City (January 1978), paper MD6.

[5] A. Yariv *et al.*: *Integration of an injection laser with a Gunn oscillator on a semi-insulating* GaAs *substrate*, Appl. Phys. Lett., Vol. 32 (1978), pp. 806-809.

[6] K. Aiki *et al.*: *Frequency multiplexing light source with monolithically integrated distributed-feedback diode lasers*, Appl. Phys., Lett., Vol. 29 (1976), pp. 506-508.

[7] C. L. Chen and J. T. Boyd: *Channel waveguide array coupled to an integrated charge-coupled device* (*CCD*), Integrated and Guided Wave Optics, Salt Lake City (January 1978), paper Ma5.

[8] R. V. Schmidt and L. L. Buhl: *Experimental* 4×4 *optical switching network*, Elect. Lett., Vol. 12 (1976), p. 575.

[9] A. Yariv *et al.*: *Chirped gratings in integrated optics*, IEEE J. of Quantum Electron., Vol. QE-13, No. 4 (1977), pp. 296-304.

[10] D. R. Scifres *et al.*: *Semiconductor lasers with integrated interferometric reflectors*, Appl. Phys. Lett., Vol. 30 (1977), pp. 585-587.

[11] D. B. Ostrowky *et al.*: *Integrated optical photodetector*, Appl. Phys. Lett., Vol. 22 (1973), pp. 463-464.

[12] A. Yariv *et al.*: *Optical guiding and electro-optic modulators in* GaAs *epitaxial layers*, Opt. Commun., Vol. 1 (1970), p. 403.

[13] I. P. Kaminov *et al.*: *Thin film* $LiNbO_3$ *electro-optic light modulator*, Appl. Phys. Lett., Vol. 22 (1973), pp. 540-542.

[14] J. C. Campbell *et al.*: GaAs *electro-optic directional coupler switch*, Appl. Phys. Lett., Vol. 27 (1975), pp. 202-205.

[15] M. Papuchon *et al.*: *Electrically switched optical directional coupler: COBRA*, Appl. Phys. Lett., Vol. 27 (1975), pp. 289-291.

[16] R. V. Schmidt and H. Kogelnik: *Electro-optically switched coupler with stepped* $\Delta\beta$ *reversal using* Ti-*diffused* $LiNbO_3$ *waveguides*, Appl. Phys. Lett., Vol. 28 (1976), pp. 503-507.

[17] R. A. Steinberg *et al.*: *New electrode design for polarization insensitive integrated optics switches*, Appl. Opt., Vol. 16 (1977), p. 2166.

[18] H. P. Hsu and A. F. Milton: *Flip-chip approach to endfire coupling*, Electron. Lett., Vol. 12 (1976), p. 404.

[19] V. Ramaswamy and M. D. Divino: *A balanced bridge modulator switch*, Integrated and Guided Wave Optics, Salt Lake City (January 1978), paper TuA4.

[20] P. W. Smith *et al.*: *Integrated nonlinear Fabry-Perot devices*, Integrated and Guided Wave Optics, Salt Lake City (January 1978), paper TuB1.

[21] M. G. F. Wilson and G. A. Teh: *Tapered optical directional coupler*, IEEE Trans. Microwave Theory Tech., Vol. MTT-23 (1975), p. 85.

[22] S. Somekh *et al.*: *Channel optical waveguide directional couplers*, Appl. Phys. Lett., Vol. 22 (1973), pp. 46-47.

[23] D. C. Flanders *et al.*: *Grating filters for thin-film optical waveguides*, Appl. Phys. Lett., Vol. 24 (1974), pp. 194-196.

[24] R. V. Schmidt: *Narrow-band grating filters for thin-film optical waveguides*, Appl. Phys. Lett., Vol. 24 (1974), pp. 651-652.

[25] H. Kogelnik: *Filter response of non uniform almost-periodic structures*, Bell Syst. Tech. J., Vol. 55 (1976), pp. 109-126.

[26] E. A. Marcatili: *Bends in optical dielectric waves*, Bell Syst. Tech. J., Vol. 48 (1969), pp. 2103-2132.

CHAPTER 3

MATERIALS, CHARACTERIZATION METHODS AND TECHNOLOGIES

by Vittorio Ghergia

As already specified in the previous Sections, in order to manufacture IO devices it has been necessary to study new materials and to establish new manufacturing technologies and special physical-chemical control methods.

3.1 Materials for active and passive components

Even if it is difficult to distinguish clearly between materials that are suitable for active and passive IO components, there is a class of semiconductor materials which can currently be considered as the most suitable for the manufacture of active components such as radiation source and detectors, and which constitute the only materials suitable for monolithic type integration. These are semiconductor materials formed by binary and ternary, sometimes even quaternary, compounds of elements pertaining to groups III and V of the periodic system.

In addition to the well-known GaAs binary compound, typical examples of the ternary compound are $Ga_{1-x}Al_xAs$ systems for visible light wavelight 0.82 μm and $Ga_{1-x}In_xAs$ for $\lambda = 1.06$ μm. A quaternary system which is widely used for $\lambda = 1.1$-1.3 μm is $Ga_{1-x}In_xAs_{1-y}P_y$, which is constituted by the two binary compounds GaAs and InP.

Table I [1], shows the lattice constant a_0, the melting temperature T_m, the energy bandgap Σ_g with the related light emission λ and the band structure type for 9 binary compounds obtained by assembling three elements of the III A group (Al, Ga, In) with three elements of the V A group (P, As, Sb). The compounds which could be obtained by assembling the remaining elements

TABLE I [2]

III-V Binary compounds.

	a_0 (Å)	T_m (°C)	E_g (eV)	λ (μm)	Band structure
AlP	5.463	2000	2.45	0.51	Indirect
AlAs	5.661	1740	2.16	0.57	Indirect
AlSb	6.138	1050	1.65	0.75	Indirect
GaP	5.451	1467	2.26	0.55	Indirect
GaAs	5.653	1238	1.42	0.87	Direct
GaSb	6.095	706	0.72	1.85	Direct
InP	5.868	1058	1.35	0.95	Direct
InAs	6.058	942	0.36	3.44	Direct
InSb	6.479	530	0.17	7.30	Direct

of group III and V are not considered, since the preparation of these materials is difficult and applications are therefore not yet known.

The main characteristic of the active elements manufactured with these materials consists, in general, in the possibility of obtaining through epitaxially grown thin films either *p-n* junctions or more complex multistructures, such as, for instance, those relating to double heterojunction semiconductor lasers. The use of ternary and quaternary compounds arises essentially from the attempt to avoid mismatching of lattice constants among the various grown layers. This fact actually gives rise to dislocations in the epitaxial structures that are particularly dangerous for light emission.

Figure 1 [1] shows wavelength light emission of the 8 most widely used ternary compounds among the 18 possible combinations of the 9 compounds. Passive component materials include those operating simply as optical waveguides and those performing functions for which electro-optic, acousto-optic and magneto-optic coefficients are important. The former require a low attenuation coefficient and a suitable refraction index n value. The refraction index of those materials varies as a function of electric field, mechanical stress and magnetic field applied. From a techno-

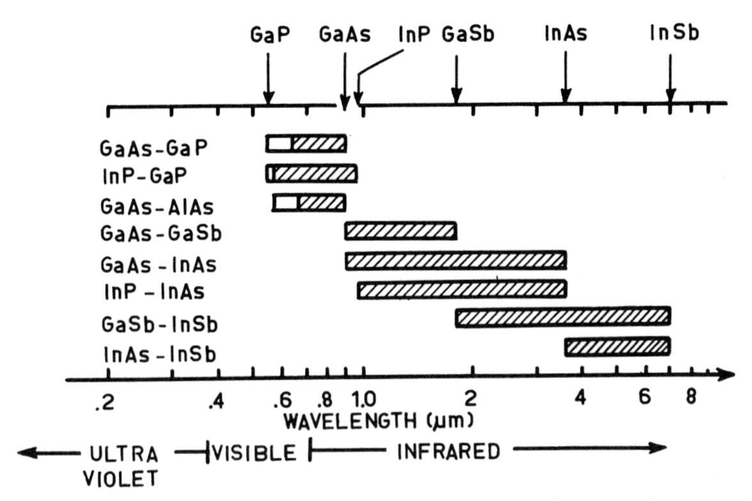

FIG. 1 – Room-temperature wavelength range covered by several commonly used III-V ternary alloys. (After [1]).

logical and preparation point of view these materials can be divided into amorphous (or semi-amorphous) and monocrystalline materials. Although amorphous materials do not present problems for the matching of the crystallographic parameters between substrate and superposed layer, they seldom allow modulation of the refraction index as indicated above. The group of amorphous materials includes glasses, photoresists, gelatines, polymeric resins and liquid crystals. For these compounds, the refraction index can be continuously varied within wide limits by changing the chemical composition and by the short-range atomic reorganization which is a function of the material preparation conditions.

It is useful to point out that intrinsic attenuation for these materials is connected to two separate phenomena: attenuation and scattering.

Light attenuation is due both to the material extinction coefficient and to small quantities of impurities which are always present in the material. As for glasses, dangerous impurities, even if in a part-per-billion concentration, are transition metal ions such as Fe^{3+}, Cr^{3+}, Ni^{2+}, etc. and oxidriles such as Si—OH.

Scattering, on the other hand, results from local variations of the refraction index of the material. This phenomenon may be caused by the presence of different materials (such as inclusions of air, dust, metal oxides, etc.), by other defects or by the presence of different phases. An example of the latter case is the deglassing of silica glass with a formation of quartz crystalline germs. Further refraction index variations can be caused by fluctuations of the material density or by local nonuniformities of the short range crystallographic re-organization of the material, which are caused by variations in the material deposition parameters.

Monocrystalline materials showing « n » modulation phenomena pertain to the no-inversion symmetry classes. Among these the most interesting are III and V semiconductors, II and IV semiconductors, as well as some dielectric crystal families. Among III and V semiconductors the $Ga_{1-x}Al_xAs$ compound is to be particularly noted. This compound was already mentioned as regards its use in active devices, and is suitable for direct integrations with sources.

Other III and V semiconductors are also interesting, such as InAs and GaSb (which have a higher refraction index than GaAs) and also the compounds

$$Ga_{1-x}In_xAs\,, \qquad GaAs_{1-y}Sb_y\,, \qquad Ga_{1-x}In_xAs_{1-y}Sb_y\,,$$

with n increasing as a function of x and y (which can operate as optical waveguides on GaAs substrata).

As already pointed out in Table I, there are problems concerning lattice constant (a_0) coupling among different materials. For these applications, however, the presence of mechanical stresses and crystallographic defects at the interfaces is not such a serious problem as in the case of sources, where they can affect the operation of the device. The danger of layer cracking can be avoided, in any case, with the right growth of quaternary compounds. II and VI semiconductors can also be utilized as optical waveguide materials for passive components. Generally they show optical properties which are less satisfactory than those of III and V materials, but, on the other hand, they have higher electro-optic coefficients. The most studied compounds of this family are ZnO and ZnS. Optical waveguides can also be obtained by diffusion of Se and Cd in CdS, ZnS and ZnSe, ZnS respectively. The most interesting insulating materials are those showing good optical properties and high electro-acusto-magneto-optic coefficients, such as $LiNbO_3$ and $LiNb_xTa_{1-x}O_3$. Optical waveguides can be obtained from these materials by diffusing Ti or Li into the substrate, since in this case a higher refraction index can be formed in a non-stechiometric compound.

Finally, a compound family that can be employed for waveguides in the $\lambda > 1$ μm range is that of magnetic garnets showing a strong magneto-optic effect. (The limit $\lambda > 1$ μm depends on the high optical attenuation coefficient of these materials in the visible region).

3.2 Material characterization methods

The analysis and characterization of materials, both for active and passive devices, is an important stage in the research and development of IO components. On one hand, they are fundamental to the preparation of materials, as control and optimization means of preparatory techniques. On the other hand, they are a necessary means of investigation on degradation and failure phenomena in materials. In the following we give a short review of the methods of analysis and characterization which are considered to be the most important for the knowledge and preparation of IO materials. We shall consider methods of chemical composition analysis and methods of crystallographic structure analysis.

3.2.1 *Methods of chemical composition analysis*

The analysis of chemical composition can be divided into two parts, namely, the definition of the stechiometry of the fundamental components of the material and the qualitative-quantitative identification of the elements which are present in the material as doping impurities.

As far as the first of the two parts is concerned, apart from chemical analysis (which is usually destructive and produces results averaged over the whole analyzed sample) we must list X-ray microanalysis (XMA), Auger electron spectrometry (AES), secondary ions mass spectrometry (SIMS), Nuclear reactions analysis (NRA), Microanalysis by proton induced X-ray (PIX) and Rutherford backscattering techniques (RBS).

TABLE II

Methods of chemical composition analysis.

Analysis method	Incident particle	Energy of the incident particle (1)	Analyzed particle or radiation	Energy of the analyzed particle (1)
XMA X-Ray Micro-Analysis	electrons	2-50 keV	Characteristic X-rays emitted from sample	0-40 keV
AES Auger Electron Spectroscopy	electrons	1-20 keV	Auger electrons emitted from sample	20-2000 eV
SIMS Secondary Ion Mass Spectrometry	ions	0.1-20 keV	Ions extracted from sample	0-100 eV
NRA Nuclear Reaction Analysis	ions	0.2-16 MeV	Protons, α particles, X-rays emitted from sample	0.2-15 MeV
PIX Proton Induced X-Ray	ions	10-400 keV	Characteristic X-rays emitted from sample	0.4-30 keV
RBS Rutherford Back-Scattering	ions	1-4 MeV	Incident ions backscattered from sample	50 keV - 4 MeV

(1) Electron volt (eV) is the kinetic energy acquired by one electron when accelerated between two points with a potential difference of 1 volt; 1 MeV = 10^6 eV.

TABLE III - Performance of different methods of chemical composition analysis.

Method of analysis	Analyzed elements (²)	Elements resolution	Analyzed depth (³)	Depth resolution	Lateral resolution	Accuracy in quantitative analysis	Limit of sensitivity (⁴)	Destructive	Analysis time
XMA X-ray Micro-Analysis	$Z \geqslant 4$	$\Delta Z = 1$	0.5-3 μm	0.5-3 μm	0.2-2 μm	1-10%	100-1000 ppm	No	30″-20′
AES Auger Electron Spectroscopy	$Z \geqslant 3$	$\Delta Z = 1$	5-30 Å	5-10% of removed thickness e.g. ~100 Å after 1500 Å ~200 Å after 3000 Å	1-5 μm	5-10%	100-1000 ppm	Yes	30″-50′
SIMS Secondary Ion Mass Spectrometry	all	$\Delta A = 1$	2-10 Å	5-10% of removed thickness e.g. ~100 Å after 1500 Å ~200 Å after 3000 Å	1-5 μm	not determined	0.1-100 ppm	Yes	30″-50′
NRA Nuclear Reaction Analysis	$Z \leqslant 20$	$\Delta Z = 1$	5-30 μm	200-400 Å for deepness < 100 Å 600-800 Å for deepness > 200 Å	~1 mm	15-20%	300-1000 ppm	No	10′-20′
PIX Proton Induced X-ray	$Z \geqslant 4$	$\Delta Z = 1$	0.5-4 μm	0.5-4 μm	~1 mm	1-10%	5-100 ppm	No	1′-20′
RBS Rutherford Back-Scattering	all	for $Z > 40$ ΔZ .. 10 for $Z < 40$ ΔZ .. 1	0.1-3 μm	200-300 Å for deepness < 1500 Å 300-400 Å for deepness > 2500 Å	~1 mm	3-6%	100 ppm for $Z > 40$ in light materials; 100 000 ppm for $Z < 40$ in heavy materials	No	10′-15′

(²) Z = Atomic number; A = Mass number. (³) 1 μm = 10^4 Å. (⁴) 0.1 ppm = 0.00001% $\simeq 10^{15}$ atoms/cm³

All these analysis techniques require the use of an ion or electron beam, which interacts elastically or unelastically with the sample under analysis producing measurable interactions. The main characteristics of the different analysis techniques and their performance are outlined in Tables II and III [2]. Performance connected to lateral and deepness resolution is particularly important for IO, since integration devices are needed having lateral dimensions of the order of microns and thicknesses of the order of some thousands of Ångstrom. The best analysis technique for this kind of problem is Auger electron spectroscopy which, used together with sputter-etching, allows the best lateral and depth resolutions.

SIMS is another interesting technique, since it shows sensitivity limits which are good enough to detect doping levels of the order of 10^{14} atoms/cm^3.

As regards the second part of chemical analysis, for the qualitative identification of the elements which are present in the material as doping impurities, the most important method is atomic absorption, with a sensitivity limit depending on different elements between 0.1-1 ppm. This technique can reach, with suitable auxiliary equipment to vaporize the sample, an accuracy of one part per billion. However, it should be noted that information derived from atomic absorption, as opposed to SIMS and AES, is connected to the whole volume of the analyzed sample, without any spatial resolution.

3.2.2 *Methods of crystallographic structure analysis*

The analysis of a crystallographic structure is generally dependent on diffractometric techniques, the most popular of which is X-ray diffractometry.

With X-ray diffractometry it is possible to evaluate with a high level of precision the lattice constant parameters, the structure phases, the crystallinity degree of the material and its preferred orientation and the internal stress of films.

The obtained data refer to the whole volume of the analyzed sample (from 10^{-3} to 1 mm^3) and are therefore average values. In analyzing films there is a thickness limit in the 500-1000 Å range under which it is difficult to derive significant data, either for intensity reduction or for excessive enlargement of the diffracted peaks.

A particular technique, called X-ray topography, exists by which it is possible to display crystallographic defects such as stacking faults, grain bounders and dislocations. In addition to X-ray diffraction techniques, there are various electron diffraction techniques which are particularly suitable for analysis of very thin films, since the analyzed volume is several orders of magnitude smaller than in the case of X-ray diffraction. Therefore, it is possible to analyze material thicknesses between 5-500 Å. Table IV compares the

TABLE IV

Diffraction techniques.

Technique	Incident radiation	Incident radiation energy	Minimum analyzable thickness
XRD X-Ray Diffraction	X-rays	5-10 keV	100-1000 Å
LEED Low Energy Electron Diffraction	electrons	20-500 eV	~ 5 Å
RHEED Reflection High Energy Electron Diffraction	electrons	10-100 keV	10-500 Å
SEM-ECP Electron Channeling Pattern	electrons	10-50 keV	~ 500 Å

X-ray diffraction technique with the most frequently used electronic diffraction techniques. On the contrary, Table V compares the X-ray topography technique used for displaying crystallographic defects with the competing scanning electron microscopy (SEM) and RBS techniques.

TABLE V

Techniques for defect display.

Technique	Incident radiation	Incident radiation energy	Analyzed thickness	Lateral reso-lution
XRT X-Ray Topography	X-rays	5-10 keV	10-1000 μm	1-10 μm
SEM Electrovoltaic barrier effect	electrons	5-15 keV	1-6 μm	0.3-2 μm
SEM Cathodoluminesence	electrons	5-50 keV	1-6 μm	0.3-2 μm
RBS Channeling	ions	1-4 MeV	1-10 μm	1-5 mm

3.3 Technologies

As already mentioned, in order to develop IO devices several technologies have been established to prepare the necessary materials and to obtain the suitable geometries. The most interesting material preparation technologies are those regarding liquid-phase epitaxial growth, vapor-phase epitaxial growth, molecular beam epitaxial growth and sputtering deposition.

In order to obtain the geometries required for IO circuits it has been necessary to employ electrolithographic technology and to improve chemical and ion etching techniques. In the following we shall briefly review the basic requirements of IO regarding the above technologies.

3.3.1 *Liquid Phase Epitaxial growth* (*LPE*)

From a thermodynamic point of view two methods of LPE exist, namely, the isothermal method and the controlled cooling method. The first method, generally used for garnet growth, consists in heating the furnace crucible (Fig. 2) up to a temperature which is slightly under the equilibrium temperature of the melted solution, which becomes metastable. If a substrate is dipped into

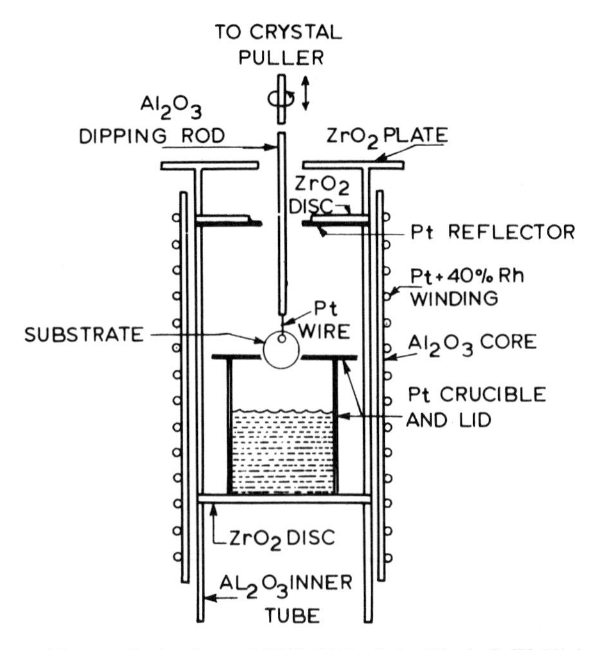

FIG. 2—Vertical furnace for isothermal LPE. (After S. L. Blank, J. W. Nielsen, Journal of Crystal Growth, Vol. 17, pp. 302-311).

the solution, an epitaxial film with the same composition as the melted solution will be grown on it. The main problem for this kind of epitaxy is to maintain thermal stability during the epitaxial growth. (To this end it is necessary to control within 1/2 degree temperatures of about 1000 °C).

Another important problem concerns the melting composition, for which a complete knowledge of state diagrams and partitioning constants is required. Liquid epitaxy with controlled temperature drop (which is largely used ofr GaAs and all III-V compounds, as well as for niobate on tantalate, germanate on titanate, etc.) is a growth technique in thermodynamic equilibrium conditions.

The substrate is dipped into a limited volume of melted solution at the saturation temperature. The temperature is then slowly and uniformly reduced. The melted solution will be epitaxially grown following the solubility curves of state diagrams with a good approximation. For this kind of growth it is necessary to have a good knowledge of state diagrams of both the main compounds and dopants.

Often, as in the case of GaAlAs compounds, it is also necessary to control the growth atmosphere, where it must be H_2 reducing. LPE with temperature drop can be obtained either in vertical furnaces, as previously described, or in horizontal multi-zone furnaces (Fig. 3). In any case, it is necessary that the temperature drop in a controlled and uniform way on the whole useful region of the furnace.

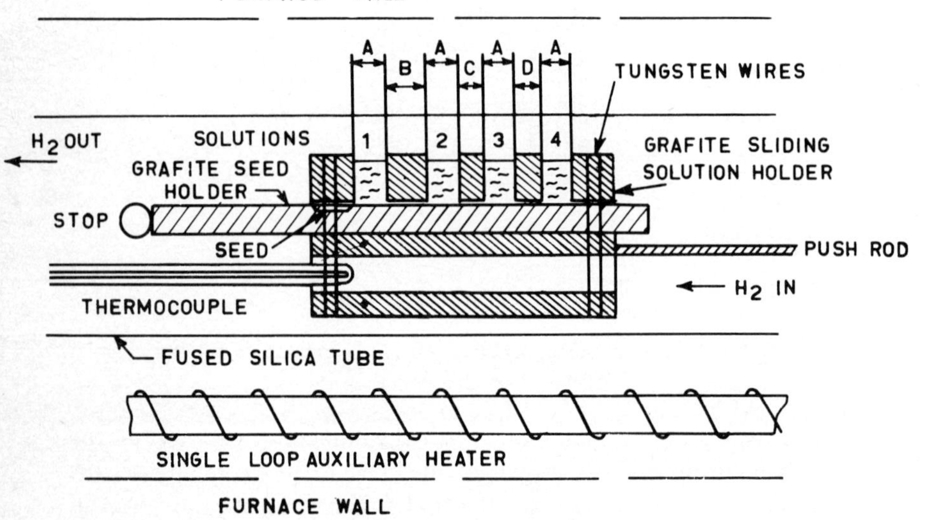

FIG. 3 – LPE Horizontal furnaces. (After [20], p. 8).

By means of the crucible shown in Fig. 3, it is possible to deposit four layers with different composition and doping by suitably moving the substrate under the various solutions.

3.3.2 *Vapour Phase Epitaxial growth* (*VPE*)

In VPE the solid materials forming the growing layer are vaporized, transferred in gas phase onto the substrate and then deposited in the form of the compound desired. There are several

VPE growth techniques according to the technical solutions adopted [3]. However, the basic equipment is a multizone furnace with well-controlled inner temperature profile and a complex system of fluxmeters and valves for the proportioning of each gas, which sometimes requires special solutions due to the reactivity and noxiousness of the gas being used.

Figure 4 shows the current gas dynamic growth basic scheme.

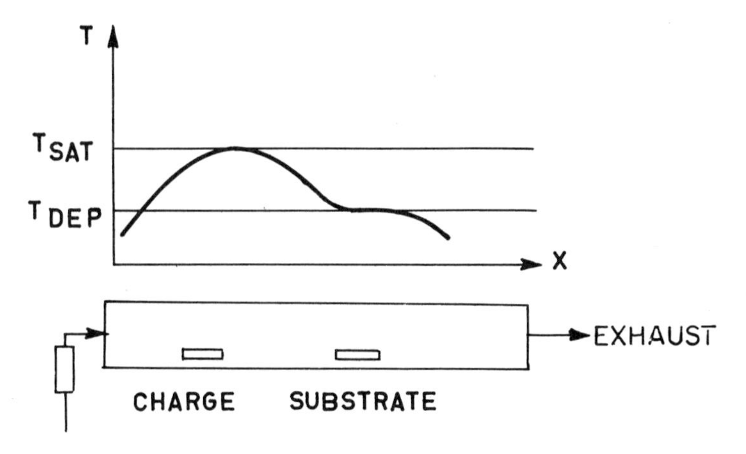

FIG. 4 – VPE current gas dynamic growth scheme.

A flux of carrier gas is saturated by the component to be deposited at temperature T_{sat} by a polycrystalline charge. At lower temperature T_{dep}, the over-saturated vapor deposits the compound epitaxially onto the substrate.

Although VPE presents considerable control difficulties in the manufacture of many layers of complex elements, it has better productivity prospects than LPE; therefore, it is more suitable for industrial growth processes.

3.3.3 *Molecular Beam Epitaxial growth (MBE)*

MBE is the most recent epitaxial growth system and is still in an experimental phase. Although the largest applications at the present time have been through LPE and VPE, MBE could become the most suitable system for obtaining particularly pure materials with perfect crystalline structures. At present, MBE is carried out in vacuum chambers as shown in Fig. 5.

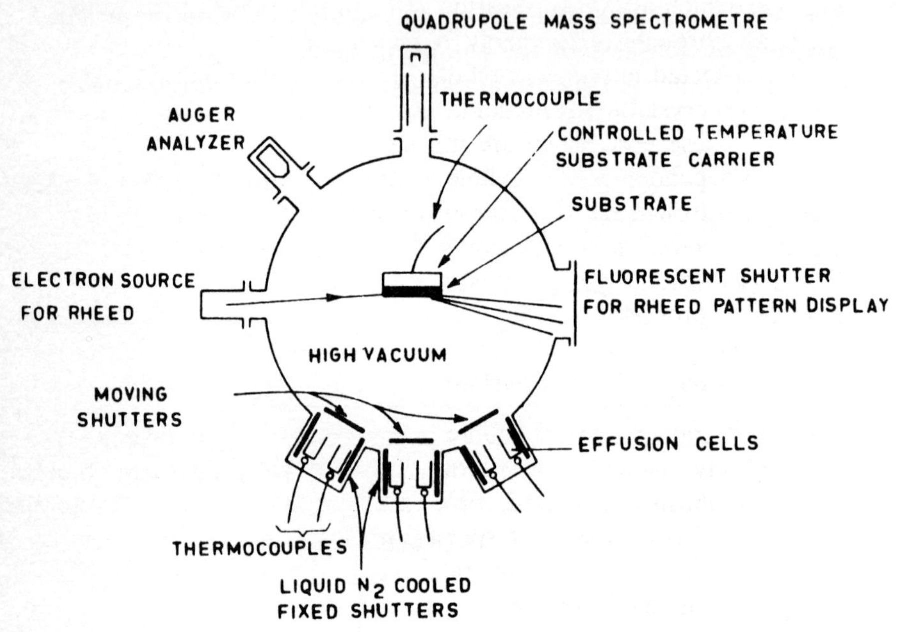

FIG. 5 – MBE system.

Vapor sources are formed by crucibles which are independently heated at controlled temperatures. Thus, for each single element a partial effusion (evaporation or sublimation) speed may be reached which is suitable for producing the right proportions in the deposited layer. By feeding several cells at the same time, different molecular beams may be obtained which, deposited on the substrate consisting of an useful monocrystal in suitable temperature conditions, can produce a composite epitaxial layer.

Fixed liquid N_2 or water cooled shutter thermically separate one source from the other. Moving shutters in front of the cells allow interruption or generation of molecular beams when several layers of different compositions have to be deposited. In MBE systems the analytical equipment installed is very important. This allows a practical on-line control of the epitaxial growth phases.

This analytical equipment normally includes: *a*) a mass spectrometre which analyzes the atmosphere of the chamber's remaining gasses; *b*) an AES analyzer which allows the analysis of layer composition at the end of each deposition phase; *c*) a RHEED analyzer which allows verification of layer crystallographic growth structure through high energy reflected electron diffractograms. All the detected parameters relate to the aim of producing one or more monocrystalline epitaxial layers with the desired composition. Since all these parameters are mutually correlated and cannot be varied independently, the whole deposition and analysis system can be controlled by an electronic computer. By means of MBE systems, heterostructure lasers including GaAs layers doped by Sn or Mg [4] and double heterostructure lasers $GaAs/Ga_{1-x}Al_xAs$ [5] have been produced.

3.3.4 *Sputtering layer deposition*

Deposition of non-crystalline layers for optical applications is obtained by means of sputtering, since evaporation, the other vacuum deposition method, produces layers with higher losses. The main characteristics of sputtering deposition are a relatively slow ($\simeq$ 30 Å/min) deposition rate and a high energy of the particles striking the substrate.

Since most of the layers which are useful for optic applications are dielectric, radio-frequency sputtering is employed. The sputtering efficiency (defined as the ratio of sputtered material to the number of ions incident on the target) is related to the relevant atomic elements. Therefore, the layer composition may differ from that of the source material.

The composition of the sputtering atmosphere can affect both the composition of the deposited layer (through compound formation or gas absorption) and sputtering efficiency. Sputtering has been employed for obtaining glass, ZnO, Ta_2O_5, TiO_2 [6], Nb_2O_5, xBaO $y$$SiO_2$ [7] layers.

3.3.5 *Mask making and electron beam lithography*

Problems regarding lithographic techniques and manufacturing of the masks required for these techniques are very complex for IO. For passive components (coupled guides, directional couplers, filters), and also for active components (stripe geometry laser or DFB laser gratings), the geometries used can require details with dimensions of 0,1 micron in intervals of 1 cm. For optical guides particularly, it is necessary to have boundary roughness lower than 500 Å to obtain attenuation lower than 1 dB/cm. It is probable that problems involved in mask making for IO will exist for a long time. But part of these problems will have to be solved for the manufacture of the next generation of integrated electronic circuits (64-100 kbit RAM memories) which are expected to be implemented in the 80's.

Mask drawing for IO devices can be of two types: periodic (as in the case of grating couplers) and non-periodic. Periodic drawings, owing to their periodicity, can be made by optical systems, through interference of two laser beams. The interferential drawing is used to expose photoresist of a photographic mask or to cover the device layer being manufactured. The minimum distances that can be obtained between the interferential fringes are of the order of 1000 Å [8].

Since we cannot count on the traditional photoreduction, limited to geometries slightly lower than one micron by the intrinsic properties of the light employed, for non-periodic drawing masks, exposure is required under energy sources with a wavelength at least one order of magnitude lower than the desired geometries.

Thus, focused laser beams, focused electron beams, or X-ray beams may be used. The exposure system by focused electron beams appears to be the most suitable, both for integrated circuit technology and for IO devices.

Either positive or negative resists which are sensitive to electrons have been developed for this kind of exposure. Although the former require slightly higher exposure energy than the latter, they present better contrast and resolution.

Considering that an electron beam cannot be deflected more than 1-2 mm (if aberrations caused by lenses and deflection systems are to be avoided), and that for optical guides larger lengths are usually required, it is necessary that the substrate to be exposed be positioned with the required accuracy. In the systems developed to date this is obtained as follows:

— a focused electron beam with a 300 Å diameter is available;

— the scanning of the beam is utilized digitally or analogally for short displacements;

— the substrate-carrier moves to allow exposures larger than those permitted by electron beam deflection, in such a way as to join together different rectangles and lines. The accuracy of substrate carrier movement is determined by interferometric methods. The finest corrections are obtained by computer-aided beam positioning.

The systems developed were initially derived from those studied for integrated circuit technology. Therefore, they are of the digital-analogue hybrids.

Most systems developed for IO (Ostrovsky, Goell, Ballantyne, Tong *et al.*) have been made for laboratory research and not for marketing. Nevertheless, there are at present several systems that can be purchased from different manufacturers, but the required investment is very high.

3.3.6 *Chemical and ion etching*

When the geometries of the parts protected by resist films are delimited by mask making, it is generally necessary to etch one or more layers of the non-protected parts.

Chemical or ion etching can be used. Chemical etching, although selective, has the disadvantage of underetching, which limits the lateral resolution considerably when a material thickness higher than 2000-3000 Å is used. Underetching can be also be used advantageously in the lift-off technique. This consists in depositing materials on a substrate partly covered by a photoresist film which is then removed (Fig. 6).

In electron sensitive positive resists the electron backscattering, from the surface under the resist, causes a line drawing which is larger at the base of the resist that at the top. This avoids connections being formed between the layer deposited on the substrate parts covered by the resist and the other parts which are not covered by the resist, thus facilitating lift-off etching (Fig. 7).

In comparison with chemical etching, ion etching has the great advantage of no underetching, since it is obtained by bombardment of ions hitting the sample surface perpendicularly. This fact allows structures having transversal dimensions lower than a micron

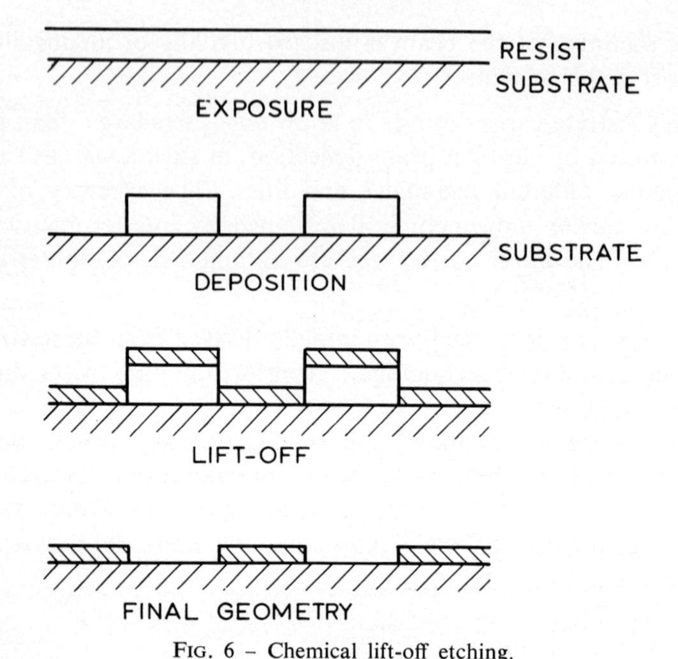

FIG. 6 – Chemical lift-off etching.

to be manufactured. In ion etching the sample is located in a high vacuum chamber. Ions are generated in a separate ionization chamber, and eventually accelerated, focused and electrically neutralized before hitting the sample.

The material not to be etched is protected by photoresists or metal masks carefully chosen in order to avoid removal during the bombardment. A metal mask must be produced using a metal

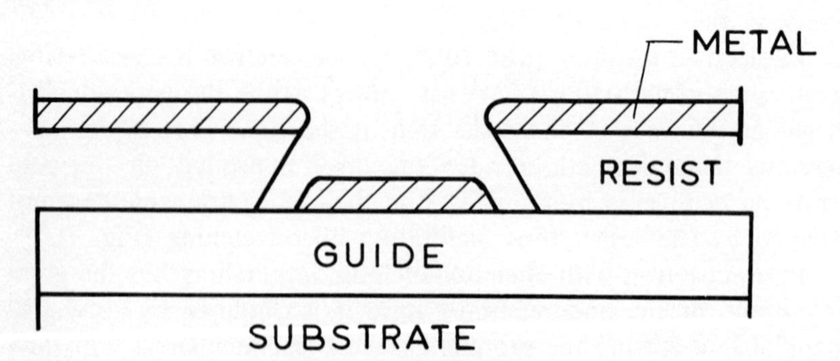

FIG. 7 – Electron beam exposure effect and advantage for lift-off technique.

having an ion etching rate lower than that of the etching layer. Moreover, this metal must not form chemical compounds with the underlying layer and must be easily removable after ion etching.

3.3.7 *Diffusion and ion implantation*

Diffusion and ion implantation are employed to obtain optical waveguides by introducing a suitable dopant and changing the refraction index under the material surface. The dopant is introduced into the substrate in two different ways.

In diffusion carried out in furnaces at temperatures of 700-1000 °C, the parameters controlling either the dopant penetration deepness into the substrate or its local concentration are the process temperature, the time required for the process, and the dopant surface concentration.

In the ion implantation process the substrate is bombarded by dopant ions accelerated at relatively high energies. The average dopant penetration depends on the acceleration energy and the doping distribution is Gaussian.

Theoretically, ion implantation can be obtained with similar criteria for a large number of doping elements to be implanted into different substrata both of the amorphous and of the crystalline type. On the other hand control of the diffusion process is critically dependent both on the type of substrate or on the dopant to be introduced. Therefore, whereas optical planar guides are produced by ion implantation of different dopant and substrate pairs (Li^+ into SiO_2, H^+ into GaAs, O^+ into GaAs; etc.), the diffusion process in IO applications was studied only for particular pairs such as Zn into GaAs and diffusion of Ti (or Nb) in $LiNbO_3$ (or $LiTaO_3$).

The refraction index in a substrate can also be changed using photon implantation. In this case the increase of the refraction index in the substrate is not caused by a doping effect, but is caused by damage in the crystal lattice caused by the proton bombardment, which introduces traps for the free carriers reducing their real concentration.

Both in ion and proton implantation the lattice damage also has a negative effect on the optical properties of the substrate; it is thus necessary to remove a part of this negative effect by annealing processes.

REFERENCES

[1] C. J. Nuese: *III-V alloys for opto-electronic applications*, J. of Electronic Materials, Vol. 6 (1977), pp. 253-293.

[2] V. Ghergia *et al.*: *Integrated optics state of art*, CSELT, Internal Report No. 76.08.143 (1976).

[3] C. Paorici: *Experimental methods in a chemical vapor deposition*, CNR, MASPEC Report (1976).

[4] A. Y. Cho and W. C. Ballamy: GaAs *planar technology by molecular beam epitaxy* (*MBE*), J. Appl. Phys., Vol. 46 (1975), pp. 783-785.

[5] M. B. Panish: *Heterostructure injection lasers*, IEEE Trans. Microwave Theory Tech., Vol. MTT-23 (1975), pp. 20-30.

[6] C. W. Pitt *et al.*: *R.F. sputtered thin films for integrated optical components*, Thin Solid Film, Vol. 26 (1975), pp. 25-51.

[7] J. E. Goell: *Rib waveguide for integrated optical circuits*, Appl. Opt., Vol. 12 (1973), pp. 2797-2798.

[8] H. I. Smith: *Fabrication techniques for surface-acoustic-wave and thin-film optical devices*, Proc. IEEE, Vol. 62 (1974), pp. 1361-1387.

CHAPTER 4

FUTURE TRENDS IN INTEGRATED OPTICS

by Vittorio Ghergia and Antonio Scudellari

As previously mentioned, IO is at present in the early phase of experimental development. The time when complete systems will be obtained with IO devices is still to come.

For this to be achieved, two types of problems must be solved relating to technological difficulties and commercial and economic applications. This latter form of difficulty now prevails over the former. It should be noted, however, that when a economic stimulus exists it is possible to solve in an short time even the most difficult technological problems, as has happened recently in the field of electronics.

As regards communications, the most promising field of application for IO is that of single mode fibre systems. Single mode fibre communications are attractive both for high bit rates and for multiterminal data busses requiring high speed switching with low crosstalk [1]. Nowadays, however, the market is completely dominated by multimode fibres, which present good characteristics and allow systems to be manufactured with performance levels valid also for applications in the near future.

It is expected that multimode fibres will continue to predominate over monomode fibres, which are seriously penalized by the difficult problem of their splicing in operational service.

In this case, IO is encouraged to lose its monomode character to include miniaturized and planar components of the multimode type.

At present many « thick film » devices compatible with multimode fibre optic systems are being developed [2-5]. In this new form IO will, however, suffer from considerable limitations as regards components and their performance. Moreover, not only theoretical but also technological problems will have to be solved.

From this point of view, it is sufficient to consider that, for passive optical waveguides, it would be necessary to pass from the monomode waveguide thickness of a few microns to multimode waveguides having a thickness of 50-100 μm, which cannot be obtained by means of traditional technologies such as vacuum deposition, diffusion etc. On the other hand, if we consider IO only for applications in transmitters and not in receivers, almost all IO components can be employed also in their monomode version. For instance, the WDM transmitter in Fig. 1 of Chapter 2 can be employed with multimode fibres, even though it is not easy then to extract signals from multimode fibres.

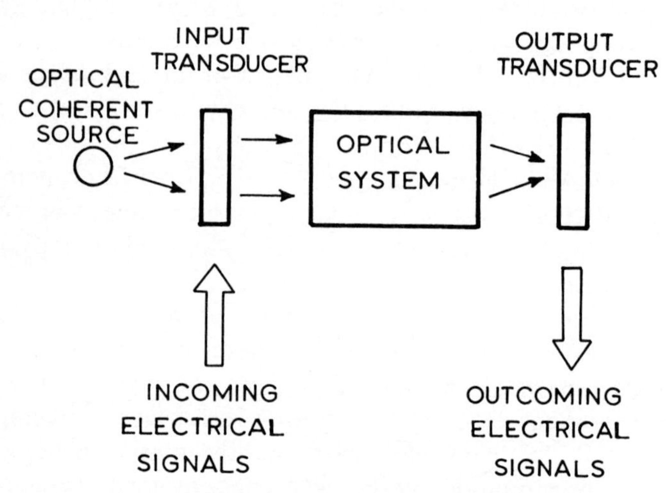

FIG. 1 – Block diagram of a typical optical processor.

On the contrary, if interest in monomode fibre were to increase IO could shortly solve many of the problems concerning monomode components.

It is worth remembering that IO could find remarkable applications in the processing of optical data.

As regards optical data processing, IO could encourage the manufacture of optical processors, which are devices where processing is performed through optical means on an electrical signal transferred to an optical coherent source (Fig. 1). An example is that of a *RF* spectrum-analyzer. Such devices produced with discrete components are employed already for some data processing systems. An IO implementatio would make them less bulky, more efficient and more stable.

The development of integrated optical spectrum analyzers is advancing rapidly [6], even if no clearly favoured material structure has yet emerged for integration. Gallium-arsenide can be used for injection laser, active waveguide components and detection circuitry, but the sensitivity to temperature is high and waveguides are lossy; silicon is the most suitable material for the detector array, but it shows poor acoustical properties and cannot be used for light generation. The best choice, therefore, seems to be that of an hybrid integration, like that shown in Fig. 2. The substrate is made of lithium niobate, which is the most appropriate substance from an acoustic point of view, and optical waveguides are hybrid-coupled to a gallium-arsenide laser and a silicon detection circuit.

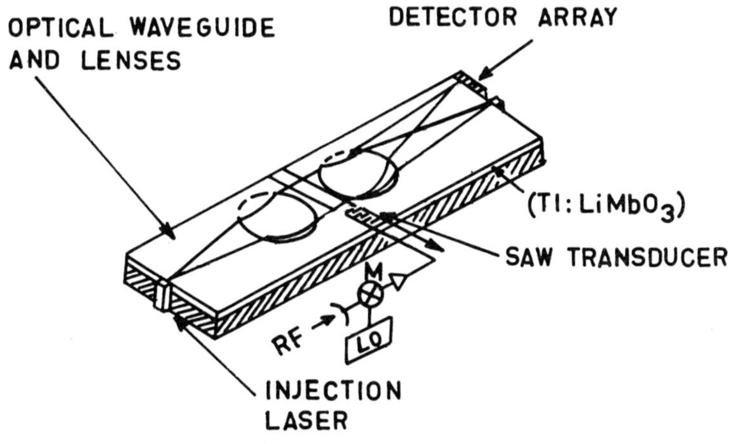

FIG. 2 – An hybrid structure, on $LiNbO_3$ substrate, for an integrated optical RF spectrum analyzer.

Another area of potential application in the field of signal processing concerns high speed *A/D* convertions [7] and the use of electro-optic modulators and switches to perform logic operations [8].

REFERENCES

[1] T. G. Giallorenzi and A. F. Milton: *Single mode fiber optics and integrated optics for use in optical communications*, Optical Fibres, Integrated Optics and their Military Applications, London (May 16-20, 1977), paper 18.

[2] M. G. F. Wilson *et al.*: *Optical power division in a multimode-waveguide intersection*, Electron Lett., Vol. 12 (1976), p. 434.

[3] A. R. Nelson *et al.*: *Electro-optic channel modulator for multimode fibers*, Appl. Phys. Lett., Vol. 28 (1976), pp. 321-323.

[4] F. Auracher: *Planar branching network for multimode glass fibers*, Opt. Commun, Vol. 17 (1976), pp. 129-132.

[5] R. Ulrich and G. Aukele: *Self-imaging in homogeneous planar optical waveguides*, Appl. Phys. Lett., Vol. 27 (1975), pp. 337-340.

[6] J. F. Mason: *Technology '79: aerospace and military*, IEEE Spectrum, Vol. 16 (1979), p. 71.

[7] D. Lewis and H. F. Taylor: *An integrated optical analog-to-digital converter*, Optical Fibres, Integrated Optics and their Military Applications, London (May 16-20, 1977), paper 24.

[8] H. F. Taylor: *Integrated optical logic circuits*, Integrated and Guided Wave Optics, Salt Lake City (January 1978), paper TuC4.

SUBJECT INDEX

SUBJECT INDEX

C

D

E

F

G

H

I

N